MW00836907

Magnetic Domains

Springer
Berlin
Heidelberg
New York
Barcelona
Hong Kong
London
Milan
Paris
Singapore
Tokyo

Alex Hubert   Rudolf Schäfer

Magnetic Domains

The Analysis
of Magnetic Microstructures

With 400 Figures, 4 in Colour

 Springer

Professor Dr. Alex Hubert †
Universität Erlangen-Nürnberg
Institut für Werkstoffwissenschaften
Lehrstuhl Werkstoffe der Elektrotechnik
Martensstrasse 7
91058 Erlangen
Germany
e-mail: hubert@ww.uni-erlangen.de

Dr. Rudolf Schäfer
Institut für Festkörper- und Werkstofforschung
Helmholtzstrasse 20
01069 Dresden
Germany
e-mail: r.schaefer@ifw-dresden.de

Library of Congress Cataloging-in-Publication Data
Hubert, Alex, 1938 – Magnetic domains: the analysis of magnetic microstructures / Alex Hubert; Rudolf
Schäfer. p. cm. Includes bibliographical references and indexes. ISBN 3-540-64108-4 (hardcover: alk. paper)
1. Magnetic materials. 2. Domain structure. I. Schäfer, Rudolf. II. Title.
QC765.H83 1998 538'.4-dc21 98-16905

Corrected Printing 2000

ISBN 3-540-64108-4 Springer-Verlag Berlin Heidelberg New York

Springer-Verlag Berlin Heidelberg New York
a member of BertelsmannSpringer Science+Business Media GmbH

© Springer-Verlag Berlin Heidelberg 1998
Printed in Germany

Typesetting: Camera-ready copies by the authors
Cover design: *design & production* GmbH, Heidelberg

SPIN: 10836233   57/3111 – 5 4 3 2 1 - Printed on acid-free paper

Preface

Magnetic domains are the elements of the *microstructure* of magnetic materials that link the basic physical properties of a material with its macroscopic properties and applications. The analysis of magnetization curves requires an understanding of the underlying domains. In recent years there has been a rising interest in domain analysis, probably due to the increasing perfection of materials and miniaturization of devices. In small samples, measurable domain effects can arise, which tend to average out in larger samples. This book is intended to serve as a reference for all those who are confronted with the fascinating world of magnetic domains. Naturally, domain pictures form an important part of this book.

After a historical introduction (Chap. 1) emphasis is laid on a thorough discussion of domain observation techniques (Chap. 2) and on domain theory (Chap. 3). No domain analysis is possible without knowledge of the relevant material parameters. Their measurement is reviewed in Chap. 4. The detailed discussion of the physical mechanisms and the behaviour of the main types of observed domain patterns in Chap. 5 is subdivided according to the crystal symmetry of the samples, according to the relative strength of induced or crystal anisotropies, and according to the sample dimensions. The last chapter deals with the technical relevance of magnetic domains, which is different for the different fields in which magnetic materials are applied. The discussion reaches from soft magnetic materials forming the core of electrical machines that would not work without the hidden action of magnetic domains, to magnetic sensor elements used in magnetic recording systems, which may suffer from domain-related noise effects.

We emphasize in this book (in Chaps. 5, 6) the analysis of actual magnetization processes, in particular of those that lead to discontinuities and irreversibilities in the magnetization curve. This approach is adequate for several important application areas, such as magnetic sensors and transformer materials.

In other areas, such as those of bulk polycrystalline soft-magnetic materials, the link between observed elementary processes and possible technical descriptions of the hysteresis phenomena is difficult to establish. Formal descriptions of hysteresis phenomena — shortly introduced by reference to actual textbooks in the last section of Chap. 6 — use abstract concepts of magnetic domains in their justification. Whether the discrepancy between the simplified assumptions of hysteresis theories and the complexities of actual domain behaviour is technically relevant remains an interesting open question.

Previous books and reviews touching on the subject of magnetic domains and magnetization processes are listed separately in alphabetic order at the end of the reference list. The few recent works among these cover important, but on the whole rather specialized aspects. One of the aims of this book is to collect the knowledge available from the investigations of many magnetic materials. Although the technical areas of electrical steels, of high frequency cores, of permanent magnets, or of computer storage media have little in common, the magnetic microstructures in most of these materials obey the same rules and differ only in quantitative, not in qualitative aspects.

Despite the preference of many magneticians for the old Gaussian system of units, standard SI units are used throughout. A reference to the old system is made only occasionally. But we prefer to reduce equations to dimensionless units anyway in all practical examples, so that results get a broader range of applicability, and at the same time the mathematical expressions become independent of the unit system.

We take one liberty, however. The magnetic dipole density of a material is given by the vector field $J(r)$ (measured in Tesla or Vs/m^2). Its official name is *magnetic polarization*, but very often and also in this book, it is called simply *magnetization*, although in the strict SI system this name is reserved to a quantity $M(r)$ that is measured in A/m and related to J by $M = J/\mu_0$. We will never use the latter quantity, so no confusion is possible. Our choice agrees with the recommendation from *P.C. Scholten*[1], except that he would prefer to use the abbreviation M instead of J, which we think might lead to confusion[2]. Anyway, in most cases we use the reduced unit vector of magnetization direction only, and we abbreviate this vector field by $m(r)$.

The careful reader will discover quite a few original contributions in this book, not published elsewhere. They are intended to improve the grasp of

---

[1] P.C. Scholten: "Which SI?", J. Magn. Magn. Mat. **149**, 57–59 (1995)

[2] The fundamental material equation is thus expressed in this book in the form $B = \mu_0 H + J$

otherwise abstract concepts and check their applicability. They also help to fill gaps in the published material. Such gaps may be too small or unimportant to justify an independent scientific publication, but they may form a serious obstacle for newcomers in their attempt to enter the field.

A few hints to the reader: Up to five text organization levels are used. Three of them (chapter, section, subsection) are systematically numbered in a decimal classification. They appear (perhaps in an abridged way) in the right-hand page headers to facilitate orientation. A further subdivision is often needed. It is marked and referred to within the same subsection by "(A), (B) etc.". In references from other subsections a small capital letter (like in "Sect. 3.4.1A") is appended. Where needed, an informal fifth level, marked "(i), (ii) etc.", is added. Parenthesis with numbers like "(3.1)" always refer to equations, while brackets like "[212]" indicate citations. Parenthesis with lower case letters like "(a)" refer to parts of a figure mentioned in the same paragraph.

Some passages in Chap. 3 on domain theory are addressed more to the interested specialist than to the general reader. They are marked by smaller print. All references are collected in the order of occurrence after the text part. Orientation within the references section is supported by an author index that comprises all authors and coauthors of all cited references (explicit text references are listed first). In addition to the regular, numbered and captioned figures we offer integrated "sketches", that are intended to facilitate reading, but which need no further explanation (thanks to John Chapman for suggesting this tool). The few tables are numbered and treated as if they were figures, which makes it easier to find them.

Erlangen, Dresden, July 1. 1998

Alex Hubert
Rudolf Schäfer

Acknowledgements

The authors want to thank all the many colleagues who helped with comments and criticism by reading the manuscript, who permitted the reproduction of their beautiful domain images, who supplied valuable samples for domain imaging, and who assisted in sample preparation, calculations, graphics, and editing of the manuscript.

Most valuable was the help of *S. Roth*, who completely and critically read the manuscript, and of *W. Andrä*, who carefully studied most of the text, supplying us with many suggestions based on his long experience. Portions of the book were read by *L. Abelmann, S. Arai, B.E. Argyle, M. Bijke, A. Bogdanov, S.H. Charap, K. Fabian, R.P. Ferrier, F.J. Friedlaender, P. Görnert, H. Groenland, M. Haast, U. Hartmann, H. Höpke, F.B. Humphrey, L. Jahn, E. Jäger, V. Kamberský, M.H. Kryder, H. Lichte, J.D. Livingston, C. Lodder, J. McCord, K.-H. Müller, M. Müller, F. Oehme, I.B. Puchalska, W. Rave, A.A. Thiele* and *P. Trouilloud*. Their criticism and suggestions are gratefully acknowledged. We also received good advice on various subjects by *D. Berkov, H. Fujiwara, O. Gutfleisch, R. Hilzinger, H. Huneus* and *K.H. Wiederkehr*.

For illustration we were permitted to show beautiful domain images and computational results by *L. Abelmann, H. Aitlamine, S. Arai, B.E. Argyle, J. Baruchel, L. Belliard, J.E.L. Bishop, H. Brückl, R. Celotta, J.N. Chapman, R.W. DeBlois, D.B. Dove, T. Eimüller, A. Fernandez, E. Feldtkeller, J. Fidler, P. Fischer, R. Frömter, J.M. Garcia, K. Goto, R. Grössinger, F.B. Humphrey, J.P. Jakubovics, A. Johnston, K. Kirk, K. Koike, S. Libovický, C.-J.F. Lin, J.D. Livingston, J.C. Lodder, F. Matthes, R. Mattheis, Y. Matsuo, S. McVitie, J. Miles, J. Miltat, K.-H. Müller, T. Nozawa, H.-P. Oepen, L. Pogany, S. Porthun, I.B. Puchalska, D. Rugar, C.M. Schneider, M. Schneider, Ch. Schwink, J. Šimšova, R. Szymczack, A. Thiaville, S.L. Tomlinson, A. Tonomura, S. Tsukahara, J. Unguris, P.A.P. Wendhausen, Y. Zheng, J.-G. Zhu, L. Zimmermann,* and *E. Zueco.* Thanks to all of them for their support.

Major contributions to this book either in computations or in experiments are due to *A.N. Bogdanov, K. Fabian, J. McCord, W. Rave* and *M. Rührig.* We are especially grateful for their help.

Special thanks are due to the many colleagues from industry and research laboratories who supplied us with valuable samples for domain imaging: *S. Arai, B.E. Argyle, W. Bartsch, R. Celotta, W. Ernst, M. Freitag, P. Görnert, B. Grieb, H. Grimm, P. Grünberg, W. Grünberger, A. Handstein, G. Henninger, G. Herzer, W.K. Ho, V. Hoffmann, A. Hütten, F.B. Humphrey, L. Jahn, J.C.S. Kools, S.S.P. Parkin, J. Petzold, B. Pfeifer, T. Plaskett,*

Ch. Polak, S. Roth, M. Schneider, R. Schreiber, R. Thielsch, W. Tolksdorf, J. Wecker, U. Wende, J. Yamasaki, and *K. Záveta.*

We are grateful to the collaborators at the University of Erlangen-Nürnberg, in particular *K. Boockmann, H. Brendel, G. Cuntze, R. Fichtner, W. Grimm, P. Löffler, H. Maier, J. McCord, M. Neudecker, J. Pfannenmüller, K. Ramstöck, K. Reber, F. Schmidt, L. Wenzel, S. Winkler* and many others, to whom we owe thanks for domain pictures which were taken with the digitally enhanced Kerr-imaging system developed there together with *P. Doslik.* Valuable assistance, for example in sample characterization and manipulation, was given by *B. de Boer, O. de Haas, A. Bürke, D. Eckert, J. Fischer, G. Große, D. Hinz, N. Mattern, S. Roth, U. Schläfer* and *A. Teresiak,* all collaboraters at IFW Dresden.

In literature research we received help in particular by *E. Dietrich, Barbara Hubert, B. Richter* and *H. Wagner-Schlenkrich.* We appreciate the support offered by the Institutes of Erlangen and Dresden by allowing the use of hardware under the responsibilities of *G. Müller, K.-H. Müller, L. Schultz, K. Wetzig* and *A. Winnacker.*

The first author is much indebted to the Magnetics Technology Center of Carnegie Mellon University, Pittsburgh, where an initial part of the manuscript was written during a sabbatical, supported by *M. Kryder.* Another part was prepared during a stay in the Institute for Physical High Technology in Jena (the famous former Magnet-Institut), supported by *P. Görnert.* And finally a sabbatical could be spent at the Institute for Solid State and Material Research (IFW) in Dresden, generously supported by *L. Schultz.*

Extremely valuable technical assistance was contributed by *W. Habel* (Erlangen) in photography and graphics, by the first author's daughter *Birgit* in the generation of the indices and the final production of the manuscript, and by *S. Schinnerling* (Dresden) in sample preparation.

Last but not least both authors would like to thank their wives *Heidemarie Hubert* and *Renate Schäfer* for their support, patience and encouragement.

The manuscript was written on the Apple Macintosh® based on the Nisus® Writer text system, with ample use of the Expressionist® formula editor, the EndNote® literature organizer, the drawing routines MacDraw® and ClarisDraw®, and the Adobe PhotoShop® Image Processing System. Any suggestions or error reports from the readers are most welcome. Updating information is intended to be collected at the website:

http://www6.ww.uni-erlangen.de/~hubert/magnetic-domains.html

Prof. Dr. Alex Hubert   Dr. Rudolf Schäfer
Institut für Werkstoffwissenschaften   Institut für Festkörper- und Werkstoffforschung
der Universität Erlangen-Nürnberg   Helmholtzstr. 20
Lehrstuhl Werkstoffe der Elektrotechnik   D-01069 Dresden
Martensstr. 7   Germany
D-91058 Erlangen   email: r.schaefer@ifw-dresden.de
Germany

Table of Contents – Overview

Table of Contents

1. Introduction

After a brief empirical definition of magnetic domains in this chapter the historical development of the knowledge of magnetic microstructure is outlined. This discussion is used to introduce the basic facts about domains.

1.1 What are Magnetic Domains?

Today it is easy to answer this question by reference to direct observation. Figure 1.1 shows the magnetic microstructure in the field-free state of different magnetic samples made visible with the help of polarization optics. In all cases, uniformly magnetized regions, so-called domains, are observed to appear spontaneously within otherwise unstructured samples.

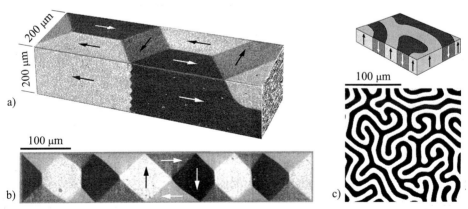

Fig. 1.1 Domains observed with magneto-optical methods on homogeneous magnetic samples. (a) Images from two sides of an iron whisker, combined in a computer to simulate a perspective view (sample courtesy *R.J. Celotta*, NIST). (b) Thin film NiFe element (thickness 130 nm) with a weak transverse anisotropy (sample courtesy *M. Freitag*, Bosch). (c) Faraday effect picture of domains in a single-crystal garnet film with perpendicular anisotropy, together with a schematic of the magnetization

The directions of magnetization in the domains of Fig. 1.1, which can be determined by additional experiments in an external field, are indicated by arrows. Example (a) shows two sides of an iron whisker (the two images were mounted together to create a three-dimensional appearance). Example (b) displays a polycrystalline thin film in which the domains are determined by a weak transverse uniaxial anisotropy. A transparent single-crystal magnetic film with a uniaxial anisotropy perpendicular to the film plane is shown in (c). The black and white areas are magnetized into and out of the image plane, as indicated schematically. Looking at such clear pictures, we can have no doubt about the reality of domains. In the microscope the situation becomes even more convincing when domains are seen moving in a magnetic field. In the beginning, when domains were first conceived, no pictures of them were available – they were discovered by theory! The development of modern understanding of domains, starting from this first theoretical postulate, is outlined in the following section.

1.2 History of the Domain Concept

1.2.1 The Domain Idea

At the beginning of the nineteenth century scientists began to realize that magnetic matter consists of *elementary magnets* in a similar sense as matter in general consists of atoms and molecules. *Ampère*'s hypothesis of elementary molecular currents (see [1]) is the best known example of such a theory. The concept of elementary magnets explains two well-known experimental facts: the impossibility of isolating magnetic south and north poles, and the phenomenon of magnetic saturation in which all elementary magnets are oriented in the same direction. In spite of the validity of this hypothesis, no progress in the understanding of magnetic behaviour was achieved until 1905 when *Langevin* [2] developed a theory of paramagnetism by using the methods of statistical thermodynamics. He showed that independent molecular magnets at room temperature lead to weak magnetic phenomena only, and he concluded that *strong* magnetism must be due to some interaction among the elementary magnets. Only two years later *Weiss* [3] elaborated this idea, following *van der Waals*' treatment [4] of the condensation of gases (which is caused by an attractive interaction among the molecules of the gas). In analogy to the "internal pressure" of van der Waals' theory, Weiss introduced a *molecular field* to model the average effect of the magnetic interaction in a tractable way. Weiss' famous theory succeeded in deriving the general shape of the temperature dependence of magnetic saturation. Adjusting the strength of the interaction so that the experimentally observed Curie temperature is reproduced, Weiss formally obtained a very large "molecular field". Only much later *Heisenberg* [5] identified the nature of this field in the quantum-mechanical *exchange* effect.

The Weiss theory also predicted that the state of magnetic *saturation* is the thermodynamic equilibrium state at all temperatures sufficiently below the Curie point. This is true because the value of the molecular field is much larger than the internal or external magnetic fields occurring in practice. External magnetic fields have almost no influence on the value of the saturation magnetization in the Weiss theory. However, since the Weiss molecular field always follows the direction of the average magnetization, the magnetization vector is fixed only in its *magnitude*, while its *direction* remains arbitrary. This feature of the Weiss theory explains the fact that a piece of iron can appear non-magnetic at room temperature, far below the Curie point: the magnetization vectors in different parts of the sample only have to cancel each other. Of course, there are infinitely many possibilities for such a macroscopically non-magnetic state. In his original work Weiss just mentioned the possibility that part of a crystal is magnetized in one direction, and part in the opposite direction. He did not introduce a name for the magnetic substructure in this article. The now almost universally adopted term *domain* structure for the subdivision into uniformly magnetized regions inside a crystal was introduced later [6 (p. 162f.)]. It still reflects the initial uncertainty about its nature, meaning something that is known only vaguely [7 (p. 120)].

1.2.2 Towards an Understanding of Domains

It was still a long way from the domain idea to a theory of magnetic hysteresis and the very high permeabilities found in ferromagnets (a piece of soft magnetic iron can have a one million times higher permeability than vacuum!). Some hints from experiment were necessary before the theory could proceed. A first confirmation of the domain concept was found by *Barkhausen* [8]. He discovered that the magnetization process is often discontinuous, giving rise to a characteristic noise when made audible by an amplifier. Originally, Barkhausen jumps had been interpreted as domain *switching*. Although this interpretation is not considered valid today, the further pursuit of the Barkhausen phenomenon led to a decisive discovery. Experimentalists had tried to find specimens in which, instead of the complicated Barkhausen noise, some simpler process took place during magnetization reversal. Certain stressed wires showed in fact only one giant jump leading immediately from one saturated state to the opposite one [9, 10]. The analysis of the dynamics of this process led *Langmuir* (see [11]) to the conclusion that such jumps could occur only by a spatially inhomogeneous process, namely by the propagation of a *boundary* between domains of opposite magnetization. This hypothesis was soon confirmed by the famous experiments of *Sixtus* and *Tonks* [11] who followed the propagation of the domain boundary in a stressed wire by electronic means. It inspired *Bloch* [12] to analyse theoretically the transition between domains, finding that the walls must have a width of several hundred lattice constants due to Heisenberg's exchange interaction that opposes an abrupt transition. Wide domain walls effectively average over local inhomogeneities, such as point defects; this result explains why domain walls can be so easily moved as shown in the Sixtus-Tonks experiment.

In a parallel development, the effects of anisotropies, magnetostriction and internal stresses on the magnetic microstructure were investigated by many prominent authors, such as *Akulov* [13], *Becker* [14] and *Honda* [15]. The textbook of *Becker* and *Döring* [16] summarizes this work. The most important results (as far as they apply to our subject) may be stated as follows:

- Crystal *anisotropy* (the preference of the magnetization vector to align with so-called easy crystal axes) and *magnetostriction* (the spontaneous deformation of the crystal related to the magnetization direction) are independent material properties that cannot be derived from the Weiss theory of ferromagnetism or from Heisenberg's exchange interaction. They are connected to spin-orbit coupling effects and can be determined by experiments on single crystals. Because of fundamental symmetry relations they do not distinguish between a magnetization direction and its opposite.

- As a consequence of anisotropy, the magnetic microstructure consists of domains that follow the easy axes of the anisotropy functional. A magnetization process can be either based on the displacement of domain walls, or on the rotation of the magnetization vectors inside the domains. The attempt to explain observed hysteresis curves of soft magnetic materials and in particular their high permeability by *rotation* processes alone fails, since the measured anisotropies are generally too large.

- If the cubic anisotropy of iron favours the ⟨100⟩ directions, inhomogeneous stresses inside a piece of iron will induce domains magnetized along more than one of the easy axes, thus generating *90° walls* (a 90° wall is a wall in which the magnetization rotates by 90° from domain to domain).

It is difficult, however, to reconcile the observed permeabilities with 90° wall motions alone, since their positions are bound to stress inhomogeneities. Also compatible with the anisotropy are *180° walls*. If they are present, their position is not determined by stresses, and they will therefore be mobile, explaining the high permeabilities. From anisotropy and stress considerations alone, however, there is no reason why these 180° walls should exist, except occasionally for continuity reasons. We postpone a graphic representation of these arguments to Fig. 1.3 for reasons that will become immediately apparent.

One element of a complete theory of magnetization processes was missing at this point. Some authors (*Frenkel* and *Dorfman* [17], *Bloch* [12], *Heisenberg* [18]) had already perceived the missing link: the magnetic *dipole* interaction, also known as magnetostatic energy or stray field energy. For a long time this interaction had been more or less forgotten after Weiss had proved it to be much too weak to explain ferromagnetism as such. The dipolar interaction was used in a crude way to derive the macroscopic magnetization curve of a finite body from the hysteresis of an infinite or ring-shaped body, using the demagnetizing factor and the shearing transformation. It was well understood that a uniformly magnetized single crystal carries an excess energy – the demagnetizing energy – which can be large compared to the usual anisotropy energies. Heisenberg [18], for example, still assumed that to minimize this demagnetizing energy, tiny threadlike domains had to develop. Again the experiment had to give a hint to the solution. In 1931 *v. Hámos* and *Thiessen* [19] and independently *Bitter* [20, 21] showed the first pictures of magnetic micropatterns obtained with the help of an improved powder method. Even if none of the observed structures could be understood in detail at the time, the pictures demonstrated three important features: domains were static, they could be rather wide, and they frequently had a periodic and regular appearance.

Probably stimulated by such observations and by their first theoretical analysis by Bloch (see the footnote at [12 (p. 321)]), *Landau* and *Lifshitz* [22] presented the solution in 1935: domains are formed to minimize the total energy, an important part of which is the stray field energy. And the stray field energy can be avoided by *flux-closure* type domains as shown in Fig. 1.2. (The basic idea of such closed flux patterns had been put forward already by *Zwicky* [23]). If the magnetization follows a closed flux path everywhere, the stray field energy is zero and therefore even smaller than in the hypothetical thread domains. Landau and Lifshitz proved for the first time that a domain model like that of Fig. 1.2 has a lower energy than the uniformly magnetized state.

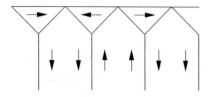

Fig. 1.2 The first realistic model of magnetic domains by *Landau* and *Lifshitz* [22]

The stray-field-free model structure of Fig. 1.3 is thought to represent a part of an extended domain pattern. It contains both 90° and 180° walls and demonstrates that both wall systems can be displaced without violating the constraint of flux-closure. Note, however, that the motion of the 180° wall system (b) is compatible with the grown-in stress pattern (which is thought to favour different axes on both sides of the diagonal), in contrast to a displacement of the 90° wall system (c).

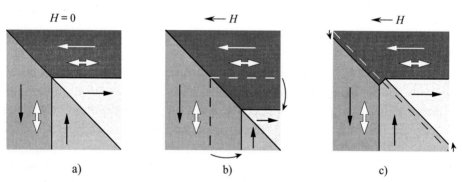

Fig. 1.3 The different behaviour of 90° walls and 180° walls in an external field *H* for a crystal subject to an inhomogeneous stress pattern (double arrows). A 180° wall motion (b) would be compatible with the stress pattern, but the independently possible 90° wall motion (c) would generate a conflict with the intrinsic stresses

Landau and Lifshitz also gave the answers to a number of questions that were still debated at that time:

- The exchange interaction tends to align the neighbouring dipoles and causes them to act together. In most cases the correspondence principle therefore permits treating the average magnetization as a classical *vector* field rather than as a quantum-mechanical spinor field.

- Thermal agitation plays a role only in small particles or at temperatures close to the Curie point. Under normal circumstances the equilibrium magnetic microstructure must be considered *athermal.*
- Macroscopically, the effect of exchange interaction can be expressed to a sufficient accuracy by a *stiffness* term, a quadratic form of the first spatial derivatives of the magnetization vector. This stiffness energy favours uniform magnetization, particularly on a microscopic scale.
- Even inside domain walls the Weiss postulate of a constant value of magnetization should be valid (in contrast to Bloch's treatment who still assumed a ferromagnet to become paramagnetic in the middle of the walls). The magnetization *rotates* in passing through the wall.
- Domain structures are a consequence of the finite dimensions of magnetic bodies. The domain size increases with the specimen size. A uniform infinite or toroidal body may have *no* domain structure in equilibrium.

1.2.3 Refinements

The model of Landau and Lifshitz proved to be too simple to explain actual observations. Starting from their basic ideas, refinements and extensions were contributed in succeeding years. In the early articles the fundamental difference between the domains in uniaxial crystals, like cobalt, and in cubic crystals, like iron, was poorly understood.

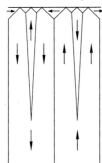

For example, in an article presented by *Lifshitz* in 1944 [24] a theory of domain *branching* was introduced (see sketch), a feature that was well known from experiments on uniaxial crystals [25]. Lifshitz' article was meant to apply to iron, a cubic material (it contained among other contributions also the first correct calculation of the 180° wall in iron). But Lifshitz failed to see the additional degrees of freedom of the domain structures of cubic crystals with their multiple easy directions. *Néel* [26, 27] in his independent work made full use of these possibilities, predicting a number of remarkable domain structures. A famous example among these, the Néel *spikes*, can be used to estimate the coercivity connected with large inclusions in iron crystals. When they were later observed experimentally [28], this was considered a striking success of domain theory.

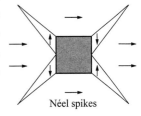

Néel spikes

Landau and Lifshitz as well as Néel had studied large crystals with weak anisotropies in which the assumption of completely flux-closed domain structures ("pole avoidance") is well justified (Fig. 1.4a, b). Thus the explicit calculation of the stray field energy was not necessary. (Lifshitz [24] did, however, calculate the energy of the internal fields in his branched structures). In small specimens or in uniaxial crystals with large anisotropy open structures as in Fig. 1.4c, d are expected; they were first calculated by *Kittel* [29, 30].

Meanwhile, experimental methods had improved considerably. Powder had been replaced by finer *colloids* [31]. Arbitrary samples were replaced by well oriented crystals, and after preparing an undamaged crystalline surface it became possible to obtain meaningful pictures. In the famous article by *Williams, Bozorth* and *Shockley* of 1949 [32] the identity between the domains of domain theory and the observed magnetic microstructure was convincingly demonstrated. In the same year Kittel [30] reviewed domain theory and experiments, and this review became the generally accepted reference for domain research. What followed can be called *application* of the established theory, stimulated by improved methods of domain observation, by the preparation of materials with surprising new magnetic structures, and by applications based on the properties of domains. This process is continuing and will be discussed in detail in this book.

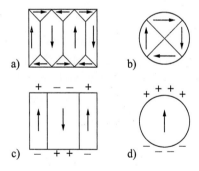

Fig. 1.4 The more or less flux-closed patterns of low-anisotropy cubic particles (a) and (b), compared to the open domain structures for high-anisotropy uniaxial particles (c) and (d)

1.3 Micromagnetics and Domain Theory

Micromagnetism is the continuum theory of magnetic moments, underlying the description of magnetic microstructure. The theory of Landau and Lifshitz is based on a variational principle: it searches for magnetization distributions with the smallest total energy. This variational principle leads to a set of differential equations, the *micromagnetic equations*. They were given in [22] for one dimension. Stimulated again by experimental work [25] and its analysis, *W.F. Brown* [33, 34] extended the equations to three dimensions, including fully the stray field effects (see [35, 36]).

The micromagnetic equations are complicated non-linear and non-local equations; they are therefore difficult to solve analytically, except in cases in which a linearization is possible. However, a number of problems in domain research needs micromagnetic methods for their adequate treatment:

- The investigation of the magnetic behaviour of small particles that are too small to accommodate a regular domain structure, but too large to be described as uniformly magnetized (Sect. 3.3.3A).
- The calculation of the internal structure of domain walls (Sect. 3.6).
- The investigation of finely divided surface magnetization patterns of samples, where the magnetization is subject to conflicting influences (Sect. 3.3.4).
- The description of rapid dynamic magnetization reactions (Sect. 3.6.6).
- The calculation of the magnetic stability limits (switching; Sect. 3.5).

To treat such problems, work on numerical solutions of the micromagnetic equations is increasingly pursued. It appears utopic, however, to apply micromagnetic methods to large-scale domain structures. The gap between the size of samples for which three-dimensional finite element calculations are possible (at most up to perhaps a micron cubed), and the scale of well defined domain patterns (often reaching millimetres and centimetres) is simply too large.

For most problems we have to rely on *domain theory*, a theory that combines discrete, uniformly magnetized domains with the results of micromagnetics for the connecting elements, the domain walls and their substructures. We will see that in many cases, reliable guidance in domain analysis can be obtained from domain theory.

Repeating some arguments, we show in Fig. 1.5 how domain theory is embedded into more general descriptions of magnetic materials. Five levels [37] are distinguished, which are connected with characteristic scales.

Level three of the scheme representing the centre of interest of this book is analogous to classical metallography. The continuum theory of micromagnetics appearing in level two forms the basis of domain analysis. Both levels together represent a *mesoscopic* approach in the description of a magnet. Level four corresponds to the discussion of a material in terms of phase diagrams, and is an important limiting aspect of domain analysis. It ignores the detailed arrangement of domains and focuses on their volume distribution. Level five, the phenomenology of technical magnetization curves, is touched in so far as connections between hysteresis and domain phenomena can be established.

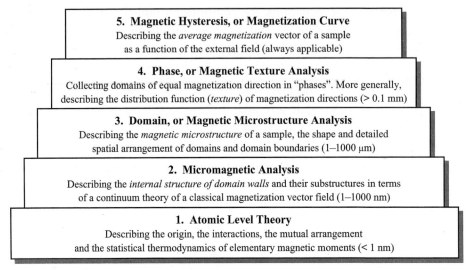

Fig. 1.5 The hierarchy of descriptive levels of magnetically ordered materials. The values in parenthesis indicate the sample dimensions for which the different concepts are applicable

The atomic foundation (level one), the question how to explain the observed magnitudes of magnetic moments, crystal anisotropies, or magneto-elastic interactions, is completely excluded. Level one also deals with the spin structure of a magnetically ordered material, the arrangement of spins on the crystal lattice sites. From the mesoscopic viewpoint of micromagnetics and domain theory it does not matter whether a material is ferromagnetic or *ferri*magnetic. In this book we consider ferrimagnets generally to be included in discussing ferromagnets for short. An interesting hybrid between atomic description and micromagnetics averages over the spins only along one or two dimensions, and retains the atomic description otherwise (see e.g. [38]).

It should be mentioned that the term "micromagnetic" is getting fashionable also outside the range of its original description. Almost every magnetic investigation that touches microscopical aspects is found to be called "micromagnetic" (some recent examples are [39–41]). We stick (in agreement with still the vast majority of authors) with the classical definition of *W.F. Brown* [34], restricting the term micromagnetics to the *continuum theory of magnetically ordered materials,* to the second level in Fig. 1.5.

2. Domain Observation Techniques

In this chapter methods to observe the magnetic microstructure are critically examined. For each method the principle, the basic experimental procedure, an outline of the theory, special techniques, and the potential are reviewed, followed by a discussion of merits and shortcomings of the method. A final section is devoted to a comparison of the various techniques in tabular form.

2.1 Introduction

A magnetic domain structure is specified if the vector field of magnetic polarization $J(r)$ is known, where r is the position vector in the sample. The magnetic polarization is connected with the magnetic flux density B and the magnetic field H by the relation:

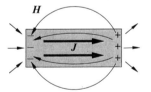

$$B = \mu_0 H + J \ . \tag{2.1}$$

Maxwell's equation div $B = 0$ yields:

$$\mu_0 \mathrm{div}\, H = -\mathrm{div}\, J \tag{2.2}$$

which means that any divergence of $J(r)$ creates a magnetic field. If no external field is present, then H is just the stray field mentioned in Sect. 1.2.2. Some methods of domain observation are sensitive to the stray field. The classical Bitter method and modern magnetic force microscopy belong to this class.

Other techniques are sensitive directly to the direction of magnetic polarization, as, for example, the magneto-optical methods, and electron polarization methods. Transmission electron microscopy is in most cases sensitive to the total flux density B which causes the Lorentz force deflecting the electrons.

Finally, there are X-ray and neutron diffraction methods that react to the weak lattice distortions connected with magnetism. The different domain observation techniques therefore yield different information, and they also have different limitations as will be discussed in detail.

Previous reviews and overviews on domain observation techniques can be found in references [42−51]. None of the available methods achieves the stated goal, namely to determine $J(r)$ throughout a sample. The search for improved experimental methods is actively being pursued, and considerable advances are continuously being made.

2.2 Bitter Patterns

2.2.1 General Features

The stray fields above a domain pattern can be decorated by a fine magnetic powder as convincingly demonstrated by *Bitter* [21]. Usually colloidal magnetite particles are used for this purpose. The historical significance of the powder method has been mentioned in the introduction. Even today, it remains an important method on account of its simplicity and sensitivity.

Progress in the art of obtaining good Bitter patterns was closely connected with the ability to prepare fine, sensitive and stable magnetic colloids. (In *Andrä*'s review [47] examples of this development are shown). Magnetic colloids were first prepared in connection with domain imaging [31], but for this purpose only tiny quantities are needed. This is perhaps why a commercial Bitter solution never appeared on the market. Every researcher had to prepare his own liquid, with varying success. Over the years, however, magnetic colloids have found other applications. They can be used as magnetic fluids for many purposes, from vacuum seals to damping devices. Other applications are connected with the technique of magnetic separation. Some of the commercial products developed for such applications also proved to be excellent Bitter colloids, such as water-based *Ferrofluids*® [52] and the low-cost magnetic fluid *Lignosite FML*® [53]. Good preparations consist of roughly isometric magnetite particles of about 10 nm diameter, covered with various surfactants.

Examples of Bitter patterns obtained with these colloids are shown in Fig. 2.1. One of them (Fig. 2.1a) shows domains in an implanted layer on a garnet epitaxial film that have never been made visible with other means. The two other examples show domain patterns on a nickel-iron alloy. These high permeability materials produce only weak stray fields, and the fact that usually no domain patterns can be observed on them with the powder technique has puzzled researchers for a long time. The alloy composition and crystal orientation in Fig. 2.1b, c are chosen so that the details of the domain pattern can be made visible, provided that a good colloid and the right auxiliary field (as explained below) are used. Often such efforts are not successful.

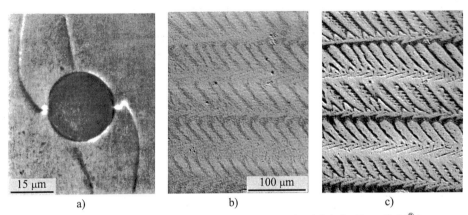

Fig. 2.1 (a) Domains in an ion-implanted garnet layer made visible by Ferrofluid® (courtesy *D.B. Dove*, IBM Yorktown Heights [54]). The circular shape is unimplanted. (b, c) Stress-induced domains on a $Ni_{55}Fe_{45}$ crystal revealed by Lignosite FML®. A perpendicular field of 1.5 kA/m was applied in (c) to improve the contrast of the same pattern. The NiFe crystal has $\langle 100 \rangle$ easy directions and a near (100) surface

From experimental experience a good Bitter colloid is characterized by the following features:

- The fine colloidal particles permit a resolution up to (and beyond) the limit of optical microscopy. (The coarse emulsions used in non-destructive testing of steel are not suitable for high resolution domain observation purposes).
- Sensitivity to weak fields of a few hundred A/m, as shown in the examples of Fig. 2.1. (Only rather weak fields appear on top of the domains and walls of soft magnetic materials).
- Reversibility of the image formation. Rearranging the domain pattern in an applied field changes the observed image. (This is the only reliable way to separate domain images from structurally induced colloid concentrations).

In addition, long-term stability and reproducibility are needed. All these requirements are fulfilled by the mentioned products. In particular, stability is very good as long as drying of the colloid is prevented. But the basic mechanism of Bitter pattern formation is poorly understood for a long time, and only in recent years a better understanding has begun to emerge.

2.2.2 Contrast Theory

The fundamental theory of magnetic particle accumulation in a field gradient has been presented by *Elmore* [55] and by *Kittel* [56] decades ago. They used a thermodynamic argument, assuming that particles accumulate in a field gradient under the simultaneous action of magnetic forces and thermal agitation. We follow this assumption since we are mainly interested in reversible particle accumulation. (Alternatively, the particles may be considered as viscously moving in the gradient until they get caught [57]. This point of view has been studied especially in connection with magnetic separation in high gradients [58, 59], but it is more complicated and may not be adequate for the low gradient situations encountered in domain research).

Two cases have to be distinguished: in the first case, the colloid particles are small enough to behave as single-domain particles with a fixed magnitude of the magnetic moment $q_c = JV$, where J is the magnetization and V is the volume of the particle. The statistical treatment of the motion and rotation of these particles in the presence of a magnetic field is related to Langevin's theory of paramagnetism. The result [55] is a colloid particle concentration p_c as a function of the magnitude H of the field $\mathbf{H}$ at a given temperature T:

$$p_c(H) = p_c(0) \sinh\left(q_c H / kT\right) / \left(q_c H / kT\right) \tag{2.3}$$

where k is Boltzmann's constant.

In the second case, the particles are larger than the single-domain limit so that they carry a magnetic moment only in a magnetic field (in domain imaging only particles with low crystal anisotropy are used). Now the result of thermodynamics is:

$$p_c(H) = p_c(0) \exp\left(q_c H / 2kT\right) \quad \text{with} \quad q_c = \chi \mu_0 H V \tag{2.4}$$

where χ is the relative susceptibility of the colloid particle and V is its volume. Both functions are plotted in Fig. 2.2.

The product $q_c H$ must exceed a value of some kT in both cases to generate an acceptable contrast. In the following section we will see that with simple assumptions it is not easy to meet this condition in the weak fields connected with many domain patterns.

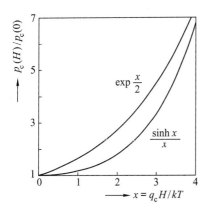

Fig. 2.2 The two functions (2.3) and (2.4) describing the expected accumulation of a colloid in a magnetic field. From the diagram the value of x needed for a required difference in the colloid concentration can be derived. Note that the quantity q_c in the definition of x has different meanings in the cases (2.3) and (2.4). Plotted explicitly as functions of H, both functions $p_c(H)$ would rise with H^2 for small fields

2.2.3 The Importance of Agglomeration Phenomena in Colloids

The problem is that — with simple assumptions about the properties of a magnetic colloid — pictures like Fig. 2.1 cannot be explained. For small single-domain particles the critical field according to (2.3) is too high. Typical isometric magnetite colloid particles with a saturation magnetization of 0.6 T and a diameter of 10 nm need a field of about 10 kA/m to obtain a particle concentration enhancement by a factor of 1.5. To obtain a sensitivity that is closer to the observed one (a few hundred A/m), larger particles would be required. But such large particles are (i) difficult to stabilize as a colloid, (ii) not single-domain particles any more (the critical single-domain diameter of magnetite is about 70 nm [60]).

Larger, spherical multi-domain particles have ideally an effective susceptibility χ of 3 because of their demagnetizing factor. Inserting this value into the predicted density (2.4), and asking for a particle density enhancement by a factor of 1.5 in a field of 500 A/m requires particles of more than 150 nm in diameter. Colloidal particles rarely exceed some 10 nm. Suspensions with big particles are used in non-destructive testing of materials for hidden flaws, but rarely in domain studies.

Some authors did see these difficulties. *Bergmann* [61] tried to resolve them by assuming unrealistically high stray fields above Bloch walls. *Garrood* [62] postulated an unrealistic remanence in the large particles to evade the consequences of (2.4). In reality, a special and unexpected phenomenon seems to be responsible for the sensitivity of good Bitter colloids. These colloids are mostly dispersed in an aqueous phase and treated so that they are stable in the absence of an external field, either by charging them electrostatically, or by increasing their effective diameter by a surfactant. In a weak magnetic field, however, the particles of some water-based colloids form needle shaped *agglomerates* constituting a "condensed phase" of particles. Removing the field, the agglomerates decay again. Both the formation and the decay of the agglomerates are time dependent. Small agglomerates form rapidly, larger ones may take seconds or minutes to form.

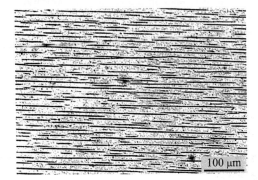

Fig. 2.3 The agglomerates forming in a magnetic field, demonstrated on Lignosite FML® in a field of 560 A/m

This kind of reversible agglomeration was first observed microscopically by *Hayes* ([63]; a similar observation can be seen in Fig. 2.3). Later workers [64–68] contributed many details to the properties of the agglomerates. Remarkable is the finding in [68] that the *largest* particles out of the particle distribution in a colloid play a decisive role in the agglomeration process. Still, the exact origin of this interesting property remains uncertain. Probably, the magnetic field induces a phase transition from a "gaseous" dispersed phase of the particles to a "liquid", condensed phase under the combined influences of magnetic and van der Waals forces. Mean-field theories of such a transition have been elaborated in [69, 70]. Further developments can be found in [71–75] and in the respective cited references.

Accepting the reversible agglomeration as a given experimental fact, Bitter pattern formation in weak stray fields is readily understood. The volume of the agglomerates is much larger than that of the individual particles, and their susceptibility is rather large because of their dense packing and their elongated shape. Take as an example an agglomerate of magnetite particles of 1 μm in length and 100 nm in diameter, and assume an effective susceptibility of 30 [67]; then (2.4) indicates that a field of only about 500 A/m is needed for a reasonable particle accumulation, a value that is more realistic than the previous estimates. Since there seems to be no limit to the size of the agglomerates, it is hard to define a limit to the sensitivity of Bitter colloids.

The new picture of Bitter pattern formation explains qualitatively observations, some of which have been known for a long time:

- The sensitivity of Bitter colloids can be enhanced by an external field perpendicular to the surface of the specimen [55] as demonstrated in Fig. 2.1 (such a field usually does not disturb the domain pattern in soft magnetic samples). Traditionally, this fact was explained with the nonlinearity of the sensitivity functions (2.3) and (2.4). Another and probably stronger contribution to this effect stems from the fact that the superimposed field induces the formation of needle-shaped agglomerates, which may then be drawn into the regions of the strongest field.

- Water-based colloids were found to be better suited for domain observation than colloids dispersed in other liquids. The reason may be that mostly water-based fluids show the reversible agglomeration phenomena [76].

- Magnetic fluids for engineering applications are usually not expected to agglomerate in a magnetic field. The gradual improvement of these materials therefore tends to suppress the agglomeration effect, thus increasing the stability, but reducing the applicability for domain observation purposes. Older water-based colloids of Ferrofluidics Corp. are better for our purpose than later developed, more stable ones. On the other hand, in magnetic separation, a tendency towards agglomeration may even be an advantage, and the properties of colloids tailored for this application may remain more favourable for domain detection in the long run.

- The tendency to agglomerate will depend on the density of the liquid. The more concentrated the colloid, the higher the sensitivity. But solutions that are too concentrated get black, and the domains become invisible through the black liquid. The optimum density has to be carefully determined empirically. Two articles demonstrate ways to circumvent this limitation and thus to increase the sensitivity: *Pfützner* [77] first applied concentrated colloid and then added *ox gall* as a detergent, which causes the colloid to agglomerate and to settle, decorating the magnetic pattern. The remaining liquid was rinsed off or removed with a piece of blotting paper, leaving a dry pattern behind. The method is sensitive enough to permit domain observations on coated transformer steel without auxiliary fields. *Rauch* et al. [53] added a strippable white coating to the colloid and allowed it to dry on the sample. After stripping, the Bitter pattern becomes visible on the side that was in contact with the sample. Both procedures are restricted to the observation of static images, but they are sensitive and can be used in routine material characterization.

- The process of the formation and the decay of large agglomerates involves the flowing of a rather viscous phase. It may therefore be slow and sometimes take minutes till equilibrium is reached [67, 76]. The weaker the fields to be detected, the larger the agglomerates needed, and the longer it takes for the colloid to react.

- The formation of needle-shaped agglomerates leads to optical anisotropy (dichroism and birefringence; [78−83]). This effect itself can be used for domain imaging (Fig. 2.4) in a polarization microscope.

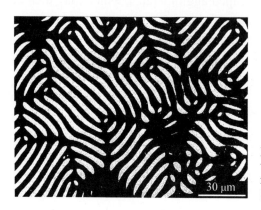

Fig. 2.4 Domains in a garnet layer made visible by the birefringence effect of magnetic colloids (Courtesy *I.B. Puchalska* [80])

The optical anisotropy of magnetic colloids in a magnetic field can also serve as a measure for the sensitivity of a colloid. Figure 2.5 shows the intensity of light transmitted through layers of different magnetic colloids between crossed polarizers as a function of the strength of a magnetic field oriented at 45° to the polarizers. In sensitive colloids, anisotropy effects are observed even in fields of a few A/cm. It was qualitatively confirmed that the colloids with the strongest optical anisotropy effects also had the highest sensitivity in domain observation.

Strong optical anisotropy in weak fields and sensitivity for domain detection are based on the same effect: the formation of needle-shaped agglomerates. The potential of the direct use of optical anisotropy for domain imaging is not completely explored. The dichroism of the particle agglomerates indicates the direction of the acting stray field: light that is polarized parallel to the field and to the submicroscopic agglomerates is absorbed more strongly than light polarized perpendicular to this direction. The contrast of domain walls is thus polarization dependent. One might consider using optical anisotropy to determine the magnitude of fields, but the nonlinearity of the effects of agglomeration and of particle accumulation will probably thwart such efforts.

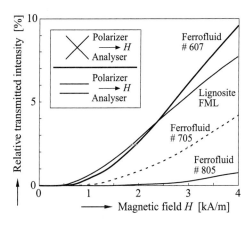

Fig. 2.5 Optical anisotropy of magnetic colloids, defined as the ratio indicated in the inset, as a function of the magnetic field. Ferrofluid #607 is a less stable (old type) magnetic colloid, #705 and #805 are more stable, but less sensitive variants. (Together with H. Mayer, Erlangen)

2.2.4 Visible and Invisible Features

Bitter patterns are traditionally thought to decorate domain walls as a result of their stray field. This point of view is debatable for several reasons [84]:

- Strong domain boundary contrasts are usually caused by a misorientation of the crystal that is connected with surface charges on the domains. Surface

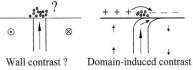

charges are the magnetic "poles" generated by vertical magnetization components at the surface. Such charges generate a maximum in the horizontal stray field component over the wall, and this component becomes visible according to the contrast laws (2.3, 2.4) which do not depend on the field direction. Bloch walls on ideally oriented surfaces are virtually invisible in low-anisotropy material, because the walls modify their near-surface structure in a largely stray-field-free way, as shown in Fig. 3.82.

- So-called V-lines, two subsurface domain walls that meet at the surface, are mostly invisible in the Bitter technique. Figure 2.6 demonstrates that these domain boundaries, which are magneto-optically apparent (b), are absent or only diffusely visible in the Bitter image (a).

- In some cases three subsurface domain walls meet at the surface. Such structures, which might be called "Ψ-lines", tend to be quite pronounced in Bitter images, but they cannot always be seen magneto-optically. The Ψ-line

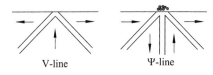

structure can dive below the surface without violating flux continuity conditions. It then becomes invisible in the Kerr image, but remains visible in

Bitter contrast. At the apparent end of the Ψ-line in transformer steel often a characteristic colloid accumulation is observed (the "tadpole head"; Fig. 2.6c, e). This structure is not visible in the Kerr image (d).

- In high-permeability materials such as Permalloy (80% Ni-Fe) or cobalt-rich metallic glasses, no magnetic microstructure can be made visible with the Bitter technique, although domains are certainly present as shown by magneto-optical investigations.

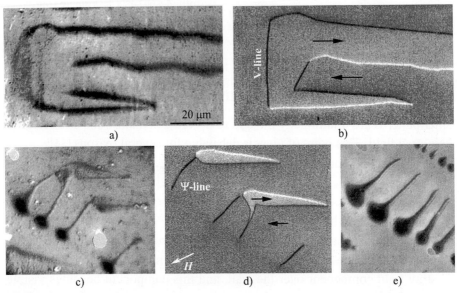

Fig. 2.6 A comparison of Bitter (a) and Kerr (b) images of similar domain patterns on a 3% Si-Fe crystal that is slightly misoriented with respect to the (110) crystal plane. The side walls of the "lancet" domains are displayed in the Bitter image, but the V-lines at the blunt ends are invisible or almost so, depending on the experimental conditions (dilution of the colloid, waiting time). In contrast, the famous "tadpole" pattern known from early Bitter pattern investigations [85] and prominently visible in (c, e) reduces to simple Ψ-lines in Kerr contrast (d) [the pictures (c) and (d) are taken from equivalent though not identical patterns]. The subsurface tadpole "head" pattern develops out of the lancet pattern in an oblique applied field that has to be carefully adjusted to get "fat" tadpoles like in (e)

- In thin films, contrast depends on the specific wall type. Asymmetric Bloch walls (Sect. 3.6.4D) which are almost stray-field-free are barely visible, whereas Néel walls, which are connected with unavoidable stray fields, show up conspicuously.

In view of all these uncertainties it must be considered dangerous to rely on the Bitter technique alone in the investigation of unknown magnetic microstructures. Within a known context, the Bitter technique may offer important information, some of which is not available with other methods.

2.2.5 Special Methods

The contrast of Bitter patterns can be enhanced with various kinds of microscopic techniques. *DeBlois* [86], for example, used dark field conditions to obtain his beautiful pictures of domains on crystal platelets. *Khaiyer* and *O'Dell* [87] introduced Nomarski's interference contrast microscopy [88] for the investigation of Permalloy thin films. Since this method enhances any surface profile it can be used with best results on samples with optically flat surfaces (as e.g. thin films). *Hartmann* and *Mende* [89] achieved high resolution with this method for the "living", liquid Bitter pattern using a *water* immersion objective lens. Another way to achieve high resolution is the use of an oil immersion objective that is corrected for a cover glass under which the Bitter colloid is confined[1]. With these refinements a resolution in the observation of mobile colloids in the optical microscope of 0.2 μm can be achieved.

A technique devised by *Wysłocki* [90] *reduces* the resolution by applying a lacquer coating as spacer between surface and colloid. This method is useful in cases in which a system of complex surface domains (Sect. 3.7.5) virtually conceals the basic domain pattern. With the spacer layer the surface domains become invisible, leaving an enhanced image of the basic domains [91].

The application of the standard Bitter methods is restricted to ambient temperatures and to optical resolution. Resolution can be increased by allowing the colloid to dry on the surface, since the particle layer will be thinner than the original colloidal solution; this "dry method" [92–95] also offers advantages in conventional optical microscopy on irregular surfaces. If some agent is added to the colloid so that it forms a strippable film after drying, resolution can be further enhanced with the help of electron microscopy (Fig. 2.7). The direct observation of the dried deposit in the scanning electron microscope [96–98] offers the double advantages of high resolution and a large depth of focus, and is therefore well suited for the investigation of small particles and rough surfaces. Also tunnelling microscopy can be used to observe Bitter patterns at high resolution [99].

If the drying of the colloid is avoided by a proper sealing of the observation chamber as in [54], live slow magnetization processes can be directly observed by the Bitter technique even for months. The amplitude of fast sinusoidal wall

---

[1] *Mark Re*, private communication

oscillations can be made visible as a double wall image marking the periodic end positions of the moving wall [100].

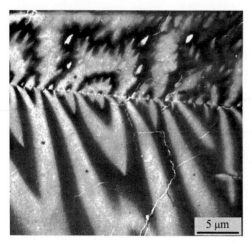

Fig. 2.7 Domains on a cobalt crystal at 4.2 K decorated with iron particles, made visible by the electron microscope replica technique [101]. The plastic appearance of the domains is caused by a twin plate that acts like a mirror for the domains. These domains are a striking manifestation of the domain branching phenomenon discussed in Sect. 3.7.5. See also Fig. 5.6c

5 μm

Extensions to higher or lower temperatures require carriers other than water. Various emulsions have been proposed [102–105]. The most versatile method uses a magnetic "smoke" which is produced either by burning a suitable substance [106], or by evaporating a magnetic material in a low pressure gas in which small particles form by condensation [107–109]. The magnetic particles are then allowed to settle on the sample. This method can be used down to temperatures of liquid helium and it can also be combined with electron microscopy. Pictures of excellent resolution can be obtained as demonstrated in Fig. 2.7. Using paramagnetic oxygen particles offers the advantage that the decoration can be observed in situ and is easily removed by heating [110]. Suitable magnetic particles can also be produced conveniently in a two-chamber sputtering process [111]. The contrast mechanism in the smoke methods is not quite the same as the one of Bitter colloids, because the smoke particles settle in a single trajectory on the surface and do not search equilibrium positions. Both methods are compared empirically in [112].

A particularly simple variant was found in the laboratories of Nippon Steel Corporation[2]. Modern printing toner is often magnetic for reasons connected with the transport mechanism in a printer or copier. This powder can be emulgated in water with the help of a household detergent. The emulsion is not a

[2] *Satoshi Arai*, private communication

colloid, but for low-resolution applications such as for domain inspection on coated electrical steels (Fig. 2.8) it offers excellent contrast and high sensitivity.

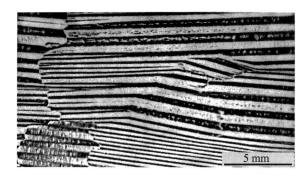

Fig. 2.8 Domains on a coated transformer steel sheet made visible by a toner powder emulsion. (Courtesy *Satoshi Arai*, Nippon Steel)

2.2.6 Summary

The main *limitations* of the Bitter method are:
- High permeability materials have only weak stray fields. If there is no stray field, Bitter patterns cannot show the domains. In other cases only certain types of domain boundaries may be visible.
- Only slow or periodic [113] domain movements can be recorded with the Bitter method. The speed of Bitter pattern movement depends on the magnitude of the available stray field.
- In strong applied fields the stray fields connected with sharp sample edges and with imperfections usually become stronger than the stray fields of domains and walls. Domain observation then becomes difficult.
- The connection between stray fields and magnetization is indirect, non-linear and non-local, making the interpretation of Bitter patterns problematic.
- The samples become rather dirty in the Bitter method.
- Although the magnetic permeability of a magnetic colloid is small, a reaction of the domains when applying the colloid cannot generally be excluded.

On the other hand there are unique *advantages* of the Bitter method:
- No special equipment is needed.
- The magnetic microstructure of misoriented crystals consists often of weakly modulated or continuously varying surface patterns. The Bitter method produces particularly clear pictures in these situations.
- The dry colloid method is the only technique suitable for at least static domain observation on rough, three-dimensional surfaces [96].

- The resolution of the Bitter technique reaches easily 100 nm and is limited by the particle size of non-agglomerating colloids to some 10 nm.
- Bitter patterns do not always ask for extreme care in surface preparation. Even coated samples can be investigated [53, 77, 114–116]. But resolution and sensitivity are higher for well polished samples.

Related to the Bitter method is the approach of using *magnetotactic bacteria* (Sect. 2.7.4) for decorating stray fields.

2.3 Magneto-Optical Methods

Magneto-optical domain observations in reflection ([117, 118]; using the magneto-optical *Kerr* effect [119]) and in transmission ([120–122]; using the *Faraday* effect [123]) have a long history. They are based on small rotations of the polarization plane of the light and become visible in the polarizing microscope. The Faraday effect can rarely be applied, as few magnetic samples are transparent. For non-transparent samples only the Kerr effect is available, but this effect was considered for a long time rather weak and difficult. Nevertheless, nice images could be obtained for favourable and well prepared samples and with a carefully optimized microscope as shown in Fig. 2.9.

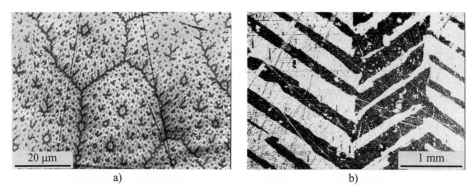

a)   b)

Fig. 2.9 Two examples of magneto-optical images from the time before the introduction of digital image processing. Branched domains on a cobalt crystal are shown in (a), taken by the polar Kerr effect in an applied field parallel to the easy axis. The overview picture (b) shows the "saw-tooth" pattern on a (110)-oriented silicon-iron crystal [124], taken with the longitudinal Kerr effect in a field perpendicular to the preferred axis of this material

For most materials the results remained disappointing, however, and only few specialists used and developed this technique. This changed with the introduction of the digital difference technique (see Sect. 2.3.7) in which the non-magnetic background image is digitally subtracted, and the magnetic contrast is enhanced by averaging and electronic manipulation. Interestingly, the same effects are also used increasingly as a magnetization measuring tool particularly for ultrathin films [125, 126], and since every technique appears to need an acronym, the old magneto-optical Kerr effect is called "MOKE". Also magneto-optical recording is based on the same effects (see Sect. 6.4.3). In view of the wide-ranging applications of magneto-optical methods we present here a rather detailed discussion.

2.3.1 Magneto-Optical Effects

Domain observation by magneto-optics is based on a weak dependence of optical constants on the direction of magnetization $m = J/J_s$ (J_s = saturation magnetization). In addition to the already mentioned Kerr and Faraday effects, the *Voigt* effect [127], which is quadratic in the components of the magnetization, can contribute to magneto-optical images. The latter effect is also known as *linear birefringence* (a birefringence for linearly polarized light) and under the name *Cotton-Mouton* effect [128]. All effects can be represented by a generalized dielectric permittivity tensor. For cubic crystals the tensor has the following form ([129–132], with the sign conventions of *Atkinson* and *Lissberger* [133]):

$$\varepsilon = \varepsilon \begin{vmatrix} 1 & -iQ_V m_3 & iQ_V m_2 \\ iQ_V m_3 & 1 & -iQ_V m_1 \\ -iQ_V m_2 & iQ_V m_1 & 1 \end{vmatrix} + \begin{vmatrix} B_1 m_1^2 & B_2 m_1 m_2 & B_2 m_1 m_3 \\ B_2 m_1 m_2 & B_1 m_2^2 & B_2 m_2 m_3 \\ B_2 m_1 m_3 & B_2 m_2 m_3 & B_1 m_3^2 \end{vmatrix} . \quad (2.5)$$

Here Q_V is the (Voigt) material constant describing the magneto-optical rotation of the plane of polarization of the light — the Faraday effect in transmission and the Kerr effect in reflection. The effect is also called *circular* magnetic birefringence, a birefringence for circularly polarized light. With this term alone the dielectric law may also be written in the form:

$$D = \varepsilon (E + i Q_V m \times E) . \quad (2.5a)$$

The constants B_1 and B_2 describe the *Voigt* effect. In isotropic or amorphous media the two constants B_1 and B_2 are equal, but in cubic crystals they are in general different. The m_i are the components of the unit vector of magnetization along the cubic axes. All constants are frequency dependent and complex in general, but the real parts of the constants Q_V, B_1 and B_2 are usually predominant.

In principle, a similar scheme applies to the tensor of magnetic permeability that contributes to optical constants as well. But it was shown [134, 135] that the "gyromagnetic" terms are about two orders of magnitude smaller than the "gyroelectric" terms at optical frequencies, so that they can be neglected. The quantities Q_V, B_1 and B_2 are not well known for the majority of materials. The constant Q_V is of the order of 0.03 in the visible range and roughly proportional to the saturation magnetization (or the sublattice magnetization) of the material as has been found for example in the Ni-Fe-system [136]. The linear magnetic birefringence was measured especially for magnetic garnets and was found there to be of the order of 10^{-4} [137]. For the purpose of domain observation the rotation effects are more important than the linear birefringence effects. The latter have been used mainly for the observation of magnetic microstructure in transparent garnets [121, 138]. Only much later, the application of this effect has been extended to regular metals [139]. The main part of the following discussion will be devoted to the Kerr effect, which is also used for most domain images in this book (especially in Chap. 5). We come back to the other effects in Sect. 2.3.12, including a demonstration of a new-found magneto-optical effect which, in addition to the classical effects (2.5), depends on certain components of the magnetization gradient tensor, not of the magnetization vector.

2.3.2 The Geometry of the Magneto-Optical Rotation Effects

The Kerr effect may be applied to any metallic or otherwise light-absorbing magnetic material with a sufficiently smooth surface, whereas the Faraday effect is restricted to transparent media. First we discuss the dependence of the two effects on the direction of magnetization, which can be rigorously derived from (2.5), Maxwell's equations, and the proper boundary conditions [129]. The symmetry of the solutions can also be described by simple arguments using the concept of a Lorentz force acting on light-agitated electrons.

Let us first assume the magnetization to be oriented perpendicular to the surface (Fig. 2.10a). Then a linearly polarized light beam will induce electrons to oscillate parallel to its plane of polarization — the plane of the electric field E of the light. Regularly reflected light is polarized in the same plane as the incident light. We call this the regular component R_N of the emerging light. At the same time, the Lorentz force induces a small component of vibrational motion perpendicular to the primary motion and to the direction of magnetization. This secondary motion, which is proportional to $v_{Lor} = -m \times E$, generates, because of Huygens' principle, secondary amplitudes: the Faraday amplitude R_F for transmission and the Kerr amplitude R_K for reflection. The superposition of R_N with R_F or R_K leads to magnetization-dependent polarization rotations.

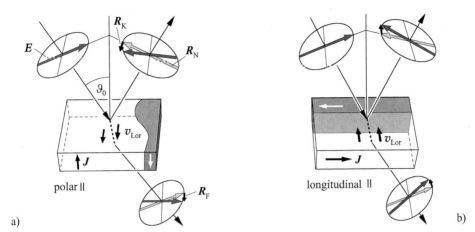

Fig. 2.10 The polar (a) and the longitudinal (b) magneto-optical Kerr and Faraday effects. R_N is the regularly reflected electric field amplitude. The magneto-optical amplitudes R_K and R_F can be conceived as generated by the Lorentz motion v_{Lor}. The polar effects would also occur for a vanishing angle of incidence ϑ_0, and they are largely independent of the direction of polarization E (chosen *parallel* to the plane of incidence). The longitudinal effects, shown here also for the parallel polarization case, increase proportional to $\sin \vartheta_0$

Figure 2.10a shows the *polar* Faraday and Kerr effect for which the magnetization points along the surface normal. This effect is strongest at perpendicular incidence ($\vartheta_0 = 0°$). As explained in the figure caption, it consists in a rotation of the plane of polarization which for $\vartheta_0 = 0°$ is by symmetry the same for all polarization directions of the incident beam.

For the *longitudinal* effect, the magnetization lies along the plane of incidence and parallel to the surface. The light beam has to be inclined relative to the surface. It yields a magneto-optical rotation both for parallel (with respect to the plane of incidence; Fig. 2.10b) and perpendicular polarization of the incoming light (Fig. 2.11a), as can be seen by inspecting the angular relations between incident light amplitude, Lorentz motion, and reflected or transmitted beam direction. The sense of rotation is opposite in the two cases. For $\vartheta_0 = 0$ the Lorentz force either vanishes (Fig. 2.10b), or points along the beam, thus not generating a detectable radiation (Fig. 2.11a).

For *transverse* orientation (Fig. 2.11b) in which the magnetization is perpendicular to the plane of incidence, no magneto-optical effect occurs in transmission [because the cross-product (2.5a) is either zero, or points along the propagation direction]. In reflection, however, light of parallel polarization will generate a Kerr amplitude, since the reflected beam has a different direction. It is proportional to $\sin \vartheta_0$ again, but the polarization direction is the same as that of the regularly reflected beam. The transverse effect therefore causes an amplitude variation of the light, which can be used for measuring purposes [140], but it will produce little contrast in a visible image. To generate a detectable rotation in the transverse case, the polarization of the incident beam is chosen *between* the parallel and the perpendicular orientation. Then the perpendicular component is not affected, while the parallel component is modulated in its amplitude, leading to a detectable polarization rotation [141].

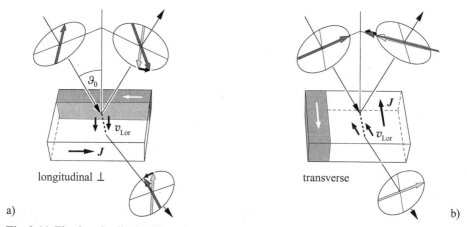

Fig. 2.11 The longitudinal effects for perpendicular polarization (a), and the transverse effect (b). The magnitude of the longitudinal effects is the same here as in the case of Fig. 2.10b, but of opposite sign. In the transverse case, only parallel polarization yields an effect, and only in reflection. In transmission there is no transverse effect for either polarization. Both effects require a non-vanishing angle of incidence

Combining these effects in a quantitative way, a general formula [142] for the Kerr effect is obtained. The light passes first through a polarizer of setting ψ_{p}, measured from the plane of incidence (Fig. 2.12). Then it is reflected from the sample, experiencing regular amplitude reflection coefficients R_{p} and R_{s} for the components parallel and perpendicular to the plane of incidence. The Kerr amplitudes R_K^{pol}, R_K^{lon} and R_K^{tra} for the polar, the longitudinal and the transverse cases are excited, depending on the magnetization components m_{pol}, m_{lon} and m_{tra}. Finally the light passes through an analyser (with the setting α_{s} measured from the axis perpendicular to the plane of incidence), leading to the total signal amplitude relative to the incident amplitude:

$$A_{\text{tot}} = -R_p \cos \psi_p \sin \alpha_s + R_s \sin \psi_p \cos \alpha_s +$$
$$R_K^{\text{pol}} \cos(\alpha_s - \psi_p) m_{\text{pol}} + R_K^{\text{lon}} \cos(\alpha_s + \psi_p) m_{\text{lon}} - R_K^{\text{tra}} \cos \psi_p \sin \alpha_s m_{\text{tra}} . \quad (2.6)$$

The regular reflection coefficients R_p and R_s can be derived from the optical constants and the angle of incidence, using Fresnel's formulae. Similar expressions also exist for the Kerr coefficients; we will present them below in (2.11–2.13), including the effects of additional dielectric interference layers.

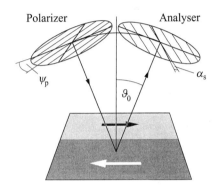

Fig. 2.12 Polarizer and analyser in a Kerr setup with definition of the respective angles

Only small values of α_s and ψ_p are needed to adjust the polar and longitudinal effects for optimum contrast. The transverse effect yields the largest modulation in the amplitude A_{tot} after the analyser if $\alpha_s \approx \psi_p \approx 45°$, provided that the two regularly reflected amplitudes R_p and R_s are about equal.

Equation (2.6) can also be used for a general magnetization direction different from the conventional polar, longitudinal and transverse orientations. In part of the literature other traditional names for the different cases are still used: the longitudinal effect is also called *meridional* effect and the transverse effect is also known as *equatorial* effect. In the next section we will study the consequences of (2.6) for domain observation in more detail.

2.3.3 Magneto-Optical Contrast in Kerr Microscopy

For two domains with opposite magnetization, the Kerr amplitudes differ only in sign. Writing (2.6) in the form $A_{\text{tot}} = A_N \pm A_K$, where A_N represents the regular part [the first line in (2.6)], and A_K is the effective Kerr amplitude [the second line in (2.6)], we may define the *Kerr rotation* as the (small) angle $\varphi_K = A_K/A_N$. We start with the analyser $\alpha_s = \varphi_K$ setting, which extinguishes exactly the light coming from one of the domains. (If this should not be possible because of an elliptical polarization, we may add a *compensator*; but let us ignore this complication for the moment). As a result, one domain appears dark, and the other domain appears more or less bright.

It is better, however, to rotate the analyser beyond the extinction point $\alpha_s = \varphi_K$, as explained in the following. If the phases of regular and magneto-optical amplitude are equal, then A_K and A_N can be taken as real numbers, and the intensity of the "dark" domain becomes relative to the incident intensity:

$$I_1 = A_N^2 \sin^2(\alpha_s - \varphi_K) + I_0 \cong (A_N \sin\alpha_s - A_K \cos\alpha_s)^2 + I_0 \qquad (2.7a)$$

where I_0 is a background intensity, and the definition $\varphi_K = A_K/A_N$ for $\varphi_K \ll 1$ is inserted again. The relative intensity from the other domain is:

$$I_2 = A_N^2 \sin^2(\alpha_s + \varphi_K) + I_0 \cong (A_N \sin\alpha_s + A_K \cos\alpha_s)^2 + I_0 \; . \qquad (2.7b)$$

The relative magneto-optical signal is the difference between the two intensities:

$$S_{mo} = I_2 - I_1 = 2\sin(2\alpha_s) A_K A_N \; . \qquad (2.8)$$

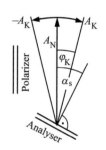

The signal is a linear function of the Kerr amplitude A_K and therefore of the respective magnetization components. It can be increased by increasing the analyser angle α_s, but this does not always offer an advantage. For example, the best contrast C_{mo} for visual observation is found by optimizing $C_{mo} = (I_2 - I_1)/(I_2 + I_1)$ with respect to the angle α_s, yielding:

$$\tan\alpha_{opt}^C = \sqrt{\frac{A_K^2 + I_0}{A_N^2 + I_0}} \; , \quad C_{opt} = \frac{A_K A_N}{\sqrt{(A_N^2 + I_0)(A_K^2 + I_0)}} \approx \frac{A_K}{\sqrt{A_K^2 + I_0}} \qquad (2.9)$$

where the last expression is valid for $I_0 \ll A_N^2$ (see [143] for equivalent expressions). For large A_N the optimum contrast depends on the background intensity and on the magneto-optical amplitude A_K only; not on the regular amplitude A_N or on the Kerr *rotation* $\varphi_K = A_K/A_N$. Contributions to the background intensity are determined by imperfect polarizers, surface imperfections, and a finite illumination aperture as explained in Sect. 2.3.6.

The best visibility of domains is, however, rarely given by the criterion of maximum contrast. If, for example, the image is too dark, larger analyser angles α_s are preferred. With increasing angle α_s the intensity increases with α_s^2. This feature is important if the signal is processed electronically for example in video systems. Then good domain visibility primarily requires a large signal-to-noise ratio r_{SN}. Weak contrast that was originally a problem can easily be enhanced in an electronic system provided r_{SN} is large enough.

Three sources of noise have to be taken into account:

1. Shot noise based on the quantized nature of light. This unavoidable noise contribution varies with the square root of the photon number in the image.

2. Electronic noise that is usually independent of the image intensity, depending only on the detection electronics.

3. Fluctuations in the light source, in the optical path, and in the sample ("media noise") which will be proportional to the image intensity.

The ideal case is that of unavoidable shot noise only. In this case the optical noise may be written as $N_{shot} = \sqrt{\frac{1}{2} F_{inc} (I_1 + I_2)}$, where F_{inc} is the incident number of photons. With the absolute signal $S_{mo} = F_{inc} (I_1 - I_2)$ we obtain $r_{SN} = S_{mo} / N_{shot} = \sqrt{F_{inc}} (I_1 - I_2) / \sqrt{\frac{1}{2} (I_1 + I_2)}$. Inserting (2.7) and minimizing with respect to the analyser angle α_s we obtain:

$$\tan \alpha_{opt}^{SN} = \sqrt[4]{\frac{A_K^2 + I_0}{A_N^2 + I_0}} \quad , \quad r_{SN}^{opt} = \frac{4 A_K A_N \sqrt{F_{inc}}}{\sqrt{A_N^2 + I_0} + \sqrt{A_K^2 + I_0}} \approx 4 A_K \sqrt{F_{inc}} \ . \quad (2.10)$$

The maximum value of r_{SN} is determined by the Kerr amplitude and the illuminating number of photons, but again not by the Kerr rotation. At $\alpha_s = \alpha_{opt}^C$ only 70–90% of this maximum are reached, depending on the background intensity I_0. Opening the analyser beyond $\alpha_s = \alpha_{opt}^C$ improves the signal to noise ratio by up to 30%, or it saves up to 50% of the necessary illuminating intensity when r_{SN} is sufficient and kept constant.

Adding electronic noise and/or fluctuations reduces the signal-to-noise ratio, but does not affect the basic features of the calculation presented above: the optimum r_{SN} is still found somewhere beyond the analyser angle (2.9) for best contrast α_{opt}^C, and this ratio is for a given Kerr amplitude A_K virtually independent of the amplitude of the regularly reflected light A_N. This statement, which can be verified by inserting reasonable values for the parameters, confirms the role of the Kerr amplitude A_K as the meaningful material quantity in magneto-optics. It is also called the "figure of merit" [144–146] and often expressed somewhat indirectly in the form $\varphi_K \sqrt{I}$.

For a phase shift between A_K and A_N, it suffices to remove the resulting ellipticity of the darker domains by a phase shifting *compensator* (such as a rotatable quarter wave plate). Once this is achieved, the formulae (2.8, 2.9) for the signal and the contrast remain valid if the magneto-optical amplitude is replaced by its absolute value [143]. Using a compensator is not always necessary, however, since at oblique incidence phase shifts can be introduced by rotating the polarizer slightly away from the symmetry positions. In practice, the polarizer and the analyser are adjusted "simultaneously" until an image of satisfactory contrast and brightness is obtained. If a compensator is available, the polarizer can be left fixed and compensator and analyser are adjusted in this way. When using a video system, it is often advisable to perform this adjustment first directly in the microscope, and then to "open" the analyser as far as necessary to obtain best results on the video screen.

2.3.4 Interference and Enhancement by Dielectric Coatings

Dielectric anti-reflection coatings can enhance the often rather weak Kerr images ([147, 148, 143], see [148] for a review of still earlier work). They reduce the regularly reflected light while enhancing the Kerr component.

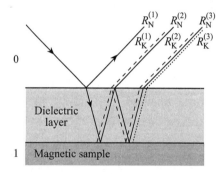

Fig. 2.13 A simple view of the enhancing effect of a dielectric layer [147]. The layer thickness is adjusted so that the normal reflectivities R_N cancel. It can be shown that in this case the Kerr amplitudes R_K (*dashed*) add up by constructive interference

The effect can be explained qualitatively with the help of Fig. 2.13. The dielectric coating thickness is tailored so that the phase of $R_N^{(2)}$ is shifted by 180° with respect to $R_N^{(1)}$. The phase difference between $R_N^{(2)}$ and $R_N^{(3)}$ is then 360° (including the 180° phase shift at the reflection from inside the dielectric layer). Under the condition that the amplitude of $R_N^{(1)}$ is equal to the combined amplitude $R_N^{(2)} + R_N^{(3)} + \ldots$, the total regular reflection will be zero, whereas the Kerr amplitudes are enhanced. This optimum case can be realized with a sufficiently high refractive index coating. Another valid interpretation is that anti-reflection coatings cause all of the incident light intensity to be absorbed in the magnetic medium rather than being uselessly reflected [143, 146].

Theory confirms these qualitative arguments [143]. The Kerr coefficients to be inserted into (2.6) for an arbitrary system of non-absorbing dielectric layers are for the three cases of the three magnetization orientations [143, 146]:

$$R_K^{\text{pol}} = \frac{iQ_V n_1}{4 n_0 \cos\vartheta_0} T_p T_s \; , \tag{2.11}$$

$$R_K^{\text{lon}} = \frac{i Q_V \sin \vartheta_0}{4 \cos \vartheta_0 \cos \vartheta_1} T_p T_s \ , \tag{2.12}$$

$$R_K^{\text{tra}} = \frac{-i Q_V \sin \vartheta_0}{2 \cos \vartheta_0} T_p^2 \ , \tag{2.13}$$

where n_0 is the refractive index of the environment, n_1 is that of the magnetic substrate, Q_V is its magneto-optical constant, ϑ_0 is the angle of incidence (measured from the surface normal) and ϑ_1 is the (complex) angle of incidence in the magnetic medium, to be calculated from ϑ_0 by Snell's law. T_p and T_s are the regular amplitude transmission coefficients from outside into the magnetic layer, for parallel and perpendicular polarization directions, respectively.

The rigorous expressions (2.11–2.13) may be simplified for the case that the cosine of the angle of incidence inside the material can be set equal to 1, which is a good approximation in most practical cases. Then the absolute values of the Kerr amplitudes (which are the values important for usable magneto-optical signals) can be expressed in terms of the easily accessible regular *intensity* reflectivities r_p and r_s:

$$|R_K^{\text{pol}}| = \left(|Q_V|/4n_1 \right) \sqrt{n_1^2 + k_1^2} \ \sqrt{(1 - r_s)(1 - r_p)} \ , \tag{2.11a}$$

$$|R_K^{\text{lon}}| = \left(|Q_V|/4n_1 \right) n_0 \sin \vartheta_0 \sqrt{(1 - r_s)(1 - r_p)} \ , \tag{2.12a}$$

$$|R_K^{\text{tra}}| = \left(|Q_V|/2n_1 \right) n_0 \sin \vartheta_0 (1 - r_p) \ , \tag{2.13a}$$

where $n_1 + ik_1$ is the complex refractive index of the magnetic medium. If the regular reflectivities r_p and r_s are reduced to zero, the Kerr amplitudes R_K assume their largest values. The problem of enhancing the Kerr effect is thus equivalent to finding an effective dielectric *anti-reflection* coating for the particular material. Since large angles of incidence are hardly used, it is sufficient in practice to minimize the reflectivity at normal incidence.

For metals a suitable dielectric is ZnS, and for magnetic oxides MgF_2 or SiO_2-layers can be applied. The material is evaporated under optical control until the reflectivity for the wavelength intended to be used in the observation is minimized. For green light the surface then displays a dark blue colour. Further possibilities including multiple layers with reduced depolarization properties are discussed in [143].

Equations (2.11a–2.13 a) also show that the polar Kerr effect is stronger than the other two effects, not only by the factor $1/\sin \vartheta_0$, but also by the factor

$\sqrt{n_1^2 + k_1^2}$ in (2.11a). This is a consequence of Snell's law that reduces the angle of incidence inside the magnetic medium.

The contrast in thin *transparent* magnetic films can also be enhanced by interference if needed. Equations (2.11–2.13) are equally valid for arbitrary thin film systems if the corresponding transmission or reflection coefficients are inserted. In principle, very thin film systems can even yield a larger magneto-optical amplitude than bulk material [149–151]. To achieve this, the film is first deposited onto a mirror, with a dielectric spacer layer of appropriate thickness. The system is then covered with an anti-reflection system as above. Such a mirror system can yield theoretically large improvements although its practical implementation may be difficult. Thin film mirror systems can in principle overcome the disadvantage of longitudinal versus polar magnetization [146, 152].

The reason for the somewhat surprising result that an ultrathin magnetic layer can produce more signal than bulk material was studied in [151]. In an optimized mirror system all energy is eventually absorbed in the magnetic medium just as in a bulk sample with an ideal anti-reflection coating. From this point of view the same efficiency as for bulk material would be plausible.

An advantage of the thin film system shows up when the *phase* of the magneto-optic amplitude is considered. In bulk material the total magneto-optical signal can be seen as a superposition of contributions from different depths, which differ in phase according to the complex amplitude penetration function $\varphi_{ap}(z) = \exp(4\pi i n_1 \cos\vartheta_1 z/\lambda_0)$ ([146, 151, 152]; the

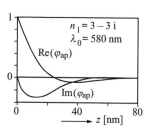

light wave has to travel into a certain depth z and out again, hence the factor 4π instead of 2π). The idea that the Kerr effect displays a depth sensitivity according to the (intensity) light penetration depth as sometimes proposed (explicitly for example in [153]) is erroneous. For bulk material the contributions from different depths differ in phase, thus reducing the overall efficiency. For an optimized ultrathin film system, however, all of the magneto-optic amplitude is generated with the same phase. This advantage becomes the more effective the smaller the absorption in the magnetic medium, because then the contributions from different depths in bulk material may even differ in sign, leading to very weak overall signals. For the highly transparent garnets, for example, no domain observation truly based on the Kerr effect has been reported (domains in garnets are observed in transmission by the Faraday effect, perhaps using a mirror for reflection observations).

The different phases of contributions from different depths can be exploited in Kerr microscopy. Using a rotatable compensator, the phase of the Kerr amplitude can be adjusted relative to the regularly reflected light amplitude. In this way light from selected depth zones can be made invisible if their Kerr amplitude is adjusted out of phase with respect to the regular light. This kind of depth-selective Kerr microscopy was demonstrated in [154, 155] for Fe/Cr/Fe sandwich samples in which the domains in the two iron layers could be imaged separately. The possibilities of such observations can be extended by using different light wavelengths [156, 157].

2.3.5 Kerr Microscopes

The most common Kerr microscopes are sketched in Fig. 2.14. The first setup is recommended for low resolution applications in the longitudinal or transverse effects to obtain an overview of the domain pattern of larger samples. The advantage of this setup is that no optical elements other than the sample exist between polarizer and analyser, so that contrast conditions are optimal. The objective lens is tilted to increase the range of focus and to reduce the image distortion [158]. Resolution in this arrangement is practically limited by the achievable numerical aperture and by aberrations in the tilted objective to about 2 μm. Microscopes of this type have to be custom-made. They can be based on a suitable stereo microscope [159].

Observations at high resolution up to the limit of optical microscopy are possible with the second setup [160] which is usually based on a conventional polarizing microscope. Using oil immersion for a high numerical aperture and blue light for a short wavelength, a resolution of 0.3 μm has been achieved for the polar effect, thus showing domains as narrow as 0.15 μm [161, 162]. Contrast is reduced to some extent since the beam passes through the mirror element and twice through the objective between polarizer and analyser. Strain-free, polarization-quality optics are mandatory. Berek prisms are the preferred mirror elements for the longitudinal effect, since they cause no light loss and introduce little depolarization. On the other hand, they restrict the viewing aperture and therefore the resolution in one direction. If light intensity poses no problem, a sheet reflector offers optimum resolution and high flexibility in the observation mode. The best solution could be a narrow reflector tongue placed in the back focal plane. Such a tongue reflector was used in the microscope of *Prutton* (see [45]), but its potential for increased resolution was not utilized there because an aperture stop was placed behind the objective lens. No such illumination problem occurs in transmission [163].

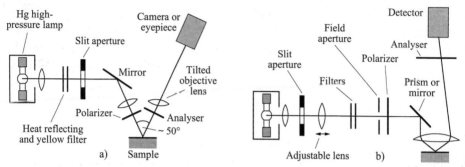

Fig. 2.14 Two Kerr microscopes. (a) A low resolution and high sensitivity version. In this case a wide angle objective lens, which can be tilted to reduce distortion, works best. (b) A high resolution, distortion-free Kerr microscope. To avoid depolarization, the objective lens and the mirror element must be strain-free

A high pressure mercury lamp with suitable spectral filters to select the green and yellow mercury lines is recommended as the best light source in most cases. It offers sufficient brightness and a suitable colour spectrum, which can be used not only in the yellow-green but also in the blue range. The disadvantage of arc lamps is their insufficient stability and lifetime. A more stable alternative can be laser illumination. Solid state green or blue lasers will be interesting when they become available and sufficiently reliable. Good results were obtained with blue argon ion lasers and with dye lasers for pulse applications. The advantages of laser illumination are (i) virtually unlimited intensity, (ii) the possibility to stabilize the output power of the laser, which is impossible for arc sources, and (iii) the use of short laser pulses for high-speed and stroboscopic microscopy.

A problem with laser illumination (as with any narrow aperture illumination) is the tendency to show strong interference fringes and speckle. Using a multimode dye laser and inserting a spinning glass plate in the illumination were shown to reduce this problem [164]. Digital image subtraction (Sect. 2.3.7) in principle eliminates interference effects, but a very high stability of all components would then be needed. A good technique consists in feeding the laser light through a vibrating multimode glass fibre to destroy the coherence ([165]; and also to reduce the potential danger connected with lasers) and, in addition, to vibrate the tip of the fibre, so that its image covers a suitable area in the back focal plane to avoid diffraction fringes. Still, the quality of the image in visual observation based on available stabilized argon lasers is not as good as with regular incoherent illumination. Satisfactory results are achieved only in photographs averaging away residual, instability-related laser effects.

Recent developments of high-intensity light-emitting diodes offer a further promising alternative [166]. If fed into an optical fibre[3] and perhaps combined into a bundle, they might well represent the ideal microscope illumination in the future, combining sufficient spectral width to avoid speckle and interference fringes, with the inherent stability of a solid state light source.

2.3.6 The Illumination Path

The illuminating aperture in a Kerr microscope should be neither too small nor too large. Too small an aperture (parallel light) leads to disturbing diffraction fringes especially around sharp defects on the sample surface. A large aperture reduces the contrast by generating a background intensity because of depolarization effects. The reason for this is sketched in Fig. 2.15a.

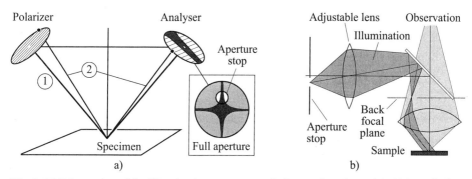

Fig. 2.15 Schematics of the illuminating aperture, polarizer and analyser (a). *If* the polarizer is set parallel to the central plane of incidence, the central beam (1) is reflected without a phase shift from any metallic surface. This is not true for an off-centre beam (2) with its different plane of incidence. This beam is reflected in an elliptical and rotated polarization state in general, and is thus not fully extinguished by an analyser oriented perpendicular to the central plane of incidence. The zone of extinction in the full aperture, observable in the back focal plane of a microscope (b), is indicated in the inset (a) for this case. An effective aperture stop is chosen to select an illumination with a good extinction ratio

Consider all light beams hitting one point of the sample. We call the plane defined by the central beam of the illuminating bundle and the surface normal the *central* plane of incidence. The positions of polarizer and analyser are defined with respect to this central plane. For all beams not lying in this plane

---

[3] *Theo Kleinefeld*, private communication

the effective polarizer angles are different. If the polarizer settings were optimized for the central illumination plane, they are no more optimized for other
beams. This can be demonstrated by adjusting polarizer and analyser for
maximum extinction and looking at the diffraction plane (the so-called conoscopical image) of the microscope (by using a built-in Bertrand lens, or an
auxiliary telescope replacing the eyepiece, or, less satisfactory, by simply
looking into the tube after removing the eyepiece). The zone of maximum
extinction is in general cross shaped in the full aperture (Fig. 2.15a, inset).

Polarization-dependent reflection and transmission effects on curved lens
surfaces and other optical elements within the microscope add to the depolarization by the sample. This is particularly true for the strong objective lenses
used at high magnification. All these depolarization effects could in principle
be reduced by suitably tailored anti-reflection coatings on lenses and sample.
No practical solution in this sense is available, however.

To obtain best contrast conditions, the illumination should be restricted to
the extinction zone in the conoscopical image. For the polar effect, a central
circular diaphragm is placed in the illuminating beam. For the longitudinal
effect, a displaced *slit* aperture oriented parallel to the plane of incidence is
preferable [167], while, for the transverse effect, a displaced slit oriented perpendicular to the plane of incidence would be the best solution.

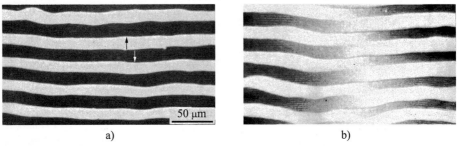

a)   b)

Fig. 2.16 A domain image with a correctly placed illuminating aperture (a) and with an
incorrectly adjusted aperture for comparison (b). The domain pattern is a so-called stress
pattern in 3% SiFe transformer steel (see Sect. 5.3.3)

The diaphragm should be effective uniformly for the whole observation
field by placing its optical (not the actual) position into the back focal plane
of the objective (Fig. 2.15b). This is achieved by adjusting the lens indicated
in this figure and in Fig. 2.14b according to the properties of the respective
objective. If this condition is not fulfilled, an inhomogeneous image up to a
contrast inversion across the image develops as demonstrated in Fig. 2.16. It

should be mentioned that the resolution of the microscope is not seriously affected by the illumination with a narrow aperture, as long as the actual diaphragm is outside the viewing light path (as shown in Fig. 2.15b). If a sheet reflector is used instead of a prism, the plane of incidence can be chosen freely by moving the image of the aperture diaphragm in the back focal plane of the objective. In this case the true transverse Kerr effect can be replaced by the longitudinal Kerr effect with a transverse plane of incidence.

2.3.7 Digital Contrast Enhancement and Image Processing

Magneto-optical observations often have to cope with surface imperfections and irregularities that produce strong "non-magnetic" contrast when polarizers are nearly crossed. In [168] a photographic method was proposed to subtract the contrast of non-magnetic origin. The results were not satisfactory because of the nonlinearity of the photographic process.

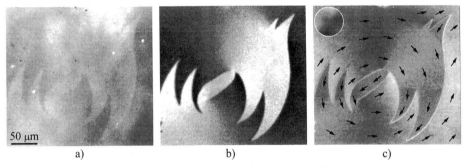

a)   b)   c)

Fig. 2.17 The same domain pattern on an amorphous iron-nickel alloy photographed before (a), after a digital contrast enhancement (b), and after a quantitative evaluation (see Sect. 2.3.8) obtained by combining two digital images (c). In the last case, a colour code was used to indicate the magnetization direction at every point in addition to the arrows[4]

The idea was revived using electronic methods [169–171, 165]. In the standard procedure, a digitized video image of the magnetically saturated state, the reference image, is first stored in a digital memory, and then subtracted from the pictures containing magnetic contrast. It is advantageous if this subtraction process is performed in "real time" at video frequencies. Live observation of magnetization processes is possible in this way, as long as the microscope settings remain unchanged, because the same reference image can be used for each of the incoming images. For recording, the resulting difference image

---

[4] See Colour Plate

can be improved by averaging and other techniques of digital image processing. Necessary is a high optical, mechanical, electrical and thermal stability of the microscope system. Figure 2.17a,b demonstrates the dramatic improvement possible with such a system. Instead of the saturated state an averaged a.c. state can be used as the reference, with the advantage that forces on the sample may be smaller in an alternating field than in a high saturating field. Small displacements between reference and actual image, as they may occur by thermal effects, can be corrected by interactively adjusting sample or camera before the image is finally recorded.

It is often desirable to study the same domain pattern in different aspects. This is possible within the digital difference procedure by *combination* techniques [142, 139]. Two different views of the same domain pattern can be recorded if one reference image is taken before the pattern in question is investigated and the second reference image, relating to a modified microscope setting, is recorded at the end of the sequence. For the different aspects exists a wide range of possibilities: (i) If the plane of incidence is changed, different magnetization components can be made visible (the quantitative image Fig. 2.17c is obtained with such a technique, see Sect. 2.3.8). (ii) At perpendicular incidence the polar component, the Voigt effect or the magneto-optic gradient effect (see Sect. 2.3.12) can be observed and correlated with a conventional longitudinal image. (iii) The sample can be shifted and different parts of the same pattern can be explored. (iv) By rotating the compensator, the depth sensitivity can be modified as explained in Sect. 2.3.4. (v) By changing the objective lens, a portion of a picture can be investigated in different magnifications. (vi) Even the rear side of a sample can be inspected and compared with the front image, if it is accessible. Complex three-dimensional domain patterns can be analysed much more reliably, if images from different sides of a sample are available (see [172] and Fig. 1.1a).

The standard digital difference procedure sometimes still contains structural or non-magnetic contrast contributions. Coarse-grained polycrystalline and non-flat samples often make such problems. An enhanced method that normalizes the standard difference image by a "saturation difference image", thus removing these artefacts, was developed and demonstrated in [173].

Electronic contrast enhancement opened a wide area of new possibilities. For the first time even domain walls in metallic soft magnetic materials could be visualized in all detail [171], as demonstrated in Fig. 2.18a. Here, the surface structure of Bloch walls in silicon-iron is depicted with the longitudinal Kerr effect, while the domains show almost no contrast except for a few domains magnetized along the plane of incidence. When the plane of incidence is rotated in the sense of a combination picture as explained above, the visibility conditions for domains and walls are exchanged (Fig. 2.18b).

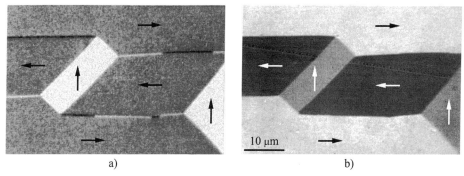

a)   b)

Fig. 2.18 Domain and wall images from a (100)-oriented silicon-iron crystal. With a vertical plane of incidence (a) the structure of the horizontal walls becomes visible. In (b) the sensitivity axis at oblique incidence is chosen horizontally, displaying mostly domain contrast and only indications of the wall substructure pattern

Another way of digital image improvement was demonstrated in [174]. Here the digitally recorded image was Fourier transformed, and those parts of the spectrum not related to the magnetic image were removed. The resulting magnetic image of maze domains in a garnet film showed an improved quality. This possibility is interesting but probably not generally applicable. Further possibilities are discussed in [175].

2.3.8 Quantitative Kerr Microscopy

Once a picture is digitized, further operations become possible like the quantitative determination of magnetization directions [142, 176, 177] and the quantitative evaluation of domain pattern characteristics and domain wall parameters. In *quantitative* Kerr microscopy two aspects of the identical pattern, obtained e.g. with two different planes of incidence, are recorded sequentially and combined in the computer.

The quantitative method is applicable to soft magnetic materials for which the observable magnetization lies basically parallel to the surface and polar components are negligible. Still it needs calibration experiments to determine the sensitivity directions of the two combined images because adjusting a microscope for a predetermined sensitivity direction is unreliable in practice.

The calibration experiments are performed before and after the actual investigation, i.e. before the first reference image, and after the second one. In [142, 177], recorded image intensities in saturating fields of varying directions were used for this calibration. Problems arise by parasitical Faraday rotations in the lenses and by polar magnetization components induced in the sample (for example at grain boundaries and sample edges) in high magnetic fields [176]. Better results are obtained if the magnetization direction

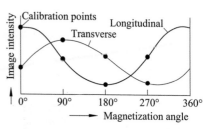

in a number of places is known a priori, for example based on crystal anisotropies, or for domains near thin film sample edges. Then this knowledge can be extended to the whole viewing area by interpolation [177]. Sometimes it is possible to generate simple patterns in a sample for the purpose of calibration, which can then be used for the more complex pattern to be explored quantitatively [178].

Another possibility of quantitative Kerr microscopy was proposed in [179]: here the sample is rotated by 90° and images are taken before and after the rotation. The method has the advantage of needing little calibration and being applicable to arbitrary patterns. It may be difficult, however, to correlate the two images precisely. There will also be problems if the sample is not completely flat and exactly oriented perpendicular to the rotation axis.

It needs special care to separate the different contributions to magneto-optic pictures if polar magnetization components are present. The polar components alone can be made visible by an experiment at normal incidence since then the in-plane components do not contribute to the image. To separate these components is more difficult if sizable polar components are present; it can be achieved by a combination experiment with different polarizer and analyser settings [180] based on (2.6).

Quantitative results can be favourably displayed with a colour code ([142]; Fig. 2.17c). The wheel of saturated colours maps the in-plane magnetization directions, with the additional option of using black and white for polar components if they should be known. The colour code can be supported by arrows at selected points in the image. A regular array of arrows tends to generate a textured impression because of the arrow arrangement. This idea has also been applied to other quantitative domain observation techniques and to the results of micromagnetic simulations (as for example in [181]).

2.3.9 Dynamic Domain Imaging

Domain dynamics can be observed visually as fast as the eye can follow. Photographing periodic processes of any frequency with long exposure times yields zones of intermediate contrast that may give valuable information at least on the amplitudes of wall motion [182, 183]. Digital image subtraction techniques (Sect. 2.3.7) add further options, such as the distinction between irreversible and reversible wall displacements (Fig. 2.19).

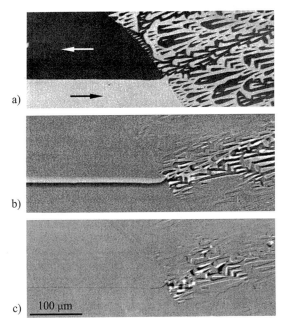

a)

b)

c) 100 μm

Fig. 2.19 Studying the kinetics of a domain pattern in a transformer sheet. The starting configuration (a) is subtracted from the image of a state in which the sample is subjected to an alternating field (b). Those parts of the domain pattern that do not move stay gray in the difference image, while the reversibly moving walls become visible by a black and white contrast in the dynamically averaged image. After switching off the alternating field, some walls remain displaced irreversibly as can be seen in the static difference image (c)

More detailed investigations also of non-periodic processes call for time-resolved high-speed photography. *Houze* [184] recorded 5000 pictures/sec with a xenon flash lamp as illumination. Only low magnification pictures are possible with this technique because of the limited intensity.

The light source is the principal problem in all high-speed investigations (even static magneto-optical observations require the brightest available sources as mentioned before). Periodic magnetization processes can be observed stroboscopically either by a pulsed light source [185, 186] or by a triggered video camera [187]. Single-shot pictures at high resolution can be taken with triggered lasers [164, 49, 188–191], for example with dye lasers pumped by a nitrogen ion laser. The laser pulses are a few nanoseconds long and can be triggered with precise timing relative to a field pulse. Sweeping the delay time yields a series of corresponding pictures of periodic or quasi-periodic processes.

Figure 2.20a shows an example of a picture taken with a 10 ns laser pulse [189]. Since this is a single-shot photograph, optical interference effects were not suppressed. This would need digital subtraction, but laser sources with reproducible shot intensities are not available. For stroboscopic imaging of periodic processes with a triggered laser, image subtraction has been demonstrated to be useful [192, 193]. An example of a picture obtained in this way is shown in Fig. 2.20b. Here the averaged domain states at different phase positions are subtracted to emphasize wall motion.

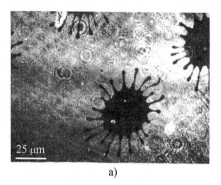
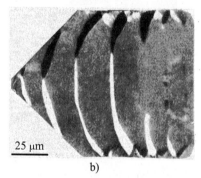

a)   b)

Fig. 2.20 (a) A picture of "exploding" magnetic bubbles in a fast negative bias pulse field applied to a transparent garnet film. The exposure time using a dye laser triggered by a nitrogen laser was 10 ns. (Courtesy *F.B. Humphrey*, Carnegie Mellon University). (b) Stroboscopic image of domain wall motion in the yoke of a thin film head, taken as a difference picture between the average over two series of 5 ns pulses, where the pulses of the second series are triggered somewhat later in the 1 MHz magnetization cycle than the pulses of the first series. The black and white contrasts indicate the range of the periodic wall motion between the two trigger points. (Courtesy *B.E. Argyle*, IBM Yorktown Heights [193]])

A record in high-speed photography was set by *Chetkin* et al. [194] in the investigation of the supersonic motion of domain walls in orthoferrites. Single shot dye laser pulses of 1 ns length were used in these investigations. In an optical difference technique part of the laser beam was split off and, after a detour leading to a delay of a few nanoseconds, was fed back into the microscope. A polarizer in the second beam was set so that the resulting contrast was opposite to that of the original beam. So contrasts cancel unless the investigated domain wall moves between the two shots. In this way local wall velocities up to 20 km/sec could be recorded. The technique took advantage of the very high magneto-optic contrast of properly prepared orthoferrite samples.

2.3.10 Laser Scanning Optical Microscopy

Using a scanning optical microscope (or "laser scanning microscope") instead of a conventional parallel-illumination setup has a number of advantages: lasers are available with a higher power than conventional light sources, and the signal can be processed and analysed point by point, offering more flexibility than a conventional photographic plate or a video camera. Variations in the laser intensity can be easily compensated in a scanning device by measuring them separately. Since the spots on the sample are illuminated sequentially, diffraction fringes are less important and can be eliminated by using a second, confocal aperture in the detection system [195]. Such confocal microscopes should also display an enhanced resolution (by a factor $\sqrt{2}$) which has, however, not yet been verified in magneto-optics.

With lock-in techniques weak variations of the optical signal can be detected and a map of permeabilities or other local properties, for example on a thin film recording head, can be generated [196, 191]. Another example of a successful application of the laser scanning technique was the first direct visualization of Bloch lines in bubble materials [197], which was based on a weak diffracted light image (asymmetric dark field microscopy). (In this case, subsequent developments showed that a corresponding method in a conventional micro-scope can lead to the same result [198–200]). In [201] amplitude modulation using the transverse Kerr effect was demonstrated. The detection of this small contrast is probably impossible in conventional imaging. The technique that works without an analyser has the advantage of being simple and sensitive only to one magnetization component. Using lock-in techniques clear scanned pictures can be obtained. Further possibilities are explored and reviewed in [202–204].

The main disadvantages of laser scanning microscopes are their high price and their slow speed compared to regular imaging microscopes. Most laser scanners are rather slow. Fast scanning instruments [205] offer advantages but are still not fast enough for live observations of domains at video frequencies. Often a cheaper but slower variant is preferred, where the sample is scanned mechanically under a fixed laser beam. A simple one-dimensional scanning device can yield valuable information in special cases such as in the investigation of Bloch walls [206, 207].

A problem lies in local heating of the sample by the laser that may influence the domains. According to [201], the stationary heating of an illuminated spot on a bulk sample is given by $\Delta T = 3P/(\pi \lambda_{th} d_{sp})$, where P is the absorbed intensity, λ_{th} the thermal conductivity and d_{sp} the diameter of the spot. With a laser power of a few milliwatts, as needed in Kerr microscopy, there results a local heating of a few degrees that may affect the details of domain patterns in soft magnetic materials. If very fast scanning systems were available, the local heating problem could be avoided by repeated scanning.

We conclude that scanning magneto-optical microscopes fall short of replacing conventional (wide field) imaging microscopes. However, they can image other quantities than simply the magnetization, and in these applications lies their particular value.

The additional possibility to raise the resolution beyond the diffraction limit by using *near-field* scanning techniques is discussed in Sect. 2.6.2.

2.3.11 Sample Preparation

The information depth of light (about 20 nm in metals) determines the quality of sample preparation necessary in Kerr microscopy. In soft metals, mechanical polishing is usually insufficient as the damaged layer is too thick in this case (an example of the domains appearing on mechanically polished, stressed iron-based samples is shown in Fig. 2.22 d–f). A mild heat treatment after polishing (such as e.g. at 800°C in vacuum for iron alloys) allows the surface to be restructured and yields satisfactory results. Hard metals like the rare-earth permanent magnet alloys can be prepared by diamond polishing alone without heat treatment. The damaged layer seems to be thin enough in this case. Electrolytic polishing is always satisfactory if applicable. A problem is the limited flatness of the surfaces after electrolytic polishing. Polishing in a laminar stream or jet of electrolyte is to be preferred. In low magnification observation the unevenness of the surface results in a strong background image. It can be largely suppressed by digital image processing, but if this technique is not available, mechanical polishing may be better for low magnification. For high-resolution observations electrolytic polishing is preferred if it is possible. As a rule, the thinning and polishing procedures successful in transmission electron microscopy give good results in Kerr microscopy too.

Some samples need no surface preparation at all: thin films and melt-spun metallic glasses already have a sufficiently perfect surface if they have not been touched before domain observation. With digital image processing the domains of many uncoated soft magnetic materials can be observed without polishing. No preparation is necessary (and possible) if domains are observed through a transparent substrate. If a glass substrate is chosen sufficiently thin

(< 0.1 mm), domains can be observed at highest resolution using conventional oil immersion objectives that are corrected for a cover glass.

Oxidic magnets can be polished by a colloidal suspension of amorphous SiO_2 [208] which is known from silicon technology [Syton® (Monsanto), OP–S® (Struers), or similar products]. Surprisingly, these products work on regular metal samples like silicon-iron as well. The polishing mechanism combines some chemical and mechanical effects. No further surface stress annealing is necessary for Kerr microscopy if polishing with a colloidal silica medium was successful.

In every case the application of a dielectric anti-reflection coating (Sect. 2.3.4) improves the contrast considerably, protects the sample, and is not known to have any adverse effects.

2.3.12 Other Magneto-Optical Effects

(A) The Faraday Effect. For transparent magnetic materials, such as most oxides and in particular the garnets, the Faraday effect offers the best method of domain observation, especially as the Kerr effect is usually weak in these materials (see Sect. 2.3.4). Even domain walls can be easily seen in such samples by the Faraday effect ([121, 209, 210]; Fig. 2.21).

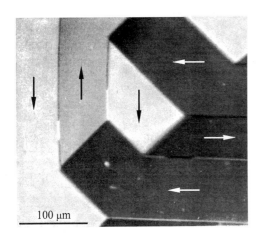

Fig. 2.21 A transmission image of a modified YIG garnet plate. Because of a small cobalt content, the easy axes in this garnet are ⟨100⟩. We see mainly the Voigt effect in the domains differing by 90°, and the polar Faraday effect in the 180° domain walls. The illumination is almost perpendicular, but a slight tilting (along the vertical axis in the picture) generates some longitudinal Faraday contrast between the 180° domains, too. (Together with *M. Rührig*, Erlangen. Sample: courtesy *W. Tolksdorf*, Philips Hamburg)

100 μm

Usually no digital image processing is necessary in transmission experiments because contrasts are strong and intensity is high. In oxides infrared radiation can be used if the crystal is too thick and the material absorbs too much light in the visible range [211]. The advantage of looking through a sample turns into a problem, however, when complicated, depth-dependent structures have to be investigated. A combination of transmission pictures with surface-sensitive methods, such as the Bitter technique, is extremely useful [211, 212]. Another option would be the use of the Faraday effect at long wavelengths where an oxide is transparent, and of the Kerr effect at short wavelengths.

In transmission experiments a problem arises if the samples are optically anisotropic (birefringent). Then samples must be cut precisely along an optical axis since otherwise the magnetic contrast can be very weak. The mentioned high-speed investigations on orthoferrites [194] were only possible with samples prepared in this way.

(B) The Use of the Faraday Effect in Indicator Films. The Faraday effect finds another important application: thin indicator films can be used to map stray fields above superconductors, hard magnetic materials, recording materials or active integrated circuits. The stray fields induce a polar magnetization component in the active layer of the detector which is recorded with the polar Faraday effect. Usually the indicator films are coated with a mirror layer on one side, so that the light passes through the film twice, thus enhancing the sensitivity.

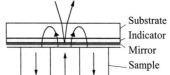

The first indicator films were paramagnetic solutions [213] or glasses [214]. The introduction of evaporated paramagnetic films (mixtures of EuS and EuF_2; [215, 216]) improved the resolution decisively. The sensitivity of these paramagnetic films is limited, however. It reaches acceptable values only close to the ordering temperature (the Curie point) of the indicator material.

The sensor material should of course not influence the sample to be investigated and should in itself be structureless. For these reasons, attempts to use transparent bubble garnet films with their characteristic maze domain pattern (see Fig. 1.1c) did not lead to very satisfactory results [217, 218]. High sensitivity and high resolution can be achieved with bubble garnets if the detection task allows the application of an in-plane field that is just strong enough to turn the magnetization into the plane. This method of exploiting a "critical state" (see Sect. 3.4.4) was applied in [219] for the detection of magnetic inclusions in a non-magnetic substrate (Fig. 2.22a).

Much better results for the regular case (in which such an applied field is not allowed) were obtained with the introduction of garnet films with in-plane anisotropy [220]. These films do not generate stray fields by themselves, and their in-plane domain pattern is largely invisible in polar observation. The method and its application on superconductors are reviewed in [221]. Another application is displaying magnetic microfields in non-destructive testing [222]. Its role for the observation of regular magnetic domains is limited, but under favourable conditions it can be useful as shown in Fig. 2.22 c, e, f.

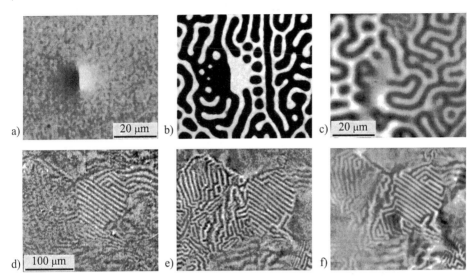

Fig. 2.22 Examples of using indicator films in magnetic domain studies. (a) Decoration of a buried magnetic defect via the critical state of a perpendicular-anisotropy garnet film in an applied field parallel to the surface. In (b) the same defect modifies a remanent domain pattern in the garnet indicator film. (c) Basic domains on a $Nd_2Fe_{14}B$ coarse-grained material made visible by an in-plane-anisotropy garnet film. Only the coarse interior domains and not the finely divided surface domains become visible, and only in parts of the sample where the indicator film is in good contact with the sample. The maze pattern on mechanically polished silicon-iron (d–f) is caused by residual polishing stresses. The longitudinal Kerr effect (d) is not very sensitive in this case. A coating filled with garnet particles {[223]; (e)} displays the domain pattern more reliably than a separate garnet film (f). Together with *R. Fichtner* (a, b) and *K. Reber* (d–f)

(C) The Voigt Effect and the Gradient Effect. The Voigt effect (linear magnetic birefringence) plays a well established role in the investigation of transparent crystals. At perpendicular illumination in-plane domains show up only by this effect, while the domain walls and perpendicular domains become visible by the stronger Faraday effect (Fig. 2.21). Rotating the sample by 90° demonstrates the difference: Voigt effect contrast is inverted since the Voigt effect is quadratic in the magnetization components. Faraday effect contrast stays the same for the polar effect, or disappears if it is caused by the longitudinal effect and some misalignment. Thus a convenient interpretation of observed pictures becomes possible.

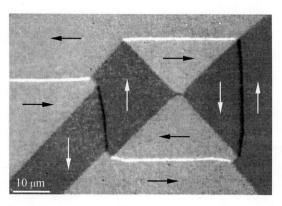

Fig. 2.23 Contrast observed at perpendicular incidence on a (100)-oriented silicon-iron crystal. The polarizer and analyser are oriented at ±45° relative to the horizontal direction, so that the Voigt effect shows up between the 90° domains, while the magneto-optic gradient-contrast becomes visible on the 180° domain boundaries

The Voigt effect was also observed in reflection with the help of the digital difference technique (Fig. 2.23; [139]). One more interesting effect can be seen in Fig. 2.23. Some domain boundaries show an additional strong contrast. This observation has been identified as being caused by a birefringence effect that depends linearly on certain components of the magnetization *gradient* [139, 224–226]. Independent of the true origin of this effect [227, 228] it may be useful in domain analysis [155]. An interpretation of an in-plane domain pattern such as the one shown in Fig. 2.23 is possible with a single perpendicular-incidence image using the contrast laws of Voigt and gradient effects. An overview over these possibilities is given in [229].

(D) Second Harmonic Generation in Magneto-Optics. A new field of magneto-optic phenomena was opened by investigating an effect of non-linear optics [230–233]. In this regime of optics the dielectric displacement contains a significant contribution from the square of the electric field, in contrast to the conventional linear dielectric law that is the basis of all regular magneto-optic effects. If light of frequency ω_L shines with high intensity on a material, the non-linear effect generates a second harmonic (SH) which can be described by a dielectric law $D_i^{(\mathrm{SH})}(2\omega_L) = \sum_{r,s=1}^{3} \varepsilon_{irs}^{(2)} E_r(\omega_L) E_s(\omega_L)$, where $\boldsymbol{\varepsilon}^{(2)}$ is the SH dielectric coupling tensor. The excited non-linear light amplitude can be easily separated from the incident light with spectroscopic methods. The coupling tensor contains both non-magnetic and magnetic contributions. Some elements of the tensor $\boldsymbol{\varepsilon}^{(2)}$ depend linearly on the magnetization. The detailed symmetry conditions are quite complex as elaborated in [230].

Symmetry analysis reveals that significant effects from the bulk of regular, high-symmetry materials are forbidden. Symmetry is broken at the surface, however, because the secondary electron motion is stopped at the surface. The observed SH amplitudes therefore originate mainly from the first atomic layers. The question of additional volume contributions in magnetic materials is discussed in [234, 235]. Also the possibility that a magnetic inhomogeneity itself can excite the generation of a SH amplitude is elaborated in the latter article. The main effect is certainly caused by the surface, however. In contrast, the conventional, linear Kerr effect displays information from the amplitude penetration depth as discussed in Sect. 2.3.4.

Strong magnetic effects in second harmonic generation were demonstrated in [236]. Laser pulses of 6 ns pulse duration and 532 nm wavelength were applied on the (110) surface of an iron crystal at an angle of 45° and parallel polarization. The specularly reflected second harmonic light was found to be magnetically modulated by 25%. The total intensity of the second harmonic was not enough in this experiment for any kind of microscopy, however, but this might be improved with shorter laser pulses because then the effective electric field increases for constant average power, thus leading to a stronger second harmonic effect. Large effects were also reported in [237].

If SH microscopy can yield good quality images in reflection, it will compare favourably with the equally surface sensitive electron polarization methods (Sect. 2.5.4), since the metal surface need not be ultra-clean in this technique. It may be covered by a transparent coating (provided this coating does not quench the ferromagnetism in the top-most atomic layer [236]). A scanning microscope based on the SH effect would offer a useful addition to available techniques. First magnetic SH images were obtained in transmission from stressed epitaxial magnetic garnet films [238, 239].

The frequently raised claim that second harmonic imaging is based on a much stronger effect than the conventional linear Kerr effect, only because SH rotations are stronger, has not to be taken seriously. We know from the analysis of the conventional Kerr effect (Sect. 2.3.3) that magneto-optical *rotations* are not related to image quality or signal-to-noise ratio. Decisive is the relative excited amplitude, and this is unfortunately still very small.

2.3.13 Summary

The standard Kerr technique of domain observation has many advantages:

- The magnetization can be observed directly without ambiguity.
- The samples are not destroyed or damaged during observation. The shape and size of the sample are largely arbitrary.

- The process of observation does not influence the magnetization (if the heating effect of the illumination is suppressed; as another exception, there are certain photo-induced magnetic anisotropy changes in magnetic oxides).
- Dynamic processes can be observed at high speed.
- The sample may be manipulated easily during observation. High or low temperature, mechanical stress or, most importantly, arbitrary magnetic fields may be applied.
- The same effects that are used for imaging may also be used for the magnetic characterization of the material, measuring local hysteresis properties. These possibilities are naturally available in the scanning variant of Kerr microscopes. They can also be added to conventional Kerr microscopes by replacing the camera with a photoelectric detector or adding a micromagnetometer to the microscope [159].

The drawbacks of magneto-optical methods are:
- Samples have to be prepared so that they are reasonably flat and smooth on a scale exceeding the chosen resolution.
- Some equipment is necessary, especially if low contrast conditions call for electronic enhancement.
- The resolution is limited to magnetic domains larger than about 0.15 µm, corresponding to an optical resolution of about 0.3 µm.
- On metals, only the surface magnetization in a layer of the penetration depth of some 10 nm can be seen (which needs not necessarily be a disadvantage as discussed in more detail in [146, 151]).

Most magneto-optical domain observations are performed in classical polarization microscopes, using either photographic plates or video cameras as detectors. The alternative laser scanning technique offers advantages in the flexibility of defining the "signal" (which can be, for example, a dynamically defined permeability or coercivity). Optical scanning techniques suffer, however, from the problem of low scanning speeds making it much more difficult to follow dynamic processes.

The Faraday effect is preferred if the sample is sufficiently transparent. The Voigt effect and the magneto-optic gradient effect can be used as additional options. The new technique of second harmonic magneto-optical effects may open the way for true surface observations with magneto-optical means. The related method of near-field magneto-optical microscopy is discussed in Sect. 2.6.2. X-ray spectroscopic methods that are treated in Sect. 2.7.3 may be considered magneto-optics at much shorter wavelengths. However, both the detailed physical background and the experimental conditions of "X-ray magnetic circular dichroism" and their kin differ so much from conventional magneto-optical methods that they deserve a separate discussion.

2.4 Transmission Electron Microscopy (TEM)

2.4.1 Fundamentals of Magnetic Contrast in TEM

Domain observation in the electron microscope [240–245] was stimulated by the technical interest in thin magnetic films. The early work has been reviewed, for example in [246–249]. An introduction into modern concepts can be found in the textbook of *Reimer* [250]. The review by *Chapman* [251] includes a thorough theoretical analysis and a survey of basic experimental techniques. A review of classical Lorentz microscopy techniques is found in [252], for new methods with an emphasis on holography see [253].

In an electron microscope electrons are accelerated to energies of 100–200 keV in conventional microscopes and up to 1000 keV in high voltage electron microscopes. These electrons have both particle- and wave-like properties. As particles, they have a velocity that is approaching the velocity of light (164000 km/s for 100 keV, 282000 km/s for 1 MeV). As waves, they have a wavelength that is much smaller than atomic distances (0.0388 Å for 100 keV, 0.0123 Å for 1 MeV). The interaction between the electrons and the magnetic induction can be described in both pictures. In the particle representation, corresponding to classical beam optics, the electrons are deflected by the Lorentz force:

$$F_{\mathrm{L}} = q_{\mathrm{e}} (v_{\mathrm{e}} \times B) \tag{2.14}$$

where q_{e} and v_{e} are electron charge and velocity, and B is the magnetic flux density. Only the components of B which are perpendicular to the electron beam are effective. The Lorentz force has to be integrated over the whole electron path, not only over the trajectory through the sample. Stray fields outside the sample will contribute to the contrast as well. In unfavourable cases they can even compensate the effects of the magnetization in the sample.

Of the three basic arrangements of bar-shaped domains (Fig. 2.24) only one is visible in an untilted sample. For the other two the net Lorentz deflection cancels to zero. In-plane domains as they typically occur in soft magnetic films are normally of the well visible type (Fig. 2.24a), whereas the problematic modes (b) and (c) occur in magnetic recording.

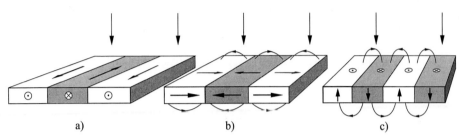

Fig. 2.24 Three infinitely long domain patterns, of which only (a) yields a net deflection of the electrons if illuminated perpendicularly. In case (b) the deflection by the magnetization is cancelled by the deflections due to the stray field above and below the sample if the pattern is infinitely wide (this can be proven using Gauss' theorem); for a finite width contrast will arise at the lateral edges. In case (c) the Lorentz force vanishes inside the sample because the electrons run parallel to the magnetization at perpendicular incidence, and the deflections by the stray fields above and below the sample cancel. (There may be a contrast from the domain walls in the latter case)

The difficulty indicated here is common to all modes of Lorentz microscopy. In some cases it can be circumvented by tilting the sample. If the sample in Fig. 2.24c is tilted around the axis perpendicular to the bands, the electrons see a magnetization component pattern as in (a). Also in Fig. 2.24b the visibility of the domains can be improved by a large angle tilting around the same axis if the width of the pattern is finite. The electrons then pass partially through the weaker lateral fields, and the deflections are no more fully compensated as elaborated in [254]. By systematic tilting and image processing it may be possible to completely separate the localized magnetization and the non-localized stray field, which is, however, not easy in an electron microscope.

Microscopy based on the Lorentz force relies on small changes in the electron direction. Those electrons that undergo strong deflections by collisions with nuclei in the sample, or which are inelastically scattered from the sample core electrons, contribute little to magnetic contrast, but generate a disturbing background. These effects also set a limit to the maximum sample thickness (some 100 nm) which can be used in Lorentz microscopy.

2.4.2 Conventional Lorentz Microscopy

(A) Defocused Mode Imaging. Even if there is a net deflection by a sample, it will not generate any contrast in the regular bright field *focused* image.

Out-of-focus, however, shadow effects (Fig. 2.25 a) delineate the boundaries of uniformly magnetized domains at least for regular stray-field-free domains (Fig. 2.24 a). Because this method of obtaining magnetic contrasts is based on the Lorentz force, it is called defocused mode Lorentz microscopy, or more specifically *Fresnel*-mode Lorentz microscopy because of the similarity of the mechanism with that of the optical Fresnel interference experiment.

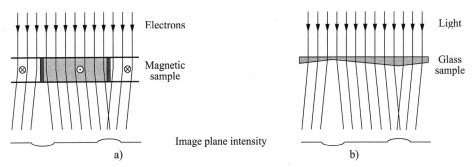

Fig. 2.25 Electron deflection in the shadow, or "Fresnel" mode (a). The microscope is focused some millimetres below the sample. (b) Optical analogue of (a)

In the discussion of the defocused mode the wave-optical aspect is particularly useful. In this point of view the magnetization influences the *phase* of the electron wave. Defining B_0 as the average component of B perpendicular to the beam, and s_b as the beam direction, the phase variation can be written in the simple form [255, 256]:

$$\text{grad}\varphi_e = (2\pi q_e D/h)\, B_0 \times s_b \tag{2.15}$$

where φ_e is the phase of the electron wave, D is the film thickness and h is Planck's constant. A uniformly magnetized domain thus corresponds in optics to a glass plate with a linearly varying thickness, as shown in Fig. 2.25b. Such an optical model reveals an equivalence between the classical aspect (refraction by the prism) and the wave-optical aspect (the phase object). In the wave-optical formulation quantitative calculations of contrast for a given magnetization distribution, including electron interference effects, can be performed [257].

Electron diffraction fringes appear, for example, in the "convergent" wall image of the Fresnel mode if a well localized, coherent electron source is used (Fig. 2.26). Such a complicated pattern carries only indirect information about the *wall* structure between two domains. (In the optical model of Fig. 2.25b, the wall structure corresponds to the shape of the prism edges). If models of walls or other micromagnetic structures are available, the comparison of observations with wave-optical calculations can test their validity [258–260], but in general it is not possible to *see* micromagnetic features immediately in the defocused mode of Lorentz microscopy.

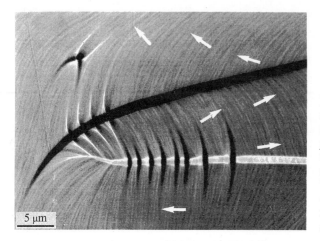

2 μm

5 μm

Fig. 2.26 Lorentz picture of a polycrystalline Permalloy film showing the streaky *ripple* texture perpendicular to the mean magnetization in the domains and *cross-tie* walls. The convergent wall displays diffraction fringes as shown in the enlargement. (Courtesy *S. Tsukahara* [261])

Perhaps the only exception is the asymmetry of Bloch walls that shows up in the "divergent" wall image of single-crystal films as demonstrated first by *Tsukahara* (Fig. 2.27; in the convergent wall image the asymmetry is less obvious. Single-crystal pictures contain, in addition to the magnetic contrast, other crystallographic contrasts (extinction contours, dislocation lines, etc.) which can, however, be readily identified by the experienced observer.

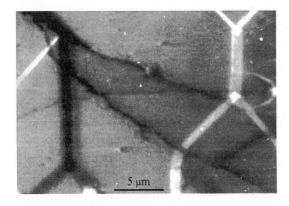

5 μm

Fig. 2.27 A defocused mode Lorentz picture from a (100)-oriented single-crystal iron sample of unknown thickness. Note the asymmetric profile of the divergent (black) wall which reflects the internal structure of the asymmetric Bloch wall (Sect. 3.6.4D). The irregular lines are lattice contrasts. (Courtesy *S. Tsukahara* [261])

In general, standard Lorentz microscopy is well suited to the observation of domains in thin films, but not to high resolution investigations of walls and their substructure. Even in domain observation there is a limitation since uniformly magnetized domains do not indicate the direction of magnetization. This can be derived, however, by several indirect methods:

- The reaction of walls to magnetic fields of various directions gives clues to the magnetization direction.
- In polycrystalline samples a characteristic fluctuation of the magnetization, the magnetization *ripple* (Sect. 5.5.2C), is always oriented perpendicular to the average magnetization direction of a domain ([243]; Fig. 2.26).
- In separate experiments *small angle diffraction* from a selected area records the direction of the Lorentz deflection and thus the average magnetization direction in this area.

(B) Classical In-Focus Domain Observation Techniques. From the beginning of magnetic domain observation in the electron microscope there were attempts to find alternate modes of Lorentz microscopy which permit a more direct image interpretation.

Some in-focus approaches used the interaction between the Lorentz deflection and Bragg diffraction in crystalline samples [262–264]. These methods did not find wider acceptance, probably because they depend on the crystal perfection of the specimen. They are useful, however, for studies of the interaction between magnetism and lattice defects.

The analogy to optical phase microscopy stimulated the approach of *Boersch* and *Raith* [241, 242] of using a diaphragm in the diffraction plane obstructing half the aperture (Fig. 2.28a). This "schlieren" or *Foucault* method shows regular domain contrast that is easy to interpret, but the details depend critically on the exact position and nature of the diaphragm edge, which are in general not known. For a long time this method was considered to offer little advantages compared to the Fresnel method. It was revived with improved technology [265] and demonstrated its value particularly in the analysis of small thin film elements (Fig. 2.28 b,c). The pictures resemble Kerr images, but they surpass optical resolution by more than an order of magnitude. This technique is well suited for the investigation of domain patterns, less so for domain wall studies because of the mentioned fundamental difficulties.

Two further in-focus methods — discussed in the following sections — offer truly quantitative information in magnetic microscopy. In the first method a *scanning* transmission electron microscope is used with a special detector employing difference images. The other method is based on electron *holography* with optical or numerical reconstruction.

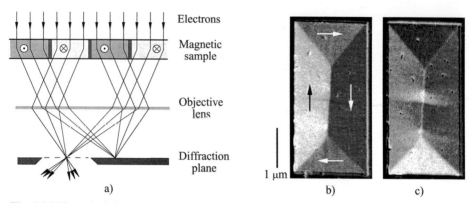

Fig. 2.28 The principle of the Foucault technique of Lorentz microscopy (a) and an example of two aspects of a domain pattern obtained with this technique on a thin Permalloy element of 24 nm thickness, displaying the longitudinal (b) and the transverse magnetization component (c). (Courtesy *K.J. Kirk* and *J.N. Chapman*)

2.4.3 Differential Phase Microscopy

(A) The Standard Scanning Technique. This method is a variant of *phase* microscopy that is based on a scanning transmission electron microscope [266–272]. The sample is scanned with a fine electron beam. In the diffraction plane of the microscope, a special split detector converts the diffracted beam into electrical signals that may be displayed on a video screen or digitized and stored in a memory (Fig. 2.29a). If the detector consists of two halves, the difference between the two signals is proportional to the magnetic deflection of the beam and therefore to the average magnetization component parallel to the axis along which the detector is split. The technique is known as *differential phase contrast* (DPC) microscopy.

The result is an in-focus picture with domain contrast as in Kerr or Faraday effect images (Fig. 2.29b), but with a much higher resolution, which can reach values better than 10 nm. It was proven that the linearity of the imaging process holds even including diffraction effects [270, 271], deviations being expected only for domain wall widths in the nanometre range.

Rotating the detector, the microscope becomes sensitive to the other component of magnetization as demonstrated in Fig. 2.30. In practice, a quadrant detector is used, the signals of which can be combined in various ways. If stray fields outside the sample can be neglected (as is typical for soft magnetic materials), the combination of two such images leads to a quantitative determination of the magnetization directions [273]. For high resolution, the electrons must be carefully collimated. The illumination system has a function equivalent to the objective lens in the conventional electron microscope. Thus the electron optics of the two microscopes do not differ fundamentally [251].

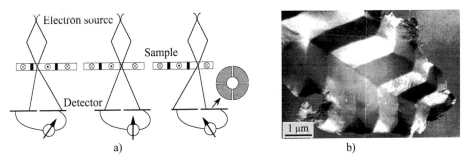

Fig. 2.29 The principle of the differential phase contrast in a scanning transmission electron microscope (a). The high resolution attainable with this method is demonstrated in (b) showing domains in an iron foil. (Courtesy *J.N. Chapman*)

The differential phase microscope offers two further advantages: the magnification may be freely adjusted by changing the scanning amplitude, and the imaging process is expected to be insensitive to inelastic scattering of the electrons as far as these deflections cancel in the difference operation. Therefore the method is in principle applicable to thicker specimens than otherwise possible in electron microscopy.

A difficulty connected with the DPC method has to be mentioned: not all structural features produce a symmetrical scattering pattern which would cancel in the difference procedure. Therefore, pictures often show a more or less pronounced phase contrast from the grain structure and from the sample edges, in addition to the magnetic contrast. The effect tends to be stronger for thinner samples. Using a ring-shaped detector, where electrons close to the central beam are excluded, largely suppresses this side-effect for samples with planar magnetization as shown in [272, 274, 275] and in Fig. 2.30. Non-magnetic *amplitude* contrasts that may become visible in the differential image because of a slight defocus can be compensated by digitally subtracting a differentiated and weighted bright field image as demonstrated in [276].

Apart from the difficulty with non-magnetic contrast, the differential phase method offers high resolution and quantitative information about the average in-plane flux density of a sample. As discussed in connection with Fig. 2.24, even this information may not be sufficient to derive the full three-dimensional magnetization vector field. *Beardsley* [277] proved that the final goal of a complete determination of the magnetization vector can be achieved if — in addition to a DPC picture — the stray field in a plane above and below the sample is recorded. Some of the methods discussed in Sect. 2.5 open the perspective of solving this fundamental problem.

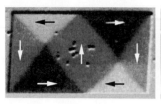 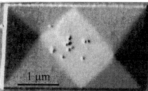

Fig. 2.30 Differential phase images showing orthogonal magnetization components of a Permalloy element of 60 nm thickness. (Courtesy *S. McVitie* and *J.N. Chapman* [273])

(B) Differential Phase Contrast in a Conventional TEM. An interesting variant of differential phase imaging was proposed and demonstrated in [278–280]. A series of Foucault images is digitally combined to obtain a quantitative image. The technique is related to the standard scanning DPC method by a duality relation. Instead of scanning the electron beam over the sample, the angle of incidence of the illuminating electron beam is systematically varied in a quadrant arrangement in the aperture plane. The recorded images in each quadrant are digitally averaged, and differences are formed between these average images as in the scanning DPC procedure. The advantage of this method is that it can be set up in a conventional electron microscope.

2.4.4 Electron Holography

Electron holography [256, 281–284] is generally a very powerful technique. It is based on recording an interference pattern from which amplitude and phase of an object can be reconstructed.

(A) Off-Axis Holography. In the standard technique of off-axis holography (Fig. 2.31) imaging is performed in two steps: first, an electron hologram is generated with the help of a coherent electron beam and then magnetic information is extracted by an optical or electronic reconstruction process. The standard holographic method is restricted to samples that do not fill the object plane completely, so that an undiffracted reference beam can pass by the sample. The two beams are reunited by an electrostatic system consisting of a charged wire and two plates (a so-called biprism from its optical analogue) to form an interference pattern, the hologram.

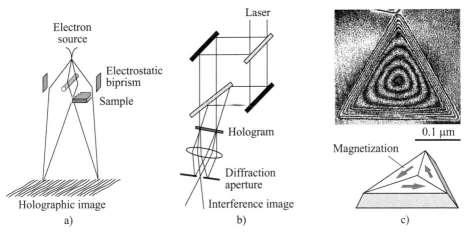

Fig. 2.31 Principle of conventional electron holography (a), optical reconstruction using a *Mach-Zehnder* interferometer (b), and the resulting image (c) showing the lines of magnetic flux (as averaged along the electron trajectories) inside a triangular cobalt platelet. The narrow lines at the sample edge are caused by the regular topographical phase shift. (Courtesy *A. Tonomura*)

The magnetic phase image can be reconstructed from the hologram by an optical interferometer of the Mach-Zehnder type [282] with the help of a split laser beam (Fig. 2.31b). The illumination of the holographic plate with coherent light produces two images of the original object, the regular and the conjugate image. The regular image of the one laser beam is combined with the conjugate image of the second beam to form a visible image. Interference produces maxima where the phase difference of the two beams is zero or a multiple of 360°, and minima in between. If we would obstruct one of the two branches in the interferometer, we would get no image of the magnetic phase.

The resulting lines of constant phase are simple to read if the phase shifts are caused by magnetic fields only: according to (2.15) the phase gradient is perpendicular to the average planar component of B. The lines of constant phase are therefore parallel to this component B_0, and the separation of the lines in the phase interferogram is given by the condition that the flux between two lines must be equal to the flux quantum h/q_e. Figure 2.31c shows an example of a picture from a cobalt platelet. To obtain a good hologram requires extreme stability of the microscope, special high-coherence electron sources, and favourable samples. Incoherent electron scattering tends to blur the hologram and limits the useful specimen thickness to well below 0.1 µm.

In a variant of the optical reconstruction process the photographic holograms are first copied on *hard* photographic film, enhancing the contrast so that even higher-order images can be combined, thus leading to a higher density of flux contours. The same effect can be achieved with modern techniques of digital image processing that are reviewed in [284] and analysed in [285]. Another advanced technique transmits the recorded hologram as a video signal to a liquid crystal display, which is simultaneously illuminated by a laser beam, thus offering the possibility to observe the reconstructed image in real time [286, 287]. It is also possible to generate a reconstructed image with electron-optical means right in the microscope, as demonstrated in [288].

(B) Differential Holography. Variants of holographic techniques are reviewed and demonstrated in [253] for the scanning transmission microscope. In one of these variants, which is called the "differential mode", domain wall profiles of remarkable quality with a resolution in the 5 nm range can be reconstructed [289, 290]. The method evaluates the interference between two beams passing through the sample displaced by a small amount. The reconstruction thus contains information about the phase difference between the two beams. The resulting image is no more dominated by interference contours in the domains as in Fig. 2.31c, but displays a uniform domain contrast just as Kerr microscopy or the Foucault technique. The reason is that the magnetization is proportional to the phase gradient according to (2.15), so that a method that displays only the differential of the electron phase has to be considered the natural choice for magnetic imaging.

Differential holography can also be implemented in a conventional (field emission) transmission electron microscope as reported in [291]. Here the two interfering beams are generated by a biprism as in conventional off-axis holography, but the hologram is recorded close to the sample image position. In the mentioned article a computer reconstruction result is presented that is equivalent to differential phase microscopy as discussed in Sect. 2.4.3, with the additional possibility of removing all kinds of image distortions and artefacts in the computation. An important advantage of differential holography is that also extended thin films can be investigated because both interfering beams pass through the sample.

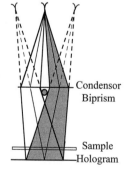

(C) Coherent Foucault Technique. Another new technique, which is not strictly holographic but yields similar pictures as standard holography, was described by Chapman et al. [292–294]. The method is based on a highly coherent microscope and a special aperture filter. As this aperture filter is similar to the one used in the Foucault technique, this new method can be called a *coherent Foucault* technique. The best results are obtained with a phase shifting filter, which shifts by π all magnetically deflected beams, as well as exactly one half of the central, undiffracted beam. By this slicing of the central beam, the electron wave that passes the sample is diffracted to interfere with the magnetic information. Clear interference patterns can be obtained with this procedure, obviously much easier and from a wider range of samples compared to conventional electron holography. An example is shown in Fig. 2.32. As in holography, it remains a necessary condition that the sample is smaller than the field of view, so that some electrons can pass it to allow interference. The resolution is determined by the fringe spacing.

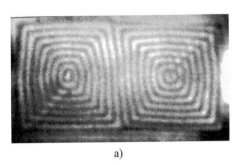
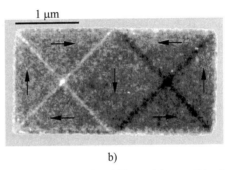

a)   b)

Fig. 2.32 Coherent Foucault image (a) compared with a conventional Fresnel image (b) of a thin film element displaying a flux-closed domain structure. The image was obtained at coherent imaging conditions, with a $\lambda/2$ phase plate in the aperture that contained a hole for the central beam, positioned so that the central beam is cut in two halves by the edge of the hole. (Courtesy *A. Johnston* and *J.N. Chapman*)

(D) Critical Assessment of Holographic Techniques. Traditional electron holography (Fig. 2.31) is often claimed to be the only quantitative Lorentz microscopy technique. In fact, the amount of flux (integrated along the electron trajectories) between two points can be quantitatively derived from a holographic image as elaborated above. But in ferromagnetism, the absolute value of the magnetization is a material constant and normally known or readily available. The sample thickness can be measured more directly than by a complex electron interference experiment. The only unknown in the investigation of magnetic microstructures is the magnetization *direction*. A quantitative domain observation technique has to be able to determine this magnetization direction, and many electron-optical and other methods can compete in this field. The additional information about the size of the magnetic induction offered by traditional holography is useful in the investigation of magnetic fields outside a sample [295, 296], but less so in magnetic domain imaging.

The traditional reconstruction mode of displaying phase contours thus offers no real advantages in magnetic investigations. Important is only the phase gradient according to (2.15). In principle, this vector field can be reconstructed from any hologram. But the new modes of differential holography directly focus on this quantity and are less restrictive in the sample shape than off-axis methods. Progress in these techniques certainly deserves attention even if they do not reach record values in resolution.

2.4.5 Special Procedures in Lorentz Microscopy

Sample preparation is particularly important in TEM. Bulk material has to be thinned, preferably by electropolishing or ion beam thinning. The surface must be about as smooth as in magneto-optical observations. The useful film thickness is limited to $\leqslant 100$ nm in conventional and to a few hundred nanometres in high voltage microscopes. The differential phase technique, if applied to a high voltage microscope, will widen the useful thickness range.

One modification of the conventional electron microscope is always required: the strong objective lens produces a large axial magnetic field that would destroy the domain pattern to be observed in most cases. Therefore this lens must either be switched off, or the sample must be moved sufficiently away from the lens. This requirement limits the resolution in Lorentz microscopy. Favourable is the solution to add to the conventional "immersion" objective lens of the electron microscope another pair of "Lorentz lenses" which produce no field at the ideal sample position [297, 292]. This weaker objective does not reach the resolution of conventional lenses, but a resolution in the nanometre range is mostly sufficient for magnetic investigations.

Most domain observations in Lorentz microscopy are performed with standard 100–200 kV microscopes. High voltage microscopes [298, 299] offer a clearer picture since the Lorentz deflection decreases more slowly than inelastic electron scattering with increasing voltage. The optimum voltage may be 300–500 kV. Another advantage is the higher penetration depth. Clear domain images from single crystals of up to 0.5 μm thickness have been observed with a 1 MV electron microscope[5]. Inelastically scattered electrons can also be eliminated by energy filtering as demonstrated in [300].

---

[5] *Takao Suzuki*, private communication (1985)

It is possible to apply weak magnetic fields perpendicular to the microscope axis. Special magnetization stages built for this purpose compensate the influence on the electron beam with one or two additional coils [301]. The simplest approach uses the weakly excited objective lens, tilting the sample according to the desired field direction. Field components of up to about 100 kA/m parallel to the film plane can be obtained in this way. Along the microscope axis, objective lens fields of several hundred kA/m are available.

Tilting stages, which are useful for many purposes in electron microscopy, help to distinguish magnetic from structural contrast. The only way to get access to the magnetization component perpendicular to the surface (Fig. 2.24c) is to tilt the sample. Investigations of such structures in thin films used tilting angles up to sixty degrees [302].

If the picture is sufficiently bright, even dynamic processes may be recorded, although this is easy only in the defocused or Fresnel mode. The experiment by *Bostanjoglo* and *Rosin* [303] was unique observing the resonance of Bloch wall substructures (so-called Bloch lines, see Sect. 3.6.5C) stroboscopically at frequencies up to 100 MHz, using a gated and synchronized image amplifier.

Intensity is often a problem, in particular if high resolution is required. Frequently the pictures cannot be observed on the fluorescent screen but can only be recorded by photography with long exposure times. Special electron sources like the field emission gun are advantageous, and are almost mandatory for the holographic and the differential phase methods. This poses a severe practical problem: the field emission cathode can only survive under ultra high vacuum conditions. *McFadyen* [304] managed to implement the differential phase method in a conventional microscope, but he had to make concessions in resolution and he had to use digital image acquisition to make up for the reduced brightness of the electron source.

2.4.6 Summary

The unique features of TEM are:
- High resolution down to the nanometre range is available with modern techniques such as electron holography and differential phase microscopy.
- High contrast and sensitivity even to small variations of the magnetization can be obtained.
- Electron microscopy can observe directly the interaction between domain walls and lattice defects.

These positive aspects have to be weighed against the limitations:

- The expensive equipment can in general not be used for other purposes after being adjusted for magnetic investigations.
- The range of sample thickness is limited (to some hundred nanometres).
- Sample preparation is difficult.
- The field of view is restricted to at most a few tenths of a millimetre.
- Applying magnetic fields or mechanical stresses is difficult.
- Magnetization and stray field may compensate each other.

The relative merits of the different modes of transmission microscopy are:

- Domain observations and also observation of magnetization processes with good but limited resolution and excellent contrast also of fine details are best performed in the defocused or Fresnel mode.
- The somewhat more subtle Foucault method is able to produce medium-resolution, but quantitative *domain* images.
- Quantitative, very high resolution domain and wall images — also from thicker specimens — can be obtained with differential phase microscopy.
- Electron holography offers quantitative information about the magnetic flux distribution particularly in its differential mode. It requires, however, an elaborate technique, and it is restricted to samples thinner than 100 nm.
- If, in addition to a quantitative differential phase image, the stray field above and below the sample can be recorded, the complete three-dimensional magnetization distribution can in principle be derived mathematically.

2.5 Electron Reflection and Scattering Methods

2.5.1 Overview

Electron *mirror* microscopy [305, 306] was an early attempt to use electrons to display magnetic structures of bulk specimens. The method suffered from distortion and limited resolution and was largely abandoned. The situation changed with the widespread introduction of the *scanning* electron microscope (SEM) in which the surface is scanned by a fine electron beam. The scattered or re-emitted electrons are collected, their intensity is processed electronically and displayed on a video screen. In the SEM the electrons actually hit the sample with energies in the 10–100 kV range, as opposed to the mirror microscope, where the electrons are reflected above the sample surface.

Two kinds of re-emitted electrons are distinguished: (i) Some electrons are *scattered back* from the nuclei of the atoms in the sample. Their energy ranges from the energy of the primary electrons (elastic scattering) to 10–20% lower (inelastic scattering). (ii) Other electrons are emitted from atoms that have been excited by the electron beam. These *secondary* electrons have energies of a few up to about 50 eV. If the sample is magnetic, all these electrons are somehow deflected, so that magnetic information can be extracted with the help of collectors that are sensitive to the electron direction. Since the effects on the different classes of electrons are different, energy-selective collectors are used. The low energy *secondary* electrons are quite sensitive to magnetic stray fields above the sample. The high-energy backscattered electrons are primarily influenced by the magnetization inside the sample. In addition, the state of *polarization* of the secondary electrons depends on the magnetization direction. An important method is based on this effect. Since the total number of re-emitted electrons will in general be different from the incident electron flux, the sample must be sufficiently conductive to avoid electrostatic charging. Insulating materials must therefore be coated with a metal film before they can be studied in a reflection electron microscope.

Previous reviews of the many possible modes of domain observation with the SEM can be found in [307–310].

2.5.2 Type I or Secondary Electron Contrast

A typical setup for observing stray field contrast in the SEM [311–313] is shown in Fig. 2.33 a. The sample is oriented perpendicular to an electron beam of less than about 10 keV. The secondary electrons are collected in an asymmetric arrangement. Their intensity depends on the magnetic field component $H_y(r)$, which deflects the electrons towards or away from the collector. In a quantitative analysis as reviewed in [308], the magnetic signal for a given detector geometry, which is tilted about the y axis as in Fig. 2.33a, depends on the integral over an electron trajectory:

$$S_y(x, y) = \int_0^\infty H_y(r) \, dz . \tag{2.16}$$

Since this integral is a smooth function of the coordinates x and y, even if the underlying domains have sharp boundaries, type I images are diffuse, displaying mostly the fundamental harmonic of the magnetic pattern. Figure 2.33 b shows a typical example.

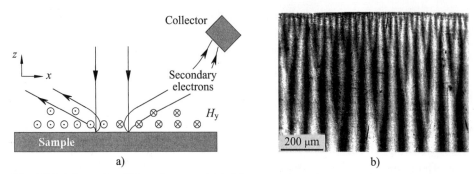

Fig. 2.33 Schematic of Type I contrast (a). The secondary electrons of the two beams see different fields and are deflected into different directions. (b) Domains on a cobalt crystal (edge view) as a typical example for this technique. (Courtesy *J. Jakubovics* [312])

With a regular large collector the maximum contrast is calculated to:

$$C = 8\,\mu_0 q_e\, S_{\max}/(\pi\, m_e v_e) \tag{2.17}$$

where $S_{\max}$ is the maximum value of the integral S_y, and q_e, m_e, and v_e are charge, mass and velocity of the secondary electrons. The contrast thus increases with decreasing electron energy. It can be enhanced by a careful adjustment of the collector shape, and by energy filtering. Empirically, images have been obtained when the integral S_y ranged between 0.2 and 0.02 A, depending on the smoothness of the sample [308]. For domains of 2 μm period, for which the integral will extend to about the same distance from the surface, stray fields of the order of 10 to 100 kA/m are necessary to keep the integral within the required limits. Therefore the method does not reach the sensitivity of the Bitter technique. An advantage is the quantitative connection with the stray field. If a model of the domain pattern exists, the electron contrast can be calculated, giving access to parameters of the model. The analysis can be refined by varying the collector position, thus displaying different components of the horizontal stray field and different magnetic periods [308].

O.Wells [314, 315] invented an interesting possibility to fully reconstruct the perpendicular field component at the sample surface from blurred type I images. His method starts from $\operatorname{div} \boldsymbol{H} = \partial H_x/\partial x + \partial H_y/\partial y + \partial H_z/\partial z = 0$ which is valid outside the sample. Integrating $\partial H_z/\partial z$ from zero to infinity, and inserting (2.16), the component H_z at the surface becomes:

$$H_z(x, y, 0) = \partial S_x(x, y)/\partial x + \partial S_y(x, y)/\partial y \; . \tag{2.18}$$

The procedure thus requires the differentiation and combination of two images S_x and S_y obtained with two perpendicular detector tilting axes. From the value of H_z at the sample surface, the field vector at every point above the surface can be calculated by potential theory. No practical realization of this possibility has been published so far.

Another "tomographic" way to determine the field above a surface using grazing angle electron reflection was demonstrated ([316]; for earlier applications of similar techniques see [315]). These methods could be used together with differential phase microscopy (Sect. 2.4.3) to determine the full three-dimensional magnetization field of a thin sample as mentioned in Sect. 2.4.3A. For further developments in stray field tomography see [317, 318].

The secondary electrons used in type I contrast may be excited also by ultraviolet light from a mercury arc source. The electrons are then called photoelectrons, but the contrast mechanism is the same as demonstrated by *Mundschau* et al. [319]. Because scanning a light beam is more difficult than electron scanning, the image was formed in a photo emission electron microscope (PEEM) rather than by a scanning technique as in regular type I contrast. The same kind of microscope is used in X-ray spectroscopic methods which will be discussed in Sect. 2.7.3.

2.5.3 Type II or Backscattering Contrast

A stray field contrast as discussed in the last section is not expected for soft magnetic materials with their low anisotropy and small stray fields. So it was a surprise when *Philibert* and *Tixier* [320] discovered clear and well-defined domain contrasts on silicon-iron transformer steel by selecting the *backscattered* electrons and choosing a tilted sample orientation.

The mechanism behind these pictures was clarified later [321–323]. Figure 2.34a shows schematically the experimental arrangement. The electrons are deflected on their path through the tilted sample by the magnetic induction — either towards the surface, thus enhancing the backscattering yield, or away from the surface with the opposite effect. The symmetry of the phenomenon can be described by:

$$S_B = S_0 + F_I(\vartheta_0, E_0) \, \boldsymbol{B} \cdot (\boldsymbol{k} \times \boldsymbol{n}) \tag{2.19}$$

where S_B is the backscattering intensity, S_0 its background value, $\boldsymbol{B}$ the magnetic induction, $\boldsymbol{k}$ the primary electron propagation direction, and $\boldsymbol{n}$ the surface normal. The factor F_I depends on the angle of incidence ϑ_0 and the primary electron energy E_0. The sensitivity is largest for a magnetization perpendicular to the plane of incidence, as in the transverse Kerr effect.

The angular dependence of the factor F_I leads to maximum contrast at about $\vartheta_0 = 40°$. The effect increases with the primary energy E_0 approximately with the 3/2 power. In any case the contrast is weak. It varies typically from tenths of a percent for conventional SEM energies of 30 keV to one percent in high voltage instruments of 200 keV. High voltage also reduces the disturbing structural and topographical contrasts relative to the magnetic signal. With electronic contrast enhancement clear pictures can be obtained in the latter case (Fig. 2.34b). Unfortunately, high voltage SEMs are not as readily available as conventional instruments. Scanning *transmission* microscopes operated in reflection mode can be used for this purpose [324]. For dynamical investigations, unwanted non-magnetic contrast can be suppressed by applying an alternating magnetic field and detecting the backscattered electrons with a lock-in amplifier [325–327]. Also digital image subtraction as in Kerr microscopy was demonstrated to solve the problem of detecting a weak magnetic contrast superimposed onto a strong structural background [328]. As even weak external fields lead to a displacement of the obtained image, the two images to be subtracted have to be carefully registered before the final difference operation is performed.

When comparing pictures such as Fig. 2.34b with corresponding Kerr effect pictures, the results of the two methods might be considered largely equivalent. But they differ in two decisive aspects, namely in resolution and in the sensitivity to magnetic surface structures. Resolution of the backscattering electron method is fundamentally limited by the path of the electrons inside the sample. At high electron energies, which give the clearest pictures, the scattering range of the electrons is largest. Measurements indicate a resolution limit of the order of 1 μm at 100 keV and about 3 μm at 200 keV [329]. This is clearly inferior to the magneto-optical techniques. On the other hand, the penetration ability of the electrons offers also advantages. At 200 keV the electrons reach a depth of about 15 μm, and the depth for maximum backscattering is estimated to be about 9 μm [329]. The method is therefore insensitive to thin surface layers, or to the quality of polishing; even the insulation coatings of electrical steels that are typically 3–5 μm thick do not disturb the magnetic image substantially [329, 330].

Since the penetration depth depends on the electron energy, domain structures can be scanned at various depths with this method [331, 327]. While the Kerr effect "sees" effectively the top 20 nm of a metal sample, the electron backscattering method sees depths between 1 μm and 20 μm depending on electron energy. But using a high penetration depth introduces a danger, too. Information about shallow surface domains may be lost. For example, the "grey" zones in the article of *Nozawa* et al. ([332]; Fig. 2.34 b) appear to be domains magnetized in a hard direction; in reality they stand for a shallow system of closure domains revealed in Kerr effect pictures of corresponding samples [333].

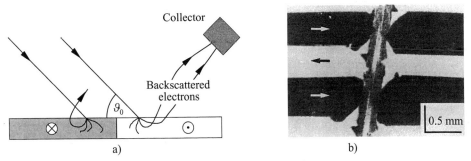

Fig. 2.34 Backscattering contrast in the SEM (a) with some typical electron paths inside the sample, and an image of a silicon-iron transformer steel sample (b) showing domains near a scratch introduced for the purpose of domain refinement. (Courtesy *T. Nozawa* [332])

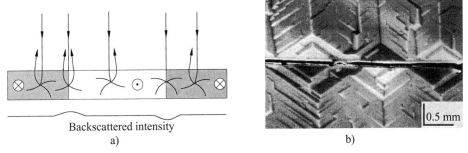

Fig. 2.35 Domain boundary contrast formation in the SEM at oblique incidence with the plane of incidence parallel to the domain walls (a). The backscattering probability is larger near the left than near the right wall. (b) Boundary contrast on a (100) SiFe crystal, together with domain contrast from closure domains near a scratch. (Courtesy *J.P. Jakubovics*)

The domain contrast mechanism discussed so far relies on the vertical deflection of the electrons, towards or away from the surface. There is another (non-local) effect that is based on an accumulation of electrons scattered from both sides of a domain wall (or the opposite effect). This leads to a domain *boundary* contrast [334–336] as in defocused-mode Lorentz microscopy, when the sample is tilted with the plane of incidence parallel to the domain wall (Fig. 2.35 a). The image width of the boundary contrast in (b) is not related to the wall width but is a measure of the scattering range, and therefore of the resolution of the method. Its sign is determined by the electron path after scattering, as indicated in Fig. 2.35 a. This interpretation was confirmed in [336] by statistical ("Monte Carlo") calculations of the scattering process.

An attractive mode of stroboscopic domain imaging [337] can be applied to all slow scanning methods. If domain walls are made to oscillate while the image lines are written, they appear meandering in the picture (Fig. 2.36). The amplitude and wave shape in the image reflect the dynamic wall behaviour.

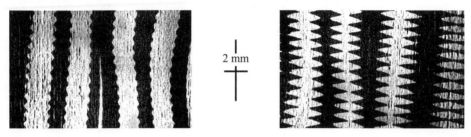

2 mm

Fig. 2.36 Stroboscopic type II SEM pictures, taken from a metallic glass while the walls were excited at 60 Hz with two different amplitudes. (Courtesy *J.D. Livingston*, Schenectady)

A variant of backscattering contrast was demonstrated in a largely unnoticed early article [338] and was discussed again in connection with observations on metallic glasses [339]. It is based on a magnetically induced *angular* asymmetry of backscattered electrons, which is detectable with a sectored backscattering detector (Fig. 2.37a).

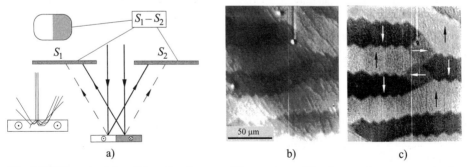

a)   b)   c)

Fig. 2.37 The principle of the direction-sensitive variant of backscattering contrast in the SEM (a) and an example of an observation with this technique (b, c) taken on (100)-oriented silicon-iron sheet (in collaboration with *L. Pogany* [340]). A saturated image was digitally subtracted from the domain image to reduce the contrast of non-magnetic origin (c)

Split detectors are standard in most SEMs, and are used for surface topography contrast. In [339], this angular contrast was not demonstrated because the authors used the sum signal of the split detector, thus getting only domain boundary contrast as in Fig. 2.35. Using the *difference* signal instead, clear domain contrast pictures at relatively low voltages can be obtained ([340]; Fig. 2.37c). An attractive feature of this difference technique is that it suppresses much of the background contrast, so that lower voltages than in the standard SEM technique are possible. This means that any standard SEM equipped with a split backscattering detector can be used. The signals are weak, however, and even with the added digital difference technique, demonstrated in Fig. 2.37c, it is difficult to completely separate magnetic from other contrasts in this mode. To fully explore the potential of this method a high-intensity, field emission instrument would have to be employed.

2.5.4 Electron Polarization Analysis

Secondary electrons emitted by a magnetic sample have one more important property in addition to their energy and their direction: they are also spin-polarized, with the magnetic moment parallel to the magnetization direction at the point of their origin [341–344]. *Koike* and *Hayakawa* [345–347] first demonstrated how a domain image can be obtained by measuring this polarization. The principle of their arrangement is shown in Fig. 2.38a.

The secondary electrons are collected and accelerated to 100 keV. The electron polarization is measured as the difference signal in a "Mott detector": the scattering of the accelerated polarized electrons by a gold foil is asymmetric because of spin-orbit coupling effects. Figure 2.38b shows a domain pattern of an iron crystal that is similar to a Kerr picture. There are several advantages of this method that is often abbreviated as SEMPA (scanning electron microscopy with polarization analysis; [348–351]):

- Since only the normalized difference signal is recorded, the image is insensitive to non-polarizing structural features.
- The method has the potential of high resolution, limited only by the electron beam width (Fig. 2.38c). In [352] an impressive correspondence between measured wall profiles and micromagnetic calculations within 10 nm accuracy was presented.
- The polarized secondary electrons see the magnetization in the top nanometre of a sample only. The method is therefore more surface-specific than Kerr microscopy that probes about 20 nm in metals.
- The connection between the electron spin and the surface magnetization vector is very direct. The signal from a pair of detectors as in Fig. 2.38 is proportional to the magnetization component perpendicular to the drawing

plane. The two in-plane magnetization components can be measured simultaneously and independently with four detectors. To measure the third, polar component, the secondary electron beam can be deflected electrostatically by 90°. Also a special spin rotator, which relies on the simultaneous action of electric and magnetic fields, can be employed [353].

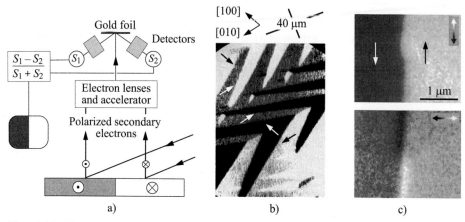

a)   b)   c)

Fig. 2.38 (a) Detecting polarized electron contrast in an ultra high vacuum SEM. The magnetic information is extracted by a signal processing unit connected to a polarization detector. (b) Domains on a (100) SiFe crystal. (Courtesy *K. Koike*). (c) High-resolution picture of the transition in the surface structure of a domain wall in an iron whisker, shown with two different sensitivity directions. (Courtesy *J. Unguris* and *R. Celotta*)

Electron polarization analysis therefore offers direct quantitative domain analysis, a possibility that is available in magneto-optical methods only after careful calibration and processing (Sect. 2.3.8). Figure 2.39 demonstrates this ability on an example in which a new feature of a classical domain pattern became visible for the first time. The texture of the in-plane magnetization component in the flower pattern of the cobalt basal plane (Fig. 2.39b) had not been seen before in Bitter (Fig. 2.7) and Kerr effect (Fig. 2.53) investigations which are dominated by the polar magnetization component. (Once it was established, it could be confirmed by special magneto-optic techniques [180]).

There are, however, limitations connected with electron polarization methods. Most important is the low intensity that leads to long exposure times because of the low efficiency of the polarization detectors. Depending on the number of picture elements and the required noise suppression it may last from 10 sec to several minutes. Improved low-energy polarization detectors [348, 349] enhance the practicability of the method to some extent.

Fig. 2.39 Polar (a) and in-plane (b) magnetization components of a cobalt crystal cut parallel to the basal plane, made visibly with electron polarization techniques. A colour code is used in the representation of the in-plane components [6] (Courtesy *J. Unguris*)

Another drawback lies in the difficult experimental technique. Since the electron polarization would be destroyed by any scattering of the electrons on their way from the sample to the collector, ultra high vacuum conditions and a correspondingly clean surface are needed. Only conducting materials can be investigated, since a non-magnetic metal coating on an otherwise insulating sample would destroy the electron polarization. A way out is a thin ferromagnetic coating, which is exchange-coupled to the substrate and which can be applied even for oxides [354]. Such a coating may also enhance the electron polarization effect of metals as demonstrated with 10–15 atomic layers of iron on a cobalt-rich recording medium [355]. The magnetization of the coating just follows the underlying magnetization pattern, thus displaying with a better polarization efficiency the domains of the underlying material.

2.5.5 Other Electron Scattering and Reflection Methods

(A) Low Energy Electron Diffraction (LEED). A modern surface imaging technique is based on the LEED method. This classical method of surface diffraction studies can be extended to a microscopic technique [356, 357]. The method is based on a *photo emission* or *surface* electron microscope, an immersion type instrument that gathers extremely low-energy electrons emitted from a surface, and which can reach a resolution in the 50 nm range [358]. Using polarized electrons instead of regular unpolarized electrons for illumination, the diffracted intensity becomes sensitive to the spin pattern of magnetic materials [359–361]. Note that on the imaging side, the electron polarization is

---

[6] See Colour Plate

irrelevant in this method so that no analyser is needed. The technique that is called SPLEEM (= spin polarized low energy electron microscopy) offers high resolution and good quality images because of its inherent efficiency (much more efficient electron polarizers than analysers are available). Its main attraction is its potential for simultaneous growth studies and surface analysis as demonstrated for example by *Duden* and *Bauer* [362] who correlate the in-plane and perpendicular magnetization components of an ultrathin Co film with the atomic step pattern of these films.

(B) Further Techniques. The interaction between Bragg diffraction and magnetic deflection of the electrons was explored by several authors [307, 363, 364]. These aspects are particularly useful for studies of the interaction between crystal features and magnetic structures. Crystal diffraction effects always contribute to the details of the observed magnetic contrasts in crystalline material. Another effect was discovered by *Balk* et al. [365] who used an acoustic detector instead of an electron collector to investigate a silicon-iron crystal. The primary electron beam is modulated with a high frequency, thus generating thermal and acoustic waves. If the detector is made sensitive to the second harmonic of the electron beam frequency, a diffuse magnetic image appears that may be caused by some reaction of the Bloch walls with the elastic wave. The acoustic waves penetrate through bulk crystals and, by monitoring their phase, information about the origin of the sound waves (probably coming from magneto-elastically active 90° walls) can be obtained.

To circumvent the intensity problem in the standard polarized electron methods *Kay* and *Siegmann* [366] propose the use of a scanning laser beam to generate polarized photo electrons. Much higher intensities are available in this way, so that even a high-speed observation of magnetic structures seems possible if the laser beam can be scanned at sufficient speed, but the resolution will be limited to about 0.5 μm.

Further methods use the imaging of optically excited "photoelectrons" rather than scanning. If regular white light is used to generate the photoelectrons, the emerging electrons are deflected by the magnetic fields above the sample as in secondary electron contrast (Sect. 2.5.2). *Spivak* et al. demonstrated in early work [367] that photoelectron microscopy of domains reaches high resolution, comparable to that of optical microscopy and better than in the later developed type I scanning electron microscopy. The reason why this method was not pursued can probably be attributed to the widespread availability of ordinary scanning electron microscopes.

2.5.6 Summary

Here are the advantages and drawbacks of domain observation methods using reflected electrons. The positive features are:

- Small sensitivity to surface conditions in the high-energy backscattered electron method makes this technique useful for industrial conditions.
- With SEM methods, one can look through a non-magnetic insulation layer and to some extent through surface domains into the bulk.
- With secondary electron methods, inhomogeneous stray fields above a magnetic sample can be registered.
- The electron polarization method offers high resolution and quantitative images of the surface magnetization.
- Domain observations may be combined with the powerful micro-analytical tools of scanning electron microscopy.

Negative aspects are:

- The standard methods suffer from poor resolution, in contrast to general expectations connected with electron microscopical techniques.
- The equipment becomes expensive especially if high-voltage microscopes or ultra high vacuum conditions are needed.
- The possibilities to handle the sample, to apply magnetic fields or elastic stresses are limited by the conditions of the electron microscope.
- Dynamic processes can be observed only by stroboscopical techniques since the image build-up requires a few seconds.
- Non-conducting samples require a metallic coating.

The methods using electron polarization analysis offer high resolution far beyond the optical limit, but they suffer from insufficient efficiency leading to severe noise problems. These problems are not of a fundamental nature, however, so that a decisive breakthrough may occur anytime. Efficiency is not a problem for the SPLEEM method. Its technology is rather complex, however, having to fight with disturbing magnetic fields, surface contamination and stability problems, so that its potential impact can not yet be evaluated.

2.6 Mechanical Microscanning Techniques

Reports about attempts to scan magnetic materials for stray magnetic fields can be found in early sources [45]. The field experienced a new impetus with the introduction of the scanning tunnelling microscope [368], the operation principle of which forms the basis for most of the presently used techniques. Another name for all these techniques is "scanning probe microscopies". Their possible applications for magnetic microstructures are reviewed in [369].

2.6.1 Magnetic Force Microscopy (MFM)

The magnetic force microscope is a variant of the scanning (or "atomic") force microscope [370], which is itself a spin-off of the scanning tunnelling microscope mentioned above. It records the magnetostatic forces or force gradients between a sample and a small ferromagnetic tip. Starting from pioneering work [371−374] the method has found a remarkably widespread acceptance, particularly in the field of imaging magnetic patterns in magnetic recording media. The two most prominent advantages of the technique contributing to this success are its potential insensitivity to non-magnetic surface coatings and relief, and good resolution down into the nanometre range [375]. Reviews of the principles and methods of MFM can be found in [376, 377] and more compact introductions in [378, 379].

(A) Experimental Procedures. In force microscopy, forces are measured by the deflection of a flexible beam (the so-called cantilever), which carries the tip-shaped probe at its free end. It can be adjusted by a piezoelectric actuator and its position can be detected by a wide variety of sensors — from a tunnelling tip over piezoelectric, dielectric to optical transducers. Optical methods using, for example, the optical interference between the tip of a fibre and the cantilever as the controlling signal [371] are favoured because of their simplicity. The control signal can be used to operate the scanning microscope in different modes. One option is to run it at constant force (equivalent to a constant deflection of the cantilever) and use the height necessary to obtain this state as the imaging information. Another mode consists in operating the elastic tongue at a frequency close to its mechanical resonance, and to detect any change in the resonance amplitude or shift in phase. Since a magnetic force gradient is equivalent to an additional contribution to the spring constant of the cantilever, profiles of constant force gradient can be recorded in this way.

A particularly elegant method was developed by Digital Instruments, the makers of the commercial Nanoscope® instrument. Here the sample is scanned twice. First the surface profile is recorded by intermittently measuring repulsive forces; then in the second run the force gradients or forces are recorded at an adjustable distance above the previously measured topographic profile. The advantage of this procedure is automatic subtraction of surface features from the total image, leaving primarily magnetic information. A constant height mode relative to the surface profile can be reached also by other means. In [380, 381] this preferable operation mode is achieved by employing a modulated electrical voltage between probe and electrically conducting sample. Electrostatic forces vary with the square of the voltage. The force signal correlated with the second harmonic therefore depends only on the distance to the surface and can be used to control the probe height.

The cantilevers are nowadays fabricated in integrated silicon technology, including a mounting bar and the tip [382]. The tip is formed by etching into a piece of silicon from all sides until reaching a sharply pointed residue. Coating the tip with a suitable magnetic film by sputtering or evaporation produces excellent probes [383]. The advantage of such tips is twofold: (i) silicon technology allows batch fabrication and thus economic and reproducible production, and (ii) the thin magnetic films generate much less far-reaching stray field than formerly used tips made of electrolytically polished wires, thus reducing the danger of unwanted interaction with the sample. The preferred coatings are thin film recording materials such as CoCr. There is no best coating for all applications. For reasons discussed below the magnetic coating has to be optimized for every class of material to be investigated.

Silicon etching techniques reach tip radii in the 10 nm range. The probe resolution can be further enhanced by growing a thin needle on top of the silicon tip in an electron microscope. An electron beam, interacting with residual gases in an oil diffusion pump vacuum chamber can lead to such a finely pointed outgrowth that consists mostly of carbon [384–386].

With respect to speed, magnetic force microscopy cannot compete with optical or electron techniques. This is no disadvantage in the investigation of information storage patterns. In regular domain observation it is advisable, however, to integrate MFM with Kerr microscopy to get an overview and to localize the points of interest [387–390]. In this way the slow scanning technique can be used economically and possible artefacts (see below) can be recognized.

(B) Interaction Mechanisms. At first sight a description of the force between magnetic tip and sample seems straightforward: the force is the gradient of the interaction energy, which can be written in two equivalent forms:

$$E_{\text{inter}} = -\int_{\text{tip}} \boldsymbol{J}_{\text{tip}} \cdot \boldsymbol{H}_{\text{sample}} \, dV = -\int_{\text{sample}} \boldsymbol{J}_{\text{sample}} \cdot \boldsymbol{H}_{\text{tip}} \, dV . \tag{2.20}$$

The two versions are in a sense reciprocal [391]. The first integral supports the conventional interpretation of magnetic force microscopy: the magnetization distribution of the tip, which is assumed to be known, interacts with the stray field of the sample. The interaction energy, including all its derived quantities (such as the force or the force gradient), should therefore offer information about the stray field. Different magnetization patterns with identical stray fields cannot be distinguished. The second integral in (2.20) tells us that two tips with different internal magnetization distributions, but identical stray fields, are also equivalent. A large number of articles was written elaborating in detail the consequences of this interaction integrals. We believe that another form offers an even better insight [392]. To see this, we transform the interaction integral (2.20) by partial integration:

$$E_{\text{inter}} = -\int_{\text{surface}} \lambda_{\text{sample}} \, \Phi_{\text{tip}} \, dV - \int_{\text{surface}} \sigma_{\text{sample}} \, \Phi_{\text{tip}} \, dS \tag{2.20a}$$

where $\lambda_{\text{sample}} = -\text{div} \boldsymbol{J}_{\text{sample}}$ is the volume charge, $\sigma_{\text{sample}} = \boldsymbol{n} \cdot \boldsymbol{J}_{\text{sample}}$ is the surface charge of the sample, while Φ_{tip} is the scalar potential of the probe stray field obeying $\boldsymbol{H}_{\text{tip}} = -\text{grad} \, \Phi_{\text{tip}}$. Depending on the degree of localization of Φ_{tip} the interaction is thus determined by a more-or-less smoothed charge pattern. A localized potential can be expected if the probe is magnetized "vertically", towards the sample or away from it. A horizontal magnetization parallel to the sample generates a zero potential in front of the tip and effectively produces a differentiated charge image that will be rarely useful.

10 μm

Fig. 2.40 Written longitudinal recording track pattern observed in a magnetic force microscope, displaying the magnetic charges at the transitions. (Courtesy *D. Rugar*, IBM Research)

To obtain the force instead of the interaction energy, Φ_{tip} has simply to be replaced by $d\Phi_{\text{tip}}/dz$, and if the microscope is operated in the force gradient mode, the corresponding second derivative has to be inserted into (2.20a). The connection between charge and magnetization is most straightforward in the presence of a polar magnetization component m_{pol}. In this case, we just have $\sigma_{\text{sample}} = m_{\text{pol}}$ and the force image should closely resemble the image obtained with the polar Kerr effect. In other cases the connection between magnetization and charges is more indirect but always instructive; for example, for a longitudinal recording pattern the charges appear at the bit transitions (Fig. 2.40).

In all variants of the interaction energy (2.20) it must be taken into account that the magnetization pattern of the sample and all its derived quantities may be influenced by the probe, or vice versa. Let us first focus on the case in which such interactions are weak.

(C) Negligible Interactions: Charge Contrast. In this case neither the sample is modified by the probe nor the probe by the sample. Experimentally, it was demonstrated that convincing domain images can be obtained even from soft magnetic materials, if strong reactions in the sample are avoided by using weak, magnetically hard tips and a large tip-sample separation (Fig. 2.41).

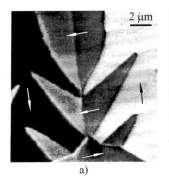 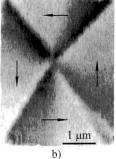

Fig. 2.41 MFM images of domains on soft magnetic materials: (a) bulk silicon-iron and (b) FeTaN thin film element of 30 nm thickness. (Courtesy *L. Belliard* and *J. Miltat* [393, 394])

The numerical simulations by *Tomlinson* and *Hill* [395, 396] offer insight into this case and its limits. The authors achieve results which closely resemble experimental observations, and they also demonstrate in detail the nature of the contrast and possible artefacts. Starting with a numerical calculation of the domain pattern of a square Permalloy thin film element, the forces between a magnetically rigid tip and this pattern are calculated for three cases: (i) that of a *very* weak tip that generates a negligible reaction in the sample (the case of interest in this section), (ii) that of a moderately weak tip that causes only reversible magnetization excursions, and (iii) that of a stronger tip that produces irreversible reactions and strongly distorted, unrecognizable images. The result for the first case is shown in Fig. 2.42a. Based on the numerical simulation of the thin film domain pattern the derived charge pattern is shown in Fig. 2.42b. In fact, it corresponds closely to the computed force image in Fig. 2.42a (and also with observations, as in Fig. 2.41b).

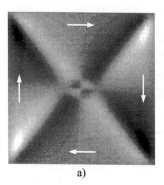 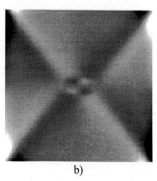

a)   b)

Fig. 2.42 A simulated magnetic force image (a) for a $3 \times 3 \ \mu m^2$ Permalloy element of 40 nm thickness (Courtesy *S. L. Tomlinson* [396]), compared with the charge distributions (b) calculated from the same micromagnetic model as used in (a) [392]

In thin film elements the charge pattern reflects characteristic properties of the extended tails of Néel walls (see Sect. 3.6.4C). This case demonstrates that magnetic force images are in fact more closely related to the *charge* pattern (an interpretation supported also in [397]), rather than to the magnetization, to some stray field component, or to the absolute value of the stray field as in Bitter contrast (2.3). Magnetic force microscopy relies according to (2.20a) always on the interactions with the magnetic charges in the sample. The charges can be induced by the presence of the tip, but in the limit of weak interactions MFM senses the original charge pattern. This method of *charge microscopy* differs from all other magnetic imaging methods. Whether the conditions for pure charge contrast are fulfilled can be checked experimentally: if the probe is magnetized along the opposite direction, the image should be inverted. Sometimes this is in fact true but more often not, indicating a further, superimposed contrast mechanism.

(D) Reversible Interactions: Susceptibility Contrast. Neither the probe potential nor the sample charges can be always considered rigid. Magnetic materials can have a large permeability, mostly because of the displacement of domain walls, but also because of magnetization rotations in low-anisotropy materials. Thus the sample magnetization can be influenced by the probe stray field, and vice versa.

The second possibility occurs in particular if a hard magnetic material is scanned with a soft magnetic tip as shown for example in [398]. If a reasonably hard probe is used, the first possibility is more often encountered. In the example of Fig. 2.43 a branched domain pattern of a cobalt crystal is represented in force microscopy images (a, b) with two different tip polarities and compared with the (interaction-free) Kerr technique (c). The general correspondence of both techniques is apparent. Because in this case the magnetization is primarily directed perpendicular to the surface, the magnetization pattern visible in the polar Kerr effect is identical with the (surface) charge distribution visible in the magnetic force image. But while the Kerr image shows black and white circular "flower" patterns, all flowers are dark and embedded in a lighter matrix in the MFM images. This means that the environment is more strongly attracted by the tip than the flowers, irrespective of the charge of flower and environment and also of the polarity of the tip. The long-range pattern in Fig. 2.43b is in fact reversed relative to (a), but the matrix around the flowers is again strongly attracted.

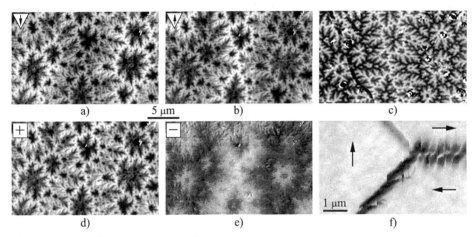

Fig. 2.43 Comparing magnetic force images for opposite tip polarities (a, b) of the branched domain pattern on the basal plane of a cobalt crystal with a Kerr image of the same pattern (c). The MFM images display a better resolution than the optical image, but are not identical because of the superposition of two contrast mechanisms, that of charge and of susceptibility imaging, which can be separated in digitally generated sum and difference images (d, e) as explained in the text (together with *W. Rave* and *E. Zueco*, Dresden [390]). The domain walls in the Fe thin film element [(f); 30 nm thickness] show hysteretic distortions occurring if the tip sample interaction is stronger than the local coercivity. (Courtesy *M. Schneider*, Jülich)

The obvious explanation of such observations is that reversible reactions in the sample lead to an attractive interaction, in the same sense as a soft magnetic material is always attracted by a permanent magnet. It is well conceivable that different parts of a branched domain pattern are susceptible to such reversible reactions in a different degree. Similar reversible reactions in the tip sample system were first clearly identified in [399] by experiments on domain walls with tips of different polarity (see also [400]).

If the average image calculated from two force images with opposite polarity probes is not uniformly grey (Fig. 2.43d), and if no irreversibilities can be identified, then this is an indication of a reversible reaction in the tip sample system leading to what we may call a *susceptibility* image. No theory of the susceptibility contrast is available and formulating one is certainly not easy, as the local susceptibility of a magnetic microstructure contains both rotation and local wall displacement effects. Numerical simulation can be applied for small objects as demonstrated in [396]. The calculated difference image between pictures taken with oppositely magnetized tips (Fig. 2.43e) will represent the charge image discussed in (C) if the imaging procedure is sufficiently linear, which depends on details of the experimental setup. If the interaction is rather strong, second-order susceptibility effects can occur, which probably play a role in observations of ultrathin perpendicular films [394, 401] as discussed in [392]. In any case, studying these sum and difference images as in Fig. 2.43d, e is very useful and offers complementary information.

Extremely non-linear but reversible probe-sample interactions were demonstrated in a most ingenious experiment by the Orsay group [394, 401]. They observed easily displaceable domain walls in garnet films by force microscopy and at the same time through the transparent substrate by the Faraday effect. The grossly differing images of the same band domain pattern in the garnet film shown in Fig. 2.44a,b are revealed in (c) to be caused by an artefact of the MFM technique: the tip expands locally the dark domains during the scanning process, thus generating the impression of overall wide dark and narrow white domains in the scanned image (b). Obviously magnetic force microscopy must be applied with extreme care on low coercivity materials.

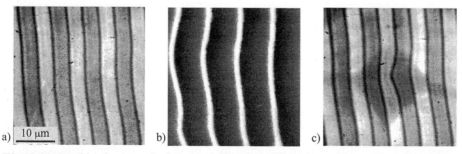

Fig. 2.44 Combined magnetic force and Faraday effect observations on a transparent magnetic garnet film with perpendicular anisotropy. (a) Undisturbed band domain pattern as seen by the Faraday effect. (b) The same pattern as observed in MFM contrast. The discrepancy between (a) and (b) is explained in (c), where the local distortion of the domain wall caused by the MFM tip is observed simultaneously in the Faraday effect image through the transparent substrate [the pyramidal shape of the probe can be seen as a shadow in (c)]. (Courtesy *J. Miltat* and *L. Belliard*, Orsay)

(E) Strong Interactions: Hysteresis Effects. If the stray field of the probe is too strong, irreversible reactions in the sample magnetization are observed [402, 403], leading to all kinds of artefacts. A domain wall that is unpinned by the tip at one point in the scanning process can be easily recognized by a resulting discontinuity in the image. But there is also the possibility that a wall is dragged along by the tip in every pass, until it breaks off to return to its original position. Confusingly, both positions, the initial and the final one, then become visible. The fine and sharp features visible in Fig. 2.43f or reported for example in [404] must be attributed to such effects. Images that are influenced by some kind of irreversible magnetization process may be summarily classified as *hysteresis* images.

The possibility of stimulating magnetization processes with a localized probe is interesting in itself [405–408]. In the first step of exploring a pattern, one should try, however, to avoid any irreversibilities by employing weak probes at a sufficient distance (which may, of course, limit the available resolution and sensitivity). An interesting proposal that would avoid these difficulties was presented in [409]. Instead of a ferromagnetic layer, a reversible, superparamagnetic film is sputtered onto the tip. Such material is attracted to magnetic stray fields, leading to a pure susceptibility contrast as demonstrated on the gap field of a recording head. It was also successfully applied in domain observation on sintered NdFeB permanent magnet material [410].

A special method that *exploits* hysteresis in the tip was demonstrated by *Proksch* et al. [411]. In this technique a weak alternating perpendicular field is superimposed, which is just strong enough to switch the soft magnetic probe, but weak enough not to affect the sample. The resulting signal is evaluated as in a fluxgate magnetometer. If the sample stray field is different from zero, the probe is magnetized in an asymmetric way by hysteresis, giving rise to second harmonics, the amplitudes of which are a linear function of the sample stray field to be measured. In this way even a quantitative evaluation of the magnetic stray fields above recording media becomes possible.

(F) Summary. Three contrast mechanisms are identified in magnetic force microscopy:
- Charge contrast for magnetically hard probing tips that produce only weak, well localized stray fields.
- Susceptibility contrast that is superimposed if reversible reactions in the sample or in the probe are relevant. Within this contrast mechanism linear and non-linear, but still reversible, effects have to be distinguished [392].

- Hysteresis contrast indicating irreversible reactions either in the sample or in the probe which may lead to nice but hardly useful images. As a rule such effects must be avoided; a requirement that limits sensitivity and resolution.

Only the first mechanism usually applies to magnetic recording media. If the magnetic hardness of the probe matches that of the investigated medium, interaction problems are negligible. Magnetic force microscopy then images the magnetic poles or charges that are directly responsible for the stray field picked up by a read head in magnetic recording. The technique is therefore the method of choice for the investigation of recording tracks (Fig. 2.40).

2.6.2 Near-Field Optical Scanning Microscopy

The concept of near-field optical microscopy circumvents the diffraction limit of optical imaging by forcing the light through a submicroscopic aperture. When this aperture is scanned over the sample, an image can be generated with ten to fifty times better resolution than with conventional optical imaging. The principle has been known for a long time and was applied, for example, in microwave imaging. It received a new stimulus in optics by the advent of scanning probe techniques [412–414]. The prospect of enhancing the resolution decisively, while keeping all the advantages of optical techniques, is fascinating.

Near-field optical techniques were also tried for observing magnetic domains based on the magneto-optical effects. This proved to be quite feasible in transmission using the Faraday effect. Convincing results were obtained by *Betzig* et al. [415], who used a monomode optical fibre which is heated and pulled to the needed thickness. Coating it on the outside with aluminium leaves the desired aperture at the tip that can be brought into close proximity of the sample. The fibre can be used either to locally collect transmitted light, or to locally illuminate the sample, collecting the light in a regular microscope. The laws of near-field optics require the aperture to be placed about as close to the sample as the desired resolution 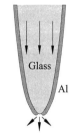 given by the diameter of the aperture. Polarization proved to be conserved in monomode fibres, although somehow modified by the specific state of the fibre. This modification of the polarization state can be compensated by phase shifters and analysers, so that magnetic information can be derived as in conventional magneto-optic techniques.

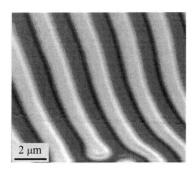

Fig. 2.45 A domain image obtained with a near-field optical scanning microscope based on the monomode fibre concept. The maze pattern in a transparent garnet film of 0.8 μm thickness is shown in transmission. All the light emitted by the submicron aperture is collected in this experiment. (Courtesy *F. Matthes* and *H. Brückl*, IFW Dresden)

Unfortunately, transmission observations are possible only for few materials. If they are possible, high quality pictures can be generated as demonstrated in Fig. 2.45. The application of the fibre concept to the more interesting reflection geometry meets difficulties, in particular for in-plane magnetization requiring oblique illumination. First examples for imaging polar magnetization components in reflection can be found in [416, 417].

The traditional near-field optical setup discussed above uses a submicroscopic aperture on the illumination side and collects all scattered light. In an alternate concept the sample is illuminated with a broad beam, and only the light scattered from the sharp tip of a chemically thinned optical fibre [418] or the sharp corner of a gallium arsenide photodetector [419] is recorded. In a further variant a tiny scat-

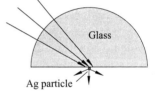

tering silver particle (as in photographic emulsions) replaces the aperture of conventional near-field microscopy [420]. The particle is placed in the zone of frustrated total internal reflection of a glass hemisphere. The laser illumination through the hemisphere does not reach the sample except near the particle. Polarization can be detected as demonstrated experimentally.

All these techniques are not yet really developed. The difficulties of controlling the submicroscopic apertures or scattering centres are considerable, as is the elimination of non-magnetic background signals. Convincing high quality reflection images with a resolution that surpasses that of conventional Kerr microscopy have yet to be demonstrated. In transmission, things are easier and good images have been taken as shown in the example. However, near-field optical writing and read-out techniques may one day enhance the capacity of magneto-optical recording (and of conventional compact discs) by several orders of magnitude. This prospect will stimulate further developments and good domain observation methods may result as a by-product.

2.6.3 Other Magnetic Scanning Methods

(A) Electron Spin-Dependent Scanning Microscopy. A spin-sensitive scanning tunnelling microscope would be the ultimate microscopical magnetic investigation tool as it would make it theoretically possible to investigate individual spins at the surface of a sample. The probe should be "non-magnetic" in the sense that it does not generate a long-range magnetic field. This is necessary to avoid a magnetostatic influence between probe and sample, which would certainly be too large at tunnelling distances. The implementation of such a device has not yet passed the concept stage. *Johnson* et al. [421] still considered a ferromagnetic tip. Evidence for spin-dependent effects in tunnelling between antiferromagnets and ferromagnets was presented in [422, 423]. Several possibilities of the antiferromagnetic-ferromagnetic option are discussed in [424]. *Jansen* et al. [425] propose a semiconductor tip that is "magnetized" by illumination with circularly polarized light. These and further options for spin-sensitive tunnelling microscopy are discussed in [426].

In such a tunnelling microscope atomic resolution should be available as well as the compatibility with large magnetic fields, which would be necessary to eliminate the non-magnetic background by subtracting the tunnelling image of a saturated state. *Shvets* et al. [424] point to the difficulty that in the close distance of a tunnelling probe magnetostrictive deformations would be measured in addition to the desired spin interactions. This should be considered welcome additional information rather than a complication.

Another possibility of a high resolution spin-dependent technique was proposed by *Allenspach* et al. [427]. They performed tunnelling experiments with a sharp tungsten tip in the field emission mode and proved, with the help of a Mott detector, that secondary electrons emerging from a magnetic sample retain some of their polarization as in conventional spin-dependent SEM (Sect. 2.5.4). Resolution in the nanometre range would be possible in this way, with the advantage of a high electron intensity.

(B) Magnetic Field Sensor Scanning. Scanning the surface of a magnetic sample with field detectors to record information about the magnetic microstructure was often tried [428–432, 115] even before the introduction of scanning probe microscopy methods. Hall probes and vibrating pick-up coils were applied, and useful results could be obtained with these techniques in measuring stray fields at defects and grain boundaries. But the demonstrated resolution in domain observation remained rather poor. Nevertheless, the important problem of detecting the basic domain structure in coated transformer steel seems to be accessible with such methods [433, 115], at least for favourably oriented grains.

Modern magnetic field sensors, as they are used in magnetic recording and in sensor technology, tend to become smaller and smaller making use of microfabrication techniques. At the same time the available scanning methods allow much smaller distances between probe and sample. Using such sensors therefore will become an increasingly interesting option.

In [434] submicron resolution was demonstrated with a miniature integrated Hall probe. A corner of the slightly tilted chip carrying the Hall circuit is used as the "tip" of a scanning tunnelling setup. A field sensitivity in the range of one A/m was achieved. The field can be measured quantitatively and without disturbing the sample magnetically. In [435] high resolution was achieved by a mathematical deconvolution of the Hall mapping of the stray field of an MFM tip. In other examples [436–438] commercial magnetoresistive recording heads are used as the basis of a scanning microscope. Such heads have still (track-) widths of several micrometres, along the track the resolution is already in the 0.1 μm range.

2.7 X-ray, Neutron and Other Methods

The methods to be discussed in this section differ from conventional methods, in part by the size of the equipment used. Instead of optical or electron microscopes, synchrotrons and nuclear reactors enter the scene. This introduction of "big science" was justified because only these methods are in principle capable of looking *into* the domain structure of bulk metallic samples. In practice, as will be shown, the results have not been encouraging enough to lead to a widespread acceptance of these methods in domain analysis. An increasing activity in X-ray spectroscopic methods can be recorded, however, which also rely on synchrotron radiation sources. This is mainly due to the unique element-specific imaging potential of these methods.

2.7.1 X-ray Topography of Magnetic Domains

(A) Lang's method. Figure 2.46a shows a schematic of the most commonly used procedure to obtain an X-ray image of magnetic structure [439]. Ideally, a monochromatic and plane parallel beam is directed onto a crystal plate oriented so that Bragg's condition is fulfilled for some set of lattice planes. Lang's method is based on conventional X-ray sources [in contrast to synchrotron systems to be discussed in (B)]. To approximate a parallel, monochromatic beam the characteristic radiation from a suitable cathode is used, and adequate measures are necessary to be able to select, for example, the $K_{\alpha 1}$ radiation only. In addition, a slit aperture limits the illuminating beam to a narrow strip on the sample to restrict the beam divergence. The diffracted beam, selected by another slit, is recorded by a high resolution and high sensitivity photographic plate (a so-called nuclear plate). Both the crystal and the plate are slowly advanced synchronously, thus scanning the sample.

A perfect crystal would generate a uniform image. Crystal imperfections disturb the process of Bragg reflection, leading to an image of these defects. Lang's method is sensitive enough to display isolated dislocations. If such structural contrasts are largely absent, the weak magnetostrictive strains and lattice rotations in magnetic crystals can be made visible [440, 441].

Figure 2.46b shows the (exaggerated) lattice orientation change connected with a 90° domain wall (see Sect. 3.2.6G for a systematic treatment). This rotation amounts only to about 10^{-5} radians in ferromagnets and is thus smaller than the usual divergence of X-ray beams. Nevertheless, a contrast develops at those positions, where either the orientation or the spacing of the lattice *changes*. These contrast phenomena are described by the dynamical theory of X-ray diffraction and they are connected with interference effects between direct and diffracted beams in the adjacent crystal regions.

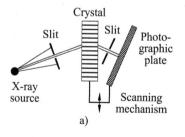 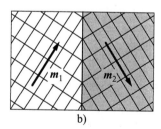

Fig. 2.46 Schematics of the Lang method of X-ray topography (a) and schematics of the lattice rotations responsible for X-ray contrast (b)

Figure 2.47 shows an example of a Lang topogram of a nearly perfect silicon-iron sample together with a line drawing of the corresponding domain structure. We see 90° walls, showing black or white line contrasts depending on the sense of magnetization rotation. Junctions of three walls show a special kind of black-and-white "butterfly" pattern. The 180° walls are invisible in this image. The domain pattern shown here — the so-called fir tree pattern — consists of shallow surface domains, which are connected with a slight misorientation of the crystal relative to the (100) surface (see Sect. 3.7.1 for the theory of such patterns and Sect. 5.3.4 for magneto-optical observations). More complicated contrast phenomena including interference fringes are observed if the X-rays pass through more than one domain [442, 443].

Fig. 2.47 X-ray topogram of a slightly misoriented (100)-oriented Fe-3% Si crystal (thickness~ 0.1 mm), with an explanation of the observed domain structure. (Courtesy *J. Miltat*, Orsay [444])

0.5 mm

X-ray topography contrast has been discussed extensively in the literature. The thorough analysis of a given observation involves two non-trivial problems: (i) The calculation of the magnetostrictive strains and lattice rotations for a given domain configuration, taking into account compatibility relations and surface relaxation (including elastic anisotropy) as indicated in Sect. 3.2.6. (ii) The calculation of the wave field in the dynamic theory of X-ray diffraction. In the few cases in which both problems have been solved, convincing agreement between theory and experiment was achieved [445–447, 444].

For 90° walls *Polcarová* and *Kaczér* [448] demonstrated a simple rule for the visibility of these walls: let m_1 and m_2 be the magnetization vectors in the two adjacent domains, and g_L the reciprocal lattice vector responsible for the Bragg reflection (i.e. a vector oriented perpendicular to the refracting planes and with length $1/d$, if d is the lattice spacing of these planes). Then a 90° wall is invisible in X-ray contrast if the condition:

$$g_L \cdot (m_1 - m_2) = 0 \tag{2.21}$$

is fulfilled. This rule may be checked by looking at Fig. 2.48 and it may be understood qualitatively by considering the lattice rotations (Fig. 2.46b): the magnetization *difference* vector of a Bloch wall is oriented parallel to the wall. Lattice planes that are parallel to this direction (so that their g_L vector is perpendicular to it) are not rotated, and thus invisible in X-ray topography. All other lattice planes are bent, giving rise to an interference contrast effect.

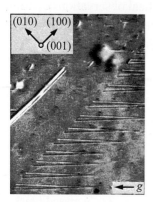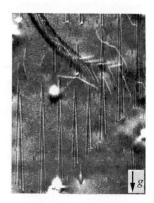

Fig. 2.48 The same type of fir tree pattern (see Fig. 2.47) viewed under three different diffraction conditions. Depending on the diffraction vector g_L, one or the other set of branches is invisible. (Courtesy *J. Miltat* [444])

There is an interesting connection between rule (2.21) and the rules governing the orientation of Bloch walls. For a given transition m_1 to m_2 only those wall orientations are allowed in which the magnetization difference vector lies parallel to the wall. [The allowed wall orientations form a cylinder around the $(m_1 - m_2)$ axis; see (3.114) ahead]. Rule (2.21) states, therefore, that if in a given experiment one wall is invisible, all other allowed walls between the same domains are invisible, too. This prediction is confirmed by the observations shown in Fig. 2.48. The shallow fir tree domains are bounded by half-tube-like walls (see Fig. 3.9), which occur in all permitted orientations around the circumference of the tube. Depending on the chosen diffraction vector, the fir tree branches are in fact either fully visible or completely invisible.

Contrast depends also strongly on the thickness D of the sample relative to the absorption length L of the X-rays. Samples that are thinner than L show simple black lines for all 90° walls [448]. Thicker samples entering the range of anomalous transmission show complicated contrast effects.

180° walls are largely invisible in X-ray diffraction. As elaborated in more detail in Sect. 3.2.6, magnetostriction does not depend on the sign of the magnetization vector, so that the two domains separated by a 180° wall carry equal deformations. The wall with its rotated spins is under stress, but the *strain* inside the wall must be the same as in the domains because of elastic compatibility conditions. Only near a surface will the stress partially relax, giving rise to some strain inhomogeneity [449]. A weak and almost undocumentable black contrast has been attributed to this surface effect [441]. In another observation, "180° walls" became visible near the tip of an iron whisker under conditions of an applied field [450]. Certainly, in this case the field produced a deviation from the 180° wall situation.

(B) Synchrotron Radiation Topography. Exposure times of more than a day are necessary in conventional Lang topography using the $CuK_{\alpha 1}$ or the $MoK_{\alpha 1}$ radiation. Alternatively, the more intense synchrotron radiation can be used for topography, permitting drastically reduced exposure times because synchrotron radiation is almost parallel, and a wide beam can be used illuminating the whole sample simultaneously.

In a kind of double crystal technique, the synchrotron radiation is made monochromatic by a structurally perfect germanium crystal. An image of the magnetic crystal can then be produced in one step, 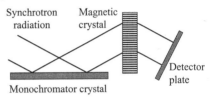 without the need of a lengthy scanning procedure as in Lang's technique. Even stroboscopic observations of dynamic magnetization phenomena become possible [451, 452], as shown in Fig. 2.49.

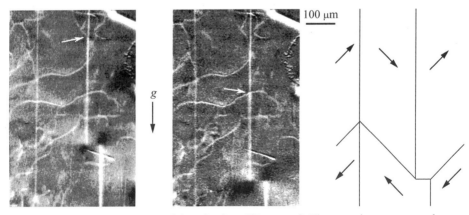

Fig. 2.49 Stroboscopic images of domains in a SiFe crystal. The two pictures were taken at subsequent times in an oscillatory field. Note the change in contrast of a 90° wall at its intersection with dislocations (arrows) indicating a direct interaction between these two elements of microstructure. (Courtesy *J. Miltat* [451])

The resolution limit in X-ray topography has not been pursued systematically. It is determined by the properties of the photographic plate and by the

shot noise of the radiation. Resolution achieved in practice is generally in the
5 μm range but values down to 1 μm have been demonstrated.

2.7.2 Neutron Topography

The wavelengths of thermal neutrons cover the same range as those of X-rays.
X-ray images of the lattice deformations can therefore in principle be reproduced
by neutron topography. The contrast in neutron topography differs, however,
from that of X-ray topography, because direct magnetic interactions caused
by the spin of the neutron are present in addition to nuclear interactions. The
spin interaction can be employed if polarized neutrons are used. They lead to
pictures analogous to polarized light images, with black and white contrast
between domains (Fig. 2.50b).

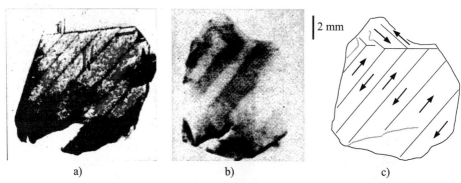

a)   b)   c)

Fig. 2.50 Domain images on a silicon-iron sample obtained with neutron topography: a
double crystal topogram showing domain boundary contrasts {(a); [453]}. In another technique a polarized neutron beam is used, leading to images with domain contrast (b). The
domain structure is sketched in (c). (Courtesy *J. Baruchel*, Grenoble)

The resolution is low in any case. It is limited by the contrast mechanism,
but also by the detection method, and by the available beam intensities, which
are too low even from special high-flux reactors [454, 455]. In the case of
unpolarized neutrons only the domain walls become visible as shown in Fig.
2.50a. An interpretation of this contrast effect was given in [453], based on a
total reflection on the domain wall. Some polarization and direction components
of the neutron beam are reflected at the domain boundary, which acts like a
discontinuity in the refractive index of the neutrons. These reflected beams
add to those beams that are deflected into the same direction by Bragg diffraction,
thus leading to an enhanced intensity near the wall.

2.7.3 Domain Imaging Based on X-Ray Spectroscopy

Effects analogous to the conventional magneto-optic effects also exist at shorter, X-ray wavelengths. The imaginary part of the Faraday effect can be seen as a magnetization-dependent absorption of circularly polarized light. The analogous effect for X-rays is called X-ray circular dichroism [456, 457]. It is related to the electron core levels of the investigated material and the radiation-induced transition of electrons into unoccupied or free states. In both cases magnetic effects can be found. For reviews and more detailed discussions see [458, 459].

As analysers for X-rays are not available, X-ray spectroscopy has to rely on dichroic effects rather than on polarization rotation effects as in optics (Sect. 2.3). The effects are element specific, depending on the atomic absorption lines. By exploiting a fine structure of these lines even different chemical states can be distinguished.

X-ray absorption can of course be measured directly by X-ray detectors [456, 460]. More elegant and better suited to microscopic applications is the detection of the excited photoelectrons. No polarization analysis, energy selection or directional selectivity is in principle necessary for these experiments. Photoemission electron microscopes, which are available for general surface studies, can thus be used to image magnetic domains, based on the magnetization-dependent polarized X-ray absorption [461]. The X-rays are usually obtained from synchrotrons, which can deliver high-intensity polarized beams particularly with the help of special devices, the so-called magnetic undulators. The magnetic contribution can be separated from other effects by subtracting the images associated with the L_2 and the L_3 core lines, because the magnetic effects for these two lines are usually equal and opposite. The achieved resolution lies in the μm range. It is anticipated that photoemission microscopes, which employ some energy filtering or chromatic aberration correction, will reach resolutions in the submicron range [462–464]. Energy filtering will also adjust the information depth of the method, which may be useful in the investigation of magnetic multilayers.

Progress in X-ray spectroscopic domain imaging techniques is remarkable. An example of a high-quality domain image is shown in Fig. 2.51a. The picture is computed as the normalized difference between photoemission images from the iron L_3 shell (corresponding to about 700 eV) for left and right circularly polarized synchrotron radiation entering at an angle of incidence of 65°. It shows mainly the horizontal magnetization component in a domain pattern on the (100) surface of an iron whisker. The resolution reaches already in this first example that of Kerr microscopy; even the narrow domain walls (V-lines in this case) of iron become clearly visible (Fig. 2.51b).

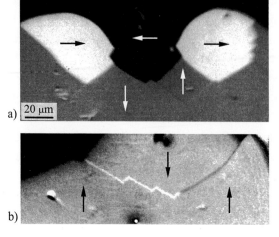

Fig. 2.51 Domain image on a mechanically stressed iron whisker, generated by circularly polarized soft X-rays from the ESRF synchrotron (Grenoble) entering in the picture from the right. The produced photoelectrons (of several eV energy) are collected by an advanced photoemission electron microscope [465]. (a) An overview of the domain pattern. (b) Details of V-line walls. (Courtesy *R. Frömter* and *C.M. Schneider* [459])

There is a large variety of possible procedures in X-ray spectroscopic imaging . In an alternative scheme higher-energy (Auger) electrons are specifically selected in a "spectromicroscope" [466], leading to stronger effects, which makes it possible to image the magnetization distribution even in single antiferromagnetic coating layers on a ferromagnetic material [467], as yet unfortunately with reduced resolution. Also linearly polarized X-rays are used in an experiment, which resembles the transverse Kerr effect [468]. Instead of an electron microscope, the use of an X-ray microscope (based on zone plate) is feasible in X-ray spectroscopy as demonstrated by *Kagoshima* et al. [469] with a scanning method, and by *Fischer* et al. [470] with a high resolution X-ray imaging approach. The latter authors reach a spatial resolution of better than 100 nm and also demonstrate the compatibility of the method with large external magnetic fields. Figure 2.52 shows an example of a magnetization process observed with this technique [471].

All variants are element specific, which is a unique property among domain imaging techniques. This feature makes it possible to look through cover layers to a specific magnetic layer if it contains a suitable element. Probably the most exciting aspect of this new possibility is the use of thin magnetic "tracer" layers embedded into a multilayer system, which consist of a different atomic species than the rest of the multilayer [472]. If the coupling between the tracer layer and the regular layers is well understood, detailed information from any depth in a multilayer can be derived from an element-specific micro-scopy method. Certainly important developments are to be expected, although the methods rely on extremely expensive radiation sources (which should rather be enhanced, as in Fig. 2.51, instead of simplified to achieve high image quality, which was still strongly affected by statistical noise in earlier demonstrations).

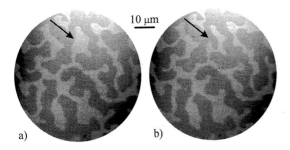

a) b)

Fig. 2.52 Domains on an amorphous $Fe_{72}Gd_{28}$ film of 60 nm thickness and perpendicular anisotropy, made visible using X-ray dichroism on the $Fe L_3$ edge. Each picture was re-corded in 3.5 sec, (b) 100 sec later than (a), thus displaying the process of domain creeping. (Courtesy *P. Fischer* and *T. Eimüller*, Augs-burg; [471])

2.7.4 Magnetotactic Bacteria

A remarkable method for domain imaging was demonstrated in [473]: quite a few anaerobic *bacteria* (among them one named *Aquaspirillum magnetotacti-cum*) living in muddy water orient themselves with the aid of the earth's magnetic field to move away from the oxygen-rich surface [474, 475]. This is achieved with the help of a tiny chain of single-domain magnetite particles of diameter 50–100 nm inside the cell of the bacterium, which is spontaneously saturated in a certain direction and held in place by being attached to the cell wall. Since the earth's magnetic field is pointing downwards when looking northwards in the northern hemisphere (as an example), all bacteria from this part of the world are north seeking. They act like tiny self-propelling compasses in the earth's field. The right polarity is chosen by selection and inherited by the progeny by sharing chain fragments, which are then restored by the synthesis of new particles. The bacteria survive in a sealed sample of their natural habitat for at least two years without additional nutrients. With the help of a magnetic field they can be extracted and collected. Putting them on a magnetic sample, they will move along the field lines of the stray field until they reach — within a few seconds — the northern end of a field line at the surface.

The method is similar to the Bitter method, but the contrast mechanism is different. Bitter patterns indicate the positions of highest absolute values of the field on the surface (Sect. 2.2.2), but the bacteria indicate the northern footpoints of the field lines. These positions need not coincide, especially if tangential fields are responsible for Bitter contrast as is often true in slightly misoriented surfaces. Therefore images from magnetotactic bacteria often resemble more a magneto-optical domain contrast than the typical wall contrast of Bitter patterns. They look similar to Bitter pictures obtained with an auxiliary field perpendicular to the surface.

An advantage of the technique is its high sensitivity: since the bacteria have to rely on the earth's field in their natural environment, they react to fields of the order of 0.5 Oe (40 A/m), and are therefore roughly an order of magnitude more sensitive than Bitter colloids. Even when dead, the magnetite chains of magnetotactic bacteria represent most sensitive magnetic particles. The resolution is limited somehow by the size of the bacteria, which are typically larger than a micrometre. Since the method is fast and sensitive it may find applications in the routine inspection, for example, of transformer steels. Further applications are explored in [476]; the contrast mechanism as well as the sensitivity of the method are discussed in [477].

2.7.5 Domain-Induced Surface Profile

In rare cases magnetic domains can generate a surface profile that may be detected with a suitable microscopic method, like Nomarski's interference contrast, or with scanning probe microscopic methods. Such an effect is occasionally observed after electrolytic polishing of uniaxial materials [478, 479] where the magnetic stray fields present during the process somehow influence the polishing rate (Fig. 2.53).

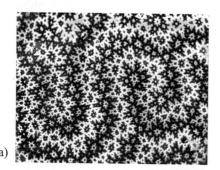 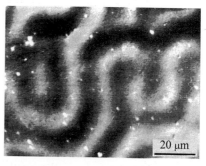

a)   20 μm   b)

Fig. 2.53 Domains on an electrolytically polished cobalt crystal made visible by the polar Kerr effect (a) and by the simultaneously present surface relief (b) shown by differential interference contrast. The sample was mounted on a magnet during polishing

Also during mechanical polishing magnetic stray fields and probably magnetic particles from the abrasion process can produce surface profiles in hard magnetic samples. Another source of a surface profile is magnetostriction, but usually this effect is too small in ferromagnets to lead to an observable topographic contrast. An exception was found in certain rare earth alloys with an extremely strong magnetostriction [480–482].

2.7.6 Domain Observation in the Bulk?

Investigating the interior domain structure of oxidic materials poses no problem in principle, as these materials are transparent at least to infrared light. Figure 5.7 displays an excellent example of these possibilities. Doing the same for ordinary metallic magnets proves to be virtually impossible. Simply slicing a sample to look for the domains in the interior is meaningless in magnetism because the cut will produce a new domain pattern. (This was tried e.g. in [483] on a thin film recording head, and in fact closure domains were found at the cutting edge that had not been there before).

Of all methods reported so far only with X-ray and neutron methods we are able to look into the interior of bulk metallic samples without cutting. But in spite of the impressive work done in clarifying the contrast phenomena in X-ray topography, there have been few original contributions to the knowledge of magnetic domain structures. Most domain patterns were known before from conventional surface observation methods. In [447] an internal zigzag folding of a 90° wall was identified from X-ray images, but even this feature was known from theory, and might not have been derived from the X-ray topogram without this knowledge. There is one notable exception: in 1993 *A.R. Lang* reports about observations he made in 1962 [484 (Fig. 8)], showing subsurface fir tree patterns in thin (110)-oriented silicon-iron sheets, a pattern that apparently was never seen or identified otherwise.

The limited magnetic information derivable from X-ray topography is surprising in view of the important contributions of this technique in other fields of material science. The reason lies in the specific laws of magnetic domains: only nearly perfect and relatively thin crystals are allowed in X-ray topography. For such crystals magnetic domains are determined by the surfaces and all domains should touch the sample surface somewhere. With surface observations and the rules derived from theory the interior domains in thin perfect crystals can be inferred. This would be impossible for inhomogeneous or deformed samples, but for such specimens the X-ray experiments do not work either.

Another unique feature of X-ray topography is the simultaneous visibility of domains and lattice defects. But again the results are somewhat disappointing. Only in few experiments an interaction between 90° walls and dislocations could be observed as in the example of Fig. 2.49. The interactions with the more important, but unfortunately largely invisible 180° walls (see Sect. 1.2.2) are not accessible. And interactions of 90° walls with structural features of less perfect crystals (such as dislocation agglomerates or inclusions) can rarely be studied, since such crystals are not suitable for X-ray topography. Synchrotron radiation topography offers a somewhat better opportunity; with its continuous spectrum, slightly bent crystals can be imaged in the Laue technique as in [485] where the interaction between 90° walls and a dislocation bundle in a plastically deformed iron whisker is documented.

The relatively poor resolution and the indirect nature of X-ray contrast are further reasons for the comparatively limited impact of X-ray methods on the investigation of magnetic microstructure. The principal merit of this method was the stimulation of important work on magnetostrictive deformations and elastic interactions of domains [447, 444, 486]. With neutron topography the magnetization pattern inside metallic samples can be directly observed, but the poor resolution renders useful applications unlikely. This argument does not hold for antiferromagnetic domains where other methods are rare [455, 487].

An interesting proposal aims at a *tomographic* evaluation of "neutron depolarization" measurements [488, 489]. The principle consists in measuring the rotation of the spin of differently polarized neutrons along many trajectories and combining the information in the manner of computer X-ray tomography for medical applications. It appears promising, but it has not yet been tested on actual samples and nothing can be said about the possibly resolution. X-ray microscopy with *hard* X-rays based on the circular magnetic dichroism effect (see Sect. 2.7.3) is another possibility, the sensitivity and resolution of which has still to be demonstrated.

A unique special method that permits the detailed investigation of the domains inside a bulk metallic sample was discovered by *Libovický* [490]. A silicon-iron alloy (Fe 12.8 at.% Si) undergoes an irreversible structural transition at about 600°C, namely the formation of ordered submicroscopic precipitates, platelets oriented along the local magnetization direction by elastic interactions. At room temperature, this "texture" gives rise to a birefringence effect after a suitable etching treatment, which can be made visible in polarized light. All three magnetization axes are distinguished and even 180° walls are visible at high magnification. By polishing away the surface successively, the deeper layers of the domain structure are revealed. Two of Libovický's pictures are shown in Fig. 2.54. This method is destructive, it permits a one-time investigation of the domains present at the reaction temperature. But in view of its unique potential to reveal the "anatomy" of a complex domain state, the technique clearly deserves further attention.

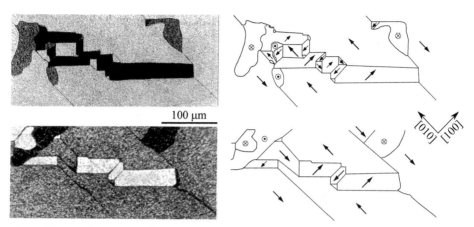

Fig. 2.54 Domain pictures obtained with the help of submicroscopic precipitates formed at a 580°C heat treatment in a SiFe alloy. In polarized light the domain structure that was present during the precipitation process becomes visible. The two pictures, each with its interpretation, show different depth levels of the same three-dimensional structure; also the polarizer setting was changed. (Courtesy *S. Libovický*, Prague [490])

2.8 Integral Methods Supporting Domain Analysis

All integral magnetic measurements, which depend on the magnetic domain microstructure, can somehow be used to help in domain analysis. Some of these integral experimental methods provide the possibility of a direct quantitative comparison with a domain model, and are thus useful in cases in which domain observation is impossible or inconclusive. Others, like integrating neutron depolarization techniques (reviewed for example in [491–493]), have to rely heavily on theory and model calculations and are research areas of their own. Here we want to discuss primarily the potential of the more direct methods, namely:
- Magnetization measurements
- Torque measurements
- Magnetostriction measurements
- Magnetoresistance measurements

They all yield a spatial average of some properties of the magnetic microstructure. The mean value of the magnetization vector $m(r)$ can be determined by *magnetization* measurements. In an applied field, the measured *torque* is proportional to the average transverse magnetization and perpendicular to the applied field and the torque axis. It is always useful to compare a domain model with the measured average magnetization in all three spatial directions.

Magnetostriction offers additional information on average domain properties. This effect is quadratic in the components of the magnetization vector. Crystal symmetry defines a spontaneous "free" deformation of every domain, depending on the magnetization direction. Although this deformation may be influenced by interaction between the domains (see Sect. 3.2.6F), the interactions cancel in the spatial average. This means that the elongation of a crystal is a linear function of the "phase" volumes v_i of domains magnetized in the various directions, where a phase volume collects the volumes of all domains magnetized along a certain direction. The motion of 180° walls has no effect on magnetostrictive strains, but the phase volume of domains magnetized for example at 90° relative to the "basic domains" can be directly measured [494]. The basic idea of the method is illustrated in Fig. 2.55.

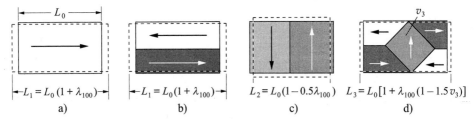

Fig. 2.55 The derivation of the volume of 90° domains from magnetostriction experiments on an iron crystal. The states (a) and (b) do not differ with respect to magnetostriction. The relative volume v_3 of the transverse domain in (d) may be derived rigorously from the measured lengths by the simple formula $v_3 = (L_1 - L_3)/(L_1 - L_2)$. Only the average elongation of the sample is indicated in (d), ignoring the inhomogeneous distortion connected with such a domain state, which cancels in the average. L_0 is the length of the sample in a hypothetical non-magnetic state

As all components of the average magnetostrictive strain tensor are in principal measurable, several domain phase volumes can be determined from magnetostriction measurements along different axes. The averaged strain tensor is a symmetric, traceless 3×3 tensor, thus characterized by five independent components. Together with the condition that the sum of all phase volumes must be the sample volume, up to six phase volumes can be measured.

The available information is not complete, however. As mentioned, domains with antiparallel magnetization do not differ in their magneto-elastic strains. But even when combining the phase volumes of antiparallel domains into "axial phases", it is not always possible to determine the phase volumes of such axial phases with global magnetostriction measurements alone. This can be easily verified for the hypothetical example of coplanar domains magnetized along the four axes [100], [010], [110] and [1$\bar{1}$0]. In contrast, if the domains are magnetized along the four $\langle 111 \rangle$ axes (as occurring in nickel crystals, see Sect. 3.2.3A), a complete axial phase analysis based on magnetostrictive measurements along all space diagonals is possible.

Even less complete information is available for materials without well-defined easy axes, as for example for polycrystals or for a metallic glass. Continuous functions describing the "magnetic texture" (the volume fraction as a function of magnetization direction) would be needed for a complete phase characterization. Magnetostriction measurements cannot provide these functions, but only certain integrals of them.

The *anisotropic magnetoresistance* effect (AMR) — the variation of the electrical resistance as a function of the magnetization direction relative to the current direction — can replace magnetostriction measurements. The directional dependence of the spontaneous magnetostrictive elongation and of this classical magnetoresistance effect are equivalent, so that both methods may be used according to convenience. Bulk samples can be measured more easily with the magnetostrictive effect, whereas thin films deposited on a substrate are more easily tested by magnetoresistance measurements [495].

The same general limitations on the derivable information as elaborated for magnetostriction applies to *Mössbauer* measurements [496, 497]. The ratio of the intensity of the second to the first Zeeman line in the Mössbauer spectrum depends on the square of certain magnetization components [498, 499]. Mössbauer measurements therefore offer information similar to magnetostriction and magnetoresistance measurements. They may have the advantage of being applicable to parts of the sample and to small samples. In particular, volume and surface of a sample are accessible separately, the latter with conversion electron Mössbauer spectroscopy (CEMS), which detects secondary electrons instead of the γ-rays in conventional Mössbauer spectroscopy.

The mentioned integral methods are powerful tools in *supporting* domain observations, not replacing them. We repeat the argument for a general but symmetric example, such as the analysis of a piece of grain-oriented transformer steel. From integral magnetization measurements we obtain the three integrals

$\int m_x \, dV$, $\int m_y \, dV$ and $\int m_z \, dV$. Magnetostriction measurements (and equivalently magnetoresistance or Mössbauer measurements) add three further expressions, $\int m_x^2 \, dV$, $\int m_y^2 \, dV$ and $\int m_z^2 \, dV$, one of which carries no information because of $m_x^2 + m_y^2 + m_z^2 = 1$. Assume we have analysed surface domain observations, leading to a hypothesis about the magnetic microstructure $\boldsymbol{m}(\boldsymbol{r})$ in the volume. We may now check whether the additional information from the five volume integrals is compatible with the assumed model $\boldsymbol{m}(\boldsymbol{r})$. To invert the problem, i.e. to derive the continuous vector function $\boldsymbol{m}(\boldsymbol{r})$ from the five measured numbers, is obviously impossible. Integral methods therefore offer *necessary* but by no means sufficient information for the analysis of magnetic micro-structure.

2.9 Comparison of Domain Observation Methods

Figure 2.56 indicates the limits of the most important techniques in three areas: in spatial resolution, in the recording time and in the information depth of the method. The last quantity determines also the quality of the necessary surface preparation. The recording time limits the dynamic capabilities of the methods, which may, however, be extended by stroboscopic experiments. The limits in this figure are guidelines, depending on the conditions of the experiment.

The table in Fig. 2.57 lists a number of qualitative criteria for the usefulness of domain observation techniques. Markedly positive features are highlighted. In the third column, the "quantitative" potential of the various methods is characterized with the following key:

Indirect: A method that permits the determination of the magnetization distribution only via a model calculation.

Direct: A method directly showing the (surface) magnetization vector.

Quantitative: A method permitting to evaluate magnetization vector components quantitatively either at the surface, or averaged over the sample thickness.

The abbreviations used in both tables are:

MFM: Magnetic force microscopy.

MO: Magneto-optic method.

SEM: Scanning (reflection) electron microscopy.

TEM: Transmission electron microscopy.

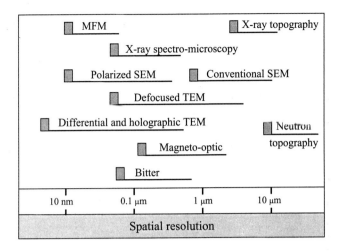

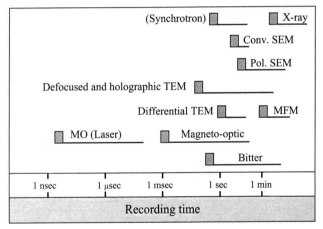

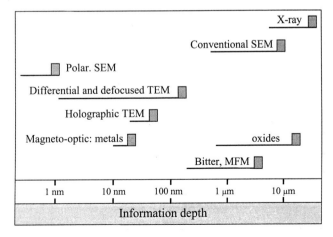

Fig. 2.56 Comparison of different domain observation techniques. Indicated are the estimated limits of the properties and their approximate range, depending on the experimental conditions

Method of domain observation	Sensitivity to small variations in magnetization	Evaluation of the magnetization vector	Allowed magnetic field range	Sample preparation quality requirements	Necessary capital investment
Bitter	very good	indirect	100 A/cm	moderate–low	low
Magneto-optic	fair	direct	any	high	moderate
Digital MO	good	quantitative	any	moderate	high
Defocused TEM	very good	indirect	3000 A/cm	high	high
Differential TEM	good	quantitative	1000 A/cm	high	very high
Holograph. TEM	good	quantitative	100 A/cm	very high	very high
Secondary SEM	poor	indirect	100 A/cm	low	high
Backscatt. SEM	poor	rather direct	300 A/cm	moderate–low	high
Pol. SEM	good	quantitative	100 A/cm	very high	very high
X-Ray topography	poor	indirect	any	moderate	extremely high
Neutron	poor	indirect	any	low	extremely high
MFM	good	indirect	3000 A/cm	low	moderate

Fig. 2.57 Qualitative comparison between different domain observation methods

3. Domain Theory

Domain observation and domain theory must go together in the study of magnetic microstructure. This chapter introduces domain theory. The focus is on application-oriented procedures rather than on the well-established fundamentals. A more detailed analysis is offered where it illustrates characteristic mechanisms. In this case intermediate steps are presented as a guideline to encourage the reader to reproduce the line of arguments. In Sect. 3.2 the free energy of a magnet is presented. The principle that this energy must be a minimum leads to the micromagnetic equations and their dynamic generalizations. A discussion of the origin of domains from the viewpoint of theory (Sect. 3.3) is followed by two sections, which deal with cases for which no knowledge of domain walls is necessary. Phase theory (Sect. 3.4) is adequate for the calculation of the reversible magnetization curve of large, bulk crystals. Small particle switching theory (Sect. 3.5) applies to small particles that are too small to contain regular domain walls. For all intermediate cases domain walls become important. Their properties are summarized in Sect. 3.6, which also includes a discussion of substructures of domain walls and of wall dynamics. In the last part of the chapter (Sect. 3.7) a number of commonly occurring domain features — like domain branching or dense stripe domains — are discussed from a theoretical point of view. This section ends with a thorough discussion of a seemingly simple, but actually rather complex model case: the Néel block.

3.1 The Purpose of Domain Theory

As discussed in the introduction (Sect. 1.2.2) the principles of domain theory go back to the famous article of *Landau* and *Lifshitz* [22]. Its presently accepted form can be found in *Kittel*'s reviews [30, 42]. There are no serious doubts about its validity or its foundations today. Domain theory is an indispensable tool in domain analysis, complementary to domain observation. In the last chapter we had to conclude that for most specimens there is no way to directly observe the interior domains. Only by theory are we able to infer the inner domains from surface observations. An obvious procedure to do this is as follows: firstly, find reasonable domain models that are compatible with the observed surface pattern. Calculate their energies and choose the best for further analysis. Then, vary the parameters of the chosen model (angles, lengths etc.) continuously, again looking for the lowest energy. If the calculated configuration is still consistent with observation, the model may be assumed to be correct; if not, other models must be checked for lower energy and better agreement with the experiment. This seemingly involved procedure is necessary because a direct numerical solution of the micromagnetic equations (see Sect. 1.3) is impossible for all but extreme cases such as submicron thin film elements. In all other cases only domain theory can support domain observation, and the continuum theory of micromagnetics can only help in supplying the necessary elements.

This chapter is intended to support such investigations with guidelines and a "toolbox" of expressions for particular cases. For background reading and earlier overviews see [34, 43–45, 37, 49] and [500–512]. *Miltat*'s review [513] gives an up-to-date introduction to the theory of magnetic microstructures (also amplifying some of the subjects of this book). The recent textbook by *Aharoni* [514] contains a careful analysis of the foundations of micromagnetics and a review with a focus on the extensive contributions of the author to this subject.

3.2 Energetics of a Ferromagnet

3.2.1 Overview

Micromagnetics and domain theory are based on the same variational principle which is derived from thermodynamic principles, as established initially in [22] and as reviewed for example in [37, 503]. According to this principle, the vector field of magnetization directions $m(r) = J(r)/J_s$ is chosen so that the total (free) energy reaches an absolute or relative minimum under the constraint $m^2 = 1$. Here we ignore the rare cases in which this constraint is not valid (meaning that the saturation magnetization J_s varies inside a micromagnetic configuration). A necessary consequence of the minimum energy principle and of the constant magnetization constraint is that at each point the *torque* on the magnetization as derived from the energy by variational calculus has to vanish. These torque conditions are known as the micromagnetic equations, which are discussed in detail in Sect. 3.2.7. Thus the magnetization-dependent contributions to the energy are the starting point both of domain theory and of the micromagnetic torque equations.

We must distinguish between local and non-local magnetic energy terms. The local terms are based on energy densities, which are given by the local values of the magnetization direction only. Their integral value is calculated by a simple integral of the form $E_{\mathrm{loc}} = \int f(\boldsymbol{m}) \mathrm{d}V$ over the sample, where the energy density function $f(\boldsymbol{m})$ is an arbitrary function of the magnetization direction $\boldsymbol{m}$. Examples are the *anisotropy* energy, the *applied field* (Zeeman) energy and the *magneto-elastic* interaction energy with a stress field of non-magnetic origin. The exchange or stiffness energy is also local in a sense, since it is calculated by an integral over a function of the derivatives of the magnetization directions.

The two non-local energy contributions are the *stray field* energy and the *magnetostrictive self-energy* — the energy connected with elastic interactions between regions magnetized along different axes. These energy terms give rise to torques on the magnetization vector that depend at any point on the magnetization directions at every other point. Non-local energy terms cannot be calculated by a single integration. For example, the stray field energy may be calculated by the following procedure (see Sect. 3.2.5): firstly, a scalar magnetic potential is derived from an integration over so-called magnetic 'charges', the sinks and sources of the magnetization vector field. A second integration over the product of charges and potential then leads to the total energy. Similar procedures are available for the magnetostrictive self-energy. Altogether two spatial integrations are necessary for non-local energy terms. The non-local terms make micromagnetics and domain analysis both interesting and complicated.

In the conventional approach indicated above, the stray field and the elastic deformations are integrated for every given magnetization distribution, looking for the minimum energy as a function of the magnetization field only. There are mathematical techniques to avoid the complication of non-local energy contributions, but at a price. In these alternatives, a modified total energy functional is varied with respect to the magnetization field as before and with respect to different potentials of the stray field (and of the elastic deformations) as additional variables [34, 37, 501]. In this case the functional depends on the variables and their derivatives only, and no further integrations are needed. The final result is the same as in the conventional approach. There seem to be advantages of the alternative approach in connection with computer solutions

[515–518], but this is not clear, as some authors abandoned it later again [519, 520]. We prefer limiting the number of independent variables as in the conventional procedure to obtain a better intuitive understanding of a situation.

A remark on notation: we generally use upper-case letters for total energy terms (E_x etc.), and lower-case letters for (volume or surface) energy *densities*.

3.2.2 Exchange Energy

(A) Volume Exchange Stiffness Energy. The fundamental property of a ferromagnet (or a ferrimagnet, see Sect. 1.3) is its preference for a constant equilibrium magnetization direction. Deviations from this ideal case invoke an energy penalty, which can be described by the "stiffness" expression [22]:

$$E_x = A \int (\mathbf{grad}\, m)^2 \, dV \tag{3.1}$$

where A is a material constant, the so-called exchange stiffness constant (dimension J/m or erg/cm), which is in general temperature dependent. Its zero-temperature value is related to the Curie point T_c [$A(0) \approx k T_c / a_L$, a_L = lattice constant, k = Boltzmann's constant]. The integrand of (3.1) can be written more explicitly:

$$e_x = A \left[(\mathbf{grad}\, m_1)^2 + (\mathbf{grad}\, m_2)^2 + (\mathbf{grad}\, m_3)^2 \right] \tag{3.2}$$

$$= A \left[m_{1,x}^2 + m_{1,y}^2 + m_{1,z}^2 + m_{2,x}^2 + m_{2,y}^2 + m_{2,z}^2 + m_{3,x}^2 + m_{3,y}^2 + m_{3,z}^2 \right]$$

where $m_{1,x}^2 = (\partial m_1 / \partial x)^2$, etc. This formula is derived by a Taylor expansion of the isotropic Heisenberg interaction $s_1 \cdot s_2$ between neighbouring spins. It imposes an energy penalty on any change in magnetization direction. The exchange energy (3.1) is called isotropic because it is independent of the direction of the change relative to the magnetization direction. Even if the Heisenberg interaction between localized spins is not applicable (as in metallic ferromagnets), (3.1) still describes phenomenologically the stiffness effect to first order. Only the interpretation of the exchange constant has to be changed.

In cylindrical coordinates the exchange energy density (3.2) is:

$$e_x = A \left\{ m_{\rho,\rho}^2 + m_{\varphi,\rho}^2 + m_{z,\rho}^2 + m_{\rho,z}^2 + m_{\varphi,z}^2 + m_{z,z}^2 \right.$$
$$\left. + \frac{1}{\rho^2} \left[(m_{\rho,\varphi} - m_\varphi)^2 + (m_{\varphi,\varphi} + m_\rho)^2 + m_{z,\varphi}^2 \right] \right\} \tag{3.2a}$$

A systematic analysis of the stiffness term can be found in Döring's review of micromagnetics [37], where he derives a generalized expression:

$$E_x = \int \sum_{i,k,l} A_{kl} \frac{\partial m_i}{\partial x_k} \frac{\partial m_i}{\partial x_l} \, dV \; . \tag{3.3}$$

The symmetric tensor A degenerates to a scalar for cubic or isotropic materials. Hexagonal or other lower symmetry crystals require more than one exchange stiffness constant. In practice, the isotropic formula is used throughout; at least no experimental determination of anisotropic exchange stiffness coefficients has been recorded.

The exchange energy (3.1) may be written in a different form that was found by *Arrott* et al. [521]. It is derived from the identity:

$$(\mathbf{grad}\,m)^2 = (\mathrm{div}\,m)^2 + (\mathbf{rot}\,m)^2 - \mathrm{div}\left[m \times \mathbf{rot}\,m + m\,\mathrm{div}\,m\right] \tag{3.4}$$

which is valid for $m^2 = 1$. The total exchange energy can therefore also be expressed as:

$$E_x = A \int \left[(\mathrm{div}\,m)^2 + (\mathbf{rot}\,m)^2\right] dV + A \int \left[m \times \mathbf{rot}\,m + m\,\mathrm{div}\,m\right] dS \tag{3.5}$$

where the first integral extends over the volume and the second over the surface. This formula contains only standard vector analytical expressions that are easy to handle, for example for a coordinate transformation. The additional surface integral is essential, however, and cannot be omitted [522]. It must be stressed that this surface integral is not at all related to surface anisotropy (see Sect. 3.2.3C) as falsely claimed in [514 (p. 138)].

The fact that the integrands in the volume integrals in (3.1) and (3.5) are different, and that part of the exchange energy density in (3.5) is transferred to the surface integral, does not support the intuitive understanding of a situation. Take a "cross" vortex with $m_1 = ay$, $m_2 = ax$, $m_3 = \sqrt{1 - m_1^2 - m_2^2}$, where a defines the length scale of the configuration. In the central point at $x = y = 0$ the integrand of (3.1) is $2a^2$, whereas the integrand in the volume integral of (3.5) vanishes, although the magnetization is not stationary around this point and the intuitive picture of a stiffness energy would predict a non-vanishing value there. Also the "refinement indicator" in the adaptive finite element algorithm of [523] would fail to detect a vortex here. We do not rely on (3.5) in the following.

Applying the gradient operator twice on the identity $m^2 = 1$, one may derive another equivalent form of the exchange energy density [30]:

$$e_x = A(\mathbf{grad}\,m)^2 = -A\,m \cdot \Delta m \tag{3.6}$$

where $\Delta = \mathrm{div}\,\mathbf{grad}$ is the Laplace operator. When the magnetization direction in (3.2) is expressed in polar coordinates, we obtain another convenient form:

$$m_1 = \cos\vartheta \cos\varphi \ , \ m_2 = \cos\vartheta \sin\varphi \ , \ m_3 = \sin\vartheta \tag{3.7a}$$

$$e_x = A[(\mathbf{grad}\,\vartheta)^2 + \cos^2\vartheta\,(\mathbf{grad}\,\varphi)^2] \ . \tag{3.7b}$$

It is this last formula (or a specialization of it) which is most frequently used in practice. [Throughout this book we prefer polar coordinate systems in which $\vartheta = 0$ at the equator, because symmetries around the equator are easier seen in this way. If you prefer $\vartheta = 0$ at the north pole, you may exchange sines and cosines of ϑ in (3.7)].

(B) Exchange Interface Coupling. If one ferromagnet is in contact with another (such as in multilayer thin films), an exchange interaction may couple the two media. In general, the coupling strength can not be derived from the volume properties of the two partners; depending on the exact nature of the interface, it can be weakened in comparison with bulk exchange, or even inverted as first discovered for iron-iron interfaces with a thin chromium interlayer [524]. In the same system, a further discovery was made: for certain chromium thicknesses a coupling favouring a non-collinear relative orientation between the two layers was found [525].

Phenomenologically, all these cases can be described by the following surface energy density expression:

$$e_{\text{coupl}} = C_{\text{bl}}(1 - \mathbf{m}_1 \cdot \mathbf{m}_2) + C_{\text{bq}}[1 - (\mathbf{m}_1 \cdot \mathbf{m}_2)^2] \ . \tag{3.8}$$

Here $\mathbf{m}_1$ and $\mathbf{m}_2$ are the magnetization vectors at the interface, C_{bl} is the bilinear and C_{bq} is the "biquadratic" coupling constant. If C_{bl} is positive, it favours parallel orientation of the magnetization in the two media ("ferromagnetic coupling"). If it is negative, an antiparallel alignment is preferred ("antiferromagnetic coupling"). A negative value of C_{bq} may lead to a 90° relative orientation if C_{bl} is small. According to present understanding the origin of the two coupling coefficients is quite different. The bilinear term is closely related to the corresponding volume exchange stiffness effect and is derivable from the same quantum mechanical foundations (for a review see [526]). In contrast, the biquadratic term is attributed to various microscopic spatial fluctuation ("extrinsic") mechanisms (related, for example, to interface roughness), as reviewed by *Slonczewski* [527]. For the purpose of micro-magnetism and domain theory the question of the nature of the coupling effects can be ignored. Important is, however, that the extrinsic fluctuation mechanisms always lead to negative values of C_{bq}.

In principle, higher-order terms and non-local expressions could be added to (3.8). The latter possibility may play a role in rare earth metal multilayers with their long-range Rudermann-Kittel interaction. A non-analytical variant of the biquadratic term was proposed by Slonczewski [527], which appears to be supported by certain experiments with antiferromagnetic interlayers.

3.2.3 Anisotropy Energy

The energy of a ferromagnet depends on the direction of the magnetization relative to the structural axes of the material. This dependence, which basically results from spin-orbit interactions, is described by the anisotropy energy. We distinguish between *crystal* anisotropies of the undisturbed crystal structure, and *induced* anisotropies describing the effects of deviations from ideal symmetry as for example because of lattice defects or partial atomic ordering. Shape effects do not belong to the anisotropy terms. They are part of the stray field energy that is treated in Sect. 3.2.5B. Whatever the origin of these anisotropies is, they must conform to the symmetry of the situation. Therefore, expansions in terms of spherical harmonics are used to describe the most important contributions. Rarely more than the first two significant terms have to be considered, since thermal agitation of the spins tends to average out the higher-order contributions. Only at low temperatures, significant higher-order terms have been observed, caused by interactions between the spins and the often highly anisotropic Fermi surface [528, 529].

Here only the lowest-order terms for cubic, uniaxial and orthorhombic systems are given. Expansions to higher orders and for other symmetries, and a discussion of the origins of anisotropy can be found in [530−533].

(A) Cubic Anisotropy. The basic formula for the anisotropy energy density of a cubic crystal is:

$$e_{\mathrm{Kc}} = K_{c1}\left(m_1^2\, m_2^2 + m_1^2\, m_3^2 + m_2^2\, m_3^2\right) + K_{c2}\, m_1^2\, m_2^2\, m_3^2 \qquad (3.9)$$

where the m_i are the magnetization components along the cubic axes. The material constant K_{c2} and higher-order terms can mostly be neglected. The constant K_{c1} assumes values in the range of $\pm 10^4$ J/m^3 for different materials. The sign of K_{c1} determines whether the $\langle 100\rangle$ or the $\langle 111\rangle$ directions are the easy directions for the magnetization (see Sect. 3.4.3A for more details). Figure 3.1 gives a graphic representation of anisotropy energy contributions, offering an impression of the local energetic environment the magnetization 'feels' in a ferromagnet. In polar angles (3.7a) the cubic anisotropy becomes:

$$e_{\mathrm{Kc}} = \left(K_{c1} + K_{c2}\sin^2\vartheta\right)\cos^4\vartheta\,\sin^2\varphi\,\cos^2\varphi + K_{c1}\sin^2\vartheta\,\cos^2\vartheta \quad . \qquad (3.9a)$$

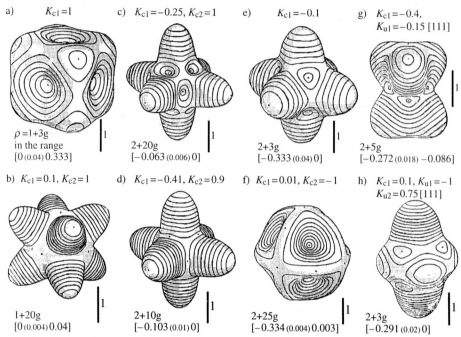

a) $K_{c1}=1$ c) $K_{c1}=-0.25, K_{c2}=1$ e) $K_{c1}=-0.1$ g) $K_{c1}=-0.4,$
                                                                                                 $K_{u1}=-0.15\,[111]$

$\rho=1+3g$
in the range
$[0\,{(0.04)}\,0.333]$

$2+20g$
$[-0.063\,{(0.006)}\,0]$

$2+3g$
$[-0.333\,{(0.04)}\,0]$

$2+5g$
$[-0.272\,{(0.018)}\,-0.086]$

b) $K_{c1}=0.1, K_{c2}=1$ d) $K_{c1}=-0.41, K_{c2}=0.9$ f) $K_{c1}=0.01, K_{c2}=-1$ h) $K_{c1}=0.1, K_{u1}=-1$
                                                                                                                  $K_{u2}=0.75\,[111]$

$1+20g$
$[0\,{(0.004)}\,0.04]$

$2+10g$
$[-0.103\,{(0.01)}\,0]$

$2+25g$
$[-0.334\,{(0.004)}\,0.003]$

$2+3g$
$[-0.291\,{(0.02)}\,0]$

Fig. 3.1 "Energy surfaces" defined by $\rho = A + B \cdot g$, where $g(\vartheta, \varphi)$ is a general anisotropy functional, such as (3.9) or (3.11) or a combination of these, and A and B are suitable scaling factors. Also plotted are contours of equal energy in the range $[C\,(S)\,D]$, where C is the minimum value of $g(\vartheta, \varphi)$, D is the maximum and S is the distance between contours. The anisotropy coefficients as well as the scaling factors and scales can be found below and above the figures. (a, b) Cubic anisotropy with $\langle 100 \rangle$ easy directions. (c, d) Cubic with $\langle 110 \rangle$ easy directions. (e, f) Cubic with $\langle 111 \rangle$ easy directions. (g) Cubic, and uniaxial anisotropy along the [111] axis superimposed. (h) Cubic, and conical anisotropy along the [111] axis superimposed

Equation (3.9a) is useful for the investigation of domains or walls in a $\langle 100 \rangle$-based coordinate system as one would use it in samples with (100) surfaces. For samples with a (110) surface or for 90° walls (Fig. 1.3), the anisotropy energy density may be expressed in another polar coordinate system:

$$\boldsymbol{m} = \cos\vartheta \cos\varphi\,(0,0,1) + \sqrt{\tfrac{1}{2}}\cos\vartheta \sin\varphi\,(1,\bar{1},0) + \sqrt{\tfrac{1}{2}}\sin\vartheta\,(1,1,0)\ ,$$

$$e_{Kc} = K_{c1}\cos^2\vartheta \cos^2\varphi\left(\cos^2\vartheta \sin^2\varphi + \sin^2\vartheta\right)$$
$$+ \tfrac{1}{4}\left(K_{c1} + K_{c2}\cos^2\vartheta \cos^2\varphi\right)\left(\cos^2\vartheta \sin^2\varphi - \sin^2\vartheta\right)^2\ . \tag{3.9b}$$

For walls in negative-anisotropy materials, a polar coordinate system of still another orientation may be used to express the crystal anisotropy energy:

$$\boldsymbol{m} = \sqrt{\tfrac{1}{2}}\cos\vartheta \cos\varphi\,(1,\bar{1},0) + \sqrt{\tfrac{1}{6}}\cos\vartheta \sin\varphi\,(1,1,\bar{2}) + \sqrt{\tfrac{1}{3}}\sin\vartheta\,(1,1,1)$$

$$e_{Kc} = K_{c1}\left[\tfrac{1}{4}\cos^4\vartheta + \tfrac{1}{3}\sin^4\vartheta - f_1(\vartheta, \varphi)\right]$$
$$+ \tfrac{1}{108}K_{c2}\left[2\cos^6\vartheta \sin^2(3\varphi) + \sin^2\vartheta f_2^2(\vartheta) + 6f_1(\vartheta, \varphi)f_2(\vartheta)\right] \qquad (3.9c)$$

with $f_1(\vartheta, \varphi) = \tfrac{1}{3}\sqrt{2}\sin\vartheta \cos^3\vartheta \sin(3\varphi)$ and $f_2(\vartheta) = 2\sin^2\vartheta - 3\cos^2\vartheta$

For arbitrary surface orientation, tensor calculus can be used to derive corresponding formulae. If the cubic anisotropy energy (3.9) is expressed as:

$$e_{Kc} = K^{(0)} + K^{(1)}_{ijkl}m_i\,m_j\,m_k\,m_l + K^{(2)}_{ijklrs}m_i\,m_j\,m_k\,m_l\,m_r\,m_s \ , \qquad (3.10)$$

the tensors $K^{(1)}$ and $K^{(2)}$ can be transformed to a new coordinate system by the standard procedure (for example $\widetilde{A}_{ijkl} = A_{mnrs}T_{im}T_{jn}T_{kr}T_{ls}$ for a fourth-order tensor A, where T is the transformation matrix).

(B) Uniaxial and Orthorhombic Anisotropy. Hexagonal and tetragonal crystals show a uniaxial anisotropy, which up to fourth-order terms becomes:

$$e_{Ku} = K_{u1}\sin^2\vartheta + K_{u2}\sin^4\vartheta \qquad (3.11)$$

where ϑ is the angle between anisotropy axis and magnetization direction. The same formula applies to a uniaxial induced anisotropy. A large positive K_{u1} describes an easy axis, a large negative K_{u1} an easy plane perpendicular to the anisotropy axis. For intermediate values, i.e. under the condition $0 > K_{u1}/K_{u2} > -2$, the easy directions lie on a cone with the angle Θ relative to the axis given by $\sin^2\Theta = -\tfrac{1}{2}K_{u1}/K_{u2}$. In short, the three cases are called 'uniaxial', 'planar' and 'conical' magnetic anisotropies. Uniaxial anisotropies can be much stronger than cubic anisotropies, reaching some 10^7 J/m^3 for rare earth transition metal permanent magnet materials.

Sometimes a generalized second-order anisotropy has to be considered. This applies to crystals of lower than tetragonal or hexagonal symmetry, or cases when several uniaxial anisotropies are superimposed. The energy density of this *orthorhombic anisotropy* is written as:

$$e_{Ko} = \sum_{i,k} K_{ik}m_i\,m_k \qquad (3.12)$$

where K is a symmetric tensor of rank two. Also anisotropies induced by elastic stress and some shape effects will reduce to this formula. The energy expression e_{Ko} defines three orthogonal axes, an easy, a hard and an intermediate axis. In a coordinate system, which is oriented along these eigenvectors of K, the orthorhombic anisotropy (3.12) reduces to:

$$e_{Ko} = K_1 m_1^2 + K_2 m_2^2 + K_3 m_3^2 \qquad (3.12a)$$

where the K_i are the eigenvalues of K, one of which is magnetically insignificant because of $m_1^2 + m_2^2 + m_3^2 = 1$.

An *induced* anisotropy in a cubic crystal can take the form of an orthorhombic anisotropy [531], as described by the second-order expression:

$$e_{\mathrm{Ki}} = F_{\mathrm{ind}}\left(m_1^2\,a_1^2 + m_2^2\,a_2^2 + m_3^2\,a_3^2\right)$$
$$+ 2\,G_{\mathrm{ind}}\left(m_1\,m_2\,a_1\,a_2 + m_1\,m_3\,a_1\,a_3 + m_2\,m_3\,a_2\,a_3\right)\;. \qquad (3.12b)$$

Here a is the magnetization direction during the annealing process at which some structural change induces the anisotropy, and F_{ind} and G_{ind} are two material parameters that are in general temperature and time dependent. Equation (3.12b) describes an orthorhombic anisotropy the easy axis of which does not necessarily coincide with the annealing axis. In the special case when F_{ind} and G_{ind} are equal, the orthorhombic anisotropy (3.12b) reduces to a uniaxial anisotropy with the axis along a. This is also true for polycrystalline or amorphous materials. For evaporated Permalloy films, for example, the uniaxial anisotropy coefficient becomes typically $10^2\,\mathrm{J/m^3}$.

(C) Surface and Interface Anisotropy. Another anisotropic energy term applies only to the surface magnetization and was introduced by *Néel* [534]. It is called magnetic *surface anisotropy* and is described by additional phenomenological parameters. In a structurally isotropic medium, magnetic surface anisotropy can be expressed to first order as:

$$e_{\mathrm{s}} = K_{\mathrm{s}}\left[1 - (m \cdot n)^2\right] \qquad (3.13)$$

where n is the surface normal. The expression must be integrated over the surface, so the dimension of the coefficient K_{s} is $\mathrm{J/m^2}$. For positive K_{s} the surface energy density e_{s} is smallest for perpendicular orientation of the magnetization at the surface ($m \parallel n$). The order of magnitude of K_{s} (between 10^{-3} and $10^{-4}\,\mathrm{J/m^2}$ [535]) is often considerably larger than the value one obtains by multiplying regular anisotropy energy constants with the thickness of one atomic layer. The effect is attributed to the reduced symmetry of the atomic environment of surface atoms. For ordinary bulk samples, however, the effects of surface anisotropies are negligible because the surface magnetization is coupled by exchange forces to the bulk magnetization. They become important for very thin films and multilayers of such films. Surface anisotropy also affects the micromagnetic boundary conditions, as will be discussed in Sect. 3.2.7D.

In cubic crystals, surface anisotropy is described to first approximation by two independent quantities [34]:

$$e_s = K_{s1}\left(1 - m_1^2 n_1^2 - m_2^2 n_2^2 - m_3^2 n_3^2\right)$$
$$- 2K_{s2}\left(m_1 m_2 n_1 n_2 + m_1 m_3 n_1 n_3 + m_2 m_3 n_2 n_3\right) \quad . \tag{3.13a}$$

If K_{s2} equals K_{s1}, this expression reduces to the isotropic formula (3.13). *Gradmann* et al. [535, 536] reviewed experiments on the magnitude of the constants K_{s1} and K_{s2}, indicating the "anisotropy of the surface anisotropy" $K_{s1} - K_{s2}$ to be small but measurable. It causes, for example, a switch in the easy direction of very thin (110)-oriented iron films from [001] to [$\bar{1}$10] as a function of thickness. The same kind of energy also occurs at *interfaces* between ferromagnetic and non-magnetic media. Higher-order terms may also become relevant particularly for uniaxial materials [537].

Experimentally, surface anisotropies are derived, either from surface-related modes in magnetic resonance [538], or from the thickness dependence of total thin film anisotropies. In the latter kind of measurements, a possible complication has to be considered. Stress anisotropy, which may be caused by misfit in epitaxial films or by deposition stresses in polycrystalline films, is primarily part of the volume anisotropy. Such stresses can gradually relax beyond a critical thickness, thus imitating surface anisotropy [539]. For the purpose of micromagnetic analysis, both effects have to be separated by additional elastic and magneto-elastic measurements, as reviewed in [540].

(D) Exchange Anisotropy. All anisotropy expressions discussed so far are *even* functions of the magnetization vector, not changing their value upon reversal of the magnetization. This is a necessary symmetry condition for chemically and crystallographically uniform samples. Non-uniform samples in which exchange-coupled ferromagnetic and antiferromagnetic phases exist side by side often show asymmetrical magnetization curves that appear displaced along the field axis. Typically the external field is only able to switch the ferromagnet. The antiferromagnetic phase may be given a specific order by cooling it through its Néel temperature under the influence of the coupled saturated ferromagnet. The result is a preference of the magnetization for this direction also in the ferromagnet, and a displaced hysteresis.

Formally, the exchange coupling between both phases may be described by an effective field H_{XA} or an equivalent "unidirectional" anisotropy constant $K_{XA} = H_{XA} J_s$ in the ferromagnet. This leads to an energy of the form:

$$e_{XA} = -K_{XA} \cos \Theta \tag{3.14}$$

where Θ is the angle between the magnetization direction in the ferromagnet and the preferred direction of the exchange anisotropy. Similar effects are observed when a soft magnetic material is exchange coupled to a hard magnetic phase that does not react in the applied field. In sufficiently high fields, it is always possible to obtain a symmetric hysteresis loop.

Even seemingly uniform materials may show exchange anisotropy on account of microscopic antiferromagnetic or magnetically hard inhomogeneities (micro-precipitates, ordering regions, stacking faults, crystalline layers on top of amorphous materials, etc.). The complex phenomena connected with exchange anisotropy are reviewed in [541–543]. The effect also finds technical applications in the field of sensors, where it pins the magnetization vector in the adjacent ferromagnetic layer and suppresses unwanted domains. This important application leads to a continuing search for new systems and better understanding of exchange anisotropy phenomena [544, 545].

3.2.4 External Field (Zeeman) Energy

The magnetic field energy can be separated into two parts, the *external field energy* and the *stray field energy* (see e.g. [37, 501, 514]). The first part, the interaction energy of the magnetization vector field with an external field H_{ex}, is simply:

$$E_{\text{H}} = -J_{\text{s}} \int H_{\text{ex}} \cdot m \, dV \quad . \tag{3.15}$$

For a uniform external field this energy depends only on the average magnetization and not on the particular domain structure or the sample shape.

3.2.5 Stray Field Energy

(A) General Formulation. The second part of the magnetic field energy is connected with the magnetic field generated by the magnetic body itself. Starting from Maxwell's equation $\text{div}\, B = \text{div}\,(\mu_0 H + J) = 0$, we define as the stray field H_{d} the field generated by the divergence of the magnetization J:

$$\text{div}\, H_{\text{d}} = -\text{div}\,(J/\mu_0) \quad . \tag{3.16}$$

The sinks and sources of the magnetization act like positive and negative *"magnetic charges"* for the stray field. The field can be calculated like a field in electrostatics from the electrical charges. The only difference is that magnetic charges never appear isolated but are always balanced by opposite charges. The energy connected to the stray field is:

$$E_d = \tfrac{1}{2}\mu_0 \int\limits_{\text{all space}} H_d^2 \, dV = -\tfrac{1}{2} \int\limits_{\text{sample}} H_d \cdot J \, dV \quad . \tag{3.17}$$

The first integral extends over all space; it shows that the stray field energy is always positive, and is only zero if the stray field itself is zero everywhere. The second integral is mathematically equivalent for a finite sample. It is often easier to evaluate, since it extends only over the magnetic sample. Equations (3.16) and (3.17) completely define the stray field energy. Methods to handle these equations are described in the following.

A general solution of the stray field problem is given by potential theory. The reduced volume charge density λ_v and the surface charge density σ_s are defined in terms of the reduced magnetization $m(r) = J(r)/J_s$:

$$\lambda_v = -\operatorname{div} m \quad , \quad \sigma_s = m \cdot n \tag{3.18}$$

where n is the outward directed surface normal. If the body contains interfaces separating two media 1 and 2 with different values m_1 and m_2 at the interface, then the interface charges 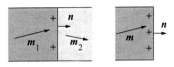 $\sigma_s = (m_1 - m_2) \cdot n$ are formed (assuming that the interface normal n points from medium 1 to medium 2). The surface charges of (3.18) are a special case of interface charges where the second medium is non-magnetic.

With these quantities the potential of the stray field at position r is given by an integration over r':

$$\Phi_d(r) = \frac{J_s}{4\pi\mu_0} \left[\int \frac{\lambda_v(r')}{|r-r'|} \, dV' + \int \frac{\sigma_s(r')}{|r-r'|} \, dS' \right] \quad , \tag{3.19a}$$

from which the stray field can be derived by $H_d(r) = -\operatorname{grad} \Phi_d(r)$. Another integration immediately yields the stray field energy:

$$E_d = J_s \left[\int \lambda_v(r) \, \Phi_d(r) \, dV + \int \sigma_s(r) \, \Phi_d(r) \, dS \right] \quad . \tag{3.19b}$$

The integrations $\int dV$ and $\int dV'$ extend over the volume of the sample, and $\int dS$ and $\int dS'$ over its surface. The stray field energy calculation therefore amounts to a six-fold integration if volume charges λ_v are present. Although the integrand diverges at $r = r'$, the integrals remain finite.

(B) Simple Cases. Sometimes the calculation of the stray field presents no difficulties at all. An example is an infinitely extended plate in which the magnetization direction depends only on the z coordinate (Fig. 3.2). In this one-dimensional case the differential equation (3.16) is readily integrated, yielding the demagnetizing or stray field H_d and its energy density e_d:

$$H_d = -\left(J_s/\mu_0\right) m_3(z)\, e_3 \;,\quad e_d = -\tfrac{1}{2} H_d \cdot J = \left(J_s^2/2\mu_0\right) m_3^2(z) \;. \tag{3.20}$$

(The integration constant in the stray field H_d has to be zero, since otherwise the field would extend to infinity). An analogous rigorous solution applies to cylindrically symmetric magnetization distributions [546].

The energy density e_d has the form of a uniaxial anisotropy energy. It is convenient to introduce an abbreviation for the coefficient in this equation:

$$K_d = J_s^2/2\mu_0 \;. \tag{3.21}$$

The stray field energy density is just K_d for a plate that is uniformly magnetized perpendicular to its surface. As this is a particularly unfavourable situation, the quantity K_d is a measure for the maximum energy densities which may be connected with stray fields (although there may be rare situations where even this value can be exceeded, as for example in coarse-grained hard magnetic materials in which a grain may be magnetized opposite to its environment). Even in cases where a stray field energy is much more difficult to calculate than by the simple formula (3.20), it will scale with the material parameter K_d. The quantity K_d becomes $2\pi M_s^2$ in the Gaussian system; this means that if any calculation of a stray field energy presented in the old system is to be transferred to the SI system, the coefficient M_s^2 has to be replaced by $K_d/2\pi$.

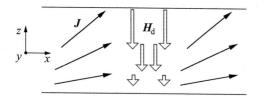

Fig. 3.2 Stray fields in a one-dimensional magnetization distribution. The figure shows the cross-section through a plate that is infinitely extended along the (x, y) plane. For such a plate the external stray field is zero

An often used dimensionless quantity is the ratio $Q = K/K_d$, where K is an appropriate anisotropy coefficient. This material parameter plays a role in many discussions in this book. If not obvious from the context, the type of anisotropy is specified in defining Q, as in $Q_u = K_u/K_d$ for uniaxial materials.

Another classical example is that of a uniformly magnetized ellipsoid. The demagnetizing field of an ellipsoid is uniform and linearly related to the (average or uniform) magnetization J by the symmetrical demagnetizing tensor N:

$$H_d = -N \cdot J/\mu_0 \;. \tag{3.22}$$

For the general ellipsoid with the axes (a, b, c), the demagnetizing factor along the a axis is given by the integral:

$$N_a = \tfrac{1}{2} abc \int_0^\infty \left[(a^2 + \eta)\sqrt{(a^2 + \eta)(b^2 + \eta)(c^2 + \eta)} \right]^{-1} d\eta \ . \qquad (3.23)$$

Analogous expressions apply to N_b and N_c. The sum of all three coefficients is always equal to one. Figure 3.3 shows the demagnetizing factors of the general ellipsoid calculated numerically from (3.23).

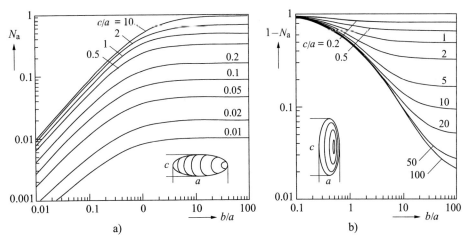

Fig. 3.3 The demagnetizing factor N of the general ellipsoid as a function of its shape. (a) applies more to prolate ellipsoids, while (b) shows $1 - N$ for oblate ellipsoids

For rotation ellipsoids of dimensions (a, c, c), explicit formulae are available:

$$N_a = \frac{\alpha^2}{1 - \alpha^2} \left[\frac{1}{\sqrt{1 - \alpha^2}} \operatorname{arcsinh}\left(\frac{\sqrt{1 - \alpha^2}}{\alpha} \right) - 1 \right] \ , \quad N_c = \tfrac{1}{2}(1 - N_a) \qquad (3.23a)$$

for prolate (cigar-shaped) ellipsoids with $\alpha = c/a < 1$, and

$$N_a = \frac{\alpha^2}{\alpha^2 - 1} \left[1 - \frac{1}{\sqrt{\alpha^2 - 1}} \arcsin\left(\frac{\sqrt{\alpha^2 - 1}}{\alpha} \right) \right] \ , \quad N_c = \tfrac{1}{2}(1 - N_a) \qquad (3.23b)$$

for oblate (disc-shaped) ellipsoids with $\alpha > 1$, the two cases joining at $N_a \approx \tfrac{1}{3} - \tfrac{1}{15}(\alpha - 1)$ for nearly spherical shapes ($\alpha \approx 1$).

The demagnetizing energy of an ellipsoid of volume V is obtained from the general stray field energy formula (3.17):

$$E_d = (1/2\mu_0)\, V\, \mathbf{J}^T \cdot \mathbf{N} \cdot \mathbf{J} = K_d V \mathbf{m}^T \cdot \mathbf{N} \cdot \mathbf{m} \ . \qquad (3.24)$$

Numerical calculations of the demagnetizing energy for arbitrary bodies can be checked by comparing them with the rigorous result for a suitable ellipsoid. The demagnetizing factor of compact bodies is often well approximated by that of the inscribed ellipsoid.

(C) Three-Dimensional Finite Element Calculations. Except in favourable cases as discussed in (B), stray field calculations are complex and regularly call for numerical evaluation. The problem is that the direct computation of the sixfold integral (3.19) would be forbiddingly time consuming. One should therefore try to perform as many steps as possible analytically. One approach that may be traced back to [547] replaces the integrations by a summation over self-energy and interaction energy terms of *finite elements*. These are volume or surface elements in which the magnetic charges are assumed constant. If the elements are of rectangular shape, all integrations can be reduced by substitution to multiple integrals of the core function $F_{000} = 1/r$. The integrals can be evaluated analytically and are listed below, using the abbreviations:

$$u = x^2 , \quad v = y^2 , \quad w = z^2 , \quad r = \sqrt{u + v + w} \quad ,$$

$$L_x = \operatorname{arctanh}(x/r) = \tfrac{1}{2} \ln\left[(r + x)/(r - x)\right] \quad \text{etc.} \; ,$$

$$P_x = x \arctan\left(yz/xr\right) \quad \text{etc.} \; ,$$

$$L_x = 0 \quad \text{and} \quad P_x = 0 \quad \text{for } x = 0 \quad \text{etc.} \; ,$$

$$F_{ikm} = i, k \text{ and } m\text{-fold indefinite integral along } x, y \text{ and } z \; . \tag{3.25a}$$

Since only differences between the integrals will be used, all integration constants are arbitrary (remember that a "constant" in an integration along, say x, may be an arbitrary function of y and z, and may become a more complicated function in the course of further integrations). In the following expressions some integration constants were retained such that the expressions become (i) symmetric if applicable, and (ii) easily verifiable by differentiation operations. We therefore begin with the source function F_{000} and continue with integrals of increasing order, omitting integrals that follow by simply exchanging the variables x, y and z:

$$F_{000} = 1/r \; ,$$

$$F_{100} = \int F_{000} \, dx = x L_x - r \; ,$$

$$F_{200} = \int F_{100} \, dx = x L_x - x \; ,$$

$$F_{110} = \int F_{100} \, dy = y L_x + x L_y - P_z \; ,$$

$$F_{210} = xy L_x + \tfrac{1}{2}(u - w) L_y - x P_z - \tfrac{1}{2} y r \; ,$$

$$F_{111} = xy L_z + xz L_y + yz L_x - \tfrac{1}{2}(x P_x + y P_y + z P_z) \; ,$$

$$F_{220} = \tfrac{1}{2}\left[x(v - w) L_x + y(u - w) L_y\right] - xy P_z + \tfrac{1}{6} r(3w - r^2) \; . \tag{3.25b}$$

Further, less frequently needed integrals are:

$$F_{211} = xyzL_x + \tfrac{1}{2}z(u-\tfrac{1}{3}w)L_y + \tfrac{1}{2}y(u-\tfrac{1}{3}v)L_z - \tfrac{1}{6}uP_x - \tfrac{1}{2}x(yP_y+zP_z) - \tfrac{1}{3}yzr \ ,$$

$$F_{221} = \tfrac{1}{2}z\left[x(v-\tfrac{1}{3}w)L_x + y(u-\tfrac{1}{3}w)L_y\right] + \tfrac{1}{3}\left[uv-\tfrac{1}{8}(u+v)^2\right]L_z$$
$$-\tfrac{1}{6}xy(yP_y+xP_x+3zP_z) + \tfrac{1}{24}zr\left[2r^2-5(u+v)\right] \ ,$$

$$F_{222} = \tfrac{1}{24}\left[xL_x(6vw-v^2-w^2) + yL_y(6uw-u^2-w^2) + zL_z(6uv-u^2-v^2)\right]$$
$$-\tfrac{1}{6}xyz(xP_x+yP_y+zP_z) + \tfrac{1}{60}r\left[r^4-5(uv+uw+vw)\right] \ . \qquad (3.25c)$$

To see how these integrals are applied, consider the interaction between
two parallel lines, each of which is uniformly charged and extending parallel
to the x axis from x_1 to x_2 and from x_3 to x_4,
respectively. The y and z positions shall be
$(0, 0)$ and (y_0, z_0). Then according to (3.19), an
integral of the following form must be evaluated
to calculate the stray field energy:

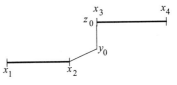

$$I_d = \int_{x_1}^{x_2}\int_{x_3}^{x_4}\left[(x-x')^2+y_0^2+z_0^2\right]^{-1/2}dx\,dx' \ . \qquad (3.26)$$

With the abbreviation $F(x) = F_{200}(x, y_0, z_0)$ it evaluates to:

$$I_d = [F(x_4-x_1)-F(x_3-x_1)] - [F(x_4-x_2)-F(x_3-x_2)] \ . \qquad (3.27)$$

Using such multiple differences, the total stray field energy can be calculated
as a sum over interaction energy expressions of the line segments.

Cells with surface or volume charges
are treated similarly. Consider the interaction between two rectangular bodies,
as sketched, with reduced volume charges
λ_1 and λ_2. Their interaction energy can
be expressed by a multiple difference as:

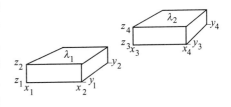

$$E_{inter} = \tfrac{1}{2\pi}\lambda_1\lambda_2 K_d\left[F_2(z_4-z_1)-F_2(z_4-z_2)-F_2(z_3-z_1)+F_2(z_3-z_2)\right] \quad (3.28)$$

with $F_2(z) = F_1(y_4-y_1,z)-F_1(y_4-y_2,z)-F_1(y_3-y_1,z)+F_1(y_3-y_2,z)$

and, using $F = F_{222}$:

$$F_1(y, z) = F(x_4-x_1,y,z)-F(x_4-x_2,y,z)-F(x_3-x_1,y,z)+F(x_3-x_2,y,z) \ .$$

Self-energy terms are special cases of interaction energy terms. As an
example, consider the self-energy of the first cell in the sketch:

$$E_{\text{self}} = \tfrac{2}{\pi}\lambda_1^2 K_d \big[\, F(x_2 - x_1, y_2 - y_1, z_2 - z_1) - F(x_2 - x_1, y_2 - y_1, 0)$$

$$- F(x_2 - x_1, 0, z_2 - z_1) - F(0, y_2 - y_1, z_2 - z_1)$$

$$+ F(x_2 - x_1, 0, 0) + F(0, y_2 - y_1, 0) + F(0, 0, z_2 - z_1)\big] \qquad (3.29)$$

with $F = F_{222}$. This expression can be derived from the general expression (3.28). It is one half of the interaction energy E_{inter} of a cell with an identical copy (two equal, coinciding cells: $x_1 = x_3$, $x_2 = x_4$, etc.). Using this relation we may write the total stray field energy of a set of n cells in the following simple way:

$$E_d = \tfrac{1}{2}\sum_{i=1}^{n}\sum_{j=1}^{n} E_{\text{inter}}(i, j) \ . \qquad (3.30)$$

If the interaction energy between two arbitrary cells is known, the total energy of a set of many cells can therefore be computed. The rectangular cells may be of any size, position and charge if the cell edges are all parallel.

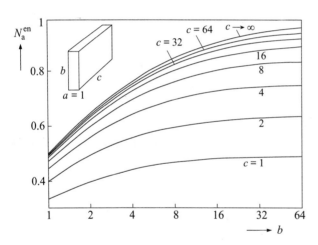

Fig. 3.4 The energy-demagnetizing factor of uniformly magnetized rectangular bodies as a function of their shape. As in the case of ellipsoids (3.23), the demagnetizing factors along the three axes obey $N_a^{\text{en}} + N_b^{\text{en}} + N_c^{\text{en}} = 1$

The well-known function of *Rhodes* and *Rowlands* [547] is a special case of F_{220}. It can be applied to high-anisotropy uniaxial block-shaped samples with surface charges as sketched. Along x the sample is normalized to unity, so that the double x difference can be performed in advance. The energies can then be expressed as a function of:

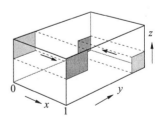

$$F_{\text{RR}}(y, z) = 2\,[F_{220}(1, y, z) - F_{220}(0, y, z)] \ . \qquad (3.31)$$

The integral F_{220} can also be used to calculate the demagnetizing factor of uniformly magnetized rectangular bodies. We define the *energy*-demagnetizing factor N^{en} by $E_d = N^{en}K_d V$, where E_d is the stray field energy of the particle saturated along the axis in question, and V is the volume [compare (3.24)]. Calculating the self- and interaction energies of the resulting surface charges we obtain the demagnetizing factors of rectangular bodies as shown in Fig. 3.4.

(D) General Features and Numerical Techniques. A number of general properties of the stray field interaction and possibilities for its evaluation can be most easily demonstrated with the help of discretized formulations like (3.30), although they could also be proven within the continuum approach presented in (A).

• The total stray field energy can be expressed in a particularly convenient way, if the cells form a periodic lattice. Let the n cells be given by their dimensions (a, b, c), their positions r_i and their charges λ_i. Then the total energy can be written in terms of an interaction matrix W:

$$E_d = \frac{1}{2\pi} K_d \sum_{i=1}^{n} \lambda_i \sum_{j=1}^{n} \lambda_j W_{ij} \quad \text{with} \quad W_{ij} = F(r_i - r_j) \tag{3.32}$$

where the interaction function F follows immediately from (3.28) by inserting the new coordinates. Because the interaction matrix depends only on the distance between two cells $r_i - r_j$ and not on their absolute position, its coefficients can be stored in a much smaller memory [of the order $O(n)$ compared to $O(n^2)$ for a general, non-periodic cell pattern].

• The mathematical form of the inner sum in (3.32) amounts to a discrete *convolution*, and such a convolution can be evaluated by highly efficient Fast Fourier Transform techniques [548]. An analogous possibility, based on the magnetization vectors, was first introduced in micromagnetic computations in [549]. The more efficient scalar charge version (3.32) was used first in [550] and elaborated in more detail in [60]. The advantages of the new techniques, which can reduce the computation time by several orders of magnitude compared to traditional techniques, were again demonstrated in [551]. The properly applied FFT technique does not introduce any approximations as checked and explicitly stated in [550]. For a given lattice, the direct summation of the stray field energy and the evaluation via FFT agree within numerical accuracy, meaning to 10^{-8} for typical problems and standard, double precision arithmetics. The doubts cast on this method by Aharoni [514 (p. 257)] are groundless.

The huge savings in memory and computation time thus possible with a periodic discretization may still be outweighed by the advantages of an adapted discretization, which uses a fine mesh only where necessary. Which of the two approaches is preferable can only be decided for the particular situation. An ingenious method to combine irregular discretization with calculations on a periodic lattice was developed in [552].

• The interaction energy between distant elements may be calculated by replacing volume or surface charges by line or point charges. Ultimately, calculations based on point charges will be numerically more accurate for large distances than the rigorous expressions for volume charges, which consist of multiple differences of large numbers. To illustrate the convergence of the different approaches, Fig. 3.5 shows the interaction between two elements as a function of their distance, together with the approximate expressions calculated with planar, line and point charges centred in the elements, each having the same total charge.

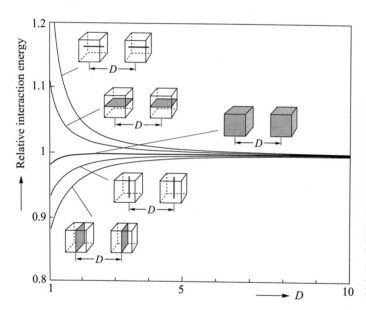

Fig. 3.5 The interaction energy of two cubes of unit size carrying uniform volume charges as a function of their distance D (*fat curve*), together with approximate expressions for planar, line and point charge interactions. The interactions are normalized by the point charge interaction

• Scaling up a configuration simply leads to an increase of the total stray field energy by the third power of the scale (i.e. proportional to the sample volume). This property is reflected in corresponding "homogeneity" properties for our versions of the integrals F_{ikm} as e.g. for F_{220}:

$$F_{220}(cx, cy, cz) = c^3 F_{220}(x, y, z) \ . \tag{3.33}$$

In stray field calculations it is always possible to reduce one dimension to unity and to apply the scaling factor at the end.

- Often it is possible to simplify a stray field energy calculation by replacing the original charge pattern $s^{ch}(r)$ by "components" s_1^{ch} and s_2^{ch}, so that s^{ch} is a linear superposition of the components $s^{ch}(r) = s_1^{ch}(r) + s_2^{ch}(r)$. The components can be either simple subsets of the charge elements, or other patterns fulfilling the superposition condition. The total stray field energy (3.30) can then be reformulated according to:

$$E_d = \tfrac{1}{2} \sum_{i,j \in s_1^{ch}} E_{i,j} + \tfrac{1}{2} \sum_{i,j \in s_2^{ch}} E_{i,j} + \sum_{i \in s_1^{ch}, j \in s_2^{ch}} E_{i,j} \tag{3.30a}$$

where $E_{i,j}$ stands for the interaction energy $E_{\text{inter}}(i,j)$. The net charge of both components shall be zero ($\sum \lambda_i v_i = 0$ for $i \in s_1^{ch}$, as well as for $i \in s_2^{ch}$, where the v_i are the volumes or other integration weights of the elements i). Then the first sum in (3.30a) can be called the self-energy of component s_1^{ch}, the second sum is the self-energy of component s_2^{ch}, and the third expression is the interaction energy between both components. The reformulation (3.30a) is useful if the interaction energy between the components is zero or negligible. Such situations of non-interacting charge patterns occur for example if (i) the stray field from one component is oriented at right angles to the magnetization pattern of the second component [for a proof of this statement see (3.17)], or (ii) if the spatial scale of one component is so small that the variation of the stray field potential in (3.19b), which is caused by the first component, can be neglected on the scale of the charge variation of the second component.

This possibility can be exploited in the analysis of complex domain patterns [553]. The sketch illustrates an orthogonal arrangement in which the total number of elements in the components is smaller than the number of elements in the original (checkerboard) pattern. The procedure can also be applied iteratively.

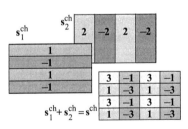

(E) Two-Dimensional Finite Element Calculations. A similar scheme as for the three-dimensional stray field calculations can be applied to two-dimensional problems, for which the charge densities are independent of the third dimension. Such problems arise for example in the calculation of domain walls in infinitely extended thin films, in which the wall structure depends on two dimensions only.

It is possible to use formally the concept of self- and interaction energy terms again, although the real self-energy of an infinitely extended charged element is infinite even per unit length. Since in magnetism the opposite charges always balance exactly, the diverging parts in the individual self- and interaction energies cancel. The sum of self- and interaction energies of a balanced system remains finite (per unit length of the system).

A suitable core function, analogous to the source function $F_{000} = 1/r$ in (3.25), can be derived as follows: calculate, as in (3.27), the interaction between two parallel line charges of length L along the z direction (the cylinder axis of our 2D problem). Expand the interaction energy per unit length in terms of $1/L$ and omit terms that do not depend on x and y and which will therefore cancel in the difference operations. Finally go to the limit $L \to \infty$. The result for the *relevant* part of the line interaction is $G_{00} = -\ln(x^2 + y^2)$. From this core function, the integrals applicable to surface and volume interactions can be calculated as in the three-dimensional case.

We use the following abbreviations for the integrals:

$$B_x = \arctan(x/y) \ , \quad B_y = \arctan(y/x) \ , \tag{3.34a}$$

$G_{ik} = i$-fold integral of G_{00} along x, and k-fold integral along y .

The source function and its most important integrals are:

$$G_{00} = -\ln(x^2 + y^2) \ ,$$

$$G_{10} = \int G_{00}\,dx = x\,G_{00} - 2yB_x + 2x \ ,$$

$$G_{11} = \int G_{10}\,dy = xy\,G_{00} - x^2 B_y - y^2 B_x + 3xy \ ,$$

$$G_{20} = \tfrac{1}{2}\left[(x^2 - y^2)\,G_{00} - 4xyB_x + 3x^2 + \tfrac{7}{6}y^2\right] \ ,$$

$$G_{21} = \tfrac{1}{2}\left[\tfrac{1}{3}y(3x^2 - y^2)\,G_{00} - \tfrac{2}{3}x\left(3y^2 B_x + x^2 B_y\right) + \tfrac{1}{6}y(22x^2 + y^2)\right] \ ,$$

$$G_{22} = \tfrac{1}{24}\left[(6x^2 y^2 - x^4 - y^4)\,G_{00} - 8xy(x^2 B_y + y^2 B_x) + 25x^2 y^2\right] \ . \tag{3.34b}$$

In the functions G_{10}, G_{20} and G_{21}, those additive parts that depend on x or y only (such as the term '$2x$' in G_{10}) may be omitted because they will cancel in the difference operations needed in applying these integrals. In the presented form (3.34b) the functions are mutually related as derivatives and indefinite integrals, which permits an easy verification and may be occasionally useful.

As an application, consider the interaction energy between two parallel, infinitely long bands that extend from x_1 to x_2 and from x_3 to x_4, with a distance q between the two planes:

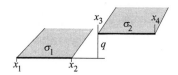

$$E_{inter} = \tfrac{1}{2\pi}\sigma_1\sigma_2 K_d[(G_{20}(x_4 - x_1, q) - G_{20}(x_4 - x_2, q))$$

$$- (G_{20}(x_3 - x_1, q) - G_{20}(x_3 - x_2, q))] \ , \tag{3.35}$$

and the formal self-energy per unit length of an infinitely long band with surface charge σ, extending along the width dimension from x_1 to x_2:

$$E_{\text{self}} = \tfrac{1}{2\pi}\sigma^2 K_{\text{d}}\, G_{20}(x_2 - x_1, 0) \ . \tag{3.36}$$

Furthermore, we study the energy gain of a band-shaped domain of opposite magnetization in a film of perpendicular anisotropy (see inset of Fig. 3.6). This calculation can be performed in an interesting manner: starting with a uniformly magnetized plate, we superimpose an oppositely magnetized band of double magnetization strength (to compensate the original magnetization and establish an opposite one). The self-energy of this band can be calculated in a straightforward manner. The interaction energy between the band and the plate leads to an energy gain. If the plate is infinitely extended, its stray field is uniform (3.20) and the interaction energy is simply $-4K_{\text{d}}V$ (V is the volume of the band). This result can also be derived by applying the formalism for finite plates and by going to the limit of an infinite plate length.

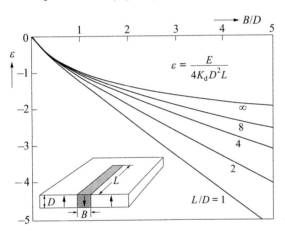

Fig. 3.6 The energy gain ε per unit length of an oppositely magnetized band-shaped domain as defined in the inset

We compare the energy gain per unit band domain length in Fig. 3.6 for domains of finite length, using the function F_{220}, with the case of an infinite extension, using G_{20}. The graph demonstrates that the curves for long band domains converge towards the curve for infinite domains. Combining the indicated explicit expressions for the stray field energy with applied field and wall energy contributions, a complete theory of rectangular "bubble" and band domains can be formulated (which is a nice exercise). Formulae that are directly applicable to more realistic cylindrical bubbles can be found in Sect. 3.7.3D, as well as in the specialized literature on that topic [49; 510].

(F) The μ^*-Method. Using the above formulae, the stray field problem may be considered solved if the magnetization field of a ferromagnet is known. The solution is analogous to electrostatic field calculations from known charges and polarizations, with the only difference that the absence of true magnetic charges simplifies calculations as shown above. Unfortunately, the situation in magnetism is seldom such that the magnetization distribution is known in advance. The stray field generated by a given magnetization in general exerts a torque on the magnetization vectors, which tends to rotate them away from the initially assumed direction. For materials with very high anisotropy, the assumption of a rigid magnetization along an easy axis is a good approximation. If the anisotropies are weaker but still large enough such that deviations from the preferred directions are small, the stray field can be calculated by a perturbation procedure: the so-called μ^*-method [32]. In this method the magnetization is first assumed to be exactly parallel to the easy directions. The starting volume and surface charges are derived from this initial magnetization distribution $\boldsymbol{m}_0$ according to (3.18):

$$\lambda_v^0 = -\operatorname{div}\boldsymbol{m}_0 \; ; \quad \sigma_s^0 = \boldsymbol{m}_0 \cdot \boldsymbol{n} \; . \tag{3.37}$$

Then possible deviations from the easy directions are taken into account by calculating the potential from these charges for a medium with an effective permeability tensor $\boldsymbol{\mu}^*$, which may be calculated from the second derivatives of the anisotropy functional with respect to rotations away from the easy directions. Along the easy axis the permeability is always equal to one; perpendicular to the axis it may have two different values. If g_1 and g_2 are the eigenvalues of the mentioned tensor of second derivatives, the diagonal elements of the effective permeability tensor become in a suitable coordinate system $(1, 1 + 2 K_d/g_1, 1 + 2 K_d/g_2)$.

In important cases, the two derivatives g_1 and g_2 are equal, leaving only one relevant value of μ^*. This is true in particular for uniaxial crystals where we get $\mu^* = 1 + K_d/K_u$. For cubic materials with positive anisotropy we obtain $\mu^* = 1 + K_d/K_{c1}$, and for negative anisotropy $\mu^* = 1 - \frac{3}{2} K_d/K_{c1}$ [expand (3.9c) around $\vartheta = \frac{\pi}{2}$ to derive this formula]. The larger the anisotropy of the material the smaller is the deviation of μ^* from the vacuum value 1.

In the next step the permeability tensor is introduced into the magnetostatic equations determining the magnetic field:

$$\mu_0 \operatorname{div}(\boldsymbol{\mu}^* \boldsymbol{H}_d) = J_s \lambda_v^0 \qquad \text{in the volume ,}$$

$$\mu_0 \boldsymbol{n} \cdot (\boldsymbol{H}_d^{(o)} - \boldsymbol{\mu}^* \boldsymbol{H}_d^{(i)}) = J_s \sigma_s^0 \qquad \text{at the surfaces ,} \tag{3.38}$$

where $H_d^{(o)}$ and $H_d^{(i)}$ are the values of the stray field outside and inside the sample boundaries. If the stray field determined by (3.38) is known, the stray field energy can simply be calculated from (3.17) with the starting magnetization distribution m_0:

$$E_d = -\tfrac{1}{2} J_s \int m_0 \cdot H_d \, dV \ . \tag{3.39}$$

This holds true although the μ^*-method implies deviations of the magnetization from its starting value by $\mu_0(\mu^* - 1) \cdot H_d$. The additional anisotropy energy and the modifications in the stray field energy connected with these deviations are included in (3.39), as was shown explicitly in [32]. The method is valid, as long as the resulting rotations are small enough to stay within the quadratic approximation of the anisotropy function in the neighbourhood of the easy directions. In terms of the potential function $\Phi(r)$ the stray field energy is given by (3.19b) when σ_s^0 and λ_v^0 are inserted as magnetic charges.

The μ^*-method has many applications in domain theory. In domain models of low-anisotropy materials the magnetization inside the domains is usually assumed to be oriented parallel to an easy direction, and the domain walls are not allowed to carry magnetic charges. Under these assumptions the interior of the sample is charge free. Charges may appear only on the surfaces, if the easy directions are inclined to the surface. Such charges generate stray fields that exert torques on the domain magnetization. As a result the magnetization in the domains will deviate from the easy directions, the initial surface charges are reduced and additional volume charges are generated. The μ^*-method permits the calculation of all these reactions in a simple way. Often this calculation reduces even to a simple scaling transformation.

An example is the case of an arbitrary one-dimensional charge pattern on the surface of an infinite plate. Let the material have a rotational permeability $\mu_\perp^*$ perpendicular to the surface, and $\mu_\parallel^*$ parallel to the surface and perpendicular to the lines of constant charge (Fig. 3.7a). The charge pattern on the surface may have a sufficiently fine scale, so that interactions with the other surface can be neglected. If $\Phi_0(x, y)$ and E_{d0} are potential and energy of the rigid model ($\mu_\perp^* = \mu_\parallel^* = 1$), respectively, then the corresponding quantities for the permeable medium are given by [554]:

$$\Phi(x, y) = a \, \Phi_0(x, \beta y) \quad \text{inside} , \quad \Phi(x, y) = a \, \Phi_0(x, y) \quad \text{outside the sample;}$$

$$E_d = a \, E_{d0} , \quad \text{with } a = 2/(1 + \sqrt{\mu_\parallel^*/\mu_\perp^*}), \ \beta = \sqrt{\mu_\parallel^*/\mu_\perp^*} \ . \tag{3.40}$$

The effective permeability thus causes both the stray field and the stray field energy to be reduced by the factor a. For a material with uniaxial perpendicular anisotropy ($\Theta = 0$) we obtain $\left[\mu_\perp^* = 1, \mu_\parallel^* = \mu^*, a = 2/\left(1 + \sqrt{\mu^*}\right)\right]$, and for an arbitrary material with a small deviation of the easy axis from the surface ($\Theta \approx 90°$) the solution is $\left[\mu_\perp^* = \mu_\parallel^* = \mu^*, a = 2/\left(1 + \mu^*\right)\right]$. The latter case plays a central role in the discussion of domains in soft magnetic materials where μ^* becomes large. With values such as $\mu^* = 40$ for iron or $\mu^* = 43$ for nickel the μ^*-"correction" a $= 2/(1 + \mu^*)$ is never negligible. Soft magnetic materials thus tend to develop closed-flux magnetization patterns, even if the easy axes are misoriented with respect to the surface.

For uniaxial films with an easy axis perpendicular to the surface ($\Theta = 0$), the μ^*-correction (3.40) is even valid for arbitrary two-dimensional surface charge patterns. If the easy axis is tilted by an angle Θ (Fig. 3.7a), the effective permeabilities are $\mu_\parallel^* = \mu^*$ along the x axis and $\mu_\perp^* = 1 + \left(\mu^* - 1\right)\sin^2\Theta$ perpendicular to the surface [555].

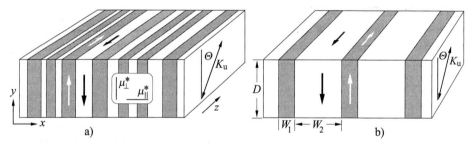

Fig. 3.7 (a) A uniaxial plate with a tilted easy axis. In the case of the μ^*-"correction" in (3.40), the one-dimensional charge distribution may be non-periodic, but it must be narrow compared to the thickness. In contrast, the expression (3.41) applies to a periodic, but arbitrarily wide pattern (b)

(G) Periodic Problems. Another class of tractable problems deals with one- or two-dimensional periodic surface charge distributions. The method of solution consists in expanding the charge distribution as well as the potential in a Fourier series, fulfilling the boundary conditions for every element in the series. The μ^*-method may be incorporated in this calculation as was first demonstrated by *Kooy* and *Enz* [554], leaning on earlier results [556] in which the μ^*-correction was still neglected. The following formula is applicable to periodic one-dimensional domains of alternating surface charge density in an infinite plate of thickness D and tilted easy axis (Fig. 3.7b):

$$E_d = K_d \cos^2\Theta \left\{ \left(D m^2/\mu_\perp^*\right) + \left(8P/\pi^3\right) \sum_{n=1}^{\infty} n^{-3} \sin^2\left[\tfrac{1}{2}\pi n(1+m)\right] \cdot \right.$$
$$\left. \cdot \sinh(\pi n g)/\left[\sinh(\pi n g) + \sqrt{\mu_\perp^* \mu_\parallel^*} \cosh(\pi n g)\right]\right\} \quad (3.41)$$

with $P = W_1 + W_2$, $m = (W_1 - W_2)/(W_1 + W_2)$,

$\mu_\parallel^* = 1 + K_d/K_u$, $\mu_\perp^* = \cos^2\Theta + \mu_\parallel^* \sin^2\Theta$, $g = (D/P)\sqrt{\mu_\parallel^*/\mu_\perp^*}$,

D = thickness , W_1 , W_2 = domain widths ,

K_u = anisotropy constant , $K_d = J_s^2/2\mu_0$, J_s = saturation magnetization ,

Θ = angle between easy direction and surface normal .

For large thicknesses $(D \gg W)$ and zero net magnetization $(W_1 = W_2 = W$, $m = 0)$, equation (3.41) simplifies to the formula of *Kittel* [30]:

$$E_d = \tfrac{1}{2\pi} 1.705\, K_d\, W \cos^2\Theta\left[2/\left(1 + \sqrt{\mu_\perp^*\, \mu_\parallel^*}\right)\right] . \tag{3.42}$$

The hyperbolic functions in (3.41) describe interactions between the charge patterns on both surfaces, which are a function of the ratio D/P in the parameter g. As a rule of thumb, these interactions become insignificant (smaller than 10^{-4} in relative magnitude) if the thickness gets larger than three times the average domain width. The Kooy and Enz formula [(3.41); neglecting the μ^*-effect $(\mu^* = 1)$] is generalized in [557] to a periodic multilayer system.

The more general case of a twofold periodic surface charge distribution (as for a bubble lattice or for an undulating band pattern) can be treated in the same way: let $\sigma_s(x, y)$ be an arbitrary charge distribution on the surface of a plate of thickness D, with periods P_x , P_y along the x and y axes, respectively. Opposite charges are located on the bottom surface of the plate. The stray field energy for $\mu^* = 1$ is then:

$$E_d = D\, K_d\left[C_{00}^2 + \sum_{r,s}' \left[\frac{C_{rs}\, C_{-r-s}}{(2\pi g_{rs})}\right]\left(1 - e^{-2\pi g_{rs}}\right)\right] , \tag{3.43}$$

with $C_{rs} = \int_0^1 \int_0^1 \sigma_s(\xi, \eta)\, \exp[-2\pi i(r\xi + s\eta)]\, d\xi\, d\eta$,

$\xi = x/P_x$, $\eta = y/P_y$, $g_{rs} = D\sqrt{(r/P_x)^2 + (s/P_y)^2}$, and

$\sum'$ = summation over integers r and s from $-\infty$ to ∞, except for $r = s = 0$.

Original references and applications of this formula to high-anisotropy materials can be found in [507]. It would be interesting to include the μ^*-effect.

In conclusion, stray field calculations are rarely easy. If analytical integrations are possible, they should be considered even if they do not offer a complete solution. For periodic problems, Fourier transformations as employed in (3.41–43) are an obvious choice. But even for aperiodic problems, Fourier techniques can decisively reduce the computation time as discussed in (D).

3.2.6 Magneto-Elastic Interactions and Magnetostriction

(A) Overview: Relevant and Irrelevant Effects in Magneto-Elastics. Up to this point we used only the magnetization vector to describe a magnetic body. Elastic effects introduce a new degree of freedom. A magnetic body will deform under the influence of a magnetic interaction, and this deformation is described by an (in general) asymmetric tensor of elastic *distortion* $p(r)$. The distortion consists of a symmetric part, the tensor of elastic strain ε, and an antisymmetric part, the tensor of lattice rotations ω. In a systematic approach, all parts of the free energy are expanded with respect to p. Because magneto-elastic effects are small in ferromagnetism, these expansions only have to include the smallest non-vanishing order. The elements of zero order form the energy of the undeformed lattice. The first-order elements define the magneto-elastic interaction energy, while second-order terms describe the elastic energy, which usually can be assumed to be independent of the magnetization direction.

The latter assumption is not adequate for extremely stressed materials. Here a further term becomes important, which may be formulated as a magnetization-dependent elastic tensor, or, alternatively, as a magneto-elastic coupling the coefficient of which is itself strain dependent. In other words, in strongly strained materials magnetostriction differs from the unstrained state. The phenomenon was first identified in Ni-Fe alloys [558] and in metallic glasses [559], but it is important for thin films, too [560, 561]. As such strong stresses are always of non-magnetic origin, they may be considered to be a part of the material properties. Just remember that magneto-elastic coefficients in a strongly strained state may differ from the standard values.

Neglecting third-order terms in the elastic strain components means linear elasticity characterized by Hooke's law. This is certainly a very good approximation for magnetostrictive effects alone, which deal with magnetization-induced strains of typically 10^{-6}–10^{-3} only. If stronger deformations of non-magnetic origin are superimposed, one may again consider this strained state as the reference by using the linear elastic constants applicable for this state. Treating magneto-elastic effects in the standard framework of linear elasticity and strain-independent magneto-elastic coefficients is therefore adequate. A more general treatment of magneto-elasticity may be found in [562].

All parts of the free energy of a magnet can somehow depend on the distortions and therefore contribute to the magneto-elastic interaction energy. The most important contribution stems from the crystalline anisotropy and leads to a local expression, which is quadratic in the components of the magnetization vector and linear in the lattice strain. The dependence of the stray field energy on deformations produces the so-called *form effect* of magnetostriction, which is important near saturation but less important in connection with domain effects. It is needed in Sect. 4.5.2 discussing measurements of magnetostriction constants, but is ignored here. Little is known about the dependence of the exchange energy on lattice strains. It gives rise to the so-called volume magnetostriction, an isotropic change in the lattice constant on magnetic ordering at the Curie point. Whether there is also a measurable dependence of the exchange stiffness energy on the lattice strains, which would influence the domain wall structure, is unknown. *Brown* [562] and *Kronmüller* [503] discuss the symmetry of such effects without being able to follow the matter further without experimental data for the relevant material parameters. Even if an access to the material parameter of exchange-induced magnetostriction should be found, these effects are probably negligible in domain theory, as opposed to micromagnetics.

There is also a dependence of the free energy on lattice rotations: if we assume that the direction of the magnetization is fixed in space, an assumed free rotation ω of the lattice would change the angle between the magnetization m and the easy axis and thus the anisotropy energy density. A magneto-rotation interaction energy may be derived by calculating the first derivatives of the anisotropy function with respect to the magnetization direction [562, 563]. In [509] this effect was included in a systematic way in the calculation of planar domain walls. But the magneto-rotation effect appears to be small even in domain walls where strong deviations from the easy directions occur. A sizeable contribution of the lattice rotation effects (~25%) was found for domain patterns in a uniaxial crystal under the action of a strong external field perpendicular to the easy axis. We refer to the original literature [161] with respect to this rather special case. In domain theory the effect will be negligible because the domains tend to be magnetized along easy directions for which the magneto-rotation energy vanishes.

We are thus left with just one important mechanism in magneto-elastics: the anisotropic magnetostriction, which is derived from the magneto-crystalline anisotropy and which is quadratic in the components of the magnetization. This effect is important for three reasons: (i) It gives rise to the stress sensitivity of magnetic materials; this applies not only to bulk materials, but also to thin films. (ii) It adds a volume energy term to magnetization configurations that do not consist of antiparallel domains only. In bulk cubic materials, this energy contribution can become predominant together with the energy contributions of domain walls. (iii) The magnetostrictive deformations, although small in scale, play an important technical role. The noise of common trans-formers has its origin in these effects. Because it is difficult to find a compre-hensive overview of magnetostrictive effects at other places, we have to go into some detail in their discussion.

(B) Magnetostriction in Uniformly Magnetized Cubic Crystals. In this section the formulae for the magneto-elastic interaction energy, for the elastic energy and for the resulting magnetostrictive interaction are collected for cubic crystals. The symmetric strain tensor ε is defined here relative to a fictitious non-magnetic state. From symmetry we obtain the following expres-sion for the magneto-elastic interaction energy (to the lowest significant order in powers of the magnetization components [16, 30]):

$$e_{\text{me}} = -3\,C_2\lambda_{100} \sum_{i=1}^{3} \varepsilon_{ii}\left(m_i^2 - \tfrac{1}{3}\right) - 6\,C_3\lambda_{111} \sum_{i>k} \varepsilon_{ik} m_i m_k \ , \tag{3.44}$$

where $C_2 = \tfrac{1}{2}(c_{11} - c_{12})$ and $C_3 = c_{44}$ are convenient abbreviations for the two shear moduli of the cubic crystal (where $c_{11} = c_{1111}$, etc., are the elastic tensor coefficients in Voigt's notation). The constants λ_{100} and λ_{111} are independent, dimensionless material parameters indicating the strength of the magneto-elastic interaction (λ_{100} and λ_{111} also measure the spontaneous magnetostriction as shown below).

The magneto-elastic energy is balanced by the elastic energy, which for cubic crystals may be written in the following form:

$$e_{\text{el}} = \tfrac{1}{2}C_1\left(\sum_{i=1}^{3} \varepsilon_{ii} \right)^2 + C_2 \sum_{i=1}^{3} \varepsilon_{ii}^2 + 2C_3 \sum_{i>k} \varepsilon_{ik}^2 \ , \tag{3.45}$$

where C_2 and C_3 are defined as above and $C_1 = c_{12}$ is another elastic constant. For a uniformly magnetized body with free surfaces, the spontaneous magne-tostrictive deformation is computed by minimizing the sum of the energy contributions (3.44) and (3.45) with respect to the components of ε. We call the result the tensor of *free* or spontaneous or "quasi-plastic" deformation ε^0:

$$\varepsilon_{ii}^0 = \tfrac{3}{2}\lambda_{100}\left(m_i^2 - \tfrac{1}{3}\right) \qquad \text{for } i = 1\ldots3 \ ,$$

$$\varepsilon_{ik}^0 = \tfrac{3}{2}\lambda_{111} m_i m_k \qquad \text{for } i \neq k \ . \tag{3.46}$$

Characteristically, the trace $\sum \varepsilon_{ii}^0$ of the spontaneous magnetostrictive defor-mation is zero, in contrast, for example, to the strain state caused by a tensile stress where the trace depends on Poisson's ratio. The elongation of the crystal along the direction of the unit vector a is given by $\delta l/l = \sum a_i a_k \varepsilon_{ik}^0$. We thus obtain the well-known formula for the *spontaneous* magnetostrictive elongation of a cubic crystal:

$$\frac{\delta l}{l} = \tfrac{3}{2}\lambda_{100}\left(\sum_{i=1}^{3} a_i^2 m_i^2 - \tfrac{1}{3}\right) + 3\lambda_{111}\sum_{i>k} m_i m_k a_i a_k \ . \tag{3.47}$$

Inserting the deformation $\boldsymbol{\varepsilon}^0$ into (3.44) and (3.45) yields the total energy of cubic crystals, subject neither to external forces nor internal incompatibilities:

$$e^0 = -\tfrac{3}{2}C_2\lambda_{100}^2 - \tfrac{9}{2}\left(C_3\lambda_{111}^2 - C_2\lambda_{100}^2\right)\left(m_1^2 m_2^2 + m_1^2 m_3^2 + m_2^2 m_3^2\right) \ . \tag{3.48}$$

Equation (3.48) has the same symmetry as the first-order cubic anisotropy (3.9). Since anisotropy measurements are in most cases performed with free crystals, this energy contribution is already included in the measured values of K_{c1}. Strictly speaking, it has to be removed from measured constants, but for most materials this correction is negligible.

In the opposite case of a "clamped" crystal, the crystal is subject to a magneto-elastic *stress* $\boldsymbol{\sigma}^0$ — a "quasi-plastic stress" in the nomenclature of *Kröner* [564] — which is calculated by Hooke's law for cubic crystals:

$$\sigma_{11} = c_{1111}\varepsilon_{11} + c_{1122}(\varepsilon_{22} + \varepsilon_{33}) \quad [= 2\,C_2\,\varepsilon_{11} \text{ if } \textstyle\sum \varepsilon_{ii} = 0 \text{ as true here}] \ ,$$

$$\sigma_{12} = c_{1212}(\varepsilon_{12} + \varepsilon_{21}) = 2\,C_3\,\varepsilon_{12} \ .$$

Applying these equations to the free deformation $\boldsymbol{\varepsilon}^0$ defined in (3.46) leads to:

$$\sigma_{ii}^0 = -3\,C_2\lambda_{100}\left(m_i^2 - \tfrac{1}{3}\right) \qquad \text{for } i = 1\ldots3 \ ,$$

$$\sigma_{ik}^0 = -3\,C_3\lambda_{111}m_i m_k \qquad \text{for } i \neq k \ . \tag{3.49}$$

The same stress is also derived from the magneto-elastic energy (3.44) simply by $\sigma_{ik}^0 = \partial e_{\mathrm{me}}/\partial \varepsilon_{ik}$. The coefficients of the strain components in e_{me} can be interpreted as the components of the stress tensor $\boldsymbol{\sigma}^0$. Equation (3.44) may therefore also be written as:

$$e_{\mathrm{me}} = \sum_{i,k} \varepsilon_{ik}\sigma_{ik}^0 \ . \tag{3.44a}$$

(C) Isotropic Materials. For elastic and magnetostrictive anisotropy (amorphous or polycrystalline materials) the equations (3.44–49) simplify by using the isotropic shear modulus $G_{\mathrm{s}} = C_2 = C_3$ and the isotropic magnetostriction constant $\lambda_{\mathrm{s}} = \lambda_{100} = \lambda_{111}$. For the magnetostrictive free deformation $\boldsymbol{\varepsilon}^0$ we get:

$$\varepsilon_{ik}^0 = \tfrac{3}{2}\lambda_{\mathrm{s}}\left(m_i m_k - \tfrac{1}{3}\delta_{ik}\right) \tag{3.50}$$

where δ_{ik} is Kronecker's symbol. For the elongation along $\boldsymbol{a}$ we obtain:

$$\delta l/l = \tfrac{3}{2}\lambda_{\mathrm{s}}\left[(\boldsymbol{m}\cdot\boldsymbol{a})^2 - \tfrac{1}{3}\right] \tag{3.51}$$

which depends only on the angle between the magnetization direction $\boldsymbol{m}$ and the measuring direction $\boldsymbol{a}$.

(D) Hexagonal Crystals and Uniaxial Materials. Hexagonal crystals behave to lowest meaningful order elastically and magneto-elastically like simple uniaxial materials, meaning that neither the elastic nor the magneto-elastic energy expressions depend on the azimuthal magnetization angle. Magneto-elastic effects in uniaxial materials are rarely important. If the domains are magnetized parallel to the axis, their elastic deformations are identical. Inside domain walls, an additional magneto-elastic energy appears, but it is usually negligible compared to the uniaxial anisotropy. An exception is the case when a strong field perpendicular to the easy axis causes the magnetization to deviate from it. Then the domains are dominated by magneto-elastic effects as analysed in [161, 565] (see Sect. 5.2.2B). We will not repeat this analysis in this book and it may therefore suffice to mention that standard uniaxial or hexagonal materials require five elastic constants instead of three for cubic crystals, as well as four instead of two magneto-elastic coefficients [532].

(E) Interaction with Stresses of Non-Magnetic Origin. The magneto-elastic interaction energy (3.44) assumes a different meaning if the elastic strain tensor $\boldsymbol{\varepsilon}$ is replaced by the corresponding stress tensor $\boldsymbol{\sigma}$, and, conversely, the magneto-elastic stress tensor $\boldsymbol{\sigma}^0$ is replaced by the corresponding strain tensor $\boldsymbol{\varepsilon}^0$:

$$e_{\text{me}} = -\sum_{i,k} \sigma_{ik}\varepsilon_{ik}^0 \qquad \text{(general)} , \qquad (3.52)$$

$$e_{\text{me}} = -\tfrac{3}{2}\lambda_{100}\sum_{i}\sigma_{ii}\left(m_i^2 - \tfrac{1}{3}\right) - 3\lambda_{111}\sum_{i>k}\sigma_{ik}m_i m_k \qquad \text{(cubic)} , \qquad (3.52a)$$

$$e_{\text{me}} = -\tfrac{3}{2}\lambda_{\text{s}}\sum_{i,k}\sigma_{ik}\left(m_i m_k - \tfrac{1}{3}\delta_{ik}\right) \qquad \text{(isotropic)} . \qquad (3.52b)$$

In this *stress* formulation the magneto-elastic energy describes the interaction of magnetization with a stress $\boldsymbol{\sigma}$ of non-magnetic origin. The latter may be an *external* stress, or a non-magnetic *internal* stress resulting from dislocations, or from inhomogeneities in temperature, structure or composition. This coupling energy is comparable to the interaction of the magnetization with an external magnetic field (3.15), and it has the form (3.12) of an orthorhombic anisotropy. For isotropic material and a uniaxial stress along the axis of a unit vector $\boldsymbol{a}$, the magneto-elastic coupling energy becomes:

$$e_{\text{me}} = -\tfrac{3}{2}\lambda_{\text{s}}\sigma\left[(\boldsymbol{m}\cdot\boldsymbol{a})^2 - \tfrac{1}{3}\right] \qquad (3.52c)$$

which describes a uniaxial anisotropy along the stress axis with an anisotropy constant of $K_{\text{u}} = \tfrac{3}{2}\lambda_{\text{s}}\sigma$.

(F) Magnetostrictive Self-Energy in Inhomogeneously Magnetized Bodies.
In analogy to the magnetic stray field energy that appears if the magnetization
pattern does not follow a closed flux path, there exists a magnetostrictive
self-energy if the spontaneous magnetic deformations of the various parts of a
domain pattern do not fit together. Its calculation in the general case is difficult
like many elasticity problems; for a review see [566]. There are at least three
equivalent approaches to this problem and it is useful to consider all of them
and to understand their relative merits. The goal is in any case to calculate the
elastic state of the sample and its energy for a given magnetization pattern.

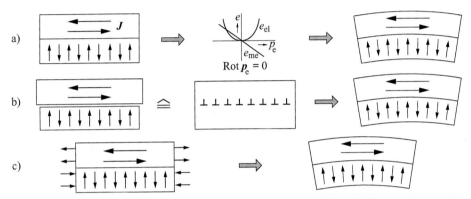

Fig. 3.8 Schematic view of the three aspects of magneto-elasticity. (a) The total energy is
minimized under the constraint of compatibility. (b) The parts of the sample can deform
freely and incompatibilities are compensated by quasi-dislocations. (c) The deformation of
the sample is calculated under the influence of the appropriate forces. The drawings are for
positive magnetostriction, i.e. elongation along the magnetization axis

- The first approach (Fig. 3.8a) starts from an undeformed contiguous "non-
magnetostrictive" body. If magnetostriction is switched on, small continuous
deviations from the non-magnetostrictive state take place, which are described
by a distortion tensor p_e. The distortions must fulfil the laws of elastic compatibility, leading to the constraint $\mathbf{Rot}\,p_e = 0$. (We use the capital letter notation
for the differential operators if they apply to a tensor instead of a vector; see
[564]). This condition implies that the distortion tensor is the gradient of a
potential, the displacement vector field. Inserting the elastic strain ε_e — the
symmetric part of p_e — into the elastic (3.45) and the magneto-elastic energy
(3.44) terms, the total energy is minimized under the compatibility constraint.
This method is particularly useful if the compatibility condition can be fulfilled
in advance by a general ansatz as in one-dimensional problems (see below). If
this method is applicable at all, it can be easily extended to crystals of arbitrary

symmetry including even the magneto-rotational effects [509]. The magneto-strictive self-energy e_{ms} is defined here as the difference between the total elastic and magneto-elastic energy (as resulting from the minimization) and the energy e^0 of the stress-free state (3.48). It is always possible to calculate an upper bound for the magnetostrictive self-energy with this approach by minimizing the energy with respect to the parameters of a suitable ansatz for the displacement vector.

- In the second approach (Fig. 3.8b), uniformly magnetized parts of the magnetic body are first considered independent and deformed according to their respective equilibrium. This deformation ε^0 is then analysed with respect to its compatibility. The incompatibilities or gaps between the parts

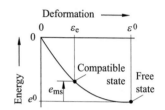

are considered as generated by a quasi-dislocation density α, which can be calculated by the vector-analytical operation $\alpha = \mathbf{Rot}\,\varepsilon^0$. To restore a contiguous body, we add a compensating distortion $p^{cp} = \varepsilon^{cp} + \omega^{cp}$ such that $\mathbf{Rot}\,p^{cp} = -\alpha$, and such that the corresponding magnetostrictive stress tensor $\sigma^{cp} = c \cdot \varepsilon^{cp}$ fulfils the equilibrium conditions ($\mathrm{Div}\,\sigma^{cp} = 0$ in the volume and $\sigma^{cp} \cdot n = 0$ at the surfaces). The problem is thus equivalent to finding the stresses or strains for a given, compensating (quasi-) dislocation density $-\alpha$ and for given boundary conditions. The stress and strain fields of dislocations are known in some important cases. This approach automatically concentrates on the origins of internal stresses and ignores stress-free deformations of elastically compatible domains. Once the additional strains are known, the magnetostrictive self-energy can be obtained by evaluating the elastic energy connected with these strains.

- The third approach (Fig. 3.8c) starts again from the undeformed body. If magnetism is "switched on" without permitting deformations, the magneto-elastic interaction acts as a "quasi-plastic" stress σ^0. In the second step, elastic deformations and balancing stresses σ^{bal} are added such that both stresses together are in equilibrium, which means that their divergence (the forces) must vanish: $\mathrm{Div}\,(\sigma^0 + \sigma^{bal}) = 0$ in the volume and $n \cdot (\sigma^0 + \sigma^{bal}) = 0$ at the surfaces. These compensating stresses cause a (compatible) deformation of the body, which must be equal to the sum $\varepsilon^0 + \varepsilon^{cp}$ in the second approach. Again the problem is reduced to a purely elastic problem. Once the quasi-plastic forces, $\mathrm{Div}\,\sigma^0$, are known, the magnetic nature of the sample can be ignored and the deformations caused by these forces can be calculated as in the general elasticity theory. The needed tools are the stresses and deformations produced by bulk and surface forces under appropriate boundary conditions (as for example for an infinite plate or for a semi-infinite body). The procedure is conceptually similar to the calculation of stray fields from their sources, the divergence of the magnetization, as discussed in Sect. 3.2.5.

In the following we restrict ourselves to a few typical examples relating to domain theory — using all of the systematic approaches sketched above. The examples can be transferred to general cases only approximately. This may be acceptable since magnetostrictive self-energies are important, but not predominant energy contributions, and it may be sufficient in practical cases to estimate their contributions to within 10–20 %. For details we refer to the specialized literature, which is reviewed e.g. in [567].

(G) One-Dimensional Calculations. As for stray field calculations, magneto-elastic problems, which depend on one spatial coordinate only, can be readily integrated and reduced to local problems. Here we present two special, but important cases, and a general formula. Both special cases apply to pairs of domains, which are separated by a domain wall, and we search for the additional magneto-elastic energy in the elastically coupled domains as a function of wall orientation. Method (a) of the approaches in Fig. 3.8 could be directly applied, but in one-dimension the strains can be calculated explicitly as demonstrated for the first example.

Here two domains are magnetized along [100] and [010] in a cubic material. If the wall plane is the symmetry plane (110), characterized by an angle $\psi = 90°$ between wall and 'magnetization' plane, the pair of otherwise free domains is compatible and there is no

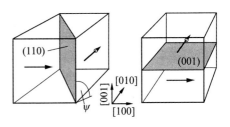

magneto-elastic self-energy. If, however, the wall is oriented parallel to (001) ($\psi = 0°$), the free deformations of the domains are incompatible, and the domains are strained according to the mean value of the two free strains. The three diagonal values of the strain matrices appearing in this problem are:

$\varepsilon_{\mathrm{I}} = \left\{1, -\tfrac{1}{2}, -\tfrac{1}{2}\right\} \lambda_{100}$ for the free deformation of the first domain,

$\varepsilon_{\mathrm{II}} = \left\{-\tfrac{1}{2}, 1, -\tfrac{1}{2}\right\} \lambda_{100}$ for the second domain,

$\bar{\varepsilon} = \left\{\tfrac{1}{4}, \tfrac{1}{4}, -\tfrac{1}{2}\right\} \lambda_{100}$ for the average of the free deformations,

$$\Delta\varepsilon_{\mathrm{I}} = \bar{\varepsilon} - \varepsilon_{\mathrm{I}} = \varepsilon_{\mathrm{II}} - \bar{\varepsilon} = \left\{-\tfrac{3}{4}, \tfrac{3}{4}, 0\right\} \lambda_{100} \tag{3.53}$$

for the deformation of the first domain relative to its free state.

Inserting this deviation from the free state into the elastic energy formula
(3.45) we get the magnetostrictive self-energy density of two incompatible 90°
domains:

$$e_{ms} = \tfrac{9}{8} C_2 \lambda_{100}^2 \ . \tag{3.54}$$

If the domain wall has an arbitrary orientation ψ between (001) and (110), the
magnetostrictive self-energy can be derived by optimizing the total energy
with respect to allowed forms of the strain tensor, leading to:

$$e_{ms} = \tfrac{9}{8} C_2 \lambda_{100}^2 \cos^2 \psi \ . \tag{3.55}$$

A similar calculation for negative-anisotropy cubic materials (easy directions
⟨111⟩) yields for the magnetostrictive self-energy of domains separated by
109° and 71° walls:

$$e_{ms} = C_3 \lambda_{111}^2 \cos^2 \psi \tag{3.56}$$

where the angle $\psi = 0$ means a wall parallel to (110). The coefficients $C_2 \lambda_{100}^2$
and $C_3 \lambda_{111}^2$ in these equations represent reference energy densities, just as the
quantity K_d (3.21) measures the stray field energy density. For most materials
these quantities are at least two orders of magnitude smaller than K_d.

A general solution for every one-dimensional magneto-elastic problem can
be derived [509] by minimizing the sum of the elastic energy and the magneto-
elastic coupling energy with respect to the components of a displacement
vector from which the compatible distortion p_e (see Fig. 3.8a) is derived. The
final result, which is presented here without proof, is quite simple in its
structure. Let n be the only direction along which variations of the magnetization
and of lattice distortions are permitted (in the case of a wall n would be the
wall normal), and let u be the spatial variable along n. The quasi-plastic stress
$\sigma^0[m(u)]$ is known, for example, from (3.49) for cubic crystals. Antisymmetric
contributions from the magneto-rotation terms may also be included if necessary.
The spatial average value of σ^0 is denoted by $\bar{\sigma}$. Then the magnetostrictive
self-energy density is:

$$e_{ms}(u) = \tfrac{1}{2} \sum_{iklm} \left[\sigma_{ik}^0(u) - \bar{\sigma}_{ik} \right]\left[\sigma_{lm}^0(u) - \bar{\sigma}_{lm} \right]\left(s_{iklm} - n_i n_l F_{km}^{inv} \right) \ , \tag{3.57}$$

where F^{inv} is the inverse of the square matrix F defined by $F_{km} = \sum n_i n_l c_{iklm}$
and s is the inverse of the elastic tensor c. This formula was applied in [509]
to the calculation of the elastic stresses in Bloch walls (see Sect. 3.6.1E).

(H) Two-dimensional Problems. *Miltat* [444] analysed a case in which the
shear stresses of (3.54) extend only over a limited range rather than to infinity.

We expect a reduced energy because part of the stresses that would be present in an infinite sample can relax at the surface. The studied case was that of "fir tree branches" occurring on slightly misoriented (001) surfaces of iron (see Figs. 2.41a, 2.48, 3.9): long and shallow domains magnetized along the [100] direction, embedded in [010] and [0$\bar{1}$0] basic domains. A branch of the fir tree pattern shall have a rectangular cross-section (Fig. 3.9b), and its elastic situation is studied in a two-dimensional approach.

We use a rotated coordinate system in the cross-section as shown in Fig. 3.9b, which differs from the crystal-oriented coordinate system (Fig. 3.9a). If the embedded domain takes over the strain of the matrix, its deformation relative to the free state will be twice as large as in the case of two domains that equally share the incompatible strain. Therefore the magnetostrictive self-energy density inside the embedded domain will be four times the value of (3.54), namely $e_{ms}^{(0)} = \frac{9}{2} C_2 \lambda_{100}^2$, while the environment is stress free. This is the extreme case of a shallow domain that takes over completely the surrounding strain.

The general case of a deeper embedded domain and non-uniform strain can be treated if elastic isotropy is assumed. Then the maximum magnetostrictive self-energy, corresponding to uniform strain, becomes $e_{ms}^{(0)} = \frac{9}{2} G_s \lambda_{100}^2$, where G_s is the isotropic shear modulus replacing the cubic coefficient C_2. Based on *Kléman*'s theory [568], Miltat derived the non-uniform strains relative to the free state taking into account the surface boundary conditions correctly by means of a mirror procedure, relying on approach (c) of Fig. 3.8. The result is:

$$\varepsilon_{13}^{cp}(x,y) = -b \left(\Phi_1 + \Phi_2 \right) \ ,$$

$$\varepsilon_{23}^{cp}(x, y) = -\tfrac{1}{2} b \ln \left(R_1 R_3 / R_2 R_4 \right) \quad \text{with } b = \tfrac{3}{4\pi} \lambda_{100} \ , \tag{3.58}$$

where the geometrical quantities are defined in Fig. 3.9b. All other components of $\boldsymbol{\varepsilon}^{cp}$ are zero. The angle Φ_1 must be taken negative if $-y$ exceeds the domain depth D. For very wide domains $\Phi_1 + \Phi_2$ approaches 2π, and ε_{13}^{cp} becomes $-\frac{3}{2} \lambda_{100}$ as in a one-dimensional problem, while ε_{23}^{cp} is negligible except in the edge zones.

Integrating the elastic energy connected with the stresses (3.58) over the cross-section yields the result shown in Fig. 3.10. The degree ρ of stress relaxation depends on the aspect ratio W/D of the embedded domain. An effective relaxation is only possible near the favourably oriented lateral borders. For deep and narrow domains (small values of the width-to-depth ratio) most of the stress can relax by an expansion of the embedded domains and the energy can be strongly reduced, but for shallow domains the energy gain by the relaxation is only 10–20%.

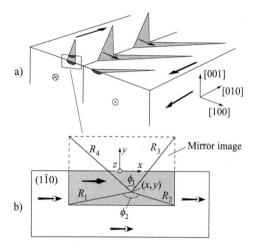

Fig. 3.9 A sketch of the fir tree pattern (a) on a slightly misoriented (001) surface. The cross-section shown in (b) is indicated by a small frame. For the calculation of the magnetostrictive stresses in a fir tree branch this cross-section is assumed to be rectangular. The calculation according to (3.58) needs various angles and distances that are indicated in (b). The slab is thought to be infinitely extended along the z direction

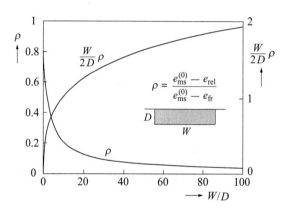

Fig. 3.10 The degree ρ of stress relaxation in fir tree branches. $e_{\mathrm{ms}}^{(0)}$ is the energy per unit volume of the unrelaxed embedded domain, e_{rel} is the energy if elastic relaxation is permitted, and e_{fr} is the energy of the free, completely relaxed state

Domains with a more realistic rounded cross-section — instead of the hypothetical one of Fig. 3.9b — will have a lower energy because they avoid the logarithmic singularities in $\varepsilon_{23}^{\mathrm{cp}}$ which occur at the edges. But since the largest contribution to the energy comes from $\varepsilon_{13}^{\mathrm{cp}}$, this will not affect the final result decisively. Figure 3.10 contains as a second curve a quantity $(W/2D)\rho$, which can be interpreted as relaxation energy gain per unit edge length. In so far as elastically anisotropic and three-dimensional calculations are not available, it is proposed to use this edge energy gain as a correction to the energy of the uniformly strained state. For the inhomogeneous strains and stresses near the edges the effects of elastic anisotropy may even cancel to some degree.

The result (3.58) was obtained in [444] using the concept of so-called *disclinations,* angular defects in continuum theory which describe the effects of the edges of dislocation walls (a continuous sheet of dislocations that may generate, for example, a small angle boundary in a lattice). For fir tree branches *twist* disclinations had to be invoked, which are both hard to describe and to understand. A more intuitive picture arises in another case of magnetic internal stresses, that of wall junctions. Even if the walls would be stress free as isolated walls between two domains, they generate stresses if they are bound together in junctions such as in the Landau and Lifshitz prototype domain structure (Fig. 1.2). The different types of such junctions for cubic materials are shown in Fig. 3.11.

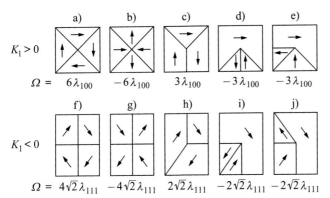

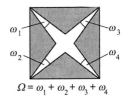

$$\Omega = \omega_1 + \omega_2 + \omega_3 + \omega_4$$

Fig. 3.11 The closure defects (see above) of different wall junctions for the two cases of positive and negative cubic anisotropy

A junction is characterized by an angular defect Ω that can be derived by geometrical considerations (Fig. 3.11): each triangular domain in (a), for example, is elongated along the magnetization vector by λ_{100} and contracted in the transverse direction by $\frac{1}{2}\lambda_{100}$. This results in a reduction of the acute corner angle by $\frac{3}{4}\lambda_{100}$ (for $\lambda_{100} \ll 1$). Summing over all eight angles we obtain an angular defect of $6\lambda_{100}$, as shown in Fig. 3.11a. Note that the angular defect Ω can be either positive or negative, depending on the sign of the magnetostriction constants and on the details of the arrangement.

The stresses are mainly determined by the angular defect Ω and less by the geometrical details. They concentrate in each case around the junction line, decreasing inversely with the distance from it. The junctions have the character of *wedge* disclinations in this case.

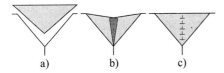

Fig. 3.12 The equivalence between disclinations (a) and a suitable array of edge dislocations (c) along a hypothetical cut (b) as used by *Pryor* and *Kramer* [569]

Energy and stress fields of wedge disclinations can be calculated by integrating the stress fields of rows of dislocations as in small angle grain boundaries. This point of view was introduced by *Kléman* and *Schlenker* [563]. A particularly useful version is due to *Pryor* and *Kramer* [569], who used *Head*'s [570] dislocation stresses for an isotropic half space, thus relying on the approach (b) in Fig. 3.8. They rearranged the incompatibility as shown in Fig. 3.12 which is allowed for small strains and rotations. The dislocation fields are easily accessible only for elastic isotropy. For the rotationally symmetric disclination stress field the effects of elastic anisotropy will cancel to some degree, so that the assumption of elastic isotropy may be tolerable.

The calculation of Pryor and Kramer leads to formulae for the magnetostrictive self-energy and interaction energies of disclinations, which, slightly generalized, are:

$$E_{\text{self}} = \tfrac{1}{2} C \Omega^2 R^2 \ , \tag{3.59a}$$

$$E_{\text{inter}} = C \Omega_1 \Omega_2 \left\{ R_1 R_2 - \tfrac{1}{4}\left[(R_1 - R_2)^2 + Y^2\right] \ln \frac{(R_1 + R_2)^2 + Y^2}{(R_1 - R_2)^2 + Y^2} \right\} \ , \tag{3.59b}$$

where $C = G/[2\pi(1-\nu)]$, G is the shear modulus, ν is Poisson's constant and Ω is the angular defect. The depths R_1 and R_2 of the disclinations and their distance Y are defined in the inset of Fig. 3.13, where these equations are evaluated. The parameter δ weighs the difference in depth of two interacting disclinations.

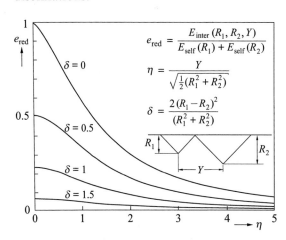

$$e_{\text{red}} = \frac{E_{\text{inter}}(R_1, R_2, Y)}{E_{\text{self}}(R_1) + E_{\text{self}}(R_2)}$$

$$\eta = \frac{Y}{\sqrt{\tfrac{1}{2}(R_1^2 + R_2^2)}}$$

$$\delta = \frac{2(R_1 - R_2)^2}{(R_1^2 + R_2^2)}$$

Fig. 3.13 The reduced elastic interaction energy e_{red} between wall junctions with identical disclination strengths according to (3.59b). The distance between the junctions is measured by the variable η, differences in their depth are expressed by the parameter δ

Focusing on disclinations of equal depth ($\delta = 0$), the reduced interaction energy e_{red} as defined in the inset of Fig. 3.13 starts with the value 1 for coinciding disclinations of the same sign (in this case the interaction energy amounts to the same value as the sum of the two self-energy contributions, so that the total energy becomes four times the self-energy of a single disclination). With increasing distance η between the disclinations the absolute value of the interaction energy decreases.

Disclinations of opposite sign show an attractive interaction given by the curves in Fig. 3.13 with the opposite sign of the ordinate. This causes a tendency for disclinations to occur in pairs of opposite sign, as shown in Fig. 3.14. Particularly, wall junctions that are embedded deep inside a sample, have a strong tendency to form such dipolar pairs as discovered by Miltat in X-ray experiments (Fig. 2.49, for example, contains such a junction pair, the kink in the 180° wall). Note that the strains around disclination cores become visible as a butterfly-like black-and-white contrast in X-ray topograms.

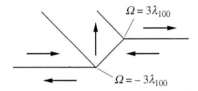

Fig. 3.14 Two wall junctions with opposite closure defect thus forming a disclination dipole. Compare with Fig. 3.11c,e

It is interesting to compare the magnetostrictive self-energy of closure domains according to (3.59) with a simple homogeneous strain model. The latter leads to an energy density of $\frac{9}{2}C_2\lambda_{100}^2$ in the closure domains as for the fir tree pattern. For elastic isotropy this becomes $\frac{9}{2}G\lambda_{100}^2$. Taking the self-energy and one half of the interaction energy for an infinite series of identical closure domains ($Y = 2nR_1$, $R_1 = R_2$), one gets [569]:

$$e_{ms} = \frac{1}{4\pi}\left(9G\lambda_{100}^2/(1-v)\right)\left\{1 + \sum_{n=1}^{\infty}\left[1 - n^2\ln(1 + 1/n^2)\right]\right\} . \qquad (3.60)$$

The infinite sum representing the repulsive interaction between the disclinations equals 0.6054, so that the magnetostrictive self-energy is $1.15\,G\lambda_{100}^2/(1-v)$ which becomes $\approx 2G\lambda_{100}^2$ for $v = 0.42$, the value of iron. This implies a reduction in energy of about 56% as compared to the simpler, uniformly strained model.

In summary, the following strategy is proposed for a reasonable estimate for the magnetostrictive self-energy of a given magnetization distribution:

1. Find a suitable one-dimensional starting model that yields an explicit upper bound for the energy (3.57), including the effects of elastic anisotropy.

2. Estimate the effects of two-dimensional stress relaxation using model calculations like Fig. 3.10 for the edges of shear stress regions (*twist disclinations*)

or Fig. 3.13 for wall junctions (*wedge disclinations*). In this step elastic isotropy has to be assumed.

For regular materials this approach should be sufficient for most problems in domain theory. Materials with giant magnetostriction like terbium-iron alloys [481] may require a more rigorous treatment, including three-dimensional and elastically anisotropic computations.

As pointed out in Sect. 3.2.6F, the calculation of the magnetostrictive self-energy for a given magnetization distribution reduces to a purely elastic problem. Work from other fields with equivalent problems may therefore be consulted. A series of articles by *Pertsev* et al. [571–573] evaluates, for example, two-dimensional problems in elastically isotropic ferroelectrics and ferroelastics (with the tools of quasi-dislocations and disclinations), which may be directly translated into corresponding magneto-elastic problems. *Fabian* and *Heider* [574] presented an approach to calculate three-dimensional magneto-elastic problems in a rigorous numerical fashion for elastically isotropic particles embedded into a non-magnetic, but elastically identical matrix (a problem applicable for example in rock magnetism).

3.2.7 The Micromagnetic Equations

The micromagnetic equations are derived by minimization of the total free energy with respect to the unit vector field $m(r)$ using variational calculus [33]. They are thus based on the same principle as domain theory. Domain theory stresses the global aspects; the differential equations of micromagnetics describe the equilibrium at every point.

(A) The Total Free Energy. Summing up all magnetization-dependent contributions of Sects. 3.2.2–6 the total energy (relative to the freely deformed state) can be written as an integral over the sample volume:

$$E_{tot} = \int \Big[A\,(\mathbf{grad}\,m)^2 \;+\; F_{an}\,(m) \;-\; H_{ex}\cdot J \;-\; \tfrac{1}{2}H_d\cdot J$$

$$\text{Total} \qquad\quad \text{exchange} \qquad\quad \text{anisotropy} \qquad\quad \text{ext. field} \qquad\quad \text{stray field}$$

$$-\sigma_{ex}\cdot\varepsilon^0 \;+\; \tfrac{1}{2}(p_e-\varepsilon^0)\cdot c\cdot(p_e-\varepsilon^0) \Big]\,dV\,. \qquad (3.61)$$

$$\text{ext. stress} \qquad\qquad \text{magnetostrictive}$$

Here, A is the exchange constant, F_{an} collects all contributions from crystal and structural magnetic anisotropies, H_{ex} is the external and H_d the stray field. The symmetric tensor σ_{ex} collects all stresses of non-magnetic origin, $\varepsilon^0(m)$ is the free magneto-elastic deformation at any given point and c is the tensor of elastic constants. The asymmetric tensor p_e is the actual distortion, the compatible deviation from the initial, "non-magnetic" state, which will try to approach the free deformation ε^0. The magnetization vector is $J = J_s m$ with $m^2 = 1$. The stray field H_d and the distortion p_e must fulfil the following conditions:

$$\operatorname{div}(\mu_0 H_d + J) = 0 , \qquad\qquad \operatorname{rot} H_d = 0 , \qquad\qquad (3.62a)$$

$$\operatorname{Div}[c \cdot (p_e - \varepsilon^0)] = 0 , \qquad\qquad \operatorname{Rot} p_e = 0 . \qquad\qquad (3.62b)$$

There are different formulations of this free energy, that choose different independent variables or employ integral relations such as the one in (3.17) or (3.5). In this book we stick to the chosen formulation that is closest to an intuitive understanding. In the magneto-elastic part it uses the approach (a) in the scheme of Fig. 3.8. See the literature [34, 37, 515] for alternatives.

(B) The Differential Equations and the Effective Field. Taking into account the constraint $m^2 = 1$, variational calculus derives, from the total free energy (3.61) and the conditions (3.62), the following set of differential equations:

$$-2A\,\Delta m + \operatorname{grad}_m F_{an}(m) - (H_{ex} + H_d)J_s - (\sigma_{ex} + \sigma^{ms})\operatorname{Grad}_m \varepsilon^0 = f_L m, \quad (3.63)$$

$$\operatorname{div}(\mu_0 H_d + J) = 0 , \qquad\qquad \operatorname{rot} H_d = 0 , \qquad\qquad (3.64a)$$

$$\operatorname{Div}[c \cdot (p_e - \varepsilon^0)] = 0 , \qquad \operatorname{Rot} p_e = 0 , \qquad\qquad (3.64b)$$

where $\Delta = \operatorname{div} \operatorname{grad}$ is the Laplace operator and f_L is a Lagrangian parameter. The magnetostrictive stress $\sigma^{ms} = c \cdot (p_e - \varepsilon^0)$ is proportional to the deviation from the freely deformed state ε^0.

In this derivation the magnetization vector m was considered the only independent variable. The stray field H_d and the distortion p_e are thought to be given by the magnetization field. For the stray field this means that H_d is derived from a potential $H_d = -\operatorname{grad} \Phi$, which is connected to the magnetic charge $\lambda = -\operatorname{div} J$ by the potential equation $\mu_0 \Delta \Phi = -\lambda$. For every distribution $J(r)$ the stray field potential can thus be computed using the tools of potential theory as demonstrated for simple examples in Sect. 3.2.5. This avenue can also be pursued in numerical solutions of the micromagnetic equations and it is in fact the most frequently chosen path. As mentioned in the introductory overview (Sect. 3.2.1) there is an alternative, namely to consider the stray field potential Φ_d (or a corresponding vector potential of the quantity $B_d = \mu_0 H_d + J$) as independent variables in the variational procedure [515]. We ignore this possibility in this book since it is less related to the intuitive arguments needed in domain theory.

This point of view is supported by the fact that in magnetostatics the solution of the potential equation can be carried out in a straightforward way. There are no problems with boundary conditions or material properties. Once the magnetization pattern is given, both the magnetic sample and its environment can be considered as magnetically equivalent to vacuum, as demonstrated in Sect. 3.2.5. For magneto-elastic interactions the analogous procedure meets some difficulties. It is again possible to derive the elastic distortion p_e from a displacement vector u by $p_e = -\mathbf{Grad}\,u$, and a kind of potential equation for this vector field follows from (3.64b): $\mathrm{Div}(c \cdot \mathbf{Grad}\,u) = -\mathrm{Div}(c \cdot \varepsilon_0)$. But it is in general difficult to solve this equation because of the complications connected with the elastic tensor c. To extract this fourth rank tensor out of the differential expressions is in general impossible, and this tensor is also material dependent, usually different for the magnetic sample and its environment. No generally applicable procedures for the solution of the generalized potential problem are available. Some accessible cases were discussed in Sect. 3.2.6F. Treating the displacement field u as an independent parameter to be optimized in a numerical procedure may be an alternative, but this has obviously not yet been tried.

The left-hand side of (3.63) can be written as $-J_s H_{\mathrm{eff}}$ with:

$$H_{\mathrm{eff}} = H_{\mathrm{ex}} + H_{\mathrm{d}} + [2A\Delta m - \mathbf{grad}_m F_{\mathrm{an}}(m) + (\sigma_{\mathrm{ex}} + \sigma^{\mathrm{ms}})\mathbf{Grad}_m \varepsilon^0(m)]/J_s \ . \quad (3.65)$$

This *effective field* offers a simple interpretation of the micromagnetic equations. In the form $J \times H_{\mathrm{eff}} = 0$, it means that the effective field must at every point be directed along the magnetization vector; the *torque* exerted on any magnetization vector must vanish in static equilibrium. The effective field is not uniquely determined by (3.65), since the torque is unaffected by J-directed components of H_{eff}. Such terms may appear, for example, when two equivalent forms of the anisotropy energy are used. In uniaxial samples the expressions $F_{\mathrm{an}}(m) = E_{\mathrm{Ku}} = K(m_1^2 + m_2^2) = K(1 - m_3^2)$ lead to two equivalent contributions to $J_s H_{\mathrm{eff}}$, namely $2K(m_1, m_2, 0)$ and $2K(0, 0, -m_3)$, the difference being $2Km$ [37].

(C) Magnetization Dynamics. Dynamically, the angular momentum connected with the magnetic moment will lead to a gyrotropic reaction if a torque $J \times H_{\mathrm{eff}}$ exists, which is described by the following equation:

$$\dot{J} = -\gamma\, J \times H_{\mathrm{eff}} \ , \quad \gamma = \mu_0 g e / 2m_e = g \cdot 1.105 \cdot 10^5 \ \mathrm{m}/\mathrm{As} \quad (3.66)$$

where γ is the gyromagnetic ratio. The Landé factor g has values close to 2 for many ferromagnetic materials.

Equation (3.66) is the starting point of any dynamic description of micro-magnetic processes. It describes a precession of the magnetization around the effective field. During this motion the angle between magnetization and field does not change. This unexpected feature is related to the fact that no *losses* have been taken into account up to this point. Losses in magnetism in general can have many origins: eddy currents, macroscopic discontinuities (Barkhausen jumps), diffusion and the reorientation of lattice defects, or spin-scattering mechanisms can all introduce irreversibilities and losses. The long-range effects of eddy currents and Barkhausen jumps cannot be treated separately from the domain structure. But even if we exclude these non-local effects by focusing on ideal, non-conducting samples and continuous changes of the magnetization pattern, there will be a remaining localized, *intrinsic* loss, which may be described by the *Landau-Lifshitz-Gilbert* equation [22, 575]. In this equation a dimensionless empirical damping factor α_G is introduced to describe unspecified local or quasi-local dissipative phenomena, like the relaxation of magnetic impurities or the scattering of spin waves on lattice defects. We thus obtain:

$$\dot{\boldsymbol{m}} = -\gamma_G \, \boldsymbol{m} \times \boldsymbol{H}_{eff} - \alpha_G \, \boldsymbol{m} \times \dot{\boldsymbol{m}} \ . \tag{3.66a}$$

The damping term allows the magnetization to turn towards the effective field until both vectors are parallel in the static solution.

The *Gilbert* equation (3.66a) is a variant of the original Landau-Lifshitz equation:

$$\dot{\boldsymbol{m}} = -\gamma_{LL} \, \boldsymbol{m} \times \boldsymbol{H}_{eff} + \alpha_{LL} \, \boldsymbol{m} \times (\boldsymbol{m} \times \boldsymbol{H}_{eff}) \ . \tag{3.66b}$$

The two equations can be converted into each other [34] by inserting (3.66b) into (3.66a) and comparing the coefficients of the vector expressions, which leads to: $\gamma_{LL} = \gamma_G/(1 + \alpha_G^2)$ and $\alpha_{LL} = \alpha_G \gamma_G/(1 + \alpha_G^2)$. For zero damping $\alpha_{LL} = \alpha_G = 0$ we have $\gamma_{LL} = \gamma_G = \gamma$ of (3.66). The choice between the two versions of the dynamic equations is often based on mathematical convenience.

At low frequencies the loss term in the dynamic equation is predominant. The gyromagnetic term has to be taken into account only in the GHz range, where it leads to quite unexpected phenomena as discussed in Sect. 3.6.6 on domain wall dynamics. The general validity of the Landau-Lifshitz equation (3.66) has been questioned in a series of articles by *Baryakhtar* and collaborators as reviewed in [576]. It is claimed that field gradient expressions have to be added to the dissipation terms, thus taking into account exchange contributions to losses. The proposed generalization predicts different relations between resonance experiments and domain wall mobility measurements. A detailed experimental confirmation of this theoretical concept, which would also give access to the material coefficients of the new damping terms, is still lacking. We therefore stick to the conventional approach in this book.

(D) The Boundary Conditions. The variational procedure also yields boundary conditions at surfaces and interfaces of a magnetic system. The surface- and interface-specific energy terms discussed in Sects. 3.2.2B and 3.2.3C only enter at this point. The boundary conditions are influenced by surface anisotropy (3.13) [37], as well as by interface coupling phenomena (3.8) if present [577], leading in the most general case to the following law [578]:

$$ m \times \left[2A \left(n \cdot \mathbf{grad} \right) m + \mathbf{grad}_m e_s(m, n) - \left(C_{bl} + 2 C_{bq} m \cdot m' \right) m' \right] = 0 . \quad (3.67) $$

Here n is the surface or interface normal, m is the unit magnetization vector at the surface or interface and m' is the interface magnetization of the adjacent medium. If the surface anisotropy e_s is zero and there is no interface coupling, the boundary equation (3.67) reduces to the requirement that the normal derivatives of all magnetization components must be zero at the surface: $(n \cdot \mathbf{grad}) m = 0$.

If the surface anisotropy has the simple form $K_s[1 - (m \cdot n)^2]$ (3.13), the second term in (3.67) becomes $-2K_s(m \cdot n)n$. In this case the derivative of the magnetization along the normal direction $(n \cdot \mathbf{grad}) m = \partial m / \partial n$ can be expressed explicitly by applying to (3.67) another cross-product with m, resulting in:

$$ A \partial m / \partial n = K_s (n \cdot m) [n - (n \cdot m) m] + \tfrac{1}{2} \left[C_{bl} + 2 C_{bq} m' \cdot m \right] \left[m' - (m' \cdot m) m \right] . \quad (3.67a) $$

Interestingly, the biquadratic coupling term (the coefficient of C_{bq}) looks formally like the first, surface anisotropy term in (3.67a), with the magnetization vector m' in the adjacent medium playing the role of the surface normal n.

This boundary condition is important in the analysis of thin films, in calculating their magnetization curves and their dynamic (resonance) behaviour. For bulk material the boundary condition is not so important, particularly as the surface anisotropy is not well known and strongly influenced by chemical surface conditions. It affects in bulk samples only a thin surface layer (of thickness $\sqrt{A/K_d}$), an effect that is better analysed separately and superimposed afterwards [579].

(E) Length Scales and Computability of Magnetic Microstructures. The micromagnetic equations are non-linear and non-local coupled partial differential equations of second order. They are non-linear in the magnetization components because of the condition $m^2 = 1$. Nonlinearities arise, in addition, in cubic crystals, where higher powers of the magnetization components appear in the anisotropy energy (3.9). The non-locality of the equations results from the stray field term $\boldsymbol{H}_d$ and from the magneto-elastic displacement vector $\boldsymbol{u}$ mentioned in (B), both of which must be calculated by a separate integration. The differential equations are coupled via the constraint $\boldsymbol{m}^2 = 1$, and via the stray field calculation. The source of the stray field is the negative divergence of the magnetization vector, which is a sum over derivatives of all three magnetization components.

Only in few cases can the micromagnetic equations be linearized and solved analytically. These cases deal with the approach to saturation and with the nucleation and switching of domains ([34]; see Sect. 3.5.4), but not with complete domain structures. Numerical techniques suffer from the wide range of *scales* appearing in magnetic problems (leading in the language of numerical analysis to "stiffness"). Figure 3.15 gives an idea of the characteristic lengths occurring in two typical materials.

Take the case of *iron*. One of the smallest lengths relevant to micromagnetics occurs in this material at places where the exchange energy is in equilibrium with the stray field energy. An example is the core of a magnetization vortex (as in a so-called Bloch line) magnetized perpendicular to the surface (see Fig. 3.27). Other examples are the already mentioned micromagnetic surface layers and the cores of *Néel* walls in thin films. The characteristic length of these features is $\sqrt{A/K_d}$, where K_d is the stray field energy constant (3.21). Still smaller structures occur at so-called micromagnetic singular points, which cannot be continuous in the micromagnetic sense (see Sect. 3.6.5D). In the core region of such a singularity, typically a few lattice constants wide, the magnetic order itself must break down. Conventional micromagnetism does not include this possibility because of the condition $\boldsymbol{m}^2 = 1$.

Next in scale among characteristic lengths is the so-called Bloch wall width parameter $\sqrt{A/K}$, where K is any anisotropy constant. The Bloch walls carry a specific energy of the order $\sqrt{AK}$. Interestingly, both quantities $\sqrt{A/K}$ and $\sqrt{A/K_d}$ are frequently called 'exchange lengths'. Micromagnetic solutions that manage to avoid the formation of stray fields (such as the classical infinitely extended Bloch wall) typically scale with the exchange length $\sqrt{A/K}$. Other configurations that cannot avoid free magnetic poles, such as the micromagnetic solutions for small particles, need for their description a resolution of the order of the exchange length $\sqrt{A/K_d}$.

Larger characteristic dimensions appear if the Bloch wall energy competes with other energies in determining the overall structure. If the competing energy is the stray field energy caused by a surface misorientation ϑ_s, a length of the form $\sqrt{AK}/K_d\sin^2\vartheta_s$ determines the character of the domain structure. If the thickness of the sample is larger than this characteristic length, the domain structure tends to be modified by the formation of supplementary domains that reduce the stray field energy at the expense of additional domain walls near the surface. For smaller thicknesses the stray field is decreased only by rotating the magnetization parallel to the surface, without the formation of additional domains and walls.

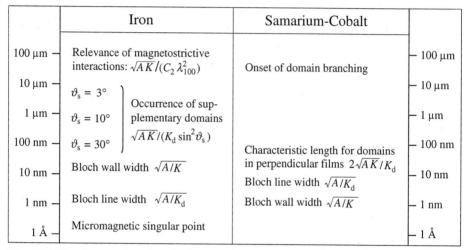

Fig. 3.15 The characteristic lengths in micromagnetics for two typical examples: for iron as a representative of the cubic, low-anisotropy materials, and for SmCo$_5$, a representative of uniaxial, high-anisotropy materials. The misorientation angle ϑ_s is defined as the angle between the nearest easy direction and the surface of the sample

Similarly, there is a characteristic thickness determining whether magnetostrictive self-energies become important. If the thickness is smaller than $\sqrt{AK}/C_2\lambda_{100}^2$, then wall energy is predominant and structures with an efficient domain wall pattern are preferred. In the opposite case, the elastic compatibility of the domains has priority, favouring stress-free arrangements.

Iron is one example showing the wide range of characteristic dimensions in micromagnetics. Another example, high-anisotropy hexagonal material (SmCo$_5$), is also included in Fig. 3.15.

The best procedure to tackle a problem in micromagnetics is to try to divide it into partial tasks according to the different scales, using reasonable approximations in each step. The occurrence of domains and walls, however, cannot be derived from such an approach. They are, instead, introduced at the beginning, supported by experimental evidence and qualitative micromagnetic arguments. This discomforting situation was most thoroughly analysed by *W.F. Brown* [34, 501] without offering a solution to the problem.

In not too complicated cases — such as in small particles and thin film elements — it is possible to numerically calculate the equilibrium structure of a given configuration. One approach consists in using the dynamical versions (3.66a) or (3.66b) of the micromagnetic equations. A selected starting configuration will relax towards the equilibrium solution with a speed depending on the damping parameter that can be chosen freely to insure best numerical convergence. With increasing computing power the boundary between numerically tractable problems and intractable ones is continually shifted upwards, while devices get smaller by modern microfabrication techniques. Supply and demand will meet somewhere, but the overlap will remain small, as is apparent from Fig. 3.15. We will come back to numerical solutions of the micromagnetic equations, such as in the area of domain walls in thin films (Sect. 3.6.4) and in the magnetization processes of small particles (Sect. 5.7.2).

3.2.8 Review of the Energy Terms of a Ferromagnet

Figure 3.16 lists the characteristic coefficients in the energy expressions discussed in this chapter. The various kinds of anisotropy are not distinguished to offer a simplified overview.

Energy term	Coefficient		Definition	Range
Exchange energy	A	[J/m]	Material constant	$10^{-12} - 2 \cdot 10^{-11}$ J/m
Anisotropy energies	K_u, K_c ...	[J/m^3]	Material constants	$\pm(10^2 - 2 \cdot 10^7)$ J/m^3
External field energy	$H_{ex} J_s$	[J/m^3]	H_{ex} = external field J_s = saturation magnetization	Open, depending on field magnitude
Stray field energy	K_d	[J/m^3]	$K_d = J_s^2/2\mu_0$	$0 - 3 \cdot 10^6$ J/m^3
External stress energy	$\sigma_{ex} \lambda$	[J/m^3]	σ_{ex} = external stress λ = magnetostriction constant	Open, depending on stress magnitude
Magnetostrictive self energy	$C \lambda^2$	[J/m^3]	C = shear modulus	$0 - 10^3$ J/m^3

Fig. 3.16 The coefficients of the energy terms discussed in Sect. 3.2.2–6 together with definitions and the order of magnitude of these terms in typical materials. A further energy term, which is connected with internal lattice rotations, was discussed on p. 135. It plays a role only in zones which deviate strongly from the easy directions and scales with $K \lambda$

3.3 The Origin of Domains

In this section qualitative arguments supporting the existence of magnetic domains are reviewed. It turns out to be impossible to assign a single origin to domain structures in all kinds of materials. Somehow the non-local energy terms, above all the stray field energy, are responsible for the development of domains. But the arguments differ considerably depending on the magnitude of the anisotropies and on the shape and size of the samples. In this chapter we limit the discussion to thermodynamically stable domains and exclude domains that are determined by non-uniform, irreversible magnetization processes, as they may be found in recording materials or in permanent magnets.

3.3.1 Global Arguments for Large Samples

From a global point of view domains are a consequence of discontinuities in the equilibrium magnetization curve and of the demagnetizing effect. We first explain the origin of jumps in the magnetization curve and then how they lead to domains. These arguments apply to extended samples for which the energy of the domain walls is negligible, in the same sense as an alloy may be discussed based on its phase diagram, neglecting the energy of grain boundaries. In magnetism, we also neglect the localized stray field and anisotropy energy contributions of so-called closure structures at surfaces and interfaces, which will be discussed in detail in Sect. 3.7.4.

(A) Magnetization Discontinuities. Discontinuities in magnetization curves are explained with the following example: a thin (100)-oriented single-crystal iron film with in-plane magnetization shall be sufficiently extended to neglect demagnetizing effects. Cubic anisotropy [see (3.9a) with $\vartheta = 0$] and a superimposed uniaxial anisotropy (3.11) with [010] as the easy axis shall be present. For both anisotropies only the first-order constants are taken into account. We are interested in the equilibrium magnetization curve for a field H applied along a chosen direction η_h relative to the [100] axis. Then the total energy density is:

$$e_{tot}(\varphi) = K_{c1} \sin^2\varphi \cos^2\varphi + K_{u1} \cos^2\varphi - H J_s \cos(\varphi - \eta_h) \tag{3.68}$$

where φ is the magnetization angle inside the film starting at the [100] axis. We will show that the magnetization curve can be derived in a straightforward way from the reduced effective anisotropy energy g as a function of the magnetization component m along the field, with g and m being defined as:

$$g = Q_{c1}\sin^2\varphi\,\cos^2\varphi + Q_{u1}\cos^2\varphi\ ,\quad m = \cos(\varphi - \eta_h)\ , \tag{3.69}$$

with $Q_{c1} = K_{c1}/K_d$, $Q_{u1} = K_{u1}/K_d$. Plotting $g(m)$ as a parameter plot based on the angle φ as parameter we obtain the result shown in Fig. 3.17 (for certain values of the constants given in the caption). Unimportant information has been omitted from this plot: for every angle φ there exists a mirror angle $(2\eta_h - \varphi)$ with the same magnetization component m. The two angles lead to different effective anisotropies g. As we are interested in the equilibrium solution, we anticipate that only the lower branch of the energy will be relevant, skipping the upper branch.

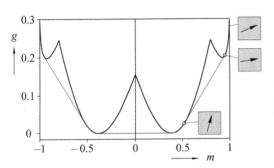

Fig. 3.17 The effective anisotropy energy g as a function of m, the magnetization component along the angle η_h. Of the two possible branches, only the lower energy branch is plotted for every value of m. The chosen parameters are $Q_{c1} = 1$, $Q_{u1} = 0.2$ and $\eta_h = \pi/8$

To derive the magnetization curve $m(h)$ we go back to (3.68). Inserting (3.69) we get the reduced energy density $\varepsilon_{tot} = e_{tot}/K_d = g(m) - 2hm$, where $h = \mu_0 H/J_s$. Minimizing ε_{tot} with respect to m leads immediately to the implicit relation $2h = g'(m)$, which has in general, however, more than one solution. The lowest-energy branches can be selected by a graphical procedure known from the analysis of metallographic phase diagrams. This procedure is demonstrated in Fig. 3.18a: the equilibrium solutions are related to the "convex envelope" $\hat{g}(m)$ of the anisotropy energy functional, which consists of the convex parts of $g(m)$, plus the tangents bridging the "concave" parts.

To see that in fact the derivative of the convex envelope $\hat{g}'(m)$ yields the magnetic field h of the magnetization curve describing the thermodynamic equilibrium states, we have to prove a few relations. First look at the common tangent A to B in Fig. 3.18a, which repeats one part of Fig. 3.17. The two points A and B belong to the same field because $g'(A) = g'(B) = 2h$ at both tangential points. In addition, the total energy $e_{tot} = g(m) - 2hm$ in both points is identical because the connecting line has the same slope:

$$2h = g'(A) = [g(B) - g(A)] / [m(B) - m(A)] \text{ , leading to}$$

$$e_{tot}(A) = e_{tot}(B) \text{ ,} \tag{3.70}$$

which is derived using $e_{tot}(A) = g(A) - 2hm(A)$ and $e_{tot}(B) = g(B) - 2hm(B)$. We now have to prove that any state that belongs to a point between A and B does not belong to the equilibrium magnetization curve. Consider two states C and D, as in Fig. 3.18a, selected to belong to the same field, meaning $g'(C) = g'(D)$. Then the tangential slope is larger than the slope of the connecting line:

$$h = \tfrac{1}{2} g'(C) > \tfrac{1}{2} [g(D) - g(C)] / [m(D) - m(C)] \text{ , leading to:}$$

$$e_{tot}(D) < e_{tot}(C) \text{ .} \tag{3.71}$$

The "inside" point C is thus disfavoured compared to a point on the envelope D. The same can be shown for all magnetization values located between the two touching points A and B of the common tangent. The equilibrium magnetization curve is therefore determined exclusively by the convex envelope of the anisotropy functional. It jumps over concave parts from one tangential point to the other, as shown in Fig. 3.18b. For our chosen example the magnetization curve displays three such jumps.

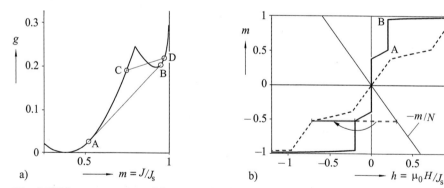

Fig. 3.18 The construction of the magnetization curve (b) based on the anisotropy functional (a). The jumps in (b) are related to the common tangents cutting off concave parts of the anisotropy functional in (a). The *dashed curve* in (b) is the sheared magnetization curve resulting from a demagnetization effect as explained in (C)

If only an external field is present and the magnetization can point into any direction without an additional magnetostatic energy, a discontinuity in the magnetization curve is not expected to lead to domain states. As long as the field is smaller than the discontinuity field, the lower branch of the magnetization curve is preferred. On exceeding this field the magnetization jumps to the upper branch. Multi-domain states *at* the discontinuity field will not occur because the interface energy, even if it is small, makes them unfavourable.

(B) Magnetization Control. As indicated in the introduction to this section and as elaborated further down in (C), it would take a demagnetizing effect to induce domains. There is a less frequently occurring but conceptually simpler case in which domains are induced also if the sample is closed by an ideal soft magnetic yoke: let a certain average magnetization be enforced in the yoke by a feedback mechanism — as opposed to the previous assumption where the magnetization direction was freely variable. If a magnetization value is enforced that lies within the range of one of the discontinuities, one of the intermediate higher-energy states could possibly be occupied. An energetically advantageous possibility is, however, that of a *mixed state* in which the two states at the endpoints of the jumps are mixed in a certain volume ratio so that the enforced magnetization is reached. The energy of such a mixed state would lie on the straight connecting lines in the $g(m)$ plot *if* a simple mixing law applies.

Because only the relative volumes of the constituent states (the states at the tangential points) enter the free energy in this construction, we may also call the domains *phases*, as in the analogous case of phase diagrams in thermodynamics. The phase volumes for an arbitrary state on the connecting line in Fig. 3.18a are determined in the usual way by the two segments a_1 and a_2. To repeat the argument: instead of a uniform transverse magnetization connected with an enhanced (anisotropy) energy, inhomogeneous lower energy states are preferred, as long as the domain wall energy and other extra energies of an intermediate state can be neglected.

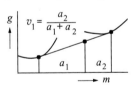

Two conditions must be met to make this simple argument valid: (i) The sample must be sufficiently large, so that the interface or wall energies are negligible as emphasized before. (ii) The postulated domains must be *compatible* in the sense that they are not connected with long-range stray fields or stresses that would add volume energies as opposed to the surface energies of the domain walls. Two adjacent domains are *magnetically* compatible if the normal components of the magnetization vectors of both domains with respect to their common domain wall are equal. This condition is fulfilled by all domain walls that contain the difference between the domain magnetization vectors (see Sect. 3.6.1D). *Elastic* compatibility between two domains adds two further conditions: the tangential components of the free deformations (3.46) must be equal in both domains for the given wall orientation. There is always such a domain boundary for magnetostrictive deformations, but it is not necessarily true that the elastically compatible boundary belongs to the manifold of walls allowed by magnetic compatibility.

It turns out that for all standard domain walls in cubic or uniaxial crystals (see Sects. 3.6.2–3) magnetic and elastic compatibility can be achieved simultaneously even if external fields are added. Magnetic domain states are thus possible without additional volume incompatibility energies. Incompatible domain patterns may occur for non-standard cases such as asymmetrical or composite anisotropies. The domains connected with the secondary jump [(A–B) in Fig. 3.18b] are in general elastically incompatible, depending on the values of the magnetostriction constants. Another example for incompatible domains will show up in the case of the [111]-anomaly in iron (Sect. 3.4.3C). We ignore these complications in the present discussion because magnetostrictive effects are often weak and negligible in a first analysis.

We have thus justified the existence of domains in magnetization-controlled circuits. Machines with a predominantly inductive load on a rigid voltage (such as idling transformers) are examples for such circuits. For this case the construction of Fig. 3.18a indicates the lowest energy configurations, including domain states representing points on the tangential "bridges". In the following, the more obvious case of finite samples, in a given external field will be discussed.

(C) The Effect of the Demagnetizing Field. At this point finally *open* samples generating a demagnetizing field are considered. For the example defined in (A) the infinitely extended film is replaced by a narrow strip oriented perpendicular to the applied field. The demagnetizing energy adds a quadratic term $NK_d m^2$ [see (3.24) and curve $\langle 2 \rangle$ in the sketch] to whatever free energy was present before. The tangents in Fig. 3.18a ($\langle 1 \rangle$ in the sketch) are thus converted to convex parabolic sections $\langle 1+2 \rangle$, yielding a unique stability field for each magnetization value m.

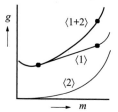

In the magnetization curve this is expressed by the *shearing* transformation (Fig. 3.18b): for a given magnetization value m, the external field H_{ex} must be augmented by the demagnetizing field $-H_d = N J_s m / \mu_0$ to reach the same state. Thus a discontinuity is transformed into a finite-slope segment with a unique magnetization value for every field value. The shearing transformation can be derived in a few lines: let $g(m)$ be the free energy without the external field and demagnetizing terms. Then the total free energy is:

$$e_{tot}(m) = g(m) - hm + NK_d m^2 . \tag{3.72}$$

From $de_{tot}/dm = 0$ follow the solutions:

$$h^{(0)} = g'(m) \qquad \text{(for } N = 0 \text{)} \text{ and}$$

$$h^{(N)} = g'(m) + 2NK_d m \qquad \text{(for } N \neq 0 \text{)} . \tag{3.73}$$

The solution for $N \neq 0$ then results from the solution for $N = 0$ if for every m the field $h^{(0)}$ is replaced by:

$$h^{(N)} = h^{(0)} + 2NK_d m \quad \text{or} \quad H^{(N)} = H^{(0)} + NJ_s m/\mu_0 . \tag{3.74}$$

If a jump from m_1 to m_2 in the unsheared magnetization curve is transformed into a straight segment with a finite slope, the states on the slope can only represent "intermediate" or domain states. These domain states are composed of domains magnetized along the boundary states m_1 and m_2 in volume ratios again determined by the lever rule applied to the $g(m)$ curves. For macroscopic samples for which the shearing transformation is considered adequate in describing the demagnetizing effects, all discontinuities in the unsheared magnetization curve thus lead to thermodynamically stable domain states after shearing. This situation is comparable to the occurrence of inhomogeneous states in the thermodynamics of closed systems. The contents of a closed vessel half filled with water is in a certain temperature range separated into two phases, the high-density liquid and the low-density vapour phase. The two-phase state is induced by thermal equilibrium under the condition of a fixed total quantity of water. A magnet closed by yokes, in which every value of the average magnetization is possible without an energy penalty, is comparable to an open vessel. The demagnetizing energy of an open magnet favours the non-magnetic state, or, in connection with an external field, a certain average magnetization. It thus has a similar effect as the lid on a vessel.

The unsheared magnetization curve depends in general on the direction of the applied field relative to the easy axes of the sample. For example, there is no discontinuity in the magnetization curve for a field along the hard axis of a uniaxial crystal. Therefore no domain structure is expected following the one-dimensional global argument elaborated in this section. But if the sample is of finite dimension along its easy axis perpendicular to the field, it will be broken up into transverse domains even without a discontinuity in the regular, longitudinal magnetization curve. The reason is that then an energy penalty is connected with a net transverse magnetization. A one-dimensional analysis cannot account for complexities like these, which are systematically treated in Sect. 3.4 in the context of *phase theory*.

Magneto-elastic interactions with the environment of a crystal may also lead to domains if a crystal is embedded in a way that is not compatible with free magnetostrictive strains of the saturated state. Domains may form to accommodate better to the elastic environment. This effect is important in

antiferromagnetic and ferroelectric substances with their often large spontaneous deformations, but it most probably also plays a role in polycrystalline ferromagnetic materials.

The global arguments are valid for large samples for which the domain wall energy and the details of possible magnetization patterns play only a minor role. For *small* samples more detailed arguments are necessary as presented in the following. These arguments depend on the relative strength of magnetic anisotropy that is measured by the material parameter $Q = K/K_d$. Here K is the absolute value of the predominant anisotropy parameter (Sect. 3.2.3) and $K_d = J_s^2/2\mu_0$ is the stray field energy constant (3.21). We will treat the three cases $Q \gg 1$, $Q \ll 1$, and moderate Q separately.

3.3.2 High-Anisotropy Particles

In this section we discuss samples of uniaxial materials with $Q = K_u/K_d \gg 1$. Since anisotropy is the predominant energy contribution, the magnetization will naturally be aligned along one of the two opposite easy directions. For large samples the arguments of the last section remain valid: in a sample of infinite length and uniform cross-section saturated easy axis states are energetically preferred; no domains are expected. The same is true for a toroidal sample with a circumferential easy axis, or for a body that is inserted into a high-permeability yoke. If, however, the sample is finite along the easy axis, so that a uniform magnetization leads to surface charges, then a domain structure may appear. In the following exploratory calculations we further assume that the domain walls are very thin and always parallel to the easy axis. The latter assumption is generally not valid. Depending on the symmetry of the domain pattern, domain walls will be bent by the effects of the stray field [511] leading to magnetic charges on the walls. Ignoring this complication makes calculations much easier and suffices for our purpose.

(A) General Relations. Focusing on *small* particles, one may try to find the energy gain by a domain structure relative to the uniformly magnetized sample, and below which sample size L domain states become unstable, taking into account the energy of domain walls. Although a micromagnetic approach is necessary for a thorough investigation, we can obtain at least a qualitative answer by assuming a *thin* wall having a size-independent specific energy γ_w. In standard uniaxial materials the specific wall energy is also independent of the wall orientation as long as the wall plane contains the easy axis. Every domain structure in such a uniaxial particle gives rise to a stray field energy E_d that can be calculated with the tools of Sect. 3.2.5C. It can be measured by the dimensionless parameter $\varepsilon_d = E_d/(VK_d)$, where V is the volume of the particle. The parameter ε_d is independent of the particle size as long as shape and domain pattern stay similar in the mathematical sense. For uniformly magnetized particles ε_d is simply the demagnetizing factor N, while for multi-domain particles it becomes much smaller.

A domain pattern may also carry a net magnetization. Its average reduced value measured along the easy axis is called m. Furthermore, every domain configuration is connected with a wall area F_w, measured by the dimensionless parameter $f_w = F_w/V^{2/3}$. The total energy density ε_{tot} in units of K_d can be expressed in terms of the three reduced quantities ε_d, f_w and m:

$$\varepsilon_{tot} = e_{tot}/K_d = \varepsilon_d + \left(f_w/l_p\right)\left(\gamma_w/K_d\right) - 2hm \ , \tag{3.75}$$

where $h = \mu_0 H/J_s$ is the reduced field applied along the easy axis, and the effective particle size l_p is defined by $l_p = \sqrt[3]{V}$ (which is the edge length of a cube with the same volume). The stray field energy and the external field energy stay constant per unit volume with changing scale, while the wall energy $\gamma_w F_w$ divided by the volume V decreases in importance with increasing particle size l_p. The anisotropy energy as well as the exchange energy enter only indirectly over the specific wall energy γ_w.

The energy of a given domain pattern can thus be extended to particles of other sizes l_p, and in other fields h. Figure 3.19a shows this energy for a number of different domain patterns for magnetically uniaxial *cubes* at zero field. As N is the reduced energy density of the uniformly magnetized particle, a domain state has a lower energy at zero field if it exceeds the "single-domain size" l_{SD} which is found by equating N with (3.75) for $h = 0$:

$$l_{SD} = \left[f_w/(N - \varepsilon_d)\right]\left(\gamma_w/K_d\right) \ . \tag{3.76}$$

The single-domain size is thus determined by the interplay between wall energy and stray field energy. It scales with the ratio γ_w/K_d, which is a characteristic material length. The coefficient of this characteristic length depends on the configuration. Beyond the threshold l_{SD} a domain state is thermodynamically stable. The domain configuration with the smallest l_{SD} will be the first to appear if only the lowest energy states are considered, followed by other patterns that are more favourable for larger sizes. The transition between two competing patterns is calculated in the same way:

$$l_{1,2} = \left[\left(f_w^{(2)} - f_w^{(1)} \right) / \left(\varepsilon_d^{(1)} - \varepsilon_d^{(2)} \right) \right] \left(\gamma_w / K_d \right) . \tag{3.76a}$$

The "single-domain diameter" for a *sphere* was estimated by Kittel [30] as $D_{SD} \approx 9\,\gamma_w/K_d$. A more precise value was derived by Néel [27] by calculating rigorously the stray field energy of a two-domain ("split") sphere ($\varepsilon_d = 0.1618$), leading according to (3.76) with $N = \frac{1}{3}$ and $f_w = \pi\,(4\pi/3)^{-2/3} = 1.209$ to a single-domain size $l_{SD} = 7.048\,\gamma_w/K_d$. This is equivalent to a single-domain volume of l_{SD}^3, and thus to a single-domain diameter of $D_{SD} = 8.745\,\gamma_w/K_d$.

(B) Cube-Shaped Particles. Comparing many possible states for cubic particles (Fig. 3.19), the simple two-domain state proves to have the lowest critical thickness. With $\varepsilon_d = 0.1707$ and $f_w = 1$ we obtain from (3.76) $l_{SD} = 6.15\,\gamma_w/K_d$. Because the surface charge pattern consists of simple rectangles, the tools of Sect. 3.2.5C can be directly applied. The value $\varepsilon_d = 0.1707$ can be expressed with (3.31) as $\varepsilon_d = \frac{1}{\pi}\,[4F(\frac{1}{2}) - 3F(0) - F(1)]$ with $F(x) := F_{RR}(x, 0) - F_{RR}(x, 1)$.

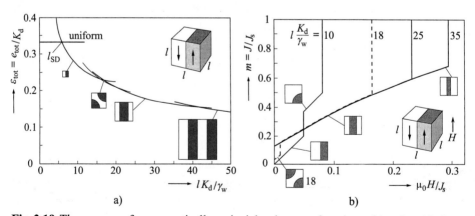

a)   b)

Fig. 3.19 The energy of a magnetically uniaxial cube as a function of its size (a) for a number of different domain patterns. Below the critical size l_{SD} the uniform magnetization is the most favourable. The corresponding lowest energy magnetization curves are shown in (b) for four different values of the cube size. The field is aligned along the easy axis. In this analysis configurations with two to five bands, one, two and four quarter circles and a centred circle were compared

For larger cubes beyond the single-domain limit we find that a sandwich-type three domain state takes over above a size of $19\gamma_w/K_d$, but in a narrow range $l_p = (15.2\dots19)\,\gamma_w/K_d$ another three-domain state with quarter-circles is energetically favoured (Fig. 3.19a). Interestingly, the latter configurations carry a non-zero remanent magnetization in thermodynamic equilibrium. For the three-plate state it is $m = 0.124$ opposite to the central domain. The stray field energy factor of this state is $\varepsilon_d = 0.10826$. The ensuing four-domain state carries no net magnetization again at zero field.

It is interesting to study the "idealized" magnetization curves for a number of configurations. Figure 3.19b shows these curves for cubes of four selected sizes resulting from energy minimization including the field term. We observe various discontinuities where one pattern gives way to another. Even a quarter-circle domain state has a stability range around $l_p = 10\gamma_w/K_d$ and $h = 0.07$.

We still have to prove that the results of these considerations are compatible with the assumption of a *thin* wall compared to the particle size. To do this we use two well-known properties of the 180° wall in uniaxial crystals (which will be derived in Sect. 3.6), namely their energy $\gamma_w = 4\sqrt{AK_u}$ and their width $W_L = \pi\sqrt{A/K_u}$. The first critical thickness for the stability of domains in the cubic particle can then be written in the form:

$$l_{SD} = 6.15\,\gamma_w/K_d = 7.83\,Q\,W_L\,, \tag{3.77}$$

which is in fact much larger than W_L for high-anisotropy materials because of $Q \gg 1$!

(C) Oblong and Oblate Particles. The calculated critical sizes depend strongly on particle shape. In Fig. 3.20a the length of block-shaped particles, for which the uniform state is energetically equal to the two-domain state, is plotted as a function of the aspect ratio. The size of thermodynamically stable single-domain particles can become large if they are elongated. For very long particles the simple two-domain state will not be the first stable multi-domain state, however. Beyond an aspect ratio of 14.65 : 1 it is replaced by a state with wedge-shaped domains at the ends, leading to a change in the slope of the single-domain size limit. In calculating these configurations with wedge domains, additional magnetic charges at the domain walls have to be taken into account. They become favourable for elongated shapes because the wall area in simple plate domain patterns becomes too large in comparison. Although the average magnetization of the wedge states is close to saturation, it will differ drastically from a truly saturated state in its stability, i.e. in its coercivity, because the wedges are efficient nuclei for the switching process.

For oblate shapes ($L/W < 1$) the plate width W_{SD} is the more meaningful quantity and is therefore added in Fig. 3.20a (for a given aspect ratio L/W the single-domain size may be given either by the length L_{SD} or by the width W_{SD}). The quantity W_{SD} displays a minimum value at about $L/W = 0.46$ and for shorter samples the single-domain width limit diverges roughly with $1/L$. Very thin perpendicular-anisotropy plates can form energetically stable domains only if they are very extended. In the range of oblate particles we also find the particle with the smallest volume at its single-domain limit in the considered class at about $L/W = 0.2$.

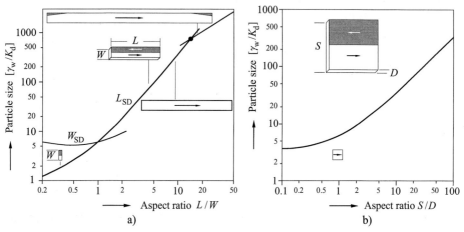

a)   b)

Fig. 3.20 (a) The thermodynamic stability limit for the length (L_{SD}) and the width (W_{SD}) of elongated single-domain high-anisotropy particles of different shapes. For particles larger than this limit multi-domain states are energetically favoured. The particles have dimensions (W, W, L) and the easy axis is along the third dimension. Beyond a value of $L/W = 14.65$ the two-domain state is replaced by a state with wedge domains at the particle ends, as found by a detailed analysis. (b) The single-domain limit for flat particles with the anisotropy axis parallel to the surface

For the sake of completeness we also present the case of high-anisotropy particles of the same shape but with the anisotropy axis oriented perpendicular to the particle symmetry axis (Fig. 3.20b). This case comprises thin films with large in-plane anisotropy and needles with a strong transverse anisotropy. The relative energies of the single-domain and the two-domain states determine the phase boundary. High-remanence states with wedge or dagger domains (as sketched) are metastable, but they cannot compete energetically with the simple two-domain state at zero field.

(D) The Significance of the Calculated Single-Domain Limits. The calculations presented in this section yield no information about how the *transition* from a saturated state to a domain state occurs (or the transitions between different domain patterns). The question of spontaneous switching between the saturated state and other, low remanence states is analysed in Sect. 3.5. The thermodynamically stable domain state will most likely be realized if the sample is cooled from above its Curie point at zero field ("thermal demagnetization"). So at least for this experimental procedure the existence of domains is justified by the simple energy considerations of this section.

3.3.3 Ideally Soft Magnetic Materials

(A) Small Particles. Here we discuss materials that have a negligibly small magnetic anisotropy (i.e. $Q \ll 1$). The only energy contributions to be considered then are the magnetic field energy terms, the exchange energy and perhaps the magnetostrictive self-energy. Firstly, if a particle is too small, no domain structure will develop. The argument is the same as for high-anisotropy materials, with the exception that the wall width cannot be considered small compared to the sample size at the critical thickness. Therefore a three-dimensional micromagnetic analysis is necessary to calculate the critical single-domain particle size in soft magnets. Such calculations have been undertaken [580–582, 60, 583] and we report here some results.

Most micromagnetic computations are based on a specific anisotropy, but different calculations for materials with small anisotropy agree in the basic result: above a certain size an inhomogeneous magnetization state with a low average magnetization takes over. This state is no more continuously related to the uniformly magnetized state. The initial high-remanence or single-domain configuration is called the "flower" state [581] because the magnetization in the corner spreads outward like the petals of a flower (Fig. 3.21a). The low remanence state, which has a lower energy beyond the single-domain limit, is called the curling or vortex state (Fig. 3.21b). The single-domain size depends on the particle shape and scales with $\Delta_d = \sqrt{A/K_d}$. For cubes, values between $l_{SD} = 6.8\Delta_d$ for relatively weak uniaxial anisotropy ($Q = 0.02$) [581, 582] and $l_{SD} = 7\Delta_d$ for stronger negative cubic anisotropy were found [60]. Other authors [583] found $l_{SD} = 7.27\Delta_d$ for $Q = 0.0053$. Low anisotropy spheres of different anisotropy show values around $l_{SD} = 4\Delta_d$ [584]. For uniaxial cobalt ($Q \approx 0.4$) $l_{SD} = 10.8\Delta_d$ resulted from numerical micromagnetic computations [585].

In the same way as for high-anisotropy particles (Fig. 3.20), the single-domain limit rises with increasing elongation of the particle, increasing, for example, for magnetite from $7\Delta_d$ for a cube to about $19\Delta_d$ for a particle twice as long as wide. For cubes and block-shaped particles not only the thermodynamic "phase boundary", but also the stability limit of the high-remanence flower state was investigated [60]. Beyond the thermodynamic transition l_{SD} the flower state was still (meta)stable, but it became unstable at about twice this length for constant aspect ratio (at which point it switched in the simulation into another state with intermediate remanence as shown in Fig. 3.21c). More about these "switching" processes will be discussed in Sect. 3.5.

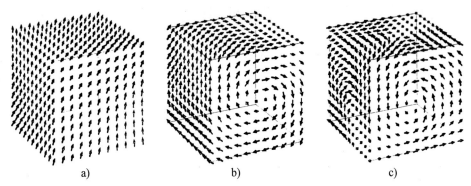

a)   b)   c)

Fig. 3.21 Micromagnetic states of small cubic particles. (a) The high-remanence flower state that continuously develops out of the uniformly magnetized state. (b) The vortex or curling state modified by trying to stay parallel to the cube edges. (c) A complex state found at larger particle sizes, which is related to a three-domain state. The calculations were done for the material parameters and anisotropies of magnetite [60], but the influence of anisotropy on these patterns is weak

To summarize, low-anisotropy particles display a thermodynamic transition from a high-remanence, near saturation state to a low-remanence vortex state in a size range of about $7\Delta_d$. In the following section we explore the formation of domains in samples that are much larger than this exchange length.

(B) Two-Dimensional Thin Film Elements. If the dimensions of the specimen are much larger than the single-domain limit, and if at the same time anisotropy is absent or negligibly small, closed-flux magnetization configurations with vanishing stray field energy are favoured if they are possible. The question is: do we expect real domain structures with identifiable domains in the absence of anisotropy, or rather a continuously flowing divergence-free magnetization pattern comparable to the velocity field in hydrodynamics? The conventional answer was that real domains must be connected with anisotropy. Surprisingly, apparently ordinary domains may appear in thin film elements in the complete absence of anisotropy.

Consider a rectangular thin-film element of large size ($\gg\sqrt{A/K_d}$), vanishing anisotropy, and zero applied field for which a completely stray-field-free magnetization pattern may be postulated if it should be possible. In such a pattern the magnetization vector field $m(x, y)$ must (i) lie parallel to the film surface ($m_z = 0$), it must (ii) be divergence-free in the interior ($\mathrm{div}\,m = \partial m_x/\partial x + \partial m_y/\partial y = 0$) and at the edges ($m \cdot n = 0$, n = edge normal), and it must (iii) have a constant length $|m| = 1$.

Intuitively, some smoothly varying vector field such as in sketch (a) might be expected. It turns out, however, that no continuous planar pattern can fulfil the three conditions simultaneously. This would be no problem with any two of the three conditions, but with all three conditions only patterns containing discontinuities ("domain walls") as in (b) are possible.

A comprehensive analysis of such thin film elements of arbitrary shape (Fig. 3.22) was achieved by *van den Berg* [586–588]. He proved that the conditions explained above can only be met if the magnetization stays parallel to the edges on every point along an edge normal as long as no other edge interferes. If an edge is straight, then a *domain* with uniform magnetization results in a certain neighbourhood of this edge.

An ingenious generalization of this result can even predict the possible range of these domains and determine the natural position of walls which separate the edge-induced domains. It also introduces a natural extension of the domain concept for curved edge shapes. The geometrical recipe is as follows ([586]; Fig. 3.22):

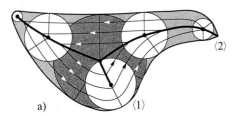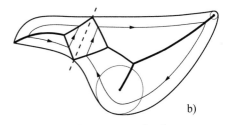

Fig. 3.22 (a) Van den Berg's construction of a stray-field-free planar magnetization pattern in a simply connected soft magnetic film element. (b) A more complicated variant obtained by introducing a virtual cut along the *dashed line*

- Take *circles* that touch the edge at two (or more) points and lie otherwise completely within the figure. The centres of all such circles form the *walls* in the pattern.
- In every circle the magnetization direction must be perpendicular to each touching radius. The walls are then stray-field-free.
- If a circle touches the edge in more than two points, its centre forms a junction (dark circle in Fig. 3.22a).
- If the touching points fall together (osculating contact, ⟨1⟩ Fig. 3.22a) the wall ends at the centre of the circle, and this point is the centre of a zone of concentric magnetization rotation.
- If the shape contains an acute corner ⟨2⟩, a boundary runs into this corner.

The domain structure thus obtained is not unique. All magnetization directions may be inverted by 180°. More complicated structures are obtained by virtually cutting the figure and applying the algorithm to the parts (Fig. 3.22b). In this way multiple wall junctions at the edge are generated, which were called "edge clusters" by van den Berg. The basic structure probably has the smallest energy for simple and compact shapes. Other patterns of the indicated class may be favoured for very elongate thin film elements or as a consequence of anisotropy or magnetostriction. Many of them may be present as metastable solutions depending on magnetic history, particularly when the edge clusters are pinned at the sample edge.

The geometry of domain wall edge clusters in thin films is not arbitrary. The condition of stray field avoidance imposes certain rules for the allowed angles and the allowed number of participating walls. These geometrical rules were explored by van den Berg [589]. They are demonstrated in Fig. 3.23 for two- and three-wall clusters. Higher-order clusters behave analogously but are practically irrelevant. Note that for three-wall clusters on a smooth edge the magnetization on both sides of the cluster is parallel. For antiparallel magnetization an even number of domain boundaries is necessary.

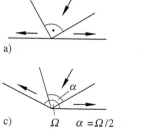
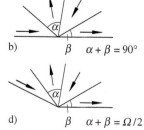

a)

b) $\beta \quad \alpha + \beta = 90°$

c) $\Omega \quad \alpha = \Omega/2$

d) $\beta \quad \alpha + \beta = \Omega/2$

Fig. 3.23 Possible edge clusters for a straight edge (a, b) and in a corner (c, d). The indicated laws for the wall angles follow from the assumption of planar magnetization and stray field avoidance

Domain boundaries are treated here as linear discontinuities in the two-dimensional magnetization pattern. In this approach van den Berg's model contains no intrinsic scale, as the construction scales with the sample size. In reality, real thin-film domain walls (Sect. 3.6.4) have to be inserted. The wall energy, its orientation and wall angle dependence, as well as the interactions between these domain walls, will modify the ideal van den Berg pattern as discussed in the experimental part (Sect. 5.5.4).

The important point of this discussion is that domains and walls can develop even without anisotropy. They can be induced simply by the boundary conditions and the principle of pole avoidance. Regions with a non-constant magnetization induced by a curved edge are a natural extension of the domain concept.

Although the above algorithm can be applied to any film shape, it is tailored for "simply connected" films. Multiply connected films (i.e. films with holes in them) can be magnetized in a stray-field-free manner without walls, as is obvious for a simple torus. The circular magnetization pattern has a lower energy in a torus than the structure obtained by the van den Berg algorithm, which carries a circular wall as sketched. In structures of generalized toroidal or infinitely extending shape a band of flux can travel around or along [587], accompanied by closed-flux eddies in the remaining niches. These possibilities are demonstrated for two examples in Fig. 3.24.

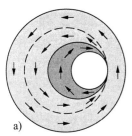
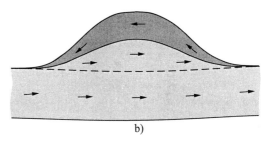

Fig. 3.24 Generalized stray-field-free configurations favourable in toroidal (a) or infinite strips (b). The pattern in (a) is constructed so that first the largest possible torus is formed and then van den Berg's basic pattern is constructed in the remaining (inner) area. Analogously, a van den Berg eddy occupies in (b) only the *bight* outside the main stream. The (*dashed*) boundaries of the continuous band of magnetization are introduced only for the purpose of interpretation and do not exist in reality

Another interesting problem is how these domain structures behave in applied fields. The topic was addressed by *Bryant* and *Suhl* [590] starting from the idea that the magnetic *charges* arising in an ideally soft magnetic material in an applied field should be distributed as in the analogous electrostatic problem. The charges on a conducting body are known to reside on the surfaces only. For a magnetic thin film element in an external field the charges are therefore expected at the edges of the element, but also at the top and bottom surfaces of the thin film. Once the charges are known, the unit magnetization vector field can be integrated mathematically. Generalized domain patterns that start from van den Berg's solution at zero field can be calculated numerically in this way for any field. Interesting is a circular element for which the van den Berg solution degenerates into a simple concentric pattern (Fig. 3.25a). Applying a field, Bryant and Suhl predicted a domain wall to develop out of the central vortex at zero field. This prediction was indeed verified experimentally (Fig. 3.25b,c).

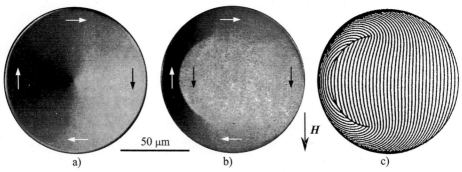

a)   b)   c)

Fig. 3.25 Kerr images [591] displaying the modification of a completely stray-field-free pattern in a circular low-anisotropy thin film element (a) by an applied field that induces a domain wall (b). This observed behaviour corresponds to the prediction of Bryant and Suhl [590] shown in (c). Sample: see Fig. 3.29

(C) Compact Three-Dimensional Bodies. It might seem straightforward to extend van den Berg's construction to truly three-dimensional bodies. But there are good arguments that its significance is not the same in this case. For the two-dimensional body van den Berg proved that a stray-field-free solution is in general only possible if one-dimensional mathematical discontinuities, the *domain walls*, are admitted. By postulating domain walls he justified the existence of domains, although not all his patterns fit into the classical domain image of regions of constant magnetization direction (remember the "rotation segments" in Fig. 3.22).

A generalization of van den Berg's construction to three dimensions is possible by replacing touching circles by touching spheres. Two-dimensional discontinuities (domain walls) are formed by this process. The following argument demonstrates that such discontinuous walls are not necessary in stray-field-free configurations in three dimensions: start with a not too thin plate for which the basic 2D van den Berg pattern has been constructed. Now consider this object as a three-dimensional body. It is known that Bloch walls in thick films have a vortex-like, two-dimensional structure that can avoid stray fields altogether (see Sect. 3.6.4D). Replacing van den Berg's linear discontinuities by these stray-field-free walls we arrive at a continuous stray-field-free magnetization pattern without any compromise with respect to the starting assumptions! This is not possible in two dimensions because continuous planar domain walls (so-called Néel walls; see Sect 3.6.4C) are always connected with stray fields, so that magnetic poles could only be avoided by discontinuous, abrupt transitions from one domain to the other.

If there is thus mathematically a difference between very thin and thick film elements, this does not mean much in the physical sense. After all, van den Berg's construction is only meant as a guideline to the construction of possible patterns; his wall discontinuities should always be replaced by real continuous walls. In all film elements these walls remain localized, determined by an equilibrium of exchange and stray field energy for thin films (Sect. 3.6.4C), and by the film thickness for thick films (Sect. 3.6.4D). If the elements are wide compared to these length scales, the general character of the domain pattern will follow van den Berg's principles in both cases. The mathematical difference becomes significant in the transition to real bulk samples, for which the thickness is comparable to the lateral dimensions. In the ideal case of zero anisotropy, domain walls in such samples are not well-defined any more. If their width scales — as in thick films — with the sample thickness, a distinction between domains and walls becomes meaningless.

This is best demonstrated by a very instructive example due to *Arrott* et al. [592] for an ideally soft magnetic finite cylinder. Arrott's model is stray-field-free and the magnetization is continuous everywhere except at two surface singular points at the top and the bottom of the cylinder. It is based on an approach that was first introduced in the construction of stray-field-free Bloch walls [579]. In cylindrical coordinates (Fig. 3.26b) the magnetization field m is derived from the φ component of a vector potential A_φ as follows:

$$m_\rho = -\partial A_\varphi / \partial z \ , \ m_z = (1/\rho) \, \partial(\rho A_\varphi)/\partial \rho \ , \ m_\varphi = \sqrt{1 - m_\rho^2 - m_z^2} \ . \qquad (3.78)$$

This vector field obeys the two conditions $\mathrm{div} m = 0$ and $|m|^2 = 1$. If we choose the function A_φ so that it vanishes at the cylinder surface, then the third condition $m \cdot n = 0$ for a stray-field-free micromagnetic configuration is also fulfilled. The following relatively simple function (basically from [592]) can lead to a low-energy configuration:

$$A_\varphi = \tfrac{1}{2} \rho u(z) f(\rho) \Big/ \sqrt{g(\rho, z) \rho^2 + u(z)^2} \ , \tag{3.78a}$$

with $f(0) = 1$, $f(R) = 0$, $u(\pm L) = 0$.

The functions $f(\rho)$, $u(z)$ and $g(\rho, z)$ are expressed in a power series and the coefficients can be determined by minimizing the exchange stiffness energy, which is the only energy term left [522]. The total exchange energy of this model for $L = R$ becomes $E_x = 39.7 A R$, when R is the cylinder radius. Most probably this configuration is very close to the lowest energy state of the ideally soft magnetic cylinder of any size. Although it is perfectly demagnetized, it differs conspicuously from the conventional picture of a domain structure.

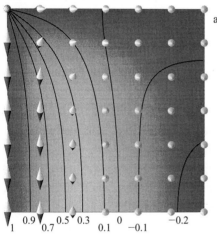

a)

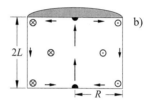

b)

0.9 0.5 0.3 0 −0.2
1 0.7 0.1 −0.1

Fig. 3.26 Arrott's stray-field-free cylinder. (a) The vector potential indicated by a *grey-shade plot* for a case of equal diameter and length, superimposed by *arrows* indicating the magnetization directions, and *contour lines* for constant vertical magnetization components. Only one quarter of the cross-section (b) is shown. The singularity in the upper left corner of (a) will in reality be replaced by a *swirl* (Fig. 3.27)

The exchange energy density in Arrott's model is finite everywhere except in two singular points — in contrast to van den Berg's models for thin film elements that contain even singular domain walls. To avoid even these singular points is impossible because of a mathematical (topological) theorem that is popularly known as the "hedgehog combing" theorem: you cannot comb a curled-up hedgehog so that all his spikes lie flat, parallel to the surface. According to the same theorem the weighed sum of all singularities for a sphere and equivalent bodies must be two (the topological weight depends on the magnetization rotation sense relative to the circulation sense of a path around a singularity. It is +1 for circular vortices as on the front surfaces of Fig. 3.21b,c, −1 for the "cross" vortex on the top surface of Fig. 3.21c).

Arrott's model proves that even for a body with edges (although without corners), such as the finite cylinder, the minimum number of singularities can be realized in a micromagnetic configuration. No well-defined magnetization direction can be ascribed to the singular point itself. The exchange energy density approaches infinity in the neighbourhood of the singularity although the integral over this infinity remains finite. In Sect. 3.6.5D we will further discuss micromagnetic singularities that may also occur inside a ferromagnet.

Based on these considerations and similar arguments [593] we suggest that singular surfaces (domain walls) and singular lines in magnetic microstructures can generally be avoided even in strictly stray-field-free three-dimensional bodies, leaving only a few unavoidable point singularities. This idea is expressed in the following mathematical conjecture:

- For all non-pathological bodies that we define as bodies with only a finite number of corners, there exist magnetization distributions $m(r)$ with $|m| = 1$, div $m = 0$ and $m \cdot n = 0$ (n = surface normal), which are continuous and differentiable everywhere except in a *finite* number of singular points.

An extension of this conjecture:

- All these singular points can be placed at the surface of the body.

If such solutions with a finite number of point singularities exist, they are probably energetically favoured compared to hypothetical solutions with line singularities or with an infinite number of point singularities, and this will probably be true also for regular domain patterns governed by anisotropy, as discussed in Sect. 3.3.4. No proof or counter-proof of this conjecture is available so far. Arguments for it were presented in [593]. See [594] for a general mathematical discussion of the problem.

If the body carries corners, micromagnetic singular-ities must necessarily reside in all corners for the stray-field-free limit. Depending on the magnetization direction in the three edges leading to the corner, two types of corner singularities can be distinguished [593]: a "saddle" and a "tripod" type as sketched. The tripod carries the  topological weight +1, the saddle singularity does not count as a topological singularity (weight 0) because it could be replaced by a continuous configuration if the corner is rounded. Whether a stray-field-free configuration with *only* one singularity in the corner suffices is also still unknown.

An important final remark in this context: when singular points are considered at the surface of a sample, these singularities are a mathematical concept within the (simplifying) assumption of completely stray-field-free patterns, not an actual structure. Since the exchange forces are locally stronger than the dipolar forces in ferromagnets, the surface singularity will be replaced by a small magnetization *swirl* in which the magnetization turns perpendicular to the surface, thus avoiding the singularity (Fig. 3.27). The width (of the order of $\sqrt{A/K_d}$) and the optimum structure of swirls are discussed in Sect. 3.6.5C. Also corner singularities will in reality be replaced by continuous patterns.

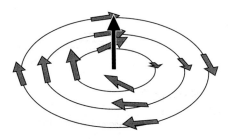

Fig. 3.27 The blown-up micromagnetic structure of a swirl, replacing the surface singularities in Fig. 3.26. On approaching the centre the magnetization rises continuously out of the plane, so that the configuration generates a strong local stray field. The structure is continuous and non-singular. It is strongly constricted at the surface and widens in the interior of the sample as shown in more detail in Fig. 3.92

(D) Infinitely Extended or Toroidal Three-Dimensional Bodies. Completely stray-field-free magnetization patterns without any singularities are possible in non-compact bodies, such as infinitely extended plates, or toroidal bodies with a constant cross-section. This was first demonstrated in [579] for domain walls in thick films, and we will discuss these solutions in more detail in Sect. 3.6.4D. The construction is easy if the magnetic structure depends on two spatial variables only as for the mentioned domain walls. The rotationally symmetric cylinder discussed above also belongs to this class. If this cylinder is hollow, a slightly modified vector potential (3.78a), which is zero also at the inner radius of the cylinder, generates a completely stray-field-free and non-singular vector field.

Infinitely extended prismatic bodies of arbitrarily shaped constant cross-sections were studied analytically in [595]. The following amusing recipe offers solutions for the zero-field case: prepare a plate of the shape of the cross-section. Cover it with the maximum amount of dry sand. According to the laws of soil mechanics the sandpile $S(x, y)$ will ideally form a roof-like body with a constant slope, i.e. with a constant absolute value of the gradient. Turn the gradient vector $(\partial S/\partial x, \partial S/\partial y)$ by 90° within the (x, y) plane and identify this rotated gradient with the planar magnetization field $(m_x = -\partial S/\partial y,$ $m_y = \partial S/\partial x)$. This 2D vector field is immediately verified as stray-field-free, obeying $\mathrm{div}\, \boldsymbol{m} = (\partial m_x/\partial x + \partial m_y/\partial y) = 0$. It reproduces van den Berg's solution for the thin plate, including his discontinuous (singular) domain walls, which correspond to the sharp ridges of the sandpile. A non-singular solution is generated by agitating the pile a bit, thus rounding the ridges. This modified sandpile displays slopes either of the standard value of the material, or smaller values (on the rounded ridges). For the magnetization field this means $m_x^2 + m_y^2 \leq 1$. We only have to set $m_z = \sqrt{1 - m_x^2 - m_y^2}$ to end up with a solution that fulfils all conditions. The rounded ridges generate smooth domain walls instead of discontinuous boundaries. The degree of rounding determines the wall width. Examples demonstrating this construction for a number of cross-section shapes are shown in Fig. 3.28.

Fig. 3.28 The construction of stray-field-free vector fields for prismatic bodies of constant cross-section, based on the slope of smoothed sandpiles. The top views simulate Kerr images, while the lateral profiles display the continuous "domain wall" transitions

Whether similar stray-field-free and non-singular vector fields with a constant modulus also exist for irregular extended or toroidal bodies, such as rods with varying cross-sections, is not known. The situation may be analogous to the corresponding thin film case in which "eddies" and extended flux systems are combined, as demonstrated for two dimensions in Fig. 3.24. Surface point singularities can probably not be avoided if closed-flux eddies have to exist. For a three-dimensional pattern in an infinitely extended thin film see [596].

3.3.4 The Effect of Anisotropy in Soft Magnetic Materials

Magnetization patterns structured only by a few singular points (or swirls) and lacking conventional domain walls can hardly be classified as domain patterns. The calculated structures shown in Fig. 3.21b,c, as well as Arrott's model (Fig. 3.26), are examples of such free-flowing magnetic microstructures in three dimensions. The question is which role such "anomalous" magnetization patterns might play in larger samples in which the magnetic microstructure is primarily dominated by anisotropies.

We consider materials with well-defined anisotropies, but these anisotropies shall be small compared to the stray field energy constant K_d (i.e. $Q \ll 1$). In order to receive guidance from experiment, let's look at what happens if anisotropy is superimposed onto a previously free-flowing magnetization pattern. Such an experiment can be done with low-anisotropy thin film elements. If the material is magnetostrictive, a superimposed strain (induced by bending or compressing the substrate) can introduce an adjustable anisotropy, as shown in Fig. 3.29. Increasing the effective anisotropy, the volume of those zones shrinks in which the magnetization is disfavoured by anisotropy, and conventional domains are formed in the rest of the sample. The transition from freely flowing to conventional, regular domains will depend on the element size. The larger the element, the less anisotropy is needed to induce a regular pattern. For small samples there exists a transition from a simple, vortex-like state to a well-defined domain state, which will depend on the ratio between the sample size and the Bloch wall width parameter $\sqrt{A/K}$.

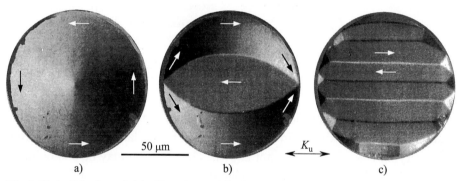

a)   b)   c)

Fig. 3.29 A disk-shaped thin film element displays an isotropic magnetization pattern (a) following van den Berg's principles. A compressive stress introduces anisotropy (b, c) [591]. The domain patterns were brought to equilibrium by demagnetization in an alternating field oriented parallel to the stress-induced easy axis (b) and perpendicular to it (c). Although the resulting domain pattern depends on the demagnetization history, both (b) and (c) can be classified as domain patterns in the classical sense, in contrast to (a). The sample is a nanocrystalline iron-Permalloy multilayer system of 300 nm total thickness behaving like a single film [597]. The images were taken with the magneto-optical Kerr effect with a vertical sensitivity direction

Quite instructive is the opposite approach: consider a magnetic crystal with an anisotropy $K \ll K_d$ and of a size such that all dimensions are very large compared to the Bloch wall width parameter $\sqrt{A/K}$. Then in the interior a domain structure will be formed that occupies easy directions only, and these domains are joined so that no magnetic stray fields are generated. Near misoriented surface zones, however, the two requirements of using only easy directions and avoiding stray fields may be incompatible when a surface does not contain an easy direction. In such a case the sample tries to find a compromise by using a domain branching scheme, a phenomenon discussed in more detail in Sect. 3.7.5. The principle can be seen in Fig. 3.30. Here the magnetization follows easy directions everywhere, and no stray fields or strong deviations from the easy directions are generated, except in the domain walls and in a very thin surface zone!

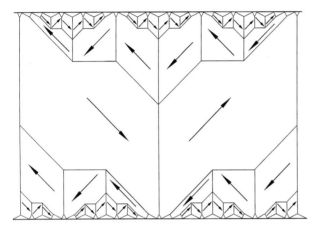

Fig. 3.30 The principle of domain refinement towards an unfavourable surface, demonstrated on the example of the "echelon pattern" in cubic crystal plates. For details consult Sect. 3.7.5C

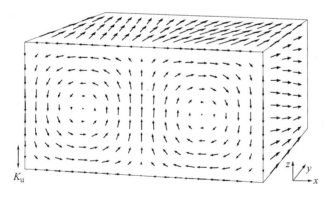

Fig. 3.31 The sinusoidal pattern (dense stripe domains, see Sect. 3.7.2) occurring above a critical thickness in a uniaxial film with weak perpendicular anisotropy. The film is thought to extend infinitely along x and y. Projections of the magnetization vector onto the three surfaces of the section are shown

Right at the surface, the magnetization must lie essentially parallel to the surface because $Q \ll 1$ (it needs not be uniform, however). Underneath, in a certain depth, it must be connected to the uppermost generation of the branching pattern. Although no rigorous calculation of such a surface zone is known, we may get an idea of the structure of the postulated "irregular" surface zone by comparing it with a thin film with perpendicular, weak anisotropy. Rigorous perturbation theory (Sect. 3.7.2A) of the nucleation mode in such films yields the pattern of Fig. 3.31 beyond a critical thickness of $D_{cr} = 2\pi\sqrt{A/K_u}$, which is twice the classical wall width. Because such a thin film can be seen as two surface zones joined together, it appears plausible to assume a continuously varying surface zone also under a misoriented surface with a thickness of the order of the wall width parameter $\sqrt{A/K_u}$.

Experimental observation indicates that the argument in favour of a continuous surface zone does not apply to *slightly* misoriented surfaces. Here refined regular domains — so-called supplementary domains — extend up to the surface. We need a continuous micromagnetic model of the surface only when the surface domain width of the supplementary domains approaches the scale of the wall width. The complexities of slightly misoriented surfaces are treated again in Sect. 3.7.1 (theoretical analysis) and in Sect. 5.3.4 (experimental observations). Apart from this special case, the concept of a layer of improper domains beneath "misoriented" surfaces appears plausible, although it cannot be considered as really confirmed. It will be discussed in more detail in Sect. 3.7.4 under the heading "Closure Domains".

3.3.5 Résumé: The Absence and Presence of Domains

The discussion of the last sections may be summarized in the following picture: continuous, "flowing" magnetization patterns will fill soft magnetic bodies only for a certain size range that depends on the material parameters. The element should be comparable in size or smaller than the wall width parameter $\Delta = \sqrt{A/K}$ (where K is a residual anisotropy that will never be exactly zero), while it should be large compared to the characteristic length of the stray field $\Delta_d = \sqrt{A/K_d}$. If the latter condition is not met, a uniform magnetization direction is enforced. If a sample is much larger than Δ it will contain a classical domain pattern in most of its volume. Continuously flowing, anomalous magnetization structures may be found near unfavourably oriented surfaces in a length scale between the two exchange lengths Δ_d and Δ.

Let us try to get an overview of the behaviour of the magnetic microstructure at least for one special case, that of cubical particles with a uniaxial anisotropy oriented parallel to one of the cube edges. Two parameters characterize this problem, as shown in Fig. 3.32: the reduced magnetic anisotropy $Q = K_u/K_d$ and the reduced particle size $\delta = D/\Delta_d$. Qualitatively, we expect three kinds of micromagnetic states: the single-domain state (SD) for small particle sizes, the regular multi-domain state (MD) for large particles, and for low-anisotropy particles an intermediate, continuously flowing state which we may call the vortex state (V). To find the V-MD-boundary we define (somewhat arbitrarily) a multi-domain state to consist of at least *three* domains. The vortex state and multi-domain states are clearly different in character for small Q: In a vortex state stray field energy is largely avoided at the expense of exchange stiffness energy, while anisotropy plays no role. In multi-domain states the much weaker anisotropy energy begins to become important so that these states resemble ordinary domain states. This difference in character vanishes for large Q, but all states can be continuously followed towards rising Q.

The phase diagram shown in Fig. 3.32 is based on finite element micromagnetic calculations [598]. For large Q we found earlier in Sect. 3.3.2B the SD–MD limit at $l_{SD} = 6.15\, \gamma_w/K_d$, which, after inserting $\gamma_w = 4\sqrt{AK_u}$, leads to a phase boundary at $\delta_{SD} = 24.6\sqrt{Q}$. For small Q the phase boundary between the single-domain (or "flower") state and the vortex state was found in [581] (cf. the correction in [582]) in agreement with new results at about $l_{SD} = 6.8\,\Delta_d$ or $\delta_{SD} = 6.8$ for $Q = 0.02$.

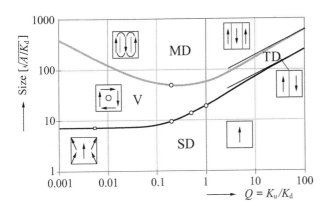

Fig. 3.32 Simplified phase diagram for the presence or absence of domains, applicable to cubical particles with a uniaxial anisotropy: SD single-domain state, V vortex state, TD two-domain state, MD multi-domain state. Based on micromagnetic simulations with *W. Rave*, IFW Dresden and *K. Fabian*, Univ. of Bremen. See [598] for details

As indicated, the V-MD-boundary in the low anisotropy case is driven by anisotropy. The demagnetizing energy is very small in the vortex state and will hardly be further reduced by splitting the domain pattern. The gain in anisotropy energy has to be paid by an extra wall or exchange energy. We therefore suspect that this boundary will scale with the classical domain wall width and will probably be found near $l_{MD} = (10-20)\Delta$ or $\delta_{MD} = (10-20)/\sqrt{Q}$. For high-anisotropy particles ($Q \gg 1$) the transition from the two-domain state to the first three-domain state was found in Sect. 3.3.2B at $15.2\gamma_w/K_d$ corresponding to $\delta_{MD} = 60.4\sqrt{Q}$. Data points for $Q \approx 1$ were computed numerically and they turn out to be compatible with the expectation derived from domain theory for small and for large Q. This kind of a phase diagram will eventually have to be extended to other particle shapes and anisotropy functionals, to external fields and stresses, and to different magnetic environments. The important point is that a wide gap exists for low-anisotropy (small Q) materials between the single-domain limit that scales with Λ_d, and the regime of regular domain patterns, the boundary of which scales with Δ. In this gap, regular, classical domains are expected to give way to continuously flowing magnetic microstructures and there are arguments as discussed before, that this is true not only for small particles as analysed in Fig. 3.32, but also for surface zones or other parts of bulk material in which regular domain patterns cannot exist because of the conflicting influences of anisotropy and the pole avoidance principle.

To conclude, domains are a widespread phenomenon in ferromagnets, but they do not appear to have a common origin. They may exist to reduce the stray field energy, or to adapt to local anisotropies or to the sample shape — depending on the material constants and the size of the sample. Magnetic domains are not a universal feature of ferromagnetic materials. Small particles contain no domain pattern even if they are commonly called "single-domain particles". In low-anisotropy materials there is an intermediate range in which continuous micromagnetic vortex states, rather than classical domains, prevail.

Even if anomalous, continuous magnetic microstructures do not look like domains, this does not mean that domain models, in which continuous transitions are replaced by domains and walls, are useless. Such models offer a first understanding for a complex situation and can be expected to be very useful if properly selected and evaluated, as will be demonstrated in Sect. 3.7.4

In all cases in which classical domains are present, they obey the same basic laws. One of the goals of this book is to elucidate the common properties of magnetic domains in different materials.

3.4 Phase Theory of Domains in Large Samples

We met equilibrium magnetization curves for small, high-anisotropy particles in Fig. 3.19b. In calculating these curves only the specific energy of the domain walls was needed, not their particular structure. In another limiting case, that of large, soft magnetic samples, the properties of domain walls become completely negligible. Only the phase volumes of different, coexisting domain classes are important, the particular arrangement of the domains is irrelevant. "Phase theory" describes these phenomena and was used in a simplified, one-dimensional version in Sect. 3.3.1 justifying the formation of domains. Here this approach is developed systematically, taking fully into account the three-dimensional character of magnetization processes. The result is a reversible, vectorial magnetization curve for extended samples of largely arbitrary shape. This magnetization curve is expected to agree with the "idealized" *anhysteretic* magnetization curve of an extended sample. It is obtained by superimposing decreasing alternating fields during the measurement, thus removing the hysteretic, irreversible contributions to the magnetization curve. In the general case of hysteretic behaviour of a material, the idealized magnetization curve acts as a reference on which domain-wall-related irreversible magnetization phenomena are superimposed.

3.4.1 Introduction

Magnetization phases are defined in a similar sense as phases in metallography or in thermodynamics. All domains magnetized in the same direction are gathered into a *phase*, which is characterized only by its volume and its magnetization direction. Interface (wall) energies are neglected. This approximation is valid if the sample is sufficiently large at least in one dimension. Stray field and magneto-elastic interactions are considered only globally in the sense of demagnetizing fields and their elastic analogue. Neglecting internal stray fields is reasonable for *soft* magnetic materials, if the domain walls can occupy their equilibrium orientation. Neglecting internal magneto-elastic interactions is generally possible except for special cases as will be indicated.

Both the external field and the demagnetizing field are assumed to be uniform. The latter is justified if the sample shape is ellipsoidal or at least approximately so, including the limiting cases of infinite plates and cylinders. It may be possible to generalize the treatment to more complex shapes by applying phase theory to several interacting parts of a body, but we will not pursue this possibility. Finally we postulate that the phase volumes can freely reach their optimum values, thus ignoring any coercivity or irreversibility effects. Phase theory therefore applies to *extended, ellipsoidal, homogeneous* and *soft* magnetic materials. The detailed geometrical arrangement of domain patterns is ignored, but the spatial periods of the implied domains must be small compared to the sample size in all dimensions.

Magnetization-induced elastic stresses have to be considered only if a crystal is embedded in a matrix or in an environment of other crystals. This is important, for example, for polycrystalline materials. We will ignore this case here and refer to the literature [599].

Phase theory dates back to work by *Néel* [600] and by *Lawton* and *Stewart* [601]. The phase concept has also been discussed in great detail by *Träuble* [504, p. 203 – 232], by *Pauthenet* et al. [602, 603], by *Birss* et al. [604, 605], and by *Baryakhtar* et al. [606]. The mathematical foundations of phase theory were analysed by *de Simone* [607]. The same thermodynamic theory applies to superconductors, antiferromagnets and many other material classes, a point of view elaborated in a general manner in [606].

3.4.2 The Fundamental Equations of Phase Theory

The only variables entering phase theory are the magnetization directions $m^{(i)}$ and the relative volumes v_i of a finite number of phases $i = 1 \dots n_{\mathrm{ph}}$. We assume $|m^{(i)}|^2 = 1$ for all i, and $\sum v_i = 1$. Then the total energy per unit volume of the sample can be written in the following form:

$$e_{\mathrm{tot}} = \sum_i v_i\, g\!\left(m^{(i)}\right) - J_{\mathrm{s}} H_{\mathrm{ex}} \cdot \overline{m} + K_{\mathrm{d}}\, \overline{m} \cdot N \cdot \overline{m} \ , \qquad (3.79)$$

where $g(m)$ is the generalized anisotropy energy including, if present, induced anisotropies and external stress energy contributions, J_{s} is the saturation magnetization, H_{ex} is the external field, K_{d} is the stray field energy constant $J_{\mathrm{s}}^2/2\mu_0$, N is the demagnetizing tensor and $\overline{m} = \sum v_i m^{(i)}$ is the average magnetization.

(A) Infinite or Flux-Closed Bodies. In the first step we disregard the demagnetizing field. We may call this the case of the infinite body. To realize such a condition in practice is not impossible. A cube with interleaving yokes short-circuiting all three magnetization components would be a possibility (Fig. 3.33). Theorists use the concept of "circular" boundary conditions for this purpose.

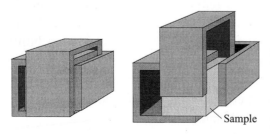

Sample

Fig. 3.33 A simulation of an infinite magnetic body. The three yokes (partially removed *on the right*) consist of ideally soft magnetic material, with sufficient thickness to allow saturation in all three directions. Coils can be wrapped around the yokes to apply fields

For a vanishing demagnetizing field the last expression in the energy (3.79) is zero and $H = H_{ex}$. Minimizing the energy e_{tot} with respect to $m^{(i)}$ under the constraints $|m^{(i)}|^2 = 1$, we obtain a vector equation for every phase:

$$T(m) - J_s H = f_L m \text{ with } T(m) = \text{grad}_m[g(m)] , m = m^{(i)}, i = 1...P, \quad (3.80)$$

a specialization of the micromagnetic equations (3.63). The torque $T(m)$ is derived from the anisotropy function $g(m)$, and f_L is a Lagrangian parameter.

The equilibrium equation (3.80) says that the magnetization vector in every phase is parallel to the total effective field in this phase, and these phases are independent of each other and not coupled, their volumes being indeterminate. Eliminating f_L by a scalar multiplication of (3.80) with m, and using the condition $m^2 = 1$, we obtain an implicit equation for m in a given field H:

$$T(m) - J_s H = m[m \cdot T(m) - J_s m \cdot H] , m^2 = 1 . \quad (3.80a)$$

For every field H one or more magnetization directions m fulfilling (3.80a) may be found. These are the possible phases, but according to our assumptions (Sect. 3.4.1) only the phase with minimum energy can exist in equilibrium. Normally, this will be just one unique magnetization direction, the (field-dependent) generalized easy direction. If, however, there is more than one magnetization direction with the same lowest energy, then they are allowed to coexist. These exceptional cases, which indicate first-order phase transitions between distinct phase states, turn out to be the important cases leading to domain formation in phase theory.

By which general method can the possible domain states be identified? Assume a field of a certain absolute strength but of arbitrary direction. For every field direction we determine the easy direction(s) from (3.80a) and sort the solutions into classes. Starting from one solution, we assign to the same class all solutions that can be reached *continuously* from the starting solution by changing the field direction. If we plot the classes on a map (Fig. 3.34a), they occupy areas in field space that are delineated by boundaries, at which two easy magnetization directions have the same energy. In passing through the boundary a vectorial magnetization jump occurs from one easy direction to the other if only uniform states are permitted. The magnetization may, however, also vary continuously through such a transition by using mixed phase states, as explained in Fig. 3.34b. The two-phase boundary lines of Fig. 3.34a form *interfaces* in field space if we allow the field strength to vary. There are also triple *lines*, in which the areas of three classes (or three two-phase boundaries) meet. In special cases more than three classes may meet at such a line. An example is the [100] field direction in negative-anisotropy cubic materials, where four adjacent $\langle 111 \rangle$ directions are energetically equivalent. But apart from such special-symmetry cases, four magnetization classes will meet only at a special field *point*. In general, the four (or more) magnetization directions that are in equilibrium at such a field point will not lie in one plane. An example is the point $H=0$ in multiaxial materials where all (zero-field) easy directions are equivalent.

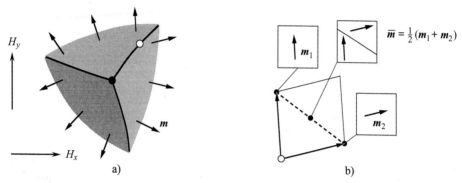

Fig. 3.34 (a) Map of magnetization directions as a function of field directions for a fixed absolute field value. The magnetization vectors in a class are continuous. At the boundaries, members of different classes have the same energy. The average magnetization of mixed states lies on the line connecting the tips of the magnetization vectors of the pure states (b). This representation applies to two dimensions, but it can be readily generalized to the three-dimensional case in which the phase boundaries are surfaces in field space

The class boundaries must not always end in a triple line meeting other boundaries. They may also end in a critical line, at which the distinction between the two phases vanishes. More on this subject is presented in Sect. 3.4.4.

(B) Finite, Open Samples. Up to this point the phase volumes v_i are indeterminate since they do not appear in the internal equilibrium equations (3.80). If a point in field space is thermodynamically compatible with more than one magnetization direction, any combination of these magnetization directions yields the same energy. This is no more true in the general case of (3.79) which includes the demagnetizing effect. Minimizing the energy (3.79) with respect to the magnetization direction we get:

$$T\bigl(m^{(i)}\bigr) - J_{\mathrm{s}} H_{\mathrm{ex}} + 2 K_{\mathrm{d}} \overline{m} \cdot N = f_{\mathrm{L}} \, m^{(i)} \ . \tag{3.81}$$

By the definition of the average magnetization $\overline{m}$ these equations couple the magnetization directions $m^{(i)}$ and the volumes v_i of the phases. Minimizing (3.79), this time with respect to the v_i, we get a second set of equations:

$$g\big(m^{(i)}\big) - J_s H_{ex} \cdot m^{(i)} + 2K_d\big(\bar{m} \cdot N\big)m^{(i)} = f_L^{(2)} \tag{3.82}$$

where the second Lagrange multiplier $f_L^{(2)}$ stems from the constraint $\sum v_i = 1$.

Although equations (3.81–82) look more complicated than (3.80), it is possible to reduce the general case to the simple case without demagnetizing field. This is achieved by introducing the *internal* field defined by:

$$H_{in} = H_{ex} - \big(J_s/\mu_0\big)\bar{m} \cdot N \ . \tag{3.83}$$

Inserting H_{ex} from its definition (3.83) into the first general equilibrium condition (3.81) and using the definition of $K_d = J_s^2/2\mu_0$, we end up again at the internal equilibrium equation (3.80), now interpreting the field H as the internal field! According to (3.80) every possible phase must be in equilibrium with the common internal field derived for the "infinite" body. Moreover, we may insert the definition (3.83) into the other equilibrium equations (3.82), yielding:

$$g\big(m^{(i)}\big) - J_s H_{in} \cdot m^{(i)} = f_L^{(2)} \ . \tag{3.84}$$

·All phases coexisting in equilibrium must have the same "internal" energy value $f_L^{(2)}$. A two-phase domain state can therefore only be stable if the internal field (3.83) lies on one of the mentioned boundaries in field space. These boundaries were derived for the closed-flux samples for which demagnetizing effects were absent. But this rule also applies to open samples of arbitrary ellipsoidal shape. In the same sense, three-phase domain states require the internal field to lie on one of the triple lines in field space. And non-degenerate four- or multiphase states are only stable if the internal field lies along the prescribed value (as e.g. the value $H_{in} = 0$ in multiaxial crystals).

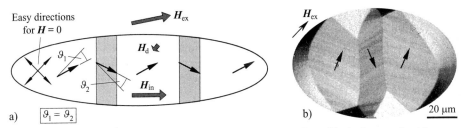

Fig. 3.35 A schematic picture of a domain structure in a flat elliptical crystal with easy axes as indicated (a). In spite of the asymmetry of the situation, the two angles ϑ_1 and ϑ_2 must be identical. An experimental observation demonstrates this behaviour (b): note the symmetric orientation of wall and magnetization directions in the basic domains in spite of the oblique applied field. (Sample: Permalloy element of 240 nm thickness with transverse easy axis; courtesy *M. Freitag*, Bosch)

Consider, as a demonstration, a regular symmetrical crystal in which the coexisting phases are crystallographically equivalent (Fig. 3.35a). If a moderate external field points into a direction that does not coincide with a symmetry axis, the domain boundaries are shifted until the developing demagnetizing field together with the external field results in an internal field, which lies on a two-phase boundary in field space. Here, the two phases coexist in a stable equilibrium. As a consequence, the magnetization rotation angles in coexisting stable phases are equal in magnitude and independent of particle shape and external field direction. This fact helps in discussing complex magnetization processes and in interpreting domain observations in an applied field.

(C) The Classification of Magnetization Processes within Phase Theory. We return to the problem of determining the magnetization phases and the average magnetization for a given external field and a given ellipsoidal sample shape. The above discussion of the properties of stable phases does not immediately help, since the unknown internal field can only be determined from the equally unknown average magnetization. Particular solutions of this problem are possible in cases of special symmetry [600–605]. Generally helpful is a classification of magnetization processes within phase theory as introduced by Néel [600].

We first note that any two domain states that comprise the same phases and that produce the same average magnetization $\bar{m}$ cannot be distinguished from the viewpoint of phase theory. To see this we insert H_{ex} from (3.83) into the energy expression (3.79) and make use of the definition $\bar{m} = \sum v_i m^{(i)}$, arriving at:

$$e_{tot} = \sum v_i \left[g(m^{(i)}) - J_s H_{in} \cdot m^{(i)} \right] - K_d \bar{m} \cdot N \cdot \bar{m} \quad . \tag{3.85}$$

As shown above in (3.84), the expression in the square bracket must be identical for all coexisting phases and the last energy term is only sensitive to the average magnetization. So it is sufficient to characterize the possible domain states by their average magnetization and the common internal field. Graphically, all states can be represented by a point inside or on the surface of the unit sphere of average magnetization vectors (Fig. 3.36). The saturated states lie on the surface of this sphere. All demagnetized states are collected in the centre. The magnetized states possible at zero internal field occupy a polyhedron that is spanned by the (zero-field) easy directions of the sample in analogy to the construction of Fig. 3.34b. In iron this is an octahedron. In nickel (negative cubic anisotropy) we find a cube. Since, in general, more than four energetically equivalent easy directions exist at zero field, any average magnetization inside the polyhedron may be realized in different ways. Phase theory is not able to distinguish between the different possible realizations.

In the magnetization curves of phase theory different modes are distinguished. Magnetization processes that occur within the polyhedron by shifting the domain phases (i.e. the walls separating the domains) without changing the internal field, are classified as belonging to *mode I*. If there exists another particular point in field space in which four or more non-planar magnetization directions are energetically equivalent, then these states define another special volume in the unit sphere of average magnetization. All states in this volume can exist at the same internal field in the same sense as all states in the polyhedron can exist at $H_{in} = 0$. The corresponding magnetization processes are again classified as belonging to mode I. As magnetization processes in mode I occur by definition at constant internal field, they are connected with a jump in the magnetization curve $J(H_{in})$ as in Fig. 3.18b. For the same reason only wall displacement processes occur in mode I, while the magnetization directions of all participating phases stay constant.

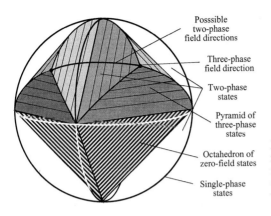

Posssible
two-phase
field directions

Three-phase
field direction

Two-phase
states

Pyramid of
three-phase
states

Octahedron of
zero-field states

Single-phase
states

Fig. 3.36 The sphere of possible average magnetization directions of domain states for a cubic material with positive anisotropy (simplified). Below the equator only the octahedron of field-free states is shown

For large fields only the saturated states at the surface of the unit sphere are stable. The space between the polyhedron and the surfaces is filled with three- and two-phase domain states. First consider the three-phase states, or, more generally, those states for which the tips of the magnetization vectors of three or more equilibrium phases lie in a plane (coplanar states). Any average magnetization denoted by a point on such a plane in the area between the possible magnetization directions can be realized by suitable combinations (domain states). With increasing strength of the internal field along some predetermined direction (the $\langle 111 \rangle$ axes in the example of Fig. 3.36), a new set of (generalized) easy directions and therefore a new plane of possible domain states is defined. All the planes together fill a volume in the sphere of average magnetizations, which becomes a spherical pyramid for iron (Fig. 3.36). (Actually, in iron, the tip of the pyramid is occupied by one of the mentioned secondary four-phase volumes, in which the three easy directions are in equilibrium with the saturated state in the [111] direction. We come back to this refinement in the next section). For four coplanar easy directions (as for nickel when the field is applied along a $\langle 100 \rangle$ direction) everything remains the same, except that spherical pyramids with a quadrilateral base, instead of a triangular base, are formed. Magnetization processes employing coplanar magnetization phases are classified as belonging to *mode II*. Wall motion and magnetization rotation processes occur simultaneously in mode II. An increasing net magnetization usually needs larger fields and rotation processes.

In principle, three-phase states might be possible even inside the mode I polyhedron of regular field-free multiphase states. But in that case an "inward" component of the internal field, opposite to the mean magnetization, would be necessary. Equation (3.85) shows that a state with such an opposite internal field has always a higher energy than a field-free state if both have the same average magnetization (the last term is the same, the first term is slightly larger in a "field state", but also the middle term is positive since $\boldsymbol{H}_{\text{in}} \cdot \overline{\boldsymbol{m}}$ has the wrong, negative sign, leading altogether to an increase in energy rather than to an energy gain). This means that in small and moderate external fields, mode I magnetization processes with zero internal field have a preference over higher modes as long as the zero field processes are possible. States with a reverse field are compatible with the equilibrium equations (3.80), but they are thermodynamically unstable.

The space between the pyramids and the surface of the sphere is filled with two-phase states. For every allowed value of the field, a line of domain states connecting the two respective easy directions is obtained. Since a two-dimensional manifold of fields is allowed as discussed above (namely all field vectors pointing to one of the *boundaries* in field space), the remaining volume in the sphere of average magnetization can also be filled. The corresponding segments of the magnetization curve are defined to belong to *mode III*. By the same argument as before, two-phase states are not stable if a three-phase state can produce the same average magnetization. In other words, mode II has a preference over mode III if both are possible for the same field.

Thus we have a one-to-one mapping between all possible and distinguishable domain states, which are represented by the points inside the sphere of average magnetization, and the points in field space required for these states. With increasing field a sample passes in general through a sequence of magnetization modes: mode I using multiphase field-free states inside the polyhedron, mode II using three-phase states in the pyramids, mode III with the two-phase states in the prisms, and *mode IV* based on single-phase states on the surface of the sphere. Depending on field direction and crystal symmetry some modes may be absent, as in magnetically uniaxial crystals, where only modes III and IV can occur.

(D) Calculating (Anhysteretic) Magnetization Curves. Based on the classification of magnetization processes discussed in (C), the magnetization curves within phase theory can be calculated in a systematic way: in mode I, the case of multiple, energetically equivalent magnetization directions spanning a volume in the unit sphere at a constant field H_0; this calculation is simple and explicit, leading to a linear segment in the magnetization curve. Firstly, we have to calculate the polyhedron of magnetization states available for this magnetization mode. Let $m^{(1)}$, $m^{(2)}$ and $m^{(3)}$ be any three of these directions at the given field. Then the vector $\bar{m} = v_1 m^{(1)} + v_2 m^{(2)} + v_3 m^{(3)}$ with $v_i \geq 0$ and $v_1 + v_2 + v_3 = 1$ defines a surface of the polyhedron. Combining all possible equilibrium phases in such sets of three components we obtain a three-dimensional body, the polyhedron introduced in (C).

For positive cubic anisotropy (as for iron), where the easy directions are mutually perpendicular, the condition defining the polyhedron simplifies to a nice analytic expression given by *Becker* and *Döring* [16]: let a_i be the components of a unit vector relative to the cubic axes, then the maximum average magnetization along the direction of a is given in the field-free state by $m \cdot a = 1/(|a_1| + |a_2| + |a_3|)$.

Returning to mode I we have to test the compatibility of a given external field with a zero internal field state. To this end we invert (3.83) assuming $H_{in} = H_0$ and check if the average magnetization $\bar{m} = \mu_0 (H_{ex} - H_0) N^{-1}/J_s$ lies within the polyhedron (N^{-1} is the inverse demagnetizing tensor). In the field range where this is true, the calculated value of $\bar{m}$ may be accepted, and in this field range $\bar{m}$ is a linear function of H_{ex}.

Mode IV, the single-domain rotation of the magnetization vector, can be calculated by writing down the total energy (3.79) for uniform magnetization:

$$e_{tot} = g(m) - J_s H_{ex} \cdot m + K_d m \cdot N \cdot m \ . \tag{3.86}$$

If the magnetization is expressed in polar coordinates, the angles ϑ and φ can be obtained numerically from the equilibrium equations $\partial e_{tot}/\partial \vartheta = 0$ and $\partial e_{tot}/\partial \varphi = 0$.

Similar equation systems can be formulated for the remaining modes. In mode II the internal field varies along a line in field space and can be characterized by its strength h alone. For every value of this field the corresponding equilibrium magnetization directions have to be determined. They are given by as many equilibrium equations as there are unknown magnetization angles. The phase volumes v_1 to v_3 are further unknowns. The corresponding equations are (3.83) (three equations), the condition $\Sigma v_i = 1$ and the equilibrium equations (as above for mode IV). Solutions of this set of equations are valid if all phase volumes are positive and if the total "internal" energy (3.84) is smaller than that of competing modes.

In the case of mode III the internal field varies along a two-dimensional manifold. Possible states can be characterized by the absolute internal field value and a field angle. The two field variables together with the two phase volumes of mode III and the magnetization angles in the allowed phases form a set of variables again. They are determined by an equivalent set of equations as for mode II and are examined along the same lines.

In this way the anhysteretic magnetization curves resulting from phase theory can be systematically calculated. This task is facilitated if the range and symmetry of the multiphase points, lines and surfaces in field space are known in advance, as demonstrated for cubic materials in the next section.

3.4.3 The Analysis of Cubic Crystals as an Example

(A) The Magnetic Classification of Cubic Crystals. Because of their high-symmetry cubic crystals have many particularly simple features and are accessible to analytic evaluation. They show some surprising properties, which give an idea of the complexities that have to be expected in general systems.

Depending on sign and size of the anisotropy coefficients, different types of cubic materials have to be distinguished. To see this, we start from the anisotropy functional (3.9) and the field energy (3.15). As discussed, we first ignore the demagnetizing energy. Reduced units are introduced according to:

$$\kappa_{c1} = K_{c1}/K_q, \ \kappa_{c2} = K_{c2}/K_q, \ h = HJ_s/2K_q, \ \varepsilon = e/K_q, \tag{3.87}$$

$$\text{with } K_q = \sqrt{K_{c1}^2 + K_{c2}^2}.$$

The quantity K_q cannot become zero except in a trivial case and is thus well-suited for normalization. The total reduced energy density then becomes:

$$\varepsilon_{\text{tot}} = \kappa_{c1}\left(m_1^2 m_2^2 + m_1^2 m_3^2 + m_2^2 m_3^2\right) + \kappa_{c2}\, m_1^2 m_2^2 m_3^2 - 2h \cdot m. \tag{3.88}$$

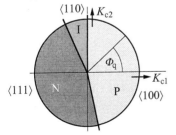

Fig. 3.37 The easy directions of cubic crystals as a function of the two anisotropy constants. The angle Φ_q can be used to characterize a material

There are three principal cases: if κ_{c1} is positive and $\kappa_{c2} > -9\kappa_{c1}$, then the $\langle 100 \rangle$ directions are favoured. If κ_{c1} is negative and $\kappa_{c2} > -\frac{9}{4}\kappa_{c1}$, then the $\langle 110 \rangle$ directions are the easy directions. In all other cases, the $\langle 111 \rangle$ directions are preferred. In the degenerate case of $\kappa_{c1} = 0$ and $\kappa_{c2} > 0$, all directions in all {100} planes are equally favoured. The other two boundary cases lead to multiple but discrete easy directions: for $\kappa_{c1} < 0$, $\kappa_{c2} = -\frac{9}{4}\kappa_{c1}$, all $\langle 111 \rangle$ and $\langle 110 \rangle$ directions are energetically equivalent, and for $\kappa_{c1} > 0$, $\kappa_{c2} = -9\kappa_{c1}$, this is true for all $\langle 100 \rangle$ and $\langle 111 \rangle$ directions. Figure 3.37 shows a sector plot of the easy directions as a function of the "cubic anisotropy angle" Φ_q defined by $\cos\Phi_q = \kappa_{c1}$ and $\sin\Phi_q = \kappa_{c2}$.

In the positive-anisotropy region P, the "polyhedron" of field-free average magnetization directions is the mentioned octahedron. In region N the $\langle 111 \rangle$ directions form a cube. In region I a 14-faced polyhedron is formed by the $\langle 110 \rangle$ directions (the surface consisting of six squares centred at the $\langle 100 \rangle$ directions and of eight equilateral triangles centred at the $\langle 111 \rangle$ directions). The polyhedron corresponding to the boundary between regions P and N in Fig. 3.37 is a spherical octahedron. The other two limiting cases lead to bodies with multi-faceted surfaces.

(B) Magnetization Curves. In Fig. 3.38 the internal field magnetization curves along the high-symmetry directions for various cubic anisotropy angles Φ_q are compiled. We can use these curves to derive the saturation fields and critical fields needed in phase theory.

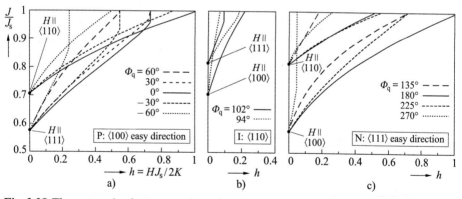

Fig. 3.38 The magnetization curves (as a function of the internal field) along the highly symmetric directions for the three different types of cubic materials. Parameter is the cubic anisotropy angle Φ_q defined in Fig. 3.37

To calculate the magnetization curves we apply the expressions for the anisotropy energy (3.9a–c), add the field term (3.15), and minimize the total energy with respect to the polar angle ϑ for given reduced fields h, choosing the appropriate azimuthal angle φ for each case. This leads to implicit equations for the magnetization curves. Note the discontinuities in the approach to saturation visible in Fig. 3.38a for $H \parallel \langle 111 \rangle$. This feature, which was noted by Becker and Döring [16 (p.118)], is discussed in more detail in (C).

The magnetization curves along the symmetry directions yield most of the information needed to calculate the possible domain states. They render immediately the simple three-phase (coplanar) states, such as the ones with the field along [111] in Fig. 3.38a or along [100] in Fig. 3.38c. The calculation of the collinear or two-phase states is facilitated by symmetry considerations. Figure 3.39 shows results for iron, for which the field directions for two-phase states lie between the [110] direction ($\eta_h = 0$) and the [111] direction ($\eta_h = 35.2°$). The two magnetization angles ϑ and φ [see (3.7a)] are plotted with the field strength h and the field elevation angle η_h as parameters.

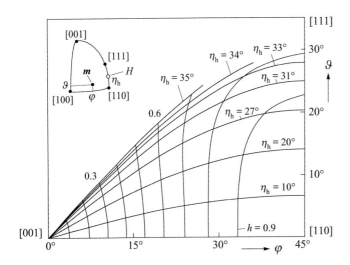

Fig. 3.39 The magnetization angles ϑ and φ as a function of the field angle η_h and the field magnitude $h = HJ_s/2K$ for the two-phase states of a positive-anisotropy material like iron ($K_{c2} = 0$). The angles apply to one of the two equilibrium states, the other one is symmetrical with the same polar angle ϑ and the azimuth $90° - \varphi$

(C) The [111]-Anomaly. In Figs. 3.36 and 3.39 the details near saturation along [111] were omitted. As shown in Fig. 3.38a the magnetization curves along the [111] direction display a discontinuity. The jump occurs at a reduced field $h = 0.7328$ (for $K_{c2} = 0$) between magnetization states, which are derived from the $\langle 100 \rangle$ easy directions (15.6° off [111]) and the [111] saturated state. Between the $\langle 100 \rangle$-derived states and the [111] state exists an energy barrier, if we admit uniform magnetization only. The height of this barrier (to be precise, of the lowest saddle point on this barrier) is calculated as $5.2 \cdot 10^{-4} K_{c1}$. For K_{c2} not equal to zero we find a qualitatively similar situation, although the numbers differ. As long as $|K_{c2}|$ is smaller than $9K_{c1}$, a magnetization jump is still formed as shown for some examples in Fig. 3.38a.

We expect domain states to appear in connection with every magnetization jump under proper conditions (see Sect. 3.3.1). This must also apply to the magnetization discontinuity close to saturation discussed above. To calculate the boundaries of these domain states, we cannot rely on symmetry relations because the participating ⟨100⟩-derived phases do not occupy any special symmetry directions. We have to perform a general analysis, beginning with locating the three- and two-phase boundaries in field space in the neighbourhood of the [111] saturation field point. If the field vector ends on one of these phase boundaries, the corresponding ⟨100⟩-derived phase is in equilibrium with the [111]-derived phase. It turns out that these boundaries are confined to a field region of about one degree around [111].

Figure 3.40 shows a field map as discussed in Sect. 3.4.2A, indicating the possible domain states. To calculate a point on a two-phase boundary, altogether five variables must be determined simultaneously, namely two magnetization directions (four angles) and one field angle. Note that for the required oblique field direction we know neither the magnetization direction of the [111]-related phase nor that of the [100]-related phase in advance. The corresponding five equations are: (i–iv) the equilibrium conditions for the two magnetization angles in both phases, and (v) the condition of equal energy. Symmetry allows the calculation to be limited to a field azimuth range of 60°.

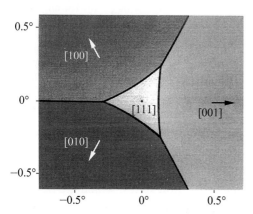

Fig. 3.40 Field map for the neighbourhood of the [111] direction, showing the stability range of the [111] phase. The *grey shades* give an impression of the variation of the magnetization direction as a function of the field angle for a given value of the size of the field ($h = 0.744$). The *arrows* indicate that the magnetization directions corresponding to fields in the three areas are derived from the respective crystal axis. Only a small interval of field directions ($\pm 0.5°$) around [111] has to be shown

Similarly, for a point on a triple line eight simultaneous equations must be solved. The map shown in Fig. 3.40 was obtained by a numerical search procedure, followed by a Newton's method solution of the applicable set of equations. The possible field directions for anomalous two-domain states are given by the boundaries of the triangle in this diagram, anomalous three-domain states need a field along the corners. More details about these states follow in the next section in connection with the discussion of critical points.

There is a possibility that could suppress the anomalous domain states. The [111] phase on the one hand and the $\langle 100 \rangle$-derived phases on the other hand are elastically incompatible for all allowed charge-free wall orientations. To find out whether this elastic incompatibility can suppress the formation of domains, the magnetostrictive self-energy in these domains has to be determined. Evaluating (3.57) with the material constants for Fe3%Si at room temperature $C_2 = 4.2 \cdot 10^{10}\,\mathrm{N/m^2}$, $C_3 = 16 \cdot 10^{10}\,\mathrm{N/m^2}$, $\lambda_{100} = 21 \cdot 10^{-6}$, $\lambda_{111} = -9 \cdot 10^{-6}$, we obtain a minimum magnetostrictive self-energy of $1.3\,\mathrm{J/m^3}$. This is smaller than the energy barrier for uniformly magnetized states calculated above to $5.2 \cdot 10^{-4} K_{c1} = 18\,\mathrm{J/m^3}$ (with $K_{c1} = 3.5 \cdot 10^4\,\mathrm{J/m^3}$). For this material and at room temperature, the anomalous domains near the hard direction should therefore be observable, provided experiments with a sufficient accuracy in field direction adjustment are performed. The predicted near-saturation domains should not be suppressed by internal magnetostrictive stresses for iron or silicon-iron. There are certainly materials or temperatures, for which the magnetostrictive energy is relatively stronger, suppressing the domains. For iron we may neglect magnetostrictive effects also in the analysis of critical points, where we come back to the near-saturation domain effects in Sect. 3.4.4B.

3.4.4 Field-Induced Critical Points

For a complete description of the phase behaviour of a material in an applied field not only the directions of the allowed fields for multi-domain behaviour, but also their range must be known. By increasing the field in a direction for which more than one phase can exist in equilibrium, one will eventually reach a maximum value at which the number of phases changes discontinuously. If the magnetization vectors in the phases move continuously towards each other to coalesce at the maximum field point, this point is called a *critical point* — as in general thermodynamics and in the Landau theory of phase transitions [608]. No time-dependent fluctuations ("critical phenomena") at micromagnetic critical points are expected because of the long-range nature of some of the micromagnetic interactions, but the character of magnetic microstructure near a critical point may be different, making phase theory inapplicable (see Sect. 3.7.3). Here we ignore these complications in order to get an overview. We first discuss ordinary, highly symmetric critical points in (A), and then analyse in (B) a case in which symmetry arguments are not available.

(A) The Ordinary Critical Points of Iron. Figure 3.41 shows magnetization curves along some directions of a cubic crystal. Among the curves there are two with critical points ([110] and [332]); in the curve with H along [110] the critical point coincides with saturation. In the other case (field along [332]) the magnetization process beyond the critical point consists of a continuous, single-phase approach to saturation. The other two curves in Fig. 3.41 display no critical points: the curve along [112] converges asymptotically towards saturation, the curve along [111] shows a first-order, discontinuous transition, the already discussed [111]-anomaly. In connection with this anomaly we will meet critical points that are not predetermined by symmetry conditions and we consider it instructive to analyse such a situation.

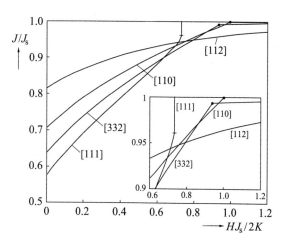

Fig. 3.41 The magnetization curves along several crystal directions for an infinite or flux-closed sample of a cubic material with $K_{c2} = 0$. Along the [332] and the [110] directions we observe a critical point, but only along the [110] direction saturation is reached at that point. The curves were calculated by optimizing the sum of anisotropy and field energy using a two-dimensional Newton method for the two spherical coordinates of the magnetization vector

How can critical fields be calculated? Let us first discuss the general case of a two-phase internal field direction. Slightly below the critical point the magnetization directions of the two equilibrium phases must be distinct but close together. As shown before in (3.84), they must have the same "internal" energy, and, since this must hold for all field values up to the critical point, it is sufficient to calculate the critical point in terms of the *internal field*. Once identified, it can be transformed into the corresponding external field for finite bodies by (3.83).

The Newton method, used for example in the calculation of Fig. 3.39, would fail in calculating the critical points, because the Jacobian to be inverted in this method becomes singular at a critical point. With standard procedures relying on the matrix of second-order derivatives of the energy functional, critical points can only be extrapolated from calculations at non-critical points. Here we are looking for a direct calculation of the critical point.

Focusing on iron-like materials the field for two-phase states must lie along the great circles connecting $\langle 110 \rangle$ and $\langle 111 \rangle$ directions (Fig. 3.36). At the critical point the magnetization must lie on the same great circle, although not necessarily at the same elevation angle ϑ compared to the field angle η_h if we use the polar

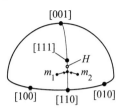

coordinate system of (3.7a). Above the critical point a single azimuthal angle φ must be stable, for example at $\varphi = \pi/4$. Therefore the second derivative of the total energy with respect to φ must be positive (the first derivative vanishes by symmetry). Below the critical point two minima appear when the second derivative becomes negative. At the critical point (h_{cr}, η_{cr}) the second derivative must be zero, leading to:

$$\partial^2 e_{tot}(\vartheta, \varphi)/\partial\varphi^2 \Big|_{\varphi = \pi/4} = 0 \ , \quad \partial e_{tot}(\vartheta, \varphi)/\partial\vartheta \Big|_{\varphi = \pi/4} = 0 \ ,$$

with $e_{tot}(\vartheta, \varphi) = \cos^4\vartheta \cos^2\varphi \sin^2\varphi + \sin^2\vartheta \cos^2\vartheta$
$$-2 h_{cr} \left[\cos\vartheta \cos\left(\varphi - \tfrac{\pi}{4}\right) \cos\eta_{cr} + \sin\vartheta \sin\eta_{cr} \right] \ . \tag{3.89}$$

A general solution of this system of equations by Newton's method would employ third-order derivatives. For cubic crystals the result can be derived analytically in terms of two implicit equations connecting the field angle η_{cr}, the field magnitude h_{cr} and the magnetization angle ϑ:

$$h_{cr} \cos\eta_{cr} = \cos^3\vartheta \ , \quad h_{cr} \sin\eta_{cr} = \tfrac{1}{2} \sin\vartheta \, (3 - 5 \sin^2\vartheta) \ . \tag{3.90}$$

The connection between the field angle η_{cr} and the magnetization angle ϑ can be written more explicitly as:

$$\tan \eta_{cr} = \tan\vartheta \, (\tfrac{3}{2} - \tan\vartheta) \ . \tag{3.90a}$$

Figure 3.42 shows the critical curve in field space, i.e. the connection between the two variables h_{cr} and η_{cr}. If the field is oriented along a $\langle 110 \rangle$ direction ($\eta_{cr} = 0$), the reduced critical field is $h_{cr} = 1$. This field is reduced for higher values of η_{cr}. It would reach the value of $\tfrac{2}{3}$ along the $\langle 111 \rangle$ direction ($\eta_h = 35.26°$) according to (3.90). In both cases (field along $\langle 110 \rangle$ and $\langle 111 \rangle$) the field direction and the magnetization direction fall together; in between they differ as described by (3.90a). However, the behaviour in the neighbourhood of the $\langle 111 \rangle$ directions has been simplified in this discussion. This point will be investigated in more detail in the following.

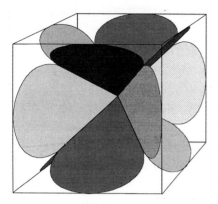

Fig. 3.42 The two-phase boundaries and critical curves in field space for positive-anisotropy cubic materials. The cubic frame is oriented along the unit cell of the material and the field vectors originate in the centre of the frame. The diameter along the diagonal [110] is 2 in units of the anisotropy field. The inside of the clover leaves indicates possible fields for two-phase states, their intersections mark possible fields for three-phase states, and their edges represent critical fields for the transition to uniform states

(B) General Equations for Critical Points. A more complicated set of two-phase boundaries exists in the neighbourhood of the [111]-anomaly, for which neither the field direction nor the magnetization directions are determined by symmetry. A critical point is then defined by four unknown variables: two magnetization angles and two field angles when the absolute value of the field is given. What are the four equations defining such a point? Two equations are obvious since the critical point must be in equilibrium with respect to an arbitrary rotation of the magnetization vector:

$$\partial e_{tot}/\partial \vartheta = 0 \ , \ \partial e_{tot}/\partial \varphi = 0 \ . \tag{3.91a}$$

Again a second derivative must vanish. It is not known in advance, in which direction the two phases develop, but one eigenvalue of the tensor of second derivatives must be zero at the critical point. This leads to the condition:

$$\left(\partial^2 e_{tot}/\partial \vartheta^2\right)\left(\partial^2 e_{tot}/\partial \varphi^2\right) = \left(\partial^2 e_{tot}/\partial \varphi \, \partial \vartheta\right)^2 \ . \tag{3.91b}$$

The last condition for the critical point involves the third derivatives. A point that fulfils all conditions (3.91a,b) can still be unstable if the third derivative along the "soft" axis, the one with the vanishing eigenvalue, is different from zero. A third-order term leads to instability because a decrease in energy can always be achieved by rotating the magnetization into that direction in which the third-order term in the Taylor expansion becomes negative. In general coordinates the condition that this should not happen becomes:

$$a^3\left(\partial^3 e_{tot}/\partial \vartheta^3\right) + 3\,a^2 b\left(\partial^3 e_{tot}/\partial \vartheta^2 \, \partial \varphi\right) + 3\,a\,b^2\left(\partial^3 e_{tot}/\partial \vartheta \, \partial \varphi^2\right) + b^3\left(\partial^3 e_{tot}/\partial \varphi^3\right) = 0 \ ,$$

with $a = \left(\partial^2 e_{tot}/\partial \varphi^2\right) - \left(\partial^2 e_{tot}/\partial \vartheta \, \partial \varphi\right) \ , \quad b = \left(\partial^2 e_{tot}/\partial \vartheta^2\right) - \left(\partial^2 e_{tot}/\partial \vartheta \, \partial \varphi\right) \ , \quad$ (3.91c)

in which the vector (a, b) points into the direction of the non-vanishing eigenvalue.

The conditions (3.91a−c) are necessary but not sufficient for a stable critical point. There exist conditions (stability limits) on the values of the remaining third derivatives and on the fourth derivatives, but if these conditions are not met, completely different magnetization states become stable. These will become evident in an initial global search procedure.

Figure 3.43 shows the critical lines in the neighbourhood of the [111] direction of Fig. 3.42 in more detail as calculated by a numerical solution of (3.91a−c). A multitude of domain states is predicted in the neighbourhood of the [111] saturation point. Although they are a straightforward consequence of the well-accepted anisotropy functional of cubic crystals, they have not been observed experimentally yet. One reason may be that the anisotropy functional has to be refined in the neighbourhood of the hard axis (the second-order anisotropy constant does not remove the discontinuity, however, as can be seen from Fig. 3.38). The other possible complication connected with magnetostrictive interactions was discussed in Sect. 3.4.3C and found not to be significant for iron. Because the angular range for the magnetic field in which these phases are expected is small, the absence of such observations may also have experimental reasons.

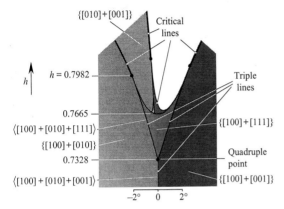

Fig. 3.43 The network of critical and triple lines for field vectors near the [111] axis omitted in Fig. 3.42. For some of the two-phase boundaries and triple lines the corresponding magnetization directions are schematically indicated. The notation [100] + [111] is meant to indicate that on this surface in field space a phase with a magnetization direction derived from [100] can be in equilibrium with a phase magnetized close to [111]. The horizontal scale indicates the deviation of the field direction from the [111] axis

Finally another interesting subtlety may be addressed: the critical lines in Fig. 3.43 form a network in which the lines are connected by branching points, the so-called tricritical points. One might expect a triple line to end in a double critical point, which would be critical with respect to both magnetization angles. But in such a point both eigenvalues of the second-order derivatives tensor and *all* third-order derivatives would have to vanish, which would lead altogether to nine conditions (two for the first derivatives, three for the second derivatives and four for the third derivatives). With only five variables (two magnetization angles and three field components) these nine conditions can only be fulfilled under special symmetry conditions. The point at $h = 2/3$ along the [111] direction in the simple positive-anisotropy material would be such a point *if* it were stable. As it is, materials like iron do not possess a double critical point. In contrast, simple negative-anisotropy materials ($\kappa_{c1} < 0$, $\kappa_{c2} = 0$) show such a point along the [100] direction at $h = 1$.

In iron the triple lines end first in a quadruple point, branching into further triple lines, and then end in tricritical points branching into critical lines. The tricritical points are given by a condition for the higher derivatives. In our case, in which the tricritical points lie on a symmetry plane, the condition is $\left(\partial^2 e_{tot}/\partial \vartheta^2\right)\left(\partial^4 e_{tot}/\partial \varphi^4\right) = 3\left(\partial^3 e_{tot}/\partial \vartheta\, \partial \varphi^2\right)^2$.

(C) Applications. Critical points are significant for domain studies. Imagine a domain state developing by reducing a field from saturation. If the field passes through a critical point, the domains will unfold at this point continuously. In all other cases a second phase can only be created by

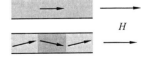

nucleation and growth. Often the final domain structure at zero field still displays this difference. Domains that developed by a continuous process can be finer and more regular than domain patterns developed by nucleation and growth. So the knowledge of the critical directions allows in some cases the control of the domain patterns in a way not otherwise possible. Conversely, if the observed domains are characteristic of their history, they can be used to determine anisotropy constants in an easy way. The remanent domain structure is observed as a function of the field direction in which it developed. If the domain structure that passed through the critical point can be distinguished from others which did not touch it, the critical directions and from it the character of the anisotropy functional can be derived. In [609] this method was applied to the analysis of a garnet film in which several anisotropies (uniaxial, cubic and a tilted orthorhombic anisotropy) existed simultaneously (the effect is demonstrated in Fig. 4.8). Other examples of domain observations related to critical points can be found in Chap. 5.

Critical points can also be measured directly. If a static field is applied to a sample and the *transverse* permeability is measured in a small superimposed alternating field, the permeability shows a peak when the d.c. field meets a critical condition. These peaks are characteristic for the anisotropy [610]. Furthermore, a magnetic sample in a critical state is very sensitive to small perturbations. In [219] a garnet film, transferred into such a state by an external field, was used to image magnetic impurities in non-magnetic ceramics.

3.4.5 Quasi-Domains

The phase concept can be extended in a useful although not perfectly rigorous way when strong demagnetizing effects restrict the allowed average magnetization directions to a subset of all possible directions. Take for example a thin soft magnetic crystal plate with surfaces that do not contain an easy direction. In the absence of an external field, every domain structure with a net magnetization component normal to the surface bears a high energy penalty and is virtually prohibited. It is thus permitted to require that the average normal magnetization component has to vanish locally in areas that are wide compared to the plate thickness. This means that only certain combinations of the easy directions are allowed in the fine structure of the domains. These combinations can be treated as new phases — the quasi-phases — from which the overall domain structure can be combined.

Every quasi-phase has an average magnetization vector oriented parallel to the surface. Its internal structure consists of subdomains magnetized along the true easy directions. The composite domain pattern made up of quasi-domains with net surface-parallel magnetization resembles regular domains. This concept was derived from experimental observations [86, 611].

Figure 3.44 shows the situation for a negative-anisotropy cubic crystal plate with (100) surfaces. The easy axes are along $\langle 111 \rangle$. Combining, e.g. [111] and [11$\bar{1}$] phases with equal volumes yields $\overline{m} = \sqrt{\frac{1}{3}}$ [110]. Another class of quasi-domains is generated by combining, e.g. [111] and [1$\bar{1}\bar{1}$], leading to $\overline{m} = \sqrt{\frac{1}{3}}$ [100]. The first class is energetically favoured by the small internal wall angle. The second class of quasi-domains is better adapted to the crystal edges for the given geometry. Using both classes, one may construct a domain structure of quasi-domains (Fig. 3.44) fulfilling all boundary conditions.

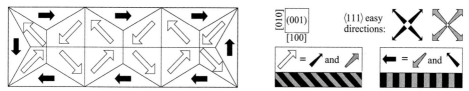

Fig. 3.44 A composite domain structure conceivable for a negative-anisotropy cubic crystal plate consisting of two types of quasi-domains explained schematically in the right part

The concept of quasi-domains can be applied to surfaces of arbitrary misorientation. The case of Fig. 3.44 is special in so far as the subdomains have equal weight to avoid a net flux out of the surface. For other surfaces the weight of the subdomains has to be adjusted accordingly.

To derive a phase theory of quasi-domains, the internal energy of the different quasi-domains must be known. This is not possible without knowledge of the micromagnetic details. For this reason the concept of a phase theory of quasi-domains cannot be followed further at this point. Examples of domain patterns composed of quasi-domains can be found in Sect. 5.5.6B.

3.5 Small Particle Switching

3.5.1 Overview

Small particle switching occurs under conditions opposite to those of phase theory. It happens in particles that display uniform magnetization or only a few domain walls (see Sect. 3.3.2). Phase theory applies to large crystals that contain many domain walls.

If a small particle is first saturated in a high field and the field is then reduced (and reversed, if necessary) it will in general switch at some point to the opposite magnetization direction or to some other state closer to equilibrium. The equilibrium states of small particles were discussed in Sects. 3.3.2 and 3.3.3A. Blocked transitions between these equilibrium states lead to *hysteresis*. The (almost) saturated state is usually metastable on approaching the equal energy limit, it cannot switch without some excitation as long as there is an energy barrier. The barrier may be overcome by thermal activation, depending on its height and shape. This leads to magnetic viscosity ("disaccommodation") and thermally induced loss of magnetization ("superparamagnetism"). In this book we focus on the idealized, athermal hysteresis and refer with respect to thermal effects to the literature [612–616]. The ideal hysteresis is in itself interesting and a prerequisite to an evaluation of the thermal effects.

The switching field is defined as the point of instability of the near-saturation state. It is also known as the *nucleation* field although not in all cases anything particular is nucleated at or close to this point (see the discussion in [514 (p. 184)]). The term "switching field" is more general and is commonly used in the fields of magnetic recording and of hard magnetic materials.

If the particles are small in units of the exchange length $\sqrt{A/K_d}$ (see Sect. 3.3 and in particular Fig. 3.32), then assuming a uniform magnetization in the particle under all circumstances can be considered a good approximation. Such particles are called single-domain particles. Only one or at most two angular variables for the average magnetization are needed and all switching processes in such particles can be analysed in closed form (Sect. 3.5.2) as first demonstrated by *Stoner* and *Wohlfarth* [617].

There are different switching modes that can be found within the Stoner-Wohlfarth regime. We investigate the properties of these modes in Sect. 3.5.3 to arrive at a classification of switching processes in larger particles that do not fulfil the conditions of uniform magnetization rotation.

A particular class of model calculations considers only samples and field directions for which the magnetization stays uniform and constant up to the instability point, where this uniform state decays spontaneously in general in a non-uniform mode. If these conditions are fulfilled, rigorous solutions can be obtained by a linearization of the micromagnetic equations around the initial state [618, 619]. This is in principle possible even for large samples, but the condition that the magnetization remains uniform up to the switching point will only rarely be fulfilled for larger particles. Nevertheless, this case dominated the discussion of nucleation phenomena for a long time. The reason was initially a technical one [36]: sufficiently powerful computers and algorithms for dealing with more general situations that are closer to experimental conditions were not available. Now that they are, the analytical solutions for the ideal particles still play a role as benchmarks for more general investigations. They are reviewed for this reason in Sect. 3.5.4.

For larger particles of non-ellipsoidal shape or for arbitrary field directions, numerical micromagnetic investigations are the only possibility. The magnetization pattern then tends to become non-uniform already before switching. How to derive the instability point from such numerical studies in a valid way is analysed in Sect. 3.5.5. A special treatment is required for the case of spontaneous symmetry breaking in second-order phase transitions. They are shortly discussed in Sect. 3.5.6.

3.5.2 Uniform Single-Domain Switching

(A) General Formulation. We consider an arbitrary small single-domain particle in which the (uniform) magnetization direction can be described by two polar angles ϑ and φ. The total energy density can be written as the sum of a generalized anisotropy functional $g(\vartheta, \varphi)$ and the external field energy (3.15). The functional g may include magneto-elastic interactions with an external stress. Since we explicitly exclude the possibility of a breakdown into domains, it is also allowed to include the stray field or demagnetizing energy as a "shape anisotropy" in the generalized anisotropy functional. (This is possible for small particles only. When larger particles switch in a non-uniform mode, which avoids the demagnetizing field, a shape anisotropy does not exist).

In very small particles the deviations from a uniform magnetization will be negligible. Under these conditions the shape of the particle can be arbitrary. It is not necessary to restrict the discussion to ellipsoidal particles in this section. Shape anisotropy is in general defined as the stray field energy of the uniformly magnetized particle as a function of the magnetization direction.

Consider first a planar problem, in which only one angle φ and two components of the field have to be taken into account. Then the total energy density of a general uniaxial particle (Fig. 3.45a) becomes:

$$e_{\text{tot}} = g(\varphi) - H_{\parallel} J_s \cos\varphi - H_{\perp} J_s \sin\varphi . \tag{3.92}$$

$H_{\parallel}$ and $H_{\perp}$ are the parallel and perpendicular components of an external field in the φ plane, "parallel" meaning a field parallel to the easy axis of the particle. In static equilibrium, the magnetization direction must fulfil:

$$\partial e_{tot}/\partial \varphi = g'(\varphi) + H_{\parallel}J_s \sin\varphi - H_{\perp}J_s \cos\varphi = 0 \ . \tag{3.93}$$

The second derivative must be positive to guarantee a stable equilibrium. For small particles it is not enough, however, to consider the absolutely stable minimum. Also metastable states have to be taken into account. Of particular interest is the stability limit, which is defined by:

$$\partial^2 e_{tot}/\partial \varphi^2 = g''(\varphi) + H_{\parallel}J_s \cos\varphi + H_{\perp}J_s \sin\varphi = 0 \ . \tag{3.94}$$

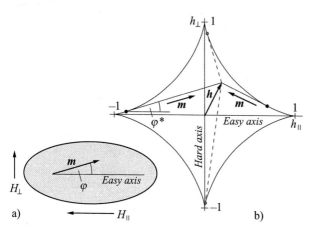

Fig. 3.45 The switching curve ["astroid"; (b)] of an ellipsoidal, uniaxial particle (a) under the influence of different fields, together with Slonczewski's construction of the equilibrium magnetization directions *m* compatible with a given field *h*. The fields are measured in reduced units $h = HJ_s/2K$. The *dashed lines* indicate unstable solutions.

Reaching the stability limit, a formerly metastable state becomes unstable and switches to a new stable state. Combining (3.93) and (3.94) leads to an implicit set of equations for the stability limit or 'switching curve':

$$H_{\parallel}^* J_s = -g'(\varphi)\sin\varphi - g''(\varphi)\cos\varphi \ ,$$

$$H_{\perp}^* J_s = g'(\varphi)\cos\varphi - g''(\varphi)\sin\varphi \ . \tag{3.95}$$

For a simple second-order uniaxial anisotropy we obtain with:

$$g(\varphi) = K\sin^2\varphi \ , \quad g'(\varphi) = 2K\sin\varphi\cos\varphi \ , \quad g''(\varphi) = 2K(\cos^2\varphi - \sin^2\varphi) \tag{3.96}$$

the switching curve in terms of the reduced field $h = HJ_s/2K$:

$$h_{\parallel}^* = -\cos^3\varphi \ , \qquad h_{\perp}^* = \sin^3\varphi \ , \tag{3.97}$$

which is the famous *Stoner-Wohlfarth* astroid ([617]; Fig. 3.45b).

(B) The Magnetization Curve. Following *Slonczewski*[620], the astroid defines not only the stability limit for equilibrium magnetization directions, but also helps to determine graphically the possible metastable magnetization directions for any given field $(H_{\parallel}, H_{\perp})$ and thus the magnetization curve.

The procedure and its proof run as follows: draw a tangent to the astroid from the endpoint of an arbitrary field vector to a tangential point with parameter φ^*. The slope of this tangent then fulfils the conditions:

$$\frac{H_\perp J_s - g' \cos\varphi^* + g'' \sin\varphi^*}{H_\parallel J_s + g' \sin\varphi^* + g'' \cos\varphi^*} \overset{!}{=} \frac{\partial H_\perp^* / \partial\varphi}{\partial H_\parallel^* / \partial\varphi} = \tan\varphi^* \ . \tag{3.98}$$

The left-hand side is derived by connecting the endpoint of the field vector and the point on the stability limit (3.95) belonging to the parameter φ^*; the centre part is the slope of the astroid at this point, which after evaluation with (3.95), simplifies to the expression $\tan\varphi^*$ on the right. Combining the left-hand and the right-hand sides of (3.98) leads, after some reordering, to the equilibrium condition (3.93) for the angle φ^* ! Thus two theorems have been proven:

- The parameter value φ^*, corresponding to the footpoint of the tangent, is an equilibrium magnetization angle for the given field.
- This magnetization direction is identical with the tangential vector, directed from the tangential point towards the tip of the field vector.

This surprising theorem is related to the mathematical theory of envelops [620], which also deals with the manifold of tangents to a curve.

Among all tangent points possible for a given field, some yield (meta-) stable and some unstable magnetization directions. For the astroid (3.97) we find four tangents if the field vector lies inside the astroid (Fig. 3.45b) and two otherwise. In the first case two tangent solutions are stable, in the second case there is only one stable solution. The stable solutions are those with the smaller angle relative to the easy axis.

Evaluating these solutions for a series of field values, one may derive complete hysteresis curves as shown in Fig. 3.46 for the simple uniaxial particle. Here the longitudinal magnetization curves, as well as the simultaneously occurring transverse magnetization excursions, are shown for different transverse bias fields. Note the strong influence of the bias field on coercivity, which decreases from $H_K = 2 K/J_s$ to zero as the transverse bias field increases from zero to H_K. (Plotting the coercive field as a function of the bias field we obtain again the astroid Fig. 3.45b).

Another important magnetic property is the transverse susceptibility $\chi_\perp$, the change in transverse magnetization $J_s \sin\varphi$ with a small field $H_\perp$ perpendicular to the easy axis. This quantity depends on the applied d.c. field and on the magnetization angle φ. It can be derived from the equilibrium condition (3.93), and, using the equation for the stability limit (3.94), it can be written as:

$$\chi_\perp = \frac{J_{\rm s}\cos^2\varphi}{\left(H_\| - H_\|^*\right)\cos\varphi + \left(H_\perp - H_\perp^*\right)\sin\varphi}\ .\tag{3.99}$$

Here $H_\|^*$ and $H_\perp^*$ are points on the switching curve (3.95). Note that the transverse susceptibility diverges on approaching the switching curve. Since the switching field scales with the anisotropy constant, information about the anisotropy distribution of a particle assembly can be derived from transverse susceptibility measurements.

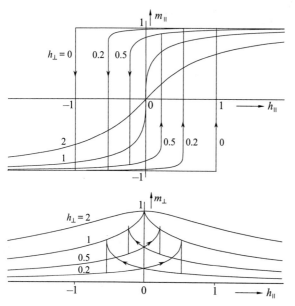

Fig. 3.46 Combining the stable solutions shown in Fig. 3.45b, hysteresis curves of a uniaxial particle may be derived. Here longitudinal $(m_\|)$ and transverse $(m_\perp)$ magnetization components are shown as a function of the longitudinal field $h_\|$ (along the easy axis of the particle) for different values of a transverse bias field $h_\perp$. As in Fig. 3.45b, the external field is expressed in terms of the anisotropy field $H_{\rm K} = 2K/J_{\rm s}$

The generalization of this analysis to three-dimensions, in which both magnetization angles ϑ and φ have to be considered, is straightforward. The equilibrium condition (3.93) is replaced by two such equations, while the condition for the stability limit is replaced by the condition of a vanishing determinant of the Jakobi matrix of second derivatives [see (3.91b)]. From these three equations the three field components for the switching 'surface' can be derived. For a simple uniaxial particle there results a surface that is generated from the astroid by rotating it around the easy axis. Also Slonczewski's tangent method of constructing magnetization curves appears to remain valid in three dimensions. The ordinary tangent of the two-dimensional problem (Fig. 3.45) is replaced by a tangential plane. Connecting the touching point of this tangential plane with the tip of the field vector gives a magnetization direction in equilibrium with the given field. (This theorem has been checked "experimentally" on a computer and found to be valid for different, quite arbitrary anisotropy functionals, field and magnetization directions).

3.5.3 Classifying Switching and Nucleation Processes

The manifold of switching and nucleation phenomena described by the small-particle Stoner-Wohlfarth case deserves a careful analysis. We will see that this elementary model can serve as a guideline to more complex switching phenomena in larger samples with a non-uniform magnetic microstructure. We choose for this analysis a simple uniaxial anisotropy, obtained by inserting (3.96) into (3.92). In reduced quantities ($h = HJ_s/2K$, $\varepsilon_{tot} = e_{tot}/K$) this leads to the following total energy expression:

$$\varepsilon_{tot} = \sin^2\varphi - 2h_{||}\cos\varphi - 2h_\perp\sin\varphi , \tag{3.100}$$

which is now analysed in the neighbourhood of various switching transitions as they are described by the astroid (Fig. 3.45).

(A) Asymmetrical, Discontinuous Switching. In an arbitrary transition somewhere on the astroid (not along one of the symmetry axes) we obtain a behaviour as shown in Fig. 3.47 with the following features:
- The transition is discontinuous and asymmetrical.
- The equilibrium magnetization angle φ_0 rotates continuously and progressively towards the switching point.
- The "susceptibility" $\chi = \partial\varphi_0/\partial h_{||}$ diverges on approaching the switching point as can be seen from the following calculation:

The equilibrium condition derived from (3.100) for constant $h_\perp$ is:

$$h_{||} = (-\sin\varphi_0 + h_\perp)\cot\varphi_0 . \tag{3.101}$$

This leads to the inverse susceptibility $1/\chi = \partial h_{||}/\partial\varphi_0$:

$$1/\chi = \left(\sin^3\varphi_0 - h_\perp\right)/\sin^2\varphi_0 , \tag{3.102}$$

which vanishes at the switching point because of (3.97).

Exploring the character of this zero we find that the inverse susceptibility approaches the switching point with a vertical tangent, as shown in Fig. 3.48a. Analytically this is confirmed by evaluating the derivative $\partial h_{||}/\partial(1/\chi)$ from (3.101) and (3.102), which is also zero at the switching point. The square of the inverse susceptibility $1/\chi^2$ approaches linearly the zero of the transition point, as demonstrated in Fig. 3.48a, so that the switching point can be extrapolated, even if the second derivative is not available. In the same way, the value of φ_0^{sw} at the switching point can be obtained by plotting the equilibrium angle φ_0 versus $1/\chi$ and extrapolating towards $1/\chi = 0$, as shown in Fig. 3.48b. This behaviour is generally found except in the special case of $h_\perp = 0$, also for general one-dimensional anisotropy functionals $g(\varphi)$.

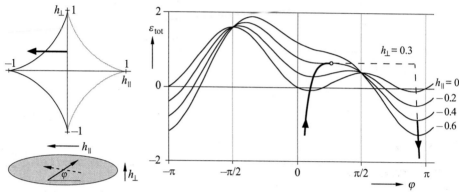

Fig. 3.47 A Stoner-Wohlfarth transition in the general, non-symmetric case, here for a fixed transverse field. The reduced total energy is plotted for several longitudinal field values as a function of the magnetization angle [617]. The *thick line* indicates the calculated energy minima approaching the instability after starting at positive saturation ($\varphi = 0$), and jumping to an angle close to $\varphi = \pi$ at the instability point. Approaching this point the magnetization angle shows a strong excursion, which permits the calculation of the switching point from the minimum positions alone, as explained in the text

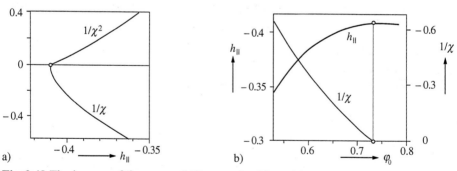

Fig. 3.48 The inverse of the susceptibility $\chi = \partial\varphi_0/\partial h_\parallel$ and its square plotted as a function of the applied field (a) for the example of Fig. 3.47. (b) The magnetization angle at the switching point is determined by extrapolating the inverse susceptibility $1/\chi$ as a function of the magnetization angle φ_0 towards $1/\chi = 0$. Also $\varphi_0(h)$ can be found in (b)

(B) Symmetrical Switching and Nucleation. In the past, a rather special case attracted the principal interest. Here the field is applied exactly along the easy axis of the particle. The behaviour of the energy profiles is shown in

Fig. 3.49. Up to the switching point the magnetization vector remains constant and it is not possible to anticipate the switching event by following the equilibrium magnetization angle as in (A). A micromagnetic stability analysis, based on the second derivative of the energy functional, is one option.

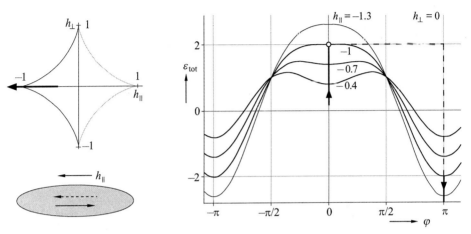

Fig. 3.49 Symmetric first-order switching in a field exactly along the easy axis. In this particular case the magnetization angle stays fixed both before and after switching

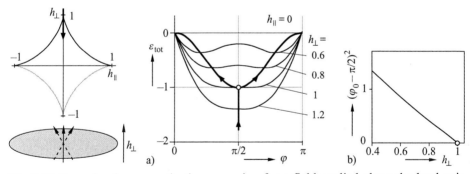

Fig. 3.50 Second-order magnetization excursion for a field applied along the hard axis direction (a). In this case the transition point can be obtained by extrapolating the square of the magnetization excursion $\varphi_0 - \pi/2$ towards the critical point from the low-field side, as demonstrated in (b)

Another special case can also be studied in the Stoner-Wohlfarth scenario: that of a continuous, second-order phase transition. Such a transition can be found in a field applied exactly perpendicular to the easy axis. The behaviour of the energy profiles for this case is shown in Fig. 3.50a. Here the magnetization angle stays constant for large fields. Below the critical point $h_\perp = 1$ the equilibrium magnetization angle φ_0 rotates continuously in a square-root-like fashion towards either side. Plotting the square of the angular deviation from the transverse direction versus the applied field, the critical point can be extrapolated "from below", as demonstrated in Fig. 3.50b.

In the following we will see that the general features of the three prototypes examined for the Stoner-Wohlfarth model find their correspondences in larger particles, in which the assumption of a rigid magnetization is no more valid.

3.5.4 Classical Solutions

A rigorous analytical treatment of the nucleation problem is possible when the magnetization stays uniform and constant up to the nucleation or switching point. In such a case the micromagnetic equations can be linearized around the starting configuration and subjected to a systematical stability analysis [618, 619]. The approach requires the following conditions to be met:

- The particle must be ellipsoidal in shape — including the limits of an infinite plate and an infinite cylinder of elliptical cross-section.
- The anisotropy must be at least orthorhombic, with the axes coinciding with sample axes.
- The field must be applied exactly along one of the axes of the ellipsoid.
- The sample must be magnetically homogeneous.
- The particle must be isolated so that interactions with other particles are negligible.

The very small single-domain particles discussed in Sect. 3.5.3B belong to this class, but here the exchange stiffness effect suppresses an inhomogeneous switching so that even non-ellipsoidal small particles can be included in the discussion. For larger particles the ellipsoidal shape is a necessary requirement for the application of analytical micromagnetic tools to be discussed here.

Most attention was paid to the switching field of ellipsoids of *revolution* with the field applied along the symmetry axis (for reviews see [34, 514]). Two principal nucleation modes were found: that of *coherent rotation* for small samples, and the "*curling mode*" for larger samples. For the coherent rotation mode the switching field is identical with that of the Stoner-Wohlfarth model. In other reduced units it may be written as:

$$h_{\mathrm{nucl}}^{\mathrm{coh}} = -Q + N_{\parallel} - N_{\perp} , \tag{3.103a}$$

where $h = \mu_0 H/J_s$ is the reduced applied field, the anisotropy constant is represented by the reduced quantity $Q = K_u/K_d$ for uniaxial anisotropy, $N_{\parallel}$ is the demagnetizing factor along the symmetry axis and $N_\perp$ is the demagnetizing factor along the transverse direction. The switching field is thus equal to the anisotropy field, corrected for shape effects. The same formula is valid for cubic anisotropy if Q is understood as $Q = K_c/K_d$. [To see the identity of (3.103a) with the Stoner-Wohlfarth result, remember that shape anisotropy was included in the total effective anisotropy in Sect. 3.5.2].

For larger particles the inhomogeneous "curling" mode takes over. This mode largely avoids the stray field which has to be overcome in the uniform rotation process. The transverse demagnetizing factor $N_\perp$ in (3.103a) determines the stray field in uniform rotation. In the curling mode, it is replaced by an exchange stiffness term that depends on the transverse particle radius. The result becomes in our units:

$$h_{\text{nucl}}^{\text{curl}} = -Q + N_{\parallel} - q^2/\rho^2 \,, \tag{3.103b}$$

where ρ is the particle radius in units of the exchange length $\rho = R/\sqrt{A/K_d}$ and q is a numerical constant that depends on the aspect ratio ($q = 2.115, 2.082, 1.841$ for an infinite plate, a sphere, and an infinite cylinder, respectively [621]). The rotationally symmetric curling mode is indicated in the sketch.
The magnetization components perpendicular to the particle axis grow in the indicated fashion towards the particle perimeter.

In most cases these two modes are the only switching modes that can occur. Only in a narrow-sized interval of elongated particles was a further mode identified, the so-called buckling mode [619]. We ignore it here because of its minor importance.

The critical size for the transition from coherent rotation to the curling mode is calculated by equating (3.103a) with (3.103b):

$$\rho_c = q/\sqrt{N_\perp} \quad \text{or} \quad R_c = (q/\sqrt{N_\perp})\sqrt{A/K_d} \,. \tag{3.104}$$

These rigorous results are interesting especially for high-anisotropy particles with $Q \gg 1$, because the switching field h is then according to (3.103b) close to $-Q$ in our reduced units. Although for larger samples domain states have clearly a lower energy at zero fields (see Sect. 3.3.2), they cannot be reached for an ideally ellipsoidal sample if one starts from saturation along an easy direction. The nucleation field is identical with the coercivity, and this technically important quantity should therefore approach the theoretical limit $H_c = 2K_u/J_s$, even for large particles. Because this result is at variance with all experimental findings, it is known as *Brown's paradox*. The solution to the seeming contradiction between theory and experiment lies in the non-ellipsoidal shape of real samples. Any deviation from the ellipsoidal shape (in particular the presence of protrusions and corners), any modulation in the anisotropy strength or direction (as around defects or inhomogeneities) can lead to premature nucleation and switching. If early nucleation happens, a low-energy domain state is occupied and magnetization continues via domain wall displacement processes. The details of these processes are still largely unknown. How they can in principle be treated is indicated in the next section.

3.5.5 Numeric Evaluation in the General Case

A number of early authors tried to explore the detailed mechanisms of the nucleation process for non-ellipsoidal particles [622, 623], although without really convincing results. Even with the advent of powerful computers and efficient algorithms the calculation of a switching field is far from trivial. The cavalier approach is to leave it up to the computer algorithm to decide when to switch from one configuration to another [581, 624, 625]. Such a procedure would only be acceptable if the independence of the results of the discretization grid is explicitly demonstrated. Otherwise numerical artefacts cannot be excluded. Unfortunately, most authors were unable to follow this postulate because they were already at the limits of memory and computation time. The sceptical attitude expressed by Aharoni [514 (p. 259 –265)] with respect to such results is fully justified. On the other hand, his postulate to include a stability check into every numerical finite element program [514 (p. 263)] is unrealistic. If an instability point for one variable is defined by the vanishing of a second derivative, this criterion corresponds for many variables to the vanishing of just one eigenvalue of the matrix of second derivatives [626]. If a particle has to be discretized with $100 \times 100 \times 100$ cells, with two angular variables in each cell, this would mean that the absence of negative eigenvalues of a matrix of $2 \cdot 10^6 \times 2 \cdot 10^6$ elements has to be confirmed.

However, there is an alternative, the basic idea of which was indicated in Sect. 3.5.3A. In the general case the magnetization does not stay constant on approaching the switching field. Either the whole magnetization field (for small particles) or just some nucleus will deviate progressively towards the instability point. As in Sect. 3.5.3A, some extrapolation procedure might be able to localize the switching point without a full-scale stability analysis. No attempt of mathematical proof is undertaken here. We tested the method on a number of examples and just demonstrate its feasibility here for one case.

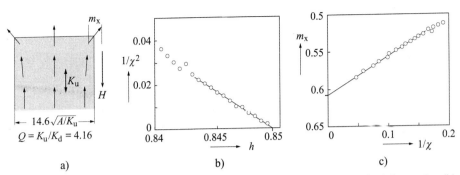

Fig. 3.51 (a) Nucleation at the edge of a prismatic, high-anisotropy uniaxial sample. (b) The switching field is obtained as in Fig. 3.48 by extrapolating the square of the reciprocal susceptibility χ, where the susceptibility is defined as $\chi = dm_x/dh$ for the edge magnetization vector. The excursion of the edge magnetization angle at the switching point results from the extrapolation (c). Further details about these micromagnetic finite element calculations can be found in [627]

A high-anisotropy particle with a rectangular cross-section is subjected to a field opposite to the initial magnetization direction [627]. Approaching the switching field, a localized excursion of the magnetization in the corners is observed, which scales with the exchange length $\sqrt{A/K_d}$. We define a "susceptibility" as the derivative of the transverse magnetization component in the corner with respect to the applied field. Plotting the square of the reciprocal susceptibility as a function of the field as in Sect. 3.5.3A yields in fact a critical field (Fig. 3.51), which becomes independent of the discretization as soon as the discretization cells are smaller than the exchange length $\sqrt{A/K_d}$. This procedure avoids a calculation at the critical point itself, which would be quite difficult.

The method sketched here and elaborated in more detail in [628] can be applied also indirectly in analysing symmetric, catastrophic switching events as they were discussed in Fig. 3.49 and in Sect. 3.5.4. If the field is tilted relative to the symmetry axis, precursors that permit the extrapolation to the critical point will be found, and these critical points can be extrapolated "along the astroid" towards the symmetrical field axis. This method might be applied to the *general* ellipsoid, which has not yet been treated analytically. In [628] another symmetric example was explored, in which the configuration did not stay uniformly magnetized on approaching the switching point, but no precursors were observed anyway. The switching mode in this case turned out to be a kind of "incoherent rotation" which can be analysed by examining the behaviour in tilted applied fields.

3.5.6 Continuous Nucleation (Second-Order Transitions)

The other special case, discussed before in Sect. 3.5.3B, is that of a symmetric, but continuous domain formation connected with a second-order phase transition and spontaneous symmetry breaking. A high-symmetry configuration, stable at large fields, decays continuously at the critical field into either of several degenerate lower-symmetry phases, which can be characterized by an "order parameter" such as the deflection angle $\varphi - \frac{\pi}{2}$ in Fig. 3.50b.

General cases can only be treated numerically by focusing on the low symmetry configuration. Increasing the field, the order parameter of the low symmetry phase decreases on approaching the critical point and can be extrapolated if plotted in the right way. This technique was extensively used in the exploration of the phase diagram of domain walls in thin films [629]. See also the discussion of dense stripe domains in Sect. 3.7.2.

The phenomenon also plays a role in small particles, not only for hard axis fields as demonstrated in Sect. 3.5.3C, but also for easy axis fields discussed in the last two sections. For example, *Ehlert* et al. [630] calculated numerically the switching field of cylindrical particles. They found that for oblate cobalt cylinders of sufficient size the nucleation field (after positive saturation) remains positive and is followed by a continuous curling process rather than by a discontinuous switching. Since curling can occur in either rotation sense, this transition must be of the postulated type. A similar second-order phase transition, which was found in a cube-shaped particle of low anisotropy, is shown and explained in Fig. 3.52.

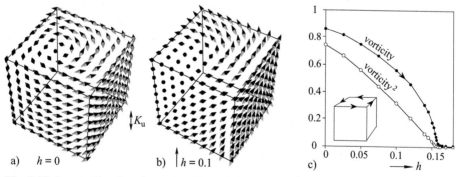

Fig. 3.52 A second-order phase transition occurring in a cubical particle of low uniaxial anisotropy ($Q = 0.001$, $L = 10\sqrt{A/K_d}$). A zero-field vortex state (a) reduces its vorticity in a field applied along the symmetry axis (b). Plotting the square of the vorticity as a function of the applied field the critical point can be extrapolated (c). The reduced field is defined as $h = \mu_0 H/J_s$. The magnetization must be understood to point towards the dark ends of the double cones in (a) and (b). (Together with *W. Rave*, IFW Dresden [628])

3.6 Domain Walls

The analysis of magnetic microstructure between the realms of phase theory for large samples and that of small particle switching needs some knowledge of the structure of domain walls, and this is the place to discuss them in detail. The calculation of domain wall structure is by far the most important contribution of micromagnetics to the analysis of magnetic domains. This is true for two reasons: experimentally, domain walls are difficult to access because they change their properties at surfaces where they can primarily be observed. Also, it is in most cases difficult to isolate a single wall from its neighbours to measure its properties. Usually domain walls interact in a complicated network. The theoretical approach, on the other hand, is straightforward and well founded. The *calculation* of domain walls is regularly the method of choice, rather than trying to determine their energy or structure experimentally. This is particularly true for walls that are sufficiently extended and flat, so that they may be considered planar and one-dimensional. Such domain walls can be calculated in a relatively easy way using the methods of variational calculus, as first demonstrated by *Landau* and *Lifshitz* [22] in their pioneering work. The treatment of two- and three-dimensional walls as they occur, for example, in thin films is more difficult but still possible. The material presented here is in large parts an abridged and updated recapitulation of the earlier textbook by the first author on this subject [509].

3.6.1 The Structure and Energy of Infinite Planar Walls

(A) The Simplest 180° Wall. Let us start with the simplest of all domain walls, a planar 180° wall in an infinite uniaxial medium with negligible magnetostriction, separating two domains of opposite magnetization (Fig. 3.53).

If the wall plane contains the anisotropy axis, the domain magnetizations are parallel to the wall and there will be no global magnetic charge, meaning that the component of magnetization perpendicular to the wall is the same on both sides of the wall. If, in addition, the magnetization rotates *parallel* to the wall plane (Fig. 3.53a), there will be no charges inside the wall, either. The stray field energy then assumes its smallest value, namely zero. This wall mode, first proposed and calculated by Landau and Lifshitz [22], is generally called a *Bloch* wall in honour of *Felix Bloch* [12], who first conceived a continuous wall transition.

To prove that for the considered 180° wall in bulk material the stray-field-free wall has the lowest energy, we look at the sphere of possible magnetization directions (Fig. 3.54a). For the purpose of this argument let us assume that the anisotropy axis coincides with the axis of the polar coordinate system of Fig. 3.54b. The magnetization in the domains then points to the north and south poles. Inside the wall the magnetization rotates from one pole to the other, moving along what we call the *magnetization path*. Later we will see that once we know the magnetization path and the anisotropy energy along this path, we are able to calculate the wall profile. So let us first look for arguments to select the magnetization path.

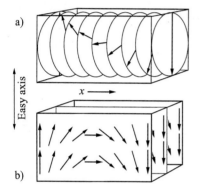

Fig. 3.53 The rotation of the magnetization vector from one domain through a 180° wall to the other domain in an infinite uniaxial material. Two alternate rotation modes are shown: the optimum mode, which is called the *Bloch* wall (a), as compared to the *Néel* wall (b), which is less favourable here but can be preferred in thin films and in applied fields. For both modes the opposite rotation is equally possible

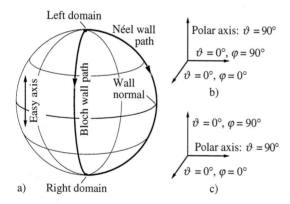

Fig. 3.54 The sphere of possible magnetization directions (a) with the magnetization paths corresponding to the two rotation modes shown in Fig. 3.53. Two alternative spherical coordinate systems used in the discussion of domain walls are defined in (b, c)

The anisotropy energy (3.11) does not depend on the azimuthal angle. Therefore all paths that differ only in the azimuth are equivalent from the point of view of anisotropy. We may safely assume that the magnetization path forms a meridian, since any deviation from this shortest path would bring no advantage. If all meridians are equivalent with respect to path length and anisotropy, those which do not produce a stray field are favoured. Possible magnetization paths of the Bloch wall are the two meridians parallel to the wall plane. No difference can be seen between the two mirror-symmetric Bloch wall paths — their energies are degenerate.

To calculate the Bloch wall energy and the Bloch wall structure we switch to a polar coordinate system with the wall *normal* as axis (Fig. 3.54c). The angle ϑ is then zero, while φ rotates in the wall from 90° to –90°. We neglect the second anisotropy constant and denote by K the first constant. Let x be the coordinate perpendicular to the wall and φ' the derivative of the magnetization angle with respect to x. Then the specific wall energy γ_w — the total energy per unit area of the wall — is an integral over the expressions (3.7b) and (3.11):

$$\gamma_w = \int_{-\infty}^{\infty} \left[A\varphi'^2 + K \cos^2\varphi \right] dx \ , \ \ \varphi(-\infty) = \tfrac{\pi}{2} \ , \ \ \varphi(\infty) = -\tfrac{\pi}{2} \ . \tag{3.105}$$

Variational calculus leads to the function $\varphi(x)$, which minimizes γ_w under the boundary conditions. The solution is derived from *Euler*'s equation:

$$2 A\varphi'' = -2K \sin\varphi \cos\varphi \ , \tag{3.106}$$

which, by multiplication with φ' and (indefinite) integration with respect to x leads to a *first integral*:

$$A\varphi'^2 = K \cos^2\varphi + C \ . \tag{3.107}$$

For an isolated wall in an infinite medium, the derivative φ' at infinity must vanish, leading with $\cos\varphi_\infty = 0$ to $C = 0$. The first integral (3.107) then tells us that at every point in the wall the exchange energy density equals the anisotropy energy density. At positions where the anisotropy energy is high, the magnetization rotates rapidly leading to a large exchange energy.

The final result is obtained by solving (3.107) for dx:

$$dx = \sqrt{A/K} \, d\varphi/\cos\varphi \ , \tag{3.108}$$

which, inserted together with (3.107) into the total wall energy (3.105), yields:

$$\gamma_w = 2 \int_{-\infty}^{\infty} K \cos^2\varphi \, dx = 2\sqrt{AK} \int_{-\pi/2}^{\pi/2} \cos\varphi \, d\varphi = 4\sqrt{AK} \ . \tag{3.109}$$

By integrating (3.108) we obtain the functional dependence $\varphi(x)$ (see Fig. 3.55):

$$\sin\varphi = \tanh \xi \ ; \ \ \xi = {}^x\!/\!\sqrt{A/K} \ . \tag{3.110}$$

[Alternate forms of this relation are $\cos\varphi = 1/\cosh\xi$ and $\tan\varphi = \sinh\xi$. The wall with the opposite rotation sense results from the opposite root in (3.108) and is described by $-\sin\varphi = \tanh\xi$].

For large x the magnetization angle approaches the boundaries in (3.105) exponentially. At a distance of $5\sqrt{A/K}$ the hyperbolic tangent differs from its asymptotic value by less than 10^{-4}. This is why the solution for an infinite medium is a good approximation even if the integration range does not extend to infinity. Rigorous solutions for a finite range can be obtained by choosing a different integration constant C in the first integral (3.107). This leads to periodic functions describing interacting parallel walls. But closely packed domain walls never occur in regular bulk ferromagnets. Closely packed walls do occur in thin films or plates in the form of dense stripe domains (Sect. 3.7.2), for which the one-dimensional theory does not apply, however. There are also materials with an intrinsic tendency towards a modulated structure [631, 632] that can sometimes be described by such solutions. In regular ferromagnets the consideration of the periodic solutions is unnecessary.

(B) Domain Walls of Reduced Wall Angle. 90° walls and other walls with a magnetization rotation of less than 180° appear naturally in materials with multiaxial anisotropy and in an applied field perpendicular to the easy axis of a uniaxial material. The proper procedure to calculate such domain walls is quite instructive and is discussed here on an elementary level.

In static equilibrium, magnetic fields acting on a wall must be oriented perpendicular to the axis of the uniaxial material. In the simplest case the field points parallel to the wall and parallel to the magnetization direction in the wall centre. The total generalized anisotropy functional is then:

$$G(\varphi) = K_{u1}\cos^2\varphi - HJ_s\cos\varphi \ . \tag{3.111}$$

In the domains, the functional $G(\varphi)$ must be stationary with respect to φ, which means that the derivative $\partial G/\partial\varphi$ must be zero. This leads to the following implicit equation for φ_∞, the magnetization angle in the domains:

$$HJ_s = 2K_{u1}\cos\varphi_\infty \ . \tag{3.112}$$

An example: $h = \frac{1}{2}HJ_s/K_{u1} = \sqrt{1/2}$ means $\varphi_\infty = \frac{\pi}{4}$, or a wall rotation angle $2\varphi_\infty$ of 90°. The energy density in the domains with (3.112) is $G_\infty = -K_{u1}\cos^2\varphi_\infty$. In calculating the wall we have to subtract this background energy from the free energy (3.111). Inserting (3.112) we thus obtain the extra energy of the wall:

$$G(\varphi) - G_\infty = K_{u1}(\cos\varphi - \cos\varphi_\infty)^2 \ . \tag{3.113}$$

This new generalized anisotropy energy density behaves quite analogously to the function $K_{u1}\cos^2\varphi$ in (3.105). It displays a maximum in the centre ($\varphi = 0$), and both the function values and their derivatives are zero at the values of φ marking the domains [$\varphi = \pm\frac{\pi}{2}$ in (3.105), $\varphi = \pm\varphi_\infty$ in (3.113)]. From it the

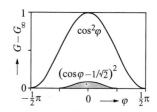

domain wall structure and energy can be calculated as in the elementary case treated in (A). We demonstrate this in connection with a more general class of domain walls in uniaxial material in Sect. 3.6.2B.

(C) Wall Widths. As demonstrated for example in (3.110), domain walls form a *continuous* transition between two domains. For this reason there can be no unique definition of a domain wall width. The classical definition introduced by *Lilley* [633] is based on the slope of the magnetization angle $\varphi(x)$, as shown in Fig. 3.55. Its value is $W_L = \pi \sqrt{A/K}$ in our example. In another definition, also indicated in this figure, the slope of the magnetization component $\sin\varphi$ in the origin is considered. It leads to $W_m = 2\sqrt{A/K}$.

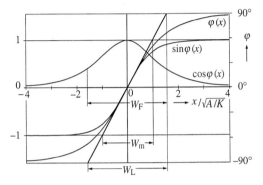

Fig. 3.55 The calculated wall profile of a 180° Bloch wall and different wall width definitions (see text)

A third definition of wall width is based on the total wall flux and reads $W_F = \int_{-\infty}^{\infty} \cos\varphi(x)\,dx$. In our example W_F coincides with Lilley's wall width W_L, since $\int_{-\infty}^{\infty} 1/\cosh x\,dx = \pi$. A fourth definition advocated by Jakubovics [634] amounts in our case to the evaluation of $W_J = \int_{-\infty}^{\infty} \cos^2\varphi(x)\,dx$, which leads to $W_J = W_m$ for the simple wall profile (3.110). It shares with W_F the advantage of being defined by an integral expression that can be determined experimentally with better reliability than definitions that are based on a single point in a profile.

Which of the four wall width definitions should be used, depends on the particular situation. Lilley's definition W_L is most commonly used. For a comparison with transmission electron microscopy observations the quantity W_m is best suited, because electrons are deflected primarily by certain magnetization components (depending on the geometry of the experiment). The integral width W_F is connected with the contrast in Bitter pattern experiments, because it determines the net magnetic flux over a wall. Jakubovics' definition W_J is sometimes preferable to W_F. For example, walls that would transport some flux towards a surface tend to compensate this flux in low-anisotropy materials by forming a surface vortex (domain walls with such properties are treated in Sect. 3.6.4D; see Sect. 2.2.4 for a qualitative discussion). This results in $W_F \approx 0$ near the surface, rendering the quantity W_F useless.

All these definitions can be applied also to domain walls with a reduced wall angle, as discussed in (B). In the tangent definitions the limiting angles $\pm\frac{\pi}{2}$ have to be replaced by $\pm\varphi_\infty$. If a metastable state is included in the anisotropy functional, more than one steepest tangent is found in the profile. How to proceed in this case is demonstrated in Sect. 3.6.2A. The integral definitions work in the same way if the proper generalized anisotropy functional as in (3.113) is used.

(D) General Theory of Classical Bloch Walls. The solution of the introductory example can be systematically generalized. We do this in a stepwise fashion, extending the range of applicability, and demonstrating how these more general cases can be reduced to the basic case treated first. This starting problem is the general classical Bloch wall without stray fields and without magneto-elastic self-energy contributions. We will see that its solution closely follows the elementary example presented in (A).

Firstly, we study the preconditions that must be fulfilled for such a wall to exist. If the wall shall be in static equilibrium, it must separate two domains of equal energy (otherwise energy could be gained by displacing the wall). This imposes a restriction on the allowed field directions, as discussed before in Sect. 3.4.2A on phase theory.

Let $m^{(1)}$ and $m^{(2)}$ denote the magnetization vectors in the two domains. The requirement that no stray fields are produced introduces (i) a restriction for the allowed wall-normal directions and (ii) a condition for the rotation mode. The allowed wall normals n are derived from demanding zero net magnetic charge on the wall (see Sect. 3.2.5A):

$$n \cdot (m^{(1)} - m^{(2)}) = 0 \ . \tag{3.114}$$

The difference vector between the two domain magnetization directions has to lie in the wall plane, as shown in Fig. 3.56 for 90° walls in iron. It defines an axis around which the wall can rotate without generating stray fields. We define the wall orientation angle ψ as the (smaller) angle between the wall plane and the $(m^{(1)}, m^{(2)})$ plane (this definition is not possible for 180° walls, for which an ad hoc definition of a wall orientation angle is necessary). Inside the wall the normal component of the magnetization must not change. In our polar coordinate system (Fig. 3.54c) the polar angle ϑ is constant, and the rotation is described by the azimuthal angle φ only. The "magnetization path" forms a *parallel* to the equator.

Taking into account these considerations, the total excess energy of a general wall with respect to the domain energy has the following form:

$$\gamma_w = \int_{-\infty}^{\infty} \left[A \cos^2 \vartheta_0 \, \varphi'^2 + G(\vartheta_0, \varphi) - G_\infty \right] dx, \quad \varphi(-\infty) = \varphi^{(1)}, \quad \varphi(\infty) = \varphi^{(2)} \quad (3.115)$$

where $G(\vartheta, \varphi)$ is a generalized anisotropy function that may include regular anisotropy, applied field energy, and/or "external" stress energy (from stresses of non-magnetic origin). G_∞ is the value of this function in the domains. The angles ϑ_0, $\varphi^{(1)}$ and $\varphi^{(2)}$ belong to the (easy) magnetization directions in the domains. The coordinate x points perpendicular to the wall.

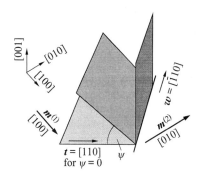

Fig. 3.56 The allowed wall orientations for a 90° wall in iron. The difference vector $w = m^{(1)} - m^{(2)}$ has to lie in the wall plane to avoid far-reaching stray fields. The tangential vector t is arbitrary. Walls with different orientation angles ψ are charge free, but differ in their specific wall energy and in their magneto-elastic properties

The calculation proceeds essentially as in (A). The first integral, the equivalent of (3.107), becomes:

$$A \cos^2 \vartheta_0 \, \varphi'^2 = G(\vartheta_0, \varphi) + C . \quad (3.116)$$

The integration constant C is derived from the boundary condition at infinity: $C = -g(\vartheta_0, \varphi^{(1)}) = -G(\vartheta_0, \varphi^{(2)}) = -G_\infty$. This leads to a formula for dx:

$$dx = d\varphi \sqrt{A \cos^2 \vartheta_0 / [G(\vartheta_0, \varphi) - G_\infty]} , \quad (3.117)$$

which can be integrated to yield implicitly the wall shape $x(\varphi)$. In the same way we get the wall energy as the definite integral:

$$\gamma_w = 2\sqrt{A} \cos \vartheta_0 \int_{\varphi^{(1)}}^{\varphi^{(2)}} \sqrt{G(\vartheta_0, \varphi) - G_\infty} \, d\varphi . \quad (3.118)$$

The calculation of classical Bloch walls is thus reduced to plain integrations, which often can be performed analytically and which are always easy to evaluate numerically.

(E) Including Internal Stresses. In the next step the effects of magnetostrictive stresses in the wall are included. Two possibilities must be distinguished (Fig. 3.57): in the simpler case, the additional elastic energy is confined to the interior of the wall. To secure this, the two domains have to be elastically *compatible* with respect to the wall plane, meaning that the free deformations of the two domains do not differ in any of their components tangential to the wall. Examples of compatible walls are all 180° walls, as well as the (110) 90° walls shown in Fig. 3.57a. In the latter case, the two domains fit together as indicated by the stress ellipses for the domains. The elastic compatibility conditions (in a one-dimensional situation) enforce constant tangential strains also throughout the wall. The spontaneous deformations inside the wall (also indicated by ellipses in Fig. 3.57a) are incompatible with those of the domains into which the wall is embedded, leading to internal stresses and to an additional contribution to the domain wall energy [27, 24].

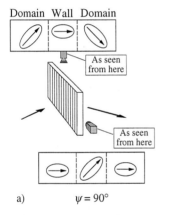

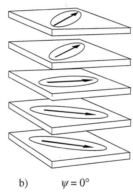

Fig. 3.57 Sketch of internal stresses in and around domain walls. In a (110) 90° wall [(a); $\psi = 90°$, see Fig. 3.56], the free strains of the domains are compatible and stresses are localized in the wall. (The strains are indicated by exaggerated ellipses for $\lambda > 0$). This can be seen from the two projections. For a (100) 90° wall (b), the strains are incompatible and the induced stresses extend to infinity

The second possibility is that of domains which are in themselves incompatible as shown in Fig. 3.57b for a (100) 90° wall. It leads to magneto-elastic stresses that extend to infinity and require a special treatment of the wall boundary conditions. We will return to this subject in connection with magnetically charged walls in (G).

In the first case of local stresses (compatible domains), we may use the general formula for one-dimensional magneto-elastic problems (3.57):

$$e_{ms}(x) = \tfrac{1}{2} \sum_{iklm} \left[\sigma^0_{ik}(x) - \bar\sigma_{ik}\right]\left[\sigma^0_{lm}(x) - \bar\sigma_{lm}\right]\left(s_{iklm} - n_i n_l F^{inv}_{km}\right) . \quad (3.119)$$

The tensor F^{inv} is the inverse of $F_{km} = \sum n_i n_l c_{iklm}$ and depends only on the elastic coefficients and the wall normal n. The components of the magnetization vector inside the wall are inserted into (3.49) to yield the quasi-plastic stress tensor σ^0. The tensor $\bar{\sigma}$ is the mean value of σ^0 in the two domains. If the domains are compatible, then the extra energy (3.119) vanishes in the domains, and its contribution $e_{ms}(x)$ inside the wall can be included into the function $G(\vartheta_0, \varphi)$, thus reducing the problem to the general theory (3.115–3.118).

Magneto-elastic effects usually amount to relatively small corrections that scale with $G_s \lambda^2$, where G_s is a shear modulus (which may be orientation dependent), and λ is a magnetostriction constant (see Sect. 3.2.6). According to Fig. 3.16 this energy density is small compared to the anisotropy energy density in most materials. Particularly in uniaxial materials, magnetostrictive contributions to the domain wall energy can usually be neglected.

However, sometimes the consideration of magnetostrictive energies is indispensable. Consider a (100) 180° wall in iron in which the magnetization rotates from the [001] direction to its opposite. This wall plane is magnetostatically allowed according to criterion (3.114). The magnetization vector in the wall must rotate parallel to the (100) plane, and in the centre of the wall the magnetization therefore meets the [010] or the [01̄0] easy direction. Thus the anisotropy energy in the centre is the same as in the domains. According to (3.117) this would lead to an infinite wall width. In other words, the (100) 180° wall tends to split into two separate (100) 90° walls by inserting a [010] or [01̄0] domain, which is, however, elastically incompatible with the outer domains. Including the magnetostrictive self-energy solves this problem, as was shown explicitly in the classical articles of *Néel* [27] and *Lifshitz* [24]. The magnetostrictive self-energy leads to a contraction of the wall, thus removing the divergence of the wall width. The resulting wall width still exceeds that of ordinary domain walls, especially for materials with small magnetostriction such as magnetic garnets (see Sect. 3.6.3A).

(F) Internal Stray Fields in Walls (Néel Walls). In the next step we permit internal stray fields in the wall, allowing the angle ϑ (Fig. 3.54c) to deviate from its domain value ϑ_0. We still retain the constraint of zero net charge of the wall. The complication now is that we have to deal with two variables, ϑ and φ. So the exchange energy must be generalized according to (3.7):

$$e_x = A(\vartheta'^2 + \cos^2\vartheta\, \varphi'^2) \ . \tag{3.120}$$

The additional stray field energy from the variation of ϑ is, in analogy to the one-dimensional stray field energy calculation (3.20):

$$e_d = K_d(\sin\vartheta - \sin\vartheta_0)^2 \ . \tag{3.121}$$

If the normal magnetization component is constant everywhere ($\vartheta = \vartheta_0$), the stray field energy vanishes. We expect deviations from this condition, for example if the magnetization path along the stray-field-free parallel to the equator is longer than the shortest path between the domain directions. Figure 3.58 demonstrates such a situation for a wall in a uniaxial material under the influence of an applied field perpendicular to the easy axis.

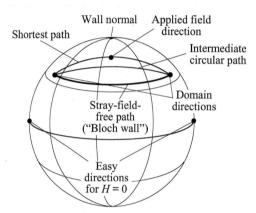

Fig. 3.58 The circular magnetization path approach for walls in a uniaxial material under the influence of an applied field. If the magnetization vector follows a circle on the orientation sphere, the wall structure can be calculated in an elementary way. Note that the orientation sphere is rotated by 90° in this figure compared to Fig. 3.54a

There are several possibilities for handling domain wall calculations needing two angular variables. A simple, approximate method assumes a *circular* magnetization path other than the stray-field-free parallel to the equator. Between the stray-field-free parallel and the shortest possible path (Fig. 3.58) there will be an optimum circular path that leads to a good approximation for the wall energy and the wall width, even if the real magnetization path is more complicated. The largest deviations occur in the vicinity of the boundary points where the influence of the precise path on the energy and width of the wall is small. Technically, the circular path approximation can be reduced to a standard one-variable problem by a proper coordinate transformation.

An advantage of the method of the circular magnetization path is that arbitrarily large deviations from the stray-field-free Bloch wall mode can be treated. If the deviation from the stray-field-free mode is small, there is another method that is based on an expansion of the functional $G(\vartheta, \varphi)$ for small $\vartheta - \vartheta_0$ [635, 509]. This assumption is usually justified for soft magnetic materials for which the stray field energy constant K_d is large compared to the anisotropy contributions. We may then neglect the exchange energy connected with the variation in ϑ and end up with the following approximate expression:

$$\vartheta - \vartheta_0 \cong -F_1/F_2 \qquad \text{with} \quad F_1(\varphi) = G_\vartheta(\vartheta_0, \varphi) - 2\,G(\vartheta_0, \varphi)\tan\vartheta_0$$

$$\text{and} \quad F_2(\varphi) = 2\,K_d\cos^2\vartheta_0 + G_{\vartheta\vartheta}(\vartheta_0, \varphi) - 2\,G(\vartheta_0, \varphi)(1 - \tan^2\vartheta_0)\ , \tag{3.122}$$

where G_ϑ denotes the derivative of the functional G with respect to ϑ, etc.

The correction $\vartheta - \vartheta_0$ entails a corresponding energy gain of the generalized wall relative to the completely stray-field-free ("Bloch") wall:

$$\gamma_w - \gamma_w^0 \cong -\frac{1}{2}\int \left(F_1^2/F_2\right)\left(\cos\vartheta_0/\sqrt{G(\vartheta_0, \varphi)} - G_\infty\right)d\varphi \quad . \tag{3.123}$$

If F_1 vanishes, as for example by symmetry, the Bloch wall is the exact solution. Deviations are induced either by asymmetries in $G(\vartheta, \varphi)$, or by the tendency of the magnetization path to shorten its length, which is reflected in the second term in F_1. The deviations are mainly limited by the second derivative of the stray field energy term (3.121), the first term in F_2. Walls in soft magnetic materials in small external fields can always be treated in this way. In large external fields stronger deviations from the stray-field-free path may occur, which can be explored roughly by the method of the circular magnetization path, or rigorously by a direct numerical integration of the two-dimensional Euler equations.

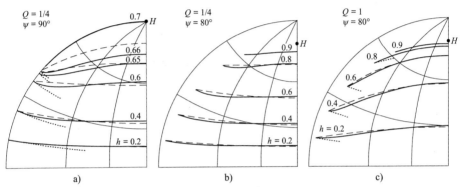

Fig. 3.59 Examples of rigorous calculations of magnetization paths of planar domain walls for a uniaxial material considering two values of the material constant $Q = K_u/K_d$ and two different directions of an applied field. The angle $\psi = 90°$ corresponds to the field direction perpendicular to the wall. Only one quarter of the orientation sphere in Fig. 3.58 is shown. The solid curves represent exact, two-variable numerical calculations. The optimum circular magnetization path is indicated by the *dashed curves*. *Dotted* are the asymptotic magnetization paths near the boundary points (representing the domains). Only when the field is parallel to the wall normal (a) a discontinuous transition from the Bloch mode to the Néel mode is observed near a reduced field value of $h = HJ_s/2K_u = 0.67$. In a high-anisotropy material the tendency towards a stray-field-free solution is less pronounced (c)

Figure 3.59 compares rigorous calculations to the approximate methods presented here to give an impression of the applicability of the latter. We see that at least in soft magnetic materials ($Q \ll 1$) the deviations from the Bloch path stay small up to external fields of about half the anisotropy field. Note that for the field orientation $\psi = 90°$ the solutions near the end points of the magnetization path either stay parallel to the stray-field-free Bloch wall mode, or they start off perpendicular to it (at large fields). An analysis of the boundary behaviour [509] confirms this observation. If we expand $G(\vartheta, \varphi)$ in the vicinity of the domain directions up to the second order, the first derivatives must vanish since the domain magnetizations must be in equilibrium. The perturbation procedure sketched above then proves that only the eigen directions of the second-order derivative tensor are allowed as starting directions of the magnetization path. These directions are indicated in Fig. 3.59 as dotted lines. A discontinuous transition from one eigenvalue to the other eigenvalue is observed at a critical field. This transition may be called a transition from a Bloch to a Néel wall (see Fig. 3.53). No discontinuous transition occurs at other field orientation angles. The circular magnetization path approximation is generally a very good one, except near the Bloch-Néel transition. Later (Figs. 3.62, 3.63) we will see that just this transition is avoided by a macroscopic wall folding.

(G) Charged Walls and Walls with Long-Range Stresses. If condition (3.114) is not fulfilled, a *net* charge appears on the wall, leading to magnetic fields extending to infinity on both sides of the wall. This is connected with a diverging stray field energy, if the wall is embedded in an infinite medium. But if the saturation magnetization of the material is small and if the actual extension of the domains is limited, such charged walls may occur if some other energy gain in the overall domain structure is achieved [636]. As will be shown, the long-range field and the associated energy in the domains can be separated from the wall, so that the wall can be calculated independently.

Charged walls possess *different* domain magnetization angles ϑ_1 and ϑ_2 in the coordinate system Fig. 3.54c. Expression (3.121) must then be generalized:

$$e_{\mathrm{d}}(\vartheta) = K_{\mathrm{d}} \left[\sin \vartheta - (v_1 \sin \vartheta_1 + v_2 \sin \vartheta_2) \right]^2 , \tag{3.124}$$

where v_1 and v_2 are the relative volumes of the domains ($v_1 + v_2 = 1$). This expression for the stray field energy density inside the wall couples the two domains. Firstly, the equilibrium domain magnetization directions for the given wall normal must be computed. The total energy of the coupled domains, including the contribution $v_1 e_{\mathrm{d}}(\vartheta_1) + v_2 e_{\mathrm{d}}(\vartheta_2)$ from (3.124), is evaluated to:

$$E_{\mathrm{tot}} = v_1 g(\vartheta_1, \varphi_1) + v_2 g(\vartheta_2, \varphi_2) + K_{\mathrm{d}} v_1 v_2 (\sin \vartheta_1 - \sin \vartheta_2)^2 . \tag{3.125}$$

Varying this energy with respect to the four variables ϑ_1, φ_1, ϑ_2 and φ_2 we obtain a set of four equations for these unknowns:

$$g_{\varphi}(\vartheta_1, \varphi_1) = 0 , \qquad g_{\varphi}(\vartheta_2, \varphi_2) = 0 ,$$
$$g_{\vartheta}(\vartheta_1, \varphi_1) + 2 K_{\mathrm{d}} v_2 (\sin \vartheta_1 - \sin \vartheta_2) \cos \vartheta_1 = 0 ,$$
$$g_{\vartheta}(\vartheta_2, \varphi_2) + 2 K_{\mathrm{d}} v_1 (\sin \vartheta_2 - \sin \vartheta_1) \cos \vartheta_2 = 0 . \tag{3.126}$$

Solving them, for example by Newton's method, leads to equilibrium magnetization angles in the domains that are influenced by the long-range stray field created by the net charge. On the other hand, the stray field energy (3.124) has to be added to the generalized anisotropy functional $g(\vartheta, \varphi)$ of simple Bloch wall theory, yielding:

$$G(\vartheta, \varphi) = g(\vartheta, \varphi) + K_{\mathrm{d}} \left[\sin \vartheta - (v_1 \sin \vartheta_1 + v_2 \sin \vartheta_2) \right]^2 . \tag{3.127}$$

This functional is again stationary in the new boundary points $(\vartheta_1, \vartheta_2)$ defined by (3.126). [To prove this, form the derivatives of $G(\vartheta, \varphi)$ treating ϑ_1 and ϑ_2 as constants, and compare the result with (3.126)]. The generalized function $G(\vartheta, \varphi)$ therefore behaves exactly like a conventional anisotropy function, and the wall shape and its additional energy can be calculated as before. Long-range stray fields require more care in the evaluation of the (coupled) boundary conditions. The wall energy covers by definition the additional energy connected with the wall transition from Bloch to Néel walls. The energy of the long-range stray field must be evaluated separately and is part of the domain energy.

The other non-local interaction, the magneto-elastic energy, gives rise to an analogous effect. The magnetostrictive self-energy (3.119) corresponds to the stray field energy (3.124). [Remember that the expression $(v_1 \sin \vartheta_1 + v_2 \sin \vartheta_2)$ in (3.124) is the spatial average of $\sin \vartheta$, corresponding to the average stress tensor $\bar{\sigma}$ in (3.119)]. The quasi-plastic stress tensor σ^0 is (in general) a function of both magnetization angles ϑ and φ, so that all four equations for the boundary conditions contain contributions from the long-range stresses, not just the two ϑ derivatives as in (3.126). Otherwise magnetostrictively "charged" walls behave as magnetically charged walls and can be calculated in the same way [509]. These walls with long-range stresses occur more frequently than magnetically charged walls, since magneto-elastic energies are usually smaller than the stray field energy and need not be avoided to the same degree. An example of this refinement will be discussed in connection with Fig. 3.70.

3.6.2 Generalized Walls in Uniaxial Materials

In this and in the following sections we present frequently occurring applications of the foregoing general theory. We begin with analytical solutions for infinite uniaxial materials — beyond the elementary case presented in Sect 3.6.1A.

(A) Higher-Order Anisotropy. Firstly, we admit a second-order anisotropy constant as in (3.11). With $\kappa = K_{u2}/K_{u1}$ we may write the generalized anisotropy function and the boundary conditions in the coordinates of Fig. 3.54c as:

$$g(\varphi) - g_\infty = K_{u1}(\cos^2\varphi + \kappa \cos^4\varphi) \ , \quad \varphi(-\infty) = -\tfrac{\pi}{2} \ , \quad \varphi(\infty) = \varphi_\infty = \tfrac{\pi}{2} \ . \quad (3.128)$$

Integrating (3.117) and (3.118) leads to the Bloch wall profile and energy:

$$\tan\varphi = \sqrt{1+\kappa}\,\sinh\left[x/\sqrt{A/K_{u1}}\right] \ , \tag{3.129}$$

$$\gamma_{\mathrm{w}} = 2\sqrt{AK_{u1}}\left[1 + \frac{1+\kappa}{\sqrt{\kappa}}\arctan\sqrt{\kappa}\right] = \sqrt{AK_{u1}}\left[1 + \frac{1+\kappa}{\sqrt{-\kappa}}\operatorname{arctanh}\sqrt{-\kappa}\right] . \tag{3.130}$$

The first version of (3.130) avoids complex numbers for positive κ, while the second version is easier applicable for negative κ. The profiles are shown in Fig. 3.60.

The parameter κ must be larger than -1, because otherwise the two domains are not stable (the magnetization with $\varphi = 0$ would have a lower anisotropy energy than $\varphi = \pm \frac{\pi}{2}$). Approaching $\kappa = -1$ the wall shows a tendency to decay into two 90° walls, as shown in Fig. 3.60 for the parameter $\kappa = -0.999$. Such walls contain a zone, which is magnetized perpendicular to the easy axis and parallel to the wall plane, indicating a metastable state in between the two stable domain states. If the effective anisotropy is modified, as for example by a superimposed field or mechanical stress, such walls may split into two halves, leading to the formation of a new domain, a process that was analysed in more detail for antiferromagnets in [637]. In Sect. 3.6.3A we will see that such widened walls are common in cubic materials in which a small effective uniaxial anisotropy is superimposed on the cubic anisotropy. For $\kappa = 0$ the solutions become identical with the previous results (3.109) and (3.110).

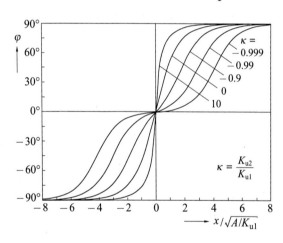

Fig. 3.60 The profiles of 180° walls in uniaxial materials as a function of the relative magnitude of the second-order anisotropy constant. The same type of wall profiles applies also to 180° walls in cubic crystals with positive anisotropy. Note the divergence of the wall width for $\kappa \to -1$

The wall width evaluation depends on the peculiarities of the wall profiles. For $\kappa \geq -0.5$ the profiles have only one point of inflection in the centre, and Lilley's wall width (Sect. 3.6.1C) is derived from the slope at $x = 0$:

$$W_L = \pi \sqrt{A/K_{u1}} \Big/ \sqrt{1+\kappa} \quad \text{for } \kappa \geq -0.5 \;. \tag{3.131a}$$

For $\kappa < -0.5$ the anisotropy functional $g(\varphi)$ contains an intermediate minimum that leads to a widened wall and to three points of inflection in the wall profile (sketch). The wall width must in this case lean on the tangents in the outer inflection points, leading to:

$$W_L = 2\Big\{ \big[\pi - 2\arctan\sqrt{-1-2\kappa} \big] \sqrt{-\kappa} + \operatorname{arctanh}\sqrt{\frac{1+2\kappa}{\kappa}} \Big\} \sqrt{A/K_{u1}} \;. \tag{3.131b}$$

(B) Applied Fields. In static equilibrium, magnetic fields acting on a wall must be oriented perpendicular to the domain magnetizations and therefore to the axis of the uniaxial material as elaborated in Sect. 3.6.1B. We denote the field orientation by the angle ψ, with $\psi = 90°$ perpendicular to the wall. In the simplest case the field is directed parallel to the wall ($\psi = 0°$). The total generalized anisotropy functional is then:

$$G(\varphi) = K_{u1}(\cos^2\varphi + \kappa\cos^4\varphi) - HJ_s\cos\varphi \ . \qquad (3.132)$$

In the domains the functional $G(\varphi)$ must again be stationary with respect to φ ($G_\varphi = 0$), leading to the following implicit equation for the domain magnetization angle φ_∞:

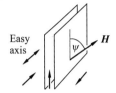

Easy axis

H

ψ

$$HJ_s = 2K_{u1}c_\infty(1 + 2\kappa\,c_\infty^2) \ , \quad c_\infty = \cos\varphi_\infty \ . \qquad (3.133)$$

The resulting φ_∞ must be checked whether it represents a stable, non-trivial solution, which is skipped here, however. Inserting (3.133) into (3.132), we obtain the extra anisotropy energy of the wall:

$$G(\varphi) - G_\infty = K_{u1}(\cos\varphi - c_\infty)^2\left[1 + \kappa\left(\cos^2\varphi + 2\cos\varphi\,c_\infty + 3\,c_\infty^2\right)\right] \ . \qquad (3.134)$$

The case $\kappa = 0$ can be treated analytically, yielding:

$$(1 - \cos\varphi\cos\varphi_\infty)/(\cos\varphi - \cos\varphi_\infty) = \cosh\left(x\sin\varphi_\infty/\sqrt{A/K_{u1}}\right) \ ,$$

$$\gamma_w = 4\sqrt{AK_{u1}}\left(\sin\varphi_\infty - \varphi_\infty\cos\varphi_\infty\right) \ . \qquad (3.135)$$

Figure 3.61 shows the energy and wall width of these solutions together with numerically integrated results for $\kappa \neq 0$. Note the discontinuities in wall energy for negative K_{u2}, which indicate discontinuous transitions to saturation, i.e. to the trivial solutions $\varphi_\infty = \pm\frac{\pi}{2}$ of (3.133).

For general field orientations $\psi > 0$ the field will also influence the angle ϑ, so that both polar coordinates ϑ and φ have to be considered. Neglecting K_{u2} we obtain for the generalized anisotropy functional:

$$G(\vartheta, \varphi) = K_{u1}(1 - \sin^2\varphi\cos^2\vartheta) - HJ_s(\cos\varphi\cos\vartheta\cos\psi + \sin\vartheta\sin\psi)$$
$$+ K_d(\sin\vartheta - \sin\vartheta_\infty)^2 \ , \qquad (3.136)$$

where the stray field energy term (3.121) had to be included to account for a variable angle ϑ. Introducing the equations $HJ_s\cos\psi = 2K_{u1}\cos\vartheta_\infty\cos\varphi_\infty$ and $HJ_s\sin\psi = 2K_{u1}\sin\vartheta_\infty$, valid for equilibrium in the domains as in (3.133), the generalized anisotropy functional becomes:

$$G(\vartheta, \varphi) - G_\infty = -K_{u1}(\cos\varphi\cos\vartheta - \cos\varphi_\infty\cos\vartheta_\infty)^2 + K_d(\sin\vartheta - \sin\vartheta_\infty)^2 \ . \qquad (3.137)$$

Numerical solutions for the optimum magnetization paths resulting from variation of the total energy were shown in Fig. 3.59.

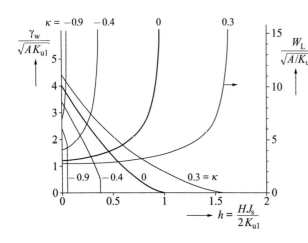

Fig. 3.61 Wall energy (decreasing with rising field and decreasing wall angle) and wall width according to Lilley (rising curves) in uniaxial materials under the influence of an applied field parallel to the wall and perpendicular to the easy axis. Parameter is the ratio $\kappa = K_{u2}/K_{u1}$

The wall energy as a function of ψ is shown in Fig. 3.62. Because the wall assumes for $\psi \neq 0$ a partial Néel character, the solutions depend on the ratio $Q = K_{u1}/K_d$, chosen here as $Q = \frac{1}{4}$. An interesting feature of this wall is its tendency to undergo a zigzag folding. For a given external field the wall parallel to the field ($\psi = 0$) has the lowest energy. If a wall is fixed in its overall orientation otherwise, it tends to reorient locally to decrease its total energy. Let the mean orientation of the wall be $\overline{\psi} = 90°$. To take into account the increased wall surface, we have to study the function $\gamma_w(\psi)/\sin\psi$, which shows in fact a minimum at an angle $\psi_0 < 90°$ (Fig. 3.62). It turns out that in equilibrium all walls with an average orientation $\overline{\psi} > \psi_0$ are composed of segments of orientation $\pm\psi_0$ only. Only the segment lengths with $\psi = \pm\psi_0$ are different if the average angle $\overline{\psi}$ differs from 90°.

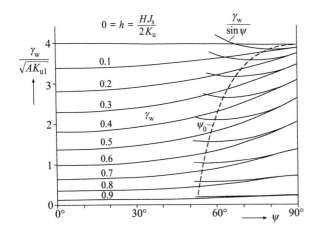

Fig. 3.62 The wall energy in a uniaxial material with $Q = \frac{1}{4}$ as a function of the orientation ψ of an applied field for different values of this field. A wall that is held at an average orientation of 90° folds into segments of orientation $\pm\psi_0$. This equilibrium zigzag angle is given by the minimum of the function $\gamma_w(\psi)/\sin\psi$

The optimum zigzag angle is shown in Fig. 3.63 for materials with different material constants Q. This zigzag folding was observed experimentally [638], including the remarkable feature that for materials with $Q < 1$ the zigzag shape persists to saturation, whereas for materials with $Q > 1$ the incipient zigzag folding vanishes at higher fields again when Néel walls are formed, which show no tendency for a zigzag folding. In contrast, the Néel walls shown in Fig. 3.59a for materials with $Q < 1$ will never be realized according to these calculations, as long as a zigzag folding is geometrically possible.

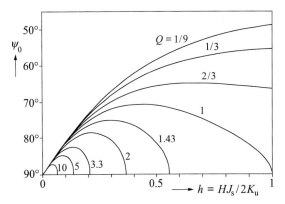

Fig. 3.63 The equilibrium zigzag angle in uniaxial materials as a function of an applied field perpendicular to the wall plane for different values of $Q = K_{u1}/K_d$

3.6.3 Walls in Cubic Materials

Walls in cubic materials appear in many varieties even without an applied field [633]. There are walls of various angles like 180° and 90° walls, and they may be oriented in different ways relative to the crystal. Magnetostriction could be neglected for walls in uniaxial materials, but it often plays a decisive role in cubic materials. We only treat the two main magnetic classes of cubic ferromagnets (see Sect. 3.4.3A): "positive anisotropy" materials like iron with $K_{c1} > 0$ and $\langle 100 \rangle$ easy directions, and "negative anisotropy" materials like nickel with $K_{c1} < 0$ and $\langle 111 \rangle$ easy directions.

(A) 180° Walls in Positive-Anisotropy Cubic Materials $(K_{c1} > 0)$. We assume the magnetization within the wall to rotate from the [001] direction to its opposite. The wall normal can point in any direction perpendicular to this axis; the orientation angle ψ is measured between the wall and the (100) plane. No fields can exist in static equilibrium with a 180° wall in cubic material for the following reason: if the field contained

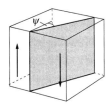

a component parallel to the [001] axis, the wall would move as in a uniaxial material. But if a field in the (001) plane would be present, it would necessarily favour one or two of the other easy directions of the cubic material and thus lead to a breakdown of the original domain pattern.

An orientation of the coordinate system like in Fig. 3.54b with the [001] axis as polar axis is most convenient to describe these walls. Then the anisotropy energy of the wall can be taken from (3.9a), replacing φ by ψ:

$$g(\vartheta) = K_{c1} \cos^2\vartheta \left[1 - \cos^2\vartheta (1 - F_\psi)\right] + K_{c2} \cos^4\vartheta \sin^2\vartheta F_\psi , \qquad (3.138)$$

with $F_\psi = \sin^2\psi \cos^2\psi$. For $K_{c2} = 0$, the functional (3.138) is equivalent to the expression (3.128) for uniaxial materials when we replace κ by $F_\psi - 1$. Substituting this in (3.130) for negative values of κ we get for the wall energy:

$$\gamma_w = 2\sqrt{AK_{c1}} \left[1 + F_\psi \operatorname{arctanh}\sqrt{1 - F_\psi} / \sqrt{1 - F_\psi}\right] . \qquad (3.139)$$

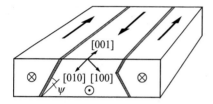

Fig. 3.64 The preferred orientation of 180° walls in (110) oriented transformer steel favours wall shapes as shown, in contrast to straight, perpendicular walls

The energy is lowest ($\gamma_{100} = 2\sqrt{AK_{c1}}$) for $\psi = 0$ or $\psi = 90°$, i.e. for the (100) or (010) walls. This is the type of wall for which magnetostrictive interactions have to be taken into account to achieve a finite wall width as shown in Sect. 3.6.1E, a correction that can be ignored in a discussion of wall energy, however. The other extreme wall orientation, (110) or $\psi = 45°$, has an energy that is about 40% larger ($\gamma_{110} \cong 2.76\sqrt{AK_{c1}}$). This energy difference has consequences for the magnetization processes in (110) oriented transformer steel. The 180° wall with the smallest area (Fig. 3.64) would be the high-energy (110) wall perpendicular to the surface. Since the wall energy is lower for {100} orientations, the (110) wall tends to rotate towards these orientations, forming tilted or zigzag walls with a lower overall energy, as shown in Fig. 3.65. This fact has to be considered especially in the discussion of dynamic processes and eddy current losses ([639]; see also Sect. 3.6.8).

The "degenerate" {100} walls show a tendency to split into two 90° walls as discussed in Sect. 3.6.1E and demonstrated in Fig. 3.60 for $\kappa \rightarrow -1$. The equilibrium distance of the two partial walls (and thus the wall width of the 180° wall) is governed by the magnetostrictive self-energy (3.119):

$$e_{ms} = \tfrac{9}{2} C_2 \lambda_{100}^2 \cos^2\vartheta - \tfrac{9}{2}\left(C_2 \lambda_{100}^2 - C_3 \lambda_{111}^2\right) \sin^2\vartheta \, \cos^2\vartheta \quad . \tag{3.140}$$

The second term is of cubic symmetry and is usually negligible compared to the corresponding K_{c1} term. The wall width is therefore determined by the first term, i.e., by the dimensionless material parameter $\rho_{ms} = \tfrac{9}{2} C_2 \lambda_{100}^2 / K_{c1}$. In iron ρ_{ms} is about 0.003, which, evaluating (3.131) with $\kappa = -1/(1 + \rho_{ms})$, leads to a 180° wall width of more than $10\sqrt{A/K_{c1}}$, much more than the sum of the wall widths of the two 90° walls (which would be $2\pi\sqrt{A/K_{c1}}$). Figure 3.66 shows the wall energy and the wall width of these walls as a function of ρ_{ms}. Note that this result applies to the hypothetical case of an infinitely extended wall. The presence of a surface may severely modify especially such "soft" walls, as will be discussed in more detail in Sect. 3.6.4D.

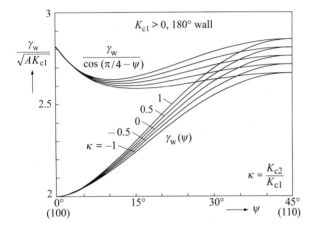

Fig. 3.65 The specific wall energy of the 180° wall in cubic materials with positive first-order anisotropy constant as a function of orientation for different values of the second-order anisotropy constant. The function $\gamma_w / \cos(\pi/4 - \psi)$ defines the optimum orientation for walls as in Fig. 3.64. The shortest wall is given by $\psi = \pi/4$

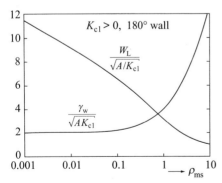

Fig. 3.66 Wall energy and wall width of {100} 180° walls in positive-anisotropy cubic materials as a function of the relative size of the magnetostrictive energy, measured by $\rho_{ms} = 9/2\, C_2 \lambda_{100}^2 / K_{c1}$. The ordinate scale is applicable to the reduced wall energy as well as to the reduced wall width

(B) 180° Walls in Negative-Anisotropy Materials $(K_{c1} < 0)$. For materials with easy directions along $\langle 111\rangle$ we use the representation (3.9c) of cubic anisotropy. It leads, neglecting K_{c2}, to an asymmetric functional:

$$g(\varphi) = -\tfrac{2}{3} K_{c1} \left[\cos^2\vartheta + c_1 \cos^4\vartheta + c_2 \cos^3\vartheta \sin\vartheta - \tfrac{1}{2} \right] , \qquad (3.141)$$

with $c_1 = -\tfrac{7}{8}$ and $c_2 = \sqrt{\tfrac{1}{2}} \sin(3\psi)$. This wall can be evaluated analytically [633]. The result for the wall energy is:

$$\gamma_w = 2\sqrt{\tfrac{2}{3} A |K_{c1}|} \left[1 + (r_2/r_0)\sqrt{\tfrac{1}{2}(r_0+1)/|c_1|}\ \text{arcsinh}\sqrt{\tfrac{1}{2}|c_1|(r_0+1)/r_2} \right.$$
$$\left. + (r_1/r_0)\sqrt{\tfrac{1}{2}(r_0-1)/|c_1|}\ \arcsin\sqrt{\tfrac{1}{2}|c_1|(r_0-1)/r_1} \right., \qquad (3.142)$$

with $r_0 = \sqrt{1 + c_2^2/c_1^2}$, $r_1 = 1 - \tfrac{1}{2}c_1(r_0-1)$ and $r_2 = 1 + \tfrac{1}{2}c_1(r_0+1)$.

This wall again tends to split into two partial walls (a 71° and a 109° wall) when the magnetization rotates through another $\langle 111 \rangle$ direction, which happens for $\psi = 30°$. The calculation of the wall width for this orientation requires the inclusion of magnetostriction. Figure 3.67 shows results for nickel, including effects of second-order anisotropy.

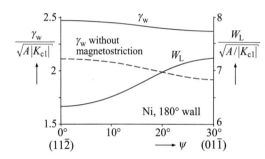

Fig. 3.67 Wall energy and wall width of 180° walls in nickel as a function of wall orientation. The second-order anisotropy is given by $K_{c2}/K_{c1} = 0.45$. The reduced magnetostriction data for nickel at room temperature are $C_2\lambda_{100}^2/K_{c1} = 0.03$ and $C_3\lambda_{111}^2/K_{c1} = 0.04$. The *dashed curve* results from neglecting magnetostriction

(C) 90° Walls. One example of this wall type in positive-anisotropy cubic materials was introduced already: just take one half of the (100) 180° wall treated before. The 90° wall of this orientation has the lowest specific wall energy among all possible uncharged orientations (see Fig. 3.65), but it describes an elastically incompatible situation. In bulk crystals the (110) orientation is preferred. It carries the highest specific energy but it avoids domain stresses, as discussed in Sect. 3.6.1E.

For the general wall we introduce an orientation angle ψ (Fig. 3.56). Except for three special orientation angles, these walls must be integrated numerically. For all orientations the magnetization in the wall must rotate from one easy direction $\boldsymbol{m}^{(1)}$ to another $\boldsymbol{m}^{(2)}$, which may be chosen as:

$$\boldsymbol{m}^{(1)} = (1, 0, 0) , \quad \boldsymbol{m}^{(2)} = (0, 1, 0) .$$

To represent the Bloch wall profile we define the unit vectors (see Fig. 3.56):

$$\boldsymbol{n} = (\sqrt{\tfrac{1}{2}} \sin\psi, \sqrt{\tfrac{1}{2}} \sin\psi, \cos\psi) , \quad \boldsymbol{w} = \sqrt{\tfrac{1}{2}}(1, \bar{1}, 0) , \quad \boldsymbol{t} = \boldsymbol{n} \times \boldsymbol{w} ,$$

where n is an allowed wall normal pointing from domain 1 to domain 2, w is a unit vector parallel to $m^{(1)} - m^{(2)}$ and t is a tangential vector perpendicular to n and w. In this system the magnetization vector can be written as:

$$m = \cos\vartheta \cos\varphi \, t - \cos\vartheta \sin\varphi \, w + \sin\vartheta \, n \ .$$

Inside the wall, ϑ stays constant and φ rotates from a negative value for $m^{(1)}$ to a positive value for $m^{(2)}$. Fields are allowed along the [110] direction only. The exact boundary values (ϑ, φ_∞) have to be calculated, including field and stress effects as discussed in Sect. 3.6.1F,G. Here are two examples for zero field and zero magnetostriction: the (001) 90° wall is given by $\psi = 0$, $\vartheta = 0$ and φ running from $-\frac{\pi}{4}$ to $+\frac{\pi}{4}$; for the (110) 90° wall we have $\psi = \frac{\pi}{2}$, $\vartheta = \frac{\pi}{4}$ and φ running from $-\frac{\pi}{2}$ to $+\frac{\pi}{2}$. The magnetization vector is inserted into the appropriate energy expressions (cubic anisotropy, applied field, stray field and magneto-elastic energies) and then (3.118) is integrated numerically. In the first step the angle ϑ is assumed to be constant, followed by a refinement using the linearization method (3.122, 3.123). Results are shown in Fig. 3.68.

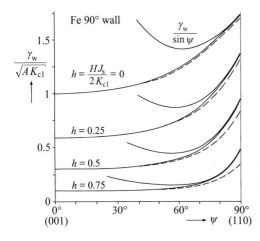

Fig. 3.68 The wall energy of 90° walls as a function of orientation. The parameter is an applied field along the [110] direction. The *dashed lines* indicate calculations that allow for deviations from the stray-field-free Bloch path

For iron the internal magneto-elastic contribution in the wall is as negligible as the stray field correction. The wall energy shows a characteristic cusp-shaped maximum at the orientation angle $\psi = 90°$ [(110) orientation]. Beyond this angle another branch describing a wall with opposite rotation sense takes over. The energy for $\psi = 90°$ is large because for this orientation the magnetization path is the longest of all orientations, and at the same time the magnetization vector rotates almost through the hard [111] direction.

A wall for which the mean orientation $\psi = 90°$ is enforced by the domain structure into which it is embedded can save energy by a zigzag folding in spite of the area increase by the factor of $1/\sin\psi$. The calculated optimum zigzag angle ψ_0 is in nice agreement with observations as was first found by *Chikazumi* and *Suzuki* [640]. Figure 3.69 shows an example of such an observation on silicon-iron. According to Fig. 3.68 the equilibrium zigzag angle of the 90° wall should be $\psi_0 = 63°$, which, when projected onto the surface, should lead to an angle Φ of 54°, as defined in the drawing of Fig. 3.69. The actually observed angle is measured as $58 \pm 3°$ in this picture. The small deviation from the predicted value is probably caused by the influence of the compressive stress that leads to the formation of this pattern by favouring the internal easy axis and that was not taken into account in the wall calculation. The "walls" visible at the surface are actually intersections of two subsurface walls, as indicated in the drawing. They are called V-lines [641] for their V-shaped internal structure. The change in wall contrast along the V-lines indicates an alternating micromagnetic configuration: the magnetization in the middle of the V-line points either parallel or antiparallel to the line.

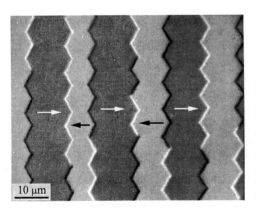
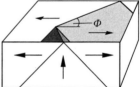

Fig. 3.69 Zigzag walls observed by Kerr microscopy at the surface intersection of closure domains on a (100) surface of a silicon-iron crystal under the influence of a planar compressive stress

In Fig. 3.69 the average wall orientation $\psi = 90°$ of the internal 90° walls is determined by the geometrical conditions of the overall domain pattern. In other cases, such as that of Fig. 1 a, the observed average wall orientation close to 90° is determined by elastic compatibility considerations. The zigzag folding induces an elastic stress field on the scale of the zigzag period. The elastic energy is balanced by the extra energy of the wall kinks.

If the average wall angle is different from 90° but lies between ψ_0 and 90°, then theory predicts wall segments to occur with orientation $\pm \psi_0$, but of different length, as discussed before in Sect. 3.6.2B. The wall segments behave effectively like equilibrium phases, as in Sect. 3.3.1.

(D) 71° and 109° Walls. These walls occurring in negative-anisotropy cubic materials separate domains magnetized along the $\langle 111 \rangle$ axes. The magnetization in a 71° wall may rotate without field

$$\text{from } \boldsymbol{m}^{(1)} = \sqrt{\tfrac{1}{3}}\,(1, 1, \bar{1}) \text{ to } \boldsymbol{m}^{(2)} = \sqrt{\tfrac{1}{3}}\,(1, 1, 1) \;.$$

Allowed wall normals and the other unit vectors are then:

$$\boldsymbol{n} = \sqrt{\tfrac{1}{2}}\,(\sin\psi + \cos\psi, \; \sin\psi - \cos\psi, \; 0) \;,\;\; \boldsymbol{m} = (0, 0, \bar{1}) \;,\;\; \boldsymbol{t} = \boldsymbol{n} \times \boldsymbol{w} \;.$$

A field can be applied along the [110] direction.

Similarly, the magnetization for a 109° wall may rotate without field

$$\text{from } \boldsymbol{m}^{(1)} = \sqrt{\tfrac{1}{3}}\,(1, \bar{1}, \bar{1}) \text{ to } \boldsymbol{m}^{(2)} = \sqrt{\tfrac{1}{3}}\,(\bar{1}, 1, \bar{1}) \;.$$

The relevant unit vectors for the wall calculation are now:

$$\boldsymbol{n} = (\sqrt{\tfrac{1}{2}}\cos\psi, \; \sqrt{\tfrac{1}{2}}\cos\psi, \; \sin\psi) \;,\;\; \boldsymbol{w} = \sqrt{\tfrac{1}{2}}\,(1, \bar{1}, 0) \;,\;\; \boldsymbol{t} = \boldsymbol{n} \times \boldsymbol{w} \;.$$

For the 109° wall a field is allowed along the [001] direction. Results for both wall types are shown in Figs. 3.70 and 3.71 for nickel.

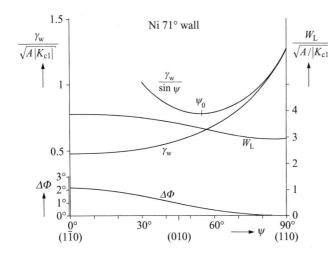

Fig. 3.70 Energy and wall width of 71° walls in nickel. The function $\gamma_w/\sin\psi$ indicates, as in Fig. 3.68, an equilibrium zigzag folding. The magnetization in the domains deviates by the angle $\Delta\Phi$ from the easy directions because of the long-range stress field of the wall. See Fig. 3.67 for the material data of nickel

Note the relatively strong influence of the magnetostrictive contribution in this material. Even the refined treatment of the boundary conditions necessary for walls between elastically incompatible domains (Sect. 3.6.1E) leads to recognizable effects. The angle $\Delta\Phi$, by which long-range stresses cause the magnetization in the domains to rotate away from the easy axes, reaches two degrees for 71° walls in nickel, as shown in Fig. 3.70.

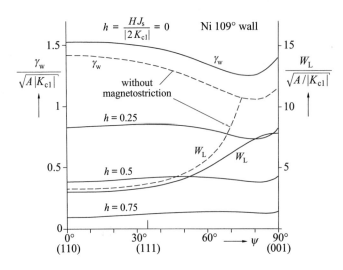

Fig. 3.71 Calculations for 109° walls in nickel. The wall energy is shown for different values of an applied field along [001]. The wall width and the *dashed curves,* calculated neglecting magnetostrictive energy, are for $h = 0$. See Fig. 3.67 for the material data of nickel

3.6.4 Domain Walls in Thin Films

Magnetic films are defined as thin, if their thickness is comparable to the Bloch wall width so that the concept of an infinitely extended wall becomes questionable. Some of the concepts that apply to domain walls in thin films also apply to the wall surface region in bulk material.

We distinguish between two basic geometries: films with *in-plane* anisotropy and films with *perpendicular* anisotropy. In "planar" films the magnetization in the domains is parallel to the film plane. The anisotropy may be either uniaxial or biaxial. We will see, however, that in such films the exact nature and size of the anisotropy is less important than film thickness and wall angle. In "perpendicular" films the magnetization in the domains is perpendicular to the film surface, which requires small saturation magnetization and large anisotropy. This is true for bubble memory carriers and for perpendicular or magneto-optical storage media. We will start with the classical in-plane anisotropy films and turn to perpendicular-anisotropy films in (H).

(A) Walls in Films with In-Plane Anisotropy — Qualitative Overview.
Néel [642] first realized that standard wall theory of Bloch walls does not hold for thin films, if the film thickness becomes comparable to the wall width. Then a wall mode using an in-plane rotation (Fig. 3.72b) has a lower energy than the classical Bloch wall mode (Fig. 3.72a).

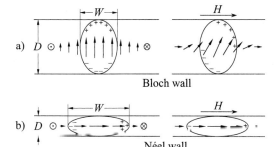

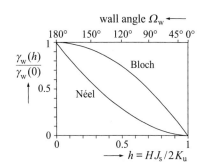

Fig. 3.72 Néel's view of domain walls in thin films. The cross-sections indicate the magnetic charges with and without an external field

Néel estimated the transition thickness by a simple argument: the wall is approximated by an elliptical cylinder of width W and height D. The demagnetizing factor of this cylinder along the vertical magnetization direction is, evaluating (3.23), for the Bloch wall $N_{\text{Bloch}} = W/(W+D)$. If W becomes larger than D, the demagnetizing energy increases and the wall prefers to flip into the "Néel wall" mode (Fig. 3.72b). The demagnetizing factor for this wall is $N_{\text{Néel}} = D/(W+D)$, which is smaller than N_{Bloch} for $W > D$. Néel allowed the wall width to be influenced by the stray field energy but kept the wall structures otherwise unchanged. He predicted a transition between the two wall modes and this transition is connected with a minimum in wall width and a maximum in the specific wall energy.

Fig. 3.73 The relative variation of the energy of Bloch and Néel walls according to Néel's model as a function of an in-plane field applied perpendicular to the wall

Figure 3.72 also indicates the modification of the Néel picture when a magnetic field is applied perpendicular to the easy axis. Such a field leads to a reduction of the wall angle $\Omega_{\text{w}} = 2\vartheta_0$ with $\cos\vartheta_0 = h = HJ_s/2K$. In a Bloch wall the vertical magnetization in the centre tilts accordingly and the corresponding surface charges decrease quadratically with the applied field. For the Néel wall, the charge reduction is much more pronounced. The integrated total charge in each half of the 180° Néel wall must be J_sD, while it is reduced to $(1-1/\sqrt{2})J_sD$ for a 90° wall. As the stray field energy varies quadratically with the charge, this predominant energy contribution is reduced by an order of magnitude. Figure 3.73 shows the resulting dependence of the total wall energy of Bloch and Néel walls on the applied field obtained by optimizing the wall width parameter of Néel's model for each field.

The 90° Néel wall has in this model only about 12% of the energy of the 180° Néel wall. This strong preference for lower-angle Néel walls has a remarkable consequence: energy can be gained by replacing a 180° Néel wall by a complicated composite wall, the *cross-tie wall* [643], which is shown in Fig. 3.74 and explained qualitatively in Fig. 3.75. Although the total length of the 90° walls within this pattern is much larger, the total wall energy is smaller.

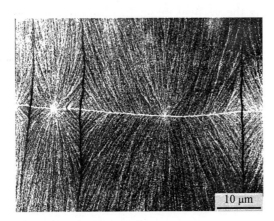

Fig. 3.74 A transmission electron micrograph of a cross-tie wall in a Permalloy film of about 35 nm thickness. The local magnetization direction can be derived from the ripple structure (Sect. 5.5.2C) which runs perpendicular to the mean magnetization. (Courtesy *E. Feldtkeller*)

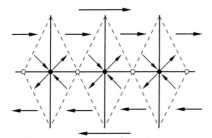

Fig. 3.75 A "domain" model of a cross-tie wall. The *dashed lines* are actually to be replaced by continuous transitions [644]. The *circular* and *cross Bloch lines* are indicated by *open* and *filled* circles, respectively. This pattern has a lower energy than the simple 180° Néel wall because it consists mainly of energetically favourable 90° walls

(B) Systematic Analysis of Domain Walls in Thin Films. By definition, domain walls in thin films need a more than one-dimensional description. We saw that in bulk material the wall structure was normally a function of a single spatial variable only. This was even true if stray fields (Sect. 3.6.1F), or magneto-elastic stresses (Sect. 3.6.1E) play a role. Only in domain walls tending to develop a zigzag folding a deviation from the one-dimensional nature had to be taken into account in bulk material. In contrast, domain walls in thin films are essentially multidimensional. We will see that the typical domain wall in very thin films, the symmetric Néel wall discussed in (C), needs basically a one-dimensional description of its magnetization structure. The stray field of this wall reaches out into the space above and below the film, however, and cannot be treated one-dimensionally. Cross-tie walls in *thin* films need at least a two-dimensional description for the magnetization and a three-dimensional treatment of the stray field. Bloch walls in thick films and their relatives are two-dimensional both in the magnetization and in the stray field. Most complex are cross-tie walls in thicker films (~100 nm for Permalloy), which need a three-dimensional description of both magnetization and field and which have never been analysed theoretically up to now.

A characteristic difference between bulk and thin film walls is connected with the occurring scales. Domain walls in bulk material generally scale with the exchange length of the anisotropy energy $\Delta = \sqrt{A/K_u}$. In thin films we will encounter domain walls that scale with the exchange length of the stray field $\Delta_d = \sqrt{A/K_d}$, or with the film thickness, or which display multiple scales within one wall profile.

Domain walls in bulk material can be computed either completely analytically, or at least explicitly up to a final, numerically trivial integration. This is no more possible for thin film walls. They can either be explored by a variational procedure based on properly chosen series expansions and test functions, or with discrete numerical procedures. Consistency tests are important in all numerical methods. Such a test was invented by *Aharoni* [645, 646]. If a computed or proposed wall ansatz or "model" fulfils the micromagnetic differential equations, then the total wall energy (3.115) or its two-dimensional generalization can be expressed in a different form based on the micromagnetic equations, the result of which has to agree with the result of the original expression. Because the differential equations have been inserted, higher-order derivatives appear in these alternative forms, not just first-order derivatives as in (3.115). The agreement between both expressions is a necessary condition for a proper solution of the differential equation, a discrepancy indicating insufficient quality of a model. Unfortunately, the criterion is boundary condition dependent and not easily generalized to new problems. It is most useful in the initial stages of an investigation, when there is still uncertainty about the nature of a solution and the adequacy of the chosen ansatz. It appears to be less useful in examining numerical inaccuracies and discretization errors in finite element calculations.

(C) Symmetric Néel Walls. The wall type of Fig. 3.72b occurs only in rather thin films (up to about 50 nm for Permalloy, for example). Therefore magnetization will be confined to the film plane and a one-dimensional description is adequate. We assume uniaxial anisotropy and an in-plane field perpendicular to the axis and parallel to the film plane. The field is given by $h = H J_s / (2 K_u)$ and the magnetization vector is represented by the angle ϑ, with $\vartheta = 0$ along the wall normal. The boundary conditions in the domains are $\cos \vartheta_\infty = h$.

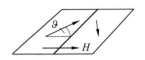

The problem resembles the case that was treated in Sect. 3.6.2B, except for the stray field energy that becomes non-local for thin films because the wall stray field extends into space above and below the film. Early attempts of computing this wall type numerically [647, 648] met numerical difficulties, which were solved, however, by *Kirchner* and *Döring* [649] who used a variable cell size. The wall profile calculated by these authors was then confirmed by many independent methods. A variational ansatz [650], which was later published as a computer program [651], reproduced the profile of Kirchner and Döring accurately in all details and extended the results to other thicknesses, wall angles, anisotropy values and anisotropy types. The ansatz leaned on earlier work of *Dietze* and *Thomas* [652] and of *Feldtkeller* and *Thomas* [653], who still did not know the exact numerical solution and therefore used an insufficient number of parameters. Small deviations from the one-dimensional structure of the symmetric Néel wall were explored in [650] and demonstrated to be insignificant. In [651] Aharoni's self-consistency parameter was confirmed for all walls to agree with unity within a few percent accuracy. It is not clear why Aharoni in his review ignores all these well-published and mutually consistent results, claiming that the exploration of the symmetric Néel wall is still "in a state of confusion" [514 (p. 155 –165)].

The problem in computing this wall type lies in its decomposition into three parts of widely different scales: a sharply localized core interacts with two extremely wide tails, which take over a good part of the total rota-

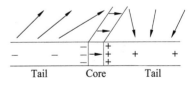

Tail   Core   Tail

tion. The core is characterized by a dipolar charge pattern, while only diluted charges of a single polarity are found in the tails. The most transparent approach to solve this problem was found by *Riedel* and *Seeger* ([654]; see also [509]). They separated the mathematical description into a differential equation for the core and an integral equation for the tail. The differential equation can be solved in the usual way, as described before. The integral equation becomes linear if the exchange energy is neglected in the tail region and can then be solved by Fourier methods. Both parts are coupled by the fields they generate in the other part, and by the condition that the two parts together must amount for the total wall rotation. Typical solutions for the wall profile are shown in Fig. 3.76. A logarithmic x scale is used to show the core and the tail in the same graph. The two parts can be clearly distinguished when the magnetization component perpendicular to the wall is plotted: the tail follows a largely logarithmic behaviour and extends from $\cos \vartheta = h$ to $\cos \vartheta = c_0$, and the core covers the range from $\cos \vartheta = c_0$ to $\cos \vartheta = 1$. The core tail boundary value c_0 can in principle lie between h and 1 but often assumes its equilibrium, minimum energy value somewhere in the middle of this interval.

As mentioned, two characteristic lengths can be identified in such a wall. The first is the core width, defined in analogy to the quantity W_{m} in Fig. 3.55. The analytical solution for the core profile yields:

$$W_{\mathrm{core}} = 2\sqrt{A\left(1-h^2\right)/\left[\left(K_{\mathrm{u}}+K_{\mathrm{d}}\right)\left(1-c_0\right)^2\right]} \ . \tag{3.143}$$

This core width turns out to be smaller than the film thickness except in ultrathin films. Because the core width scales with $\sqrt{A/K_{\mathrm{d}}}$ when K_{d} is much larger than K_{u} as in soft magnetic materials, the Néel wall core is in such materials smaller than the wall width in a corresponding bulk sample that scales with $\sqrt{A/K_{\mathrm{u}}}$. In the core the exchange energy (A) is primarily balanced by the stray field energy (K_{d}) connected with the dipolar charges.

The second characteristic length is the tail width, defined by extrapolating the logarithmic profile to $\cos \vartheta = h$ (Fig. 3.76). The value resulting from the integral equation for this quantity is:

$$W_{\mathrm{tail}} = \mathrm{e}^{-\gamma} D K_{\mathrm{d}}/K_{\mathrm{u}} \approx 0.56 \, D K_{\mathrm{d}}/K_{\mathrm{u}} \ , \tag{3.144}$$

where $\gamma = 0.577\ldots$ is Euler's constant. The Néel wall tail is determined by a balance between charge distribution (K_{d}) and anisotropy energy (K_{u}).

The stray field energy thus has an opposite effect on both parts of the wall. With increasing K_{d} the core width decreases and the tail width increases. The following reasoning may help to explain this behaviour: in the centre of the wall the flux into the wall normal direction is J_{s}, or 1 in reduced units. In the domains this flux is $h = \cos \vartheta_\infty$. In each half wall the reduced total magnetic charge $\pm(1-h)$ has therefore to be somehow distributed. Part of the charge is concentrated in the core, where it supports a low energy state by the close interaction with its counterpart of opposite polarity. This part is limited by the exchange energy, which prevents an arbitrarily narrow core width. The other part of the charge gets widely spread in the tail. The wider the tail, the smaller the charge density, thus reducing this contribution to the stray field energy. This mechanism is balanced by an increasing anisotropy energy as manifested by the formula for the tail width (3.144). The two parts of the Néel wall are coupled by two mechanisms: the part of the wall rotation taken over by the core determines the amount of charge left over to be spread in the tail, and the tail generates a positive stray field that stabilizes the core.

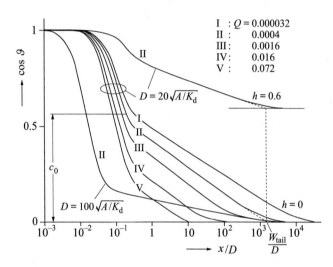

Fig. 3.76 Calculated profiles of typical Néel walls [650]. The magnetization component perpendicular to the wall is plotted as a function of the distance from the centre in a logarithmic scale. Parameters are the reduced anisotropy $Q = K_u/K_d$ and the reduced applied field $h = HJ_s/2K_u$. For the $h = 0$ curve of case II the definition of the external wall width W_{tail} and of the core tail boundary c_0 are indicated

The extended tails of Néel walls lead to strong interactions between them. Figure 3.77 shows the wall energy for four cases of parallel walls described in the caption. The interactions become important as soon as the tail regions overlap. The sign of the interaction depends on the wall

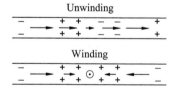

rotation sense. Néel walls of opposite rotation sense (so-called unwinding walls) attract each other because they generate opposite charges in their overlapping tails. If they are not pinned, they can annihilate. Néel walls of equal rotation sense (winding walls) repel each other. If they are pressed together by an external field disfavouring the region between the walls, they form stable pairs, so-called 360° walls. These can be annihilated only in large fields. For a Permalloy film of 50 nm thickness Néel wall interactions extend over distances of at least 100 μm! For the energy of symmetric Néel walls see Fig. 3.79.

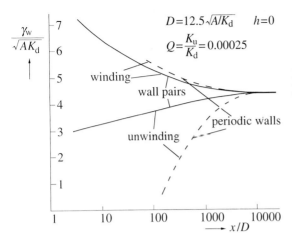

Fig. 3.77 The energy of interacting parallel Néel walls [655] for four cases: for winding and unwinding pairs, and for periodic series of the same pairs. The variable x denotes the distance between the walls. For the indicated value of the anisotropy parameter Q (applicable to Permalloy layers) the nominal tail width (3.144) is $W_{tail} = 2240\,D$. The interactions begin when the tails overlap ($x < 2\,W_{tail}$)

These long-range interactions are responsible for many interesting hysteresis phenomena in thin magnetic films, which will be discussed in more detail in Sect. 5.5.2. They certainly influence the cross-tie wall (Fig. 3.74). The basic mechanism of cross-tie formation was already discussed in connection with Néel's model (Fig. 3.73). Some energy of 180° walls is saved by replacing them with a complex pattern of 90° walls. Although the total 90° wall area is 3–4 times larger than the original 180° wall area (Fig. 3.75), the total energy decreases because the specific energy of 90° walls is smaller than that of 180° walls by a factor that overcompensates the larger wall area of the cross-tie pattern. This fundamental property of Néel walls is confirmed by more detailed calculations, such as those presented in Figs. 3.76.

An attempt to calculate the equilibrium *period* of cross-ties has to include the repulsive interaction between the main wall segments and the adjacent cross-ties. Both generate the same charge polarity in their tails according to their rotation sense, and when the tails are calculated for isolated walls, they would overlap in the cross-tie pattern. In addition, the extra energy connected with the continuous transitions between the partial walls, the so-called Bloch lines (see Sect. 3.6.5), has to be considered. No reliable estimates are available for either of these contributions, which means that a consistent theory of the cross-tie wall is still lacking. Numerical computations [656] point into the right direction but are necessarily restricted to small cross-tie periods.

(D) Asymmetric Bloch Walls. The walls to be discussed in this and the next section are stable in somewhat thicker films than those, in which the symmetrical Néel walls are stable. They are derived from the simple symmetric Bloch walls shown in Fig. 3.72. For a long time the stray field shown there was considered unavoidable. So it was a surprise when a possibility was discovered of avoiding the stray field completely by allowing an asymmetry in the wall structure with respect to the wall plane [579, 657, 658].

The solutions shown in Fig. 3.78 appear like common Néel walls on both surfaces. Despite the superficial similarity, the energetics and the internal structure of these walls are quite different from those of symmetrical Néel walls. Rather than considering them as Bloch walls with 'Néel caps', they should be called *vortex walls* if a general term is required. The difference can be most clearly seen if the charge distribution in the walls is compared. In symmetric Néel walls the dipolar charge is unavoidable and the distribution of this charge determines the wall profile as discussed in (C). In contrast, the two-dimensional eddies in the vortex walls are set up just to avoid charges. There is also a dramatic difference in scales. The Néel wall core scales, as mentioned, with $\sqrt{A/K_\mathrm{d}}$, which is about 5 nm in Permalloy, while the eddies of vortex walls are geometrically determined, scaling with the film thickness.

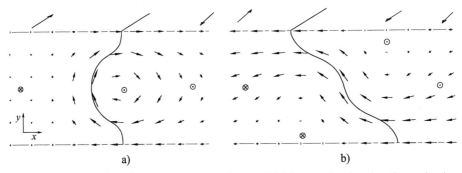

a)   b)

Fig. 3.78 Cross-sections through two types of stray-field-free walls, showing the projection of the magnetization vectors onto this cross-section for the Bloch type (a) and the Néel type (b) vortex wall. The *contour lines* indicate the "centre" of the wall, i.e. the surfaces on which the z component of the magnetization passes through zero. Based on model calculations for Permalloy films of 100 nm thickness [658]

The solution in Fig. 3.78a is called the asymmetric Bloch wall. It is the simpler of the two and favoured for small or zero field. The other structure, the asymmetric Néel wall, which will be discussed in more detail in (E) becomes more favourable in applied fields, i.e. for reduced wall angles. The Bloch wall has four equivalent orientations, since the vortex may be oriented either to the left or to the right, and, in addition, the polarization in the centre may point either upwards or downwards.

The history of the asymmetric Bloch wall is quite interesting. *Brown* and *LaBonte* [647] first put Néel's idea on a quantitative basis by calculating one-dimensional wall models with rigorous numerical methods, thus offering a reference for further developments. LaBonte then abandoned in his thesis [659] the constraint of one-dimensionality, allowing for two-dimensional excursions of the magnetization field in the wall core, which are necessary to obtain local torque equilibrium. Although he still kept the seemingly natural symmetry of the Bloch wall, the wall energy decreased considerably (by about 30%) in this step. The result was reproduced and confirmed by a variational approach in [650] within a margin of a few percent.

The first to relax the symmetry condition was Aharoni [660]. As it turned out later this was the right approach, but at this point the situation remained undecided because Aharoni's model, which was tailored to be evaluable analytically, displayed an energy higher than LaBonte's rigorously calculated two-dimensional symmetric model [659].

The decisive step was done in [657] and [579]. LaBonte [657] relaxed the symmetry requirement in his numerical approach, and surprisingly achieved a dramatic reduction in wall energy, thus proving the broken symmetry of the thin film Bloch wall. The rationale of this property of Bloch walls was displayed in the independent and simultaneous work [579], where the possibility of a completely stray-field-free Bloch wall was established by model structures *if* they were asymmetrical. Optimized stray-field-free Bloch wall configurations were quite similar to the numerically obtained solutions. The energy was lower than the best symmetrical model [659], thus proving independently the asymmetry of the Bloch wall structure in thin films. Aharoni's energetically unfavourable ansatz [660] was cited in [579] as having similar symmetry features as the final result, but it actually played no role in the discovery of the possibility of completely stray-field-free domain walls in thin films in spite of Aharoni's unsubstantiated claim [514 (p. 168)]. This also would have made no sense: how can a higher-energy model (relative to LaBonte's symmetrical solution) be a guide to a lower-energy configuration?

The asymmetric Bloch wall description of [579] was analytical in a sense. Although it needed some computer support in establishing its explicit structure and evaluating its energy, the necessary numerical effort was smaller by orders of magnitude compared to the rigorous numerical solution of the two-dimensional micromagnetic equations pioneered by LaBonte. There were extensions to other wall angles, film thicknesses and anisotropy types, as well as to interacting and moving walls [658, 661, 662, 655, 663] as reviewed in [509, 664].

Also other authors engaged in constructing more or less analytical models. Attempts to use explicit expressions led to wall structures [665–670], which were only *nearly* stray-field-free. The stray field energy has to be calculated numerically with LaBonte's method. Unfortunately, all proposed configurations display a higher energy than the stray-field-free models established in [579, 658]. *Semenov* [671, 672], on the other hand, used the approach of a completely stray-field-free model of [579], and by improved test functions he found solutions that are lower in energy than previous proposals and that come very close to LaBonte's rigorous solution.

As the approach of stray-field-free magnetization fields has proven to be quite valuable both in obtaining good models for micromagnetic configurations and for a better understanding of the mechanisms, it is outlined below for an asymmetric Bloch wall in zero external field. The construction of a strictly stray-field-free wall in two dimensions can be performed as follows (see Fig. 3.78 for the chosen coordinates): first choose a scalar function $A_z(x, y)$ (the z component of a vector potential) that is constant on both surfaces. The magnetization directions are derived from this function by:

$$m_1 = \partial A_z/\partial y \ , \ m_2 = -\partial A_z/\partial x \ , \ m_3 = \pm\sqrt{1-m_1^2-m_2^2} \ . \tag{3.145}$$

This ansatz automatically yields a stray-field-free vector field with div $\boldsymbol{m} = 0$ in the volume and $m_2 = 0$ at the surfaces. If the function $A_z(x, y)$ is symmetrical with a two-dimensional maximum in the centre of the film, the (m_1, m_2) magnetization pattern rotates around this peak and is already asymmetric. But such a magnetization pattern does not yet constitute a domain wall. To describe a 180° wall, the component m_3, which approaches -1 for $x \to -\infty$ and $+1$ for $x \to \infty$, must continuously pass the value zero somewhere in the wall. This means that the function $M = m_1^2 + m_2^2$ has to fulfil two conditions:

(i)   It must never exceed the value of one.

(ii) It must be equal to one on a continuous line $x_0(y)$ that defines the wall centre (the contour line in Fig. 3.78).

From (i) and (ii) follows that the function $M(x, y)$ must be stationary on the whole central line. This leads to the following equations for $M(x, y)$:

$$M_x[x_0(y), y] = 0 \ , \ M[x_0(y), y] = 1 \tag{3.146}$$

where M_x indicates the derivative $\partial M/\partial x$. To fulfil these conditions we introduce a new vector potential function by replacing $A_z(x, y)$ in (3.145) by $A^*(x, y)$:

$$A^*(x, y) = A_z(\xi, y) \ , \ \xi = x + S(y) \tag{3.147}$$

with an arbitrary function $S(y)$. Introducing it into (3.145) we get:

$$m_1 = \partial A_z(\xi, y)/\partial y + s(y)\, \partial A_z(\xi, y)/\partial \xi \ , \quad m_2 = -\partial A_z(\xi, y)/\partial \xi \ , \qquad (3.148)$$

where $s(y)$ is the derivative of $S(y)$. The transformed magnetization pattern is still stray-field-free because it obeys $\partial m_1/\partial x + \partial m_2/\partial y = 0$. Since $M_x = 0$ follows from $M_\xi = 0$, the transformed system $M_\xi[\xi_0(y), y] = 0$, $M[\xi_0(y), y] = 1$ as derived from (3.146) represents two ordinary equations for the two one-dimensional functions $\xi_0(y)$ and $s(y)$. By a simultaneous solution of these equations the position ξ_0 of the maximum of M is found for every y, and the derivative s that enters through (3.148) is adjusted so that this maximum reaches the value of unity. This proves to be possible everywhere by simply solving two coupled equations by Newton's method — except at special (symmetry) points at which either m_1 or m_2 vanishes. The starting ansatz $A_z(\xi, y)$ has to take care of these special points, which for the Bloch wall are in the centre and at the surfaces. Finally the function $s(y)$ is integrated to calculate the shearing function $S(y)$ and to recover the original coordinate x. In this way a stray-field-free model of a wall (or of other micromagnetic objects) can be constructed by relatively simple means — the solution of two coupled ordinary equations for as many y points as are needed in the numerical integration.

Apart from the mentioned conditions at the special points and the condition that the vector potential is constant on both film surfaces, the function $A_z(\xi, y)$ can be chosen freely as a function containing some parameters. The resulting magnetization is inserted into a variational procedure based on the total micromagnetic energy consisting of exchange, anisotropy and external field terms. The parameters of the model are determined by minimizing the energy.

It is interesting to study the relative contributions to the total energy in the resulting solutions. The stray field energy was excluded by the initial assumption of pole avoidance, and the anisotropy energy usually plays a minor role. The predominant energy term is the exchange energy. This unusual feature is caused by the geometrical constraints of a stray-field-free ansatz. The width of the vortex in Fig. 3.78a is governed by the film thickness. Even if the anisotropy energy were to allow a wider wall, the vortex could not become wider for a given thickness without increasing its energy. This is a consequence of the y dependence of the magnetization pattern. Expanding the vortex in the x direction reduces the exchange stiffness energy contributions from the x derivatives, but those from the y dependence grow because the region in which they appear increases. As long as the film thickness is smaller than the wall width of the classical Bloch wall in an infinite medium, the asymmetric Bloch wall scales essentially with the film thickness.

The structure of the asymmetric Bloch wall loses its mirror symmetry in an applied field perpendicular to the easy axis. The stability limit of the Bloch wall in such a field was studied numerically, for example in [550], but the treatment of the tail regions of these walls was not rigorous in these calculations. It is quite clear, however, that beyond a certain field value another vortex wall type, the *asymmetric Néel wall*, becomes more favourable because it is better adapted to an applied field.

(E) Asymmetric Néel Walls. This distinct wall type was first established in [579, 658] (Fig. 3.78b). It shows point symmetry in its cross-section instead of the mirror symmetry of the Bloch wall. To be precise, the asymmetric Bloch wall at zero field transforms according to $[(m_1 \rightarrow -m_1$ and $m_2 \rightarrow -m_2)$ for $(y \rightarrow -y)]$. The asymmetric Néel wall, in contrast, obeys for all fields the relation $[(m_1 \rightarrow m_1$ and $m_2 \rightarrow m_2)$ for $(x \rightarrow -x$ and $y \rightarrow -y)]$. There are two equivalent variants of the Néel wall if the horizontal component of the wall is considered to be given by an external field: the y excursion in the core can be either up or down. For comparison, there are four energetically equivalent variants of the asymmetric Bloch wall.

The magnetization of an asymmetric Néel wall points in the same direction at both surfaces, which is, as mentioned, favourable for an applied field along this direction. This property is also the reason why the wall can gain some energy by splitting off an extended tail, reducing the core energy in the field generated by the tail. For the symmetric Néel wall, as discussed in (C), the reduced dipolar charge $(1 - h)$ had to be distributed between core and tail, and often about half of this charge was displaced into the tail. In the asymmetric Néel wall most of this charge is avoided by the two-dimensional vortex pattern and only about 10 % are distributed in the tails, thereby reducing the necessary complexity and the exchange energy in the core. Numerical calculations properly including this feature were presented in [673], which is unfortunately not true for other numerical studies [674, 550, 675]. The tail part of the wall profile increases in relative importance with an applied field, so that less of the vortex structure becomes visible with decreasing wall angle. At a critical value of the applied field the asymmetry disappears in favour of a symmetric Néel wall structure (see Fig. 3.80).

(F) Comparison of Wall Energies in Parallel-Anisotropy Films. Figure 3.79 shows the calculated wall energy as a function of two variables: film thickness and wall angle [658]. The latter results from an applied field perpendicular to the easy axis and parallel to the film plane. The diagram is valid for uniaxial anisotropy K_u with the easy axis parallel to the film plane.

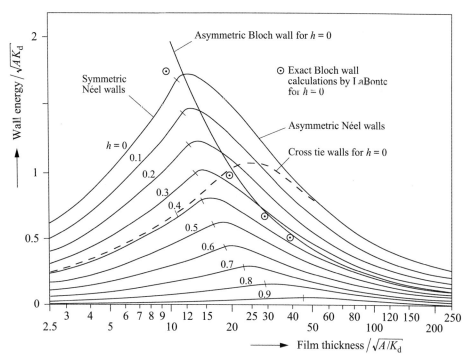

Fig. 3.79 Energy of different wall types as a function of film thickness for different values of the applied field $h = HJ_s/2K_u$. The wall energies are upper bounds, obtained by careful Ritz method calculations for the material parameters of Permalloy ($Q = 0.00025$), with the exception of a few numerically accurate points due to LaBonte [657]. The *markings* on the Néel wall curves indicate the onset of asymmetry. The *dashed line* is a crude estimate for the energy of a 180° cross-tie wall [650] based on the scheme of Fig. 3.75

As both thickness and field are expressed in reduced units, there remains only one further parameter: the material constant $Q = K_u/K_d$ for which $Q = 2.5 \cdot 10^{-4}$ was assumed, typical for Permalloy films. Since the anisotropy has only a moderate influence on the energy of walls in thin films (it influences primarily the Néel wall tails), this diagram should be typical for a wider range of low-anisotropy materials. The data stem from variational calculations and can serve only as a guideline for this reason. Rigorous numerical results are not yet available in this wide parameter range.

For larger film thicknesses, we see that at zero field, i.e. for 180° walls, the asymmetric Bloch wall always has a lower energy compared to the asymmetric Néel wall. The latter is expected to take over above about $H = 0.3 H_K$

corresponding to a wall angle of $\Omega_w = 2\arccos h = 145°$. At small thicknesses the symmetric Néel wall is the most stable configuration. 180° walls decay into the cross-tie pattern in this thickness range for fields smaller than about $0.4\,H_K$. Note that a pair of the hypothetical "Néel caps" of an asymmetric Bloch wall would have a higher energy than the Bloch wall itself, thus proving again this interpretation of vortex walls to be inadequate.

A phase diagram shows the preferred wall forms as a function of film thickness and wall angle (Fig. 3.80). Many features of this phase diagram were also seen in Lorentz microscopical investigations ([676, 664]; Fig. 5.69). Exploratory numerical simulations confirm the general picture. But numerical investigations are severely limited in the tractable computation area, and particularly Néel walls with their long-range tails are difficult to treat adequately. *Yuan* and *Bertram* [677] found the transition from the asymmetric Bloch to the asymmetric Néel wall at $h = 0.4$ for a Permalloy film of 100 nm thickness. But the calculated quantity was the instability field of the asymmetric Bloch wall, not the thermodynamic transition field at which both wall types are equal in energy. If energy considerations will be included in variable-grid algorithms such as that developed by *Miltat* and *Labrune* [673], a verification of the tentative phase diagram (Fig. 3.80) may at last become available.

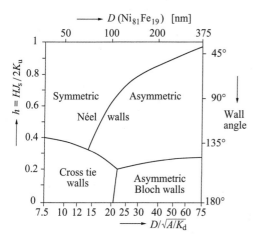

Fig. 3.80 Phase diagram of wall types derived from the previous figure and from the scheme of Fig. 3.75 for an estimate of the cross-tie wall. The calculations are based on Permalloy films ($Q = 0.00025$), but in the presented reduced form they should be a guideline for other films with low in-plane anisotropy as well

Unsolved questions still exist in the following areas:

- The shape of the cross-tie region in the phase diagram is unknown in the thin film range. For very thin films cross-tie walls become unstable according to experimental observations, apparently because of the energy of the Bloch lines (Sect. 3.6.5C), which was not included in deriving Fig. 3.80.

- The phase diagram has not been determined for different forms and values of the anisotropy. We recently performed rigorous micromagnetic calculations of two-dimensional domain walls in a narrow strip [629]. In such a strip no extended tails can be formed, so that valid, reproducible calculations are possible. Instead of the applied field the anisotropy parameter Q was varied in these calculations over a wide range. The resulting phase diagram (Fig. 3.81) shows two remarkable features: (i) a stability range for the asymmetric Néel wall at zero field (even for large Q), and (ii) a stability range of a symmetric Bloch wall for Q values exceeding a threshold of about 0.5. The latter wall type, which was first obtained in the thesis of LaBonte [659], is characterized by a two-dimensional symmetric cross-section pattern with two weak vortices on either side of the central perpendicular magnetization vector. In this study the three-dimensional cross-tie structure was excluded. Note that the Bloch-Néel transition, which was found in Fig. 3.79 (for $h = 0$) at about $12\sqrt{A/K_d}$, appears already at about $7\sqrt{A/K_d}$ for the narrow strips, because the extended tails in the Néel walls are suppressed. It would be interesting to extend such calculations to wider films, to three-dimensional wall structures and to different wall angles.

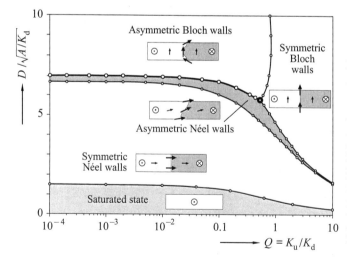

Fig. 3.81 A phase diagram of domain walls [629] in narrow strips of fixed width/thickness ratio of 4 : 1 and in zero external field. One input parameter is the anisotropy parameter $Q = K_u/K_d$ related to a uniaxial anisotropy with the easy axis parallel to the film and to the walls. The second parameter is the film thickness

- At larger thicknesses beyond the range of the phase diagram the energetical difference between asymmetric Bloch and Néel walls disappears, as indicated in Fig. 3.79. For Bloch walls some exploratory calculations were performed for large thicknesses [661]. Above a film thickness of the order of 5Δ ($\Delta = \sqrt{A/K_u}$) the vortices were predicted to become confined mainly to the neighbourhood of the surfaces. In any case, the thickness range of the (continuous) transition between two-dimensional vortex walls and mostly one-dimensional Bloch walls with surface vortices has to scale with Δ because the stray field energy plays no role in either of the two variants. The behaviour of the solutions as a function of film thickness leads to the conclusion that even for bulk low-Q materials Bloch walls should carry a surface vortex where they meet a surface, a prediction that has been widely confirmed by experimental observations (see Sects. 5.4.3, 5.5.1). The Q limit for the occurrence of a wall surface cap may be around $Q = 1$, but neither theoretical nor experimental results are available on this point.

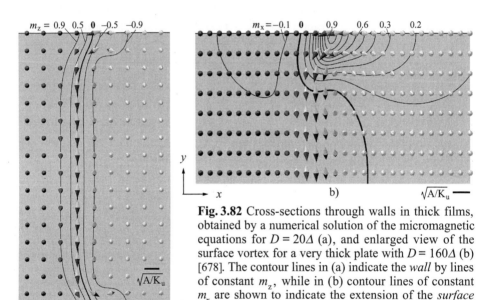

Fig. 3.82 Cross-sections through walls in thick films, obtained by a numerical solution of the micromagnetic equations for $D = 20\Delta$ (a), and enlarged view of the surface vortex for a very thick plate with $D = 160\Delta$ (b) [678]. The contour lines in (a) indicate the *wall* by lines of constant m_z, while in (b) contour lines of constant m_x are shown to indicate the extension of the *surface cap* structure

Earlier numerical calculations of Bloch walls in thick plates [352, 679, 680] confirm the general picture of the transition between the thin film solutions and the Bloch wall in infinite media. In particular, they confirm the predicted vortex-like structure of a Bloch wall intersecting a surface. The widening of the wall in the surface zone was found even more pronounced than anticipated from Ritz method investigations. Calculations including a careful elimination of discretization errors, which were not examined in detail in earlier work, were performed in [678]. In these calculations, which apply to the relatively high reduced anisotropy of $Q = 0.1$ (uniaxial in-plane anisotropy), computations were extended as far as $D = 160\Delta$. The very large surface cap width was confirmed and was found even weakly increasing up to these plate thicknesses without signs of saturation to a "bulk" value. Examples for calculated config-urations from this work are shown in Fig. 3.82.

(G) Charged Walls in Thin Films. In bulk material the avoidance of a net magnetic charge according to (3.114) represents an overriding rule, the violation of which is extremely rare. The discussion in Sect. 3.6.1G of walls not obeying this rule was introduced more for reasons of completeness than for practical purposes. In thin films, however, charged walls are frequently observed because the energy penalty is less severe: the stray field far from a line charge in a thin film falls off like $1/r$, whereas the field connected with a planar charge in bulk material stays constant. For widely separated, oppositely charged thin film walls the total stray field energy still diverges logarithmically with the distance. The degree of divergence is, however, much weaker than for planar charged walls in bulk material, which are effectively prohibited in ferromagnets.

The typical situation in which charged walls occur is shown in Fig. 3.83a. If two domains meet head-on, their separating wall develops a characteristic zigzag shape to reduce the charge density. A straight wall would have the strongest charge concentration. Increasing the zigzag angle, the charge density decreases at the expense of wall surface (it is assumed that the overall wall orientation is fixed by the magnetic environment). A quantification of this argument is difficult because of the non-local nature of the stray field energy.

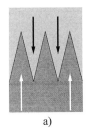 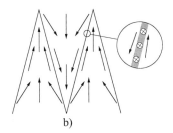

a)   b)

Fig. 3.83 A charged zigzag wall sep-arating two domains meeting head-on (a) and the formation of an uncharged core in such a wall (b)

Charged walls usually form a special structure that manages to reduce the stray field energy connected with the net charge. They develop a long-range tail, which is related to the tail of Néel walls and which carries most of the charge. The core stays largely uncharged, a concept which is sketched in Fig. 3.83b and was first put forward by *Finzi* and *Hartmann* [681]. The charge is distributed in the tails on both sides of the wall. Depending on the film thickness, the core will resemble ordinary thin film 180° walls (Bloch, Néel or cross-tie walls depending on film thickness). In the case of Néel walls, an interaction between dipolar Néel wall charges and the distributed monopolar charges has to be considered. This was done for straight periodic charged walls in [682, 683]. For the zigzag arrangement (Fig. 3.83), only the simple argument has been outlined [684] that the charge will in a good approximation somehow fill the available space uniformly.

(H) Walls in Films of Perpendicular Anisotropy. If the preferred axis of a film is oriented perpendicular to the surface and if the anisotropy is sufficiently strong, a completely different situation arises. As can be seen in Fig. 3.84 the *domains* are now connected with stray fields. In the walls the magnetization lies parallel to the surface along different directions. This is the typical situation of a bubble film or a film for a magneto-optical memory. "Bubbles" are isolated cylindrical domains that can be used for magnetic storage applications. The material constant $Q = K_u / K_d$ must be larger than unity for stability reasons. Often it is so large that one may treat $1/Q$ as a small quantity and deal with stray field effects as a perturbation.

The film thickness of bubble films is usually considerably larger than the Bloch wall width, so that these are not *thin* films in the strict sense. But the wall structure is strongly influenced by the surface, and the details of these effects have consequences for the behaviour of bubble domains.

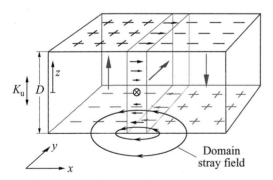

Fig. 3.84 Schematics of the stray field and the wall structure in a perpendicular-anisotropy film. The magnetization direction in the wall gets twisted by the influence of the stray field of the domains

The basic features of walls in perpendicular films were analysed by *Sloncyzewski* ([685]; see also [686]): the wall is a regular Bloch wall in the centre of the film. Towards the surfaces the stray field from the domains acts on the wall twisting the magnetization towards a Néel wall, as described in Sect. 3.6.2B. The horizontal stray field caused by the domains is almost singularly high right at the surface. Take for example a periodic arrangement of bar-shaped perpendicular domains of period $2W$ and sample thickness D, assuming infinitely thin domain walls and an infinite extension in the y direction (Fig. 3.84). Then the x component of the field in the wall centre plane can be expressed as:

$$H_x(z) = \tfrac{1}{\pi}\left(J_s/\mu_0\right)\ln\left\{\tanh\left[\tfrac{\pi}{4}(D-2z)/W\right]/\tanh\left[\tfrac{\pi}{4}(D+2z)/W\right]\right\} \quad . \qquad (3.149)$$

This field diverges at both surfaces ($z = \pm\tfrac{1}{2}D$). This divergence will be somehow smoothed and removed by the exchange stiffness effect, but large field values have to be expected nonetheless

Figure 3.63 shows that above a critical field H_{cr} the Bloch wall is converted into a pure Néel wall, recognizable by the absence of a zigzag folding so that uniformly $\psi = 90°$. This critical field can be evaluated from Fig. 3.63 where we see that for large Q it approaches $H_{cr} \approx \tfrac{2}{\pi}J_s/\mu_0$. This horizontal field may be exceeded near the surface of perpendicular anisotropy films of sufficient thickness. Then there exists a critical depth d_{cr} above which $H > H_{cr}$, thus generating a Néel wall as in bulk materials (Sect. 3.6.2B). More or less twisted Bloch walls are preferred in the small-field range deeper than d_{cr}.

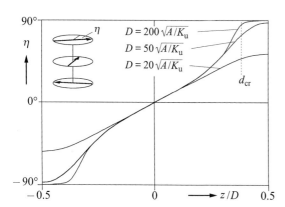

Fig. **3.85** The twisting angle of a wall in a perpendicular-anisotropy film according to a one-dimensional numerical calculation for three different film thicknesses D. For thicker films the wall approaches the Néel wall mode ($\eta = \pm90°$) at a critical depth d_{cr} below the surfaces. The calculations are based on a solution of the one-dimensional differential equation for the reduced material parameter $Q = 4$ [687]

More detailed calculations, which take into account the exchange energy from the z variation as well as a number of other effects [687], yield profiles as shown in Fig. 3.85. Here the twisting angle η is plotted as a function of z, $\eta = 0°$ for the Bloch wall in the centre and $\eta = \pm90°$ for the Néel walls at the surfaces. The transition profile from Néel wall to Bloch wall structure plays a crucial role in the dynamics of walls in films that are thick enough to show this transition (Fig. 3.96). The example $D = 200\sqrt{A/K_u}$ demonstrates the presence of Néel wall zones near the surfaces ($\eta = \pm90°$). For sufficiently thin films the exchange stiffness energy suppresses wall twisting.

3.6.5 Substructures of Walls — Bloch Lines and Bloch Points

Most domain walls can occur in two equivalent forms, differing only in their rotation sense. If the energies are equal, both modes may coexist in the same wall. The dividing lines between such wall segments are called *Bloch lines* (an attempt to classify these lines further in a similar sense as one talks of *Bloch* and *Néel* walls has largely been abandoned in view of the complex internal structure of these lines. In this book we treat the term *Néel line* as a synonym of the general term *Bloch line*).

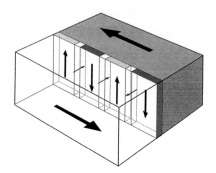

Fig. 3.86 Periodic Bloch lines in a soft magnetic material reducing the stray field energy of a Bloch wall (schematic, neglecting in particular the modification of the wall and Bloch line structure near the surfaces)

The reason for the occurrence of wall substructures is not always clear. There is experimental evidence [688, 641, 689, 690] that these subdivisions are sometimes formed for energetic reasons. Bloch walls carry some magnetic flux, and if this flux is intercepted by a surface, a subdivision reduces the extra stray field energy [691], as indicated schematically in Fig. 3.86. This is particularly true for the wide (100) 180° walls in iron (Sect. 3.6.3A).

Primarily, stray field suppression is achieved in soft magnetic materials by a surface vortex similar to asymmetric Bloch walls in thin films (Fig. 3.82). But a subdivision can support flux compensation if combined with a peculiar twisting of the wall, which deviates some of the wall flux into the domains [691]. The energies in question are tiny, however, and frequently the subdivisions will

form just by chance. Once they are formed, their annihilation may be hindered by an energy barrier. Bloch lines essentially have to be considered a part of the real wall structure that will influence its properties and behaviour. This was most thoroughly studied in connection with bubble walls, where even a memory scheme using wall substructures was proposed (Sect. 6.6.1C).

When a wall moves, the Bloch lines can move within the wall and influence the wall motion in a drastic way. In soft magnetic films Bloch lines are markedly narrower than the Bloch wall as will be discussed in (C), so that they interact strongly with pinning centres. The translation of Bloch lines may even contribute to magnetic permeabilities in cases where wall motion is impossible for reasons of symmetry. The interior structure of Bloch lines is rather experimentally inaccessible because of its small extension. It therefore has to be analysed theoretically.

A further point of interest is concerned with substructures of Bloch lines and possible transitions within Bloch lines. Bloch lines can still be formed in a micromagnetically continuous manner, whereas transitions from one to the opposite sense of rotation within a Bloch line cannot form continuously and must contain at least one micromagnetic singularity, as was first shown by *Feldtkeller* [692]. This kind of singularities will be discussed in (D).

(A) Bloch Lines in High-Anisotropy Films. For uniaxial films with perpendicular anisotropy and $K_u > K_d$, as typical for bubble domains, the calculation of Bloch lines is relatively straightforward at least in a first approximation [49]. Figure 3.87 shows a cross-section through such a transition. Two regular, Bloch wall segments (both described by ϑ varying from $-90°$ to $90°$) are separated by a transition in which the angle φ changes from $0°$ to $180°$.

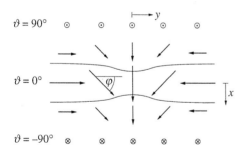

$\vartheta = 90°$

$\vartheta = 0°$

$\vartheta = -90°$

Fig. 3.87 Cross-section through a straight wall carrying a Bloch line in a perpendicular-anisotropy film. This cross-section corresponds to the mid plane ($z = 0$) in Fig. 3.84

If the transition region is wide compared to the Bloch wall width, then the use of the one-dimensional, local approximation (3.20) for the stray field energy in the Bloch line is permitted. In a reasonable approximation ϑ depends mainly on x, and φ depends only on y. The total energy can then be written as:

$$E_{tot} = \iint \left\{ A\left[\left(\frac{\partial \vartheta}{\partial x}\right)^2 + \cos^2\vartheta\left(\frac{\partial \vartheta}{\partial y}\right)^2\right] + K_u\cos^2\vartheta + K_d\sin^2\varphi\cos^2\vartheta \right\} dx\,dy \;. \quad (3.150)$$

Solving the variational problem as in standard Bloch wall theory, at first for constant angle φ, yields the following relations:

$$A\left(\partial\vartheta/\partial x\right)^2 = K_{\mathrm{u}}\cos^2\vartheta + K_{\mathrm{d}}\sin^2\varphi\cos^2\vartheta \ , \tag{3.151a}$$

$$\gamma_{\mathrm{w}}(\varphi) = 4\sqrt{A(K_{\mathrm{u}} + K_{\mathrm{d}}\sin^2\varphi)} \quad . \tag{3.151b}$$

The integral over the expression on the right-hand side of (3.151a) is one half of the wall energy (3.151b). Inserting this into the total energy (3.150) and expanding for small values of $1/Q = K_{\mathrm{d}}/K_{\mathrm{u}}$ we get:

$$E_{\mathrm{tot}} = 4L\sqrt{AK_{\mathrm{u}}} + 2\sqrt{A/K_{\mathrm{u}}}\int\left[A\left(\partial\varphi/\partial y\right)^2 + K_{\mathrm{d}}\sin^2\varphi\right]\mathrm{d}x \ , \tag{3.152}$$

in which L denotes the total length of the wall. The first term in the total energy represents the regular Bloch wall energy, while the second term is the additional energy of the Bloch line. Optimizing equation (3.152) with respect to $\varphi(y)$ represents a variational problem of the same form as the one for the standard Bloch wall. The solution is a Bloch line energy of $8A/\sqrt{Q}$ and a Bloch line width of $\pi\sqrt{A/K_{\mathrm{d}}}$. Note that the Bloch line width is larger than the Bloch wall width $\pi\sqrt{A/K_{\mathrm{u}}}$ since we assumed $K_{\mathrm{u}} > K_{\mathrm{d}}$, thus justifying the approximation of a local treatment of the stray field energy. A more general analysis, including non-local stray field and interaction effects between Bloch lines can be found in [693] and more recently in [694, 656].

In the local approximation all orientations of Bloch lines are equivalent. Actually, different orientations of Bloch lines have different properties. The Bloch lines oriented perpendicular to the wall magnetization (in bubbles these are called *vertical* Bloch lines; Figs. 3.87 and 3.95) carry a net charge that leads to a long-range interaction between Bloch lines. If they approach each other, there are two possibilities: they either can annihilate because their sense of rotation is opposite. This is called the *unwinding* case. Or they cannot annihilate because the magnetization is *winding*. Winding Bloch lines form clusters with an equilibrium distance that was calculated at approximately $2\pi\sqrt{A/K_{\mathrm{d}}}$ [693]. Vertical Bloch lines are the carriers of information in the Bloch line memory (Sect. 6.6.1C). A high concentration of winding vertical Bloch lines leads to almost immobile *hard* bubbles, which are rather useless.

The Bloch lines running parallel to the wall magnetization (*horizontal* Bloch lines in bubbles, also the straight Bloch line in Fig. 3.88c belongs to this type) do not show a long-range magnetostatic interaction. In bubble films they are strongly influenced by the domain stray fields also responsible for the twisted wall structure in such films (Fig. 3.84), and they can be generated dynamically out of the Bloch-Néel transition near the film surfaces ([49, 686]; see Fig. 3.96). Occasionally one observes bubbles with several stacked horizontal Bloch lines of the same sign. Because the Bloch lines can be shifted inside the wall, these walls exhibit high inertia in resonance experiments and are therefore called *heavy* walls.

Details of the complicated interrelations between Bloch lines and bubble dynamics and statics can be found in the extensive literature on this subject, which is reviewed in the textbook of *Malozemoff* and *Slonczewski* [49]. More recently various authors [694, 656, 695, 200] have performed rigorous numerical calculations of Bloch lines in different kinds of garnet films. In a particular case with tilted uniaxial anisotropy, for which the Bloch lines can be observed magneto-optically, good agreement between observations and calculations was achieved. Three-dimensional calculations yielded a characteristic tilting of vertical Bloch lines in bubble films, which is important for the magneto-optical observation of these lines ([696]; see Sect. 5.6.3).

(B) Bloch Lines in Bulk Soft Magnetic Materials. For low anisotropy materials (K_u, $K_c \ll K_d$) one might expect comparatively narrow Bloch lines if they would again scale with the parameter $\sqrt{A/K_d}$ derived in (A) for high-anisotropy materials. This quantity is smaller than 10 nm for typical materials, and thus much smaller than the wall width. This would imply a strong pinning of Bloch walls that contain Bloch lines (following the principle that every micromagnetic unit interacts strongest with defects that are comparable in scale). The discovery of a Bloch line structure, which avoids the strong stray field [658], is therefore of some relevance. The structure is asymmetric and thus related to the stray-field-free wall structures in thin films.

Originally, the same tools as described for asymmetric Bloch walls (Sect. 3.6.4D) were used to construct a completely stray-field-free Bloch line. These ideas were confirmed by finite element computations [550, 697]. The extra energy per unit length for a Bloch line inserted into a 180° wall in a uniaxial material is about 13 A, weakly depending on the anisotropy parameter Q. The wall width in the region of the Bloch line is reduced, but not strongly, remaining of the order of $\Delta = \sqrt{A/K_u}$ (Fig. 3.88 a).

The solutions shown in Fig. 3.88a,b represent only one possible orientation of the Bloch line in a wall, namely the one that runs parallel to the magnetization in the centre of the adjacent wall parts (the straight line in Fig. 3.88d). Here no net flux is transported towards the line, so that a completely stray-field-free Bloch line is conceivable. But even if the Bloch line is tilted with respect to this axis (the triangle in Fig. 3.88d), a stray-field-free Bloch line can be constructed if the two wall segments are displaced, so that the net flux becomes zero again (Fig. 3.88c). The charge of the Bloch line is neutralized by the domains. On iron whiskers this effect was in fact observed by *Hartmann* and *Mende* using a high-resolution Bitter method [698].

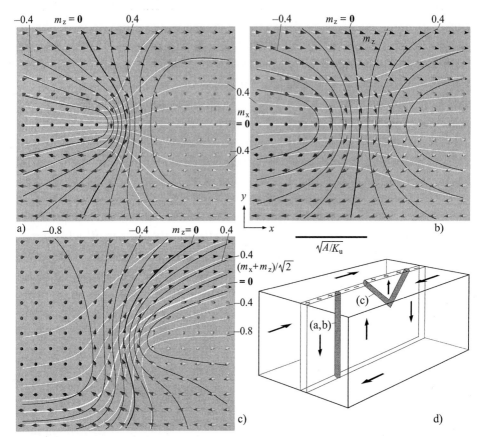

Fig. 3.88 Numerically calculated cross-sections through Bloch walls in uniaxial material containing a nearly stray-field-free Bloch line for a material with $Q = 0.01$ (a) and the simpler Bloch line (b) in a higher-anisotropy material ($Q = 0.2$). These two Bloch lines are oriented parallel to the wall magnetization. (c) Structure of a tilted Bloch line for $Q = 0.01$ and a tilting angle of 45°. Here the net wall flux is compensated by a displacement of the wall parts. The cross-sections are oriented perpendicular to the Bloch line axis and the contour lines indicate the asymmetric Bloch line shape (*black*) and the constricted wall shape (*white*). The three-dimensional sketch (d) shows the environment in which the two types of Bloch lines may occur. [Together with *K. Ramstöck* [697] (computation) and *J. McCord* (representation), Erlangen]

A stray-field-free Bloch line is not possible if the line runs perpendicular to the wall magnetization. Such a line will carry a high energy and will therefore decay into a zigzag shape. An optimum zigzag angle of 35°, measured against the perpendicular direction, was computed in [697] for a uniaxial material with $Q = 0.01$.

The fact that Bloch lines can change their direction in a wide range without a severe energy penalty should be considered in an analysis of the interaction of domain walls with lattice defects. Bloch walls are relatively stiff, at least in one dimension. For this reason they tend to average out a large part of the interaction with point defects and dislocations, and this circumstance will reduce the interaction with statistically distributed small defects. In contrast, Bloch lines may adapt to the shape and the distribution of the defects and thus finally produce a stronger interaction. On the other hand, they occupy a much smaller volume than the Bloch walls, and the energy penalty for their extension is not very large, which tends to reduce the strength of the interaction. A more detailed analysis of the net effects of Bloch lines in domain walls on wall pinning and coercivity has not yet been presented.

To conclude, Bloch lines in soft magnetic materials have a complicated two-dimensional vortex structure. Their width is of the same order of magnitude as that of the wall in which they are inserted. They should therefore be as mobile as the walls themselves except at the intersections with the surface where they get narrow. The local conditions of a Bloch line hitting a surface resemble a Bloch line in a thin film to be discussed below.

The full theoretical analysis of the three-dimensional structure of a 180° domain wall in a thick iron crystal remains a formidable theoretical problem that has by no means been solved [698]. The elements that have to be taken into account are:

(i)   the surface vortices of the domain walls near both surfaces,
(ii) the subdivision of the wall by Bloch lines,
(iii) the twisting of the wall between Bloch lines, distributing the wall flux,
(iv) the gradual transition of the wall width from bulk to surface,
(v)   constricted planar eddies or "swirls" where the Bloch lines hit the surfaces.

A consistent theory would have to couple all these elements, facing the difficulty of largely different scales ranging from hundreds of microns for the twisting phenomenon (iii) to the nanometre range expected for the swirls (see next subsection). Any theoretical analysis could rely on detailed experimental studies (see for example [699, 226] for magneto-optical studies and [89, 698] for Bitter pattern observations).

(C) Bloch Lines in Thin Films with In-Plane Magnetization. Lorentz microscopy of thin films first attracted attention to certain "singular" regions in the magnetization pattern. The magnetization rotates around these points in two different modes, called *circular* and *cross* Bloch lines (Figs. 3.75, 3.89).

Feldtkeller and *Thomas* [653] were the first to calculate the micromagnetic structure of Bloch lines in thin films. The calculation is expanded here for thicker plates, in particular in view of the already mentioned *swirls* expected at the surface of bulk soft magnetic materials (Sect. 3.3.3C, Fig. 3.27).

Common to Bloch lines in thin films and to surface swirls is the perpendicular magnetization in the centre. Conclusive experimental observations of this prediction are not known. Pictures taken by *Hartmann* [698] with a high-resolution Bitter method may indicate this feature. The narrow width of Bloch lines in thin films is supported by the observation that Bloch lines become hard to move in very thin films [644] because of interactions with small scale irregularities. Micromagnetic analysis offers the best access to these elements. The following exploratory computations may serve as a guideline.

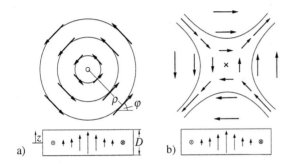

Fig. 3.89 Circular (a) and cross (b) Bloch lines in thin films and the coordinates used for their calculation

The circular Bloch line shows axial symmetry at least in its core region. Using cylindrical coordinates (Fig. 3.89) we start with a model that does not depend on φ and z. In addition, we require magnetic charges to appear at the surface only. In cylindrical coordinates the condition div $\boldsymbol{m} = 0$ reads:

$$\partial(\rho m_\rho)/\partial\rho + \partial m_\varphi/\partial\varphi + \rho\, \partial m_z/\partial z = 0 \ . \tag{3.153}$$

Our starting configuration is obtained by choosing $m_\rho = 0$, where m_ρ is the radial magnetization component. Then the remaining components m_φ and m_z are coupled by $m_\varphi^2 + m_z^2 = 1$, whence it suffices to define the function $m_z(\rho)$ to describe the Bloch line in this approximation. This function must be 1 in the centre ($\rho = 0$) and smaller than 1 elsewhere, tending to 0 at infinity. We neglect anisotropy energy influences for the core region because this energy term will be negligible compared to exchange and stray field energy contributions in this region. This assumption justifies the axially symmetrical ansatz. The exchange energy in cylindrical coordinates was given in (3.2a), and can be simplified here for cylindrical symmetry to:

$$e_x = A\left[m_{\varphi,\rho}^2 + m_{\varphi,\rho}^2 + m_{z,\rho}^2 + m_{\rho,z}^2 + m_{\varphi,z}^2 + m_{z,z}^2 + \frac{1}{\rho^2}\left(m_\varphi^2 + m_\rho^2\right)\right], \quad (3.154)$$

where $m_{\varphi,\rho}^2$ is short for $(\partial m_\varphi/\partial\rho)^2$ etc. Here only the second, the third and the seventh term in this expression are different from zero. The term m_φ^2/ρ^2 leads to a divergent integral if the integration is extended to infinity, since m_φ becomes 1 for large ρ. This feature poses no serious problem, however. It is sufficient to cut off the integration of the exchange energy at a radius at which the vertical z component of the magnetization is negligible. The omitted energy outside this cut-off radius is not coupled to the Bloch line core anyway.

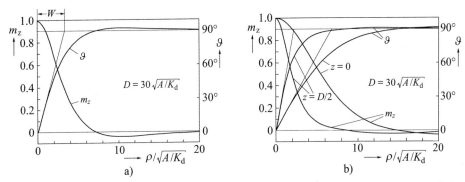

Fig. 3.90 Calculated Bloch line profiles based on the cylindrical ansatz (3.155) (a) and the two-dimensional ansatz (3.157) (b). The Bloch line width W is defined by the slope of the function $\vartheta(\rho)$ where the angle ϑ is defined by $\vartheta = \arccos m_z$

Feldtkeller and Thomas [653] chose for m_z the ansatz $m_z = \exp(-2\rho^2/b^2)$, where b is an optimization parameter. For this ansatz describing a cylindrical structure, the stray field energy can be integrated analytically. This is still true if the trial function is generalized according to:

$$m_z(\rho) = \sum c_i \exp\left(-2\rho^2/b_i^2\right), \quad \sum c_i = 1, \quad (3.155)$$

where the c_i and the additional b_i are further parameters. By generalizing the result of [653] the stray field energy can be expressed as:

$$E_d = 2\pi K_d \int_0^\infty \left[1 - \exp(-uD)\right]\left[\tfrac{1}{4}\sum c_i b_i^2 \exp\left(-\tfrac{1}{8}u^2 b_i^2\right)\right]^2 du . \quad (3.156)$$

Minimizing the total energy with respect to the parameters c_i we get profiles as shown in Fig. 3.90 a. Here six elements in the sum were used and the b_i were distributed evenly and left fixed. Note the reverse magnetization in the outer region, which compensates a part of the upwelling flux in the core. This feature could not be described with the original function employed in [653].

The Bloch line width W as defined in Fig. 3.90a is plotted as a function of the film thickness in Fig. 3.91, as the curve marked "cylindrical ansatz".

Feldtkeller and Thomas [653] derived a simple differential equation for small thicknesses D. Our multi-parameter cylindrical configuration reproduced correctly the predictions of this analysis.

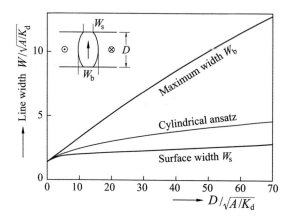

Fig. 3.91 The Bloch line width as a function of the film thickness for the cylindrical ansatz, and the more accurate two-dimensional configuration. The latter model reveals that for large thicknesses the line width at the surface is much smaller than in the interior

For larger thicknesses we try a two-dimensional model including a z dependence, which allows the Bloch line to get wider in the bulk and stay narrow at the surfaces. An ansatz with surface charges only is:

$$m_z(\rho,z) = \sum_{i=1}^{n} c_i f_i(\rho, z) \ , \quad f_i(\rho,z) = \exp\left(-2\rho^2/g_i(z)\, b_i^2\right) ,$$

$$g_i(z) = a_i\,(1 - 4z^2/D^2) \ , \quad \sum_{i=1}^{n} c_i = 1 \ , \quad \sum_{i=1}^{n} a_i c_i b_i^2 = 0 \ ,$$

$$m_\rho(\rho,z) = \frac{1}{4\rho} \sum_{i=1}^{n} c_i g_i'(z)\,[2\rho^2/g_i(z) + b_i^2]\,f_i(\rho,z) \qquad (3.157)$$

where m_ρ was obtained by integrating the condition div $\boldsymbol{m} = 0$ (3.153).

The exchange energy needs a two-dimensional integration, while the stray field energy can be calculated as before. Minimizing the energy with respect to the parameters a_i, b_i and c_i under the indicated conditions, we obtain a strongly reduced surface line width compared to the simple ansatz (Fig. 3.90 b, 3.91) if the film thickness is larger than about three times the characteristic length $\sqrt{A/K_d}$. At the same time the energy is significantly reduced.

The diameter of the magnetization "swirl" at the surface seems to approach a constant value of about $3\sqrt{A/K_d}$ for large thicknesses — in contrast to the result of the cylindrical model (the remaining slope of the surface line width W_s for large D in Fig. 3.91 is probably caused by insufficiencies of the Ritz method computations). The (quarter) cross-sections through the Bloch lines in Fig. 3.92 reveal the ton-shaped profile for larger thicknesses.

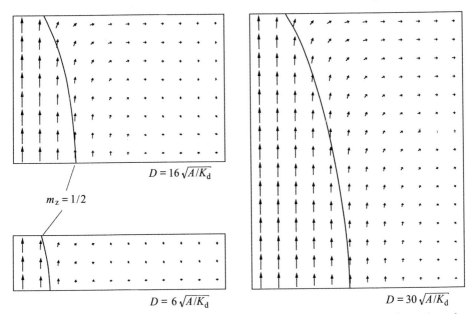

Fig. 3.92 Cross-section through calculated two-dimensional Bloch line configurations for low-anisotropy plates of three different thicknesses. Symmetry makes it possible to show only one quarter of the cross-sections

Bäurich [700] analysed also *cross* Bloch lines in thin films. His results can be summarized in the statement that the core of this structure, although a bit more complex, has the same characteristic dimensions as the circular Bloch line. The cross Bloch line necessarily differs from the circular Bloch line in the way it can be embedded: the circular Bloch line can reside in a planar, rotationally symmetric, stray-field-free magnetization pattern as long as anisotropy allows this. For the cross Bloch line a fragmentation of the pattern into domains is enforced by the stray field, as demonstrated in Fig. 3.89, and in the cross-tie pattern (Fig. 3.75).

In summary, Bloch line cores in planar thin films, as well as the intersections of Bloch lines with the surface of bulk samples, scale with the characteristic length $\sqrt{A/K_{\mathrm{d}}}$, which is small compared to the wall width in soft magnetic materials. This explains the observation that Bloch lines in very thin films cause wall pinning and act as a source of coercivity in these materials. On the other hand, $\sqrt{A/K_{\mathrm{d}}}$ is large compared to atomic distances (5 nm for NiFe, 3 nm for iron), so that a micromagnetic continuum theory of Bloch line cores is definitely still valid, in contrast to the erroneous point of view of Aharoni [701 (p. 135)]. Bloch lines in thin films are *no* micromagnetic singularities.

(D) Micromagnetic Singularities (Bloch Points). Figure 3.93 shows the sketch of a wall in a perpendicular-anisotropy (bubble) film with a Bloch line consisting of two parts of opposite sense of rotation. As was first shown by *Feldtkeller* [692], the transition between the Bloch line segments cannot be accomplished in a continuous manner. The argument leading to this conclusion is sketched in Fig. 3.93b. The magnetization directions occurring on a closed surface around the transition are transferred onto the unit sphere of magnetization directions. As demonstrated, *every* magnetization direction occurs at least once in the vicinity of the transition, while all magnetization directions together form a continuous mapping from the chosen surface onto the unit sphere. If we now try to shrink the surrounding surface towards the transition region, a theorem of topology states that there exists no *continuous* path from the surface to a single point with a well-defined magnetization direction. At least one micromagnetic discontinuity or singularity must exist inside the surface. The properties of these singularities or *Bloch points* were studied by a number of authors [49, 692, 702–705]. Essentially they consist of a small region of a few lattice constants in which the ferromagnetic order is destroyed.

The exchange energy of the singularity can be calculated in the conventional way assuming a real singular point since the exchange energy density increases as $1/r^2$ approaching the Bloch point. This function is integrable over a volume containing the origin. The minimum micromagnetic exchange energy of a singular point in the centre of a sphere of radius R is $E_{ex} = 8\pi AR$, largely independent of the actual magnetization distribution in the core that can assume many different forms [702].

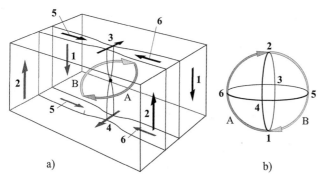

Fig. 3.93 (a) Bloch line consisting of two parts of opposite rotation sense in a perpendicular-anisotropy film. The Bloch line must contain a singular point for topological reasons. This is shown by a mapping of the magnetization directions onto the unit sphere (b), which is indicated for two example trajectories, the *round arrows* A and B

The integrated core energy of the point singularity thus remains finite even in micromagnetism where the exchange stiffness energy density becomes infinite approaching the centre. The total exchange energy of a spherical

Bloch point has no finite energy if the outer radius R of the configuration is allowed to become larger and larger. This means that a singularity has always to be analysed together with the environment into which it is embedded. A spherical surrounding is just the simplest case that is useful when the centre structure of the singularity is of primary interest. In contrast to point singularities, *line* singularities would have an infinite micromagnetic core energy and will therefore hardly occur in ferromagnets. This point of view is supported by topological arguments [706].

As the micromagnetic exchange energy expression represents an approximation presuming weakly varying spins, it is certainly not valid within a neighbourhood of the singular point having atomic scale [705]. *Reinhardt* [703] calculated the exchange energy of singular points using lattice sums instead of continuum theory. He found smaller energies compared to the micromagnetic approach, but the difference amounted only to some percent for Bloch points in regular domain walls. Lattice theory also yields different energies for different positions of the singularity in the lattice, and in particular, a preference of these points for vacancies and non-magnetic substitutional atoms. The energy differences between different sites, including interstitial positions and vacancies, turned out to be of the order of 2–3 kT. A rigorous theory of singular points and their kinetics would have to include quantum-mechanical and thermo-dynamical elements in addition to the lattice approach.

Micromagnetic singular points are not that exotic as might be thought by looking at the seemingly artificial construction of Fig. 3.93. Döring [37] pointed out that conventional spike domains (see Sect. 1.2.3) that end inside a sample have to carry at least one singularity if they are surrounded by regular Bloch walls without Bloch lines. The generation and motion of singular points are the only way to switch the rotation sense of Bloch lines, and this possibility was used in special bubble memory schemes. The technique of Bloch line switching is sketched in Fig. 3.94.

A soft magnetic capping layer, exchange coupled to the bubble film, is saturated by an external field. If the bubble occupied the lowest-energy, *unichiral* state at the beginning, this state is not topologically consistent with the saturated capping layer. A singularity is necessarily generated, tending to move along one of the Bloch lines induced by the external field. The bubble is thus converted into a different state (of *rotation number* 0) that is compatible with the saturated capping layer. The switching of a Bloch line has measurable consequences both in bubble dynamics and in the interaction between the Bloch lines. In the concept of the Bloch line memory they played a central role. Their nucleation and propagation represents the writing process. If it occurs spontaneously, information is destroyed (see Sect. 6.6.1C).

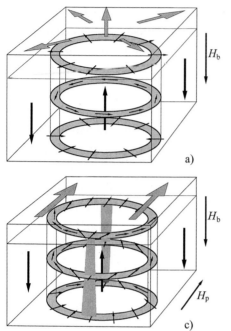

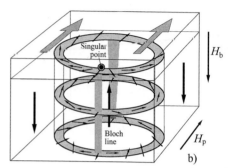

Fig. 3.94 Generating a Bloch point in a bubble wall by *cap switching*. The soft magnetic capping layer first follows the wall magnetization in the configuration (a). It is then forced into an essentially uniform magnetization by a horizontal field H_p (b), thus triggering the generation of two unwinding Bloch lines and a singular point. The singular point can move through the film and disappear, thus switching the Bloch line into the configuration (c)

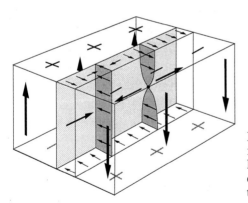

Fig. 3.95 Vertical Bloch line in a bubble film with and without a singular point. The line containing the singularity is better adapted to domain-generated stray fields and to the twisted wall structure

Micromagnetic singularities will interact strongly with point defects in the lattice. Their motion will be governed by thermal agitation. Larger defects in a ferromagnet may help in generating singular points, for example if a wall moves over them. The generation of singular points will be facilitated when a singular micromagnetic structure has a lower energy than a continuous structure.

A vertical Bloch line in thick bubble films was identified as such a case [704, 707]. Here the domain stray fields (which lead to the twisted wall structure; see Sect. 3.6.4H) produce an unfavourable situation for the Bloch line close to one of the surfaces (Fig. 3.95). Introducing a singular point in the centre reduces the total energy for film thicknesses larger than about $7.3\sqrt{A/K_d}$ in spite of the additional exchange energy connected with the singularity [707].

3.6.6 Wall Dynamics: Gyrotropic Domain Wall Motion

The dynamic behaviour of domain walls, Bloch lines and other micromagnetic objects attracted a lot of interest in connection with the high-speed motion of cylindrical domains in magnetic bubble memories (Sect. 6.6.1A). Many unexpected phenomena such as domain wall resonance effects, the skew motion of magnetic bubbles in a gradient field, or the dynamic conversion of bubble domains by Bloch line nucleation turned out to be a consequence of the gyrotropic nature of magnetization dynamics described by the Landau-Lifshitz equation (3.66b). We limit the discussion to this approach, ignoring the additional kinetic terms discussed in [576]. This section may serve as an introduction into the special literature reviewed in [49, 708, 512]. It deals with magnetic insulators in which neither eddy currents nor thermally activated processes play a role. The latter phenomena are treated in Sects. 3.6.7 and 3.6.8.

(A) **Kinetic Potential Formulation of Magnetization Dynamics.** We start from the Landau-Lifshitz-Gilbert equation (3.66) describing the dynamics of a magnetization vector in a non-conducting medium. In an arbitrary polar coordinate system ($\vartheta = 0$ at the equator) equation (3.66a) is transformed into:

$$\cos\vartheta\,\dot{\vartheta} = \left(\gamma_G/J_s\right)\delta_\varphi e_{tot} + \alpha_G\cos^2\vartheta\,\dot{\varphi}$$

$$\cos\vartheta\,\dot{\varphi} = -\left(\gamma_G/J_s\right)\delta_\vartheta e_{tot} - \alpha_G\,\dot{\vartheta}\ , \tag{3.158}$$

where $\delta_\varphi e_{tot}$ and $\delta_\vartheta e_{tot}$ are the variational derivatives of the total energy density with respect to φ and ϑ, replacing the effective field in (3.66a). The dimensionless damping factor α_G appearing in these equations is usually smaller than 0.1 in bubble garnets, but materials with values smaller than 0.01 or larger than 1 can be prepared. For garnets α_G depends on the orbital angular momentum of the rare earth constituents. Non-magnetic ions like Y^{3+}, or spherically symmetric ions like Gd^{3+}, contribute little to losses.

For static problems the time derivatives vanish, resulting in the micromagnetic equations $\delta_\varphi e_{tot} = 0$ and $\delta_\vartheta e_{tot} = 0$ applicable to the calculation of static domain wall structures. There are several ways to solve these equations for moving Bloch walls. One approach neglects the damping terms ($\alpha_G = 0$). Then the free energy can be augmented by a *kinetic potential*, so that $\delta_\vartheta(e_{tot} + p_{kin}) = 0$ and $\delta_\varphi(e_{tot} + p_{kin}) = 0$ replace the equations (3.158). This means that the dynamic equations can be expressed in the standard variational form of the static micromagnetic equations. Possible forms of p_{kin} are:

$$p_{kin}^{(1)} = \left(J_s/\gamma\right)\left(\sin\vartheta - c_{kp}^{(0)}\right)\dot\varphi \ , \tag{3.159a}$$

$$p_{kin}^{(2)} = -\left(J_s/\gamma\right)\left(\varphi - c_{kp}^{(1)}\right)\cos\vartheta\,\dot\vartheta \ , \tag{3.159b}$$

where $c_{kp}^{(0)}$ and $c_{kp}^{(1)}$ are arbitrary constants. Which of the two versions and which of the constants are preferred depends on the boundary conditions. They must be selected so that p_{kin} vanishes in the domains. Kinetic potentials are linear in the velocities, different from kinetic energies.

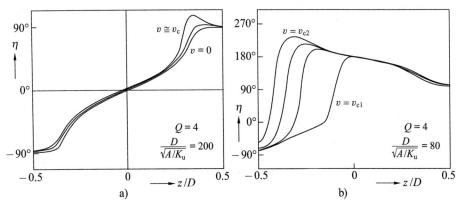

Fig. 3.96 The profiles of straight walls in perpendicular anisotropy films (using the same angle as in Fig. 3.85) moving with constant velocities. (a) The formation of a horizontal Bloch line, emerging at a critical field v_c out of the Bloch-Néel transition region. (b) The profile of a moving wall containing a Bloch line. The regular wall (a) is stable in a velocity range $0 \le v \le v_c$, while the wall containing a Bloch line (b) is stable dynamically between v_{c1} and v_{c2}, where v_{c1} is smaller and v_{c2} is larger than v_c. These solutions were obtained by a Ritz method to find starting functions for a numerical solution of the one-dimensional differential equation [687]

Based on this generalized free energy, variational (Ritz method) calculations of static micromagnetic structures can be extended to constant-velocity, *conservative* motion of the same objects. As there are no losses, no driving field is needed either. Domain walls can be calculated for a given velocity with the same methods as in statics. The kinetic potential takes care of the effects of motion. Examples of such calculations are shown in Figs. 3.96 and 3.97. The solutions can — to a good approximation — also be used for walls moving with the same velocity under the influence of a driving field and a corresponding dissipation [509], because the torques exerted by a driving field and by the dissipation mechanism are usually weak compared to the gyrotropic torques considered in the analysis of conservative motion.

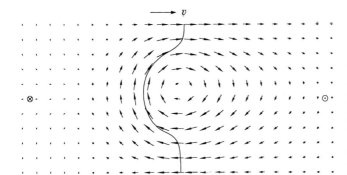

Fig. 3.97 The structure of a moving asymmetric Bloch wall, as calculated by a Ritz method for Permalloy films with a thickness of 165 nm at a velocity of 350 m/s [509]

(B) Thiele's Dynamic Force Equilibrium. An elegant method to formulate the dynamic behaviour of arbitrary micromagnetic objects was discovered by *Thiele* [709, 710]. He derived generalized dynamic *forces* that may be applied to any micromagnetic feature such as a wall, a Bloch line or a whole domain. Following [49] first the dynamic equations (3.158) are rearranged to:

$$\delta_\vartheta e_{\text{tot}} = -\left(J_s/\gamma_G\right)\left(\cos\vartheta\,\dot\varphi + \alpha_G\dot\vartheta\right) \ ,$$

$$\delta_\varphi e_{\text{tot}} = \left(J_s/\gamma_G\right)\left(\cos\vartheta\,\dot\vartheta - \alpha_G\cos^2\vartheta\,\dot\varphi\right) \ . \tag{3.160}$$

The static force f_s acting on a volume element due to the energy e_{tot} is:

$$f_s = \delta_\vartheta e_{\text{tot}}\,\mathbf{grad}\,\vartheta + \delta_\varphi e_{\text{tot}}\,\mathbf{grad}\,\varphi \ . \tag{3.161}$$

Multiplying the dynamic equations (3.160) with **grad** ϑ and **grad** φ, respectively, and adding the two equations, we obtain on the left-hand side the static force f_s. Then the right-hand side represents a dynamic counterforce $-f_d$, which can be written concisely by introducing the velocity v with which the micromagnetic structure is moving. Replacing $\dot\varphi$ by $(-v\cdot\mathbf{grad}\,\varphi)$ and $\dot\vartheta$ by $(-v\cdot\mathbf{grad}\,\vartheta)$ leads to the following expression for the dynamic force f_d:

$$f_d = -\left(J_s/\gamma_G\right)\left(g \times v + d\cdot v\right) \ , \quad g = \cos\vartheta\,(\mathbf{grad}\,\vartheta \times \mathbf{grad}\,\varphi) \ ,$$

$$d = -\alpha_G\left[(\mathbf{grad}\,\vartheta \otimes \mathbf{grad}\,\vartheta) + \cos^2\vartheta\,(\mathbf{grad}\,\varphi \otimes \mathbf{grad}\,\varphi)\right] \ . \tag{3.162}$$

The *gyrotropic vector* g generates a force perpendicular to v. All dissipation effects are taken care of by the *dyadic* d. With these definitions the vector equation (3.161) describes simply a balance of static and dynamic forces:

$$f_s + f_d = 0 \ , \tag{3.163}$$

meaning that the classical generalized force f_s must be in equilibrium with a dynamic reaction force f_d comprising gyrotropic (g) and dissipative (d) terms.

The quantities g and d can be integrated to yield the net force on a wall or some other micromagnetic entity. The integration leads for g to a remarkable result. Take for example the third component of the gyrotropic vector:

$$g_3 = \cos\vartheta\left[(\partial\vartheta/\partial x)(\partial\varphi/\partial y) - (\partial\vartheta/\partial y)(\partial\varphi/\partial x)\right] \tag{3.164}$$

for a configuration that does not depend on the coordinate z. For a given z the integral of this expression over x and y can be transformed into an integral over the orientation sphere because the parenthesis in (3.164) is the Jacobian of the mapping of $(\vartheta, \varphi) \to (x, y)$. This leads to the simple result:

$$\int g_3 \, dx \, dy = \int_F \cos\vartheta \, d\vartheta \, d\varphi \ , \tag{3.165}$$

where the angular integration extends over the area F on the unit sphere occupied by the magnetization vectors occurring inside the cross-section z. The gyrotropic reaction force of a structure thus depends only on this integral over the orientation sphere. If a micromagnetic object does not change its internal structure too much when moving, the evaluation of the dynamic reaction forces with this formalism is straightforward.

The Bloch line of Fig. 3.87 serves as an example. In this cylindrical configuration the gradients of ϑ and φ lie in the (x, y) plane, so that the gyrotropic vector g is oriented along the z direction. The magnetization component $\cos\vartheta$ extends from -1 to 1, whereas the angle φ turns from 0 to π or vice versa, so that the integral (3.165) becomes $\pm 2\pi$, depending on the sign of the Bloch line rotation. The dynamic force acting on a Bloch line moving together with a wall into the y direction therefore is $\pm 2\pi J_s v/\gamma_G$. It points perpendicular to the direction of motion, so that Bloch lines will move laterally inside a wall if the wall moves fast enough to overcome pinning (coercivity) effects. A similar argument shows that bubble domains with a net rotation in their wall magnetization differ dynamically from bubbles without such a net rotation.

Most discussions of magnetization dynamics [510, 49, 711, 712] start from (3.165). An exception are regular, one-dimensional domain walls for which (3.165) vanishes because the area F in (3.165) degenerates to a line. How this situation can be treated is shown in the example discussed in the next section.

(C) Walker's Exact Solution of Moving 180° Walls. These walls in uniaxial materials are unique in the sense that their motion can be calculated rigorously (below a critical velocity) as discovered by *Walker* [713]. This solution is quite instructive, and to generalize it numerically to other one-dimensional walls is easily possible.

A polar coordinate system is chosen with the polar axis along the easy axis (Fig. 3.54b). An external driving field is applied along the easy direction $\vartheta = 90°$. We describe the wall by the transition of the angle ϑ from $-90°$ to $90°$, with the angle φ indicating the wall character. The static wall is char-

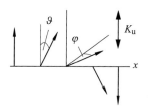

acterized by $\varphi = 0°$ for the stray-field-free Bloch wall. The dynamic forces generate a transition from the Bloch towards the Néel wall that would be reached at $\varphi = 90°$. Analytically, this is described by the dynamic equations (3.158). The free energy density to be inserted there is:

$$e_{\text{tot}} = A\left(\vartheta'^2 + \cos^2\vartheta\,\varphi'^2\right) + \cos^2\vartheta\,(K_u + K_d\sin^2\varphi) - HJ_s\sin\vartheta . \tag{3.166}$$

With $d/dt = -v\,d/dx$ the time derivatives in (3.158) can be converted into x derivatives (ϑ', φ') if the wall is assumed to move with a constant velocity v along the wall normal (the x direction). Using the abbreviations $c = \cos\vartheta$ and $s = \sin\vartheta$ this leads to the following equations for steadily moving walls:

$$-vc\,\vartheta' = \frac{2\gamma_G}{J_s}\left[-A\left(c^2\varphi'\right)' + K_d c^2\sin\varphi\cos\varphi\right] - \alpha_G vc^2\varphi' , \tag{3.167a}$$

$$-vc\,\varphi' = \frac{2\gamma_G}{J_s}\left[A\,\vartheta'' + cs\left(K_u + K_d\sin^2\varphi + A\,\varphi'^2\right)\right] + H\gamma_G c + \alpha_G v\vartheta' . \tag{3.167b}$$

One solution of these equations is a motion with constant φ throughout the wall. As has to be confirmed, the last two terms in (3.167b) — representing the driving force and the dissipation — cancel each other for this solution. Then the expression in the square brackets has to vanish identically. We thus obtain a first integral [see (3.116)], as in conventional domain wall theory, leading to:

$$\vartheta' = \cos\vartheta\,\sqrt{(K_u + K_d\sin^2\varphi)/A} . \tag{3.168}$$

Using $K_d = J_s^2/2\mu_0$, the first equation (3.167a) yields relations between the velocity and the equilibrium constant angle φ. The applied field necessary for this motion is derived from the second equation (3.167b):

$$v = -\left(\gamma_G J_s/\mu_0\right)\sin\varphi\cos\varphi\,\sqrt{A/(K_u + K_d\sin^2\varphi)} , \tag{3.169a}$$

$$\mu_0 H = \alpha_G J_s\sin\varphi\cos\varphi . \tag{3.169b}$$

The expressions (3.168) and (3.169) fulfil in fact the Landau-Lifshitz-Gilbert equations (3.167), and we confirm in particular the cancelling of driving and dissipation forces [the last two terms in (3.167b)]. The wall velocity (3.169a) has a maximum value:

$$v_p = \gamma_G \sqrt{\frac{2AQ}{\mu_0}}\, f(Q), \text{ with } Q = \frac{K_u}{K_d} \text{ and } f(Q) = \sqrt{1 + \frac{1}{Q}} - 1. \text{(3.170a)}$$

No constant velocity solution of (3.167) is possible beyond this point. For the peak velocity v_p the other variables become:

$$\sin \varphi_p = -\sqrt{Q f(Q)}\ , \quad H_p = \alpha_G H_K f(Q) \sqrt[4]{1 + 1/Q}\ , \quad H_K = 2K_u/J_s\ . \quad \text{(3.170b)}$$

Note that only a finite field H_p is needed to reach the peak velocity. Inserting typical values for a magnetic garnet ($\gamma_G = 2 \cdot 10^5$ m/As, $A = 4 \cdot 10^{-12}$ J/m, $Q = 3$) the peak velocity v_p reaches values of the order of 100 m/s.

Walker's explicit solution offers valuable insight into the mechanisms of domain wall dynamics. As apparent from (3.168) and (3.116), the place of K_u in the static solution is occupied in moving walls by $K = K_u + K_d \sin^2\varphi$. With increasing velocity the angle φ increases [see (3.169a)], causing a decrease in the wall width ($\sim\sqrt{A/K}$) and an increase in the wall energy ($\sim\sqrt{AK}$). The deviation of φ from zero produces a stray field, giving rise to the second term in the square bracket in (3.167a). Because $\varphi' = 0$ in Walker's stationary solution, this term is the only term giving rise to magnetization rotation $\dot{\vartheta} = -v\,\vartheta'$ in the moving wall. The right-hand side of (3.169a) can be interpreted as the torque acting on the central vector in the wall. The maximum velocity is thus determined by the maximum value of this torque.

In the case analysed by Walker and discussed here the torque determining the critical velocity is just caused by the stray field. It is possible, however, to enhance the maximum velocity by adding torques from an transverse external field or from an orthorhombical anisotropy [708].

Beyond the critical field no steady state is possible as already mentioned. Formally, oscillatory solutions of the original equations (3.158) can be derived in which the angle φ undergoes a more or less free precession. In these solutions the wall magnetization cycles periodically through Bloch ($\varphi = n\pi$) and Néel [$\varphi = (2n+1)\frac{\pi}{2}$] configurations, while the wall itself oscillates back and forth [712]. Figure 3.98 shows numerical solutions for these modes together with the resulting average velocities.

Figure 3.98c displays also a "negative average mobility" region beyond the peak velocity, in which the average velocity $\bar{v}$ decreases with increasing applied field. Still stationary solutions are formally possible, but these solutions are unstable. As was shown in [714], the dynamic free energy can be reduced if part of the wall advances and other parts lag back. The stationary solutions must therefore be replaced in this region by spatially inhomogeneous, "chaotic" modes [714] that are difficult to analyse in detail.

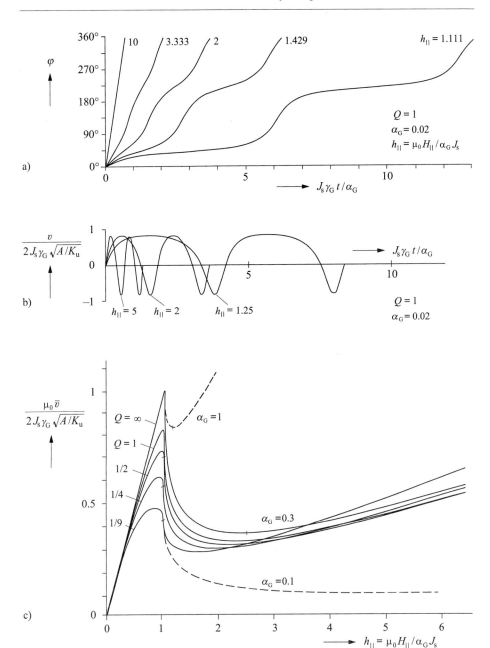

Fig. 3.98 Numerical solutions [509] of Slonczewski's oscillatory wall motion for the time dependence of the wall azimuth angle φ and wall velocity v (a, b) for different values of the driving field $H_\parallel$ along the easy axis, and for a fixed value $\alpha_G = 0.02$ of the damping parameter. Note the backward motion ($v < 0$) which occurs for certain ranges of φ between $90°$ and $180°$ and between $270°$ and $360°$. (c) Average velocities $\bar{v}$ as a function of $H_\parallel$ for different α_G and for different values of the reduced uniaxial anisotropy parameter Q

Similar considerations also apply to the oscillatory domain wall motion regime. Numerical results were published by *Schryer* and *Walker* [715], including transient and non-periodic solutions, which have to be expected in varying external fields. Any domain wall motion beyond the external field corresponding to the peak velocity has to be considered highly irregular. For this reason the maximum velocity possible for a stationary wall motion is also called the "breakdown" velocity of this wall. It determines the maximum possible speed of devices that are based on the mechanism of domain wall motion.

For thin films with perpendicular anisotropy (bubble films) the breakdown mechanism represents the generation of horizontal Bloch lines (Fig. 3.96). These Bloch lines can be converted into vertical Bloch lines and finally lead to immobile, "hard" bubbles, so called because they survive when regular bubbles collapse. It is important to avoid the peak velocity in practical devices because of the danger of hard bubble formation. For further details see [49].

(D) Wall Mass. *Döring* [635] discussed another aspect of wall dynamics: in resonance experiments the domain walls exhibit an effective *mass*, which is connected with the change in wall energy at small velocities. Take, as an example, Walker's solution of the moving 180° wall (3.168, 3.169). The wall energy can in this case be written in the form $\gamma_{\mathrm{w}} = 4\sqrt{A(K_{\mathrm{u}} + K_{\mathrm{d}}\sin^2\varphi)}$. As the angle φ begins to increase linearly with the wall velocity, the wall energy increases quadratically. Following Döring we define the effective wall mass by expanding the wall energy for small velocities according to $\gamma_{\mathrm{w}} = \gamma_{\mathrm{w}}^0 + \frac{1}{2}m^*v^2$, where $\gamma_{\mathrm{w}}^0 = 4\sqrt{AK_{\mathrm{u}}}$ is the wall energy at rest. Inserting φ from (3.169a) for small values of φ yields the 180° wall Döring mass $m^* = \frac{1}{2}\mu_0\,\gamma_{\mathrm{w}}^0/(A\gamma_{\mathrm{G}}^2)$.

The wall mass can more generally be calculated by the perturbation method discussed in Sect. 3.6.1F if we add the kinetic potential [(3.159a); with $k_0 = 0$] and the stray field energy (3.121) to the general wall energy functional. This approach yields for the general one-dimensional wall (in the coordinate system of Fig. 3.54c):

$$m^* = \frac{J_{\mathrm{s}}^2}{\sqrt{A}\,\gamma_{\mathrm{G}}^2\cos\vartheta_0}\int \frac{\sqrt{G^0(\varphi)}\,\mathrm{d}\varphi}{G_{\vartheta,\vartheta}^0(\varphi) - (1 - \tan^2\vartheta_0)\,G^0(\varphi)} \qquad (3.171\mathrm{a})$$

with $G(\varphi, \vartheta) = g(\varphi, \vartheta) + K_{\mathrm{d}}(\sin\vartheta - \sin\vartheta_0)^2$,

where $\cos\vartheta_0$ is the magnetization component parallel to the wall that rotates in the classical Bloch wall, and $G(\varphi, \vartheta)$ is the generalized anisotropy function consisting of the anisotropy and applied field expression $g(\varphi, \vartheta)$ and the stray field energy contribution (3.121). The function G^0 is G taken at $\vartheta = \vartheta_0$, the value of the domains. For low-anisotropy materials ($K_d \gg K_u$) the stray field term $G^0_{\vartheta\vartheta}(\varphi) = 2K_d\cos^2\vartheta_0$ dominates in the denominator of (3.171a), so that the integral can be expressed in terms of the total wall energy γ_w (3.118), yielding:

$$m^* = \tfrac{1}{2}\mu_0\gamma_w^0 \big/ \big(A\gamma_G^2\cos^4\vartheta_0\big) \ . \tag{3.171b}$$

The result agrees for 180° walls ($\cos\vartheta_0 = 1$) with the expression derived above for Walker's solution. By inserting the wall energy as in (3.109) we see that the wall mass is inversely proportional to the wall width parameter $\sqrt{A/K}$ for one-dimensional walls in soft magnetic materials.

The phenomena of Bloch wall motion and the occurrence of a wall mass may be discussed in a different way. Let us look at the conservative motion of a regular stray-field-free 180° Bloch wall. The external field, which is expected to move this wall, is directed perpendicular to the magnetization in the centre of the wall. According to the Landau-Lifshitz equations (3.158) the field

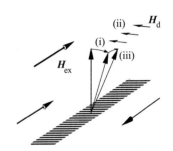

will not directly induce a rotation of the spins inside the wall that would give rise to wall motion — contrary to intuitive expectations. It will rather generate, in step (i), a rotation of the magnetization vector into the direction perpendicular to the wall, thus inducing a deviation from the Bloch wall mode [the magnitude of this deviation is given in the Walker solution by (3.169b)]. This deviation in the φ angle has two consequences: it generates in step (ii) a demagnetizing field, which will in step (iii) — according to the Landau Lifshitz equations — produce a precession in the wall plane, thus causing the wall to move. At the same time the demagnetizing field induces an increase in stray field energy giving rise to the effective wall mass.

The internal stray field in moving walls is limited. The maximum is reached when the internal magnetization vector points perpendicular to the wall, and the wall becomes unstable already before this point. The phenomenon of a maximum velocity of wall motion is connected with this limitation.

For thin films with their two-dimensional wall structures there is no simple formula for the wall mass. Explicit calculations of the energy of undamped wall motion at constant velocity (as in Figs. 3.96 and 3.97) can be used to derive the wall mass [687]. Anomalous giant mass effects are found when a wall contains Bloch lines. With respect to these complicated phenomena the reader is referred to the specialized literature [49, 512].

(E) Wall Mobility. A general result can be derived from the balance between the driving field and the dissipative term of the Gilbert equation (3.66a). For a wall moving with a constant velocity along the x direction, its mobility β_w, i.e. the ratio between its velocity and the driving field, is derived as:

$$\beta_w = \frac{2\gamma_G \sin(\Omega/2)}{\alpha_G \varepsilon_A} \quad \text{with} \quad \varepsilon_A = \frac{1}{D} \int_{-D/2}^{D/2} \int_{-\infty}^{\infty} \left(\frac{\partial \boldsymbol{m}}{\partial x}\right)^2 dx \, dy \,, \tag{3.172a}$$

where Ω is the wall angle and D is the thickness of the sample. For one-dimensional walls the integration over y (the direction perpendicular to the surface) can be omitted and $A\varepsilon_A$ becomes identical with the exchange energy in the wall, and thus with one half of the total wall energy γ_w, yielding:

$$\beta_w = 4A\gamma_G \sin\left(\tfrac{1}{2}\Omega\right)/(\alpha_G \gamma_w) \quad \text{for one-dimensional walls .} \tag{3.172b}$$

The higher the energy of a wall, the lower its mobility. This result ignores all coercivity effects as well as eddy currents and disaccommodation. Thus the conditions for experimental confirmation have to be carefully examined.

(F) Slonczewski's Description of Generalized Wall Motion. Walker's solution is the starting point for a general description of wall motion effects, which is due mainly to *Slonczewski* [712]: he integrated the two equations (3.167) over the coordinate perpendicular to the wall, assuming the functional behaviour (3.168) and its integral to be valid even when this might not be strictly true. Replacing the detailed functions $\vartheta(x, t)$ and $\varphi(x, t)$ of (3.168) and (3.169a), which describe a moving wall (in the coordinate system of Fig. 3.54b), two new variables are defined: the position of the wall centre $q(t)$ and the azimuthal angle φ of the wall, which is now denoted as $\psi(t)$. The result is a set of coupled equations (for 180° walls in uniaxial materials):

$$\partial E_{\text{tot}}/\partial q = -(2J_s/\gamma_G)\left[\dot{\psi} + \alpha_G \sqrt{K/A}\, \dot{q}\right] \,,$$

$$\partial E_{\text{tot}}/\partial \psi = -(2J_s/\gamma_G)\left[\dot{q} - \alpha_G \sqrt{A/K}\, \dot{\psi}\right] \,. \tag{3.173}$$

The total energy E_{tot} depends only on the wall position q and the internal azimuthal angle of the wall ψ. It includes the wall energy as well as the wall-position-dependent parts of the domain energy. These equations can be applied to arbitrary wall and Bloch line motions and can be extended even to gently curved walls or to the dynamics of whole domains as reviewed in [49]. In [509] a generalization based on the circular magnetization path approximation is presented with which the Slonczewski formalism can be extended to wall angles smaller than 180°.

(G) Wall Dynamics in Weak Ferromagnets. The discussion of wall dynamics dealt up to this point with regular ferromagnets and ferrimagnets for which the magnetization vector $m(r)$ remains a good representation of the system even during domain wall motion. This is not true for the so-called weak ferromagnets, which are essentially antiferromagnets with a slight canting between the almost antiparallel sublattice magnetizations. The orthoferrites, such as $YFeO_3$ and haematite $\alpha\text{-}Fe_2O_3$, are the best known representatives of this class of magnetic materials (the first mobile bubble domains were discovered in orthoferrites [510]). If the two sublattice magnetizations of a weak ferromagnet are denoted as m_1 and m_2, the resulting weak magnetization can (in the absence of a magnetic field) be described in terms of the so-called antiferromagnetic difference vector $l = \frac{1}{2}(m_1 - m_2)$ by an expression of the form $m \cong d \times l$, where d is a fixed lattice vector, the *Dzyaloshinskii* vector. It characterizes the polar nature of the respective crystal lattice. Both the static wall structure and wall dynamics are best described in terms of the antiferromagnetic vector l, deriving the ferromagnetic vector $m = \frac{1}{2}(m_1 + m_2)$ as a secondary property. The static properties of domain walls in weak ferromagnets turn out to be mostly equivalent to regular ferromagnets. This is not true for wall dynamics, however. The equation of motion for the antiferromagnetic vector differs from that of ferromagnets in decisive detail. The most important feature is a much higher limiting velocity of the order of 10^4 m/sec. In addition, the velocity limit has a relativistic character in weak ferromagnets, meaning that the critical velocity is an upper bound, which can be approached only with infinite driving fields as in the theory of relativity [remember that the Walker velocity can be reached in finite fields; see (3.170b)].

Another remarkable effect is an interaction of rapidly moving domain walls with elastic excitations when the wall velocity passes the ultrasonic barriers for the different acoustic modes. The effect can be described as a Cherenkov emission of sound waves. All these interesting properties were explored both theoretically and experimentally in an extended literature mainly by authors from the former Soviet Union. A comprehensive review of this work was given by *Baryakhtar* et al. [716]. See also [717] for a critical analysis of these phenomena from a discrete-lattice point of view.

3.6.7 Wall Friction and Disaccommodation-Dominated Motion

In addition to the intrinsic damping introduced in the last section, several other mechanisms can determine wall mobility: wall pinning can lead to a kind of friction effect, diffusion-induced disaccommodation can severely impede wall motion, and, in bulk metals, most important of all, eddy currents have to be considered. In this section we deal with static and dynamic wall friction and with the time-dependent effects known as disaccommodation or after-effect. When these effects are predominant, the gyrotropic effects discussed in the Sect. 3.6.6 can usually be neglected.

(A) Wall Friction. Domain walls interact with inhomogeneities in the material. If the defects are microscopic on the domain wall level, their effect may summarily be described by a wall-motion coercivity.

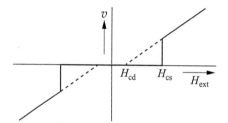

Fig. 3.99 Schematics of wall motion under the influence of intrinsic damping and friction effects. This curve defines the static wall motion coercivity H_{cs}, a threshold for wall motion, and the dynamic coercivity H_{cd} describes a displacement of the $v(H_{ext})$ curve

In analogy to mechanical phenomena, two field ranges have to be distinguished: a range of small fields, in which static pinning causes the walls not to move, and a range of larger fields, in which friction causes an energy loss per unit displacement of motion. The velocity field relation looks schematically as in Fig. 3.99. Wall motion starts at the static coercivity H_{cs}. At higher fields the effective external field appears to be reduced by the dynamic coercivity H_{cd}. The slope is determined by intrinsic damping or other dissipative processes. Little is known about the details of these phenomena, but they usually have to be considered to obtain a reasonable agreement between experiment and theory. Some empirical facts about friction effects in garnet layers are cited in [49].

In the same sense as for wall motion, special static and dynamic coercivities must be attributed to the motion of Bloch lines. Coercivity effects acting on wall substructures can lead to unexpected domain wall motion effects. Consider a series of field pulses, which would not lead to a net wall motion if applied to a system without coercivity. These could be unipolar pulses along the hard axis of a ferromagnet that do not drive the wall but rather cause some dynamic internal reactions in the wall. Assume now that the risetime and the falling time of the pulses are different. Dynamic forces on Bloch lines are proportional to the wall velocity according to (3.162), so that in one half of the cycle, when the velocity is low, a static Bloch line coercivity may be effective, and in the other half, when the velocity is high, it can be overcome, leading to a net displacement of the Bloch lines. Because Bloch line and wall motion are coupled, the effect can be a net displacement of the wall. A nonlinearity in the velocity field relation may have the same effect.

Another example is the application of a field pulse parallel to the bubble magnetization and perpendicular to the film. Under reversible conditions this field would cause the bubble to breathe, associated perhaps with some rearrangement of the Bloch lines in the bubble wall. Asymmetric pulses in connection with coercivity effects may cause a net motion of the bubble. Various phenomena of this kind are discussed in [49] under the heading "automotion".

The concept of a wall motion coercivity, which relies on individual walls with a velocity-field relation as in Fig. 3.99, cannot be applied to situations with strongly localized pinning that leads to wall deformation, Barkhausen jumps and domain rearrangements. It is only applicable to sufficiently isolated walls that maintain their identity in the magnetization process.

(B) Wall Motion and Magnetic After-Effect (Disaccommodation). One of the origins of wall friction as discussed in (A) may be thermal diffusion leading to an induced anisotropy. If the wall moves fast enough so that the induced anisotropy changes little when the wall moves by its width, the diffusion effects just add to friction-like losses. If the wall rests, however, it becomes hard to move after some time, resulting in static coercivity. In between, the magnetic after-effect results in a time dependence of permeability and wall mobility. The following is an outline of the most simple phenomena. For details we refer to the literature [613, 718−721, 509].

The classical example for such a process is the diffusion of carbon in iron. Carbon occupies octahedral interstitial positions in the body-centred cubic iron lattice. These sites have a lower, tetragonal symmetry, compared to the cubic symmetry of the lattice itself. For a given magnetization direction m the three possible interstitial sites will have three different energies $\varepsilon_i(m)$, because 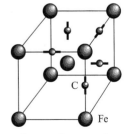 the presence of the interstitial will somehow influence the exchange interaction between the iron atoms.

To first order, the energies may be written as $\varepsilon_1 = \kappa\, m_1^2$, $\varepsilon_2 = \kappa\, m_2^2$ and $\varepsilon_3 = \kappa\, m_3^2$, because the symmetry axes of the octahedral interstitials coincide with the crystal axes. The factor κ is a phenomenological constant. In thermal equilibrium (i.e. at $t = \infty$) the occupancy $n_{i,\infty}$ of the three sites is determined by Boltzmann statistics:

$$n_{i,\infty}(\boldsymbol{m}) = \exp\left[-\varepsilon_i(\boldsymbol{m})/kT\right]\Big/\sum_i \exp\left[-\varepsilon_i(\boldsymbol{m})/kT\right] \ . \tag{3.174}$$

If we assume that the lattice defect can jump from one place to the other by a simple thermally activated process characterized by a relaxation time τ_{rel}, we obtain the following formula for the time dependence of the occupancy if we start with a random distribution $\bar{n}$:

$$n_{i,t}(\boldsymbol{m}) = \bar{n}\,\exp(-t/\tau_{\mathrm{rel}}) + n_{i,\infty}(\boldsymbol{m})\left[1 - \exp(-t/\tau_{\mathrm{rel}})\right] \ . \tag{3.175}$$

Now consider a resting Bloch wall. If the lattice defects were evenly distributed by an alternating field, they will relax to their equilibrium distribution according to the local magnetization, thus forming a potential well for the wall. This well is described by the "stabilization function" $S(x, t)$:

$$S(x, t) = S_0(x)\left[1 - \exp(-t/\tau_{\mathrm{rel}})\right] \ , \quad S_0(x) = \sum S_i^0(x) \ ,$$

$$S_i^0(x) = \int \left\{\varepsilon_i[\boldsymbol{m}(\tilde{x} + x)] - \varepsilon_i[\boldsymbol{m}(\tilde{x})]\right\} n_{i,\infty}[\boldsymbol{m}(\tilde{x})]\, \mathrm{d}\tilde{x} \ . \tag{3.176}$$

$S(x, t)$ corresponds to the energy that must be expended to move the wall to the position x. Figure 3.100 shows the stability functions $S_0(x)$ for resting (110) 90° and (100) 180° walls in iron. Characteristically, the 180° wall generates a localized potential as wide as the wall width (the chosen wall type has a particularly large wall width; see Fig. 3.66, $\rho_{\mathrm{ms}} = 0.003$). For the 90° wall the two domains differ in their magnetization axis and thus in the thermodynamic state of the point defects, resulting in an extended stabilization function.

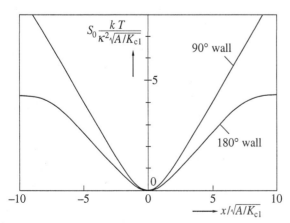

Fig. 3.100 The stabilization function $S_0(x)$ of 90° and 180° walls under the influence of a lattice defect of tetragonal symmetry

This stabilization function has a number of consequences: it defines a time-dependent (decreasing) initial permeability; it generates a threshold for the wall to break free from the potential well; and during wall motion with a constant velocity v the potential well generates an opposing field given by:

$$H_{\text{dif}} = -\left(1/\left[2J_s \sin(\Omega_w/2)\right]\right) \int_0^\infty \exp(-t/\tau_{\text{rel}}) S_0'(vt)\, dt \ , \tag{3.177}$$

where Ω_w is the wall angle.

In Fig. 3.101 two characteristic velocity field relations for disaccommodation-controlled motion are shown. In (a) the wall is caught below a critical field in its after-effect potential well and is only able to creep. Beyond the threshold, it becomes free with a mobility given by the intrinsic losses rather than by disaccommodation-related damping, so that the wall is accelerated to a much higher velocity. This behaviour is found for all 180° walls, but depending on defect symmetry also for certain 90° walls. In (b) a 90° wall is shown under the influence of a (tetragonal) lattice defect, which causes a displaced mobility curve since all the lattice defects in the domains have to be reoriented. Effectively, disaccommodation leads here to a dynamic wall friction as in Fig. 3.99 and to wall pinning for low applied fields. Without disaccommodation the $v(h)$ relation would be a straight line determined by intrinsic wall damping.

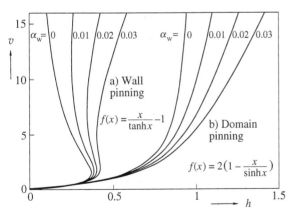

Fig. 3.101 Field dependence of wall velocity under the joint influence of disaccommodation and intrinsic damping using reduced units defined in (3.178). (a) A moving 90° wall which, because of the [111] symmetry of the responsible defect, behaves like a 180° wall and displays a localized stabilization field. (b) The same wall interacting with a tetragonal defect, leading to a far-reaching stabilization field

Figure 3.101 is based on a reformulation of (3.177), adding an intrinsic damping field that is characterized by the damping factor α_w:

$$h = (1/\tilde{v}) \int_0^\infty \exp(-x/\tilde{v}) f(x)\, dx + \alpha_w \tilde{v} \tag{3.178}$$

with $h = 2H_{\text{dif}} J_s \sin(\Omega/2)/\tau_{\text{rel}}$, $f(x) = S_0'(x)$, $\tilde{v} = v\,t$.

The processes discussed in this section are related to applications. They can be used for a study of the symmetry of lattice defects as well as for the measurement of characteristic Bloch wall parameters. If diffusion occurs only at elevated temperatures, these processes allow engineers to induce tailored anisotropies, which are frozen-in at working temperatures. For a recent analysis of the problem of disaccommodation-dominated wall motion including the effects of moving Bloch lines see [722].

3.6.8 Eddy-Current-Dominated Wall Motion

Domain wall motion in bulk metallic samples is dominated by eddy currents, an effect we ignored before. Eddy currents produce losses and limit the mobility of domain walls. Their calculation is comparable in complexity with stray field calculations. Both interactions are non-local and cannot be easily treated for arbitrary geometries. For a micromagnetic approach integrating eddy currents in numerical finite element calculations see [723].

Here we concentrate on the effects of eddy currents on the motion and shape of 180° walls in an extended (110) oriented FeSi sheet as used in transformers. The domain pattern shall be independent of the direction along the easy 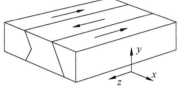 (z) axis, which would be the working axis of the transformer sheet [724–729]. Under this assumption, which represents by far the most frequently discussed case, the problem becomes two-dimensional.

Eddy current calculations start from Maxwell's equations, which lead to:

$$\Delta H_{\mathrm{E}} = \sigma \dot{B} \ . \tag{3.179}$$

Here B is the magnetic induction, which for soft magnetic materials is essentially identical with the magnetization vector, H_{E} is the magnetic field generated by the eddy currents, σ is the conductivity and Δ is the Laplace operator. A field is applied along the z axis, which sets the domain walls into motion along the x axis. Usually the magnetization change is concentrated on the moving domain walls and any permeability inside the domains can be ignored. In our case the magnetization change and the eddy current fields point into the z direction. The overall effect of the induced eddy field is a weakening of the driving field because the eddy current field is opposed to the external field, thus reducing the average wall velocity, a phenomenon known as eddy current damping. This damping effect depends, however, on sample and domain wall geometry, as has to be discussed in more detail.

The eddy current field is given by $H_E = 0$ at the two surfaces, and by:

$$(n \cdot \mathbf{grad}\, H_E)^{(1)} - (n \cdot \mathbf{grad}\, H_E)^{(2)} = 2\,\sigma v J_s \qquad (3.180)$$

at every wall element moving with the velocity v along its normal n. The indices (1) and (2) mark the field on either side of the wall, which is treated as infinitely thin in this context. The factor $2J_s$ describes the amount of flux change across a 180° wall. Since (3.179) is a linear equation, the total eddy current field can be superimposed from the fields generated by individual wall segments. The following two formulae of *Bishop* [727, 729] both describe the field produced by a wall element at position x_0, y_0 moving with velocity v parallel to x in a plate extending from $y = 0$ to $y = D$:

$$H_E(x,y) = 2\,\sigma J_s v \sum_{n=1}^{\infty} \frac{\sin(n\,\pi\, y_0/D)\,\sin(n\,\pi\, y/D)}{n\,\pi\,\exp(n\,\pi\,|x-x_0|/D)} \quad, \qquad (3.181a)$$

$$H_E(x,y) = \frac{1}{2\pi}\,\sigma J_s v \ln \frac{\cosh\big[\pi(x-x_0)/D\big] - \cos\big[\pi(y-y_0)/D\big]}{\cosh\big[\pi(x-x_0)/D\big] - \cos\big[\pi(y+y_0)/D\big]} \quad. \qquad (3.181b)$$

The first formula has the advantage that it can be easily integrated analytically to obtain the field generated from finite wall elements acting on themselves or on neighbouring elements. The second more compact form is best suited to calculate the interaction fields between distant elements, where no integration is needed. Actual eddy current calculations have normally to be performed numerically. The external field is balanced on every wall element by the eddy current field and by the specific wall energy, which acts as a wall surface tension. From the many detailed results available in the literature we present a simple formula and an interesting result of a numerical calculation.

A planar wall oriented perpendicular to the sheet surface will stay undistorted at low velocities. Its eddy-current-limited mobility is (see [729]):

$$\beta = 7.730 \,/\, (2J_s\sigma D) \quad. \qquad (3.182)$$

The mobility is inversely proportional to the sheet thickness, a result that demonstrates the non-local character of the eddy current effect.

Strong eddy current effects may lead to drastic deformations of the moving domain walls, which will influence the eddy current losses (reducing them relative to the predicted value for rigid walls). Figure 3.102 shows the velocity-induced distortion of a wall in the cross-section of a (110) oriented iron sheet. At moderate velocities the wall tends to move ahead at the surface where the eddy current field is small. It follows this tendency by utilizing the different preferred wall orientations known from Fig. 3.64.

The naturally tilted walls can easily form a dynamically favourable shape and thus produce smaller losses at small amplitudes compared to a (100) oriented sample with straight perpendicular walls. This (slight) advantage of the (110) grain-oriented transformer steel has certainly not been anticipated by the inventors. At high velocities the walls become grossly deformed to run mostly parallel to the surface (see sketch; [730]) reducing the above-mentioned effect. Eventually, magnetization processes occur mostly in a surface-parallel sheet, a phenomenon known classically as the skin effect.

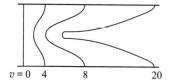

$$v = 0 \quad 4 \qquad 8 \qquad\qquad 20$$

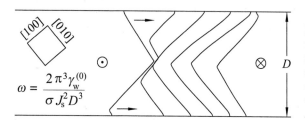

$$\omega = \frac{2\pi^3 \gamma_{\mathrm{w}}^{(0)}}{\sigma J_{\mathrm{s}}^2 D^3}$$

Fig. 3.102 Deformation of a moving 180° wall in SiFe sheet in response to eddy current fields during one half cycle of a sinusoidal motion towards the right of frequency ω. (Courtesy *J.E.L. Bishop* [727])

If there is a pinned wall in the neighbourhood of a moving wall, it will feel an increased driving field from the additional eddy field of the first wall. The second wall will therefore tend to get unpinned and to join the magnetization process. In addition, new walls may be nucleated by the eddy current field, or otherwise immobile surface and grain boundary domains may be activated and converted into regular mobile domains. The velocity of every wall can stay small if the number of walls is large for a given induction change. Therefore these mechanisms of wall bending and dynamical wall multiplication reduce the eddy current losses compared to rigidly moving walls.

These effects are also important for the experimental study of domain patterns. If one tries to obtain an equilibrium domain pattern by a.c. demagnetization, eddy currents may lead to domain multiplication, and thus to a domain width different from the true static equilibrium. To approach true equilibrium one should reduce the frequency of the alternating field until the resulting domain pattern becomes frequency-independent. For regular transformer steel of 0.3 mm thickness technical frequencies (50 Hz) are already sufficient to display eddy current effects in the demagnetized domain patterns.

3.6.9 Résumé: Wall Damping Phenomena and Losses.

The different contributions to wall damping, i.e. intrinsic damping discussed in Sect. 3.6.6, disaccommodation (Sect. 3.6.7B), and eddy currents (Sect. 3.6.8) are in principle additive. They all contribute to *losses*. The question is which effects dominate under which circumstances. Intrinsic damping and eddy currents become more effective with increasing frequency. Disaccommodation becomes important for slow processes. In metallic samples eddy currents should be checked first. As an example: for 1 µm thick alloy films in inductive recording heads eddy currents are dominant for working frequencies of 10 MHz, while for films of only 0.1 µm thickness eddy currents are still negligible. Disaccommodation (after-effect) is rarely relevant in practice for metallic materials. Because of its connection with ageing phenomena, technical materials are usually prepared so that disaccommodation becomes negligible (by removing or binding the disturbing point defects, notably dissolved carbon). This may not be true for high or low temperatures outside the usual working temperature range. At high temperatures impurities, which are immobile at working temperatures, may become mobile. An example is silicon in iron alloys. At low temperature other impurities such as hydrogen, which can move rather freely at working temperatures, may begin to "freeze", thus causing additional losses. Disaccommodation and related phenomena are also important in high-permeability ferrites that contain two- and three-valent iron ions on the same, octahedral lattice sites. Electron hopping between these ions needs little activation energy and leads to conductivity, but also to magnetic after-effect because of the asymmetry of the lattice sites. In ferrites these effects are hard to suppress and have always to be considered [731, § 52].

Intrinsic damping has to be taken into account when the other mechanisms are absent, particularly in clean insulators and in thin films. The gyromagnetic effects discussed in Sect. 3.6.6 can be observed only under these conditions.

Losses can also occur at arbitrarily low frequencies. These so-called hysteresis losses are related to discontinuous magnetization processes. They can be measured for all materials as the area of the quasi-static hysteresis and become apparent in any kind of domain observation. Examples of such quasi-static irreversible processes are presented in Chap. 5. Their theoretical analysis is complex, constituting a fascinating subject of ongoing research. For small single-domain particles hysteresis effects were discussed in Sect. 3.5.2.

To separate quasi-static losses from dynamic losses asks for frequency-dependent measurements, as shown in Fig. 3.103. Here the area of the hysteresis loop for a given induction amplitude is plotted as a function of frequency f. The value obtained by extrapolating $f \to 0$ is defined as the hysteresis loss.

In conducting media there is also a "classical" eddy current loss effect, which would be present even for an ideal, uniform magnetization process. It is calculated from Maxwell's equation under the assumption of a uniform, sinusoidal induction process along the applied field direction. The result for sheet geometry is:

$$P/f = \tfrac{1}{6}\pi^2 D^2 f B_{\mathrm{m}}^2 / \rho \ . \tag{3.183}$$

where P/f is the loss per cycle, D is the sheet thickness, B_{m} is the induction amplitude, f is the frequency and ρ is the resistivity. This classical eddy current loss can be found in Fig. 3.103 as a straight line. Frequency-related losses beyond (3.183) are called *anomalous* or *excess* eddy current losses. Based on (3.182) they would be represented by another straight line with steeper slope, the slope being determined by wall velocity. The larger the density of the walls, the smaller the velocity of every wall for a given B_{m}. That the actually measured curves in Fig. 3.103 are curved indicates that the number and activity of the domain walls changes with frequency.

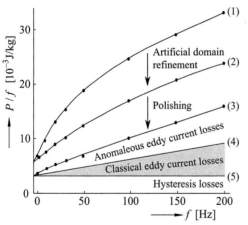

Fig. 3.103 Loss per cycle P/f and unit weight for $B_{\mathrm{m}} = 1.7$ T as a function of frequency f for specially prepared transformer steel (Fe 3.2% Si) sheets. The strongly bent curve (1) of the unmodified material (thickness 0.18 mm) is characterized by marked domain-wall-related hysteresis, and strong anomalous eddy current losses. The latter can be reduced by domain refining, as explained in Sect. 6.2.1A (2). The hysteresis losses are reduced by polishing to 0.15 mm thickness (3). Classical eddy current losses (3.183) for the polished specimen are indicated for comparison. Modified from [732]

In general, the analysis of losses as a function of frequency and amplitude helps in deciding which contribution dominates. Diffusion-related disaccommodation effects can be identified by temperature-dependent measurements. Particularly at high frequencies often the imaginary part of the magnetic susceptibility is measured, instead of the hysteresis curve and its area. Also the analysis of this quantity as a function of frequency, temperature and amplitude offers access to the basic loss mechanisms of magnetic materials.

3.7 Theoretical Analysis of Characteristic Domains

In this section we apply domain theory to some features of domain patterns that appear in one form or other in different materials and which therefore deserve a general analysis. Here is an overview:

- Ideally oriented cubic samples in which all surfaces contain at least one easy axis display only simple domains determined by the sample shape. Intricate magnetic surface patterns are induced, however, by slight misorientations. These *"supplementary domains"* are discussed in Sect. 3.7.1.

The following sections are devoted to samples with *strongly misoriented* principal surfaces of increasing thickness:

- We start with low-anisotropy thin films: the homogeneous nucleation of densely packed *stripe domains* beyond a critical thickness is characteristic of such films and is reviewed in Sect. 3.7.2.

- Thin films with a large perpendicular anisotropy are the media in which maze and *bubble domains* are formed by heterogeneous nucleation. The basic properties of these fascinating structures are introduced in Sect. 3.7.3.

- When the sample thickness increases, the surface zone of the interior domains of strongly misoriented low-anisotropy samples is transformed into *closure domains*, which are discussed in Sect. 3.7.4.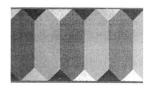

- At still larger thicknesses the basic domains become modified by *domain branching* when approaching a surface. These branching phenomena are systematically analysed in Sect. 3.7.5.

- A discussion of a seemingly simple example, the *Néel block*, follows in Sect. 3.7.6. The complexity of the actually observed phenomena in this model geometry puzzled many researchers. Its analysis sheds light on the relevance of domain theory, which can help to understand experiments, but can by no means replace them.

3.7.1 Flux Collection Schemes on Slightly Misoriented Surfaces

(A) Overview. In Sect. 3.3.4 surfaces not containing an easy direction were discussed qualitatively. We characterize such a surface by a misorientation angle ϑ_s, defined as the angle between surface and the nearest easy direction. Here we deal with slightly misoriented surfaces on soft magnetic materials for which ϑ_s amounts to a few degrees only. The intricate domain patterns occurring on such crystals fascinated observers from the beginning [32].

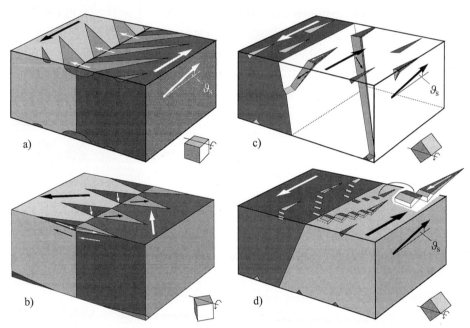

Fig. 3.104 Four examples for flux-collecting domain patterns: (a,b) Two variants of the *fir tree* pattern observed on slightly misoriented (001) surfaces of iron. Variant (b) is preferred when a crystal is misoriented by a rotation around the indicated [110] axis and when a field is applied along this axis. (c) Standard *lancets* observed on a misoriented (110) crystal. Here the flux is transported to the opposite surface or towards a 180° wall by internal "transverse" domains. The quasi-fir-tree pattern (d) is an alternative to (c) in which the flux is transported parallel to the surface to the neighbouring domains, using an undulating scheme of quasi-domains (indicated schematically)

The principle underlying the different observed patterns is the introduction of shallow surface domains collecting the net flux, which would otherwise emerge from the surface. Thus the effective domain width at the surface is reduced. The flux is transported to a suitable surface of opposite polarity of the magnetic charges and distributed again. Because this system of compensating domains is superimposed on whatever basic domains would be present without misorientation, these domains are known as *supplementary domains* [733, 734].

In the examples (Fig. 3.104, etc.) the magnetization is always assumed to follow strictly the easy directions. The effect of a subsequent application of the μ^*-correction (Sect. 3.2.5F) is not shown. In reality, the deviation of the magnetization vectors from the surface will be much smaller than given by the misorientation angle. The μ^*-correction accounts for this effect and cannot be ignored in any quantitative analysis of supplementary domains.

Figure 3.104 shows four patterns observed on iron crystals and related alloys. Two of the examples (a, b) apply to samples with an approximate (100) surface, thus displaying a four-phase domain pattern on the surface. The first variant (a) is related to 180° basic domains (this is the *fir tree* pattern discussed in Sect. 3.2.6H), and (b) shows how the flux can be collected and distributed for 90° basic domains. The other two examples (c, d) are found on crystals deviating slightly from the (110) orientation. Here all surface domains are magnetized along the only available easy axis, but internally the additional easy directions help in the flux distribution. In the first variant, the so-called lancet pattern (c), the flux is transported away from the surface, either to the opposite surface or towards the neighbouring domains. This now generally accepted model for the lancet pattern was first proposed and tested by two-sided Kerr observations in [734], to be confirmed and elaborated in much more detail in [735]. It is the standard supplementary pattern on slightly misoriented (110) iron surfaces as they typically occur in transformer sheets.

Under special circumstances an alternative structure is observed, in which the collected flux is transported laterally to the neighbouring domains. The example shown in Fig. 3.104d, the *quasi-fir-tree* pattern, is favoured for thicker crystals (thicker than ~ 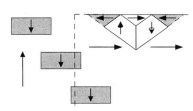 0.5 mm for iron), or when elastic stress makes a transportation of flux through the bulk unfavourable. The lateral flux transport in the quasi-fir-tree branches cannot be achieved by using the surface easy directions only. Rather, it employs a complex scheme of small surface domains connected by (invisible) internal domains, which are magnetized along the two oblique transverse axes. The quasi-fir-tree branches may be understood as quasi-domains (see Sect. 3.4.5) simulating the analogous regular fir tree branches on (100)-related surfaces.

Note that always more than one easy axis is engaged in supplementary domain patterns — either in the supplementary domains themselves or in the basic domain patterns. The generation and annihilation of supplementary domains is therefore connected with magnetostrictive noise.

Within certain limits, the strength of the misorientation influences the scale of the pattern, not the character. The larger the misorientation, the denser the pattern. Above a critical misorientation a different type of domain arrangement takes over which will be discussed in Sect. 3.7.5. Another critical value is the smallest misorientation giving rise to the type of supplementary domains discussed here. This critical misorientation is evaluated below in a quantitative analysis of the lancet pattern. We will see that in iron usually 1–2° misorientation are sufficient to give rise to supplementary domains.

(B) Quantitative Analysis of Lancet Combs. As an example of a theoretical analysis, let us choose a variant of the lancet pattern of Fig. 3.104c that occurs at a somewhat larger misorientation — typically at about 2–3°. As shown in Fig. 3.105 the lancets join in this case into *combs* using a common internal transverse domain to transport the flux to the opposite surface.

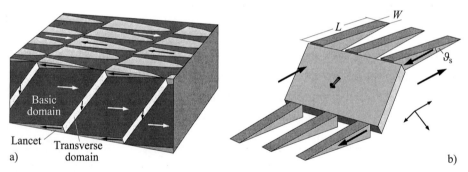

Fig. 3.105 A variation of the pattern of Fig. 3.104c in which the lancets are collected into combs. (a) A three-dimensional schematic view. (b) Simplified model used in the calculation. {How (b) drifted obviously from an early manuscript of this book into [736] is somewhat unclear}

The plate-shaped transverse domains are magnetized along the easy axes at 45° to the surface and are oriented to avoid a magnetostrictive self-energy [734]. {The observed surface orientation of the combs along the $\langle 111 \rangle$ directions supports this interpretation.

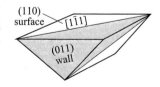

Note that the $[1\bar{1}1]$ axis is the intersection between the surface (110) and the elastically favoured (011) 90° wall of the plates}. A simple but still well-defined model of such a structure is shown in Fig. 3.105b. We assume the cross-section and the shape of the lancets to be rectangular, although the observation might rather suggest a pointed shape. In [553] a more detailed calculation based on a triangular shape was presented, but the rectangular model is simpler and more effective in reducing the stray field energy, so that it may be considered a good starting point from which refinements may be derived. We also neglect the details of the transition between the lancet combs and the transverse plate. (For single lancets these details are discussed in ⌊737⌋). In our model the magnetization in the basic domains can be assumed to be uniform, and the small misorientation angle allows the replacement of $\sin \vartheta_s$ by ϑ_s.

The model contains two parameters: the lancet width W and the lancet length L. If the depth of the lancets is $\vartheta_s L$, the base wall will be parallel to the easy axis and internal stray fields are avoided. The initial surface charge (before the application of the μ^*-correction) both of the basic domains and of the lancets is $\pm J_s \vartheta_s$. The energy of the complete lancet system per unit surface of the sheet is then for small misorientation angles ϑ_s:

$$e = \gamma_{180}\left(1 + \frac{L\,\vartheta_s}{W}\right) + C_s\,W\,\vartheta_s^2 + \frac{\sqrt{8}\,D\gamma_{90}}{L} \quad \text{with } C_s = \frac{1.705}{2\pi}\,\frac{2K_d}{1+\mu^*}, \quad (3.184)$$

where γ_{180} is the specific wall energy of the 180° walls of the lancets, γ_{90} applies to the 90° walls of the transverse domains, and D is the crystal thickness. The closure coefficient C_s is proportional to K_d for large anisotropy ($\mu^* \approx 1$), but becomes proportional to K_{c1} for low-anisotropy materials ($\mu^* \gg 1$). The factor of the 180° wall energy γ_{180} consists of two parts. The first part represents the bottom walls of the lancets which, taking top and bottom sides of the plate together, agrees in its area with one surface area in our model. The second part applies to the side walls, the area of which increases with increasing misorientation. The factor $\sqrt{8}$ in the coefficient of the 90° wall energy γ_{90} is caused by the oblique orientation of the transverse plate. An average orientation angle $\psi = 90°$ (see Fig. 3.68) has to be inserted for the specific energy γ_{90}, allowing for the zigzag folding of this wall. The estimate of the stray field energy is based on (3.41).

Minimizing (3.184) simultaneously with respect to W and L we get:

$$W_0 = \frac{1}{\vartheta_s}\left(\frac{\sqrt{8}\,\gamma_{180}\,\gamma_{90}D}{C_s^2}\right)^{1/3}, \quad L_0 = \frac{1}{\vartheta_s}\left(\frac{8\,\gamma_{90}^2\,D^2}{C_s\gamma_{180}}\right)^{1/3}, \quad (3.185a)$$

$$e_0 = \gamma_{180} + 3\vartheta_s\left(\sqrt{8}\,\gamma_{180}\,\gamma_{90}\,C_s D\right)^{1/3}. \quad (3.185b)$$

If we had optimized with respect to L alone keeping W constant, the lancet length L would have resulted proportional to $D^{1/2}$. The coupled optimization leads to a different power law. The lancet dimensions W_0 and L_0 are predicted to decrease inversely with increasing misorientation angle ϑ_s according to (3.185a), which is at least qualitatively in agreement with observations.

To decide about the stability of the lancet pattern, the energy must be compared with the energy $K_d D \vartheta_s^2 / \mu^*$ of the basic domain without a supplementary pattern. Equating both expressions, we obtain a quadratic equation for ϑ_s, the solution of which is plotted in Fig. 3.106 for silicon iron. We see that for a sheet thickness of $D = 0.3$ mm a misorientation angle of about one degree suffices to induce the lancet comb pattern in this simplified approach.

A number of refinements would be needed for a full treatment: the detailed geometry of the connection between lancets and transverse domains would have to be taken into account. Minor magnetostrictive stresses will occur (i) within the zigzag folds of the main 90° walls between basic domains and transverse plate, and (ii) around the junction between lancets and transverse domain. The lancets are certainly not rectangular in their cross-section, as assumed in Fig. 3.105. Rather a triangular shape was found theoretically in a study of this problem [738]. Also the surface appearance of the lancets will be pointed rather than rectangular as in the model (although a deviation towards a broader shape is observed particularly at higher misorientation; see [739] for a nice theoretical analysis of the lancet shape). Finally a theory of the competing isolated lancet pattern (Fig. 3.104c) would be needed. This is not the place for such a detailed analysis. The presented simple theory serves well in elucidating the basic mechanisms of supplementary domain formation.

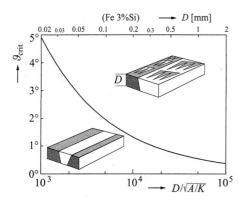

Fig. 3.106 The critical misorientation angle for the occurrence of lancet combs on (110)-oriented silicon-iron crystals as a function of crystal thickness. The data inserted to obtain the scale for silicon iron are: $K = 3.5 \cdot 10^4$ J/m^2, $A = 2 \cdot 10^{-11}$ J/m, $\gamma_{180} = 2.76$, $\gamma_{90} = 1.42 \sqrt{AK}$ as appropriate for the occurring wall orientations, and $\mu^* \gg 1$

(C) Variations. Other surface collection schemes may be treated along similar lines. For single lancets (Fig. 3.104c) the effect of magnetostrictive stresses in the transverse domains [which are avoided in the transverse plates of the lancet comb pattern treated in (B)] has to be included. The same is true for the fir tree branches as discussed in Sect. 3.2.6H.

We have distinguished between collection schemes that act parallel to the surface (Fig. 3.104a, b, d), and patterns that transport the flux through the bulk of the sample to an opposite surface (Figs. 3.104c, 3.105). Even if a surface-parallel scheme is possible at first, it may be suppressed in an applied field when the basic domains to which the flux was transported are eliminated. Figure 3.107 shows such a situation on a SiFe crystal close to the (100) orientation, which in the demagnetized state shows a regular fir tree pattern (Fig. 3.104a) employing a surface-parallel flux transport. Applying a field, the fir tree branches generate *trunks* in which their flux is collected. A trunk is joined at its base to a transverse domain leading, along a $\langle 110 \rangle$ axis because of flux continuity, to the opposite surface as for the lancet structure (Fig. 3.104c).

The switching between different flux collection schemes in the course of the magnetization cycle is a common phenomenon, which makes domain observations often rather confusing and which contributes to losses and noise in electric machines employing these materials. The humming of transformers is in fact caused by such processes as will be elaborated in Sect. 5.3.4.

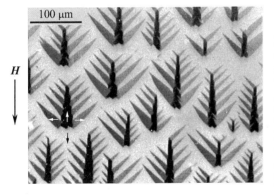

100 μm

H

Fig. 3.107 The 'true' fir tree pattern observed by the Kerr effect on a silicon-iron crystal sheet of about 0.4 mm thickness with a surface close to (100). A moderate field is applied just strong enough to suppress the basic domains. The flux from the misorientation is collected by the branches and fed into the trunk, and from its base through an invisible internal domain to the opposite surface

3.7.2 Dense Stripe Domains

In general, domain nucleation during a regular magnetization cycle along an easy direction is a discontinuous process that is dominated by defects, inhomogeneities and surface irregularities. As long as these are not precisely known, the nucleation process cannot be calculated. In the absence of defects numerical methods can be used, but these techniques are usually limited to small particles as discussed in Sect. 3.5. The situation is more favourable in the neighbourhood of a continuous, second-order phase transition. Near such a point in field space — see the discussion of critical points in phase theory (Sect. 3.4.4) — and for suitable geometrical conditions, the evolution of domains can be calculated rigorously by linearizing the micromagnetic equations. We touched upon this subject in Sect. 3.5.5. The solutions for a low-anisotropy thin film with perpendicular anisotropy have the form of closely packed stripes. They were first discovered theoretically [740, 741] in the course of domain nucleation studies. When they were later observed experimentally [742, 743], their identity with the previously derived micromagnetic structures was not understood for a long time.

In the course of experimental studies also a second class of stripe domains was discovered [744–746]. They are not formed spontaneously as regular stripes, but develop rather by nucleation and growth. For their strong contrast in the electron microscope they are called *strong* stripe domains; the regular stripe domains are called, more specifically, *weak* stripe domains. In this section we focus on weak stripes. Strong stripe domains are a characteristic feature of thicker films. They are best described by domain models and their discussion is postponed to Sect. 3.7.4.

(A) Weak Stripe Domains: Overview. The situation to be discussed is indicated in Fig. 3.108. In the limit of small thickness D a thin film with a weak perpendicular anisotropy ($Q = K_u/K_d < 1$) would be spontaneously magnetized parallel to the surface, because the anisotropy energy density K_u for in-plane magnetization is smaller than the stray field energy density K_d for uniform perpendicular magnetization and we assumed $K_u < K_d$. The chosen in-plane direction is defined as the z direction. Beyond a critical thickness the magnetization starts to oscillate out of the plane in a periodic manner to save part of the anisotropy energy. For small anisotropy the mode of oscillation is of a flux-closed character (a), while for large anisotropy the oscillation mode is essentially one-dimensional (b).

At the critical thickness the amplitude of the magnetization oscillation starts at a zero value as "dense" stripes, the half period of which typically equals the film thickness. If the thickness becomes larger, the amplitude of the magnetization modulation increases rapidly, but this process lies beyond the linearized theory of the nucleation mode. Applying an in-plane external field along the starting magnetization direction stabilizes the in-plane state and shifts the critical thickness to larger values. In the following we study the effects of thickness and field simultaneously.

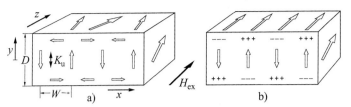

Fig. 3.108 Schematics of stripe domains in a soft magnetic layer ($Q \ll 1$) (a) and in a higher anisotropy material (b)

(B) Rigorous Theory of Stripe Domain Nucleation. The exact mathematical solution of homogeneous stripe domain nucleation was presented by *Muller* [740]. In the original work [740, 741] the available computing power was still limited and certain approximations were introduced in the evaluation, which are no more necessary.

The theory starts from the micromagnetic equations (3.63). Based on (3.64a) the stray field is expressed by a scalar potential ($\boldsymbol{H}_d = -\operatorname{grad} \varPhi_d$) defined by:

$$\Delta \varPhi_d = (J_s/\mu_0)\, \operatorname{div} \boldsymbol{m} \ . \tag{3.186}$$

The anisotropy is given by $K_u(1 - m_2^2)$, and stress effects are neglected for now. Then the micromagnetic equations (3.63) can be written in the form:

$$\boldsymbol{m} \times (2A\,\Delta\boldsymbol{m} + 2K_u m_2 \boldsymbol{e}_2 + J_s H_{ex}\, \boldsymbol{e}_3 - J_s \operatorname{grad}\varPhi_d) = 0 \ , \tag{3.187}$$

where the $\boldsymbol{e}_i$ are the unit vectors along the i directions. We limit the discussion to a two-dimensional pattern that does not depend on the coordinate z. In addition, we assume that at nucleation the components m_1 and m_2 are small. The potential $\varPhi_d$ can then also be considered small, because it depends linearly on the components m_1 and m_2 according to (3.186), and the third component $m_3 = \sqrt{1 - m_1^2 - m_2^2}$ can be treated constant ($= 1$) to first order. The micromagnetic equations (3.187) are now linearized in m_1, m_2 and $\varPhi_d$, leading together with (3.186) to the following set of differential equations:

$$-2A\,\Delta m_2 - 2K_u m_2 + J_s H_{ex} m_2 + J_s\, \partial\varPhi_d/\partial y = 0 \ ,$$

$$2A\,\Delta m_1 - J_s H_{ex} m_1 = 0 \ ,$$

$$\Delta\varPhi_d = \left(\frac{J_s}{\mu_0}\right)\left(\frac{\partial m_1}{\partial y} + \frac{\partial m_2}{\partial y}\right) \ , \quad \Delta\widetilde{\varPhi}_d = 0 \ , \tag{3.188}$$

where $\widetilde{\varPhi}_d$ denotes the potential outside the film. At the surfaces the boundary conditions [see (3.67)] are:

$$\varPhi_d = \widetilde{\varPhi}_d \ , \quad \frac{\partial m_1}{\partial y} = \frac{\partial m_2}{\partial y} = 0 \ , \quad \frac{\partial\widetilde{\varPhi}_d}{\partial y} - \frac{\partial\varPhi_d}{\partial y} = -\frac{J_s m_2}{\mu_0} \quad \text{for} \ \ y = \pm\frac{D}{2} \ . \tag{3.189}$$

Solutions of this set of linear differential equations with constant coefficients can be derived rigorously with an ansatz of the form:

$$m_1 = B\exp[i(\mu x + \kappa y)] , \qquad m_2 = C\exp[i(\mu x + \kappa y)] ,$$

$$\Phi_d = U\exp[i(\mu x + \kappa y)]|_{\text{inside}} , \qquad \tilde{\Phi}_d = \tilde{U}\exp[i(\tilde{\mu} x + \tilde{\kappa} y)]|_{\text{outside}} . \qquad (3.190)$$

From $\Phi = \tilde{\Phi}$ at the surfaces follows $\tilde{\mu} = \mu$, and $\Delta\tilde{\Phi}_d = 0$ outside the film together with the requirement that $\tilde{\Phi}_d$ must vanish at infinity leads to $\tilde{\kappa} = i\mu$ for $y > D/2$ and $\tilde{\kappa} = -i\mu$ for $y < -D/2$. The spatial frequency μ is so far arbitrary, while the values of κ are determined by the solutions of a bicubic secular equation derived from (3.188). A detailed analysis reveals that there is always one real and positive root for κ^2, which we call κ_3^2. The other two roots may be either both real and negative, or complex and conjugate.

Linear combinations of the functions (3.190) based on the altogether three roots for κ^2 (or six roots for κ) are formed to obtain real functions for m_1, m_2 and Φ_d. The solutions can be separated into two symmetry classes: (i) m_2 even, m_1 and Φ_d odd in y, and (ii) m_2 odd, m_1 and Φ_d even in y. Class (i) with m_2 even is energetically favoured, so that the other class may be ignored. If the three roots of the secular equation for κ^2 are real, the functions m_1, m_2, Φ_d and $\tilde{\Phi}_d$ assume the form:

$$m_1 = [b_1\sinh(|\kappa_1|y) + b_2\sinh(|\kappa_2|y) + b_3\sin(\kappa_3 y)]\sin(\mu x) , \qquad (3.191a)$$

$$m_2 = [c_1\cosh(|\kappa_1|y) + c_2\cosh(|\kappa_2|y) + c_3\cos(\kappa_3 y)]\cos(\mu x) ,$$

$$\Phi_d = [u_1\sinh(|\kappa_1|y) + u_2\sinh(|\kappa_2|y) + u_3\sin(\kappa_3 y)]\cos(\mu x) ,$$

$$\tilde{\Phi}_d = \tilde{u}\exp(-\mu y)\cos(\mu x) \quad \text{for } y > D/2 .$$

For large anisotropies the secular equation has one positive (κ_3^2) and two conjugate complex roots (κ_1^2 and κ_2^2). Then the solutions inside the plate can be written with $\sqrt{\kappa_2^2} = v_1 + i v_2$ in a somewhat different form:

$$m_1 = [b_1 sh_2(y)co_1(y) + b_2 ch_2(y)si_1(y) + b_3\sin(\kappa_3 y)]\sin(\mu x) , \qquad (3.191b)$$

$$m_2 = [c_1 ch_2(y)co_1(y) + c_2 sh_2(y)si_1(y) + c_3\cos(\kappa_3 y)]\cos(\mu x) ,$$

$$\Phi_d = [u_1 sh_2(y)co_1(y) + u_2 ch_2(y)si_1(y) + u_3\sin(\kappa_3 y)]\cos(\mu x) , \text{ with:}$$

$$si_1(y) = \sin(v_1 y), co_1(y) = \cos(v_1 y), sh_2(y) = \sinh(v_2 y), ch_2(y) = \cosh(v_2 y) .$$

In the next step, inserting either of these equations into the differential equations (3.188) furnishes conditions for the amplitudes b_i, u_i and $\tilde{u}$ in terms of the amplitudes c_i of m_2. Inserting further these relations into the boundary conditions (3.189) yields a set of homogeneous linear equations for the three coefficients c_i of the m_2 function (3.191). The determinant of this system of equations must vanish for a solution to exist, and this condition can be fulfilled by choosing a particular value for the film thickness D which enters through the boundary conditions (3.189). Thus the critical thickness for domain nucleation is defined for a given wavenumber μ defining the basic x period, and for a given applied field. The eigensolutions belonging to this critical eigenvalue establish relations between the three coefficients c_i. Only a common amplitude factor is left open; it can be chosen arbitrarily. A rigorous non-linear theory would be needed to determine its size.

In the last step, the wavenumber μ is optimized to find the lowest possible critical thickness in a given field. The result are sinusoidal, stripe-shaped deviations of the magnetization from the initial direction along the field. The general solution depends on three parameters: the film thickness D, the field H_{ex} and the material parameter $Q = K_u/K_d$. So we may either define a *critical thickness* D_{cr} for a given field below which the film is magnetized uniformly, or a *critical field* for a given thickness beyond which no stripes are stable.

Figure 3.109a shows the critical thickness as a function of the material parameter Q with the applied field as parameter. The inverse of the equilibrium stripe domain width W_{cr} is plotted in Fig. 3.109b.

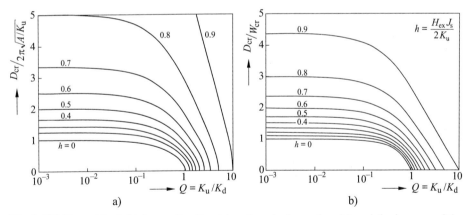

Fig. 3.109 The critical thickness D_{cr} for stripe domain formation (a) and the inverse of the optimum stripe domain width W_{cr} at nucleation (b) as a function of the material parameter Q for different applied in-plane fields along the stripe direction. For every applied field there exists also a critical value of Q. At this value Q_{cr} (which is $Q_{cr} = 1$ for zero field) the critical thickness becomes zero, while the stripe domain width diverges

The nucleation modes are shown in Fig. 3.110 for some examples. They are valid only for small amplitudes (at, or near nucleation). In the drawings the amplitudes are normalized. The solutions for small-anisotropy materials

(a) are particularly interesting. Although the driving force for the stripe domains is a uniaxial perpendicular anisotropy, the resulting modulation assumes the character of a two-dimensional flux-closed pattern.

For very small Q the solution approaches the analytical expression [579]:

$$m_1 = \frac{aD}{W} \sin\left(\frac{\pi y}{D}\right)\sin\left(\frac{\pi x}{W}\right) , \quad m_2 = a\cos\left(\frac{\pi y}{D}\right)\cos\left(\frac{\pi x}{W}\right), \quad \Phi_d = 0 \qquad (3.192)$$

leading to the critical condition:

$$D_{cr} = 2\pi\frac{\sqrt{A/K_u}}{1-h} , \quad W_{cr} = D_{cr}\sqrt{\frac{1-h}{1+h}} , \quad \text{with } h = \tfrac{1}{2}H_{ex}J_s/K_u . \qquad (3.193)$$

The cosine function in m_2 of (3.192) reaches zero at the surfaces ($y = \pm D/2$), but the derivative is not zero there, violating the boundary condition (3.189). But the rigorous solution follows the cosine of the approximate solution (3.192) over most of the thickness for small Q, deviating from it in a narrow surface layer only. The micromagnetic surface layer becomes thinner with decreasing material parameter Q. It scales with $\sqrt{A/K_d}$ and is expressed by the hyperbolic cosine contributions in (3.191), which are absent in (3.192).

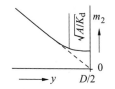

The rigorous results for the critical thickness and the optimum stripe domain width (Fig. 3.109) agree well with the results for the stray-field-free limit (3.193) for small $Q < 0.1$. In this range the critical thickness is approximately $D_{cr} = 2\pi\sqrt{A/K_u}$ in zero field. The stripe width at nucleation is practically equal to the film thickness for zero field, becoming somewhat smaller for thicker films for which nucleation occurs in an applied field. Stripe domains that are wider than the film thickness are expected only for large Q. When Q approaches the value of unity, the critical thickness vanishes and the optimum nucleation period diverges. For sufficiently large Q, stripe domains are therefore predicted even for ultrathin films. An observation that seems to confirm this theoretical prediction was reported by *Allenspach* et al. [747], who investigated ultrathin cobalt films. The Q parameter of bulk cobalt is too small for this effect ($Q \approx 0.4$), but surface anisotropy adds to the perpendicular anisotropy for very thin films, so that the effective Q becomes larger than unity. With increasing film thickness the relative influence of surface anisotropy decreases, leading for a certain thickness (of about 0.6 nm) to a value of $Q = 1$. At this point a fine although somewhat irregular domain pattern was observed. The phenomenon should be observable also for $Q > 1$ if an in-plane field is applied. According to Fig. 3.109a the value of Q, for which D_{cr} vanishes, varies with the field as $Q_{cr} = 1/(1-h)$.

For arbitrarily large thicknesses stripe domain nucleation is predicted at fields close to the anisotropy field ($h \to 1$). In this range the equilibrium period becomes much smaller than the plate thickness D, it approaches for small Q the constant value $\sqrt{2}\,\pi\sqrt{A/K_u}$ according to (3.193) independent of D.

A phenomenon that is analogous to the dense stripe domains of thin films was discovered numerically by *Hua* et al. [748] for bulk soft magnetic material. The stripe-like pattern is then limited to a surface layer and induced by a strong *surface* anisotropy rather than by a bulk perpendicular anisotropy.

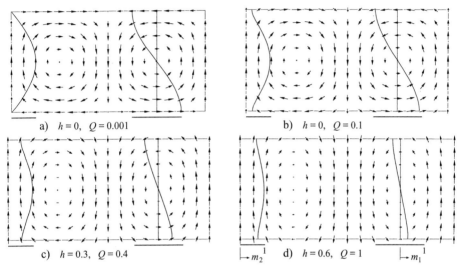

a) $h = 0$, $Q = 0.001$ b) $h = 0$, $Q = 0.1$

c) $h = 0.3$, $Q = 0.4$ d) $h = 0.6$, $Q = 1$

Fig. 3.110 Four examples of calculated stripe domain profiles for different values of the reduced applied field $h = HJ_s/(2K_u)$ and the material parameter $Q = K_u/K_d$. Cross-sections through one period of the optimized solutions are shown. The *curves* indicate the functional dependence of the magnetization components m_1 and m_2, normalized to the maximum of the vertical amplitude m_2.

(C) Magnetostriction and Stripe Nucleation. Only exchange, anisotropy, stray field and applied field energy terms entered the standard theory of dense stripe domains presented in the last section. One might not expect a weak effect like magnetostriction to contribute essentially to this solution. However, domain analysis for thicker uniaxial plates in a hard axis field had indicated a strong influence of magnetostriction on the domain orientation, particularly at large fields near saturation [161]. Figure 3.111 demonstrates this argument in a qualitative way, replacing the continuous stripe pattern by regular domains.

The orientation dependence of wall energy favours stripes along the field direction. Figure 3.62 shows that walls parallel to a field ($\psi = 0°$) have a lower energy than walls perpendicular to a field ($\psi = 90°$). But the difference between the parallel and perpendicular wall energies becomes small at large fields close to the saturation field $H_K = 2K_{ul}/J_s$. Evaluating this difference $\gamma_w(\psi = 90°) - \gamma_w(\psi = 0°)$ from Fig. 3.62 as a function of the reduced applied field h shows that it disappears quadratically approaching saturation.

While the orientation dependence of the wall energy vanishes with increasing field, the magnetostrictive interaction between the basic domains gains in relative importance. This can be seen by an argument as in Sect. 3.2.6G. Repeating this one-dimensional calculation for uniaxial symmetry and for a variable magnetization angle ϑ in the domains (where ϑ is the angle between the magnetization direction and the easy axis), results in a magnetostrictive self-energy that varies as $e_{ms} \approx \sin^2\vartheta \cos^2\vartheta$ for the configuration of Fig. 3.111a. In equilibrium the domain magnetization angle ϑ is $\sin\vartheta = h$ with $h = H_{ex}/H_K$. The value for the reduced field $h = 1$ marks saturation in the film plane. Inserting $\sin\vartheta = h$ into e_{ms} we get for the angle-dependent part the expression $h^2(1 + h)(1 - h)$, which approaches zero only linearly for $h \to 1$. The magnetostrictive self-energy therefore dominates over the influence of the wall energy at large fields, favouring the perpendicular relative to the parallel domain orientation.

The domain-theoretical argument thus predicts stripe domains to be nucleated perpendicular to the applied field in a material with non-vanishing (positive) magnetostriction because sufficiently close to saturation the magneto-elastic interaction should prevail. The initial stripe orientation tends to persist down to zero field even if at low fields the wall energy becomes predominant [161], because a reorientation of the stripes would require a collective rearrangement of the domain pattern. This qualitative reasoning explains the experimental observations of domain patterns in thick uniaxial crystals (see Sect. 5.2.2B).

The argument is not watertight, however, because real stripe domains are continuous, micromagnetic structures as in Fig. 3.110, and not simple domain patterns as in Fig. 3.111. For small thicknesses the saturation field is shifted away from the anisotropy field, thus invalidating the simple argument. *Patterson* and *Muller* [749] addressed this problem properly with a micromagnetic approach, assuming a sinusoidally varying strain field in the analysis. After introducing the compatibility relations, the resulting magneto-elastic energy was included in the nucleation analysis. The authors find a transition from longitudinal stripe domain nucleation at small film thicknesses to transverse nucleation for thick films, which depends on the relative strength of the magnetostrictive and anisotropy energy contributions. Domain theory did not predict such a threshold thickness, but experimental observation seems to support this concept showing longitudinal stripes for very thin cobalt samples [161].

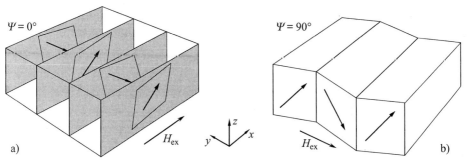

Fig. 3.111 Magnetostriction favours stripe nucleation perpendicular to the field (b) as opposed to parallel stripes (a) in which the magnetostrictive deformations are incompatible. The axes have been chosen differently compared to Fig. 3.108 to allow direct use of the equations supplied in Sect. 3.2.6G on magnetostriction

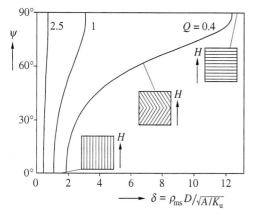

Fig. 3.112 The calculation by *Patterson* and *Muller* [749] predicts a continuous transition between the longitudinal ($\psi = 0°$) and transverse stripe domains ($\psi = 90°$) with increasing film thickness. The diagram is valid for small values of the dimensionless parameter ρ_{ms}, which is true for most materials. For cobalt ($Q = 0.4$) the reduced thickness assumes the value $\delta = 1$ for an actual thickness of $D = 0.15\ \mu m$

Decisive is the material constant $\rho_{ms} = c_{44}(4\lambda_D - \lambda_A - \lambda_C)^2/K_u$, the relative weight of magnetostrictive and anisotropy energy terms. The λ_A to λ_D are magnetostriction constants and c_{44} is a shear modulus for uniaxial symmetry. This ratio is much smaller than unity in most materials; in cobalt, for example, it assumes a value of 0.035. In Fig. 3.112 the calculated optimum stripe orientation angle according to [749] is plotted as a function of the film thickness for some values of the material parameter $Q = K_u/K_d$.

A universal behaviour is derived in terms of the reduced thickness parameter $\delta = \rho_{ms}D/\sqrt{A/K_u}$. Below a reduced thickness of $\delta_1 = \left[\pi^2/(1 + Q)^3\right]^{1/2}$ the influence of magnetostriction is negligible, leading to conventional longitudinal stripes. Beyond a second critical thickness $\delta_2 = \left[\pi^2/Q^3\right]^{1/2}$ only the transverse stripes are nucleated. In the transition range the stripes nucleate at intermediate orientations given by $\cos^2\psi = \left[\pi^2/\delta^2\right]^{1/3} - Q$. Outside the transition range the solutions $\psi = 0$ and $\psi = \pi/2$ take over. The calculation is based on the approximation $\delta > 25\sqrt{Q}\rho_{ms}$, which is mostly valid because of $\rho_{ms} \ll 1$.

3.7.3 Domains in High-Anisotropy Perpendicular Films

We have seen in Fig. 3.109 that films with a weak perpendicular anisotropy are magnetized parallel to the film plane for thicknesses below a critical thickness that depends on the anisotropy ratio Q. This is true for all materials with $Q < 1$. Even beyond this value a uniform in-plane magnetization may be enforced in an applied in-plane field, but in-plane fields are not in the centre of interest in this section, in which we are dealing with films with an anisotropy ratio $Q > 1$ and an easy axis perpendicular to the film plane. For such high-anisotropy films the state uniformly magnetized along the perpendicular axis is at least metastable.

Domains can exist in these films at zero field even for arbitrarily small thicknesses. In a field along the perpendicular easy axis the domains may be suppressed again. There exists, however, usually a threshold between domain states and the uniformly magnetized state, so that domains are generated and annihilated by discontinuous processes, not spontaneously as for dense stripe domains discussed in the last section.

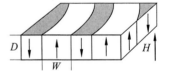

Many different domain patterns can be formed for the same values of the perpendicular field in these films, and marked hysteresis effects are connected with their generation, annihilation and transformation, which make them interesting both for fundamental studies and for memory applications. The domains are always large compared to the Bloch wall width, so that a domain-theoretical approach is adequate in their analysis rather than the micromagnetic approach needed in the analysis of dense stripe domains in low-anisotropy films.

(A) Theory of Equilibrium Parallel Band Domains. We start the discussion with the Kooy and Enz formula (3.41), which describes the stray field energy for an arrangement of parallel band domains in an infinite plate, separated by domain walls of zero thickness. We assume perpendicular anisotropy ($\Theta = 0$). A large anisotropy compared to the stray field energy parameter ($Q \gg 1$) means $\mu^* \approx 1$. Then (3.41) reduces to:

$$E_d = K_d D \left\{ m^2 + \frac{4p}{\pi^3} \sum_{n=1}^{\infty} n^{-3} \sin^2 \left[\tfrac{\pi}{2} n(1+m) \right] \left[1 - \exp\left(\frac{-2\pi n}{p} \right) \right] \right\} \qquad (3.194)$$

with: W_1, $W_2 =$ domain widths , $D =$ film thickness ,

$$m = \frac{W_1 - W_2}{W_1 + W_2} = \text{reduced magnetization} , \quad p = \frac{W_1 + W_2}{D} = \text{reduced period},$$

$J_s =$ saturation magnetization and $K_d = J_s^2 / 2\mu_0$.

Writing the stray field energy in the form $E_d = K_d D \left[m^2 + g(p, m) \right]$ and adding the wall energy and the applied perpendicular field energy, the total energy of the band pattern may be expressed in the following reduced form based on the *characteristic length* $l_c = \gamma_w / 2K_d$ ($\gamma_w =$ specific wall energy):

$$\varepsilon_{tot} = 4\lambda_c / p - 2hm + m^2 + g(p, m) \qquad (3.195)$$

with $\varepsilon = E / DK_d$, $\lambda_c = l_c / D = \gamma_w / (2DK_d)$ and $h = HJ_s / 2K_d = \mu_0 H / J_s$.

Minimizing the total energy with respect to p and m yields:

$$\lambda_c = \left(p^2 / \pi^3 \right) \sum_{n=1}^{\infty} n^{-3} \left[1 - (1 + 2\pi n/p) \exp(-2\pi n/p) \right] \sin^2 \left[\tfrac{\pi}{2} n(1+m) \right] \quad (3.196)$$

$$h = m + (p / \pi^2) \sum_{n=1}^{\infty} n^{-2} \sin[n\pi(1+m)] \left[1 - \exp(-2\pi n/p) \right] . \qquad (3.197)$$

These are two implicit equations that yield, for every pair of the reduced period p and of the reduced magnetization m, the corresponding values of the reduced characteristic length λ_c and the reduced bias field h. The inverse problem (λ_c and h given, p and m to be determined) can be solved numerically (most efficiently using mathematical transformations as shown in [750]).

A number of equilibrium magnetization curves derived in this way are shown in Fig. 3.113. The domain wall energy, which enters via the parameter λ_c, has little influence on the magnetization curve for large thicknesses (small λ_c). In accordance with phase theory the magnetization curve approaches in this limit the straight line $m = h$ as determined by the demagnetizing factor $N = 1$ of the plate. With increasing λ_c the initial slope of the magnetization curve rises and saturation is reached at fields that are smaller than the demagnetizing field. The higher initial permeability and the reduced saturation field for smaller thicknesses are a direct consequence of domain wall energy, which reduces the stability range of domain states relative to phase theory in which wall energy contributions are neglected.

The neighbourhood of saturation in Fig. 3.113 is of particular interest. The transition looks like a continuous, second-order transition. The true nature of this transition is different, however. On approaching saturation the equilibrium period $W_1 + W_2$ diverges (Fig. 3.114). At the same time the domain width of the minority domains W_2 approaches a constant value. Rather than vanishing by further reducing their width, the minority domains increase their mutual distance beyond all bounds, thus mimicking a second-order transition.

Whether the predicted behaviour can actually be observed depends on the possibility to adjust the true equilibrium domain period for every field, which might be achieved by superimposing for every static field an alternating field. If this is not done, the domains can only fold or unfold, branch or unbranch to achieve a different effective domain period. This leads to complex two-dimensional domain patterns instead of regular parallel band arrays, as discussed and demonstrated in Sect. 5.6.1.

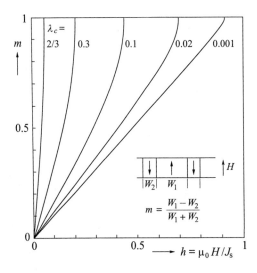

Fig. 3.113 The equilibrium magnetization curves of band domains as a function of an applied field perpendicular to the surface. Parameter is the ratio λ_c between the characteristic length $l_c = \frac{1}{2}\gamma_w/K_d$ and the film thickness D

In Fig. 3.114b the domain period P is shown in units of the material length l_c, not reduced by the film thickness as in Fig. 3.114a. The smallest domain period for $h = 0$ is found at a film thickness of about $D = 4l_c$, or $\lambda_c = \frac{1}{4}$.

When the parameter λ_c increases beyond the value of one, the saturation field becomes small and the equilibrium domain width becomes large. This means that when films become thinner than the characteristic length $l_c = \gamma_w/2K_d$, equilibrium domain patterns can hardly be formed in such films. The energy gain in the demagnetized state becomes minute for large λ_c, and any trace of wall coercivity will suppress the formation of equilibrium domains. So there is a practical limit for domains beyond $\lambda_c = 1$. Theoretically, however, there is no critical thickness for single-domain behaviour in perpendicular films [554]. There always exists a domain state that has a lower energy in an infinitely extended thin perpendicular film at zero field than the saturated state. This somewhat counter-intuitive result is caused by the long-range nature of the magnetostatic interaction. An argument in *Eschenfelder*'s book [751 (p. 31)] suggesting a discontinuous transition to the single-domain state at $\lambda_c = 1$ is not correct.

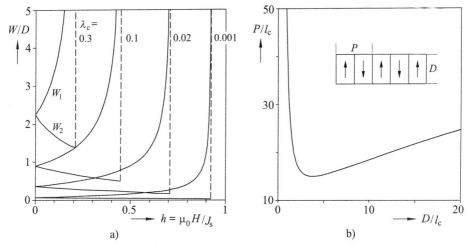

Fig. 3.114 Relative band domain width as a function of the applied field (a) and absolute band period $P = W_1 + W_2$ as a function of the film thickness (b) for $h = 0$

In the neighbourhood of the saturation field (which is shown in Fig. 3.115 together with other characteristic fields to be explained later) the bands get separated from each other as mentioned before. Their magnetostatic interaction becomes negligible, so that this characteristic field and the band width at saturation can as well be calculated for *isolated* band domains.

(B) Isolated Band Domains. The energy of a single straight infinite band domain can be calculated by using the stray field energy formulae of Sect. 3.2.5E. Starting from a saturated plate of thickness D, we superimpose a band of width W and of opposite magnetization $-2J_s$. Its stray field energy can be expressed by the self-energy of its surface charges (3.36) and their interaction (3.35). We add the interaction energy of the embedded band with the plate magnetization (as in Fig. 3.6), which amounts to $-4K_d WD$, and also the wall energy and the external field energy expressions. Finally we obtain the total energy per unit length of the band domain relative to the saturated state in the following form:

$$e_{tot} = \tfrac{4}{\pi} K_d \left[G_{20}(W, 0) - G_{20}(W, D) + G_{20}(0, D) \right]$$
$$- 4 K_d W D + 2 D \gamma_w + 2 H J_s W D \, , \tag{3.198}$$

where the function $G_{20}(x, y)$ is defined in (3.34). To calculate the equilibrium band width, e_{tot} is varied with respect to W, which yields, with the abbreviation for the relative domain width $w_r = W/D$, the implicit relation:

$$1 - h = \tfrac{1}{\pi} \left[2 \arctan(w_r) + w_r \ln(1 + 1/w_r^2) \right] \, . \tag{3.199}$$

Saturation will be reached when the total energy of the band becomes negative, because the band can then retract, thus reducing the total energy. (If the band is pinned at its ends and not able to contract, or if the field is applied only locally, it will remain metastable beyond the saturation point considered here). With $e_{tot} = 0$ and (3.198) we obtain after inserting the definition (3.34) and the equilibrium condition (3.199) a second equation for the saturation field:

$$2 \pi \lambda_c = w_r^2 \ln(1 + 1/w_r^2) + \ln(1 + w_r^2) \, . \tag{3.200}$$

Eliminating (numerically) the strip width w_r from (3.199) and (3.200) we obtain the saturation field h_{s0} as a function of λ_c (Fig. 3.115).

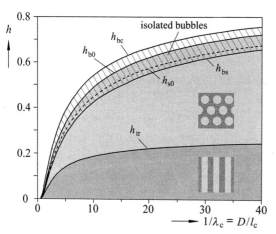

Fig. 3.115 Several characteristic fields for domain patterns in a uniaxial plate as a function of the inverse reduced characteristic length λ_c. The saturation field for the bubble lattice h_{b0} is larger than h_{s0}, the saturation field for band domains. Individual bubbles are stable between h_{bc}, the bubble collapse field and h_{bs}, the strip-out field. Finally, h_{tr} marks the boundary above which an optimized bubble lattice is energetically favoured relative to band domains. (Calculated after [752] together with *A. Bogdanov*)

(C) Bubble Lattices. The Kooy and Enz theory presented in (A) was incomplete in so far as it examined only one possible configuration, that of parallel band domains. In particular in the neigh-

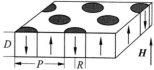

bourhood of saturation a bubble lattice has a smaller total energy than the band pattern. As a metastable configuration, bubble lattices may exist at the same values of applied fields as band arrays. Usually, the bubbles in a lattice are arranged in a close-packed, hexagonal way. In the presence of in-plane anisotropies bubbles may also form other lattices. In addition, irregular and even amorphous bubble arrangements are possible. The following analysis, which is based on [752] and [753], applies to the standard hexagonal lattice.

Let the period in a hexagonal bubble array be P and the bubble radius R. They are related to the relative mean magnetization by $m = 1 - 4\pi(R/P)^2/\sqrt{3}$. Then the reduced total energy is with abbreviations as in (3.195):

$$\varepsilon_{\text{tot}} = \left(4\lambda_c/p\right)\sqrt{(1-m)\pi/\sqrt{3}} - 2hm + m^2 + g(p, m) \ ,$$

$$g(p, m) = \left(2p(1-m)/\sqrt{3}\,\pi^2\right)\sum_{n=1}^{\infty}\sum_{k=0}^{n}\left(Z/s_b^3\right)[\,1-\exp\left(-2\pi s_b/p\right)]\,J_1^2(s_b y) \ ,$$

$$s_b = \sqrt{\tfrac{4}{3}\left(n^2 + k^2 + nk\right)} \ , \ y = \sqrt{\sqrt{3}\pi(1-m)} \ ,$$

$$Z = 6 \cdot \begin{cases} 1 & \text{if } k=0 \text{ or } n \\ 2 & \text{else} \end{cases} \ . \tag{3.201}$$

The variable s_b measures the distance between bubble centres in the hexagonal lattice and Z counts the number of equivalent bubbles for given (n, k). The sine function in (3.194) is replaced in (3.201) by the Bessel function of the first kind $J_1(x)$. The minimization of this energy leads to the implicit equations:

$$\lambda_c = \tfrac{1}{4}p^2\sqrt{\sqrt{3}/\left[\pi(1-m)\right]}\,\frac{\partial g}{\partial p} \ ,$$

$$\frac{\partial g}{\partial p} = \frac{2(1-m)}{\sqrt{3}\,\pi^2}\sum_{n=1}^{\infty}\sum_{k=0}^{n}\left(Z/s_b^3\right)\left[1-\exp\left(\frac{-2\pi s_b}{p}\right)\left(1+\frac{2\pi s_b}{p}\right)\right]J_1^2(s_b y) \ , \tag{3.202}$$

$$h = m - \left(\frac{\lambda_c}{p}\right)\sqrt{\frac{\pi}{\sqrt{3}(1-m)}} - \frac{1}{2}\frac{\partial g}{\partial m} \ ,$$

$$\frac{\partial g}{\partial m} = \frac{2p}{\pi}\sqrt{\frac{1-m}{\sqrt{3}\pi}}\sum_{n=1}^{\infty}\sum_{k=0}^{n}\left(Z/s_b^2\right)\left[1-\exp\left(\frac{-2\pi s_b}{p}\right)\right]J_0(s_b y)\,J_1(s_b y) \ . \tag{3.203}$$

The convergence of the double infinite sums is poor for large domain periods. Evaluating the two parts of the p-dependent expressions in the square brackets separately improves the convergence to some extent. A more effective mathematical approach based on Ewald's method [754] is discussed in [752].

It turns out that beyond a relative magnetization m of about 0.3 and reduced perpendicular fields h of about 0.2 the bubble lattice is energetically advantageous compared to the band array ([752]; Fig. 3.115). But the energy difference is small, and negligible as a driving force for domain rearrangement processes. The bubble lattice remains metastable at zero field and even at negative fields where it is converted into a network or "froth" structure that finally coarsens and decays, as studied in detail in [755, 756] (see also Sect. 5.6.1). The ideal

magnetization curve of a perpendicular film is shown in Fig. 3.116. As there
is no continuous path for a transition between the two domain patterns, no
spontaneous conversion between band and bubble domain patterns is observed
experimentally, and real magnetization curves follow mostly the continuous
lines for either the band domain pattern or the bubble lattice pattern.

The bubble lattice saturation field h_{b0} (Fig. 3.115) is defined as the field at
which its energy (3.201), after inserting the equilibrium conditions (3.202) and
(3.203), equals the energy of the saturated state. At this point the equilibrium
period P goes to infinity, while the bubble radius R stays finite. This field h_{b0}
is larger than the band domain saturation field h_{s0} because bubbles with their
circular domain walls are more efficient in reducing the stray field energy
close to saturation than bands. It is, on the other hand, smaller than the
collapse field h_{bc} of individual bubbles that will be derived in (D).

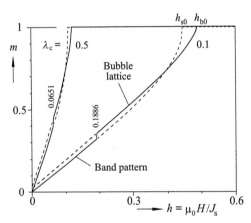

Fig. 3.116 The ideal magnetization curve
of a bubble film in a field oriented perpendicular to the surface for two selected
values of the thickness parameter λ_c. For
every field value the optimized band and
bubble lattices were computed based on
(3.195) and (3.201). Note the non-zero equilibrium remanence of the bubble lattice.
(Together with *A. Bogdanov*)

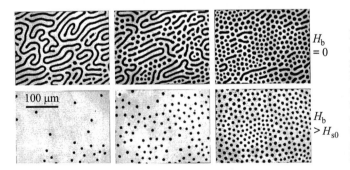

Fig. 3.117 Depending
on the nature of the re-
manent state (*top row*)
bubble states of lower
or higher bubble densi-
ty develop in a perpen-
dicular field H_b beyond
the stability limit H_{s0}
of the band domains
(*bottom row*)

Even beyond the saturation field of the band domain array when the bubble
lattice is still stable, the bands do not collapse into an equilibrium bubble

lattice, but into as many bubbles as there were independent band segments (Fig. 3.117). At this stage the bubbles can be treated as independent entities already. The properties of individual bubbles are discussed next.

(D) Isolated Bubble Domains. Individual bubbles are stable in a larger field range beyond and below the saturation field of the equilibrium bubble array. To calculate this stability range is the elementary task of bubble theory. The upper, collapse field was first calculated by Kooy and Enz [554]. *Thiele* [757] added a derivation of the lower, strip-out field at which a bubble spontaneously expands into a band domain. In the following we review the basic formulae without proof. For details see the textbooks [510, 49, 751].

We restrict ourselves to large Q, which means that we neglect the μ^*-effect again. Calculations for small Q, for which the wall width is no more small compared to the bubble diameter were presented by *DeBonte* [758]. The result was that the standard theory is well applicable for $Q > 1.5$, while bubbles become rapidly unstable for smaller Q.

In this approximation ($Q \gg 1$) the energy of an individual bubble of radius R relative to the saturated state can be written in the form:

$$E_{tot} = 2\pi K_d D^3 \left[-I(d) + \lambda_c d + \tfrac{1}{2} h d^2 \right] \quad \text{with } d = 2R/D , \tag{3.204}$$

$$u^2 = d^2/(1+d^2) \quad \text{and} \quad I(d) = -\tfrac{2}{3\pi} d \left[d^2 + (1-d^2) \, \mathsf{E}(u^2)/u - \mathsf{K}(u^2)/u \right] .$$

$\mathsf{E}(u)$ and $\mathsf{K}(u)$ are the complete elliptic integrals:

$$\mathsf{E}(u) = \int_0^{\pi/2} \sqrt{1 - u \sin^2\alpha} \, d\alpha \quad , \quad \mathsf{K}(u) = \int_0^{\pi/2} d\alpha/\sqrt{1 - u \sin^2\alpha} \quad .$$

The reduced material parameter λ_c and the reduced applied field h are defined as in the last section. The function $I(d)$ measures the stray field energy gain by the bubble. Its derivative is the so-called force function:

$$F(d) = \partial I/\partial d = -\tfrac{2}{\pi} d^2 \left[1 - \mathsf{E}(u^2)/u \right] . \tag{3.205}$$

It can be used to formulate the equilibrium condition: for a given reduced applied field h and a given ratio $\lambda_c = l_c/D$ the reduced bubble diameter $d = 2R/D$ must fulfil the condition [as derived by differentiating (3.204)]:

$$F(d) = \lambda_c + h d . \tag{3.206}$$

This implicit relation for d can be displayed graphically (Fig. 3.118). The values of λ_c and h define a straight line, which in general intersects the $F(d)$ curve twice. The intersection with the larger d value yields the stable solution d_0. Increasing the field causes the bubble diameter to decrease until at a critical field the straight line just touches the $F(d)$ curve. This is the bubble *collapse field* h_{bc}. Bubble collapse can be described by the condition:

$$\lambda_c = S_{bc}(d) = F(d) - d\,\partial F/\partial d = \tfrac{2}{\pi}\left[d^2\left(1 - E(u^2)/u\right) + u\,K(u^2)\right] \; . \qquad (3.207)$$

If $S_{bc}(d)$ is added to Fig. 3.118, the bubble diameter at collapse can be derived from this diagram. The critical field h_{bc} follows from (3.207) or graphically from the corresponding slope in the diagram.

Finally, the stability against bubble strip-out is calculated by studying the bubble energy relative to an elliptic deformation [757]. It leads to the condition:

$$\lambda_c = S_{bs}(d) = \tfrac{1}{3}\left[\tfrac{1}{2\pi}d^2\left(\tfrac{16}{3} - L_{bs}(d)\right) - S_{bc}(d)\right] \qquad (3.208)$$

with $L_{bs}(d) = 16\,d\int_0^{\pi/2}\left(\sin^2\!\alpha\,\cos^2\!\alpha\Big/\sqrt{1 + d^2\sin^2\!\alpha}\,\right)d\alpha$.

See also Fig. 3.118 for the evaluation of this quantity.

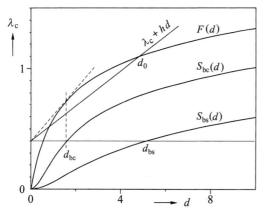

Fig. 3.118 A diagram to derive bubble stability conditions. For a given λ_c and h the (second) intersection of the straight line $\lambda_c + h\,d$ with $F(d)$ yields the bubble diameter d_0 . The intersection of the parallel to the d axis through λ_c with the curves $S_{bc}(d)$ and $S_{bs}(d)$ yields the reduced bubble diameters d_{bc} and d_{bs} at bubble collapse and strip-out. The collapse field follows from the slope of the (tangential) line connecting $(0,\lambda_c)$ and $[d_{bc}, F(d_{bc})]$, and correspondingly for the strip-out field

Another characteristic field h_{b0} is the field at which a bubble begins to carry a higher energy than the saturated film without a bubble. To find it, the total energy (3.204) is set equal to zero. In addition, (3.206) must be fulfilled. The field h_{b0} agrees with the stability limit (the saturation field) of the bubble lattice and was included in Fig. 3.115.

An individual bubble can exist in a metastable state beyond h_{b0} up to h_{bc}. Strictly speaking, a bubble is also in a metastable state below h_{b0} because a bubble lattice carries a lower energy in this field range. *De facto* bubbles can be treated as independent stable units in the field range between h_{bs} and h_{bc}. This fact is utilized in magnetic bubble memory technology. All critical fields are collected in Fig. 3.115 as a function of the parameter λ_c together with the corresponding values for band arrays. The largest stability interval $h_{bc} - h_{bs}$

exists at a value of $\lambda_c = \frac{1}{8}$, which marks therefore the preferred thickness range for bubble films.

3.7.4 Closure Domains

In this section we come back to low anisotropy materials with strong misorientation, but now of larger thickness beyond the threshold for the nucleation of dense stripe domains discussed in Sect. 3.7.2. With increasing thickness, regular domains develop out of the incipient stripes, and these domains tend to be closed at the surface for $Q < 1$. These structures are analysed here.

The driving force for the formation of closure domains is the reduction of the misorientation-induced stray field energy at the surface. The same goal is achieved for small misorientations by the μ^*-effect; see Sect. 3.7.1. Which of the two possibilities is preferred depends on the degree of misorientation. As indicated in the sketch, closure domains can avoid stray fields for arbitrarily large misorientation angles, but their formation is a non-linear phenomenon that is out of the applicability range of the μ^*-effect.

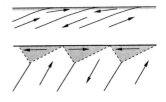

In addition, domain *refinement* towards the surface can in both cases reduce the volume in which deviations from easy directions are necessary. For small misorientations this was achieved by supplementary domains as discussed in Sect. 3.7.1. Domain refinement for strong misorientations is treated separately under the heading "domain branching" in Sect. 3.7.5.

(A) Numerical Experiments. Closure domains are magnetized by definition more or less parallel to the surface. Their internal energy must be higher than that of the basic domains, either caused by misorientation and the associated extra anisotropy energy of surface-parallel domains, or caused by superimposed stress anisotropy (Fig. 3.69) or magnetostrictive self-energy effects (Fig. 3.12). On account of this intrinsic energetic imbalance between basic and closure domains, domain walls separating these two types of domains cannot exist in *local* equilibrium, a point that was stressed in particular by *Privorotskij* [759, 511 (p. 40)]. The position and orientation of the closure walls can be derived geometrically starting from the assumption of perfect stray field avoidance, as will be explained in detail in Fig. 3.122. But their internal structure is basically unaccessible with the tools of conventional domain wall theory as treated in Sect. 3.6.

To demonstrate that closure domains are real anyway, rigorous numerical simulations were performed under various conditions. A prism of $2:1$ aspect ratio and infinite length was investigated under the influence of a weak perpendicular anisotropy. For a reduced anisotropy of $Q = 0.04$ a critical thickness of about 6Δ for the formation of dense stripe domains is expected according to Fig. 3.109 ($\Delta = \sqrt{A/K_u}$), so that for $D = 20\Delta$, well-defined domains such as those schematically indicated in the sketch are expected.

The results of the computations (Fig. 3.119) agree with these expectations. In conventional domain patterns the exchange energy density term is concentrated in the domain walls, while it is virtually zero in the domains. By plotting this energy we may therefore test whether a micromagnetic structure has a domain character or not. This is done for four anisotropy variants. In each case the exchange energy plots are supported by profiles of characteristic magnetization components along three lines (see sketch).

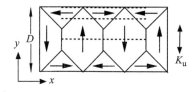

For uniaxial materials (Fig. 3.119a) true closure domains cannot be identified in the numerical results, so that in this case the domain concept can only serve as a model for a more detailed micromagnetic analysis. More convincing closure domains are found when cubic anisotropy is added (b). Interesting is case (c) in which a *planar* anisotropy induces well-defined domains on the surface, while a rather continuous transition is found in the subsurface zone. With a strong planar anisotropy (d) marked domains become visible again in analogy to ideally soft magnetic thin film elements (Sect. 3.3.3B).

Somewhat surprisingly, well-defined closure domains are also found for purely uniaxial anisotropy at larger plate thicknesses, as demonstrated in Fig. 3.119 e. This is observed, however, only if the anisotropy functional is highly stationary in the magnetization direction preferred in the closure domain. Tilting the anisotropy axis about the x axis removes these unexpected domains as documented in (Fig. 3.119 f). The same effect can be achieved by applying a moderate field along the z axis.

It would certainly not be economical to employ numerical calculations for every problem of domain analysis (this may even be impossible as discussed in Sect. 3.2.7E). We will demonstrate that *domain models* that generalize the elementary Landau structure (Fig. 1.2) are a powerful alternative in most cases. We analyse this subject in two stages. The first approach simply ignores the energy content of all improper domain walls or continuous transition zones. Relative to this model, the true micromagnetic solution will save some anisotropy energy by the continuous transition, at the expense of extra exchange and possibly stray field energy. Nevertheless, the best scale and orientation of a domain pattern can often be derived based on such simple domain models. This approach is followed throughout this subsection in (B–E), until in (F) a generalized "micromagnetic" domain theory is outlined, which systematically incorporates also improper domain walls and similar continuous variations.

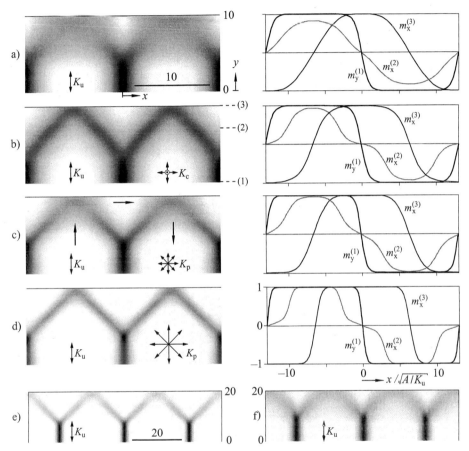

Fig. 3.119 Two-dimensional numerical modelling of closure domains in plates of 20Δ thickness. The square root of the exchange energy density (normalized to the maximum value in the basic walls) is plotted as a grey-shade image to indicate domains and walls, supported by magnetization-component contour plots (right-hand side). (a) Perpendicular anisotropy $K_u(1-m_y^2)$ only, with $Q = K_u/K_d = 0.04$, leading to a continuous closure configuration. (b) Marked closure domains are formed by adding a cubic anisotropy $K_c(m_x^2 m_y^2 + m_x^2 m_z^2 + m_y^2 m_z^2)$ with $K_c/K_u = 4$. (c) Superimposing a planar anisotropy $K_p(m_x^2 + m_z^2)$ with $K_p/K_u = 1$ generates domains at the surface, but only weak closure domains in the volume. (d) A strong planar anisotropy $K_p/K_u = 4$ forces the magnetization into the cross-section plane, with pronounced basic and closure domains. (e) Increasing the thickness to 40Δ, well-defined closure domains are found even with a purely uniaxial anisotropy as in (a). These domains disappear, however, if the anisotropy axis is tilted by $20°$ out of the cross-section plane (f). (Developed initially together with *K. Ramstöck*, Erlangen and finally with *W. Rave*, IFW Dresden)

(B) Elementary Theory of the Landau Model for Uniaxial Material. An infinite plate shall carry a *uniaxial* anisotropy K_u with the axis perpendicular to the plate surface. Its thickness D is assumed to be large compared to the Bloch wall width parameter $\sqrt{A/K_u}$, and the anisotropy energy constant K_u shall be small compared to the demagnetizing energy constant K_d. This is the case for which the Landau structure (Fig. 1.2) was tailored.

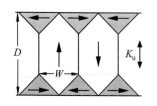

The first step in a calculation of this structure is straightforward. The domain width W results from an equilibrium between the basic 180° wall energy and the energy of the closure domains. Since the 90° walls delineating the closure domains are not well defined, they are ignored. Per unit volume, i.e. per unit length and per unit cross-section, the total energy is then:

$$e_{tot} = K_u W/2D + 4\sqrt{AK_u}\,[1/W - 1/D] \tag{3.209}$$

which, optimized with regard to W, yields the domain width:

$$W_{opt} = 2\sqrt{2D\sqrt{A/K_u}}\ , \tag{3.210a}$$

and the average energy of the domain structure:

$$e_{opt} = 2\sqrt{2K_u\sqrt{AK_u}/D} - 4\sqrt{AK_u}/D\ . \tag{3.210b}$$

The energy formula (3.209) for the Landau model is applicable only as long as the length of the resulting 180° walls stays positive, i.e. up to $W = D$, which is reached at $D = 8\sqrt{A/K_u}$. In its applicability range the average energy density (3.210b) of the Landau pattern is smaller than the anisotropy energy density K_u of the uniformly in-plane magnetized state, and much smaller than the energy density K_d of the uniformly magnetized perpendicular state.

(C) Partial Closure. For high uniaxial anisotropy ($Q = K_u/K_d$ comparable with unity) the completely closed Landau model becomes unfavourable. The open Kittel model (3.42) including the μ^*-correction is available for very large Q. It is interesting to find out which conceivable domain model yields the lowest energy and thus comes closest to reality in an intermediate range of Q values in which neither the Landau nor the Kittel model are adequate. We use mainly the tools of Sect. 3.2.5 to calculate various more-or-less closed domain models. This kind of analysis will help in understanding numerical calculations such as those presented in Fig. 3.119.

Three modifications of the Landau structure as shown in Fig. 3.120 are compared with the open Kittel structure. In (a) the surface charge density is reduced relative to the completely open structure by tilting the magnetization in the closure domains by an angle ϑ at the expense of anisotropy energy [25, 161]. Model (b) realizes partial closure by a reduced closure domain size [760]. The model in (c) is a combination of (a) and (b). In calculating the stray field energy of these models we make use of (3.43) for the stray field energy of periodic domain patterns. To superimpose the μ^*-correction (Sect. 3.2.5F) is not allowed for the structures of Fig. 3.120 because this correction can only be applied to patterns consisting of easy-axis domains. The closure domains already describe the effect of the μ^*-correction in a rough way, which can be evaluated, however, for arbitrary tilting angles ϑ up to $\vartheta = 90°$.

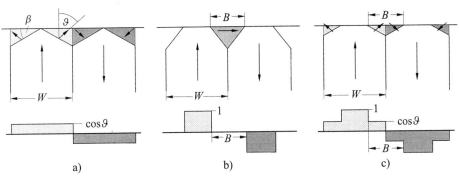

a)   b)   c)

Fig. 3.120 Generalized closure models applicable to plates with perpendicular uniaxial anisotropy with an intermediate value of the reduced anisotropy constant $Q \approx 1$. For each model the surface charge profile is also indicated. The general model (c) connects Landau's flux-closed domain pattern ($B = W$, $\vartheta = 90°$) with Kittel's open structure ($B = 0$, $\vartheta = 0°$)

The closure energy coefficient is calculated for the four models assuming a constant domain width W. The resulting coefficient may then be inserted into the total energy of a closure pattern as in (3.209).

(i) *Tilted closure.* No charges are assumed except at the surface. Then the energy terms to be considered are (taken per unit area of one sample surface) the stray field energy E_d according to (3.42) and the anisotropy energy E_K:

$$E_d = \tfrac{1}{2} K_d S_c W \cos^2 \vartheta \quad \text{with } S_c = \tfrac{1}{2\pi} 1.705 \dots , \tag{3.211}$$

$$E_K = \tfrac{1}{4} K_u W \tan \beta \sin^2 \vartheta \quad \text{with } \tan \beta = \sin \vartheta / (1 + \cos \vartheta) \ . \tag{3.212}$$

The angle β is derived from the condition of zero charge on the internal closure domain boundaries, as explained below in Fig. 3.122d. The reduced total energy density in units of K_u and for unit domain width W thus becomes:

$$e_1 = \left(S_c / Q \right) \cos^2 \vartheta + \tfrac{1}{4} (1 - \cos \vartheta) \sin \vartheta \ . \tag{3.213}$$

(ii) *Partially open model.* The stray field energy of this model can be derived from (3.41), yielding for one sample surface:

$$E_d = \left(\frac{4WK_d}{\pi^3}\right) \sum_{n=0}^{\infty} (2n+1)^{-3} \cos^2\left[\frac{\pi}{2}b_c(2n+1)\right] \quad \text{with } b_c = B/W . \qquad (3.214)$$

The anisotropy energy is simply:

$$E_k = \tfrac{1}{4} W K_u b_c^2 . \qquad (3.215)$$

This yields the reduced total energy:

$$e_2 = \left(4/Q\pi^3\right) \sum_{n=0}^{\infty} (2n+1)^{-3} \cos^2\left[\frac{\pi}{2}b_c(2n+1)\right] + \tfrac{1}{4}b_c^2 . \qquad (3.216)$$

Note that for $b_c = 0$, $\vartheta = 0$ this model becomes the open Kittel model (3.42) with $E_k = 0$ and $E_d = K_d S_c W$ [making use of $S_c = (8/\pi^3)\sum (2n+1)^{-3}$].

(iii) *Combined model.* For this model we obtain the total reduced energy:

$$e_3 = \frac{4}{Q\pi^3} \sum_{n=0}^{\infty} \frac{\left\{(1-\cos\vartheta)\cos\left[\frac{\pi}{2}b_c(2n+1)\right]+\cos\vartheta\right\}^2}{(2n+1)^3} + \tfrac{1}{4}(1-\cos\vartheta)\sin\vartheta b_c^2 . \quad (3.217)$$

It contains two adjustable parameters, the tilting angle ϑ and the opening ratio $b_c = B/W$. The energy density e_3 coincides with e_1 for $b_c = 1$; for $\vartheta = \frac{\pi}{2}$ the energy e_3 merges with e_2. The general model must therefore lead to a lower energy than either of the two "parent" models (i) and (ii) if energetically favourable and legitimate values for the parameters ϑ and b_c can be found.

(iv) *Kittel's open structure.* This model is primarily applicable to large Q, but its applicability is extended towards lower Q by the μ^*-correction. As the transverse susceptibility μ^* derived for small excursions from the easy directions underestimates the permeability at large excursions, the energy of the Kittel model can be treated as an upper bound for the true continuous micromagnetic structure. In reduced units the energy of the μ^*-model is, according to (3.42):

$$e_4 = 2S_c / [Q(1+\sqrt{\mu^*})] , \quad \mu^* = 1 + 1/Q . \qquad (3.218)$$

This expression contains no parameters. It is rigorous in the limit of large Q.

The four energies are compared numerically, optimizing their respective parameters. The results can be summarized as follows:

- The minimum energy of the tilted model e_1 is smaller than the minimum energy of the partially open model e_2 for all Q.
- The combined model obtained after minimizing e_3 yields small improvements compared to e_1 particularly in the range $Q = 0.3$–1. The resulting opening is smaller than 5% ($b_c > 0.95$) for any Q, leading to an energy gain of some 10^{-4} in the reduced energy density e_3 versus e_1. Some examples

of optimized combined structures are shown in Fig. 3.121. Surprisingly, a partial opening of the tilted model is energetically disfavoured for $Q > 1.25$ (the energy function points to an optimum beyond $b_c = 1$ which is illegitimate; the best legitimate value $b_c = 1$ means no opening). The predicted partial opening effect at lower Q may be observable in high-resolution domain imaging such as by the SEMPA method (Sect. 2.5.4, Fig. 2.39).

- The energy of Kittel's open model (including the μ^*-correction) is smaller than that of the combined closure pattern for $Q > 0.8$. It cannot compete for smaller Q. At $Q = 0.8$ the calculated excursion in the closure domain magnetization angle ϑ in model (i) amounts already to about 50°.

In summary, for $Q > 0.8$ the μ^*-corrected Kittel model offers the best estimate of the generalized closure energy of uniaxial samples. The domain models give an indication of the occurring angles in this parameter range. For smaller anisotropies (including cobalt with $Q \approx 0.4$) the tilted closure model is the best choice because it is both simple and efficient. If special features or increased accuracy are required, the combined model that allows for a partial opening of the closure domains can be used instead.

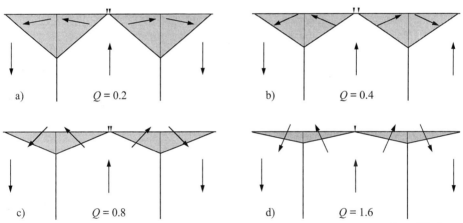

Fig. 3.121 Resulting optimized closure structures for different Q. Note the small equilibrium opening of the closure domains for intermediate Q values

(D) Models for Arbitrary Anisotropy Axes and Symmetries. As mentioned in the introduction, the concept of complete closure is applicable to arbitrary systems with weak anisotropy and strong misorientation. The first step in an analysis of such a situation is the construction of a model as in Fig. 3.122. It shows various closure domain geometries for the example of iron. The principles used in constructing such models are:

(i) The basic domains must occupy true easy directions.

(ii) The closure domains are assumed to prefer *relative* easy directions — directions of lowest energy among all directions parallel to the surface.

(iii) Finally, the condition of stray field avoidance (3.114) shall be fulfilled at all domain walls of a model. This condition determines the internal angles, and thus the volume of the closure domains as demonstrated in Fig. 3.122d.

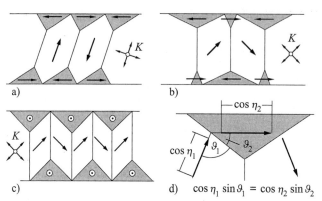

Fig. 3.122 Cross-sections through different types of closure domains in a cubic material. (a) Tilted 180° basic domains. (b, c) Two possible variants for 90° basic domains. The wall angles are determined by the principle of flux closure (d). The angle η used here is measured between magnetization vector and drawing plane

a)

b)

c)

d) $\cos \eta_1 \sin \vartheta_1 = \cos \eta_2 \sin \vartheta_2$

The total anisotropy energy in the closure domains can then be computed. Often different, non-equivalent orientations of the closure pattern are possible. These can be energetically compared to select the minimum energy orientation. In addition, a partial opening as worked out in (C) can be tested.

(E) Domain Models and Strong Stripe Domains. The tools developed in the last section are particularly fruitful in the analysis of dense stripe domains beyond their nucleation. As mentioned in the introduction of Sect. 3.7.2, rigorous analytical theory can only describe the immediate neighbourhood of the nucleation point correctly. Numerical finite element calculations can add to the understanding of the transition between the incipient dense stripe domains and regular, "ripened" domain patterns [761]. But numerical methods are not applicable to large sample dimensions, and the large efforts and expenses connected with such calculations suggest them more for typical and instructive examples than for systematic investigations as a function of multiple external parameters. Domain models are a useful alternative for exploring in a semi-quantitative way the behaviour of micromagnetic structures.

Domain models are particularly well suited for the analysis of thick films and platelets with complex tilted or multiaxial anisotropies. Experimental investigations revealed for such samples a variety of stripe-like patterns with different orientations and different contrasts in Bitter pattern or transmission electron microscopy observations. The most thoroughly investigated pattern of this kind are the *strong* stripe domains found in thicker films with tilted uniaxial or orthorhombic anisotropy [744–746]. Figure 3.123b shows a domain model of strong stripe domains as contrasted to conventional weak stripe domains (Fig. 3.123 a). The name of strong stripe domains is derived from contrast in Lorentz electron microscopy. Because the individual stripes in the strong stripe pattern differ in their longitudinal magnetization components, they become strongly visible by the Lorentz deflection. Weak stripe domains carry only unipolar longitudinal components, leading to weak Lorentz contrast.

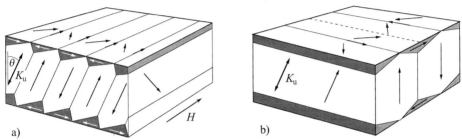

Fig. 3.123 Domain models for films with tilted uniaxial anisotropy. The weak stripe domains (a) are favoured in an applied field and at small thicknesses, and strong stripe domains (b) are favoured in zero field in thicker films. Strong stripes are preferred under the indicated conditions because both the wall length and the specific closure volume are smaller than for weak stripes. The basic domains in the strong stripe pattern are magnetized antiparallel along the tilted easy axis

The picture of Fig. 3.123 offers a qualitative argument why strong stripe domains become favoured at a certain field below the nucleation field or beyond the nucleation thickness for weak stripes. The anisotropies shall be small compared to the stray field energy constant K_d. The true easy axis of the material is inclined relative to the film normal by an angle Θ. For a *uniform* in-plane magnetization, as in a very thin film, one of the directions parallel to the film plane may be preferred. With increasing film thickness, 'weak' stripe domains develop out of this relative easy axis, as shown in Fig. 3.123a. The nucleation mode will resemble the conventional stripe domains discussed in (A), differing only in a tilting of the pattern induced by the tilted easy axis.

This pattern will, however, not remain stable at larger thicknesses or at reduced applied fields. Beyond a second transition point (when the magnetization in the basic domains approaches the true easy axis) the alternative 'strong' stripe domain pattern (Fig. 3.123b) has a lower energy for two reasons:

(i) The walls between the basic domains need not be tilted in this pattern, thus saving wall area. (ii) The closure volume for a given domain width is smaller than in the alternative pattern of Fig. 3.123a. It will therefore be preferred to the initial transverse weak stripe domain pattern. A characteristic feature of the strong stripe domain pattern is its continuous magnetization rotation at least on the surface, as indicated in the scheme. It follows naturally from the coupling of the closure domains to the underlying basic domains. In contrast, weak stripe domains show only oscillatory magnetization patterns.

The transition between both patterns is discontinuous because there can be no continuous transition between an oscillatory and a rotating magnetization pattern. It will proceed by nucleation and growth. This explanation in terms of a (simpler) domain model was first offered by *Hara* [495]. Micromagnetic calculations by *Miltat* and *Labrune* [761, 762] confirm the general picture adding many interesting details.

(F) A "Micromagnetic" Domain Model. In simple domain models as analysed in (B−E) the energy contributions of the walls delineating the closure domains are neglected, as are the torques exerted by these walls on basic and closure domains. Such models are therefore unable to describe the behaviour at small scale and in particular the transition between a domain state with flux closure, and the state of uniform in-plane magnetization, which will be preferred for very thin films and low anisotropy. Here a method is reviewed that takes into account these effects in a simple but systematic manner, describing the important aspects of the situation in a reasonable approximation within the spirit of conventional domain theory [763]. Although it needs some care in application, it is mathematically much simpler than a rigorous micromagnetic treatment.

We start with a conventional domain model (Fig. 3.124a), in which the closure domains are constructed following the recipe of Fig. 3.122d. The starting structure may thus consist of arbitrary domains separated by charge-free domain walls. It is not necessary that the domains are magnetized along easy directions, and surface charges are also permitted; the corresponding energies have to be added separately. The geometrically defined domain walls will in reality tend to spread to a certain degree into the adjacent domains (Fig. 3.124b). They will continue to spread until they either collide with the "spreading zone" of another wall, or until an increase in anisotropy energy balances the gain in exchange stiffness energy. To calculate the optimum spreading width, we simply assume that a transition gains exchange energy with increasing spreading space, i.e. volume of the spreading zones.

In detail, the procedure is as follows: in a first step we determine the magnetization direction in the geometrical wall centre, assuming the magnetization to rotate like in a stray-field-free Bloch wall. The wall rotation angle $\Delta\Theta$ is defined as the length of the magnetization path from one domain to the other on the orientation sphere (see Fig. 3.54). *One half* of this wall rotation angle is assigned to each of the spreading areas on both sides.

To estimate the exchange energy connected with a spreading area we first evaluate its depth d^{sp}, which can be expressed in the angles defined in Fig. 3.124b, as shown in (3.219). We replace the triangular spreading area by a rectangle of the same length and area (Fig. 3.124c) and assume a linear transition profile for the magnetization angle. Integrating over the rectangle we obtain for the exchange energy the expression e_x in (3.219).

In the same way the excess anisotropy energy connected with a spreading zone is calculated. We introduce the quantity $\Delta K = K\Delta\kappa$ as the anisotropy energy difference between the wall centre and the adjacent domain. Here K is the anisotropy constant of the material and $\Delta\kappa$ is a dimensionless factor. Assuming a linear variation of the anisotropy energy density and integrating over the rectangle of Fig. 3.124c, leads to the expression for e_K in (3.219). Altogether, we obtain for a closure domain belonging to a wall of length L:

$$d^{\mathrm{sp}} = \tfrac{1}{2} L/(\cot\beta + \cot\gamma) \;,\; e_x = \tfrac{1}{4} A(\Delta\Theta)^2 L/d^{\mathrm{sp}} \;,\; e_K = \tfrac{1}{2} L K \Lambda\kappa\, d^{\mathrm{sp}} \;. \qquad (3.219)$$

The dimension of the energy expressions is [J/m] because they apply to unit length in the third dimension perpendicular to Fig. 3.124. The detailed assumptions leading to e_x and e_K are indicated graphically in Fig. 3.124e. The exchange energy is proportional to the square of the wall rotation angle $\Delta\Theta/2$ and inversely proportional to the depth d^{sp}. This expression represents the integrated exchange energy of one of the transition areas. The anisotropy turns out to be proportional to d^{sp}. The factor $\tfrac{1}{2}$ in e_K stems from the integration.

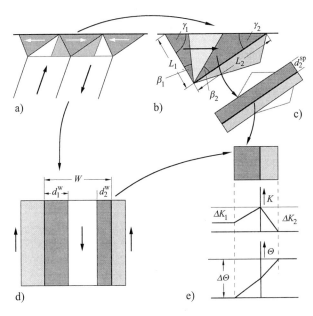

Fig. 3.124 Scheme for estimating the excess exchange and anisotropy energy contributions of continuous surface transition regions modelled by closure domains. The rotation of the magnetization in the walls in model (a) is assumed to extend into "spreading areas" (*grey*) in the adjacent domains. (b) The details of a closure domain. (c) The triangular spreading zones are replaced by rectangular regions of equal area. (d) A similar scheme can be applied to the wall between two basic domains. (e) The excess anisotropy energy ΔK and the wall rotation angle $\Delta\Theta$ are distributed linearly in the spreading zones

The following geometrical relations help in calculating the quantities $\Delta\Theta$ and $\Delta\kappa$ needed in (3.219). Let m_1 and m_2 be the magnetization directions on both sides of a wall with the normal n. Then we obtain for the stray-field-free magnetization path the wall rotation angle (= magnetization path length):

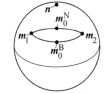

$$\Delta\Theta^{\mathrm{B}} = \sqrt{1 - c_n^2}\, \arccos\left[(c_w - c_n^2)/(1 - c_n^2)\right] \qquad (3.220\,\mathrm{a})$$

with $c_n = m_1 \cdot n = m_2 \cdot n$ and $c_w = m_1 \cdot m_2$.

The magnetization vector in the centre of this "Bloch type" wall becomes:

$$m_0^{\mathrm{B}} = c_n n \pm \sqrt{\tfrac{1}{2}(1 - c_n^2)/(1 - c_w)}\, (m_1 - m_2) \times n \;, \qquad (3.220\,\mathrm{b})$$

where $+$ or $-$ is chosen according to the shorter magnetization path. Alternatively, we may try the shortest (Néel type) path. The expressions for this case are:

$\Delta\Theta^N = \arccos c_w$ and   (3.220c)

$m_0^N = (m_1 + m_2)/\sqrt{2(1 + c_w)}$ for $c_w > -1$, $m_0^N = n$ for $c_w = -1$.   (3.220d)

The two values of the anisotropy threshold $\Delta\kappa$ of the transition zone are defined as $\Delta\kappa_1 = g(m_0) - g(m_1)$ and $\Delta\kappa_2 = g(m_0) - g(m_2)$, respectively, where $Kg(m)$ is the generalized anisotropy functional. Note that the $\Delta\kappa$ may well be negative, not for regular walls, but in closure domain walls for example of uniaxial materials. The algorithm will be able to deal with this situation, which is unfamiliar from the study of regular domain walls.

In the case of a Néel type transition (3.220c,d) an internal stray field energy term $\Delta\kappa_d = (1/Q)\left[1 - (m_0^B \cdot m_0^N)^2\right]$ has to be added (in equal parts) to $\Delta\kappa_1$ and $\Delta\kappa_2$. The assumption of a stray-field-free path (3.220a,b) is adequate for low-anisotropy situations ($Q = K/K_d \ll 1$), whereas the shortest path (3.220c,d) is preferred for $Q \gg 1$. Also possible is a compromise between the two alternatives as in Fig. 3.58, to minimize the total energy.

We now express the energy per unit length in terms of the exchange stiffness constant $\varepsilon = e/A$. The lengths and widths of the spreading zones are measured in units of the wall width parameter: $\Lambda = L/\sqrt{A/K}$ and $\delta = d/\sqrt{A/K}$. Collecting the contributions for the two spreading zones *inside* a closure domain (Fig. 3.124b) leads to:

$$\varepsilon_{cl} = \tfrac{1}{4}\left[\Lambda_1(\Delta\Theta_1)^2/\delta_1^{sp} + \Lambda_2(\Delta\Theta_2)^2/\delta_2^{sp}\right] + \tfrac{1}{2}\left[\Lambda_1\Delta\kappa_1\delta_1^{sp} + \Lambda_2\Delta\kappa_2\delta_2^{sp}\right]$$ (3.221)

with $\delta_i^{sp} = \tfrac{1}{2}\Lambda_i/(\cot\beta_i + \cot\gamma_i)$ (3.221a)

under the constraint $\beta_1 + \beta_2 \leq \pi - \gamma_1 - \gamma_2$.   (3.221b)

Minimizing the energy (3.221) with respect to the β_i is easy for large Λ and positive $\Delta\kappa_i$. We then obtain simply $\delta_1^{sp} = \Delta\Theta_1/\sqrt{2\Delta\kappa_1}$ and $\delta_2^{sp} = \Delta\Theta_2/\sqrt{2\Delta\kappa_2}$, leading with (3.221a) to the angles β_1 and β_2. These angles stay small for large Λ and positive $\Delta\kappa_i$, so that inequality (3.221b), being satisfied, has no further consequences.

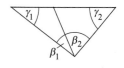

In the other case of small dimensions or negative $\Delta\kappa$, the optimization has to be performed under the constraint $\beta_1 + \beta_2 = \pi - \gamma_1 - \gamma_2$, which avoids an overlap of the spreading zones. This condition leads to a linear relation between δ_2^{sp} and δ_1^{sp}, so that δ_2^{sp} can be eliminated. Minimizing the total energy with respect to δ_1^{sp} yields a fourth-order equation for δ_1^{sp} to be solved numerically. Alternatively, the position of the spreading zone boundary can be added to the variables of the model, which are finally determined by minimizing the total energy.

A similar procedure is applied to the walls between the basic domains (Fig. 3.124d). Again two spreading areas are defined, which are given by their widths δ_1^{sp} and δ_2^{sp}. Instead of the condition for the angles (3.221b) we have $\delta_1^{sp} + \delta_2^{sp} \leq W/\sqrt{A/K}$. Everything else proceeds as for the triangular closure domains. The outlined procedure estimates the energy of domain walls to about 10 % accuracy compared to a rigorous calculation if one is possible. The advantage is that regular domain walls as well as locally unstable or asymmetrical closure walls can be treated in a unified way. None of the domains in a model configuration *must* be magnetized along an equilibrium direction, which is a necessary requirement in conventional domain wall analysis. The indicated formalism offers a reasonable estimate of transition energies for all cases.

The algorithm is thus able to simulate the micromagnetic behaviour of quite complex configurations. For every set of external parameters, such as the period and orientation of a periodic domain pattern, the appropriate conventional domain pattern is first constructed. The total energy is then given by a sum over expressions like (3.221) for all parts of the pattern. The minimum total energy is found by calculating the width of the various spreading zones. Finally the external parameters are varied in order to find the lowest energy model.

Figure 3.125 shows the result of such a calculation [763]. Beyond nucleation, the maximum perpendicular amplitude in the stripe pattern rises quickly with increasing thickness. The formation of regular domain and closure patterns for larger thicknesses is indicated by the formation of internal 180° walls. This calculation demonstrates that the completely continuous nucleation of stripe domains can be simulated in our generalized domain theory.

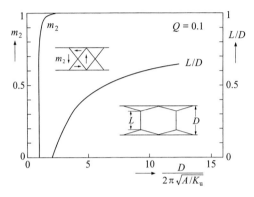

Fig. 3.125 The maximum perpendicular magnetization amplitude of stripe domains together with the length of the developing wall as a function of the film thickness, calculated at zero field based on a domain model [763]

The second application resumes the subject of Fig. 3.32. There the transition between the uniformly magnetized state and different demagnetized states of small cube-shaped particles was calculated with rigorous micromagnetic methods. The lowest energy states were calculated as a function of two external parameters: the particle size L, and the reduced uniaxial anisotropy strength Q. Our goal is to compare our "micromagnetic" domain theory with rigorous results and to extend the calculations to parameter combinations that are virtually inaccessible with rigorous methods. We analyse three configurations: the single-domain state, and two- and three-domain states. Only "two-dimensional" models that can be defined by a single cross-section drawing are considered, although three-dimensional generalizations, as they are also found in particular for the three-domain states (see Fig. 3.32), could certainly be elaborated. To cover the whole anisotropy range from soft to hard materials, we use split closure domains as elaborated in (C).

All domain models (indicated in the insets in Fig. 3.126) are defined by the closure tilting angle φ, and the common magnetization angle η leading to a uniform magnetization component $\sin\eta$ perpendicular to the cross-section plane. Large angles $\eta \to \frac{\pi}{2}$ lead to the transversely magnetized single-domain state. Complete closure is described by $\varphi = 0$, while $\varphi = \frac{\pi}{2}$ represents open domain patterns as discussed in Sect. 3.3.2 for large Q. The total energy of these models consists of three contributions: (i) The energies of the spreading zones as defined in this section. (ii) The anisotropy energy of the domains. (iii) The stray field energy for non-zero angles φ and η, which can be expressed by using the same expressions as in Sect. 3.3.2. Minimizing this total energy with respect to the angles φ and η leads to the phase diagram shown in Fig. 3.126. For the available rigorous numerical results for the phase boundaries they agree very well. The largest deviations are found in the transition range $Q \approx 1$. This is also the range in which numerical methods are best applicable. Calculating the "multi-domain boundary" for either large or small Q is virtually impossible with numerical micromagnetics. On the other hand, the reliability of the presented "micromagnetic" domain theory is particularly good there. Numerical micromagnetics and domain theory should thus be considered complementary techniques in the theoretical analysis of magnetic microstructure.

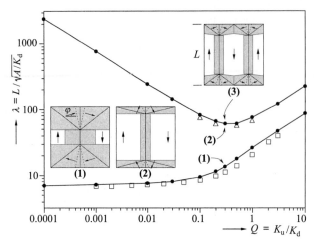

Fig. 3.126 The phase bounda-ries for two-dimensional do-main states in magnetically uniaxial, cube-shaped particles, as resulting from "micromag-netic" model calculations. In the *insets* the spreading zones of the domain walls in the mod-els are indicated by *grey shad-ing*. The isolated data points are rigorous results from Fig. 3.32 obtained by *W. Rave*, IFW Dresden by numerically solving the micromagnetic equations

(G) General Aspects of Closure Domains. The presented model-based ana-lysis of closure domains cannot replace rigorous micromagnetic computations such as the one shown in Fig. 3.119 if quantitative results for example for a comparison with experimental observations are needed. Finite element calcula-tions automatically take into account a number of subtleties that are ignored even in the enhanced domain model analysis presented in (F). Other complica-tions are even beyond the reach of available finite element techniques and can be taken into account only qualitatively. Here is a list of such problems:

- The micromagnetic properties of the domain wall junctions or knot lines are not included in any domain model analysis. These can be constructed in a stray-field-free manner with the magnetization in their centre pointing along the knot lines, adding to the torque on basic and closure domains. Finite element calculations fully describe these features and may be analysed to understand their role.

- A complete treatment of a closure structure would have to include the magnetostrictive self-energy of the closure domains. As a first approximation the results of Sect. 3.2.6H may be sufficient, but these results are not fully adequate for four reasons: (i) They neglect elastic anisotropy. (ii) They are not valid for thin plates but rather for semi-infinite bodies. (iii) The contin-uous transition in the closure domains considered in (F) would have to be part of the elastic analysis, including shear stresses in the widened transition regions. (iv) The geometry of magneto-elastic model calculations differs from the geometry of the closure domains. No example of a finite element calculation that includes these elastic effects has as yet been presented.

- For large thicknesses D the simple closure structures will give way to mostly three-dimensional branched structures, as first conceived by Lifshitz (sec Sect. 1.2.3). The details of this branching process are discussed in Sect. 3.7.5. They are out of the range of available numerical methods.
- The application of external fields adds a new dimension to the manifold of phenomena. In fields perpendicular to the surface band-shaped domains may collapse into insular patterns ("bubble lattices"; see Sect. 3.7.3C) to save wall energy, again leading to three-dimensional structures. In horizontal fields magnetostrictive interactions between the basic domains influence the preferred domain orientation relative to the field (Sect. 3.7.2C).

The refinements in treating the closure energy do not change its basic functional behaviour: its averaged value increases proportional to the basic domain period, thus favouring narrow domains at the surface. In this sense it behaves analogously to the stray field energy of an open domain model as in (3.41). Since there are all kinds of transitions between flux-closed and open patterns, it is convenient to introduce the concept of a *generalized* closure energy — the sum of regular closing energy and surface stray field energy. This energy per unit sample surface always turns out to be proportional to the domain width of the underlying domain pattern, irrespective of the details; it may therefore replace the regular closure energy in calculations such as (3.209).

3.7.5 Domain Refinement (Domain Branching)

(A) **Overview**. Domain branching is a phenomenon encountered in large crystals with strongly misoriented surfaces. Close to the surface a fine domain pattern is enforced to minimize the (generalized) closure energy, but in the bulk a wide pattern is favoured to save wall energy. The branching process connects the wide and the narrow domains in a way that depends on crystal symmetry and in particular on the number of available easy directions.

We distinguish between

- two-dimensional branching, which can be completely described in a cross-section drawing (Fig. 3.127a), and
- three-dimensional branching, which exploits geometrically the third dimension (Fig. 3.127b,c).

A further distinction is made between

- two-phase branching, which has to achieve the domain refinement with two magnetization phases only (as in Fig. 3.127), and
- multi-phase branching, which can use more than two magnetization directions in the branching process.

The classical example of two-phase branching was proposed by *Lifshitz* ([24]; Sect. 1.2.3) based on earlier considerations for superconductors [764]. Although intended to represent a good model for iron, it is now understood as applying more to uniaxial crystals. A prototype of multi-phase branching patterns, the echelon pattern (Fig. 3.30), was first presented by *Martin* [765].

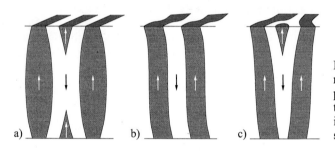

Fig. 3.127 Domain refinement and branching in two-phase systems. (a) Planar two-dimensional branching. (b, c) Three-dimensional variants

The pioneers in explaining the branching phenomenon [764, 24] recognized already the essential feature of branching: the progressive domain refinement towards the surface by iterated generations of domains (see the sketch, and as another illustration of the iteration principle, the 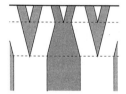 mentioned echelon pattern). But the early authors assumed the branching process to proceed indefinitely towards a zero surface domain width. This is not a realistic assumption because finally the gain in closure energy will be smaller than the additional expense in domain wall energy, as elaborated in a quantitative way in the following.

(B) Theory of Two-Phase Branching. Branching schemes with only two easy magnetization directions are natural in uniaxial materials with an anisotropy parameter $Q = K_u/K_d \gg 1$. But they occur also in cubic materials under the two-phase condition of phase theory (mode III of the magnetization curve; see Sect. 3.4.2C). With only two easy directions, and if deviations from these easy directions are forbidden, it is impossible to reduce the domain width towards the surface without generating stray fields.

The proof of this statement is based on Gauss' theorem considering an area in the cross-section extending from the centre of the plate to the surface as sketched. If the lateral flux is zero, and if practically no flux leaves the finely divided surface (which is the purpose of domain branching), the magnetization flux entering the area at the base has to be dissipated in the interior, thus creating stray fields. The argument remains, as can be seen, also valid for less symmetrical situations in which the two phases are not magnetized antiparallel to each other.

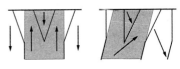

A theory of two-phase branching has to comprise three energy terms: (i) the wall energy that increases towards the surface with increasing wall density, (ii) the (generalized) closure energy depending only on the domain width of the last generation, and (iii) the energy connected with the internal stray fields. The latter may be calculated by assuming rigid magnetization directions along the easy axis and by applying the μ^*-correction (Sect. 3.2.5F). Nevertheless, an explicit calculation particularly of three-dimensional structures is rather difficult and was pursued only by few authors [766, 767].

Here we discuss a simplified approach that abstracts from less important details and concentrates on basic principles [101]. It uses as the only variable an effective domain width W that depends on the distance from the surface in a continuous manner. To define a domain width is trivial for simple patterns. In general an effective domain width can be defined by forming the ratio between an area and the integrated wall length in this area (Fig. 3.128). This definition of the domain width agrees with the ordinary definition for straight parallel walls. A method to determine it from domain images with the help of a stereological procedure is also indicated in the figure [768].

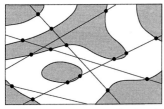

$$\text{Domain width} = \frac{2 \cdot \text{Total test line length}}{\pi \cdot \text{Number of intersections}}$$

Fig. 3.128 The domain width of an arbitrary pattern can be defined as the ratio between a test area and the total domain wall length in this frame. One procedure of evaluating the wall length and from it the domain width is based on a stereological method: intersections of the domain walls with arbitrary test lines are counted and evaluated

The total wall energy per unit volume in a thin cross-sectional slice is by definition equal to γ_w/W, where γ_w is the specific wall energy. The closure energy will be proportional to the surface domain width W_s; the proportionality factor can be determined with the tools of Sect. 3.7.4C. Finally we have to estimate the internal field energy connected with the branching process. Certainly this energy increases with the *rate* of change of the domain width. All conceivable models agree in the property that an attempt to accelerate the branching process has to be paid by increasing internal charges, and the stray field energy is proportional to the square of the charges. To describe this behaviour, the internal stray field energy is assumed to be proportional to the square of the derivative of the domain width with respect to the coordinate perpendicular to the surface. The proportionality factor can be estimated either from more or less elaborate models (the most simple one is shown in Fig. 3.129), or from experiments.

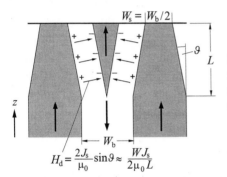

Fig. 3.129 Estimating the internal stray field for a one-generation branching process. To derive the coefficient F_i in (3.222), compare the stray field energy in one branch in the sketch $\frac{1}{2} W_b L \cdot \frac{1}{2} \mu_0 H_d^2$ with the corresponding area $W_b L$, and identify the relative decrease in the domain width $\frac{1}{2} W_b/L$ with dW/dz. After applying the μ^*-correction the result is $F_i = \frac{1}{2} K_d/\mu^*$

This leads to the following expression for the energy of a two-phase branching structure in a plate of thickness D, taken per unit surface of the plate:

$$e_{tot} = 2 \int_0^{D/2} \left[(\gamma_w / W(z)) + F_i (dW(z)/dz)^2 \right] dz + 2 C_s W_s , \qquad (3.222)$$

where F_i and C_s are factors describing the internal field energy and the closure energy, respectively, and z is the coordinate perpendicular to the plate, measured from the plate centre. The coefficients F_i and C_s will depend on the circumstances. Let us treat them for the moment as phenomenological parameters.

The optimum solution $W(z)$ will depend on the film thickness D and on the parameters γ_w, F_i and C_s. To find it we treat (3.222) as a variational problem and proceed in the same way as in domain wall calculations (Sect. 3.6.1D). The difference lies in the boundary condition. It is influenced by the contribution of the closure term, which depends on the boundary value $W_s = W(D/2)$.

We start with the Euler equation and its first integral:

$$\frac{\gamma_w}{W(z)} - \frac{\gamma_w}{W(0)} = F_i \left(\frac{dW(z)}{dz} \right)^2 \quad \text{or} \quad dz = -\sqrt{\frac{F_i}{\gamma_w}} \frac{dW}{\sqrt{1/W - 1/W_0}} . \qquad (3.223)$$

This result makes it possible to integrate the total energy, leading to:

$$e_{tot} = 2\sqrt{W_b\gamma_w F_i}\left[3\arccos\sqrt{\omega} - \sqrt{\omega(1-\omega)}\right] + 2C_s\omega W_b \quad , \tag{3.224}$$

where $W_b = W(0)$ is the basic domain width and $\omega = W_s/W_b$ is the relative surface domain width. We obtain an implicit relation between W_s and W_b by integrating the spatial dependence (3.223) and inserting $z = D/2$:

$$D = \sqrt{W_b^3 F_i/\gamma_w}\left[\arccos\sqrt{\omega} - \sqrt{\omega(1-\omega)}\right] \quad . \tag{3.225}$$

Minimizing the energy (3.224) with respect to W_b and ω under the constraint (3.225) leads to Fig. 3.130. Here the basic domain width W_b and the surface domain width W_s are plotted in reduced units as functions of the film thickness.

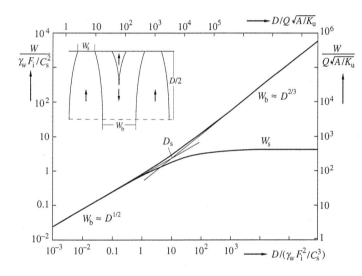

Fig. 3.130 The calculated surface and basic domain widths of two-phase branching structures as a function of the film thickness. The right-hand and upper scales use more conventional units for the case $Q \gg 1$ for which the μ^*-correction becomes negligible

We see that for small thicknesses an unbranched structure with $W = W_s = W_b$ and $W = \sqrt{D\gamma_w/2C_s}$ prevails. This is the same classical behaviour of unbranched patterns as shown before explicitly for the Landau model (Sect. 3.7.4B). Starting around a characteristic thickness $D_s = \frac{1}{8}\pi^4\gamma_w F_i^2/C_s^3$ branching with $W_s < W_b$ sets in. The basic domain width W_b increases stronger following a $D^{2/3}$ law, explicitly:

$$W_b = \sqrt[3]{(4/\pi^2)(\gamma_w/F_i)D^2} \tag{3.226}$$

as derived from (3.225) with $\omega = 0$. The surface domain width levels off for large thicknesses towards:

$$W_s = 4\gamma_w F_i/C_s^2 \tag{3.227}$$

This property of a constant surface domain width for thick samples seems to be a general feature of branched domain patterns. The transition thickness D_s is mathematically defined as the intersection point of the two asymptotic laws (3.226) and (3.227) in a logarithmic plot as indicated in Fig. 3.130.

There is a surprising rigorous relation between the surface and the interior domain width, which is valid within our model for all thicknesses:

$$(1/W_s) - (1/W_b) = (1-\omega)/(\omega W_b) = C_s^2/(4\gamma_w F_i) .\tag{3.228}$$

For small sample thicknesses the domain widths in the middle and at the surface differ only slightly. An experimental verification in this range is barely possible because the mathematical identity (3.228) is here based on a small difference between two large numbers (the reciprocal domain widths). For large thicknesses (3.228) leads immediately to (3.227) again, as $1/W_b$ becomes negligible compared to $1/W_s$.

Figure 3.131 shows the dependence of the domain width $W(z)$ on the distance from the surface in a logarithmic scale. The linear part of these curves demonstrates the *fractal* nature of the domain branching process. Deviations from this linear relation are found in the centre region because of symmetry, and in a surface layer that is comparable in thickness to the surface domain width W_s. In between, the domain width follows a $W \sim (D/2 - z)^{2/3}$ law meaning that a fractal dimension of $\frac{2}{3}$ may be attributed to the pattern.

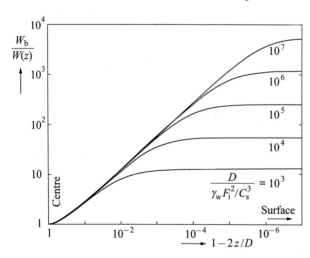

Fig. 3.131 The power law dependence of the calculated domain width of uniaxial branching patterns as a function of the distance from the surface for various sample thicknesses, indicating the fractal nature of branching

The functional dependence (3.226) for the basic domain width can be formally derived from the following energy expression:

$$e_{tot} = \gamma_w D/W + (\pi^2/8) F_i W^2/D .\tag{3.229}$$

Minimizing this energy expression with respect to W leads back to (3.226). This means that the domain wall energy is balanced by an effective closure energy, which depends on the plate thickness and on the internal branching energy penalty coefficient F_i. The real closure coefficient C_s is irrelevant for the basic domain width of thick plates. This interpretation is useful in studying sample properties that depend mainly on the basic domain width, such as eddy current losses in non-oriented electrical sheets.

To achieve a better understanding of these results let us look at actual expressions for the three coefficients in (3.222). We will do this for the most important case, namely uniaxial materials with high anisotropy, focusing on the simple two-dimensional model of Fig. 3.127a. Then the wall energy is $\gamma_w = 4\sqrt{AK_u}$ according to (3.109), the closure energy coefficient according to (3.42) is (for one surface) $C_s = 0.136 K_d \cdot 2/(1+\sqrt{\mu^*})$, and the internal field energy coefficient is estimated based on Fig. 3.129 as $F_i = 0.5\ K_d/\mu^*$. For large anisotropies ($\mu^* \approx 1$) this leads to of $D_s = 5000\ Q\ \sqrt{A/K_u}$ (remember the definitions $K_d = J_s^2/2\mu_0$, $Q = K_u/K_d$ and $\mu^* \approx 1 + 1/Q$). Two-phase branching is therefore a typical phenomenon of bulk crystals. It sets in whenever the length of the crystal along the axis reaches the mentioned threshold range of several thousand times the wall width parameter.

Three-dimensional two-phase branching models like Fig. 3.127b,c are expected not to differ qualitatively from simple two-dimensional ones. In [101] several calculations are cited demonstrating that the closure energy coefficient increases by not more than 10% if other than plate-like domains (such as corrugated plates, honeycomb or checkerboard patterns) are considered. The assumption of a continuously changing domain width also becomes more plausible for three-dimensional patterns in which wall corrugation (Fig. 3.127b) can be interpreted as an incipient domain refinement that changes its magnitude continuously. (To see this, apply the method of Fig. 3.128 to different cross-sections through the model). In fact, experimental observations regularly yield three-dimensional branching patterns that start (for increasing thickness) with corrugations and proceed with dagger domains as in Fig. 3.127c, followed by iterations of these processes. Only when an orthorhombic anisotropy or an applied field leads to a strongly anisotropic specific wall energy, three-dimensional branching becomes unfavourable and the simple plate-shaped branching scheme, which is the basis of Fig. 3.129, can be preferred, as seems to be indicated by certain observations on strongly stressed samples as in Fig. 5.9a. Nevertheless, we suppose that the ansatz (3.222) should also be applicable to three-dimensional patterns. There are arguments that three-dimensional configurations like that in Fig. 3.127c are favourable with respect to the internal field energy, but detailed investigations of this aspect are still missing. If this conjecture proves right, then primarily the coefficient F_i for the branching speed term in (3.222) must be modified for three-dimensional structures.

(C) Four-Phase Branching — the Echelon Pattern. Domain refinement in cubic materials differs from the uniaxial case because multiaxial branching can be achieved without internal stray fields. The energy connected with the branching process is thus lower, so that the critical thickness for the occurrence of branching is also much lower than for uniaxial materials.

As a simple two-dimensional example let us consider a thin long single-crystal iron plate of width L (Fig. 3.132). The surface shall be a (001) surface with edges along [110]. The thickness is assumed to be large enough to allow bulk wall structures, and thin enough to favour in-plane magnetization. Think of a thickness of 20–100 μm as a guideline.

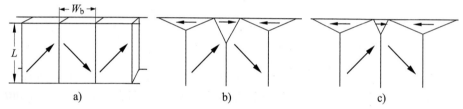

a)   b)   c)

Fig. 3.132 Elementary domain structures conceivable in an elongated cubic plate with [110] edges. (a) shows the simple open structure, (b) a general flux-closed structure, and (c) a pattern that is optimized with respect to the volume of the closure domains

The primitive domain structure shown in Fig. 3.132a would have free poles at the edges and is certainly not valid since the ratio of the stray field energy constant K_d to the anisotropy constant K_{c1} is over 30 in iron. As a starting point we therefore avoid stray fields and postulate complete closure at the surface at the expense of some anisotropy energy (Fig. 3.132b,c).

In the first step we analyse only the simple configuration of Fig. 3.132c, to compare it in the second step with branched alternatives. The smallest volume of the surface domains is obtained if both the narrow and the flat closure domains have the same depth, namely $W_b/\sqrt{8}$. The angles of the closure domains are determined by (3.114) (see also Fig. 3.122d). Then the anisotropy energy density in the closure domains is $K/4$ for the [110] magnetization direction if we abbreviate K_{c1} by K. The exchange energy connected with the transition between the basic domains and the closure domains is neglected. In calculating the total closure energy density we take into account the closure domains on both edges, and add the wall energy of the basic walls:

$$e_{tot} = W_b K/(8\sqrt{2}\,L) + (L - W_b/\sqrt{2}\,)\gamma_w/(W_b L) \;, \tag{3.230}$$

where γ_w is the specific wall energy of the 90° walls $[\gamma_w = 1.42\sqrt{AK}$ for $\{110\}$ zigzag walls; Fig. 3.68]. With reduced units:

$$\varepsilon = e/K \;,\; \Lambda = L/\sqrt{A/K} \;\;,\; \omega_b = W_b/L \;,\; \eta = \gamma_w/\sqrt{A/K}$$

the total energy becomes:

$$\varepsilon_{tot} = \omega_b/(8\sqrt{2}\,) + \eta/(\omega_b\Lambda) - \eta/(\sqrt{2}\,\Lambda) \;. \tag{3.231}$$

Optimizing this expression with respect to ω_b yields:

$$\omega_{opt} = 4\sqrt{\eta/(\sqrt{2}\Lambda)} \;,\; \varepsilon_{opt} = \sqrt{\eta/(2\sqrt{2}\Lambda)} - \eta/(\sqrt{2}\Lambda) \;. \tag{3.232}$$

In real units: the domain width W should be proportional to the square root of the sample length — as expected for a simple, unbranched pattern.

As an alternative, let us look at the echelon pattern (Fig. 3.133), a branched pattern that reduces the domain width step-wise towards the surface [765]. This reduction occurs in a two-dimensional, stray-field-free manner within a relatively narrow zone near the edges.

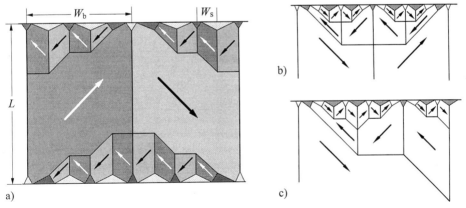

Fig. 3.133 Domain refinement in a cubic crystal according to the echelon pattern. (a) The stray-field-free staircase-like edge structure reduces the volume of the closure domains by a factor of 5 relative to the unbranched pattern of Fig. 3.132c for the same basic domain width W_b. A reduction by a factor of 7 (b) and by a factor of 13 (c) is achieved by using two echelon generations

Figures 3.133b,c demonstrate the possibility to iterate the echelon pattern using two generations of refinement. In the same way more generations may be added (a possibility that had not been mentioned by Martin [765]). Multiple-generation branching was also studied by *Privorotskij* [511 (p. 51)], who used, however, a somewhat less favourable two-dimensional model.

For a general analysis we characterize an echelon pattern by the symbol $b^{(n)}$, where n is the number of generations and b is the branching ratio per generation. For example, the pattern of Fig. 3.133c consists of $n = 2$ generations with a branching ratio of $b = 3$ each, represented by the symbol $3^{(2)}$. The total domain multiplication factor for an arbitrary pattern is evaluated as:

$$F_n = b^0 + b^1 + \ldots + b^n$$

leading, for Fig. 3.133c, to the total branching ratio $F_2 = 1 + 3 + 9 = 13$ mentioned in the figure caption. As a consequence the surface (closure) domain width is reduced to $W_s = W_b/F_n$ by the branching process.

For the total energy of the additional walls we have to sum up all walls. This leads to a recursive formula for wall multiplication factors M_i that count the extra 90° and 180° wall area elements in the echelon pattern:

$$M_1 = 1 \ , \quad M_i = M_{i-1} b + F_{i-1} \quad (i = 2, \ldots, n) \ .$$

With it the total energy of the branched structure as a function of the parameters n and b can be expressed with the reduced quantities of (3.231) as:

$$\varepsilon_{tot} = \omega_b/(8\sqrt{2}\, F_n) + \eta_{90}/(\omega_b \Lambda) + \frac{1}{\Lambda}\left[c_{90}\eta_{90} + \sqrt{2}\, c_{180}\eta_{180} - 1/(\sqrt{8}\, F_n)\right] \quad (3.233)$$

with $c_{90} = 2 g_{90} M_n/F_n$, $c_{180} = 2 g_{180} M_n/F_n$,

$$g_{180}(b) = \text{int}\left[\tfrac{1}{2}(1+b)\right] \text{ and } g_{90}(b) = \begin{cases} \tfrac{1}{4}b(b+2) & \text{for } b \text{ even} \ , \\ \tfrac{1}{4}(b+1)^2 - 1 & \text{for } b \text{ odd} \ . \end{cases}$$

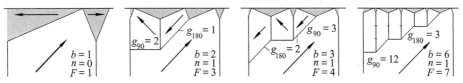

Fig. 3.134 Evaluating the total length of 180° and 90° walls in the echelon pattern

The first term in (3.233) is the remaining anisotropy energy in the closure volume, while the second term represents the energy of the basic 90° walls. The shortening of this wall area by the closure pattern is described by the last term in the parenthesis. The coefficients g represent the extra number of 180° and 90° wall elements per generation, which depends on the branching ratio b. The quadratically increasing 90° wall area for large b favours an increase in the number of generations n rather than the use of excessive values of b to reach the same total branching factor F_n. The counting procedure is indicated in Fig. 3.134 for a number of examples.

Minimizing the total energy with respect to the relative basic domain width ω_b we can compare the energies of different models (differing in the number of generations n and in the branching ratio b) and select the best one. The only parameter is the reduced sample length parameter $\Lambda = L/\sqrt{A/K}$. Results of numerical calculations are shown in Figs. 3.135–136.

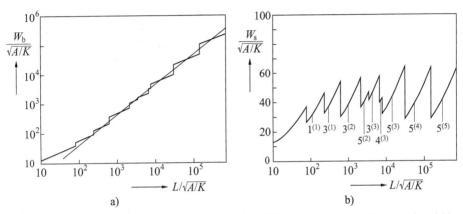

a)   b)

Fig. 3.135 The basic (a) and the edge (b) domain width as a function of the sample width. For $L < 85\sqrt{A/K}$ an unbranched domain pattern with $W_b = W_s$ is expected

The following features of the results are remarkable:

- The first branched domain pattern is found at about $\Lambda = 85$ (which means explicitly $L = 85\sqrt{A/K} = 1.8\ \mu m$ for iron), at much smaller thicknesses in reduced units than in the case of two-phase branching.
- Multiple branching — employing more than one generation — becomes favourable at about ten times larger thicknesses compared to the onset of branching. More precisely, a two-generation echelon pattern ($n = 2$) is preferred starting at $\Lambda = 740$, followed by three-, four- and fivefold patterns.
- The basic domain width increases roughly linearly with the crystal width and is described by $W_b \approx 0.35L$ over four decades in width L. A simple, non-branching model would predict a $\sqrt{L}$ behaviour (3.210), while in two-phase branching we derived a $L^{2/3}$ dependence (Fig. 3.130). Interestingly, D.H. Martin, the inventor of the echelon pattern, still predicted a $L^{3/4}$ behaviour [765] because he did not include multiple generations of echelons in his analysis. Experimentally, a proportionality of the basic domain width

with the sample width L was in fact observed in early experiments by *Carey* [769], which remained unexplained at that time.

- As in two-phase branching the edge domain width W_s stays roughly *constant*, independent of the complexity of the interior pattern. It deviates by less than a factor of two from $44\sqrt{A/K}$ ($\approx 1\,\mu\text{m}$ in iron) over four decades in L.

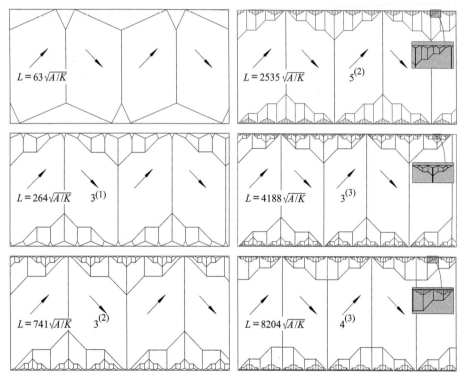

Fig. 3.136 Six of the simpler echelon patterns found by the optimization process for an increasing sample width L. The drawing resolution is not sufficient to display the third generation in the wider examples, which are therefore shown enlarged in the insets. The surface domain width is comparable in all cases (see Fig. 3.135b), while the scales differ

To summarize, multi-phase branching becomes possible in multiaxial crystals without the generation of internal stray fields. This makes branching possible at much smaller sample thicknesses than for two-phase branching. The constant surface domain width seems to be a common feature of all branching processes, but the power law for the basic domains is different: we found a $W_b \sim D^{2/3}$ law in two-phase branching at large D (Fig. 3.130), but multi-phase branching appears to lead to a linear relation between the basic domain width W_b and the sample dimension D or L (Fig. 3.135).

(D) An Analytical Description of Multi-Phase Branching. Is there a way to describe multi-phase branching with a similar theory as presented in (B) for two-phase branching? The following proposal attempts to cover the crucial point. As before, we use a continuously varying domain width $W(z)$. The total energy will contain the same elements as the total energy (3.222) for uniaxial branching, with the exception that no internal stray field will be generated in the branching process. The branching speed will still be limited, not by an energetic penalty as in two-phase branching, but by some geometrical constraint. One way of formulating such a constraint is an upper limit G_m for $dW(z)/dz$ which may be of the order of unity. There may be other ways of formulating such a constraint, but as we will see, the assumption of a maximum branching speed reproduces in fact the basic features of multi-phase branching.

We will furthermore expect that the basic domain width is limited by a secondary mechanism such as magnetostriction. The quasi-closure structures of the basic domains will carry some magnetostrictive self-energy that will be proportional to the basic domain width. Altogether this leads to the following expression for the total energy per unit surface of the plate:

$$e_{\mathrm{tot}} = 2\int_0^{D/2} [\gamma_w / W(z)]\, dz \;\; + 2\,C_s\,W_s + 2\,C_p\,W_b \;\; \text{with} \;\; W'(z) \le G_m \;, \quad (3.234)$$

where C_p is the effective closure energy coefficient for the basic domains and G_m is the geometrical constraint for the branching speed. The other quantities have the same meaning as in (B). The optimum solution is obviously one in which the domain width rises from its surface value W_s to its bulk value W_b with the maximum allowed speed G_m:

$$W(z) = W_s + G_m z \qquad \text{for } z \le (W_b - W_s)/G_m \,,$$
$$W(z) = W_b \qquad\qquad \text{for } z > (W_b - W_s)/G_m \,.$$

This ansatz is now inserted into (3.234) and integrated, leading to:

$$e_{\mathrm{tot}} = \gamma_w \left[\frac{D}{W_b} - 2\,\frac{W_b - W_s}{G_m W_b} + \frac{2}{G_m} \ln\frac{W_b}{W_s} \right] + 2\,C_s\,W_s + 2\,C_p\,W_b \;. \qquad (3.235)$$

Minimizing this expression with respect to W_b and W_s leads to:

$$D = 2\,W_b^2\,\frac{C_p}{\gamma_w} + \frac{C_s}{G_m\,C_s\,W_b + \gamma_w} \;, \qquad W_s = \left(\frac{\gamma_w\,W_b}{G_m\,C_s\,W_b + \gamma_w} \right) \;. \qquad (3.236)$$

This simple description reproduces qualitatively all inferred features of multi-phase branching:

- For strong branching ($W_b \gg W_s$) we have $G_m C_s W_b \gg \gamma_w$ and the surface domain width $W_s = \gamma_w /(G_m C_s)$ becomes independent of the bulk domain width, proportional to the domain wall energy, as for two-phase branching.

- In a certain thickness range the bulk domain width becomes proportional to the plate thickness D, as long as the "quasi-closure" coefficient C_p remains negligible: $W_b = \frac{1}{2} G_m D$. If the C_p term becomes important, a classical $W_b \sim \sqrt{D}$ behaviour takes over. The "branching depth", i.e. the zone below the surface filled with an additional network of domains, becomes proportional to the basic domain width $(G_m W_b)$ in this limit.
- In the small D limit we obtain $W_s = W_b$ and the classical $W_b \sim D^{1/2}$ behaviour (explicitly: $W_b = \sqrt{0.5 \gamma_w D/(C_s + C_p)}$.

A typical example for the general behaviour is shown in Fig. 3.137. In this example the $W_b \sim D$ range extends over about three decades.

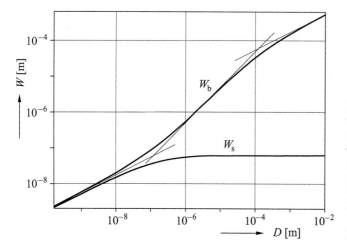

Fig. 3.137 Bulk domain width W_b and surface domain width W_s as a function of the plate thickness D as resulting from (3.236) with $G_m = 1$ and inserting parameters typical for a cubic material: $C_s = 10^4$, $C_p = 10$ J/m^3, $\gamma_w = 6.3 \cdot 10^{-4}$ J/m^2

(E) Three-Dimensional Multi-Phase Branching. If the magnetization vector is not confined to a plane, the third dimension adds new possibilities for branching that differ qualitatively from the two-dimensional solutions discussed in the last section. Not only the magnetization vector can make use of the third dimension, but also the geometry of the domain pattern can assume a three-dimensional character.

The essential point becomes clear from the example of the so-called checkerboard pattern (Fig. 3.138 [124]). This is a second-generation branching pattern applicable to an iron plate with a (110) surface, for which the in-plane [001] easy axis is disfavoured for

example by an external stress or an external field of suitable direction. In this geometry we are looking for structures for which the basic domains are magnetized along the two favoured internal axes, and only the closure domains are magnetized along the [001] axis.

The checkerboard pattern fulfils this condition. It consists of three classes of domains: the basic domains are magnetized along the two favoured easy axes. An intermediate layer of domains is magnetized partially antiparallel (and parallel) to the basic domains, but still along the two favoured easy axes. Only the shallow surface domains are magnetized along the disfavoured easy axis. Of course, the checkerboard pattern is rather complex, and there are simpler echelon type patterns with which it has to compete. A variant of the echelon pattern is used here for the extended crystal plate in which the closure domains are magnetized along the third, disfavoured easy axis, as indicated in the insets in Fig. 3.139, rather than along the hard axis necessary in thin plates discussed in (C). With respect to the magnetization directions this variant of the echelon pattern is already three-dimensional; the spatial arrangement of the domain walls is still two-dimensional, however.

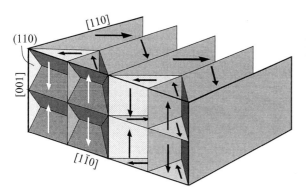

Fig. 3.138 The checkerboard pattern as an example of three-dimensional branching. In the right-hand half of the picture the surface domains were omitted for better insight. This orientation of the drawing (with the checkerboard pattern on the front side) has been chosen because it will be used again in the discussion of the Néel block (Sect. 3.7.6)

The result of a comparison between the three-dimensional checkerboard patterns and the structurally two-dimensional echelon patterns is shown in Fig. 3.139. Here the domain wall energy per unit sample surface needed for the branching configuration is compared with the necessary mean closure domain thickness or *closure depth* relative to the basic domain width. In the wall energy all necessary walls are summed up, weighted with their appropriate specific wall energy. The closure depth weighs only those shallow surface domains that are not magnetized along one of the favoured easy axes. A small relative closure depth is preferred for thick samples.

We see that relative closure depths of about 0.05 or smaller can be achieved with the three-dimensional checkerboard pattern with less average wall energy than with any conceivable variant of the two-dimensional echelon pattern. The fat continuous line in Fig. 3.139 for the checkerboard pattern corresponds to a continuous variation in the length-to-width ratio of its surface cells. Replacing square checkerboard cells by rectangular ones saves in closure volume at the expense of internal wall energy.

The checkerboard pattern is only one, relatively simple example of a three-dimensional branching structure. Once a three-dimensional pattern has turned out to be more favourable than a two-dimensional pattern, the three-dimensional branching process will continue for larger sample thicknesses. No attempt is made to give a detailed description of these processes.

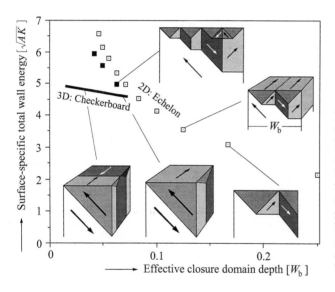

Fig. 3.139 A comparison of two- and three-dimensional branching schemes. Plotted is the extra wall energy of the branching system per unit sample surface area versus the effective closure domain depth in units of the basic domain width W_b. The specific wall energies of the different wall types are based on the example of iron

(F) Quasi-Domains in Branching. The concept of *quasi-domains* introduced in Sect. 3.4.5 offers a global description of complicated branched structures. This method is outlined in Fig. 3.140 for the example of a (111)-oriented nickel platelet. Diagrams (a,b) indicate the available $\langle 111 \rangle$ easy directions as seen from the top, and the different domain generations. We start with the cross-section (c), showing the basic domains and schematically the closure domains in the first level. We postulate that the basic domains consist of 180° domains to save magnetostrictive energy. The closure domains of these basic domains are chosen from combinations of the easy directions which result in surface-parallel net magnetization directions — the quasi-domains.

The angles of the walls between basic and closure domains are determined by the condition of stray field avoidance, as shown in Fig. 3.122 d. The only difference is that the magnetization vector of a closure domain must be taken as the mean value of the magnetization vectors of the applicable quasi-domain system. The lengths of the magnetization vectors in the basic domains correspond to the projection of the true directions onto the drawing plane. Thus the principle of flux closure can be seen directly. Which of the possible basic band orientations and the possible closure quasi-magnetizations are to be chosen depends on the energy — mainly on the basic wall energy and the necessary closure volume, and for large domains also on the magnetostrictive energy. Here an example without systematic energy optimization is shown just to demonstrate the principle.

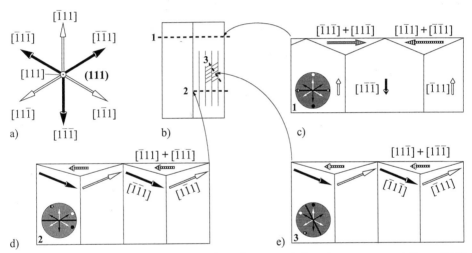

Fig. 3.140 Multiple branched structure based on quasi-domains constructed for a (111) nickel plate. This is only an example of an allowed sequence. It is still to be optimized with respect to the orientation angles and periodicities of the successive generations

In the next branching generation (d) we use the components of a quasi closure domain from (c) as the new basic domains. The best domain orientation for the second generation will in general not coincide with the first-generation domain orientation. In this special case, however, a subdivision, which is oriented along the longitudinal direction of the primary domains, seems to be a rather favourable choice. Probably energy can be saved by rotating the axis of the secondary pattern by some angle. The third generation (e) should definitely prefer an orientation different from the previous generation; a reasonable proposal is indicated in the schematic drawing. This means that the overall structure is three-dimensional. Nevertheless, we can draw each generation in a two-dimensional way as shown schematically in Fig. 3.140c–e for the first three generations. The process ends when a model with uniformly magnetized closure domains is preferred to a further generation of quasi domain, i.e. when the additional domain wall energy connected with a subdivision exceeds the anisotropy energy in the closure domains.

Also the checkerboard pattern can be described by the quasi-domain concept. In the first generation the closure domains carry a zero net magnetization, whereas in the cross-section indicated in Fig. 3.141 a regular Landau-type closure structure is revealed.

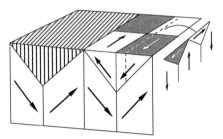

Fig. 3.141 The quasi-domain representation of the checkerboard pattern

(G) Stripe Domains and Branching. Multi-phase branching is not restricted to cubic or other multiaxial materials. Similar phenomena, which rely on three-dimensional configurations, may occur in uniaxial materials if the anisotropy is weak compared to the stray field energy ($Q \ll 1$). This is demonstrated in Fig. 3.142a where an experimental observation of a branched pattern on a metallic glass is shown. The sample carries a stress-induced anisotropy with the easy axis perpendicular to the surface. Beyond a critical stress level the basic structure consists of periodic perpendicular domains similar to the domains that develop out of the dense stripe domains discussed in Sect. 3.7.2. Quite surprisingly the closure domains of these stripe domains (visible on the left) become themselves modulated at only slightly higher stress levels (from top to bottom).

The quantitative microscopic analysis of such patterns [180, 770] displayed a continuous modulation of the magnetization direction on the surface, agreeing in character with the analytical solutions for dense stripe domains in perpendicular-anisotropy thin films (Fig. 3.110). This observation suggested a model for this phenomenon [180] as shown in Fig. 3.142b. The closure domains of a stripe pattern are magnetized in a hard direction with respect to stress anisotropy. At a certain anisotropy level they should therefore decay spontaneously into a secondary stripe pattern in about the same way as for dense stripe domains in thin films (Sect. 3.7.2A). More detailed considerations indicate that secondary stripe oscillations can be connected with the basic domains if they assume the characteristic corrugated shape visible in photograph and model. With a further increase in sample thickness or anisotropy the secondary stripe domains will grow into regular domains, the closure domains of which will decay into a third generation of stripe pattern as shown in Fig. 3.142 c.

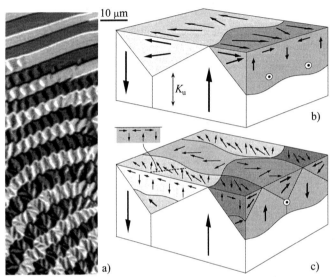

Fig. 3.142 Branching phenomena of weakly uniaxial materials, observed on a metallic glass of ~20 μm thickness with a stress-induced perpendicular anisotropy increasing in strength from top to bottom. Model (b) shows a basic Landau pattern with modulated closure domains. A second generation of branching as visible in the lower part of (a) is interpreted schematically in (c)

This principle of continuous stripe formation in the closure domains of the previous stripe generation seems to be a general feature of low-anisotropy magnetic materials. It agrees with the general ideas discussed in Sect. 3.3.4 about the possibility of a continuous closure zone superimposed onto a regular domain pattern. The micromagnetics of dense stripe domains restricted to the surface of a thick sample was computed numerically in a different context in [771, 748]. The transition from simple stripes to branched stripes as shown above occurs at critical thicknesses of the order of $100\sqrt{A/K}$, rather than at the much larger values of high-anisotropy uniaxial materials (Fig. 3.130). The phase diagram proposed in [763] based on two-phase branching alone was therefore inadequate for the range of $Q < 1$.

3.7.6 The Néel Block as a Classical Example

(A) Introduction. *Néel* [600] initiated the quantitative discussion of a seemingly simple geometry for iron crystals, which was first studied by *Kaya* [772, 773]: a long block-shaped crystal with a (100) surface and (01$\bar{1}$) sides (Fig. 3.143). Analogous shapes exist for negative-anisotropy materials like nickel.

The crystal of thickness D and width L is assumed to be very long or embedded into an ideally soft magnetic yoke, so that longitudinal demagnetizing effects can be neglected. All dimensions are assumed to be large compared to the Bloch wall width. A field is applied along the sample axis, leading ideally to a simple plate-shaped basic domain pattern. Somehow these basic domains must be closed on the side surfaces. Néel calculated the magnetization curve and the domain period based on phase theory and the simplest conceivable closure structure as shown in Fig. 3.143, hoping that these predictions might be a test for the validity of domain theory. After reassuring first experiments on thin and not-so-well-shaped crystals [774, 775], all further studies on larger and better defined crystals [776–787] resulted in strong discrepancies between calculated and observed domain periods, typically by more than a factor of two. The background of these discrepancies is discussed in the following.

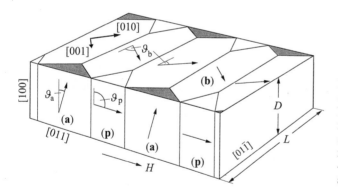

Fig. 3.143 The geometry of the Néel block and Néel's original domain model. The magnetization angle ϑ_b in the basic domains may deviate from 45° in an applied field along the sample axis

(B) Elementary Theory of the Néel Block. We start with a discussion of the theoretical boundary conditions as derived from phase theory. We assume a finite (if small) effective field may act along the sample axis, thus ignoring the completely demagnetized state in which additional 180° walls would be introduced. The *basic domain* pattern then has to be as shown in Fig. 3.143. These domains are magnetized parallel to the top and bottom surfaces of the crystal and they are also elastically compatible. The total energy, applicable to the basic domains, consists of the Zeeman energy (3.15) in the applied field along the [011] direction, and the cubic anisotropy energy that can be expressed using (3.9b) with $K_{c1} = K$, $K_{c2} = 0$ and inserting $\varphi = \frac{\pi}{2}$:

$$e_{tot} = \tfrac{1}{4}K\left(\cos^2\vartheta - \sin^2\vartheta\right)^2 - HJ_s\sin\vartheta \ . \tag{3.237}$$

Optimization with respect to ϑ leads to the following conditions for the magnetization angle ϑ_b in the basic domains (see Fig. 3.143):

$$h = \sin\vartheta_b\left(\sin^2\vartheta_b - \cos^2\vartheta_b\right) \qquad \text{for } h \le 1 \ ,$$

$$\vartheta_b = \tfrac{\pi}{2} \qquad\qquad\qquad\qquad \text{for } h > 1 \ . \tag{3.237a}$$

with the abbreviation $h = HJ_s/(2K)$. The energy $e_b = e_{tot}(\vartheta_b)$ and magnetization angle ϑ_b are plotted in Fig. 3.144 as functions of the reduced field h.

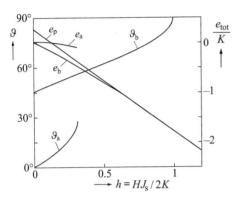

Fig. 3.144 The energies and the magnetization angles of the basic and the closure domains in the Néel block geometry as a function of the reduced applied field

To go beyond phase theory and to derive the domain period requires the consideration of wall and closure energies as functions of the applied field. The energy of the walls between the basic domains is obtained from Fig. 3.68. The mean orientation of the walls in the Néel block is $\psi = 90°$, so that we expect zigzag walls with an average energy corresponding to the minimum of the $\gamma_w/\sin\psi$ curves. (We ignore the energy of the kink regions in the zigzag wall and the magnetostrictive energy in the zigzag segments [784]).

For the closure domains Néel postulated as a first step a mixture of the two kinds of domains shown in Fig. 3.143. Both domain types are problematic at small fields: the a-domains, which are magnetized close to an easy direction are not properly closed at the top and bottom end surfaces, and the p-domains are magnetized along a hard direction. But let us first follow Néel's line of thought and calculate the energy densities and the equilibrium widths of the a- and p-domains, ignoring the stray fields of the a-domains.

As the magnetization on the side surfaces is restricted to the $(01\bar{1})$ plane, its energy can be calculated from (3.9b) with $\varphi = 0$:

$$e_{ss} = K \sin^2 \vartheta \left(\tfrac{1}{4} \sin^2 \vartheta + \cos^2 \vartheta \right) - H J_s \sin \vartheta \ , \tag{3.238}$$

leading by minimization with respect to ϑ to two solutions:

$$h = \sin \vartheta_a \left(\cos^2 \vartheta_a - \tfrac{1}{2} \sin^2 \vartheta_a \right) \qquad \text{for } 0 \le h \le \tfrac{1}{9}\sqrt{8} \ , \tag{3.239a}$$

$$\vartheta_p = \tfrac{\pi}{2} \qquad\qquad\qquad\qquad \text{for } h \ge 0 \ . \tag{3.239b}$$

The energies corresponding to these solutions are plotted in Fig. 3.144. Both modes are energetically disfavoured compared to the basic domains if the field is different from zero. But they may exist as closure domains. The p-domains may occur at any field values below saturation, whereas the a-domains can exist only at small fields $h < \tfrac{1}{9}\sqrt{8} = 0.314$.

The total energy can be optimized with respect to the volume of p- and a-domains, and to the period of the basic domains which determines the volume of the closure domains. For given surface widths w of the a- and p-domains, their volumes v, and hence their energies e_a and e_p, are:

$$v_a = \tfrac{1}{4} w_a^2 d_a D \ , \quad d_a = \frac{\cos \vartheta_b}{\sin \vartheta_b + \sin \vartheta_a} \ , \quad e_a = v_a e_{ss}(\vartheta_a) = F_a w_a^2 D \ ,$$

$$v_p = \tfrac{1}{4} w_p^2 d_p D \ , \quad d_p = \frac{1 + \sin \vartheta_b}{\cos \vartheta_b} \ , \quad e_p = v_p e_{ss}(\vartheta_p) = F_p w_p^2 D \ . \tag{3.240}$$

The factors d_a and d_p measure the depths of the prismatic closure domains (see Sect. 3.7.4D), and F_a and F_p are abbreviations for functions of the applied field, the detailed definitions of which follow from (3.238) and (3.239). The total energy per unit volume is then:

$$e_{tot} = \left[2(F_a w_a^2 + F_p w_p^2) + 2L \gamma_z - \gamma_z (w_a d_a + w_p d_p) \right] / \left[L(w_a + w_p) \right] \ . \tag{3.241}$$

Here γ_z is the average specific energy of the zigzag wall. The last term describes the correction of the wall area by the presence of the closure domains. The total energy (3.241) must be optimized numerically with respect to the closure domain widths w_a and w_p. The result is close to the following simple relations for which the last (correction) term in (3.241) is omitted:

$$w_a^{opt} F_a = w_p^{opt} F_p \ , \quad p_{opt} = w_a^{opt} + w_p^{opt} = \sqrt{(F_a + F_p) \gamma_z L / F_a F_p} \ . \tag{3.242}$$

These equations predict the dependence of the relative size of the two closure domain types in Néel's model, as well as the overall period p_{opt} of the basic domains as a function of the applied field that enters through the two functions F_a and F_p.

(C) Discussion. With known geometrical relations and energy densities these expressions for the basic domain period p_{opt} as a function of the field can be readily evaluated. What are the reasons for the failure of an experimental confirmation? Here are the most important points, as they emerged in the course of many careful studies mostly on silicon iron and on nickel crystals (for nickel the geometry of the Néel block has to be adapted to the negative cubic anisotropy of this material):

- Néel did not include the magnetostrictive energy of the closure domains which is decisive in low field situations. It was included by *Spreen* [781, 782] in a reasonable approximation. But the magnetostrictive energy gives only a minor correction for higher fields where the energy differences between closure and basic domains shown in Fig. 3.144 are predominantly caused by crystal anisotropy effects.

- Another source of discrepancies between experiment and theory was identified in a series of articles by *Ungemach* [786, 785, 787]. Continuous and smooth transitions between domain structures with different periods are not possible. To decrease the period, a threshold for the nucleation of new domains has to be overcome. It is therefore difficult to reach in an experiment the lowest energy pattern. Superimposing an alternating field of decreasing amplitude over the constant applied field helps, and the best results are obtained with a low frequency circular field as generated by an a.c. current along the crystal axis [781, 782]. Additional problems arise in some materials by induced anisotropies that favour a given domain width and resist the transition to a different period. They can be circumvented by a heat treatment in an alternating field. Finally, the attempt to generate an equilibrium configuration by the superimposed alternating field may generate a tendency towards finer domains because of eddy current effects [786]. This effect can be avoided by using low frequency demagnetizing fields of a few Hertz.

- The principal reason for the discrepancies between theory and experiment, however, lies in the actually occurring complex domain structures. In large crystals the domain pattern saves energy by *branching* (Sect. 3.7.5). Because no branching phenomena were considered in the elementary theory, small samples come closer to Néel's assumptions.

In hindsight, the limitations of Néel's model (Fig. 3.143) are not surprising. It postulates domains with strong surface charges as well as domains magnetized along a hard direction, while even completely stray-field-free patterns that use only the easy directions are possible, as shown in Fig. 3.145. Another example for such a solution is the checkerboard pattern (Fig. 3.138), which is compatible with the boundary conditions of the Néel block.

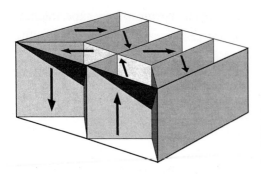

Fig. 3.145 A stray-field-free model to replace Néel's model of Fig. 3.143, which would be preferable particularly at small applied fields

The checkerboard pattern [783] and under other conditions also an echelon pattern [784] have actually been observed (Fig.3.146a). At higher fields these patterns are broken down and a refinement of the domains on the side surfaces is initiated by an enhanced folding of the zigzag walls near the surface.

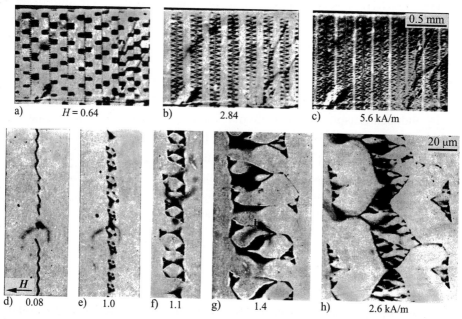

Fig. 3.146 Low resolution Kerr effect pictures (*top row*) and Bitter pattern detail observations (*bottom row*) of the closure domains on the side faces of a silicon-iron Néel block ($80 \times 4 \times 2.5$ mm^3). The checkerboard pattern (a) — invisible in the Bitter technique (d) — replaces the a-domains of Néel's model in small fields. In a field along the sample axis, complicated branched zigzag patterns grow instead of the p-domains. Their growth is shown in (b–c) and again in high resolution (f–h). Pictures (e–g) can be derived from the checkerboard pattern when its centre (zigzag) wall (d) folds into a complicated lace pattern. Pictures (b, h) mark the transition to a two-phase branching scheme with daggers inserted into the basic domains. (Photographs courtesy *Ch. Schwink*)

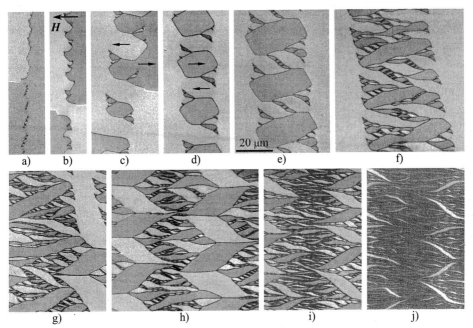

Fig. 3.147 High-resolution Kerr effect images of a 2 mm thick (110)-oriented Fe(3%Si) slab under the influence of an increasing field along the hard axis, displaying in detail the magnetization processes also observable on the side faces of a Néel block. The plane of incidence is chosen along the hard axis to display mainly wall contrast rather than domain contrast. The series starts in (a) with a domain wall of the checkerboard pattern (Fig. 3.146a), which is progressively folded in a branching process (b−f). The structure in (c, d) is related to the "cord" pattern that will be discussed in Fig. 5.23. A complete rearrangement leads then to the spectacular pattern (h) that corresponds to Fig. 3.146h and that remains unexplained in its details. Increasing the field further the pattern simplifies again towards saturation (j)

For still higher fields approaching saturation dagger-shaped domains are inserted (sketch), laced with complex closure domain patterns (Fig. 3.146). This process indicates a transition from a multi-phase branching pattern to a two-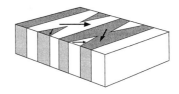phase branching pattern as discussed in Sect. 3.7.5. The appearance of parallel lines indicates the formation of p-type closure domains, still connected with a plate-like two-phase branching structure with its typical dagger-shaped domains and internal stray fields. Figure 3.146 shows overview Kerr effect and high-resolution Bitter pattern observations [784] displaying details of the branching processes in a well-prepared Néel block which are by no means fully understood. A better understanding of these fascinating patterns might be achieved by studying high-resolution Kerr effect images as they are shown in Fig. 3.147.

These pictures are taken on a thick (110)-oriented SiFe plate, not on an actual Néel block. But the surface orientation as well as the applied field direction are the same and the observed patterns are obviously very similar because they display a similar degree of branching. Studying such series as a function of the applied field and combining them with model calculations may eventually lead to a better understanding of these intricate patterns.

After more than four decades of experimental and theoretical work, many of the pitfalls in the analysis of the Néel block are now known, and a better agreement between experimental measurements and theoretical predictions might be achievable. But this would ask for laborious calculations, including a comparison of many different domain models. Carefully prepared and shaped crystals without traces of induced anisotropy would be needed on the experimental side. But all this would probably not be worth the effort. After all, Néel had devised his theory to provide a *simple* check between domain theory and experiment. After it turned out that the actual domain structure of the Néel block is complex, there is no point in forcing its analysis.

3.8 Résumé

We have to conclude that domain theory and micromagnetic analysis should in general not be used to replace experimental observation by theoretical prediction. Rather, theory should help in the analysis and interpretation of observations, providing guidelines for the construction of three-dimensional models based on surface observations. If more than one such model exist, domain theory can decide which model has the lower energy. Every interpretation of experimental observations should be tested for plausibility with the help of domain theory. But if the need for experimental confirmation of the theory is felt, it is wise to focus on details such as certain continuously variable critical angles or lengths in a confirmed model rather than on the model as a whole.

4. Material Parameters for Domain Analysis

Domain observation and domain theory are linked by a small number of intrinsic material parameters. Without the knowledge of these parameters no reasonable domain analysis can be performed. None of these material properties can be calculated from first principles to the required accuracy. In some cases there exist interpolation rules that allow the material parameters of mixtures to be derived from their components, or that predict the temperature dependence of one quantity on the basis of the temperature dependence of another. In general, however, the experimental measurement of the important parameters cannot be avoided. In this chapter we review the classical and modern methods, which are available for their measurement, as well as some theoretical relations. It must be emphasized that magnetic domains depend primarily on extrinsic parameters such as sample shape, grain structure and internal stresses. This chapter deals with intrinsic material properties only, the knowledge of which is a necessary but by no means a sufficient prerequisite in domain analysis.

4.1 Intrinsic Material Parameters

The following magnetic material parameters are needed in domain analysis of an arbitrary material:

- The saturation magnetization J_s.
- The exchange or stiffness constant A (Sect. 3.2.2).
- The gyromagnetic ratio γ and the damping constant α_G or α_{LL} (Sect. 3.2.7C).
- The crystal and other anisotropy constants K (Sect. 3.2.3).
- The magnetostriction constants λ (Sect. 3.2.6B–D).

In addition, some non-magnetic material parameters, such as the elastic constants and the electrical resistivity, have to be available.

The first three quantities in the list are essentially isotropic and can thus be measured on polycrystalline samples. No fundamental problem exists for the saturation magnetization and its temperature dependence because various magnetometers, among them the vibrating sample magnetometer (VSM) offer reliable tools for practically all materials. In contrast, the exchange constant is rather inaccessible because it refers to microscopic inhomogeneities in the magnetization. This quantity is most often derived indirectly from the Curie point or the temperature dependence of the magnetization. A direct determination of the stiffness constant is rare. The dynamic constants can usually be derived from resonance experiments.

The remaining parameters describe anisotropic properties that can reliably only be determined on single crystals. Unfortunately, many materials have never been grown in this form. In addition, when a property like anisotropy or magnetostriction depends on annealing or deposition history of a sample, this history may be difficult to reproduce for bulk single crystals. Indirect methods from which information about the anisotropic material constants can be extracted from given polycrystalline samples are therefore important, even if they may be less rigorous than direct measurements on single crystals.

Instead of discussing the measurement of the above-listed parameters in sequence we organize the discussion according to the techniques employed:
- Mechanical measurements (Sect. 4.2).
- Magnetic measurements (Sect. 4.3).
- Resonance techniques (Sect. 4.4).
- Dilatometric measurements (Sect. 4.5).
- Domain methods (Sect. 4.6).

Many of these techniques can be used to determine different basic material constants. For example, resonance experiments yield information on anisotropies, magnetostriction, the exchange constant and the dynamical parameters. Torque measurements can be used for an accurate determination of anisotropies, but also of the saturation magnetization. The same is true for VSM measurements.

The derivation of the exchange stiffness constant from thermal data is discussed in Sect. 4.7, followed by some theoretical rules in Sect. 4.8. Previous reviews of experimental methods and results can be found in textbooks such as [788, 789]. A review of the special methods applicable to the determination of the exchange stiffness constant was presented by *Döring* [37].

Naturally, there is a considerable overlap between specific methods suitable for the determination of material constants and general magnetic measurements for the characterization of technical magnetic materials. It depends on the circumstances whether fundamental constants can be extracted for example from a magnetization curve. Some results such as measurements of Barkhausen noise or magnetic losses may contain unpredictable components and are fundamentally unsuited for our purpose. In other cases there is a better chance to derive material parameters as will be elaborated in more detail in Sect. 4.3.

4.2 Mechanical Measurements

Static or dynamic measurements of mechanical forces on magnetic samples comprise both old and modern techniques. Because they require good isolation from extrinsic vibrations they are found more in fundamental research than in routine material characterization. A classification is derived from the nature of the mechanical interaction between field and sample: a uniform field H, acting on a uniformly magnetized sample of magnetization J and volume V, generates a mechanical *torque* $T_m = V H \times J$. If, on the other hand, the field is not uniform, its gradient generates a mechanical *force* $F_m = V \, \mathbf{grad}(J \cdot H)$.

4.2.1 Torque Methods

Torque measurements offer the best and most direct methods for measuring anisotropies. Such measurements require uniform, in general single-crystalline samples, preferably of spherical shape. Also circular disks can be used if spheres are not available, but corrections may be necessary in this case.

(A) Theory of Operation. The mechanical torque T_m exerted by a ferromagnetic body is defined as the derivative of its total energy E_{tot} with respect to an angular variable φ: $T_m = -\partial E_{tot}/\partial \varphi$. For a spherical, saturated sample the total energy consists of the external field energy [(3.15); in a field H along the angle η_h, see Fig. 4.1] and the anisotropy energy density $e_K(\varphi)$ (3.9, 11, etc.). The demagnetizing energy does not depend on the sample orientation for a sphere and can be ignored, and local stray fields do not occur at saturation. For the same reason exchange stiffness energy contributions do not enter. Magneto-elastic interactions can be excluded in the absence of external stresses. The effects of any kind of internal stresses must cancel if the sample is uniform and saturated. [Of course, the anisotropy measured in this way contains a magnetostrictive part as it applies to the freely deformed state (Sect. 3.2.6B)].

The total energy for a sphere of volume V is then:

$$E_{tot} = [-HJ_s \cos(\eta_h - \varphi) + e_K(\varphi)]\, V\,. \tag{4.1}$$

In static equilibrium the torque must vanish, so that the torque per unit volume from the external field just balances the intrinsic torque T_{int}:

$$T_{int}(\varphi) = HJ_s \sin(\eta_h - \varphi) \quad \text{with} \quad T_{int}(\varphi) = -\partial e_K/\partial\varphi\,. \tag{4.2}$$

If H and J_s are known, the difference $\eta_h - \varphi$ can be derived for any value of the intrinsic torque T_{int} by the relation $\eta_h - \varphi = \arcsin[T_m/(HJ_s)]$. The torque is first measured as a function of the field direction η_h. It can then be transformed to $T_{int}(\varphi)$ by the operation shown in Fig. 4.1. For large H the difference $\eta_h - \varphi$ becomes linear in T_{int}, so that the operation is analogous to the *shearing* of a magnetization curve. Then the *maxima* of the torque curves stay invariant under the shearing operation and the desired information about the anisotropy parameters may be determined from the maximal values of the torque curve alone. For a simple uniaxial anisotropy $e_K = K_{u1} \sin^2\varphi$ the maximum torque *is* the anisotropy constant K_{u1} because of $T_{int}(\varphi) = -\partial e_K/\partial\varphi = K_{u1} \sin(2\varphi)$. Exactly this case is plotted in Fig. 4.1. The same simple evaluation is possible for cubic crystals if they are rotated around [110] and if K_{c1} and K_{c2} of (3.9) suffice to describe the anisotropy functional.

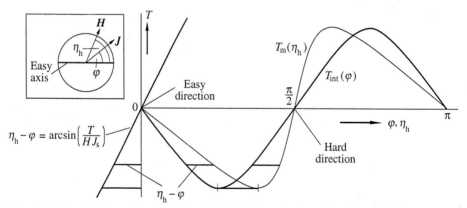

Fig. 4.1 Determination of the derivative $T_{int}(\varphi)$ of the anisotropy function from the measured torque $T_m(\eta_h)$ with the help of a shearing operation, for the example of a single crystal of simple uniaxial anisotropy

In general, the anisotropy energy can be integrated from the function $F(\varphi)$ obtained from the shearing operation even if the symmetry of the anisotropy is unknown at first. The field dependence of the shearing correction can also be used to determine the saturation magnetization of homogeneous materials from torque curves if the saturation is not known from other measurements.

(B) Experimental Setup. The first magnetic torquemeter was used by *P. Weiss* [790] in the investigation on pyrrhotite crystals which led him to his famous theory of ferromagnetism. The principle of the apparatus was the same as it is used today (Fig. 4.2): the sample is suspended on a torsion wire and stabilized either with a heavy weight or a second suspension. A strong rotatable electromagnet acts in a direction perpendicular to the torsion axis. The torsion is detected with the help of a light beam reflected from an attached mirror. In modern equipment the torque is automatically compensated by a current in a compensation coil, so that the value of the current can serve as a measure of the acting torque.

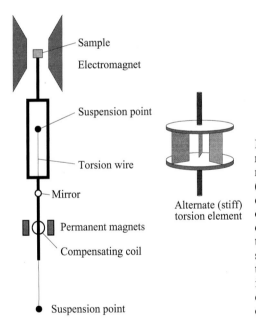

Sample

Electromagnet

Suspension point

Torsion wire

Mirror

Alternate (stiff) torsion element

Permanent magnets

Compensating coil

Suspension point

Fig. 4.2 Schematics of a torque magneto-meter. The sample is mounted on a long rod inside a magnet. The whole mechanism (except the magnet) can be put inside an oven or a Dewar for measurements at high or low temperatures. The rod is suspended on a torsion wire or on an elastic construc-tion that permits the desired torsion but is stiff against other displacements. The torque is detected optically or electron-ically (i.e. compensated by an auxiliary coil, thus measuring the torque by the nec-essary compensating current)

Most torquemeters are custom-made for the range of torques which have to be measured. For high torques the stability of the instrument is of prime importance. In one design for such applications [791] the torsion wire is replaced by a peculiar structure (Fig. 4.2), which is mechanically stiff in all directions except against torsion. The necessary sensitivity is achieved by a suitable electronic displacement sensor. In the other extreme of highest sensitivity [792] (which is necessary for measurements on thin films) the whole instrument was made of fused silica to avoid any stray magnetic forces that might be generated in the large external field if the sample holder is somehow magnetic. The stability against such magnetic forces can be increased also by a pole shape that produces a field maximum at the centre [793]. The range of torques measurable with these instruments extends from 10^{-13} to 10^{-2} Nm. The magnitude of the applied field is limited only by magnet technology. Superconducting magnets [794] permit fields in excess of 10^7 A/m. An in-depth review of the design of torquemeters was given by *Pearson* [789].

(C) Refinements. Torque magnetometry is a direct method needing only few corrections. If the magnetic field is sufficiently large to completely saturate the sample, the magnetic anisotropy can be integrated directly from the measured curve as explained in (A). This applies, however, only to the measuring plane in which the sample is rotated. How to derive the full anisotropy functional by combining measurements in different planes is discussed in [795, 796].

A refinement is necessary if the saturation magnetization cannot be considered isotropic. *Aubert* [795] investigated this correction with great care for pure nickel. He demonstrated that the measured anisotropy constants have to be corrected by linear terms in the applied field which contain corresponding coefficients of the anisotropy of *magnetization*. If K'_n is a measured coefficient, then the true anisotropy coefficient is given by $K_n = K'_n + H_{ex} J_n$, where H_{ex} is the external field and J_n is the corresponding coefficient of the orientation dependence of the saturation magnetization. By combining torque curves measured at different fields, the two contributions — crystal anisotropy and magnetization anisotropy — can be separated. In nickel a magnetization anisotropy of the order of 10^{-3} was found and had to be taken into account for a reliable determination of higher-order crystal anisotropy constants.

Another complication arises if only cylindrical samples are available. Then the *inscribed* ellipsoid will be saturated first, whereas the remaining volume needs larger fields for saturation. *Kouvel* and *Graham* [797] analysed such a situation and found that the residual edge domain structure depends in a complicated way on the field orientation. This introduces extra torques that seem to vary with $1/\sqrt{H_{ex}}$ [797]. An attempt to eliminate this effect by extrapolation to large fields did not lead to satisfactory results. Reliable measurements of the anisotropy on thick cylinders are obviously impossible without a detailed theory of the edge domain structure which is not available. Preparing spherical single crystals for which these complications do not occur is probably easier. Similar effects will be less important in thin films that are typically measured in the shape of circular disks. Here the exchange stiffness effect will suppress strong deviations in the edge zones.

(D) Torsional Oscillation Methods. These techniques [798], which are applicable to ultrathin films down to atomic monolayers, use a variant of the torque method. Instead of measuring a static balance of torques, small torsional oscillations are excited in the presence of a magnetic field. The frequency of these oscillations is given by the moment of inertia of the pendulum on the one hand, and the magnetic torques as in (4.2) on the other hand. For small oscillation amplitudes the magnetic torques are expanded around the equilibrium position of the pendulum. The oscillation frequency is thus a function of the second derivative of the free energy of the sample with respect to the rotation angle. Plotting the square of the oscillation frequency as a function of the inverse applied field [799–801] both the magnetic moment and the mentioned second derivative of the free energy can be derived. If the shape of the anisotropy functional is known beforehand, the missing anisotropy coefficients can be derived graphically; otherwise a numerical evaluation of measurements along a multitude of field directions may be necessary. The decisive advantage of the torsional oscillation method is its sensitivity that allows even fractions of an atomic monolayer to be measured. In particular surface anisotropies in ultrathin films have thus been accurately determined [536, 537].

4.2.2 Field Gradient Methods

(A) Faraday Balance. This classical method measures the static force on a sample in a magnetic gradient. It is highly sensitive and can be applied to all kinds of magnetic substances. In ferromagnetism it is best suited to measure the saturation magnetization with high precision. A uniform field is generated by an electromagnet, and the gradient field is produced by an additional coil set optimized for a uniform magnetic gradient [802]. If the sample is mounted elastically, a displacement by the force can be detected and compensated by a calibrated electromagnetic counter-force. The compensation current is a measure of the force. The only quantity needed in an evaluation of the magnetization is the sample volume.

(B) Vibrating Reed Magnetometers. These magnetometers, which were originally invented to measure tiny ferromagnetic particles [803], resemble superficially the well-known VSM to be discussed in Sect. 4.3.3B. In the latter the sample is agitated mechanically and an electric signal is derived from this motion. In contrast, a magnetically excited signal is recorded in the vibrating reed magnetometer. The sample carrying some magnetic moment is mounted on an elastic cantilever or "reed", which is excited into resonant vibration by an alternating gradient field. In the original design the vibration was recorded microscopically. The introduction of a piezoelectric pick-up system [804] enhanced the sensitivity, so that even barium ferrite particles of micrometre dimension can be measured. A more versatile and rugged modern instrument [805] includes a compensation scheme in which the magnetic moment of the sample is balanced by the moment of a coil, so that no vibration is excited any more. The compensation current is then a measure of the magnetic moment. Sweeping the external field and applying high and low temperatures becomes possible in this technique just as in a vibrating sample magnetometer. The sensitivity is two orders of magnitude higher, however.

4.3 Magnetic Measurements

4.3.1 Overview: Methods and Measurable Quantities

The well-known classical methods of measuring magnetic hysteresis loops can sometimes be used to derive information about fundamental quantities. In addition to the mechanical methods discussed in Sect. 4.2 there are three magnetic ways to measure a magnetization curve: the magnetometric, the inductive, and the optical methods.

In the *magnetometric* methods the demagnetizing field of a finite sample, which is proportional to the mean magnetization, is measured by some field sensor as a function of an external field. In the *inductive* methods the sample is surrounded by a coil in which a voltage is induced when the magnetization of the sample changes. This voltage is integrated, thus yielding a signal directly proportional to the flux change. In a variant of the inductive method the magnetization of the sample is not changed, but the sample is moved or rotated instead. *Optical* magnetometers measure the surface magnetization via magneto-optic effects (Sect. 2.3) and are particularly useful for thin films where the signals of inductive or magnetometric methods are weak.

The primary fundamental constant to be determined from a magnetization measurement is the saturation magnetization. If other quantities shall be derived, the effects of magnetic hysteresis (irreversibility) must be avoided. Reversible parts of the magnetization curve, such as the approach to saturation, can be evaluated in terms of anisotropy parameters if the sample is single crystalline and free from excessive lattice defects. Magnetostriction constants can be derived under the same conditions if the measurements are performed as a function of an external stress.

Usually single crystals are needed to derive anisotropic properties from magnetization measurements. Under favourable circumstances, however, certain anomalies in the first or second derivative of the near-saturation magnetization curve of polycrystals can be used to derive anisotropy constants.

4.3.2 Magnetometric Methods

The classical magnetometer measures the dipolar field generated by a magnetized sample with the help of a rotatable needle. Nowadays some electronic field detection device such as a Hall probe is preferred. In an arrangement as in Fig. 4.3 the difference signal of the two probes is proportional to the magnetic moment of the sample, and insensitive to the driving field.

A drawback of magnetometric methods is their sensitivity to non-uniform external fields. An advantage is that arbitrarily slow magnetization processes can be followed. The probes must be placed sufficiently far away from the sample, so that only its dipolar field is detected. This means that the samples should preferably be small and short, but apart from this condition they can be of any shape. Hard and high-anisotropy magnetic materials are therefore better suited than soft magnetic materials where demagnetization will dominate the intrinsic properties of short samples.

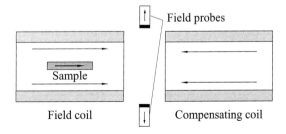

Field probes

Sample

Field coil   Compensating coil

Fig. 4.3 An example of a setup for the magnetometric measurement of the magnetization. The two field probes are connected in series and give a positive signal for fields along the arrow directions. The compensation coil is operated oppositely in series with the field coil

In the magnetometric method the sample can be easily exposed to various environmental conditions such as high or low temperature or mechanical stress. The method is inexpensive and readily implemented. A requirement is that the driving field does not change its geometry when varying the field strength, which might happen if the setup contains ferromagnetic components or impurities.

4.3.3 Inductive Measurements

Magnetization measurements based on induction integrate the voltage induced in a pick-up coil to obtain a signal that is proportional to the magnetization. The inductive effect can be produced by various methods: in the classical way by removing the sample from the coil, by rotating or oscillating it relative to the pick-up coil, or by sweeping through the magnetization curve (which must be done slowly to avoid the influence of eddy currents). The induced signal must be integrated. Classically, this was done either by the ballistic or by the creep galvanometer. Nowadays electronic fluxmeters are preferred, which can be adjusted so that integration over minutes becomes feasible without a disturbing drift. Progress in the suppression of drift in the integrating circuits determines the usefulness of the quasi-static induction methods. For this reason dynamic measurements based on sample oscillation or rotation are preferred. Not all sample shapes permit such dynamic methods, however. They cannot be applied for closed magnetic circuits, for example. We first discuss quasi-static measurements where the sample is at rest, and come back to dynamic measurements in (B).

(A) Quasi-Static Measurements. A closed magnetic circuit or an elongated sample are advisable in static inductive measurements — in contrast to the magnetometric methods, where compact, short samples can and must be used. Such a closed flux geometry is indispensable for high-permeability, soft magnetic samples.

The pick-up coil should include a compensating coil to subtract the effect of the applied field. The difference signal is then proportional to the magnetization alone. There are two possibilities: if there is enough room, two identical coils can be arranged side by side; alternatively, external layers of the winding of a coil can be connected in such a way that their winding area is equal and opposite to the core winding area. Of course, the two coils are rarely equivalent at every field strength and for all sample shapes. Also, the compensation coil may pick up some residual demagnetizing field from the sample. These circumstances limit the sensitivity and accuracy of the inductive method.

The signal induced in the compensation coil can also be used to measure the effective internal field. By virtue of Maxwell's equations the tangential component of a magnetic field must be equal on both sides of a sample surface (if no current is flowing in the sample surface). A coil placed close to the surface may therefore measure the internal field with sufficient accuracy (for precision measurements the signal has to be measured at various distances and extrapolated to zero distance from the surface). With this technique the unsheared magnetization curve can be measured even for short samples.

In static inductive methods the pick-up coils must be placed close to the sample, so that the application of temperature or stress is difficult. The time of measurement is limited by the stability of the fluxmeter. Since the zero setting of a fluxmeter is arbitrary, this method can perform only relative measurements, needing a well-defined reference state such as saturation. Another disadvantage is the limited sensitivity of integrating fluxmeters, which makes its application for thin films difficult. For *bulk soft* magnetic materials, however, the inductive technique is the method of choice.

Usually the pick-up coils are placed around the middle of a sample. To place them in front of or around the end of a short sample is not advisable since the magnetization at the ends of non-ellipsoidal samples is not representative for the whole volume. The situation is different if the sample is placed inside a yoke of soft magnetic material. Then a thin pick-up coil in a thin gap between the sample and the yoke measures only the flux density and is not influenced by demagnetizing fields. This arrangement offers the advantage of being applicable to different sample cross-section shapes.

(B) Vibrating Sample Magnetometers. The VSM [806–808] combines advantages of the magnetometric and static inductive methods. In the standard design (Fig. 4.4) the sample is placed inside a magnet and vibrated perpendicular to the field direction. The signal from a pair of pick-up coils with many windings is compared with the signal induced in a pair of reference coils by a permanent magnet. This arrangement is insensitive to *static* fields of any geometry, so that strong external fields can be applied without adverse effects. Some instruments use superconducting magnets that are able to saturate even hard magnetic materials. Since the sample is well-separated from the pick-up coils, it can be surrounded by cooling or heating devices. The sensitivity of the method is limited mainly by the noise mechanically transmitted from the vibrator to the pick-up coils, which is well suppressed in commercial models. The sample magnetization is static in the VSM, so that no eddy current effects have to be considered. The shape of the sample is largely arbitrary with the same limitations as for the magnetometric technique.

The method is rather direct. Ideally, no calibration would be needed to derive the magnetization of the sample if the pick-up coil geometry and the sample volume are known. This would be true if the magnetic field could be generated by a simple air coil. To generate the necessary field either electromagnets or superconducting magnets have to be used, however, and the magnet materials will interact with the measuring process. In electromagnets "mirror images" of the sample are formed by the presence of soft magnetic iron yokes, and these images will also induce a voltage in the pick-up coils. As the strength of the mirror images depends on the permeability of the iron yoke that in turn depends on the induction level in the magnet, the VSM must be calibrated. The best procedure is to replace the sample by a nickel sample of the same size and shape, and to rely on the accurately known saturation magnetization of nickel.

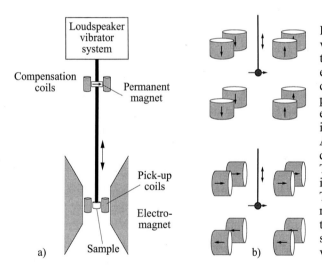

Fig. 4.4 The principle of the vibrating sample magnetometer (a). The sample is oscillated by means of a loudspeaker coil. The signal induced in the pick-up coils is compared electronically with the voltage induced in the reference coils. Alternatively, a variable capacitor setup can be used here. The sample can be enclosed in heating or cooling stages. Two alternate pick-up coil arrangements are shown in (b), the arrows indicating the sensitivity directions of the individual coils

The VSM is the best instrument to measure the saturation magnetization in any ferromagnetic material, because the measurement is not influenced by demagnetizing effects and also because high fields can be applied to reach saturation. As the method is very sensitive, it is usually possible to reduce the size of the sample, so that it is small compared to the dimensions of the pick-up coil arrangement. Then only the dipolar moment of the sample is registered and the effects of higher moments are suppressed. The apparatus is less useful for the measurement of other parameters of the magnetization curve in soft magnetic materials because of demagnetization effects — as for the magnetometric method. For high-anisotropy or hard magnetic materials, however, the VSM is the preferred instrument for many kinds of magnetic measurements.

A VSM can be enhanced by additional pick-up coils that record the magnetization component *perpendicular* to the field direction [809, 810]. According to the torque balance equation (4.2) this component $J_\perp = J_s \sin(\eta_h - \varphi)$ measures directly the torque function $T_m(\eta_h) = H_{ex} J_\perp$. Such a device is therefore equivalent to a torquemeter, permitting — with suitable single-crystal samples — precision anisotropy measurements as discussed in Sect. 4.2.1. Instead of using extra coils, transverse magnetization components can also be measured by combining a standard coil set as in Fig. 4.4b in different ways.

A variant of the VSM measures the field with a superconducting quantum interferometer (SQUID). The SQUID counts the number of flux quanta through the pick-up coil, thus providing a high-sensitivity static fluxmeter. In this magnetometer a set of superconducting pick-up windings is connected in opposition, so that uniform external fields do not contribute to the net current. When a sample vibrates between the windings, a current is induced and transferred to the SQUID circuit placed outside the range of the magnet. The SQUID magnetometer [811] is more sensitive than regular VSMs, but it is more susceptible to external perturbations that may cause spurious signals. Thermal or mechanical pulses may unpin flux lines in the superconducting material and have to be suppressed by careful design.

4.3.4 Optical Magnetometers

The magneto-optical rotation effects discussed in Sect. 2.3 are linear functions of the magnetization and therefore in principle well-suited for magnetometry. In bulk, opaque samples optical methods are not suited for our purpose because the light scans the sample surface only. Surface domains differ in general drastically from interior domains. Measuring the surface magnetization curve cannot add to the knowledge of basic material parameters for this reason and is therefore rarely useful in material characterization. This restriction is not valid for transparent materials for which the average magnetization curve can be measured optically by the Faraday effect.

For non-transparent materials optical magnetometry makes sense only for thin films for which the surface magnetization is representative for the interior. For thin films the optical method has many advantages: it is direct, it can be fast, but also quasi-static measurements can be performed. Space-resolved measurements are possible by scanning over the surface. And optical measurements can be performed on-line during preparation or treatment of a material for example inside a vacuum chamber.

Optical magnetometers may consist of a Kerr setup as in Fig. 2.14a, equipped with an optical detector instead of the camera. Laser light sources or high-intensity light-emitting diodes [812] are preferred to arc lamps because of their better stability. A useful magnetometer that is able to detect even weak signals needs some means of suppressing non-magnetic noise. One way to achieve this is to feed a split-off part of the laser light as a reference signal into the amplifier. If the polarization of the light is modulated by a spinning analyser or an electro-optical device [813–815], the magnetic signal can be detected by a lock-in amplifier, thus achieving virtually unlimited sensitivity. For no apparent reason many authors prefer to use for Kerr magnetometry the acronym MOKE (Magneto-Optical Kerr Effect).

Optical magnetometry is also able to measure directly the frequency-dependent magnetic permeability by alternating field and lock-in techniques. Transverse permeability measurements (in which the alternating field is applied at a right angle to a static field) can be useful in determining anisotropy constants via critical points in the magnetization curve [610]. The connection between critical points and anisotropy was elaborated in Sect. 3.4.4. Plotting the inverse susceptibility as a function of the static field offers even a general method of measuring anisotropy functionals of thin films as elaborated in [816].

The optical magnetometer usually needs careful calibration and adjustment. The magnitude of the signal depends on the precise settings of the polarizer and analyser and on surface conditions, i.e. on the presence or absence of surface layers. The safest procedure is therefore to normalize the optical signal with the saturation signal, taking the saturation magnetization from other measurements. Instead of a signal that becomes constant at saturation, sometimes a superimposed signal that is linear in the applied field is observed. This can be caused by the Faraday effect in the optical components of the magnetometer and is easily corrected.

Another problem lies in the exact dependence of the signal on the magnetization direction. For every setting of polarizer and analyser at oblique incidence a certain in-plane magnetization direction yields the maximum Kerr effect signal, while the perpendicular direction is inactive. The 'sensitivity direction' has to be adjusted to agree with the desired measuring direction. The difficulty of determining the sensitivity direction can be avoided by using the pure *transverse* Kerr effect. The polarizer is set parallel to the plane of incidence and the analyser can be omitted. Then only the component of the magnetization perpendicular to the plane of incidence causes a variation of the reflected intensity, which can be detected electronically.

An advantage of the transverse design is that it fits nicely into a magnet and also into a Kerr microscope [159]. In this way the local variation of material parameters can be measured in a controlled way at a micrometre resolution [206], while observing at the same time the magnetic microstructure. The transverse optical magnetometer also avoids possible influences of magneto-optical effects that are quadratic in the magnetization components and which may lead to asymmetrically distorted magnetization loops [817, 818].

4.3.5 Evaluation of Magnetization Curves

Apart from saturation magnetization, several anisotropic quantities can be determined with more or less accuracy from magnetization curves. This is possible in a relatively direct way for single crystals by comparing magnetization curves along hard and easy directions. Polycrystalline materials need a more careful evaluation before they can yield useful results.

(A) Single-Crystal Measurements. The determination of anisotropy energy terms from the magnetization curves seems to be most straightforward since magnetic anisotropy is defined as the energy differences needed for saturation along different axes. The calculated curves for cubic and uniaxial crystals (Fig. 4.5) demonstrate how to derive directly the anisotropy constants from the differences of the areas above the magnetization curves. The shearing transformation to be applied for finite samples would not change these relations. In practice, however, there are two kinds of complications that limit drastically the usefulness of this method:

(i) Hysteresis effects make the determination of the area ambiguous. Relying on the *idealized* magnetization curve (which is obtained by superimposing an alternating field of decreasing amplitude onto every d.c. field) offers a fair solution to this problem. If it is not practical, one may choose the mean value between the two hysteresis branches to get at least a rough estimate.

(ii) The second problem arises from the non-ideal behaviour in the approach to saturation. Internal stresses, inclusions and shape irregularities lead to a rounding of the magnetization curve. These effects depend on the magnetization direction because the magnetization deviations around such irregularities are influenced by anisotropy. The direct determination of the anisotropy from the magnetization curves is therefore mostly unreliable.

Highly textured materials are used as transformer steels and permanent magnets. Magnetization curves along different axes of such materials can be evaluated in a similar way as single crystals, if either the texture is known from independent measurements, or if the texture can be deduced from measurements along different directions [819, 820].

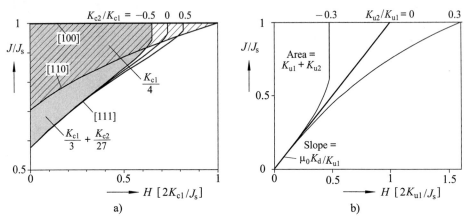

Fig. 4.5 Calculated magnetization curves along simple crystal directions for ideal cubic crystals (a) with positive anisotropy ($K_{c1} > 0$). Anisotropy constants can be derived from the areas above the magnetization curves. The curve along the [111] axis depends on the second anisotropy constant K_{c2} as demonstrated for three examples. Either the curvature of the magnetization curve or the relation between initial slope and area gives access to the second-order anisotropy constant of uniaxial crystal as demonstrated in (b)

(B) The Approach to Saturation in Polycrystals. If the deviation from saturation is measured for an untextured, polycrystalline sample over a wide field range, this deviation can be analysed in terms of inverse powers of the applied field. The anisotropy constants can in principle be derived from the evaluated coefficients [821] as elaborated below. The approach to saturation is, however, influenced by two further processes [503]: the effect of defects and localized inhomogeneities, and the so-called paraprocess, i.e. the suppression of thermal fluctuations of the spin system by the applied field. The latter effect is predicted by spin-wave theory to vary with the inverse square root of the effective field. Usually it is small in moderate fields, provided the temperature does not approach the Curie point.

The contributions from defects like pores or dislocations are complicated in nature. In some model calculations [821] a $1/H$ dependence of the deviation from saturation was derived for this effect in a wide range of moderate to high fields. At very high fields every contribution from defects has to turn into a $1/H^2$ behaviour because otherwise the magnetic energy contained in the deviations would integrate to infinity. Elaborate models lead to a more complicated field dependence [503]. The merit of all these investigations lies in the possibility to extract information on the nature and distribution of lattice defects, and less on the fundamental magnetic properties. These can only be obtained if the effect of anisotropy can be separated from the effects of the lattice defects. Anisotropies that are uniform inside each grain always generate a $1/H^2$ behaviour. If there is an intermediate field range in which this behaviour predominates, one may ascribe it to anisotropy effects. In higher fields the $1/H$ contributions of the defects will take over, which will change again their behaviour — as mentioned — at very large fields.

Provided the contributions of defects and of the paraprocess are sufficiently small in some field interval, the crystal anisotropy constants of polycrystalline materials can be derived from the approach to saturation as detailed here for cubic materials. The average effect of the arbitrarily oriented crystallites on the approach to saturation was calculated by *Akulov* [822]:

$$J \cong J_s \left[1 - \tfrac{2}{105} \left(H_K / H \right)^2 \right] \quad \text{with } H_K = 2K_{c1}/J_s . \tag{4.3}$$

But this formula is only applicable if the grains in the polycrystal would not interact magnetically as elaborated in a careful analysis by Néel [823]. If, in contrast, the grains are close-packed as in a solid body, magnetostatic interactions have drastic effects on the approach to saturation. For large fields the external field H has to be replaced by the larger Lorentz field $H_{Lor} = H + \tfrac{1}{3} J_s/\mu_0$. At moderate fields the effect is less dramatic, but amounts still to a factor of about one half in the $1/H^2$ term. Taking into account these interactions by Néel's theory may yield a reasonable estimate of the first anisotropy constant, although the uncertainties contained in the theory render this approach all but a precision method. Even the sign of the anisotropy constant would remain uncertain because H_K appears quadratic in (4.3). Another limitation arises if the grains are smaller or comparable to the Bloch wall width, so that exchange coupling effects between the grains become important. This difficulty makes it impossible to apply the approach-to-saturation method, for instance, to conventional evaporated thin films in which the grain size is usually comparable with the exchange lengths.

If the measurement of the anhysteretic or idealized *permeability* is possible, this quantity *may* offer useful information. This is not always true, however. In polycrystalline multiaxial samples that are magnetized by wall motion, the ideal initial permeability is determined by the demagnetizing effect only, thus measuring the effective demagnetization factor, but no intrinsic properties of the sample. Polycrystalline *uniaxial* samples show a complicated behaviour governed by the interaction between wall displacements and demagnetizing effects within and between the grains, globally described by an obscure "internal demagnetizing factor". No fundamental information can therefore be derived from anhysteretic magnetization curves of polycrystals.

Idealized permeability measurements are useful if a regime of purely rotational magnetization exists. Then the rotational permeability μ^* (see Sect. 3.2.5F) is directly related to anisotropy (more precisely, to a second derivative of the generalized anisotropy functional). Also *induced* and *magneto-elastic* anisotropies in polycrystalline cubic or amorphous materials can be determined in this way. Small magnetostriction constants in amorphous ribbons can be determined by measuring the magnetization curve along the hard axis under external stress [824].

The limitations discussed in this section apply to the longitudinal magnetization (the one measured along the applied field) and to polycrystalline samples. As mentioned before, the *transverse* magnetization curves of single crystals are equivalent to torque curves, thus offering anisotropy measurements in principle with the same accuracy as with the torque method.

(C) Singular Point Detection in Polycrystals. This method, invented by *Asti* and *Rinaldi* [825], detects singularities in the magnetization curve of a polycrystalline sample which are caused by singular contributions from certain grains. Take the different magnetization curves plotted in Fig. 3.38 for cubic materials. There are field orientations for which the curves are smooth. For other orientations they show characteristic kinks or even jumps (first-order magnetization transitions). In a polycrystalline sample, the kinks and jumps of accordingly oriented grains will show up as singularities (maxima) in the second derivative of the magnetization curve, while the other grains only contribute to a smooth background (Fig. 4.6). The derivative of the magnetization is recorded electronically in pulsed field experiments [826]. If a peak is detected, its field position can be related to the anisotropy field by a detailed theoretical analysis that has been provided by the authors for all the more important crystal symmetries.

The (rather involved) theory of these phenomena starts with the assumption that the grains do not interact with each other. This assumption is rarely valid in magnetic materials, and we have seen in (B) that interaction effects make the evaluation of the approach to saturation of polycrystalline samples nearly pointless because they give rise to corrections which are as large as the effects to be measured. Remarkably, the interactions do not seem to affect strongly the singular point detection method. Experiments on uniaxial single-grain powders of various densities between 1% and 100% showed no effect within experimental error [825], and the same proved to be true for different degrees of orientation of the particles in a sample. It is not quite clear why the method shows so little sensitivity to magnetic interactions, but one explanation could be that grains that are too strongly influenced by unfavourable neighbours simply do not show up in the singularity signal. This may also be the reason why the kinks in the magnetization curves of phase theory, i.e. the transitions between two-phase regions and single-phase regions, etc., have not been recorded in singularity measurements. The observed peaks in polycrystalline samples correspond rather to the true saturation points ("anisotropy fields"), which can be calculated ignoring all domain effects. Domains would always interact with neighbouring grains via exchange and dipolar forces, and their discontinuous transitions may be suppressed in polycrystals.

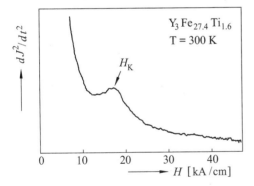

Fig. 4.6 An example of a recorded signal in a singular point detection measurement. (Courtesy *R. Grössinger*, Vienna)

The singular point detection method does not work for all materials. Fine-grained samples, in which exchange interactions between the grains dominate over the anisotropy of the grains (as in microcrystalline evaporated thin films), follow different laws. But the method has become a valuable routine tool for anisotropy measurements particularly in hard magnetic materials [826].

4.4 Resonance Experiments

4.4.1 Overview: Types of Resonance

In typical resonance experiments the sample is saturated along different directions in a strong static field. The field must be strong enough to induce a uniformly magnetized state. It is not necessary that the magnetization direction coincides with the applied field direction. To induce the resonance phenomenon a high-frequency alternating field is superimposed at right angles to the magnetization direction. Because of the alternating field standard resonance methods are not applicable to bulk metallic samples, in contrast to non-conducting oxidic materials, thin films, and powdered materials. Some variants affect only the surface of bulk (metallic) samples. Resonance phenomena in unsaturated samples are less important in connection with the subject of this chapter.

The high-frequency field stimulates a precession of the magnetization vector, which is detected by a suitable pick-up coil. If the precession is uniform in the sense that the dynamic deflection does not depend on the position in the sample, we have *ferromagnetic resonance*. At higher frequencies non-uniform precession modes may be excited. If their wavelength is comparable with the sample size so that sample-shape-dependent demagnetization effects are important, one speaks of *magnetostatic modes*. Resonance phenomena at shorter wavelengths depend on the exchange stiffness constant and are called *spin-wave modes*. Also the domain walls themselves can be excited into *wall resonance*. Finally, there are surface modes of magnetic resonance which can be excited for example by light instead of an alternating field. Then spectroscopy replaces the standard inductive detection methods.

4.4.2 Theory of Ferromagnetic Resonance

The dynamic micromagnetic equations (3.66) lead in a certain uniform field H_{res} to a resonant precession at a frequency given by $\omega_{res} = \gamma\, H_{res}$. The static field H_{res} at which resonance is observed is a measure for the stability of the equilibrium state which is tested by the resonance experiment. For small deviations from equilibrium and spatially uniform precession the effective resonance field can be derived from the generalized anisotropy functional $G(\vartheta, \varphi)$ by the formula [827]:

$$H_{res} = \frac{1}{J_s \cos\vartheta} \sqrt{\frac{\partial^2 G}{\partial\vartheta^2}\frac{\partial^2 G}{\partial\varphi^2} - \left(\frac{\partial^2 G}{\partial\varphi\,\partial\vartheta}\right)^2} \,. \tag{4.4}$$

Here the function $G(\vartheta,\varphi)$ includes — as in domain wall theory (Sect. 3.6.1B) — the external field energy, the demagnetizing energy and the various anisotropy contributions. Take as a simple example a uniaxial

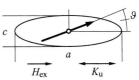

and ellipsoidal crystal with the anisotropy axis and the direction of the static applied field coinciding with the a-axis of the ellipsoid. In a polar coordinate system oriented along the c-axis (the sketch shows a cross-section with the definition of the angle ϑ only) the generalized anisotropy function becomes:

$$G(\vartheta,\varphi) = K_u(1 - \cos^2\vartheta\,\cos^2\varphi) - H_{ex}J_s\,\cos\vartheta\,\cos\varphi$$
$$+ K_d\left[\cos^2\vartheta\,(N_a\cos^2\varphi + N_b\sin^2\varphi) + N_c\sin^2\vartheta\right]\,. \tag{4.5}$$

Inserting the derivatives of this function around $\varphi = \vartheta = 0$ into (4.4) we obtain:

$$H_{res} = \sqrt{\left[\frac{2K_u}{J_s} + H_{ex} + \frac{J_s}{\mu_0}(N_b - N_a)\right]\left[\frac{2K_u}{J_s} + H_{ex} + \frac{J_s}{\mu_0}(N_c - N_a)\right]}\,. \tag{4.6}$$

Here N_a, N_b and N_c are the demagnetizing factors along the a-, b- and c-axes, respectively. For an ellipsoid of revolution ($N_b = N_c$) the effective resonance field is the sum of the anisotropy field $H_K = 2K_u/J_s$, the static external field H_{ex} and the demagnetizing field $H_d = J_s(N_b - N_a)/\mu_0$. Thus the anisotropy can be determined if the saturation magnetization J_s, the demagnetizing factors N and the gyromagnetic ratio γ are known. For other geometries see [828, 829].

From measurements along different axes of the sample and at different values of the applied field, not only the anisotropy constants, but also the saturation magnetization as well as the gyromagnetic ratio can be derived. The intrinsic damping constant α_G in (3.66a) is related to the resonance linewidth if the sample is sufficiently uniform and non-conducting. In most resonance experiments the measurement along a few symmetric directions supplies sufficient information to derive the anisotropy constants in question. But if an unknown anisotropy functional has to be determined, measurements along a continuous range of angles may become necessary. Then corrections are needed to account for the difference between the field and the magnetization direction. The corrections can be done iteratively, but this path is rarely chosen because one would lose the main advantage of resonance techniques in providing a rapid, rather accurate and direct method of material characterization.

In principle, anisotropy properties may show dispersion and depend on the measuring frequency. Resonance methods measure anisotropy constants at microwave frequencies, not quasi-statically as most other methods. This aspect is usually ignored, however.

Increased sensitivity in anisotropy measurements is achieved if the applied field is modulated by an alternating field to record effectively the derivative of the resonance signal with respect to the field. Resonance measurements of anisotropies can even detect inhomogeneities in a sample. If, for example, a thin film system consists of several layers with different magnetic properties, the resulting spectrum of resonance peaks can be analysed accordingly.

Magneto-elastic coefficients are derived from the peak shift under elastic deformation. Alternatively, the resonance can be modulated by the application of ultrasound. An advantage of the ultrasound technique is that more different stress components can be applied than under static conditions [830].

4.4.3 Spin Wave Resonance

Non-uniform magnetic resonance modes are comparable to vibrational modes of an elastic body. They are an interesting object of study, but only rarely suited to the derivation of fundamental quantities.

No demagnetizing effects are expected for a thin film with the static field perpendicular and the exciting field parallel to the surface. One gets a series of spin modes comparable to the vibrations of a string [831−833]. The excitation and exact positions of the resonances depend on the boundary condition (3.67) and thus on surface anisotropy, which is often not well-known. In practice the modes usually appear to be *pinned* at the surfaces, however. This observation has been attributed to a thin surface zone that possesses a different saturation magnetization, or a different effective anisotropy. If the bulk is in resonance, the surface zone may be out of resonance, thus leading to an effective dynamical pinning. *Wigen* et al. [834] confirmed this concept with the help of experiments at different angles of the magnetic field.

For surface pinning the effective field for resonance depends on the order n of the spin waves according to:

$$H_n = H_0 - (2A/J_s)(\pi^2 n^2 / D^2) \quad , \quad n = 1, 3, 5 \ldots \ . \tag{4.7}$$

Here H_0 is the resonance field for uniform rotation and D is the film thickness. In a uniform field only the odd modes are excited, as long as the two surfaces are equivalent. The sequence of peaks is primarily determined by the exchange stiffness constant A. The spin wave resonance method is an important method for the direct measurement of this quantity, and in particular of its temperature dependence. Unfortunately, it can only be applied to thin films, not to bulk media, and the films must be of good quality to get useful spectra.

4.4.4 Light Scattering Experiments

Since light is absorbed in a magnetic metal in a surface layer of some 10 nm, it can serve in exciting magnetic surface spin waves. These resonance effects are comparable to ordinary water waves. Their excitation results in an energy loss of the light wave that may be detected spectroscopically in the scattered light.

This is done in highly specialized experiments in which a laser beam is directed on a surface. At right angles to the reflected beam weak scattered satellites to the primary line can be detected in a spectrometer. The magnetic origin of these satellites follows from the variation of their position in the applied field. They represent the generation and absorption of surface spin waves, and by comparison with a detailed theory quantities like the exchange stiffness constant and surface anisotropies can be derived. An advantage of this method is that it offers local information from the illuminated area, not just global information. Also, the sample needs not be in a state of overall saturation to study optical scattering if only the illuminated area is uniformly magnetized. For details of this powerful method — known as Brillouin light scattering — see the extensive original literature [835–840].

4.4.5 Wall and Domain Resonance Effects

Magnetic resonance phenomena are not restricted to the oscillations of the almost saturated state. Any domain pattern in static equilibrium can be excited into oscillations where the second derivative of the total energy with respect to the oscillatory mode acts as the spring constant. The inertia is represented by the wall mass discussed in Sect. 3.6.6D.

Domain and wall resonance phenomena are rarely evaluated in terms of basic constants since the damping mechanism and the wall mass are all usually complicated functions of the particular sample geometry, domain and wall structure. The main disturbing influence comes from Bloch lines, the number and arrangement of which are mostly unknown. When a wall is set in motion, its Bloch lines move more rapidly if they are not pinned, determining completely the kinetics of the wall. Such phenomena have been studied in particular in connection with bubble films [49].

4.5 Magnetostriction Measurements

The saturation magnetostriction constants can be determined both directly or indirectly: direct measurements evaluate the elongation of a magnet depending on the magnetization direction. In indirect methods the stress sensitivity of a suitable magnetic property is analysed. For the connection between the two approaches see Sect. 3.2.6E. Direct measurements are preferred for bulk samples of sufficient size. The indirect methods are more suitable for thin films and wires. The need to obtain precise magnetostriction parameters especially on thin films led to the development of remarkable direct techniques also for thin films, however. A review on classical magnetostriction measurement techniques was presented by *Lee* in [789]. Only the magnetization-dependent *anisotropic* magnetostriction is considered in this chapter. The temperature-dependent "volume magnetostriction" is ignored.

4.5.1 Indirect Magnetostriction Measurements

Several indirect methods for the measurement of magnetostriction constants have been mentioned before: magnetization curve and resonance measurements, if performed as a function of external stress, can be evaluated in terms of the magneto-elastic coefficients. A special indirect method, the small angle magnetization rotation method, proved particularly useful for bulk metallic glasses [841, 842]. In this method the ribbon-shaped sample is subjected simultaneously to a magnetic field and to a mechanical stress, both along the ribbon axis. The field is large enough to saturate the sample. An alternating transverse field generated by a current through the sample causes a small-angle oscillation of the magnetization, which is detected as a second harmonic signal in a pick-up coil. The longitudinal field and the stress are changed simultaneously, so that the measured signal remains constant. From the field/stress ratio the magnetostriction parameter can be derived. The method is sensitive enough to measure magnetostriction values in the 10^{-9} range. A special advantage is the possibility to measure magnetostriction rapidly, even as a function of an applied stress, during a creeping process or at elevated temperatures.

Interesting details about the magnetostrictive properties of thin films were obtained in [843] relying on an indirect method. These authors employed an optical magnetometer and measured the stress anisotropy induced by bending as a function of the magnetization direction using the magnetization energy method indicated in Fig. 4.5.

4.5.2 Direct Measurements: General Procedures

Every direct determination of a magnetostriction constant requires the measurement of a change in length between two different saturated states — usually parallel and perpendicular to the measuring direction. An experiment with a field applied along only one axis of a crystal yields no useful information on the magnetostriction constants, because then only some arbitrary demagnetized state and the saturated states can be compared. As explained in Fig. 2.55, this procedure would measure the phase volumes in the demagnetized state, not the magnetostriction constants.

Different magnetostriction constants of a crystal call for measurements of the elongation along different axes (in addition to rotating the field in each case). For example, if in a cubic crystal the measurement is performed along [100], the relative change in length between the states at saturation parallel and perpendicular to this axis is $\Delta \frac{\delta l}{l} = \frac{3}{2} \lambda_{100}$ as can be derived from (3.47). Likewise, a measurement of the elongation along [111] yields $\Delta \frac{\delta l}{l} = \frac{3}{2} \lambda_{111}$. For a complete characterization of a material several differently oriented crystals may be needed.

In polycrystals only the average saturation magnetostriction λ_s can directly be determined. This value depends on texture and also on the anisotropic elastic moduli and is related in a complicated fashion to the single-crystal values. For cubic crystals and random orientation of the crystallites the following formula can be considered a good approximation [844, 845]:

$$\lambda_s = c \lambda_{100} + (1-c) \lambda_{111} \; , \; c = \frac{2}{5} - \frac{1}{8} \ln r_a \; , \; r_a = 2c_{44}/(c_{11} - c_{44}) = C_3/C_2 \; . \quad (4.8)$$

Here r_a is a measure of elastic anisotropy of the cubic material. Some textbooks still use the simpler expression with $c = \frac{2}{5}$, which is valid only for elastic isotropy $r_a = 1$. The more adequate formula (4.8) takes into account that in a polycrystal one may not simply average the strains λ_{100} and λ_{111}. It would be likewise incorrect to average the magnetostrictive stresses $C_2 \lambda_{100}$ and $C_3 \lambda_{111}$. The true mean value lies between the two possibilities. This kind of correct averaging has not yet been extended to non-saturated states. If it were available, the magnetization-dependence of the magnetostriction could, taken together with phase theory (Sect. 3.4), lead to a measurement of the anisotropic magnetostriction coefficients in a polycrystal: take an iron-like material. In the initial stage of the magnetization curve only the magnetostriction constant λ_{100} will be active because all domains will be magnetized along one of the $\langle 100 \rangle$ easy directions. Therefore this constant can be derived by comparing the magnetostrictive strains of zero internal field states for different external field directions.

The other constant λ_{111} can then be obtained from the average saturation magnetostriction constant λ_s and its theoretical estimate (4.8). A material with randomly oriented grains is required and the grain size must be large enough to make grain boundary effects unimportant.

The order of magnitude of magnetostrictive strains is 10^{-5} in most materials. It is thus comparable to the thermal expansion of solid materials in one degree. Useful magnetostriction measurements therefore require a temperature stabilization of the sample setup to at least $0.01\,\mathrm{K}$.

Since direct magnetostriction measurements have to be performed in saturation, a correction, which is called the *form effect*, cannot always be neglected. It stems from the demagnetizing energy (3.24) in which the demagnetizing factor depends on the sample shape. Conversely, a saturated sample tends to elongate along the field direction to reduce the demagnetizing energy in addition to the regular magneto-elastic effect.

The smaller the demagnetizing factor of a sample, the smaller is the form effect. This is why thin disks are easier to evaluate for magnetostriction constants than thick samples or spheres. In any case, the form effect can be calculated and subtracted. *Gersdorf* [846] compared rigorous calculations with the simplified assumption that the strain is uniform. This approach proved to be usually sufficient. It can be calculated for ellipsoids in the following way: for a general ellipsoid with the dimensions a, b, and c, the form-effect *stress* along the a-axis is given by:

$$\sigma_{aa}^{form} = -K_d \left[m_a^2 \left(N_a - \frac{\partial N_a}{\partial a} \right) + m_b^2 \left(N_b - \frac{\partial N_b}{\partial a} \right) + m_c^2 \left(N_c - \frac{\partial N_c}{\partial a} \right) \right] \qquad (4.9)$$

with analogous expressions for the other axes. The shear stresses σ_{ab}^{form}, etc., are zero in this coordinate system. Here K_d is the stray field energy constant (3.21), N_a is the demagnetizing factor along the a-axis (3.23) and m_a is the direction cosine of the magnetization along this axis. From the stresses the strains can be derived using Hooke's law. Inserting numbers into (4.9) we see that the form effect is usually negligible if the demagnetizing factors of a disk-shaped sample parallel to the disk plane are small compared to unity.

4.5.3 Techniques of Elongation Measurements

(A) Strain Gauges. Among the different available techniques strain gauges offer the advantage that they can be applied locally on a favourable small region of a single crystal or even on a grain in a coarse-grained sample. Strain gauges with an active area down to 1 mm^2 are available. They consist of a plastic foil on which structured metal films act as sensors, and they are mounted with a thin layer of a cyanacrylate-type adhesive. For bulk samples (including thin sheets) the reaction forces from the strain gauge can usually be neglected. It is always advisable to apply strain gauges on both sides of a sample, and to connect them in series to avoid an influence from sample bending induced by one-sided heating by the measuring current.

The gauge factor of a strain gauge, i.e. the ratio between the resistance change and the strain, is a material property that is supplied by the manufacturer as measured by bending experiments on calibrated elastic beams. *Lee* (in [789]) pointed to a subtle difficulty in connection with magnetostriction measurements: as the calibration is done on a conventional material with a transverse contraction (Poisson's ratio) of typically about 0.3, the gauge factor need not necessarily be correct for *magnetostrictive* strains that are characterized by a transverse contraction ratio of 0.5 as can be seen from the formula for cubic materials (3.44). The difference is probably small. To take it into account would require calibration experiments under biaxial stress.

Strain gauges are evaluated with a carrier frequency bridge that measures the changes in resistance connected with the elongation. Magnetostrictive effects of the order of 10^{-7} can be readily measured. Usually it is sufficient to use an element on an equivalent sample, which is not magnetized, as the balance in the bridge. This procedure also suppresses the effects of the unavoidable heating of the sample by the strain gauge.

Metallic strain gauges are preferable to the more sensitive semiconductor strain gauges for their better temperature stability. It is important that the sensor material is non-magnetic at the measuring temperature. Strain gauges and adhesives are available for high and low temperatures. Suitable mineral substrates and cements lead to a somewhat thicker total measuring package. In this case thicker samples should be used also to minimize reaction forces. Of course, the gauge factor may depend on temperature.

(B) Dilatometers. With non-contact dilatometric measurements one avoids many of the difficulties of the strain gauge technique only to meet other problems. A large variety of direct strain measuring techniques has been employed. With normal samples of centimetre dimension a resolution in the nanometre range is required, which can be realized with capacitive sensors and optical interferometers. Modern fibre-optical techniques offer a particularly versatile solution.

The difficulties of the dilatometric techniques lie in the necessary uniformity of strain and temperature if the measurement is performed at the sample edges. The sample should be structurally uniform and ellipsoidal because a uniform magnetization and dilatation is difficult to be achieved otherwise. An advantage of dilatometry is that there is virtually no limit in sensitivity. Even tunnelling sensors as they are used in the scanning tunnelling microscope can be used for magnetostriction measurements. With increased sensitivity, measurements on smaller samples become possible, thus reducing the uniformity problems. For dilatometric measurements on small spheres all the mentioned difficulties disappear and precise measurements can be obtained [847]. However, in this case the form effect (4.9) may have to be taken into account.

Especially for thin films there is a convenient direct method, that was introduced by *Klokholm* [848]. Here the bending of a sample-substrate-composite in a magnetic field is not avoided but detected as the primary measuring signal. In a modern variant [849] the sample is exposed to a periodically rotating field parallel to the sample plane. The magnetostrictive bending is detected optically by a reflected laser beam, a quadrant detector and lock-in techniques. If the rotating field is large enough to saturate the sample, an accurate determination of the magnetostriction constant λ can be obtained: Take a cantilever of length L, which is mounted on one end and covered on one side with the magnetic film. Properly applied theory of elasticity [850, 851] yields the deflection difference d_f of the free end of the cantilever for longitudinal and transverse magnetization. The result for a freely deformable polycrystalline (elastically isotropic) substrate material is:

$$d_f = d_\parallel - d_\perp = \frac{3 D_f L^2}{D_s^2} \frac{E_f (1 + v_s)}{E_s (1 + v_f)} \cdot \tfrac{3}{2} \lambda_s \sin^2 \vartheta \qquad (4.10)$$

where the Ds are the thicknesses, the Es and vs are Young's moduli and Poisson's ratios of the film and the substrate, respectively, and ϑ is the  magnetization orientation angle ($\vartheta = 0$ for magnetization along the cantilever). The slope at the end of the cantilever $2 d_f / L$ can be detected by the laser beam. The advantage of this direct technique of measuring the magnetostriction of thin films is that it is completely independent of other magnetic properties and of the initial magnetic state of the sample — in contrast to the indirect procedures of measuring stress-induced anisotropies, which have traditionally been preferred for thin films. *Watts* et al. [852] and *Marcus* [853] extended the elastic analysis to the case in which the cantilever is mounted rigidly at one end.

For single-crystal films (such as garnet films) sensitive X-ray diffraction techniques can be used. The change in lattice constant perpendicular to the film surface as measured in a double crystal diffractometer is recorded when the external field is rotated. Films of different crystal orientation are used to determine different magnetostriction constants [854].

In conclusion, a universal method for the measurement of magnetostriction for all kinds of materials and sample shapes does not exist. Thin films can be measured directly by bending experiments, or indirectly by magnetometric or resonance experiments. The strain gauge technique is a reasonable choice for most bulk samples. Especially for thin foils dilatometric methods may be the best alternative, but with sufficient care dilatometric methods can also achieve precision results.

4.6 Domain Methods

Under favourable circumstances material constants may be derived directly from observed domains. Such an approach requires an equilibrium situation for which a reliable theoretical treatment is possible.

4.6.1 Suitable Domain Patterns

To qualify as a suitable domain structure for a quantitative evaluation, a pattern must fulfil a number of requirements:
- It must be sufficiently simple to permit a complete analysis.
- It must contain at least one feature (such as an angle or a characteristic length) which can reach its optimum value independent of its environment.
- This feature must be sensitive to an interesting material parameter.

Let's look at some examples. Certainly branched structures (Sect. 3.7.5) are too complicated as a whole for a quantitative analysis. But the surface domain width in such patterns is well-defined by an equilibrium between wall and closure energy terms, independent of the overall configuration.

Similarly, the maze domain pattern observed in uniaxial films with perpendicular anisotropy (Fig. 1.1c) develops all kinds of random configurations depending on field history and the number and distribution of nuclei. But the band domain *width* is largely independent of the overall pattern and regularly used for the characterization of such films.

In regular soft magnetic materials the domain walls usually form a complicated network consisting of a wide range of large and small domains. In this
situation domain methods for material characterization are hardly applicable.
Only specially prepared samples with independent domain walls can be useful
in such materials.

As an example for a property that is *invariant* against material parameters
look at the period of dense stripe domains in low-Q thin films (Sect. 3.7.2A).
According to Fig. 3.109b the stripe width is just equal to the film thickness
for small fields and small Q — independent of any material parameter. In
contrast, the critical thickness for stripe formation (3.193) is directly proportional
to the wall width parameter $\sqrt{A/K_u}$.

It is difficult to outline the range of suitable domain patterns. The usefulness
of domain methods for material characterization depends on the progress in
domain and micromagnetic theory, on the knowledge available about the
sample and on the precision of experimental observation techniques. In the
following we give some examples of successful applications of domain methods.

4.6.2 Band Domain Width in Bubble Materials

A bubble material is usually characterized by its saturation magnetization J_s,
its quality factor Q, its film thickness D and the equilibrium width W of the
band domains in the demagnetized state. This fact indicates that in bubble
technology a domain measurement constitutes a reliable standard characterization method [855]. From the band domain width W the specific wall energy
γ_w and indirectly the exchange constant A and the characteristic length
$l_c = \frac{1}{2}\gamma_w/K_d$ of the bubble material are derived ($K_d = \frac{1}{2}J_s^2/\mu_0$).

Experimentally, parallel bands instead of the maze pattern are generated
by applying suitably tilted fields. The field is modulated with decreasing
amplitude until a well-defined equilibrium state is reached at zero field. The
evaluation is based on the *Kooy* and *Enz* formula (3.41) (with $m = 0$ and
$\Theta = 0$). The total energy of a band pattern, consisting of this expression plus
the wall energy per unit area of the film $\gamma_w D/W$, is minimized with respect to
the domain period $P = 2W$, which leads to:

$$\frac{\gamma_w}{2K_d D} = \frac{l_c}{D} = \frac{2P^2}{\pi^3 D^2}\sum_{n=1,3..}^{\infty}\frac{1}{n^3}\left[\frac{\sinh x}{\sinh x + \sqrt{\mu^*}\cosh x} - \frac{x\sqrt{\mu^*}}{\left[\sinh x + \sqrt{\mu^*}\cosh x\right]^2}\right]$$

with $x = \pi n \sqrt{\mu^*} D/P$, $\mu^* = 1 + K_d/K_u$.   (4.11)

The right hand side of this relation depends only on the reduced period P/D. Equation (4.11) therefore describes implicitly the dependence of the band domain width on the reduced characteristic length l_c/D on the left hand side. Figure 4.7 shows this dependence for different values of the material constant $Q = K_u/K_d$ that determines the rotational permeability $\mu* = 1 + 1/Q$. The theory relies on a Q well larger than one.

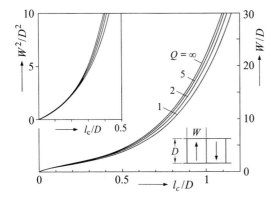

Fig. 4.7 The theoretical relation between the domain width W in the demagnetized state and the characteristic length $l_c = \gamma_w/2K_d$ of a perpendicular-anisotropy film for different values of the material parameter Q

The step from the wall energy to the exchange constant requires a reasonable theory of the twisted domain wall structure of bubble materials (see Fig. 3.84). Alternatively, one may resort to numerical calculations of band domains that have been performed for various values of Q [856] including values close to 1. The tabulated results of this study support a direct evaluation of band domain observations in terms of the nominal wall energy $4\sqrt{AK_u}$.

The direct observation of bubble collapse is also used for material characterization. With some well-known corrections this critical field can yield the saturation magnetization, thus saving a magnetometer measurement [855, 856].

A low wall motion coercivity is a characteristic property of bubble films, so that equilibrium properties of domain patterns can be easily extracted. However, domain investigations can also be applied to high-coercivity films if the effects of coercivity and irreversible behaviour are carefully taken into account. Measurements in alternating fields, at enhanced temperatures or after thermal demagnetization, can be useful. Properly analysed observations of the magnetic microstructure may help to answer fundamental questions even in such materials. We demonstrated in [162] that a single-domain experiment can decide a long-lasting controversy by proving that the most commonly investigated perpendicular recording media have a continuous, exchange-coupled nature rather than a discontinuous, granular character.

4.6.3 Surface Domain Width in Bulk Uniaxial Crystals

One feature of the theory of branched structures in bulk high-anisotropy uniaxial materials (Sect. 3.7.5B) can be exploited for material characterization: for sufficiently thick crystals the surface domain width W_s (Fig. 3.130) is predicted to become constant, independent of the sample dimensions and the basic domain width. The surface domain width depends (for large D) only on the wall energy γ_w and the stray field energy constant K_d. For large Q ($\mu^* \approx 1$) and the simple two-dimensional branching scheme (Fig. 3.127a) it becomes:

$$W_s = 108 \, \gamma_w / K_d \ . \tag{4.12}$$

The proportionality factor will be influenced by the branching mode. Three-dimensional branching, as indicated in Fig. 3.127c and as normally observed in experiments, will reduce the numerical factor because the same amount of stray field reduction can be achieved with less extra wall energy in a three-dimensional pattern. In [768] the factor was determined experimentally by comparison with a simple unbranched plate structure in thin samples which can be evaluated explicitly as shown in the last section. A value of 24.5 ± 2 instead of the factor 108 in (4.12) was obtained there. This calibration was confirmed by an independent measurement on a different material in [857]. The method of determining the wall energy from the surface domain width (see Fig. 3.128) has been accepted as a convenient procedure for the evaluation of high-anisotropy magnetic materials. An arbitrarily shaped piece of material can be examined if it contains a sufficiently large grain oriented roughly perpendicular to the investigated surface. Usually the polar Kerr effect is used, but some high-resolution technique must be employed if the domains are too fine. Magnetic force microscopy, or the Bitter technique evaluated in the scanning electron microscope [98], are useful techniques for this purpose.

4.6.4 Internal Stress Measurement by Domain Experiments

Magnetic domains in low-anisotropy materials are usually strongly influenced by long-range internal stresses. Although usually soft magnetic materials are annealed to remove stresses as far as possible, it may be desirable to measure them locally. The effects of the true internal stresses cancel in linear, integrating measuring techniques (like the torque method) because the spatial average of internal stresses must be zero in homogeneous materials within linear elasticity. There are standard methods for the measurement of internal stresses such as X-ray diffraction methods. But deformations of the order of magnetostrictive strains ($\sim 10^{-5}$), which can hardly be measured by X-rays, can completely reorder a domain pattern in a bulk sample. Therefore domain experiments represent an interesting possibility to measure the weak internal stresses relevant in magnetism.

In such experiments the reactions of domains to external stresses are studied. The internal stress at the observation point is deduced from a series of observations with different external stresses. The result may be influenced by anisotropies of other origin such as growth or field-induced anisotropies, which may be subtracted if they are known from separate global measurements.

The potential of this procedure was demonstrated in [858] on metallic glasses that are elastically isotropic. If the local easy axis lies parallel to the surface, its direction can be seen by moving a 180° wall to the place in question. The direction of a regular 180° wall indicates the preferred axis. To determine the strength of the effective anisotropy, external stresses are applied to rotate the effective easy axis. From the rotation of the wall as a function of the external stress the effective internal stress can be derived. It may be necessary to apply bending or twisting stresses [859] to achieve sufficient rotation when tensile stresses can be applied only parallel to the wall.

4.6.5 Stripe Domain Nucleation and Annihilation

The transition between a stripe-shaped perpendicular domain structure and uniform planar domains in a perpendicular-anisotropy soft magnetic plate as a function of an applied in-plane field has been analysed in Sect. 3.7.2B for samples of arbitrary thickness. A perpendicular easy axis can be induced by local compressive stresses (for positive magnetostriction) as in so-called *stress pattern* in metallic glasses resembling a fingerprint in Kerr images (see Fig. 5.46). If such a pattern is visible on a sample, it can be destroyed by an external field or by a superimposed external stress, thus yielding information about material parameters. (In actual experiments, an additional decreasing alternating field has to be applied, so that the equilibrium pattern can be observed under all conditions). For low-Q materials (like all soft magnetic materials) the critical condition for the nucleation (or annihilation) of stripe domains is given by (3.193). In thick samples such a measurement determines more or less directly the effective perpendicular anisotropy: the field necessary to eliminate the stripes is, according to (3.193), identical with the anisotropy field ($h = 1$ or $H = H_K$) apart from a correction $4\pi^2 A/K_u D^2$, which is small for thick samples. Measuring the critical field at which a stripe domain pattern disappears therefore basically measures the local value of the anisotropy field H_K at the observation point.

In the same sense the stress necessary to eliminate the stripe domains is equal and opposite to the effective stress responsible for the internal stress pattern. Applying the external stress in different directions offers access to different components of the internal stress tensor [859, 860]. The stripes will vanish as soon as one of the principal values of the total planar stress tensor becomes zero instead of negative (for a positive magnetostriction coefficient). For thin films, when the correction becomes important, it is possible to determine the exchange constant from stripe nucleation if the incipient stripe pattern can be resolved and if the anisotropy constant is known.

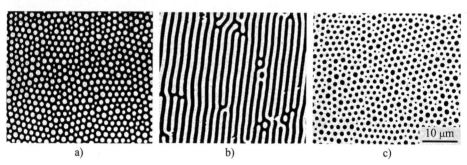

Fig. 4.8 The remanent domain structure of a bubble garnet film after nucleation from slightly different in-plane field directions. The band domain pattern (b) indicates second-order stripe nucleation, which occurs if the field was applied perpendicular to the effective easy direction (vertically, parallel to the bands). In contrast, first-order bubble nucleation (a, c) occurs when the field is tilted by few degrees towards the film normal

In [609] stripe domain nucleation was utilized to determine small extra anisotropy contributions, hardly accessible otherwise. Growth-induced perpendicular anisotropy in a garnet film is determined by the growth direction relative to the crystal axes. If the growth direction deviates slightly from a symmetrical crystal direction (like [111]), deviations from the intended anisotropy function may occur, such as a tilting of the anisotropy axis and an orthorhombical contribution. These deviations can be determined by registering the field direction necessary to induce stripe nucleation as a function of the azimuth of applied in-plane fields. Figure 4.8 shows how the *remanent* domain structure indicates sensitively the field direction at which a second-order stripe nucleation occurred. In this way the easily observable zero-field patterns can be used to determine the details of the effective anisotropy, instead of the nucleation pattern that has a small period and a small amplitude and is therefore very difficult to observe. The procedure was applied and generalized for example in [861].

In another remarkable application of stripe nucleation theory *Yang* and *Muller* [862] derived a *uniaxial* magnetostriction constant in an epitaxial garnet film, a quantity that had not been accessible before. Usually bubble films are characterized by the magnetostriction constants λ_{100} and λ_{111}, and the uniaxial induced anisotropy. The counterpart of the latter quantity, an induced magnetostriction of uniaxial symmetry, is mostly ignored. The analysis of Yang und Muller proved that it exists and that it has physical consequences.

4.6.6 Domain Wall Experiments in Soft Magnetic Materials

It is not easy to perform domain wall experiments in soft magnetic materials in which domain walls usually form complex networks, anchored at sample imperfections, corners or grain boundaries. Carefully prepared, perfect samples are needed to avoid these complications.

An ideal example is the classical *window frame* crystal in which a single wall can be studied independently. But there are also other geometries for which successful experiments that give access to the domain wall energy have been published, as in the following examples.

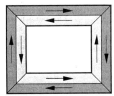

In the first experiment the displacement of a wall stabilized by a weak diffusion after-effect is studied. The width of the potential well (see Fig. 3.100) corresponds to the wall width, and measuring the wall displacement up to the point of break-off thus measures directly this quantity [863].

In another case a simple perpendicular domain wall divides a long silicon iron strip with (010) orientation of the surface into two domains. A transverse current through the strip generates an inhomogeneous

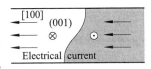

field that induces an *S*-shaped distortion of the wall [864, 865]. A critical field for a transition from a weak to a strong distortion is predicted, which can be observed from both sides of the sample. It is a sensitive measure of the specific wall energy. Alternatively, a dynamic wall bending can be induced in the same geometry by the eddy currents due to a longitudinal alternating field. Again an instability is predicted from a detailed theory. Experimentally it becomes visible as a kink in the velocity-field relation. The wall energy is again derived from the comparison of theory and experiment [866].

In these experiments the wall displacements can be recorded simultaneously for several walls by a magnetometric method, if the number of walls has been counted by direct observation and if the independence and equivalence of the walls has been confirmed. Since it is difficult to verify these conditions, the direct observation of the domain wall reactions is to be preferred.

Difficult is the derivation of the domain wall energy from observed domain periods. The basic domain width can be used only if domain branching (see Sect. 3.7.5) can be excluded. Even after checking for this complication as in [867] the observed domain period may differ from the minimum energy configuration. The experiment was performed on thin film stripes with transverse anisotropy. Characteristically, the observed domain period depended strongly on the direction of a previously applied ac-demagnetizing field. Perhaps the evaluation of surface domain widths on strongly branched domain patterns can be used after suitable calibration experiments also in cubic materials, as elaborated for uniaxial materials in Sect. 4.6.3.

4.6.7 Measurements of Domain Wall Dynamics

Wall mobility experiments on individual walls in insulators give immediate access to the damping constant in the Landau-Lifshitz-Gilbert equation if the wall profile is known [see (3.172)], thus complementing measurements of the resonance linewidth. Particularly for low-damping materials such as yttrium iron garnet, the results of wall mobility experiments and resonance experiments for the phenomenological damping constant differ systematically [49], indicating deviations from the simple Landau-Lifshitz equation. Less useful for basic material characterization are the critical velocity and the turbulent wall motion beyond this point (see Sect. 3.6.6 C), phenomena that are too involved to permit the extraction of basic material constants. In bulk metallic samples wall mobility is determined almost completely by eddy currents. It thus reflects primarily the resistivity, the sample shape and the wall shape (see Sect. 3.6.8), and does not offer any clues on other, fundamental properties.

4.7 Thermal Evaluation of the Exchange Constant

The methods to measure the exchange constant discussed so far are only applicable to specially prepared samples. A universal if somehow indirect approach is the analysis of temperature dependence of the saturation magnetization. This curve can be readily measured by the methods discussed in Sect. 4.3. It reflects the interplay of thermal excitation and exchange interaction, and several of its features give access to the exchange stiffness constant.

4.7.1 The Curie Point

The theory of the Curie point is involved and rarely accurate enough to offer a first-principles determination of the stiffness constant. The statistical treatment of the phenomena near the critical point has to include high-energy excitations of the spin system, which in a classical picture would mean large angles between neighbouring spins. The exchange stiffness constant, however, measures smooth, long-range variations with only small angles between neighbouring spins. This is why the proportionality between the Curie point T_c and the micromagnetic stiffness constant A contains a structure-dependent factor:

$$kT_c = fJ \ , \ J = a_L A/(2S_e^2) \tag{4.13}$$

where a_L is the lattice constant, S_e is the elementary spin quantum number and f is a factor that varies, depending on interaction range and character and lattice type, between 0.1 and 0.6. If for one member of a given material symmetry class and coupling scheme the stiffness constant has been determined by other methods, (4.13) can be used to extend the measurement to related materials.

4.7.2 Molecular Field Theory

The largest part of the $J_s(T)$ curve can be well described by a molecular field theory, such as Weiss' theory of ferromagnetism or Néel's theory of ferrimagnets. When two or more sublattices are coupled, the shape of the curve may differ characteristically from the standard shape of ferromagnets, yielding information about the exchange coupling between the sublattices. For ferrimagnets with two sublattices, the following equations define implicitly the saturation magnetization:

$$J_a(T) = J_a(0) \ B_{S_a}\left[(S_a g_L \mu_B/k T)(N_{aa}J_a + N_{ab}J_b)\right]$$

$$J_b(T) = J_b(0) \ B_{S_b}\left[(S_b g_L \mu_B/k T)(N_{bb}J_b + N_{ab}J_a)\right] \ , \tag{4.14}$$

where B_S is the Brillouin function for spins of magnitude S, g_L is the Landé gyromagnetic ratio, μ_B is Bohr's magneton and the N_{ij} are the molecular field parameters. Solving this system of implicit equations for the two sublattice magnetizations one gets the net magnetization curve that describes the experimental measurements with suitable interaction parameters N_{ij} [868–873]. This leads for spinel-type ferrites to the expression:

$$A = \frac{2g_L\mu_B}{J_s(0)}\frac{2N_{aa}S_a^2 + 4N_{bb}S_b^2 - 11N_{ab}S_aS_b}{16|S_a - 2S_b|}a_L^2 \tag{4.15a}$$

for the stiffness constant. Similarly, for rare-earth iron garnets one obtains:

$$A = \frac{2g_L\mu_B}{J_s(0)}\frac{5(40N_{aa} - 25N_{ad} + 15N_{dd} - 4N_{dc}S_c - 10N_{ac}S_c)}{16(6S_c - 5)}a_L^2 \tag{4.15b}$$

where Fe^{3+} sits on the a- and d-sites and S_c is the spin of the rare earth residing on the dodecahedral c-site.

No clear theoretical prediction exists about the temperature dependence of the exchange stiffness parameter. Certainly the parameter A vanishes at T_c, and a temperature variation of the stiffness parameter following the square of the saturation or sublattice magnetization has been invoked. For a discussion see [49 (p.11–13)].

4.7.3 Low Temperature Variation of Magnetization

Molecular field theory does not predict correctly the approach to absolute saturation at low temperatures. An exponential dependence of the saturation magnetization on the temperature is predicted by the mean field theory, but actually a deviation following a $T^{3/2}$ law is observed. This deviation is caused by long-range cooperative excitations called spin waves, the energy of which is determined mainly by the exchange stiffness constant. A statistical treatment of these excitations for bulk ferromagnets predicts the following law [874]:

$$J_s(T) = J_s(0)\left[1 - (T/\Theta_c)^{3/2}\right] \quad \text{with} \quad k\Theta_c = 13.25\,A(0)\sqrt[3]{g_L\mu_B/J_s(0)}, \tag{4.16}$$

meaning that the exchange stiffness constant at absolute zero $A(0)$ is proportional to the characteristic temperature Θ_c derivable from $J_s(T)$. The cubic root reflects the linear dimension of one elementary magnetic cell. As mentioned in connection with the definition (3.1) of the stiffness constant, this material parameter is also related to the Curie temperature T_c. The relation (4.16) offers a more direct access to the exchange stiffness constant at least at low temperatures because, as mentioned, the low-temperature deviation from absolute saturation $J_s(T) - J_s(0)$ is directly linked to long-range spin waves, which obey the rules of classical, continuum micromagnetics.

4.8 Theoretical Guidelines for Material Constants

Theory has derived predictions for the dependence of magnetic material constants on temperature and constitution that can be quite useful. They are not very reliable but can serve as guidelines when other information is not available.

4.8.1 Temperature Dependence of Anisotropy Parameters

The basic idea is the expectation that the elementary anisotropic spin-lattice interaction is smoothed out with increasing thermal agitation of the spins. Such considerations lead to a relation between the temperature dependence of a certain anisotropy constant and the temperature dependence of the saturation magnetization. These laws — first derived by *Zener* [875, 876] for a simple localized picture of a ferromagnet — can be formulated as follows: let n be the degree of a certain anisotropy energy or magnetostriction term as expressed in terms of the direction cosines of the magnetization. Then this coefficient is predicted to vary with temperature according to $K_n \propto J_s^{n(n+1)/2}$. As an example, for the first-order uniaxial anisotropy K_{u1} we have $n = 2$ according to (3.11); it should therefore vary as J_s^3. For the cubic anisotropy constant K_{c1} the order $n = 4$ (3.9) leads to a variation with J_s^{10}. A more systematic variant of the Zener theory was presented by *Callen* and *Callen* [877]. The result differs from Zener's formula particularly near the Curie point.

   To apply these laws properly, the energy terms should not be analysed in the conventional form as in (3.9) but rather in spherical harmonics that may generate slightly different predictions for the conventional anisotropy coefficients. Also, the magnetostrictive contribution (3.48) in the anisotropy of a free crystal should be separated in such an analysis. With these precautions a fair agreement between predictions and experiments was found. Deviations can occur in connection with lattice transformations as for cobalt [878]. This metal switches at about $420°C$ from the hexagonal to the face-centred cubic crystal structure. The exchange coupling and thus the saturation magnetization is not much affected by the subtle change in the stacking order, but the anisotropic properties are severely influenced. But even apart from such anomalies there are large discrepancies between the predictions and experiments both on metals [879] and on garnet oxides [880]. In any case, the theory describes correctly a trend of anisotropy and magnetostriction constants that depends stronger on temperature than saturation magnetization, and this is particularly so for cubic materials.

4.8.2 Mixing Laws for Magnetic Insulators

The largest contribution to magnetic anisotropy and magnetostriction in insulators stems from the interaction of the spins with the electronic configuration of its own atom (rather than its neighbours). The electronic orbitals are determined by the valence of the ion and by the symmetry of the ionic environment. Global anisotropy and magnetostriction parameters can therefore be predicted from mixing laws, taking into account the distribution of the ion species on the various lattice sites. The contributions of the ions for the different environments in the spinel and garnet crystals have been tabulated [880]. It is thus relatively easy to predict these properties for new compositions.

The method of mixing laws in oxidic materials is reliable within about 10%. Problems may arise occasionally by insufficient knowledge about the ion distributions that depend sensitively on thermal history.

4.8.3 Empirical Rules for Alloy Systems

Metallic alloys are different from magnetic insulators because the metallic interaction is inherently non-local. As a rule, the magnetic parameters of metallic systems have to be measured and cannot be predicted theoretically up to now. The behaviour of anisotropy and magnetostriction as a function of composition is not completely random, however. There is a tendency to show equivalent properties in alloy systems with equivalent total numbers of valence electrons, provided that also the structures stay the same. This tendency is apparent in the famous Bethe-Slater diagram, which orders magnetic moments as a function of the average valence electron number in 3d transition metals.

Anisotropy and magnetostriction seem to obey similar laws as exemplified by the rule of *Rassmann* and *Hofmann* [881]. A nickel-iron base alloy shows a maximum permeability if the cubic anisotropy and the average magnetostriction vanish simultaneously, which is obtained with the following recipe:

- *Take 14.5 at% Fe,*
- *Take other metals, their atomic percentage multiplied with their valence summing up to 19.5 at%,*
- *Fill up with Ni.*

Characteristically, the total number of electrons (percentage × valence) is the only variable in this rule.

5. Domain Observation and Interpretation

In this chapter a systematic overview over observed domain patterns is given, linking them with domain-theoretical concepts. After introducing a classification of magnetic materials from the viewpoint of their magnetic microstructure, the most important among these classes are presented and analysed. Special emphasis is put on the role of domains in reversible and irreversible magnetization processes.

5.1 Classification of Materials and Domains

A better understanding of the wide variability of domain phenomena is possible if a basic classification of magnetic materials with respect to their magnetic microstructure is kept in mind. In developing this classification we lean on domain theory (Chap. 3).

5.1.1 Crystal and Magnetic Symmetry

Like in optics, the many different crystal classes do not lead to as many fundamentally different magnetic materials. A natural classification distinguishes materials according to the manifold of easy directions, the magnetization directions possible in zero field. This criterion leads to the three classes shown in Fig. 5.1 based on the arguments of phase theory: we showed in Sect. 3.4 that the 'polyhedron' of possible average magnetization vectors accessible in zero internal field is spanned by the easy directions of a sample. This polyhedron is a three-dimensional body in the multiaxial class III, degenerates to a plane in the planar class II, and to a line in the uniaxial class I.

In phase theory a multiaxial material (class III) would show a large initial susceptibility, independent of the field direction, even for a polycrystal (the permeability would tend to very large values if there are no non-magnetic gaps between the grains, and if magnetostrictive interactions between the grains are small or negligible). This is the typical property of a soft magnetic material. Materials with a continuous three-dimensional manifold of easy directions, such as isotropic metallic glasses, fine-grained polycrystalline materials, or uniaxial crystals with conical anisotropy, are equivalent in this respect with cubic materials.

Uniaxial materials (class I) are preferred in all applications of the hysteresis properties of magnets, such as in permanent magnets and in magnetic recording. Planar materials (class II) are of marginal practical importance.

	Class	Easy directions	Examples
I	Uniaxial	One easy axis	Hexagonal, orthorhombic, tetragonal crystals with positive anisotropy
II	Planar	Two or more easy axes in a plane	Hexagonal, tetragonal crystals with negative anisotropy
III	Multiaxial	Three or more non-planar easy axes in space	Cubic crystals, metallic glasses, polycrystalline materials with small grains

Fig. 5.1 Classification of magnetic materials according to the manifold of easy directions

5.1.2 Reduced Material Parameters

Among the many dimensionless parameters that can be formed by combining micromagnetic energy coefficients, the ratio between the anisotropy and the stray field energy is the most important. As introduced in Sect. 3.2.5B we call this ratio Q, defined by $Q = K/K_d$, where $K_d = \frac{1}{2}J_s^2/\mu_0$ is the stray field energy coefficient, and K is the effective anisotropy constant, defined as the smallest curvature of the anisotropy functional evaluated around an easy direction. Another quantity related to this parameter is the rotational permeability $\mu^* = 1 + 1/Q$ (see Sect. 3.2.5F).

The material parameter Q determines the applicability of a material in many ways. A soft magnetic material should have a small value $Q \ll 1$. Strong permanent magnets are often based on materials with $Q \gg 1$. Recording materials, which should be both easy to switch and permanent, often have a Q

in the neighbourhood of 1. It has to be stressed, however, that the properties of high and low relative anisotropy are not directly related to magnetic hardness, i.e. to high and low coercivity. Coercivity, as an extrinsic property, is determined mainly by additional structural features such as lattice defects, grain boundaries, sample or particle size, and surface irregularities in conjunction with the magnetic microstructure. The intrinsic Q parameter determines the ease with which a desired magnetic hardness (or softness) can be achieved. In this sense Fig. 5.2 is to be understood.

	Definition	Name	Typical applications
A	$Q \ll 1$	Low anisotropy	Soft magnetic materials
B	$Q \approx 1$	Medium anisotropy	Recording media
C	$Q \gg 1$	High anisotropy	Permanent magnets, Magneto-optic recording

Fig. 5.2 Classification of magnetic materials by the parameter $Q = K/K_d$, which compares the magnetic anisotropy constant K with the stray field energy parameter K_d

5.1.3 Size, Dimension and Surface Orientation

The orientation of the surfaces and the dimensions of a sample have a decisive influence on the character of the domain patterns and thus on the magnetic properties. Particularly drastic changes occur when the sample dimensions become comparable to or smaller than one of the critical sizes listed in Fig. 3.15. The most important characteristic length is the domain wall width parameter $\Delta = \sqrt{A/K}$, where A is the exchange constant and K the anisotropy constant. Depending on the number of dimensions that are comparable to or smaller than Δ, largely different magnetic microstructures are observed. For example, in isotropic small particles that are in all dimensions small compared to the Bloch wall width Δ no regular domain pattern is to be expected as discussed in Sect. 3.3.3. Thin magnetic films are defined as materials thinner than the Bloch wall width. In such samples the domain wall structure and energy are strongly modified as discussed in Sect. 3.6.4, and this has severe consequences for the domain patterns as well. A classification of sample shapes based on these considerations is shown in Fig. 5.3.

Another important aspect is the orientation of the main surfaces with respect to the easy directions. The magnetic microstructure of samples with 'mis-oriented' surfaces, i.e. with surfaces that do not contain an easy direction, differs strongly from samples with only well-oriented surfaces. Polycrystalline bulk materials invariably belong to the class with strong misorientations and complex domain patterns because even if the polished surface of a grain is well-oriented, the internal surfaces of this grain will almost always be strongly misoriented. The aspect of surface orientation is included in Fig. 5.3 as a secondary criterion.

		Definition	Name	Typical applications
1	a	**All dimensions** $\gg \Delta$	Oriented sheets	Transformers
	b		Non-oriented sheets and bulk materials	Rotating machines, yokes
2	a	**Two dimensions** $\gg \Delta$	In-plane thin films	Soft film elements, Longitudinal magnetic recording
	b		Perpendicular	Perpendicular recording, Magneto-optical recording
3		**One dimension** $\gg \Delta$	Thin wires	Sensor elements
			Needle-shaped particles	Particulate recording media, Permanent magnets (low anisotropy materials)
4		**No dimension** $\gg \Delta$	Isometric magnetic particles	Particulate recording media, Permanent magnets (high anisotropy materials) Magnetic liquids, Geomagnetism

Fig. 5.3 Classification of magnetic samples based on their shape and size relative to the Bloch wall width parameter $\Delta = \sqrt{A/K}$. The orientation of the anisotropy axes relative to the main surface serves as a secondary criterion indicated by (a, b), where (a) applies to no or small misorientation and (b) refers to strongly misoriented bulk material and to thin films with perpendicular easy axis

The concept of well-oriented samples makes sense for thin single-crystal sheets and for films or sheets with a grain size that is much larger than the sheet thickness. Some electrical steels and in particular oriented transformer sheets belong to this class. Later we will distinguish between the cases of slight misorientation, in which the basic domain pattern is essentially the same as in ideally oriented samples (Sect. 3.7.1), and that of strong mis-orientation, leading to new, completely different domain arrangements. The considerations of Sect. 3.7.2–5 (stripe, maze, closure domains and branching) apply to strongly misoriented samples.

In the Néel block analysed in Sect. 3.7.6 the domain character changed with the value of an applied field. In zero field the domain pattern of an ideally oriented sample is observed, whereas in large fields the branched patterns of strongly misoriented samples appear on the side surfaces of the crystal. The reason was that in the applied field the easy axis in the side surface becomes disfavoured relative to the other easy directions, which are favoured in the external field but incompatible with the side surfaces. The same phenomenon can be observed in oriented transformer steel if a field is applied perpendicular to the preferred direction of this material. The property of a surface to be 'well-oriented', meaning that it contains an easy direction, may thus be field dependent. Such complications cannot be classified in a rational way and need an *ad hoc* discussion.

5.1.4 Further Aspects and Synopsis

In addition to the outlined intrinsic material properties and geometrical boundary conditions there is an important *extrinsic* aspect: the ability of a material to approach the equilibrium magnetization configuration predicted by domain theory. Hysteresis is a common property of all ferromagnetic materials, but most important in connection with domain analysis is a special aspect of hysteresis, namely the question whether domain wall motion is severely impeded or not. If the irreversibility is due to metastable domain states that differ in global aspects, such as the number of domain walls and domain wall junctions, while the domain walls themselves are free to move into their local equilibrium positions, a discussion in terms of equilibrium domain theory is possible. If, on the other hand, the *wall motion coercivity* is large compared to the effective fields of micromagnetics, no local equilibrium is expected and the observed domain pattern assumes a completely different character, usually much more irregular than in the case of equilibrium structures.

For moderate wall motion coercivities, equilibrium configurations for a given external field may still be approached by superimposing an alternating field of decreasing amplitude, a process called "idealization". Such a procedure is always recommended in domain observation, if possible. Sometimes only very general features, such as preferred domain wall densities in the demagnetized state, are related to equilibrium properties. As far as such characteristic phenomena can be identified, we will treat hard magnetic materials in the following in connection with the corresponding soft materials, concentrating on the equilibrium rather than on the non-equilibrium features. Largely immobile domains with no apparent connection to equilibrium domain theory can be found in materials for permanent magnets and magnetic recording, but also in certain multilayers with an antiferromagnetic coupling type.

Every discussion of a magnetic microstructure has to begin with a proper classification. Without the information needed for this decision no meaningful domain analysis is possible. To get started, the sample geometry, the saturation magnetization, the symmetry and size of the anisotropy, the crystal orientation and the Curie point (to estimate the exchange stiffness parameter) should be known. In addition, the degree of magnetic hardness, i.e. the ratio between the wall motion coercivity and the anisotropy field, should be known.

Even if the proposed classifications with respect to the three intrinsic criteria of Figs. 5.1–3 are accepted as sufficient, there are $3 \times 3 \times 6 = 54$ more or less distinct types of magnetic samples. In addition there are transition forms between the classes: the sample sizes, the relative anisotropy parameter Q, and the domain wall width can assume any positive real number, so that a clear-cut classification based on these quantities is impossible. Also symmetry arguments do not always permit a clear distinction. Consider a cubic material with a weak superimposed induced uniaxial anisotropy. According to the classification scheme of Fig. 5.1, this material would become immediately uniaxial because its easy axes are no more equivalent, irrespective of the size of the superimposed anisotropy. As this makes no sense, this classification should also be understood more as a guideline.

Not all possible classes have found attention in applications and deserve a systematic treatment. In the following the magnetic microstructure of the more important sample classes will be presented and discussed. Starting with uniaxial and multiaxial bulk material, we proceed with samples of reduced dimensions, the most important class of which are magnetic thin films in all their varieties. The chapter closes with a section on the special properties of magnetic samples of lower dimension (needles, wires and particles).

5.2 Bulk High-Anisotropy Uniaxial Materials

The materials to be discussed in this section are the typical raw materials of modern permanent magnets. They become magnetically hard if converted into more or less fine particles, oriented and compacted. As bulk crystals, they show the beautiful branched domain patterns discussed theoretically in Sect. 3.7.5B (Two-Phase Branching).

5.2.1 Branched Domain Patterns

Figure 5.4 shows a series of domain pictures on the basal plane of a cobalt crystal at different sample thicknesses. The images (as most other images in this chapter) were taken with the Kerr technique.

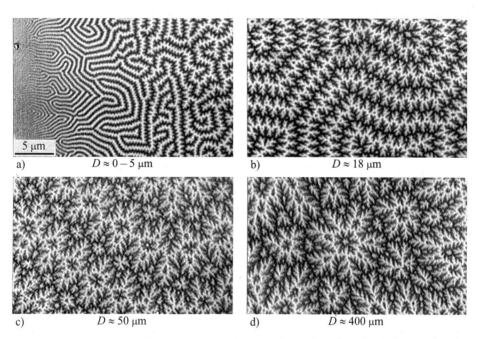

a) $D \approx 0-5 \, \mu m$ b) $D \approx 18 \, \mu m$

c) $D \approx 50 \, \mu m$ d) $D \approx 400 \, \mu m$

Fig. 5.4 Kerr effect observations on the basal plane of a wedge-shaped cobalt crystal. Bulk cobalt has a uniaxial anisotropy at room temperature so that the easy axis is perpendicular to the observation surface. With increasing sample thickness D the degree of branching increases, while the surface domain width stays essentially constant

Remarkable is that even with the Kerr effect that is sensitive to the surface magnetization, also the basic domains of the interior are still recognizable in spite of the increasing complexity of the domain structure with increasing thickness. Other methods, which are sensitive to the stray field connected with the basic domains, such as the powder technique (Fig. 2.7), magnetic force microscopy (Fig. 2.43) or even the obscure effect demonstrated in Fig. 2.53 display the basic domains even more clearly. Evaluating such series of photographs, several authors [211, 766, 882–889] confirmed the characteristic $D^{2/3}$ dependence of the basic domain width (D = sample thickness; see Fig. 3.130). As to the surface domain width, we recognize at least qualitatively its independence of the sample thickness according to (3.227).

The fine structure of the branching pattern of cobalt looks somehow different from that of regular high-anisotropy materials, as can be seen by comparison with branched patterns on NdFeB (Fig. 5.5). The reason is a different anisotropy ratio in both materials: the high anisotropy of NdFeB ($Q \approx 4$) leads to pure two-phase branching as sketched in Fig. 3.127, characterized by rather smooth surface pattern contours. This kind of pattern is found in all kinds of high-anisotropy materials, notably also in hexaferrites [211, 882, 887].

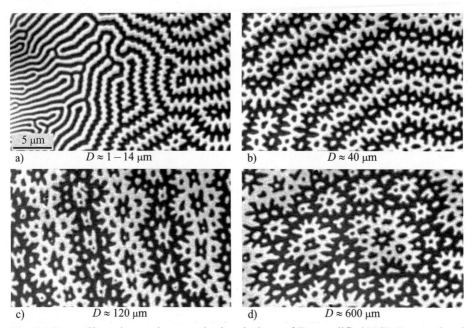

a) $D \approx 1 - 14 \, \mu m$ b) $D \approx 40 \, \mu m$

c) $D \approx 120 \, \mu m$ d) $D \approx 600 \, \mu m$

Fig. 5.5 Kerr effect observations on the basal plane of Dy-modified NdFeB crystals of various thicknesses. Compared to cobalt the surface domain pattern is coarser, but also of a slightly different, less ragged character. (Sample: courtesy *A. Handstein*, IFW Dresden)

In the lower-anisotropy cobalt ($Q \approx 0.4$) the character of the pattern is similar in the macroscopic aspects but different in the fine structure. Here the branching mode in the bulk agrees with that of high-anisotropy materials. The surface domain width is smaller because of the smaller wall energy according to (3.227). But the fine surface pattern also displays a more jagged appearance, that is even more clearly visible in its details in domain observation with higher resolution as shown in Fig. 2.7. This contrasts with the rounded shape of the branching protrusions in high-anisotropy samples such as in Fig. 5.5.

The reason for this difference was not understood before high-resolution imaging with electron polarization methods [890] concentrated on the *in-plane* components of the surface magnetization rather than on the conventional polar components (Fig. 2.39b). Similar results were then also obtained by magneto-optical means [180]. The basic reason for the different pattern of cobalt appears to be the presence of closure domains in the uppermost level of branching (see Fig. 3.121b). Obviously these closure domains are *modulated* in a dense stripe domain pattern, in analogy to the stress patterns in metallic glasses, shown and interpreted in Fig. 3.142. According to these findings, cobalt with its intermediate anisotropy forms a kind of hybrid, following the high-anisotropy two-phase branching scheme in the bulk, and a low-anisotropy multiaxial branching scheme at the surface.

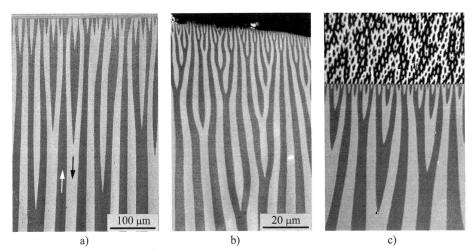

a)   b)   c)

Fig. 5.6 Domains on the side plane of a cobalt crystal (a) and of a slightly misoriented NdFeB crystal (b) demonstrating the process of domain branching towards the sample edge. A NdFeB twin crystal displays the branching process (c), where the twin boundary acts like a mirror for the domain pattern (as in Fig. 2.7). The misorientations of the twins are 13° and 52°, respectively. (Sample: courtesy *A. Handstein*, IFW Dresden)

On the side plane of a uniaxial crystal the branching mechanism can be immediately observed (Fig. 5.6). Although such pictures seem to be intuitively convincing, one should not consider them as undisturbed cross-sections through a branching pattern. The presence of the free side surface changes the energetics of the internal stray field, which will cause some relaxation of the pattern near the surface. Probably the branching "speed" is enhanced near a free surface because the internal stray field energy coefficient F_i in (3.225) will be reduced near a free surface. This becomes particularly apparent in Fig. 5.6b, where the strongly curved domain boundaries observed at the surface indicate domain walls that intersect the observation surface under an acute angle connected with a crystal misorientation. The surface appearance of the domain pattern may thus be quite different from its internal structure even on a lateral crystal surface.

In Fig. 5.6c a twin crystal displays interior and surface domains at the same time as sketched in a cross-sectional view. Because of the specific orientation of the twin pair the polar magnetization com- ponents in the two grains are opposite, explaining the contrast reversal in the Kerr image that can be seen in following domains over the boundary.

A particularly beautiful demonstration of branching is due to *R. Szymczak* [211], who investigated a magnetic oxide crystal with two complementary methods (Fig. 5.7). Such crystals are transparent in the infrared, thus displaying their internal domain pattern by the Faraday effect. The surface domains can be made visible by the Bitter technique. In spite of the striking difference in details, both patterns clearly belong together.

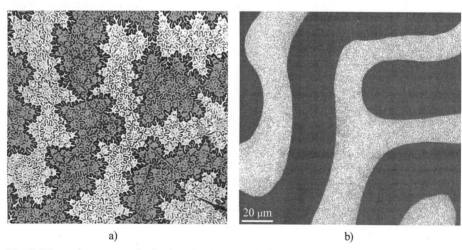

a)   b)

Fig. 5.7 Domain pattern of a barium ferrite crystal plate of 0.8 mm thickness. The surface domain pattern (a) is made visible with the Bitter technique, the basic domains (b) are observed by the Faraday effect in infrared light. The same domain pattern is shown in both views. (Courtesy *R. Szymczak*, Warsaw [211])

A systematic series of branched domain patterns on NdFeB crystals as a function of orientation is shown in Fig. 5.8. Obviously the crystal orientation can be inferred in such coarse-grained samples from a simple inspection of the domains to within 10° — except near 90° where the pattern is insensitive to the orientation angle.

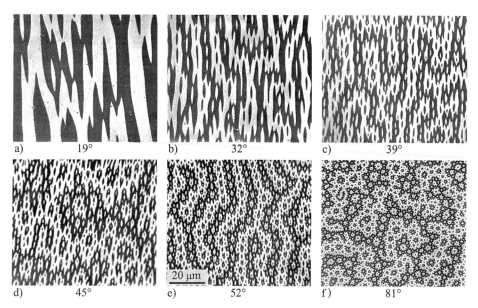

Fig. 5.8 Domain observations on NdFeB (Dy) crystal grains of increasing misorientation. (Sample: courtesy *A. Handstein*, IFW Dresden)

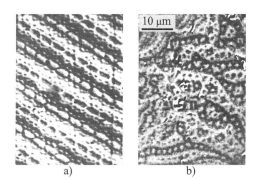

Fig. 5.9 Two-phase branching observed on cubic materials under strong uniaxial stress. (a) Domains near the surface of a iron-nickel meteorite (Fe with ~7% Ni). (b) Domains on an embedded titanomagnetite crystal, a low-magnetization spinel abundant in granite rocks. (Together with *V. Hoffmann*, München; [208])

Figure 5.9 demonstrates two examples of two-phase branching observed on basically cubic materials {an iron-base alloy in (a), a titanium-rich spinel in (b); see also [891]}. In both examples strong internal stresses make the cubic material effectively uniaxial. The different appearance of the flower pattern compared to the one of hexagonal crystals stems from the anisotropy of the magneto-elastic interaction, which produces in general an orthorhombic instead of a uniaxial anisotropy (3.12, 3.52). As in conventional uniaxial materials one axis is favoured in orthorhombical materials, but the other two (hard) axes are not equivalent, which will influence the domain wall energy. Apparently a strong anisotropy of the specific wall energy was present in these crystals.

The branched domain patterns discussed here are specific for uniaxial materials. The transition to planar anisotropy is studied in [892−894]. Examples of domain observations on a planar anisotropy material are shown in Fig. 5.10. On the basal plane (b) the domains display no general preferred direction, although at every point a preferred axis is evident, which is apparently due to internal stresses. The layered structure observable on lateral planes (a) indicates that in such materials the domain patterns are mainly closed within the basal planes and less correlated along the axis.

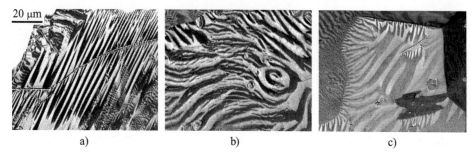

Fig. 5.10 Domains observed on planar-anisotropy Sm_2Fe_{17}. In addition to the easy-plane anisotropy, the observed domains are strongly influenced by microstructural stresses. (a) Side plane, displaying domains extending along the basal plane, as well as finely divided "stress patterns", in which local stress favours a perpendicular magnetization. (b) Domains on a somewhat misoriented basal plane, showing some branching subdivisions. (c) Domains on the basal plane, with strong stress-induced branching near the grain boundaries. (Sample: courtesy *A. Handstein*, IFW Dresden)

5.2.2 Applied Field Effects

(A) Field Parallel to the Easy Axis. The beautiful domains that can be observed on the basal plane of bulk uniaxial crystals under the action of a field parallel to the axis have never been analysed quantitatively. The picture series in Fig. 5.11, taken on barium ferrite, shows the pattern evolution depending on magnetic history. At corresponding fields we observe either a pattern of disjunct dendritic elements if coming from saturation (a), or of a branched network (c) if coming from the demagnetized state (b). A similar picture for a cobalt crystal was shown in Fig. 2.9a. Remarkable is a strong deviation of the measured bulk magnetization from the observed surface magnetization. Apparently, closure domains, as they were discussed before in Sect. 5.2.1, are used to spread the flux evenly over most of the surface, thus leading to very narrow or vanishing opposite domains at the surface and avoiding strong surface charges. A model of this mechanism is shown in Fig. 5.12a, which is at least compatible with observation (b,c).

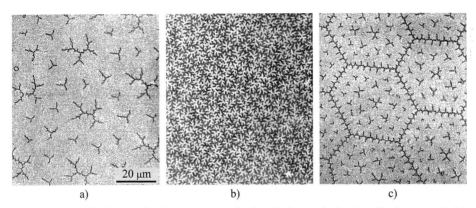

Fig. 5.11 Domain hysteresis phenomena on the basal plane of a barium ferrite crystal: (a) after nucleation starting from saturation, (b) at remanence, (c) at the same field as in (a) but starting from the demagnetized state. The topology of the domain pattern depends strongly on magnetic history. (Sample: courtesy *L. Jahn*, TU Dresden)

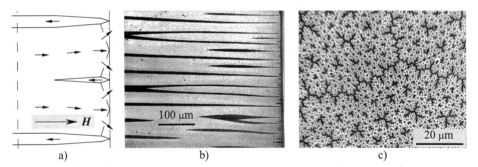

Fig. 5.12 (a) A possible magnetization pattern in a cobalt crystal magnetized 80% of saturation along the easy axis and perpendicular to the surface. The scheme explains how a uniform flux emergence from the surface is possible at a moderate expense of crystal anisotropy energy. This may be compared with observations on a side plane (b) and on the basal plane (c). Although the black domains occupy only a small fraction of the surface area, the crystal is only magnetized in this state to 60% of saturation

(B) Field Parallel to the Basal Plane. Up to this point, only the magnetostatic energy and the 180° wall energy determined the domain structure. In a strong field perpendicular to the easy axis the wall angle gets smaller than 180°. This introduces two new aspects: (i) the wall energy decreases and becomes orientation dependent (Fig. 3.62); (ii) magnetostrictive interactions between the basic domains become important.

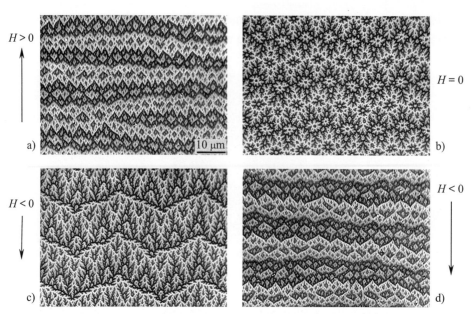

$H > 0$

$H = 0$

a)

$H < 0$

$H < 0$

c)   d)

Fig. 5.13 Applied fields parallel to the basic plane of a cobalt crystal (thickness ~180 μm). At positive fields (a) and at remanence (b) the basic domains are oriented perpendicular to the previous saturation field. In opposite fields (c) a zigzag folding saves total wall energy. In stronger fields the basic domains straighten again (d) by magnetostrictive interactions

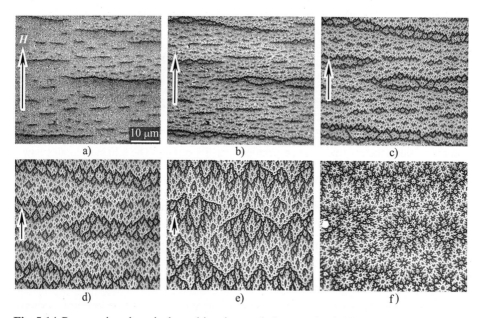

H

a)   b)   c)

d)   e)   f)

Fig. 5.14 Progressive domain branching in a cobalt crystal of thickness ~200 μm in a decreasing in-plane field. The sudden change of appearance from zigzag-shaped walls (e) to the flower state in (f) is connected with the orientation dependence of the wall energy, which disappears at zero field

The magnetostrictive energy favours a perpendicular orientation of the basic domains relative to the field (see Fig. 3.111). For large thicknesses and high fields this influence predominates [161]. At small fields, however, the wall energy favours a parallel orientation. Starting at saturation, nucleation occurs in the form of perpendicular stripes. The perpendicular domain orientation (Fig. 5.13a) is retained when the field is reduced to zero (b), while the domain width and the degree of branching increase. The maximum basic domain width is reached at zero field. Increasing the field again in the opposite direction (c), two tendencies act together: the domain width tends to decrease again, and the domain walls try to rotate parallel to the applied field.

Both effects combine in forming zigzag patterns, as first observed and analysed by *Kaczér* and *Gemperle* [895]. In higher fields the magnetostrictive interaction takes over again, leading to straightening of the basic domains (d). The effect was also found in high-magnetostriction garnet films [896].

The field-dependent features discussed so far apply to the behaviour of the basic domains and did not consider branching. The nucleation and unfolding of branching can be followed if an in-plane field is decreased from the critical field downwards, as demonstrated in Fig. 5.14. The initial picture close to nucleation (a) shows a narrow, unbranched pattern. Successive stages of branching appear in decreasing fields (b–f).

5.2.3 Polycrystalline Permanent Magnet Materials

Coarse grained permanent magnet materials usually consist of an aggregate of independent high-anisotropy grains. This case is studied in Fig. 5.130 in the context of small magnetic particles. The characteristic demagnetized domain pattern of such material is also visible in the left hand side of Fig. 5.15a. The grains in fine-grained permanent magnet materials are usually strongly correlated, and can therefore not be treated as an aggregate of independent particles. They tend to display a rather irregular magnetic microstructures in the demagnetized state as visible in the right hand side of Fig. 5.15a as well as in (b, c). These patterns are probably determined by predominantly dipolar interactions between the fine grains in the sense indicated (as a proposal) in the sketch for two grains which are coupled *only* laterally. This kind of domains was first observed in Alnico materials [897, 898]. They were also found on so-called ESD (elongated single domain) magnets [899, 900], for which the presence of any exchange interaction between the grains can be excluded.

To indicate their special nature, they were called "magnetostatic interaction domains". A review of early observations can be found in [44 (p. 187—196)].

Recently they were found again in fine-grained rare-earth magnets [398, 901—903]. Obviously also in these materials dipolar interactions between the grains dominate over the (certainly present) exchange interactions.

A characteristic feature of magnetostatic interaction domains is their elongated shape extending along the preferred axis of the material as visible in Fig. 5.15d for the example of AlNiCo. Figure 5.16 shows that the "domain boundaries" of the interaction domains can move laterally to some extent, but only in a very irregular and irreversible way.

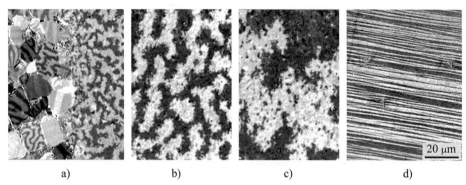

a)   b)   c)   d)

Fig. 5.15 Coarse- and fine-grained polycrystalline NdFeB material prepared by melt spinning and hot deformation. (a) The transition of the characteristic domain patterns in a sample in which grain growth occurred in the *left* hand side. The irregular domain pattern in the fine-grained part (*right*; grain size ~100 nm) is not in equilibrium. This can be seen by comparing the state obtained by demagnetizing in a suitable field sequence (c) with the thermally demagnetized state (b). (d) A view on the side face of an oriented Alnico crystal. (Together with *K.-H. Müller, W. Grünberger* and *A. Hütten*, IFW Dresden)

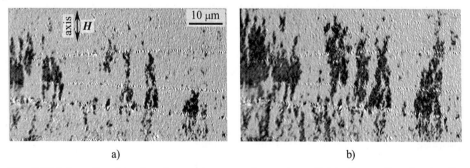

a)   b)

Fig. 5.16 Interaction domains observed on a surface cut parallel to the preferred axis of a nanocrystalline NdFeB material as in Fig. 5.15a—c. The irregular patches, that have been formed in a field of 700 kA/m opposite to a previous saturation field (a), expand primarily along the axis, but also laterally if the field is increased to 800 kA/m (b)

5.3 Bulk Cubic Crystals

Multiaxial materials generally have low relative anisotropy values ($Q \ll 1$). Exceptions are rare. Almost compensated ferrimagnets may be an example. Little is known about the domain structure of true bulk polycrystalline material in which the grains are much smaller than the smallest sample dimension. Such materials are used, for example, in the yoke of an electromagnet. Since there is no way of observing the interior domains in bulk soft magnetic materials, the discussion of their magnetic microstructure has to rely on theoretical considerations like those of phase theory (Sect. 3.4), in conjunction with the analysis of magnetization and magnetostriction measurements.

For soft magnetic sheets or plates used in electrical machines and other inductive devices the magnetic microstructure can be analysed in more detail because large grains extending through the thickness of the sheet are typical and preferred. This means that many essential features of the domain structures in such materials may as well be studied on single crystals. As long as the sheet thickness is larger than the wall width, these materials have to be classified as bulk materials even if they are only some microns thick.

The domain arrangement in bulk cubic crystals is primarily determined by the principle of flux closure. Almost as important is magnetic anisotropy in these materials, and most details of the domain patterns are therefore determined by the surface orientation relative to the easy directions. Several cases must be distinguished. From the simplest case, a surface with two easy axes, to strongly misoriented surfaces with no easy axis, the domain patterns become progressively more complicated. In positive anisotropy material such as iron the (100) surface contains two easy directions. The (110) surface in a negative-anisotropy material, such as nickel, is analogous to (100) in iron because it contains two easy $\langle 111 \rangle$-axes. Surfaces with only one easy direction are, for example, (110) for $K_{c1} > 0$ and (112) for $K_{c1} < 0$. The (111) surface contains no easy axis for positive anisotropy and the same is true for (100) and for (111) surfaces for negative anisotropy. Certainly domains in nickel-like materials are not the same as domains in iron-like materials, even if they are investigated on equivalent surfaces. They differ in the details of the allowed domain wall angles because the easy axes in iron are all mutually perpendicular, whereas this is not true for nickel-like substances. Domains in the two symmetry classes are highly analogous, however, so that we need not discuss them separately in each example.

5.3.1 Surfaces with Two Easy Axes

The simple and beautiful domain patterns of Fe (100) are well-known and
easy to interpret (Fig. 5.17a). The observed domains are often more complicated
than expected, however. Small residual stresses and possibly induced anisotropies are responsible for these deviations from the ideal structure.

At every point of a sample usually one of the crystallographic easy axes is
preferred slightly relative to the others. Figure 5.17b shows a pattern that was
obtained at the same place as the pattern in (a) after another demagnetizing
treatment. The details, in particular the magnetization *directions* differ from
those in (a), but the locally preferred *axes* agree largely in both patterns.

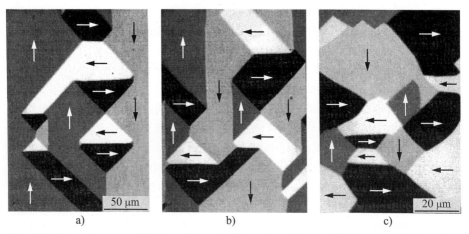

a)   b)   c)

Fig. 5.17 Domains on a (100) surface of silicon-iron (Fe 3 wt% Si, a material used as
electrical steel, magnetically equivalent to pure iron). (a) The domains are all magnetized
parallel to the surface, yielding a particularly clear pattern. (b) An alternate pattern from
the same spot after another demagnetization. In contrast, residual stresses generate "*V*-lines"
(c) which mark internal domains magnetized perpendicular to the surface

Residual stresses are also responsible for patterns in which the flux closure
principle appears to be violated (Fig. 5.17c). Here hidden internal domains hit
the surface as in the basic Landau-Lifshitz pattern (Fig. 1.2). The "domain
wall" visible at the surface consists in reality of a domain wall pair, the
components of which intersect at the surface forming a *V* [641]. As for regular
walls, one may distinguish between 90° and 180° *V*-lines, depending on the
angle between the surface magnetization directions. Figure 5.18 shows the
internal structure of the two cases, which can both be found in Fig. 5.17c.

V-lines can assume a curved overall shape because they can be rotated
about the surface normal without generating a stray field. Two different

orientations for both V-line types are shown in Fig. 5.18. There are no exceptions in this class of materials to the rule that all domain patterns are of the closed-flux type, i.e. essentially stray-field-free.

This can be proven by a comparison of a Kerr effect picture with the corresponding Bitter pattern. If any of the boundaries that we interpret as V-lines were actually charged walls, a strong colloid agglomeration would have to be expected along them. In reality, V-lines are almost invisible in the Bitter technique, as demonstrated in Sect. 2.2.4. A qualitative explanation for this phenomenon was given in [84]: the internal domains can move up or down, and also tilt a little to make up for misorientation as shown in the sketch, thus compensating any net charge from walls or domains that might be present otherwise.

Extended V-lines usually reveal the preferred orientation of the participating subsurface walls in the form of a zigzag shape (Sect. 3.6.3 C). In contrast, the geometry of regular walls, which separate two easy directions in the surface, cannot be deduced from surface observations because the difference vector in (3.114) coincides with the trace of the wall in the surface. Whatever the (allowed) internal orientation of the wall, its intersection with the surface remains the same. Only the observation of the same wall on different surfaces offers information about the interior.

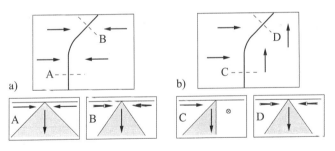

Fig. 5.18 180° V-lines (a) and 90° V-lines (b) on a (100) iron surface, both in two different orientations. The zigzag folding of the internal 90° walls is omitted in these diagrams

Figure 5.19 shows a corresponding picture from a negative anisotropy material. The 90° walls are replaced by 71° and 109° walls. Apart from these domain-wall-related details, the patterns look quite similar to the positive anisotropy case. Note that wall substructures (Bloch lines) and very wide walls show up also in 109° walls, not only in the 180° walls as in Fig. 2.21. This observation agrees with the theoretically predicted behaviour of 109° walls (see Sect. 3.6.3D).

Fig. 5.19 Domains in a (110)-oriented thin plate of an yttrium iron garnet crystal in which the easy directions are along the ⟨111⟩ axes. The domains are made visible magneto-optically in transmission at slightly oblique incidence. The domain contrast is mainly due to the Voigt effect, with some contribution from the longitudinal Faraday effect. The polar Faraday effect generates strong contrasts in the domain walls. (Together with *R. Fichtner*)

5.3.2 Crystals with One Easy Axis in the Surface

(A) Basic Domains. The basic zero-field domain structure of (110)-oriented iron sheets, as used in Goss-textured transformer material, is rather simple: it consists of slab-like domains magnetized parallel and antiparallel to the easy [001] direction (Fig. 5.20). The only peculiarity is invisible: the 180° walls are not oriented perpendicular to the surface but tilted, as sketched in Fig. 3.64 and as can be confirmed by observation from both sides. The significance of this tilting has been discussed in connection with eddy current effects (Sect. 3.6.8).

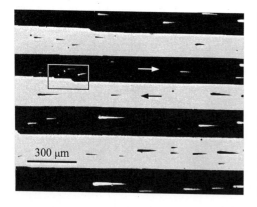

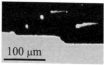

Fig. 5.20 Slab domains on a largely undisturbed (110)-oriented silicon iron crystal. The isolated lancets as well as the short kinks in the main walls (enlarged above) are connected with internal transverse domains

A rich variety of domains can be observed on this surface if for some reason the *internal* easy directions are favoured. This may be caused by defects, stresses or transverse external fields. Also demagnetizing effects at grain boundaries or the crystal edge may induce such domains. In Fig. 5.20

we detect at some places wall segments that form a large angle with the preferred axis. They are *V*-lines again, indicating internal transverse domains that are probably connected to inclusions somewhere inside the sample.

The simple bar domain pattern is disturbed by grain boundaries. Let φ_1 and φ_2 denote the angles between the projected easy axes and the boundary normal (assuming it to lie in the sheet plane). If the magnetization on both sides of the boundary follows the respective easy axes, a nominal magnetic charge of:

$$\sigma_s = \cos\varphi_2 - \cos\varphi_1 = 2\sin\left[\tfrac{1}{2}(\varphi_1 + \varphi_2)\right]\sin\left[\tfrac{1}{2}(\varphi_1 - \varphi_2)\right] \qquad (5.1)$$

is generated. We see — best from the second version of (5.1) — that this charge density is zero for identical easy directions $\varphi_1 = \varphi_2$, as well as for the symmetrical arrangement $\varphi_1 = -\varphi_2$. The charges appearing otherwise may induce grain boundary compensation or closure structures. Some examples are shown in Fig. 5.21.

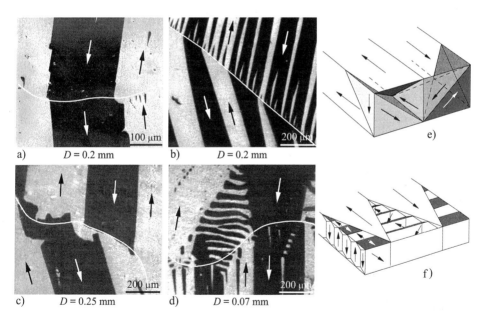

Fig. 5.21 The effects of grain boundaries on the domain pattern in (110)[001]-textured material depend on the nominal charge of the grain boundary as defined by (5.1), but also on other circumstances such as the stress state and the tilting angle of the grain boundary. Four typical examples are shown: (a) perfect domain continuity, (b) dagger-shaped compensating domains, (c) closure by interior transverse domains as outlined in (e), and (d) quasi-closure-domains for thinner plates as indicated in (f). (Together with *S. Arai*)

A complete analysis has to include a tilting of the grain boundary which can reduce the charge density. In addition, the magnetization can reduce the charge by rotation away from the easy axes, as described by the μ^*-effect (Sect. 3.2.5F). Figure 5.21a shows an example in which the domains traverse the grain boundary apparently undisturbed. With increasing nominal charge σ more complicated flux compensation domains are introduced in the neighbourhood of the grain boundary as demonstrated in (b–d). The extra domains are connected with an additional energy (which can be reduced by reducing the basic domain width — at the expense of 180° wall energy). The details of the extra grain boundary domains depend on the nominal charge σ. For small σ the flux is distributed by dagger-shaped antiparallel domains (b). For stronger charge a complicated closure pattern is generated based on invisible internal transverse domains (c; [904]). If the basic domains are wide compared to the sheet thickness, closure at an unfavourable grain boundary can only be achieved by quasi-domains as shown in (d). The closure pattern is also influenced by applied stresses. Under the influence of a tensile stress along the preferred axis the dagger-shaped domains of (b) are favoured compared to closure schemes containing large volumes of transverse domains.

(B) Transverse Applied Fields. Figure 5.22 shows a series of domain patterns observed in an external field transverse to the preferred [001] axis, together with the corresponding cross-sectional models. The models follow from the conditions that they transport some flux into the transverse direction, that only easy directions are occupied, and that no stray fields are generated. Remarkable is the *saw-tooth* pattern (Fig. 5.22a; [905, 906, 124]) which manages to transport a certain transverse flux (up to about $0.3 J_s$) without generating strong magnetostrictive stresses. This is because (110)-oriented 90° walls separate the internal transverse domains from the basic domains, which allow a stress-free relaxation of magnetostrictive strains (as in Fig. 3.105b in a more transparent geometry).

In higher fields (Fig. 5.22b) a pattern with elastically incompatible 90° walls between the completely reorganized basic domains is enforced. Increasing the field still further so that the external field is no more compensated by the demagnetizing field we get branched structures (Fig. 5.22c, d) as we met them in connection with the Néel block (Sect. 3.7.6).

If the field is applied not exactly perpendicular to the preferred axis, there appear domains of a different type, characterized by thin, tilted surface lines that can assume a cord-like appearance (Fig. 5.23). Without the field component along the preferred axis the closure configuration in this "cord pattern" would be less favourable than the one of the regular column pattern (Fig. 5.22b). On the other hand, magnetostrictive compatibility between the basic domains is better realized in the pattern of Fig. 5.23. The overall orientation of the cord pattern can be computed as an equilibrium between wall energy, magnetostrictive interaction energy, and closure energy terms [904].

In the cord pattern three internal domain walls meet at the surface (Fig. 5.23d). Slight changes in the external field can make these lines almost invisible. If the field is rotated away from the transverse direction, the whole structure dives below the surface. Alternatively, an expanded cord pattern (Fig. 5.23b) is produced for the opposite field rotation sense. The cord pattern is related to the zigzag folding of the internal 90° walls. The zigzag amplitude is enhanced near the surface, thus reducing the volume of the large closure domains. The visible cords represent a secondary closure structure of the enhanced zigzag pattern, as indicated in Fig. 5.23e,f. In higher fields, the cords grow into a complicated branched pattern (Fig. 5.23c), which we encountered previously on the side faces of the Néel block (Fig. 3.146). All the domains shown in Figs. 5.22 and 5.23 are specific for (110)-oriented samples of the standard thickness around 0.3 mm typical for transformer sheets.

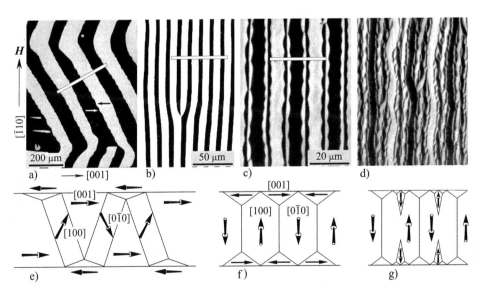

Fig. 5.22 Three types of domain patterns occurring on (110)-oriented silicon iron sheets of 0.3 mm thickness in transverse fields. The saw-tooth pattern (a, e) at about 0.3 × saturation is followed by the column pattern (b, f) near the onset of magnetization rotations (~0.7 × saturation). In higher fields (c, d) the development of branching is observed. The models represent cross-sections through the banded patterns as indicated. The model (g) shows the most simple branching pattern, approximately as in (c)

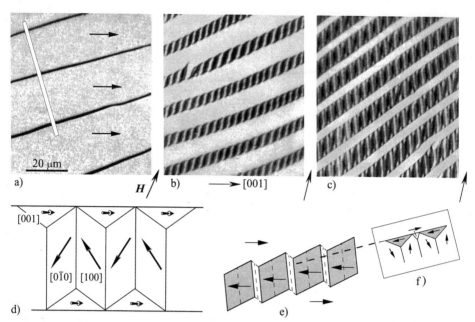

Fig. 5.23 Another pattern occurring in an oblique field in which all closure domains are essentially magnetized along the same direction. The diagram (d) shows a cross-section through the pattern in (a). If the applied field is rotated back towards the transverse direction, the volume of the closure domains tends to shrink at the expense of a folding of the internal walls, forming the characteristic cord pattern (b) which is explained in a top view in (e). The cross-section (f) shows the immediate subsurface environment of the cord pattern. In stronger fields a complicated branched pattern is generated (c)

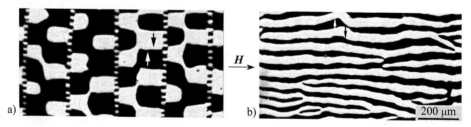

Fig. 5.24 Domains on (110)-oriented silicon iron crystals of different thicknesses in transverse fields. For thick samples (~1 mm) the checkerboard pattern replaces the saw-tooth pattern [Fig. 5.22a; (a)]. In thin samples (~0.1 mm) an irregular column pattern (Fig. 5.22b) is formed already at low field (b)

Figure 5.24 shows observations on material of different thicknesses: the checkerboard pattern occurring in a transverse field at about 1 mm thickness, and the irregular column pattern at about 0.1 mm, both replacing the saw-tooth pattern. The checkerboard structure represents a higher-order branching pattern (see Fig. 3.138) that is magnetostrictively favourable; it is therefore preferred in thicker samples. The column pattern is basically simpler than the saw-tooth pattern. Its higher magnetostrictive energy is less important in thin sheets.

Domain refinement towards a tapering edge of a thin SiFe crystal is shown in Fig. 5.25. The equilibrium domain width decreases with the thickness, and the observed structure adapts to this tendency by inserting diamond-shaped quasi-domains which duplicate the domain number in each step. What appears as a domain branching process, differs from this surface-related phenomenon (see Sect. 3.7.5) because the domain width is a function of the local thickness here, while it is a function of the distance to a surface in branching.

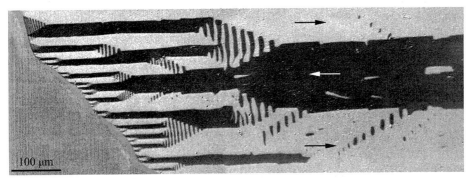

Fig. 5.25 Domains on a thinned, tapering Si-Fe plate. The domain width is reduced towards the thin edge by inserted diamond-shaped quasi domains. Note the irregularities along the 180° walls in the right, thick part of the sample. Possibly, these are traces of an "internal fir tree pattern" mentioned in Sect. 2.7.6

5.3.3 Stress Patterns

A compressive stress along the preferred axis of (110)-oriented iron-like material generates aesthetically attractive stress patterns [907–909], a selection of which is shown Fig. 5.26. The first of these, stress pattern I, develops in mild stresses (b). Resembling superficially the pattern of Fig. 5.22b, it differs by containing 180° walls between the basic domains, as indicated in Fig. 5.26g. At stronger pressure the closure domains of the simple stress pattern are subdivided, leading to the branched stress pattern II (c–e). In higher stresses (f) a further splitting of the pattern is observed. Stress pattern II occurs in differently oriented variants depending on the direction of an alternating field applied before. An equilibrium pattern cannot be identified by a.c. demagnetization. The result depends systematically and markedly on the field direction, a feature that can be qualitatively understood from a model of this pattern, as elaborated below.

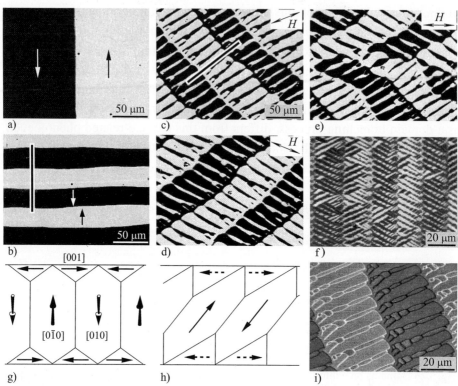

Fig. 5.26 Stress patterns produced by compressive stresses along the preferred axis which favour the two internal easy directions of a (110)-oriented silicon iron crystal. Starting from the unstressed state (a) we get stress pattern I at moderate stresses (b). Higher stresses and a.c. demagnetization fields with components along the easy axis favour the branched stress pattern II (c–e), shown here in different variants that are preferred with different directions of the alternating field indicated by arrows. Further branching occurs at still higher stresses (f). Cross-sections through (b) and (c) are shown in (g) and (h), for (h) in form of a quasi-domain representation (see also Fig. 5.27). The nice pattern of "magnetic ants" [(i); rejected by INTERMAG conference, Brighton 1990] is an enlarged version of (c) at slightly different conditions and observed in domain wall contrast (as in Fig. 2.18a). The segmentation of the "ants" can probably be understood as a third-order branching process, but no attempt of a detailed analysis is made

A model of stress pattern II (Fig. 5.27) indicates that in this variant of a multiaxial branched structure only the shallow surface domains are magnetized along the easy direction, which is disfavoured by the compressive stress [904]. The basic domains are separated by 180° walls as in stress pattern I. The inserted intermediate subsurface domains carry a net flux in the transverse direction, and they are therefore favoured under the action of transverse fields — in contrast to stress pattern I, which cannot be magnetized transversely without rotation processes. This interesting property of stress pattern II is demonstrated more clearly in the right hand inset of Fig. 5.27, in which the surface domains are ignored and the intermediate domains are represented by their average magnetization as quasi-domains. The closure regions can grow and shrink in transverse fields, thus transporting some flux along the field direction.

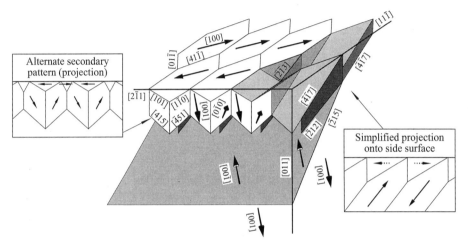

Fig. 5.27 A three-dimensional model of stress pattern II of Fig. 5.26c. The first generation quasi-closure domains indicated in the *right inset* carry a net flux along the $[1\bar{1}0]$ axis. This flux component can be used in a magnetization process. The near-surface secondary pattern is simplified in the *main picture* and shown in more detail in the *left inset*

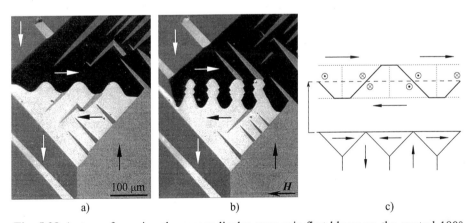

Fig. 5.28 A stress favouring the perpendicular easy axis first blows up the central 180° wall on a (100)-oriented silicon iron crystal, leading to a sinusoidal modulation of the wall (a). On applying a weak field the underlying perpendicular domains are converted into a regular stress pattern (b). The internal structure of the wavy wall pattern (a) is sketched for a straightened variant in (c). The underlying perpendicular domains are indicated by *dotted lines* and a cross-section through the mid-plane

Peculiar structures can also be observed in very weak stresses that are not strong enough to induce perpendicular domains in the volume, but which suffice to grossly modify already present domain walls. Such observations were first reported for iron whiskers [641] and for bulk magnetic garnet samples [210]. They can also occasionally be found on (100)-oriented silicon iron sheets as demonstrated in Fig. 5.28. The stress deforms the horizontal 180° domain wall into a pronounced sinusoidal pattern (a) tracing small subsurface perpendicular domains as shown schematically in (c).

Figure 5.29 shows attractive surface patterns that can be generated in transformer steel samples by applying simultaneously some mechanical stress and magnetic field. No attempt is made to interpret these patterns in detail. We leave it to the interested reader to find out how precisely they can be prepared and understood.

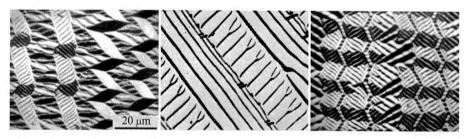

Fig. 5.29 Various patterns that can be found on (110)-oriented silicon iron when playing around with external fields and stresses

5.3.4 Slightly Misoriented Surfaces

Here we deal with crystals for which the misorientation — the angle between the closest easy axis and the surface — is less than about 5°. Then the basic domains correspond to those of the ideally oriented crystal. They are supplemented by a system of flux-collecting domains that reduce the stray field energy (Sect. 3.7.1). Figure 5.30 presents observations of the best known among these patterns, the fir tree pattern belonging to misoriented (100) surfaces [85, 910, 911]. There are two kinds of fir tree patterns, those associated with 180° walls, as explained in Fig. 3.104a, and those connected with 90° walls, as sketched in Fig. 3.104b.

Both orientations of fir tree patterns may coexist depending on the overall basic domain pattern as demonstrated in Fig. 5.31. A variant of the regular pattern, which is formed if the basis domains are removed in an applied field, was demonstrated earlier in Fig. 3.107 and called the "true" fir tree pattern.

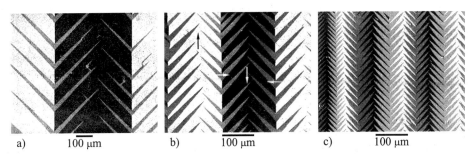

Fig. 5.30 Fir tree patterns on surfaces 2°, 3° and 4° misoriented with respect to (100). Note the different scales used in the examples

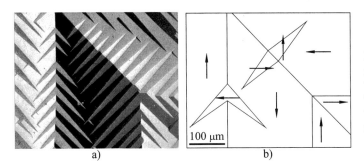

Fig. 5.31 Two variants of fir tree patterns (a) associated with different wall types in the basic domain structure as schematically indicated in (b)

Other variants of supplementary domains on the (100) surface are shown in Fig. 5.32. They are observed in connection with different basic domain patterns as they occur in applied fields and stresses. In (a) a field along the diagonal induces a two-phase basic domain pattern in which one of the phases (along the y-axis in the picture) is markedly more misoriented than the other one. The potential stray fields from this basic domain phase are compensated by dense black lancet-shaped supplementary domains. Not all black lancets are continued in narrow light-grey lancets in the favourable white basic domain. The rest of the flux is transported into the depth, forming V-lines at the surface. This interpretation is confirmed by the irregular shape of the boundary line between the basic domains.

The pattern in Fig. 5.32b is observed if a sample is almost saturated along a misoriented easy axis. As in (a), one axis is less misoriented and thus preferred. Irregular spike domains magnetized along the more favourable axis are still left close to saturation. In (c) supplementary domains are shown superimposed onto a stress pattern (see Fig. 3.69 for the basic structure). Note the V-lines at the tails of the "butterflies" and "moths" which indicate that the collected flux is transported into the interior, towards the basic domains of the stress pattern. Near the zigzag V-lines no supplementary domains are necessary as explained in connection with Fig. 5.18.

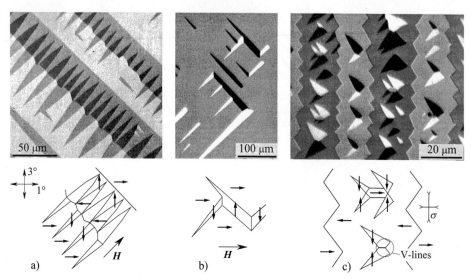

Fig. 5.32 Different types of supplementary domains on a (100)-related FeSi crystal. A field was applied along a diagonal in (a), and along a [100] axis in (b). (c) Supplementary domains inside the closure domains of a stress pattern (see Fig. 3.69). The diagrams indicate the surface magnetization direction in characteristic elements of the patterns

Corresponding patterns for the slightly misoriented (110) surface [734] are shown in Fig. 5.33 both in an overview and at high magnification. With stronger misorientation the supplementary domain system becomes increasingly complicated, often revealing the basic domains clearly only during remagnetization. Some of the details are discussed below in connection with Fig. 5.35.

In Sect. 3.7.1B we analysed theoretically the phenomenon of lancet combs in (110)-oriented sheets in which individual lancets join to use a common plate-shaped internal domain that transports the collected flux to the opposite surface. These lancet combs (see Fig. 3.105) show an interesting behaviour in applied fields. The common internal plate domains are oriented so that magnetostrictive energy is avoided. At the same time these plates carry flux into the transverse direction. A transverse field component prefers one internal easy direction against the other, and this preference can be satisfied by a switching of the combs so that their surface traces flip between the two [111] axes on the surface (Fig. 5.34).

Such domain reorganization processes play a double role: they contribute to the overall permeability, increasing in particular the *transverse* magnetizability in (110)-oriented material. At the same time they lead to excess losses because the energy content of the domain structures is lost during domain rearrangement. In transformer steels these processes are expected to become important near strongly misoriented grains and near the joints between the individual transformer yokes.

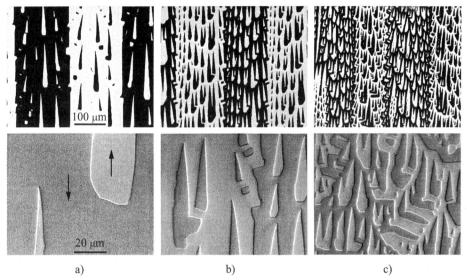

a)   b)   c)

Fig. 5.33 Some variants of the lancet pattern (see Fig. 3.104c) on (110) surfaces with 2°, 4° and 8° misorientation. High-resolution pictures (*below*) show the details of the overview pictures above with an emphasis on domain wall contrast. Weak kinks in the 180° domain walls indicate the presence of Bloch lines (see Fig. 3.86). The surface cap orientation (see Fig. 3.82) displays a fixed sign (*white* on the *right* of the darker domains, *black* on the *left*) as determined by the stray field caused by the misorientation [593]

Figure 5.34 shows the same phenomenon of lancet reorientation also for a crystal of much stronger misorientation. Here each lancet is covered with many secondary lancets that further reduce the stray field. The overall pattern is the same as for weak misorientation (top row) indicating that the complex pattern is based on the same kind of plate shaped transverse domains connecting the top and the bottom sheet surfaces.

In higher transverse fields at constant longitudinal field the lancets develop peculiar lateral protrusions, forming the so-called chevron pattern ([912]; Fig. 5.35). If the longitudinal field is increased, the chevrons contract to linear structures known from colloid investigations and called the *tadpole* pattern for a characteristic colloid concentration at the end, resembling the head of a tadpole ([911]; see Sect. 2.2.4, Fig. 2.6). Locally, the pattern in Fig. 5.35c corresponds to the "cord" structure observed for ideal (110) orientation in oblique fields (Fig. 5.23a).

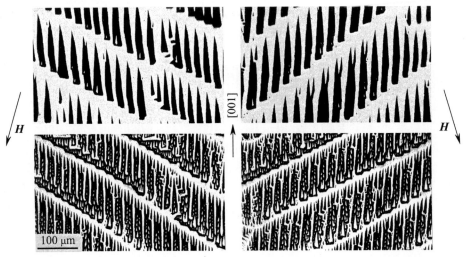

Fig. 5.34 Lancet combs switching their orientation depending on field direction. The fields are slightly rotated away from the preferred direction. (110)-oriented SiFe crystals with a misorientation of 4° (*top row*) and 8° (*bottom row*)

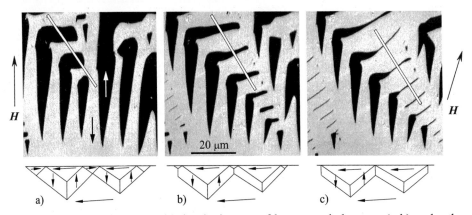

Fig. 5.35 The tadpole pattern (c) developing out of lancets and chevrons (a,b) under the influence of a transverse field component

The supplementary domains also reorder under the influence of stresses. They are enhanced under compressive stress and suppressed under tensile stress applied along the preferred axis of (110)-based material, because tensile stress disfavours the transverse domains that are attached to all supplementary domains. A uniaxial tensile stress along the preferred axis has the same magnetic effect as a twice as strong planar tensile stress as can be confirmed by evaluating the magneto-elastic coupling energy (3.52). Bending a (110)-oriented sample around its transverse axis favours the surface-parallel quasi-fir-tree pattern (Fig. 3.104d) on the compressive side as shown in Fig. 5.36. Under these conditions the transverse domains cannot reach the opposite, tensile side.

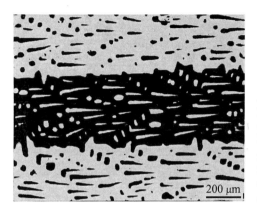

Fig. 5.36 The quasi-fir tree pattern appearing on bent (110)-related silicon iron crystals (bending axis perpendicular to the preferred axis). This pattern is formed on the compressed side, transporting magnetic flux from one domain to the neighbouring domain as explained in Fig. 3.104d. On the opposite surface supplementary domains are suppressed

The supplementary domains are often attached to the walls of the basic domains as, for example, in the fir tree pattern on (100)-related surfaces. But this may also be true for the lancet domains on misoriented (110) surfaces as indicated in Fig. 3.104c. The analysis of the characteristic magnetostriction curve of oriented transformer steel (Fig. 5.37) leads to similar conclusions. Ideally oriented (110) crystals would not show any magnetostrictive change in length during remagnetization along the [001] easy axis. The actually observed elongation (Δ_1 in Fig. 5.37) indicates the existence of transverse domains in the demagnetized state, which are just the internal domains attached to the lancets [913–915]. The initial contraction (Δ_2 in Fig. 5.37) proves the total volume of transverse domains to *increase* as the basic domains are eliminated. Obviously, many of the transverse domains were attached to the 180° walls between the basic domains. They have to extend all through the thickness to the opposite surface when the main 180° walls are eliminated. In addition, more lancets are created where the main walls had been before.

To summarize again the role of supplementary domains in magnetization processes, we note that during remagnetization, the system of supplementary domains is completely and repeatedly destroyed and rebuilt. The energy bound in the supplementary domains is lost in every cycle, thus forming an important part of hysteresis losses. Indirectly, this connection between supplementary domains and hysteresis was confirmed in [494]. In this study the total volume of transverse domains was determined by magnetostriction measurements as in Fig. 5.37. These "cross volumes" showed a marked correlation with the coercivity, i.e. with the width of the hysteresis loop.

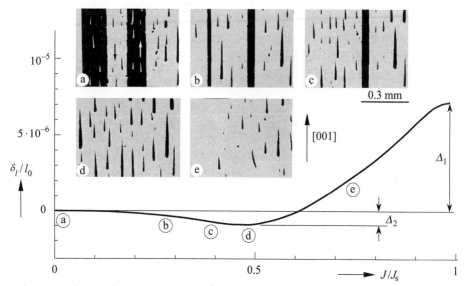

Fig. 5.37 The magnetostrictive elongation of a typical (110)-oriented transformer steel material, magnetized along the preferred axis. The length change is correlated with domain observations during this process. Note that the total surface area of supplementary domains is larger in an applied field (d) than in the demagnetized state (a), corresponding to the minimum in the magnetostriction curve

The magnetostrictive effects of supplementary domains are the main source of acoustic transformer noise. On the other hand, these domains can help to avoid stray fields at grain boundaries in polycrystalline materials and thus enhance the magnetizability of a material. To see this, look at the internal transverse domains attached to the lancet combs. As these transverse domains are reoriented (Fig. 5.34), flux along the transverse direction can be transported at little additional expense, which is not possible in the ideal slab domain structure. If a grain is misoriented laterally in a textured material, strong stray fields at its sides would develop during magnetization if the supplementary domains could not compensate the transverse flux. Suppressing the supplementary domains, for example by an applied stress, raises the danger that such grains are passed by in the magnetization process, thus increasing the burden and the losses in the better oriented grains. For this reason, a "stress coating" on grain-oriented transformer steel, which suppresses transverse domains, offers an advantage only if the lateral grain misorientations are small ([911], see Sect. 6.2.1C) .

5.3.5 Strong Misorientation

The domain patterns encountered on strongly misoriented surfaces are most involved and certainly not completely understood [32, 765, 916–918]. Surface domains on strongly misoriented surfaces are not at all representative for the underlying bulk magnetization. Measuring surface magnetization loops on such crystals makes sense only in connection with loss studies because the interesting and beautiful surface patterns certainly contribute to hysteresis losses. But the average surface magnetization is virtually independent of the average bulk magnetization. The interior domains can only be inferred from subtle features and the dynamics of the surface pattern.

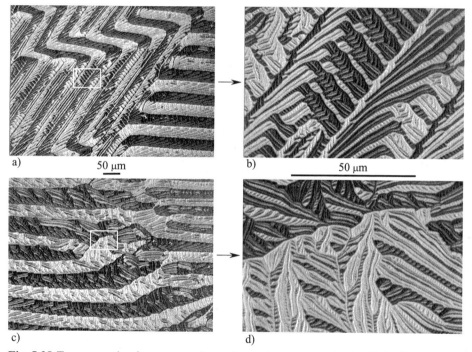

Fig. 5.38 Two examples for a comparison of overview pictures (a, c) with high-resolution images (b, d) of the same strongly branched patterns on the (111) surface of a silicon iron crystal. The area blown up in the *right column* is indicated in the *left pictures* by a *white frame*. (Developed together with *J. McCord*)

Figure 5.38 shows some observations from the (111) surface of iron. Since all magnetization directions on a (111) surface are equivalent from the point of view of the first-order anisotropy energy expression, these pictures are not easy to interpret. From the arrangement of domain boundaries alone the local magnetization directions cannot be derived as is usually true for other surfaces.

An impression of the complexity of these patterns can be obtained by displaying the same spot once in overview and once at high resolution (Fig. 5.38). The many generations of branching (four in the example) cannot be seen in a single photograph — a characteristic property indicating the fractal nature of branched patterns (see Fig. 3.131 for two-phase branching).

The variety of patterns is a challenge for a systematic analysis that has not yet been performed. Probably the best approach is to start such an analysis with high field conditions where the domains are still unbranched, and then to reduce the field leading to progressively more branched configurations. An example of an experimental study of the branching process in this sense is shown in Fig. 5.39. While (a) shows essentially a single-domain generation (with a second generation beginning to build up), two and three generations are visible in (b) and (c). The zero-field picture (d) corresponds to Fig. 5.38d, which was shown there to be part of a four-generation branching pattern.

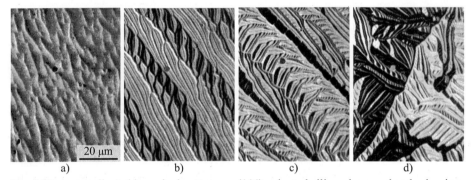

a)   b)   c)   d)

Fig. 5.39 A complicated branched pattern on (111)-oriented silicon iron as developing in a field decreasing from (a) (close to saturation along a hard axis) to (d) (remanence)

Further remarkable patterns, observed on a strongly misoriented silicon iron crystal the surface orientation of which is closer to (100), are shown in Fig. 5.40. The striking "fat walls" are no real walls but traces of internal domains as explained in the cross-sections. The explanation offered in the figure caption does not rely on a quantitative analysis; it is rather based on the well-established principles of charge avoidance and compensation.

Similar patterns are also observed on the surfaces of ideally oriented bulk crystals in a field applied perpendicular to a sample edge. This is shown in Fig. 5.41 for a gross iron whisker with {100} surfaces. Approaching the edges the characteristic patterns of increasing misorientation are observed.

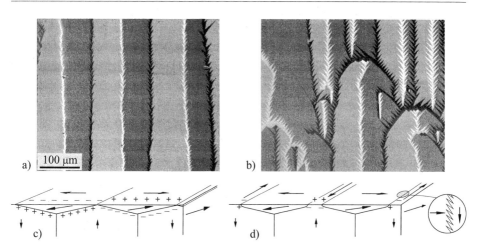

Fig. 5.40 Domain observations on a surface deviating by about 7° from both easy axes of (100). The characteristic "fat wall" pattern is explained in (c, d) in three steps. The pattern consists mainly of basic domains along one of the easy axes which are covered by shallow closure domains magnetized along the other axis. The cross-section (c) demonstrates how the charges on the surface are compensated by charges at the lower boundaries of the shallow domains. Wall energy is saved in the second step by slightly opening the basic domains (d). The charges on the steeper wall segments formed in this opening process can be distributed by a small-scale zigzag folding shown in the inset. Raw charges, not yet modified by the μ^*-effect, are shown in the *drawings*. (Developed together with *S. Arai*)

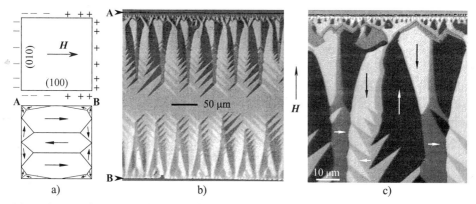

Fig. 5.41 Supplementary domains observed on an ideally {100}-oriented iron crystal of elongate shape and 0.3×0.3 mm^2 cross-section under the influence of a field of 2.6 kA/cm applied perpendicular to a sample edge. (a) The geometry of the experiment in cross section, with a possible internal domain arrangement as elaborated in the text. (b) The observed domain pattern on the top surface. (c) Magnified image of the near-edge zone

The first idea in analysing this phenomenon is that effectively misoriented surfaces are generated by the superposition of applied field, demagnetizing fields and crystal anisotropy. But this interpretation cannot be true because only one oblique direction along the field would be favoured by this mechanism, which could not give rise to complex domain patterns. The following analysis uses three steps the first two of which are indicated by sketches in Fig. 5.41a: (i) We start with the approach of Bryant and Suhl (Fig. 3.25) of calculating the surface charges in the applied field as for a material of infinite permeability. (ii) A simple domain structure is constructed that is compatible with (i). This structure includes shallow surface domains that distribute the magnetic flux but at the expense of a large wall energy. (iii) Replacing the shallow domains by a branched pattern of supplementary domains saves energy. While we interpreted supplementary domains in Sect. 3.7.1 as flux *collection* structures, we understand here that equivalent patterns can also be formed under conditions requiring flux *distribution*.

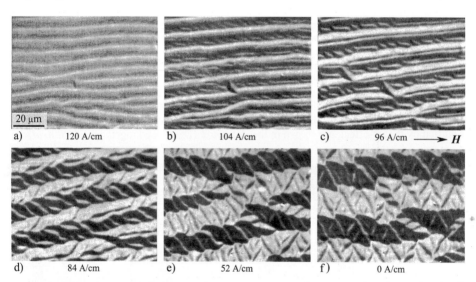

Fig. 5.42 Stress-induced domains on a nickel-iron crystal (~55 % Ni). The series (a–f) displays the unfolding of branched patterns in a decreasing field. The final pattern (f) is less complex in the finest details than the corresponding picture Fig. 5.26f for silicon iron, however, because of the smaller effective anisotropy and the smaller thickness of the NiFe sample. (Together with *J. McCord*, Erlangen)

Using another material with a smaller anisotropy is also helpful in analysing the domain configurations in strongly misoriented crystals. In iron, from which the examples shown up to this point were taken, the minimum domain size at the surface is found to be about 0.6 µm. If this minimum size scales with the domain wall width as suggested in Sect. 3.7.5C, choosing an alloy with a larger domain wall width (resulting, for example, from a smaller anisotropy) will remove resolution problems in optical microscopy. Nickel-iron alloys offer this possibility, as can be seen in Fig. 5.42. Also other domain observation techniques such as magnetic force microscopy (Fig. 5.43) can be useful in detailed studies of a particular pattern even if they do not offer the flexibility in following the evolution of a pattern. To make up, such pictures offer better resolution of the details, and they enhance different aspects of the same structure, which may help in a complete analysis.

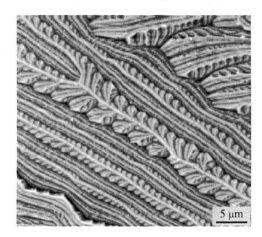

Fig. 5.43 High-resolution magnetic force image of a (1̄11)-oriented silicon iron crystal. (Courtesy *J. M. Garcia, L. Belliard* and *J. Miltat*, Orsay)

a)   b)

Fig. 5.44 Domains on a wound core of high-permeability Permalloy material. The domain pattern in (a) has to be distinguished from the crystallographic twin structure made visible in (b) as a difference picture between two oppositely saturated states. (Together with *J. McCord*, Erlangen)

Even smaller values of crystal anisotropy occur in the nickel-iron alloys around 80 % Ni. These high-permeability alloys, known under the trade names of Permalloy or Mumetal, are adjusted in their composition to reach zeros of the first-order anisotropy constant K_{c1} and of the mean magnetostriction constant λ_s. Figure 5.44 shows domains of a polycrystalline wound core of such a material in its optimized state. Regular magnetic domains are visible, which are influenced by the crystal orientation. The notion that in high-permeability materials domains are replaced by a smoothly varying magnetization is disproved by such observations, supporting the arguments of Sect. 3.3.4.

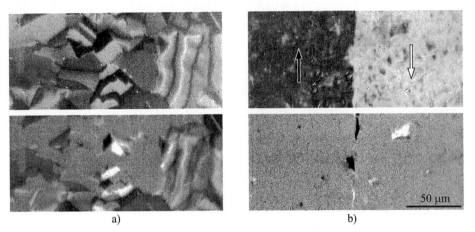

a)   b)

Fig. 5.45 Demagnetized state (*top row*) of coarse- (a) and fine-grained (b) $Ni_{81}Fe_{13}Mo_6$ material of 20 μm thickness, and the image difference between an initial demagnetized state and the state after applying an alternating field of moderate amplitude (*bottom row*). The alternating field sweeps the domain wall in (b) back and forth in a "laminar" fashion, but it virtually returns to its initial position after switching off the field. In contrast, the process in (a) makes a turbulent impression, leading to strong irreversibilities [919]

The domain patterns in high-permeability alloys depend strongly on the grain size, and these differences have direct consequences for static and dynamic loss and noise properties [919]. Figure 5.45 shows a comparison of two materials, one with grains of about 30 μm diameter (a), the other with about 13 μm grain size. In the coarse grained material the domains scale with the grain size, while in the fine-grained material macroscopic domains extend over many grains and are only modulated somehow by the local anisotropies in the grains. The static coercivity is much smaller for the coarse-grained material because the domain walls can be displaced freely inside the grains. The dynamic losses in the kHz regime are high for both materials for different reasons: in the fine-grained material the wide domain wall spacing leads to large excess eddy current losses, while for the coarse-grained material the irregularity of the domain pattern and of the magnetization behaviour appears to be responsible for high losses. This is demonstrated in the bottom row of Fig. 5.45, where the initial domain pattern is recorded as the reference image in digitally enhanced Kerr microscopy. After applying a 1 kHz alternating field of moderate amplitude all changes in the domain pattern show up, which indicates irregular and irreproducible magnetization behaviour. Very strong effects are seen in the coarse grained material (a), while similar irreproducibility effects are almost absent in (b). These findings also explain the observation that electronic noise measured on the same samples increases steeply with grain size. Almost no noise is observed for the reversible magnetization process of the fine-grained material in spite of its strongly enhanced losses. This feature is important in connection with sensor applications.

5.4 Amorphous and Nanocrystalline Ribbons

Rapidly quenched amorphous ribbons do not possess a crystal anisotropy by definition. Nevertheless, well-defined domain patterns are observed in such materials as a result of residual anisotropies. Internal stresses are the main source of these spurious anisotropies. The stresses can be the result of the manufacturing process, in particular of differences in the quenching speed between contact and free surface, and of surface irregularities from air bubbles trapped under the foils during quenching. Also crystalline inclusions and partial crystallization can lead to stresses. The magnetization is coupled to the stress by the magnetostriction constant λ_s, that need not be zero in metallic glasses. There are magnetostriction-free cobalt-rich metallic glasses, but even these materials show regular domains and sometimes even weak stress-induced patterns, probably caused by inhomogeneities in the magnetostriction constant that may be zero only in the spatial average.

In addition, metallic glasses usually carry an *induced* anisotropy that is of more or less random orientation if the material was not heat treated in a magnetic field. This induced anisotropy is based on minute deviations from random pair orientations of the components of the material, and it is induced by the magnetization pattern present during cooling through the Curie point, and possibly also to the flow pattern during quenching. The random nature of both sources of anisotropy makes it difficult to separate them. When referring in the following to stress anisotropies, this may include some unspecified contributions from random structural anisotropies.

5.4.1 The As-Quenched State of Amorphous Ribbons

Two kinds of patterns are visible in the as-quenched state (Fig. 5.46): wide curved domains with 180° walls that follow a local in-plane easy axis everywhere, and narrow fingerprint-like "stress patterns" which indicate an easy direction perpendicular to the surface [337, 920–926]. Strictly speaking, all domains visible in Fig. 5.46 are determined by stresses, but it is usual to designate only these characteristic parts as "stress patterns". Altogether, the magnetic microstructure of as-quenched material can be classified as disordered, in contrast to the more ordered states discussed in the next section.

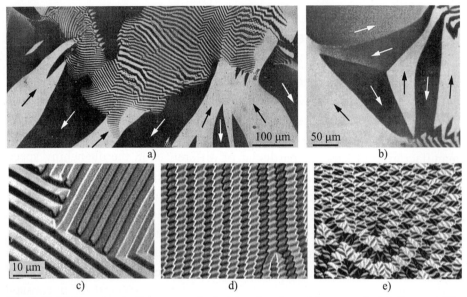

Fig. 5.46 (a) Domains in the as-quenched state of an iron-rich, magnetostrictive metallic glass ($Fe_{78}B_{13}Si_9$, thickness 25 μm) showing two types of stress-dominated domain patterns: wide curved domains with 180° walls along the principal direction of a tensile stress, and finger-print-like 'stress patterns' in the regions of a compressive planar stress. An enlarged look at the stress pattern reveals simple (c) and branched (d) patterns, and sometimes even secondary branching (e). At special points in the range of in-plane patterns the planar tensile stress tensor is degenerate, leading to walls with wall angles of less than 180° (b)

To understand the observed irregular state, we note that the spatial average of the responsible internal stresses must always be zero [920]. No uniform component of the stress tensor can be present in the absence of external forces, so that the domain configuration must be necessarily inhomogeneous if the internal stresses are the dominant source of anisotropy. A further consequence of the theory of elasticity is that because of elastic boundary conditions the stress tensor at the surface must be planar, characterized by two principal values and a characteristic angle. For a positive magnetostriction material, the wide planar domains occur in areas in which at least one of the principal stresses is positive. The local preferred direction is the direction of the largest principal stress. In certain points the two principal stresses may be positive and degenerate (= equal), leaving no effective in-plane anisotropy. In the neighbourhood of such points we observe domain patterns with less-than-180° walls as shown in Fig. 5.46 b.

If the two principal stresses are negative, they generate (for positive magnetostriction) a perpendicular easy axis leading to the mentioned "stress" pattern, which is shown in more detail in Fig. 5.46c–e. The domains visible at the surface are the closure domains of internal perpendicular domains. In strong stress conditions the stress pattern can be branched, a phenomenon explained in Fig. 3.142. The same arguments apply to negative-magnetostriction materials if compressive and tensile stress components are exchanged.

The stress pattern can appear to be superimposed onto a regular wide domain structure (Fig. 5.47). An explanation for that has to be a layered distribution of stresses, not implausible in view of the origin of the stresses in the rapid quenching process. Such layered stresses will be difficult to characterize. A destructive method would consist in analysing the domain patterns as discussed in Sect. 4.6.4, and polishing down the sample step by step. Taking into account the stress relaxation connected with removing part of the material, one could reconstruct the original stress distribution, but this theoretical possibility has not yet been verified as far as we can see.

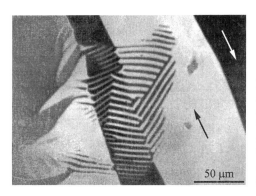

Fig. 5.47 An apparent overlay pattern (*centre*), consisting of underlying planar domains and an embedded stress pattern. The cause must be layered intrinsic stresses — compressive near the surface, tensile underneath. The domains towards the *left* edge indicate layered in-plane anisotropies with twisted easy directions

50 μm

In conclusion, the analysis of magnetic domains in metallic glasses in the native state is complicated by the fact that they are often dominated by internal stresses that are usually not well-known. The possibilities to determine the effective stresses by domain experiments (Sect. 4.6.4) do not reach far when the stresses are layered, i.e. when they are inhomogeneous through the sample. Because the stress tensor field of as-quenched metallic glasses is in general unknown and inaccessible, it appears principally impossible to attain a complete understanding of the magnetic microstructures of metallic glasses in the as-quenched state. In the following subsection we will analyse cases in which the initial stresses have been largely removed by annealing, and a uniform induced anisotropy has been superimposed. But a complete removal of the stresses — as in crystalline materials — is impossible in metallic glasses because of the danger of crystallization. Therefore some knowledge of the consequences of irregular stresses is useful.

5.4.2 Ordered Domain States of Metallic Glasses

A heat treatment below the crystallization temperature usually causes relaxation of more than 95% of the stress. At the same time a controlled induced anisotropy can be generated if the annealing is performed in an applied field. At temperatures below the Curie point the magnetization will induce some preferential atomic pair ordering of the alloy components within the amorphous state, in particular when the alloy consists of more than one species of transition metal elements. Induced anisotropies can also be generated by mechanical stress if it is strong enough to cause a plastic flow [927]. This creep-induced anisotropy can develop even beyond the Curie point, in contrast to magnetic field annealing that does not work in the paramagnetic state. The result is in both cases an ordered domain state, dominated by a single anisotropy axis. This anisotropy can extend throughout the sample, which is desired for most applications, or it can be limited only to a part of the sample volume such as a surface zone. Inhomo-geneous anisotropy patterns can lead to spectacular domain patterns. Some examples are presented in (B), following the discussion of regular, uniformly ordered states in (A).

(A) Uniform-Anisotropy States. Annealing an amorphous sample in a uni-form field (below the crystallization temperature) generates various regular, ordered domain states (Fig. 5.48). If the annealing field is applied along the longitudinal sample axis, it produces wide, almost undisturbed domains (Fig. 5.48a) which, however, lead to high anomalous eddy current losses in alternating fields. To avoid this effect, samples can be annealed in a transverse field generating a pattern of periodic transverse domains along the annealing axis with branched closure domains at the edges ([928]; Fig. 5.48b). Such closure configurations are modified or absent on wound cores, apparently because then a certain degree of flux closure between the laminations is possible. In a longitudinal field these domains are essentially magnetized by rotation, with only minor domain rearrangement processes starting from the edge configurations (see Fig. 6.12). Losses are correspondingly low.

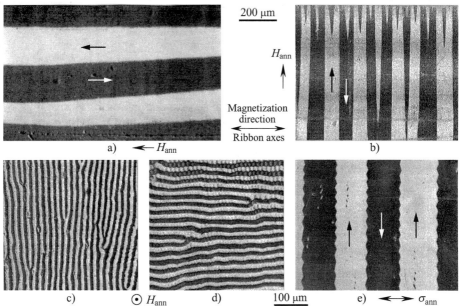

Fig. 5.48 Five patterns obtained in metallic glass strips after different annealing treatments. (a) Longitudinal domains obtained after annealing in a field H_{ann} parallel to the strip axis (usually even much wider domains are found in this case). (b) Domains obtained after a transverse field treatment on a 15 mm wide strip, showing the neighbourhood of the strip edge. (c) Narrow stripe domains induced after annealing a high-magnetostriction material in a field perpendicular to the surface. The stripes are oriented perpendicular to a previously applied alternating field. (d) The same for a low-magnetostriction material, leading to a "parallel" orientation of the narrow stripes. (e) Transverse domain pattern obtained by a stress or "creep" annealing treatment. (Sample and treatment: courtesy *K. Závěta*, Prague)

An induced anisotropy with the easy axis perpendicular to the sample surface is somewhat more difficult to produce in practice. If it succeeds, the generated pattern (Fig. 5.48c) corresponds to a domain state developing out of dense stripe domain nucleation in thin films (Sect. 3.7.2). Magnetization in longitudinal fields occurs by rotation again, as after transverse annealing. Losses can be very low — within a factor of two above the classical eddy current limit [929, 930]. But this low-loss condition is found only for materials with sufficient *magnetostriction*. In magnetostrictive material the stripes orient themselves perpendicular to the field direction (Fig. 5.48c; see also Fig. 3.111b). Longitudinal applied fields cause in this case the closure domains to "breathe", but not to switch discontinuously. In magnetostriction-free material the stripes align along the field (Fig. 5.48d; see also Fig. 3.111a). Here the switching of the closure structure in the applied field (Fig. 5.49) causes extra losses.

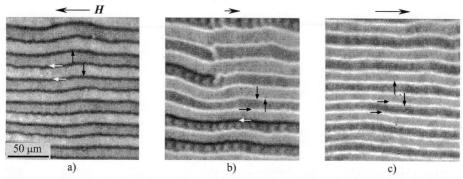

Fig. 5.49 Magnetization processes in perpendicular domains induced in a magnetostriction-free metallic glass. The stripe domains are magnetized along the sample and field axis as in Fig. 5.48d. Discontinuous reorientation processes inside the closure domains occur in low longitudinal fields and are the origin of excess losses. In (a) all V-lines are black, in (c) they have switched to white, and in (b) some are black, some are white and one shows a transition between black and white. See Fig. 3.142 with respect to the faint modulation inside the stripes

A transverse orientation of domains can also be induced by a creep annealing process, which tends to lead to a *negative* uniaxial anisotropy, i.e. a planar anisotropy with an easy plane perpendicular to the creep direction [824, 931–933]. Domain patterns induced in this way, as shown in Fig. 5.48e, often display strongly curved or zigzag domain walls [933]. They are a consequence of the *planar* anisotropy produced by the strain annealing process if for some reason the perpendicular axis is slightly favoured relative to the transverse axis. The phenomenon is analogous to the situation shown in Fig. 5.28 for an iron crystal. The slight preference of the perpendicular axis suffices to convert the internal parts of domain walls into regular internal domains [933].

The preference for the perpendicular axis can be field dependent, as shown in Fig. 5.50. This effect, which was observed in different variants of such materials, must be due to some higher-order anisotropy that is not yet fully explored. In any case, the magnetization processes observed in such samples are very different from the expected simple domain rotation process.

An almost ideal, reversible magnetization rotation behaviour has been demonstrated for perpendicular domains as in Fig. 5.48c if the stripe domains are oriented perpendicular to the working field [929]. This can also be achieved in transversely annealed ribbons (Fig. 5.48b). Often a less ideal behaviour is observed that can be related to an oblique orientation of the induced easy axis, and to anisotropy inhomogeneities that were found to be correlated with anomalies in the domain behaviour [934]. An example of a near ideal behaviour at least in a limited field range is shown in Fig. 5.51. Near saturation, very often severe domain refinement instead of a simple magnetization rotation is observed. This domain-splitting process is connected with hysteresis at least on the domain level. Its origin is not completely clear [935].

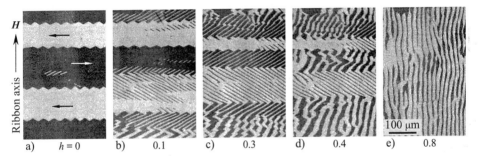

Fig. 5.50 Equilibrium domain patterns in a stress-annealed metallic glass as a function of an applied field along the tape axis. The annealing process generated a basically easy-plane anisotropy, obviously with a slight preference for the perpendicular axis. In the zero-field state (a) the domains are magnetized along the transverse axis, with strongly modulated domain walls as explained in Fig. 5.28. With rising field (b—e) a preference for internal perpendicular domains is indicated by the observed domains. The domain patterns were recorded after applying a.c. idealization for each field. (Sample: courtesy *K. Závěta*, Prague)

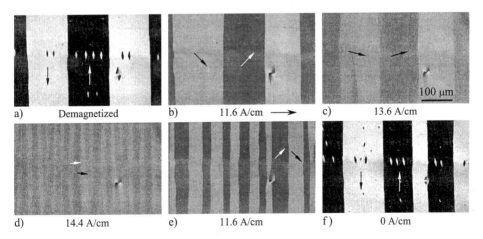

Fig. 5.51 Regular magnetization rotation in a metallic glass strip of transverse anisotropy (as in Fig. 5.48b) leading to a near-ideal soft magnetic behaviour within the field range (a –c). Only if a limiting field is surpassed (d), domains become refined and a hysteretic behaviour is observed in the domain pattern [compare (b) and (e)]. At zero field (f) the original domain width is recovered. (Sample: courtesy *G. Herzer*, Vacuumschmelze)

(B) Partially Ordered Domain States. In this section we discuss three examples of domain patterns in which the configuration observed at the surface extends not throughout the sample but is limited more or less to a surface zone. This can be demonstrated by the apparent layering of the patterns when wide basic domains are "seen" to move below a fine surface pattern. An example was already shown in Fig. 5.47 dealing with as-quenched domain patterns. Also the domains appearing in an applied field in Fig. 5.50 are probably confined to a surface layer but this is not completely clear.

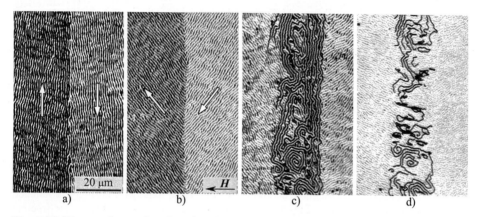

Fig. 5.52 Fine surface stripe domains (a) superimposed onto wide basic domains. The stripe pattern is induced by an effective surface anisotropy resulting from surface crystallization [936]. (b) The reaction of the dense stripes to a rotation of the magnetization direction in the underlying domains. The spectacular patterns (c, d) are formed when the basic domain wall was agitated in an alternating magnetic field

The reason for inhomogeneous effective anisotropies can be manifold. Bending stresses can be an obvious reason, but also surface crystallization or other modifications in a surface layer can have such an effect. An example is shown in Fig. 5.52. The observed dense stripe domains [936] can be attributed to oxidation and metalloid depletion, leading to the crystallization of a thin subsurface zone. The reduced volume after crystallization exerts an elastic stress on the remaining amorphous substrate, leading to an effective surface anisotropy. Strong surface anisotropy can induce dense stripe domains on the surface of bulk material in a similar way as perpendicular volume anisotropies in thin films, as demonstrated by micromagnetic modelling by *Hua* et al. [748]. The surface crystallization mechanism was confirmed by X-ray diffraction analysis and by etching experiments. The dense stripe domain nature of the observed surface domains is consistent with the observation that these stripes follow the magnetization direction in the basic domains which can be rotated by an external field (Fig. 5.52b). The fine stripes are thus closely bound to their basic domains. If the basic domain wall is displaced, the stripes may be turned over, leaving peculiar wound-up patterns as shown in Fig. 5.52c. These metastable patterns may be classified either as strong stripe domains, or as 360° walls that are difficult to destroy even in large fields (d).

A second example of inhomogeneous domain patterns in metallic glasses is shown in Fig. 5.53. These conspicuous domains are found in ribbons that are annealed in a transverse field, and which under other circumstances show simple transverse basic domains as in Fig. 5.48b. The modified patterns were first identified in [937] and were also observed in [938]. They were later studied in more detail in [939]. The transverse domains appear to be modulated by fine stripes that are oriented perpendicular to the principal magnetization direction of the transverse basic domains. This agrees with the symmetry and arrangement of *ripple* domains, well known from thin polycrystalline films (Sect. 5.5.2C). This interpretation of the observed pattern [939] is problematic, however. Ripple in thin films is induced by irregular anisotropies connected with the polycrystalline microstructure of the films. Such a source of magnetic fluctuations is absent in metallic glasses. In addition, ripple was never observed in bulk materials; it is a thin film phenomenon, intimately connected with the properties of the characteristic wall type of thin films, the Néel wall (Sect. 3.6.4C). A further argument is that the observed pattern is periodic, not resembling the typically aperiodic ripple pattern.

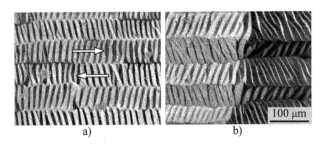

Fig. 5.53 Dense stripe domains embedded into transverse basic domains in a transversely annealed metallic glass ($Fe_{73}Ni_6B_{14.5}Si_{6.5}$) ribbon. In (b) traces of longitudinal domains can be seen. (Sample: *W. Grimm*, Erlangen)

100 μm

a)   b)

The correct interpretation must be a dense stripe domain pattern again. (Remember that volume "dense" stripe domains in the sense defined in Sect. 3.7.2 scale in their period with the film thickness. Although the modulations considered here are much wider than those of Fig. 5.52, their nature is the same, only that the effective thickness reaches further into the sample here). We saw in Fig. 5.48c and in Sect. 3.7.2C that stripe domains can be formed perpendicular to the initial magnetization direction in magnetostrictive materials. Here this modulation is embedded into the transverse domains that probably occupy half of the sample thickness and which remain visible primarily through their domain boundaries that interrupt the stripe pattern. In addition to the transverse domains also traces of wide longitudinal domains can be identified in Fig. 5.53b which appear to occupy the lower half of the sample.

The third example relates to a low-magnetostriction material. Again dense stripe domains are superimposed onto wide basic domains (Fig. 5.54). The difference to the previous case is that the stripe domains in this low-magneto-striction material are oriented parallel or oblique to the magnetization direction of the basic domains as in Fig. 5.49, not perpendicular as in Fig. 5.53.

The anomalous patterns presented here indicate the difficulties of controlling the domain structure in metallic glasses. Stresses cannot be removed completely, and beginning partial crystallization can generate unexpected results.

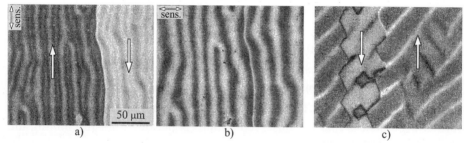

Fig. 5.54 Stripe domains, superimposed onto wide basic domains in a low-magnetostriction metallic glass. The stripes (a,b) are oriented parallel to the magnetization in the basic domains (*arrows*). They are shown with two different sensitivity directions. (c) Sometimes the stripes are formed at oblique angles. (Sample: courtesy *K. Závěta*, Prague [940])

5.4.3 Wall Studies on Metallic Glasses

The surface structure of Bloch walls in bulk materials (see Fig. 3.82) can be observed by various methods, such as magneto-optical scanning methods [941, 206], Kerr microscopy with digital image subtraction [171, 942], spin polarized scanning electron microscopy [352, 943, 944] and magnetic force microscopy [945–947]. The Bitter method can reveal many details of the deep features of domain walls but is less conclusive for the surface closure configuration [698, 89]. Magnetization processes involving substructures of domain walls are of particular interest [948–951]. The low anisotropy of metallic glasses favours wide domain walls that can be easily observed optically.

All observed walls in these materials are vortex walls, derived from the asymmetric Bloch wall in thin films (see Sect. 3.6.4). Particularly interesting

features are found after a heat treatment in the presence of a domain wall
[699]. Then the pattern of induced anisotropy follows the magnetization profiles
of domains and walls at annealing. After cooling, preferred positions for
domain walls are left, which interact in a complex manner with freely moving
walls, leading to anomalous magnetization curves that can be used in sensor
applications. Some of these processes are shown in Fig. 5.55.

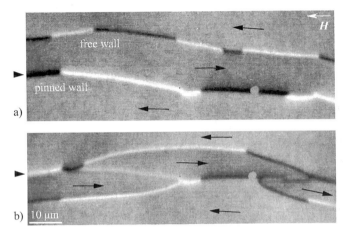

Fig. 5.55 Domain wall reactions [699] in a metallic glass sample annealed in a field-free state. In an applied field a free domain wall approaches a baked-in, pinned wall in (a). The complex interaction of the two walls (b) switches parts of the pinned wall, but leaves the transition line pattern of this wall unchanged

Figure 5.56 shows the different transitions possible inside a Bloch wall in
a metallic glass ribbon (see Sects. 3.6.4–5 for the theoretical background).
A "Bloch line" marking a change in the basic rotation sense of the wall is
connected with a kink. A "cap switch" affects the surface vortex of the wall
only. It is characterized by a change in the magneto-optical contrast, and, if
not coinciding with a Bloch line, displays a lateral displacement only. Similar
observations can be made on other soft magnetic materials.

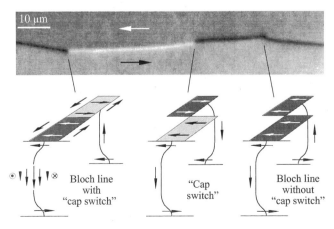

Bloch line with "cap switch"

"Cap switch"

Bloch line without "cap switch"

Fig. 5.56 High-resolution magneto-optical images of a (free) Bloch wall in a magnetostriction-free metallic glass, showing all three possible types of wall transitions

Figure 5.57 shows how wall transitions can be moved in a field perpendicular to the surface, which has virtually no effect on the domains. The field does not influence any features in the baked-in wall. In the free wall the *Bloch lines* are displaced, while the cap switch transitions stay unaffected.

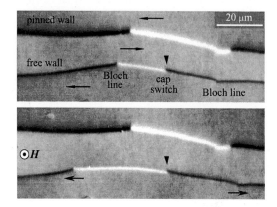

Fig. 5.57 Motion of Bloch lines in a vertical field observed on the same samples as in the previous figures. The *triangle* points to a cap switch transition that is not connected with a Bloch line and which therefore does not move in the vertical field

5.4.4 Domains on Soft-Magnetic Nanocrystalline Ribbons

High-permeability nanocrystalline materials [952] consist of about 10 nm wide crystalline grains of ferromagnetic silicon iron (in the most widely used alloy) which are magnetically coupled by a ferromagnetic metallic glass matrix. The exchange coupling between the two magnetic phases leads to a seemingly uniform magnetic material, the magnetic microstructure of which differs little from amorphous ribbons as shown in Fig. 5.58a.

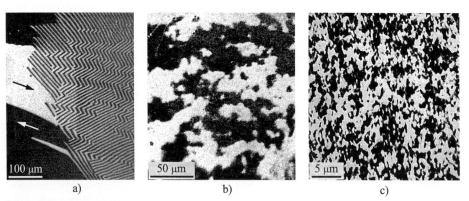

Fig. 5.58 Domains on a nanocrystalline FeCuNbSiB ribbon at room temperature (a) and at 350°C (b) beyond the Curie temperature of the amorphous component [953]. (c) The irregular magnetic microstructure of a sample that was annealed at 800°C, destroying the nanocrystalline microstructure. (Sample: courtesy *G. Herzer*, VAC Hanau)

Heating the ribbon above the Curie point of the amorphous phase at about 300°C interrupts the coupling. The coercivity rises and the magnetic domains get a dramatically different appearance (b). A state equivalent to that in (a) is recovered after cooling from these moderate temperatures to room temperature when the amorphous matrix gets ferromagnetic again. If, however, the nanocrystalline microstructure is destroyed by "overannealing" at 800°C, the soft magnetic behaviour is permanently lost, and only very hard and immobile fine domains can be seen as shown in (c). Field and stress annealing treatments [954] may produce ordered domain states in nanocrystalline materials, just as in amorphous ribbons (see Fig. 5.48).

5.5 Magnetic Films with Low Anisotropy

5.5.1 Overview: Classification of Magnetic Films

In this section we deal with thin films with $Q < 1$, which means that if no domains are present the magnetization lies essentially parallel to the surface, independent of the orientation of the easy directions. To be more precise, the lowest-energy state with a *uniform* magnetization is in the absence of a magnetic field magnetized parallel or nearly parallel to the surface. It is exactly parallel if the easy axis is parallel to the film, but also if it is perpendicular. If the easy axis is oblique, a uniform magnetization is tilted by a small angle out of the film. All this does not mean, however, that the state of minimum energy for homogeneous magnetization is also the lowest energy state. A domain state, in which the interior domains follow the true easy directions, can be more advantageous. Such a domain state requires a critical thickness as discussed in Sect. 3.7.2A. Below this thickness the films behave essentially like films with an in-plane anisotropy, even if the true easy axis points out of the plane.

In thin films the structure of the domain walls is strongly influenced by the film thickness. The domain wall properties dominate the magnetization behaviour to a larger extent than in bulk material. Particularly the distinction between *symmetric* Néel walls with their long range stray field, and the largely stray-field-free *asymmetric* Bloch and Néel walls (Fig. 3.80) is important. The following classification of magnetic low-anisotropy films, which is based on this and further arguments, helps in ordering the wide variety of phenomena:

- *Thin* and ultrathin films with in-plane magnetization both in the domains and in the walls. The domain boundaries are Néel walls with cross-ties and their characteristic long-range interactions.

- *Thick* films with uniaxial in-plane anisotropy, showing essentially stray-field-free vortex walls (asymmetric Bloch and Néel walls) which tend to be rather mobile.

- Film *elements*, the domain structure of which is determined more by the element shape than by the intrinsic properties of the material. Small elements can also be interpreted as magnetic particles.

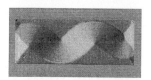

- Thick films and platelets with weak perpendicular or oblique anisotropy, forming various kinds of dense *stripe* domains (weak and strong stripe domains).

- Multiaxial *single-crystal* films and platelets with multiple easy axes. Lower-angle walls that are possible in these films largely replace 180° and cross-tie walls.

- Double and multiple films for which domain patterns and wall structures follow special rules, including antiferromagnetically coupled films and films with 90° coupling.

The classical Bitter pattern images by *Middelhoek* [955] on a series of Permalloy films with increasing thickness demonstrate the difference between thin and thick films in a beautiful way. Figure 5.59 shows a similar series obtained with the most versatile technique of today, the digitally enhanced magneto-optical Kerr effect.

An attempt was made to reach equilibrium configurations using an alternating field of decreasing amplitude along the hard axis. The basic character of the walls is reproducible in this way, and the number of visible transitions in the walls is random in particular in very thin films (a) and at large thicknesses (f–h). The boundary between thin and thick films lies for this material around 90 nm, with an interesting transition region with domain walls of the highest specific energy, the smallest overall width and the most complex three-dimensional structure.

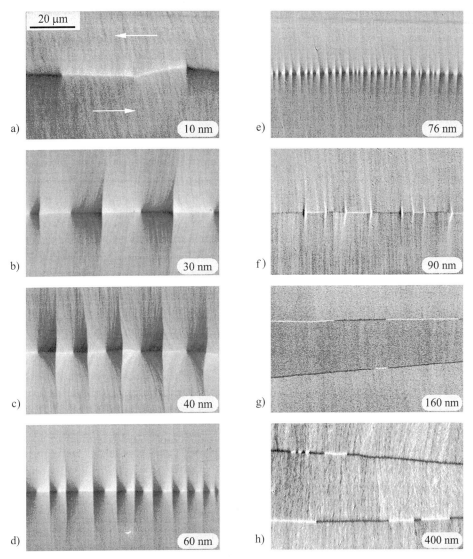

Fig. 5.59 Domain wall images of 180° walls observed on sputtered and evaporated Permalloy films with increasing thickness. The material is fine-grained polycrystalline with a weak uniaxial induced anisotropy. (a) Pure symmetric Néel wall with occasional, non-equilibrium wall transitions, separating segments of different wall rotation sense. (b–e) Cross-tie walls with increasing cross-tie density. (f) Mixed wall with cross-ties and asymmetric Bloch wall segments. (g, h) Asymmetric Bloch walls of increasing wall width with occasional, non-equilibrium transitions. (Samples: courtesy *J.C.S. Kools*, Philips Eindhoven, *U. Wende*, FHG Dresden, and *R. Thielsch*, IFW Dresden)

5.5.2 *Thin* Films

Typical representatives of this class are polycrystalline Permalloy films of a thickness below 80 nm. Such films are deposited by evaporation, sputtering or electrochemical methods. During deposition a magnetic field is applied, resulting in a uniaxial induced anisotropy. If no field is applied, a local anisotropy is generated, following the more or less random magnetization direction that happens to be built up in this process. Anisotropy can be suppressed by a rotating field or by rotating the sample during deposition, but this possibility is rarely employed.

(A) Néel and Cross-Tie Walls. The characteristic feature of thin films is the symmetric Néel wall. The conspicuous cross-tie wall in somewhat thicker films within the thin film stability range (Fig. 5.59b–e) has to be considered a variant of the symmetric Néel wall (see Sect. 3.6.4A). Both wall types can be described by an essentially in-plane magnetization distribution. As discussed in detail in Sect. 3.6.4C, magnetic charges cannot be avoided in Néel walls. In a completely planar model these charges would reside in the volume of the film. Actually, the charges are transported towards the surfaces, and to achieve this, a small deviation from the surface-parallel magnetization is added. A discussion that ignores this subtlety does not introduce a significant error in single-layer films, however. We may therefore consider the whole magnetization pattern of a thin film as two-dimensional, except in Bloch lines (Sect. 3.6.5C) where the magnetization has to point into the normal direction.

The magnetic microstructure of such films can be most readily investigated in transmission electron microscopy [247–249, 956–959]. Also the Bitter technique has proved valuable for these samples because the domain walls in thin films generate strong far-reaching stray fields [960, 955]. Under these favourable circumstances many details of their behaviour were known before they were understood micromagnetically. Some structures, such as the cross-tie wall, still form a challenge to theoretical analysis. First numerical results [656] confirm the experimentally observed picture. But the calculations are so far limited to patterns with a small period, and also the achievable discretization is probably insufficient for the fine structure of the Bloch lines.

The long-range stray field interaction between Néel walls (Sect. 3.6.4C) causes strong irreversibilities in thin film magnetization processes. Figure 5.60 demonstrates these interactions for films of different thicknesses. The interactions are especially apparent between the "cross-ties" of cross-tie walls.

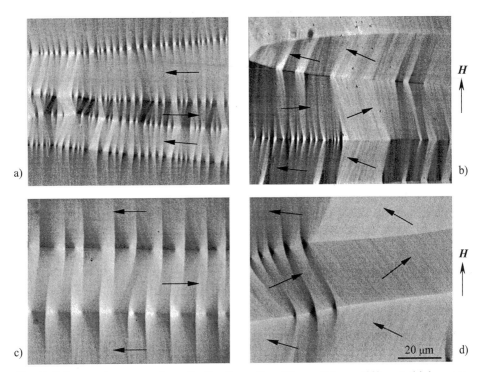

Fig. 5.60 Interacting cross-tie walls for Permalloy films of three different thicknesses. (a) 76 nm; the cross-ties interact strongly between neighbouring walls. (b) 90 nm, near the transition between cross-tie and vortex walls; cross-ties display strong interactions, while vortex walls (asymmetric Bloch and Néel walls) have a more localized character, with interactions visible only between the transitions in these walls. (c) 59 nm; classical cross-tie structures, which disappear in hard axis fields. (d) This process has happened in one half of the picture frame, while in the other half the strongly interacting cross-ties are still present. (Samples: courtesy *R. Thielsch*, IFW Dresden, *J.C.S. Kools*, Philips Eindhoven, and *S.S.P. Parkin*, IBM, Almaden)

(B) Easy Axis Magnetization Loops. If extended thin film elements are magnetized along the easy axis, domains are usually nucleated at a sample edge oriented perpendicular to the field. In low-coercivity films the nuclei propagate along the easy axis, separated from the original domain by a zigzag-shaped wall (Fig. 5.61a).

Zigzag walls in thin films are charged metastable walls that would not exist if an overall rearrangement of the domain pattern would be possible. Charged walls are allowed in thin films because the stray field can spread into the space above and below the film, thus avoiding the concentrated stray field that forbids the occurrence of charged walls in bulk soft magnetic materials. High resolution observations (Fig. 5.61b) reveal that the domain walls within a zigzag pattern may consist of cross-tie walls just as regular 180° walls in films of the same thickness. This finding confirms the concept of the internal structure of zigzag walls indicated in Fig. 3.83b, in which the charges are distributed in the space between the zigzag walls, and the walls themselves are conventional uncharged 180° walls, which decay into cross-tie walls if the equilibrium cross-tie period is small compared to the zigzag period.

The zigzag angle is characteristic for a given film, but the zigzag period depends on the circumstances. The period is reduced in a field gradient (Fig. 5.61c) which leads to a thermodynamically stable zigzag state. In thin films, which are used for longitudinal recording, the zigzag walls may separate the information bits; their irregularities are a source of unwanted noise.

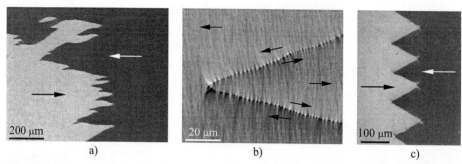

Fig. 5.61 Zigzag walls separating antiparallel domains. (a) Co film (thickness 60 nm) in which domain walls move only jerkily. (b) High-resolution image of a cobalt film (thickness 42 nm) displaying a cross-tie pattern on the individual walls. (c) Equilibrium zigzag wall generated by a gradient field acting on a Permalloy film of 50 nm thickness. (Samples: courtesy *B. Pfeifer*, Univ. Augsburg, and *W. Ernst*, IFW Dresden)

There are magnetic films that do not show a difference in the magnetic behaviour along the easy and hard axis although a well-defined anisotropy is measured by torque methods (Sect. 4.2.1). In such films the wall motion coercivity H_c is larger than the anisotropy field H_K. Before the origin of their irregular behaviour was clarified they were given the generic name "inverted" films [961–963]. The typical behaviour, which ordinary films show only in magnetization along the hard axis as discussed below, is found in inverted films for all field directions.

(C) Hard Axis Loops — Ripple and Blocking. A striking feature of hard axis loops is the large remanence even after domains have been nucleated. The remanent state consists, as shown in Fig. 5.62c, of a densely packed system of interacting low-angle Néel walls which stabilize themselves, forming a metastable, blocked or "buckling" state [964–968].

The system of blocked domains evolves out of an incipient *ripple* structure (Fig. 5.62a) which reflects the irregular polycrystalline nature of the films. The ripple phenomenon was described theoretically as the reaction of the film to a statistical perturbation by the crystal anisotropy of individual grains [969–971]. Its characteristic texture orthogonal to the average magnetization direction is caused by the stray field energy that is smaller for a longitudinal modulation than for a transverse modulation. If two laterally neighboured grains are magnetized along different angles ϑ_1 and ϑ_2 relative to the average easy axis, a "transverse" charge $\lambda_{\text{trans}} = \sin\vartheta_1 - \sin\vartheta_2$ is generated at the grain boundary.

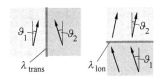

For small ϑ this is much larger than the longitudinal charge $\lambda_{\text{lon}} = \cos\vartheta_1 - \cos\vartheta_2$ appearing at the grain boundary of longitudinally neighboured grains. The stray field energy therefore suppresses lateral variations of the magnetization, but allows small longitudinal variations, particularly for the case $\vartheta_1 = -\vartheta_2$ for which λ_{lon} stays zero even for large deviations.

The transition from ripple to the blocked state occurs when, with increasing excursion angles ϑ_1 and ϑ_2, the domain wall between two oppositely deviating zones becomes important.

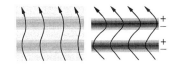

Symmetric Néel walls with a localized core and dipolar charges as discussed in Sect. 3.6.4C are formed. The stray field of one wall acts as an effective field for the next wall, thus stabilizing an array of nearly 90° walls and avoiding energetically unfavourable 180° walls. The walls in the blocked structure (Fig. 5.62c) in fact display mostly a wall angle close to 90° as can be derived from the ripple that becomes visible in the hard-axis sensitivity aspect (right hand column) of the same figure.

Qualitatively, the blocking mechanism was clearly understood already by Feldtkeller [962]. A quantitative theory of the buckling or blocking phenomenon calls for the same tools as the analysis of symmetric Néel walls [972, 973] as performed by van den Berg [974]. One obvious result of a complete theory is that the distance between the interacting Néel walls in the blocked state must be less than the range of the Néel wall tail (3.144).

The blocking breaks down in a field of opposite polarity relative to the previous saturation field, which stimulates the nucleation of Bloch lines. For the given film thickness in Fig. 5.62 this leads to the formation of cross-tie walls (d). As explained in Fig. 3.75, the cross-tie pattern achieves the replacement of 180° walls by a system of 90° walls in a different way.

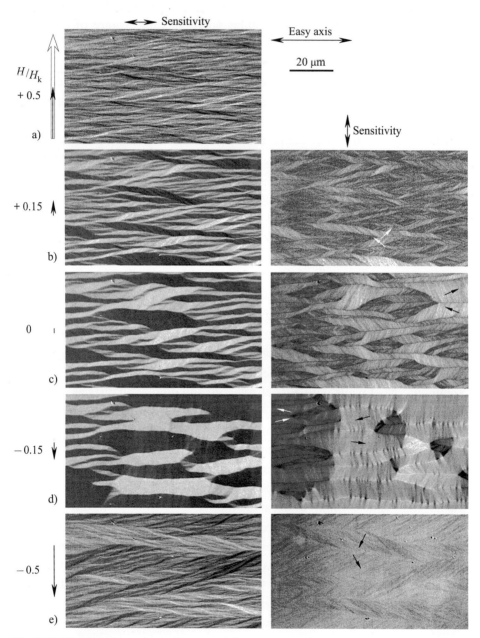

Fig. 5.62 The domain patterns observed during a magnetization cycle along the hard axis of a thin polycrystalline uniaxial cobalt film of 42 nm thickness. One branch of a hysteresis loop is shown from *top* to *bottom*, the two *columns* displaying the magnetization components along the easy axis (*left*) and along the hard axis (*right*). The local magnetization directions can be derived from the ripple pattern (a) which is not only the origin of all the structures shown here, but which can be seen locally in all patterns best on the transverse sensitivity images on the *right*. (Sample: courtesy *B. Pfeifer*, Univ. Augsburg)

Once cross-ties are generated, their segments can be shifted by applying a field along the hard axis. This process shows a considerable hysteresis connected with the pinning of the Bloch lines. The circular Bloch line is more mobile than the cross Bloch line (Fig. 5.63), which can be understood qualitatively from the environment of the two lines: the circular Bloch line is free to move inside its Néel wall segment, but the cross Bloch line is bound to the intersection of two walls. When it moves, the transverse wall has to move with it, and additional friction and pinning effects will be connected with this displacement. Pairs of Bloch lines in a cross-tie wall were proposed as information carriers [975] in a non-volatile solid state memory.

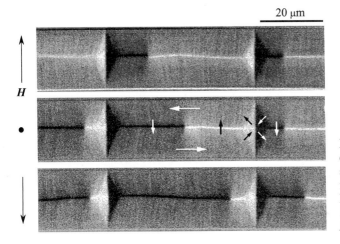

Fig. 5.63 Bloch lines in a cross-tie wall moving in small external hard-axis fields (150 A/m). (Sample: Permalloy film element of 50 nm thickness; courtesy *M. Schneider*, HL Planar)

In oblique fields a coherent magnetization rotation is expected according to the Stoner-Wohlfarth theory (see Sect. 3.5.2). Actually, an asymmetrically blocked pattern is formed, consisting of phases that would not be separately stable (Fig. 5.64). The magnetization cycle starts after saturation along the easy axis with a slightly inhomogeneous, but still almost saturated zero-field state in (a). In an opposite field this state would be switched by a 180° domain wall motion. Applying, however, an opposite field at an angle 35° off the easy axis, the net magnetization rotates towards the field, simultaneously developing a strong ripple pattern (b, c). In (d) a blocked configuration has been formed, based on stray field interactions between its components.

This texture is formed by two effects: (i) The average magnetization starts to rotate essentially as predicted by a uniform rotation model. (ii) The ripple dispersion develops with the wavevector parallel to the average magnetization,

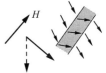

and with *symmetric* excursions to avoid charges on the ripple walls. The two subphases forming in this way carry different levels of anisotropy energy. Switching the field off (e, f), the favoured (white) phase simply turns into the nearest easy direction, while the disfavoured (dark) phase decays into a secondary pattern, forming the so-called labyrinth pattern. All these blocking and incoherent switching processes must be avoided if rapid switching is needed in device applications.

Further examples of magnetization processes in thin films will be shown in Sect. 5.5.4 dealing with microfabricated film elements.

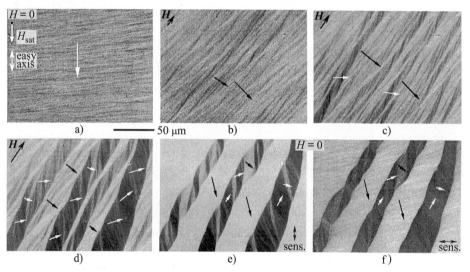

Fig. 5.64 A blocked pattern (d) forms in an oblique field of 150 A/m. Switching it off leads to the labyrinth pattern (e, f) which is shown in two different aspects. (Sample: Permalloy 40 nm; courtesy *M. Freitag*, Bosch)

(D) Very Thin Films. In films of less than about 30 nm thickness (for Permalloy) cross-tie walls are no longer observed. This is connected with the increasing energy and decreasing mobility of Bloch lines. The Néel walls in these films decrease in energy and increase in width with decreasing thickness (see Fig. 3.79), but the Bloch lines display just the opposite behaviour (Fig. 3.91). The Bloch lines are easily caught by defects, particularly by "pinholes", non-magnetic gaps in the film in which the high exchange and stray field energy concentrations of the Bloch line core can be saved. The Bloch lines remain pinned even if the domain walls generate all kinds of reactions and non-equilibrium configurations as a consequence. For example, if a 180° wall passes over such a defect and a Bloch line gets caught, a 360° wall [960] is generated connecting the wall and the defect as shown in Fig. 5.65. This metastable wall type is not found in thicker films. In very thin films it can only be eliminated in much higher fields than needed for moving regular 180° walls. Responsible is an increasing penalty on turning the magnetization perpendicular to the film plane — the only way of annihilating the winding wall pair that makes up a 360° wall. These phenomena were investigated in classical transmission electron microscopy and Bitter pattern studies [960]. The Kerr effect images shown here confirm the earlier results offering more freedom in the choice of samples and field conditions.

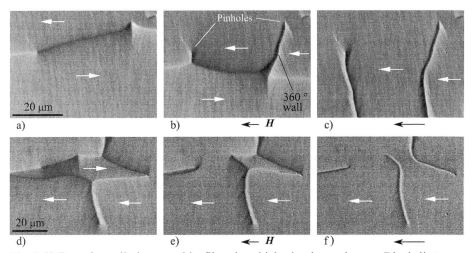

Fig. 5.65 Domain walls in very thin films in which circular and cross Bloch lines are pinned by defects in the film, observed on a sputtered Permalloy film of 10 nm thickness. (a–c) The formation of two "vertical" 360° walls out of pinned Bloch lines in a 180° wall. The second sequence (d–f) shows the annihilation of two "unwinding" 180° wall segments, and the formation of "horizontal" 360° walls out of winding 180° wall segments. (Sample: courtesy *J.C.S. Kools*, Philips Eindhoven)

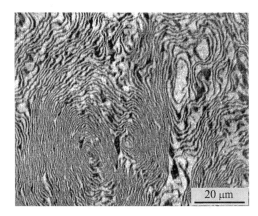

Fig. 5.66 Spiral walls generated by an alternating field consisting of densely packed 360° walls, observed on a Co film of 5 nm thickness. (Sample: courtesy *T. Plaskett*, IBM)

Spectacular spiral domain patterns [969, 976, 977] can be generated in such films by applying alternating fields as shown in Fig. 5.66. Somewhat paradoxically, alternating fields produce in this case not an equilibrium pattern but rather the contrary. The reason is that the used alternating fields here are not sufficient to activate the decisive pinning sites.

With modern ultra high vacuum thin film deposition techniques epitaxial films with much improved quality can be deposited. With these techniques the range of ultrathin films that consist of few atomic layers only becomes accessible for the first time. The problem is that few, widely separated defects suffice to pin the domain wall network. Domains in ultrathin films therefore regularly do not follow equilibrium laws. In such films domain wall coercivity is usually larger than any of the forces derived from the equilibrium energies.

Strong changes in the anisotropy conditions are reflected in the domain texture, however, as demonstrated in Fig. 5.67. Here a thickness-dependent surface anisotropy contribution causes a transition from $Q > 1$ to $Q < 1$, i.e. from a perpendicular easy axis for very thin films to a preferably in-plane behaviour beyond 4 monolayers. The domain width observed with electron polarization methods turns out to be rather small in the neighbourhood of $Q = 1$. Note that in a certain thickness zone the domains are not resolved, neither in the perpendicular (a) nor in the longitudinal observation mode (b).

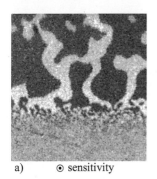 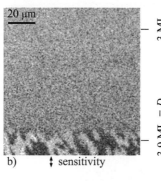

a)   ⊙ sensitivity   b)   ↕ sensitivity

Fig. 5.67 Domains in a wedge-shaped ultrathin epitaxial Co film displaying a transition between perpendicular and in-plane anisotropy. (ML: monolayers; electron polarization image courtesy H.-P. Oepen, Halle [978])

5.5.3 Thick Films with In-Plane Anisotropy

Thick films are defined as films with a thickness beyond the Néel-Bloch wall transition, which occurs typically at about 100 nm for Permalloy films or $20\sqrt{A/K_\mathrm{d}}$ in general (Fig. 3.80). The upper limit can be set at about $5\sqrt{A/K_\mathrm{u}}$ for uniaxial films (corresponding to about 1.5 µm for Permalloy), when the characteristic thin film Bloch wall gradually transforms into the classical planar wall at least inside the sample (Fig. 3.82). Within this thickness range, the specific Bloch wall energy decreases with increasing thickness (Fig. 3.79), while the wall width scales with the thickness. In spite of its large width the Bloch wall stays localized and does not extend into the domains, in contrast to the symmetric Néel wall of thin films.

(A) Magnetization Processes in Thick Films. The hard axis magnetization loop is again the most interesting and is shown for an extended film in Fig. 5.68. Coming from saturation along the hard axis, first interacting Néel walls are generated as in thin films (Fig. 5.62). Initially, they are nucleated as symmetric Néel walls, but at a lower field they are spontaneously converted into asymmetric Néel walls according to theory (Sect. 3.6.4E).

This transition is hard to observe experimentally, but the result shown in Fig. 5.68a at somewhat lower fields displays all characteristics of the asymmetric Néel wall type. Near remanence the metastable Néel wall system breaks down to be replaced by asymmetric 180° Bloch walls (Fig. 5.68b, c). The discontinuous transition between asymmetric Néel and Bloch walls, which calls for the introduction of transition or Bloch lines as in Fig. 5.62, is immediately apparent in the wall contrast images of Fig. 5.68 (see (B) for more details). Applying small alternating fields, the transition lines between the different wall types can be shifted. In large fields of opposite polarity another wall conversion to the asymmetric Néel wall takes place (Fig. 5.68d). Interestingly, not much domain widening occurs in the extended film in connection with wall conversions — in contrast to the situation in thin films (Fig. 5.62). This is due to the small energy difference between asymmetric Bloch and Néel walls.

The Bloch-Néel transition leads to unexpected phenomena. One of them, wall creeping [979, 980], contributed to the downfall of a historically promising technology, the fast random-access thin-film memory (see Sect. 6.5.3A for details), the predecessor of modern semiconductor computer memories. In this device domain walls are exposed to hard axis field pulses, which leads to a Bloch-Néel conversion. If easy axis pulses are simultaneously present, the walls are observed to creep by an amount roughly equal to the film thickness for each hard axis pulse, even if the easy axis pulses alone would not be sufficient to move the walls.

The explanation of the wall creeping phenomenon in thick films follows immediately from the domain wall properties. In every Bloch-Néel conversion one half of the wall is restructured, so that it can come free of local pinning centres [658]. The wall creeps into the direction of the applied easy axis field in a manner comparable with intermittent snail motion.

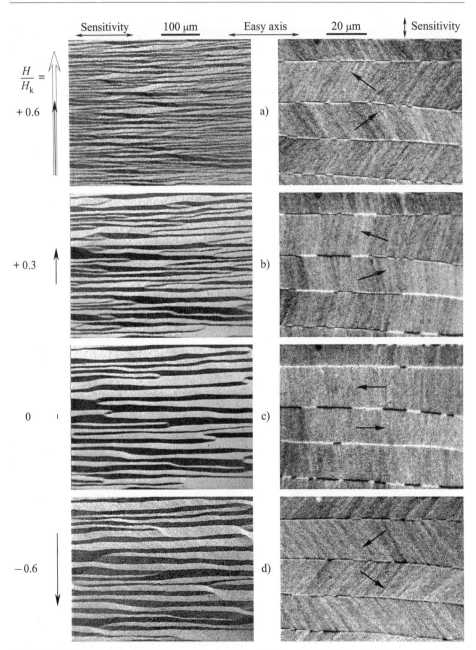

Fig. 5.68 The magnetization process in an extended Permalloy film of 460 nm thickness magnetized along the hard axis. Kerr effect overview domain images are shown in the *left column*, and magnified wall images of the same states on the *right*. (a) A system of interacting asymmetric Néel walls is formed coming from saturation. Near remanence or in the demagnetized state this is replaced by a pattern with asymmetric Bloch walls separating the principal domains (b, c). Increasing the field in the reverse direction (d) the Bloch walls are converted into Néel walls again, but the system of densely packed interacting walls is not reconstructed. (Sample: courtesy *U. Wende*, FHG Dresden)

(B) High-Resolution Domain Wall Investigations. The characteristic wall structures of thick films predicted by theory have been a favourite subject of experimental study. An overview over domain walls in thick films was shown in Fig. 5.68 and discussed in (A). The first direct observation of the asymmetric Bloch wall was achieved by *Tsukahara* [261] by conventional Fresnel mode Lorentz microscopy on single-crystal iron films (Fig. 2.27). In this mode the connection between contrast and magnetization distribution is indirect. But contrast calculations that started from theoretical models led to a convincing agreement between expected images and observations [981].

The in-focus differential phase technique (Sect. 2.4.3) gives direct access to the average in-plane component of the magnetization within a wall [267], but this is of course still short of a complete two-dimensional analysis of the wall structure. Tilting of the sample has been performed in conventional Lorentz microscopy [302] confirming the perpendicular component of both the asymmetric Bloch and Néel walls. Detailed work using conventional Lorentz microscopy made it possible to distinguish between all three wall types, the asymmetric Bloch and Néel wall and the symmetric Néel wall expected at low wall angles (Fig. 5.69a; [676, 664]). Even the theoretically predicted phase diagram (Fig. 3.80) was confirmed with reasonable accuracy (Fig. 5.69b).

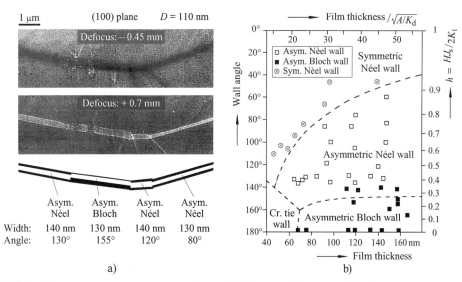

Fig. 5.69 Lorentz microscopy observations of different wall types on $Ni_{50}Fe_{50}$ single-crystal samples (a). The evaluation of the divergent and the convergent image helps to assign the wall type to each segment. The wall angle can be derived from the ripple in the domains. The observations were entered into a phase diagram (b) confirming qualitatively the theoretical predictions of Fig. 3.80. (Together with *L. Zepper*, Stuttgart [676])

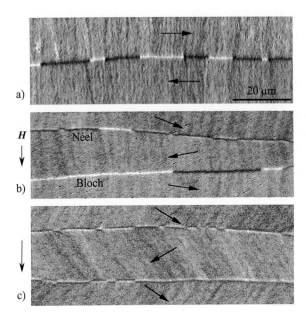

a)

H

b)

c)

Fig. 5.70 Kerr effect wall observations on a Permalloy film of 460 nm thickness. (a) 180° asymmetric Bloch wall consisting of segments of different rotations. (b) Coexisting asymmetric Bloch walls (black and white contrast) and asymmetric Néel walls (dipolar contrast) for wall angles between 160° and 180°. (c) Asymmetric Néel walls for a wall angle of about 120° with two different orientations. (Sample: courtesy *U. Wende*, FHG Dresden)

Figure 5.70 shows wall observations with the Kerr effect at higher magnification compared to Fig. 5.68, displaying clearly the different appearance of asymmetric Bloch and Néel walls. Asymmetric Bloch walls are characterized by strong black or white contrast as shown before in Fig. 5.56 for bulk amorphous material. In films the wall kinks indicating Bloch lines are less pronounced than in bulk material.

In Fig. 5.70b asymmetric Bloch walls are found to coexist with a different wall type recognizable by a weaker, dipolar contrast. Their appearance agrees with the properties of asymmetric Néel walls (Sect. 3.6.4E). As theoretically expected, asymmetric Néel walls are found in two orientations with alternating dipolar contrast. They are particularly stable at moderate applied fields, that is at wall angles of about 120° as in (c). The visualization of the dipolar contrast of asymmetric Néel walls in Kerr microscopy is quite demanding and needs a careful adjustment of the microscope as first achieved by *M. Rührig* [591].

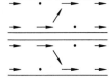

5.5.4 Thin and Thick Film Elements

The analysis of small elements is rewarding for two reasons: (i) magnetization processes in small elements can be better understood than in extended samples because the complete domain pattern can be observed, and (ii) in applications small elements of various shapes are used more often than extended films. The subject has been studied for quite some time, firstly on naturally grown platelets [641, 982, 983] to which we return in Sect. 5.5.6 on single-crystal films, later on artificially structured deposited films [588, 589, 974, 984–992], and lately also on very small elements [993–995] or "particles" (see Sect. 5.7.2). Most of the phenomena observable on small "particles" can also be found on larger elements, with the advantage of easier access.

The magnetization processes in film elements are primarily determined by their shape. Because this shape is given by the two-dimensional outline of the particle and the film thickness only, these processes can be analysed more easily than those in general three-dimensional particles. Still they display a fascinating variety of patterns, and we can only give an impression of a few typical examples, focusing in particular on hysteresis phenomena. Many features in film elements are common or at least similar in thin and thick films. This is why we treat them together here. We mainly present magneto-optical observations from somewhat larger thin film elements that can be analysed in all detail in their equilibrium configuration and in their magnetization behaviour.

For very small elements a treatment as a single-domain particle, relying on the Stoner-Wohlfarth analysis (Sect. 3.5.2), is justified. But even Co elements of 1 μm × 0.1 μm displayed an incoherent switching behaviour although having a single-domain zero-field equilibrium state [996]. A single-domain analysis makes no sense for larger elements in which domain patterns are formed spontaneously. Although this approach can be found in many traditional discussions of this subject, it cannot be considered helpful. The only feature in common between a Stoner-Wohlfarth particle and an extended thin film element is the existence of hard and easy axes.

(A) Demagnetized States. Figure 5.71 shows attempts to find the magnetic ground state on thick film elements of different shape and easy axis orientation.

The domain patterns were obtained by applying an alternating field of decreasing amplitude along different directions relative to the element axis. All obtained ground states agree with simple or modified van den Berg models (Fig. 3.22). But the patterns are not always reproducible. The most frequently appearing results are shown, and these depend on the direction of the alternating field. This finding demonstrates that even in apparently perfect samples the experimental definition of a magnetic ground state, which might be compared with theory, is quite problematic. The interaction of wall clusters with the sample edge seems to be the primary source of this ambiguity. Where it is

absent as in Fig. 5.71b, c unique and reproducible results can be obtained by
a.c. demagnetization.

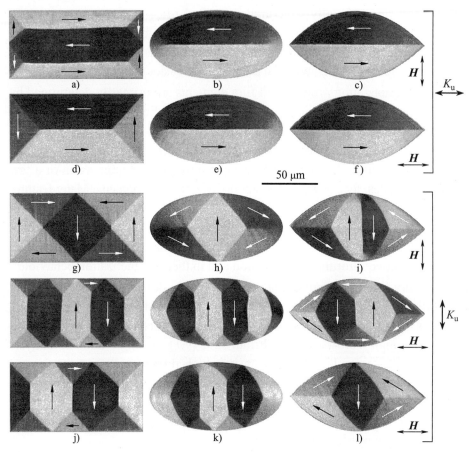

Fig. 5.71 Demagnetized states of various thick film elements (Permalloy = $Ni_{81}Fe_{19}$ of
240 nm thickness). The particles differ in their shape, and in the orientation of their easy
axis relative to the particle axis. The resulting demagnetized states depend markedly on the
alternating field axes used in the demagnetizing procedure. Interestingly, this is not true for
elliptical and pointed shapes with a longitudinal easy axis (b–e, c–f). Different patterns
can be formed under the same conditions as shown in the *last two lines*, which apply to
transverse easy axes and longitudinal a.c. fields. (Samples: courtesy *M. Freitag*, Bosch,
Stuttgart. Samples of this set are used throughout this section unless otherwise stated)

The corresponding behaviour of thin film elements is generally characterized
by a preference for 90° walls. Two alternatives are frequently encountered in
an element with a transverse easy axis after a demagnetizing treatment as
shown in Fig. 5.72: a variant of the cross-tie pattern (a, b), and a blocked
system of interacting Néel walls, preferred after a hard-axis idealization (c, d).

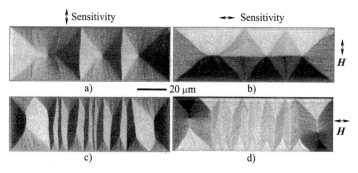

Sensitivity   Sensitivity

H

a)   —— 20 μm   b)

c)   d)

Fig. 5.72 Demagnetized states of a rectangular thin film element (Permalloy of 40 nm thickness), obtained after easy-axis (a, b) and hard-axis (c, d) demagnetization, both observed with two sensitivity axes

In very thin films it is difficult to find the lowest energy state because domain wall pinning is much stronger than the domain-energy-derived forces. This is demonstrated in Fig. 5.73 for a quadratic Permalloy element with the easy axis along the diagonal. The attempt to reach the minimum energy state by applying an alternating field of slowly decreasing amplitude leads to dramatically different results for different field directions.

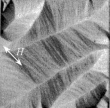

Fig. 5.73 "Equilibrium" states of a quadratic Permalloy element of 20 nm thickness obtained after a.c demagnetization along three different axes

(B) Easy Axis Magnetization Processes. Figure 5.74 shows in the top row again the lowest energy state for thick film elements of rectangular, elliptical and pointed shapes with the easy axis along the particle axis. By applying a small or moderate easy axis field on these states we observe an almost completely reversible magnetization behaviour. This is documented in the figure by difference images between positive and negative remanent states.

Because the domain wall can thus be displaced freely almost without pinning, the material behaves for easy-axis fields like an infinite (internal) permeability material. This was the requirement for the theory of Bryant and Suhl (Fig. 3.25) in which the externally measurable permeability is determined by the sample shape alone. Thick film elements are therefore the best test samples for the verification of this concept [591].

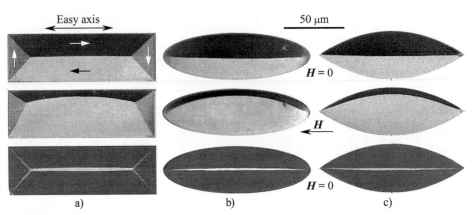

Fig. 5.74 Symmetrical, a.c. demagnetized domain patterns (*top row*) of thick film elements (Permalloy films of 240 nm thickness). The domain walls are displaced reversibly in an applied field (*second row*) and return almost to their previous position after removing the field. This is documented in the *third row* by difference images between positive and negative remanent states (taken with a zero remanence, iron-free magnet system)

A different behaviour is observed in thin film elements in which the easily mobile vortex walls of Fig. 5.74 are replaced by Néel and cross-tie walls (Fig. 5.75). Although the ground states (top row) look quite similar to the thick film case, the magnetization behaviour (middle and bottom row) is quite different. The cross-tie domain wall moves intermittently, pinned mainly by the interaction of the cross Bloch lines with small defects in the sample.

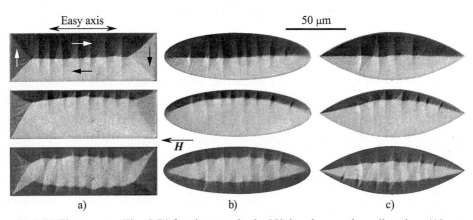

Fig. 5.75 The same as Fig. 5.74 for elements in the Néel and cross-tie wall regime (40 nm thickness). In contrast to the thick film elements, cross-tie walls cause strong irreversibilities in the wall motion. Particularly in the difference images of the *bottom row* the cross Bloch lines at the intersections between original wall and cross-ties can be seen to lag back in the wall motion process

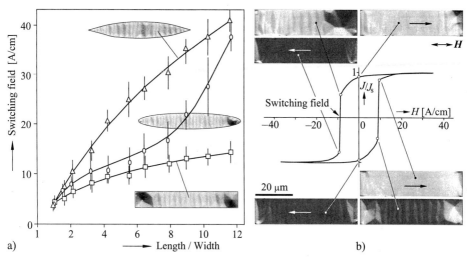

Fig. 5.76 The observed coercivity or switching field of thin film elements (a) depends strongly on the sample shape (here every data point represents an average over five independent events). The *inserted pictures* show the state just before switching for the three different shapes. (b) The illustrated easy-axis hysteresis curve of a rectangular element. (Samples: Permalloy elements of 40 nm thickness)

If a particle is first saturated in a large easy-axis field, it is interesting to observe the critical field at which the near-saturation state breaks down. Elongate elements of different shapes with parallel easy axis differ strongly in their switching behaviour, as demonstrated in Fig. 5.76. Rectangular and also elliptical shapes form a closure pattern at the blunt ends, while this is not observed for pointed shapes. The greater stability of pointed elements in a near saturation state represents an advantage in sensor applications [178, 995].

Circular elements offer the opportunity to study particularly well the interaction of domain walls with a sample edge. Often a repulsive interaction is observed, with interesting irreversibilities if this is overcome [597]. Some typical observations are shown in Fig. 5.77.

These are examples of *configurational* hysteresis, irreversible magnetization processes that are not connected with defects but with the topology of a domain pattern. A subtle difference, such as the annihilation of a wall segment between Fig. 5.77d and (e), causes the formation of completely different "backward" pattern sequences, leading in this case also to different final zero-field states (i) and (n). Similar phenomena were observed in *thin* film elements [997].

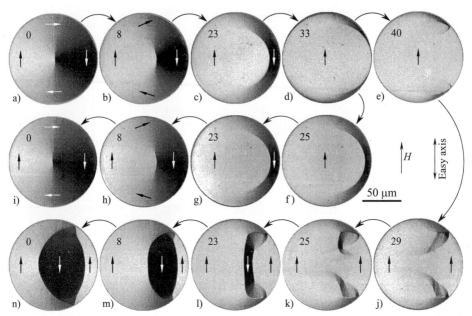

Fig. 5.77 Magnetization processes in circular thick film elements showing the interaction of a domain wall that is formed in an applied field (a—d) with the sample edge. If the repulsive interaction is not overcome, the process stays reversible (f—i). After a punch-through (e) a completely different sequence is observed (j—n). The field (which is indicated in every element in units of A/cm) starts (a) and ends in both cases (i, n) at zero. (Sample: nanocrystalline Fe-NiFe multilayers of 300 nm overall thickness; courtesy *W. Bartsch*, Siemens Erlangen)

The magnetization processes in very thin elements, as they are preferred in sensor applications, differ considerably from those in thicker films. The main reason is that domain wall displacement is more difficult in those films so that details of the switching process can be pursued that are inaccessible in thicker films. Figure 5.78 shows the switching process of an elongated rectangular element. Switching starts from closure configurations at the sample ends as shown and interpreted in different stages on the left side. The result of the switching process that consists of a discontinuous expansion and collision of the closure patterns from both sides is shown on the right. The final configuration varies statistically, but often the two typical patterns shown here are observed. The choice between both configurations depends on the relative orientation of the triangular closure domains at the sample ends. The simpler pattern (b) is found when the triangular closure domains are magnetized antiparallel, while the metastable configuration (d) appears (sometimes) for parallel closure domains. The schematic drawings are derived from quantitative Kerr images ([178]; not shown). The behaviour stays similar as long as the closure domains are not destroyed in very large fields. Remarkably, very similar patterns were also obtained by numerical simulation [998].

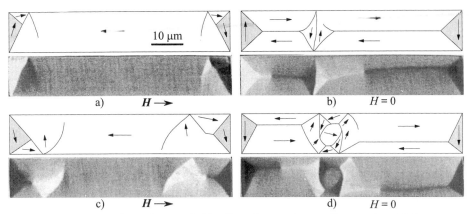

Fig. 5.78 Switching of an elongated thin film element (Permalloy, 17 nm thickness) as influenced by the polarity of the closure domains at the sample ends. The final states (b, d) are formed at the coercive field marking the coalescence of the end domain patterns, but they remain stable when the field is turned off. (Together with *J. McCord* [178])

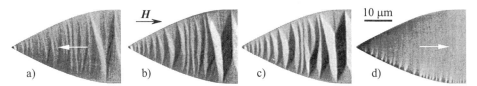

Fig. 5.79 The switching process in a pointed Permalloy thin film element of 48 nm thickness. Almost no edge-induced nucleation structure can be seen (a). For this reason the eventual switching is preceded by a buckling process in opposite fields (b, c). The result of the switching process may contain irregularities, such as a cross-tie edge wall (d). Only the left half of the element and domain state is shown. (Together with *J. McCord* [178])

Further examples of switching processes in a broader pointed thin film element are shown in Fig. 5.79. Here nucleation at the sample ends plays virtually no role (see also Fig. 5.76) and the switching field is determined by the stability of the developing "buckling" state [see for example Fig. 5.62 and (C)]. The fact that such elements are not remagnetized by nucleation and growth of end domains means that they are particularly stable in their saturated state as first established in [178].

(C) Hard Axis Magnetization Processes. Starting at the ground state of Fig. 5.74a and applying moderate hard axis fields, some nice irreversible processes can be documented (Fig. 5.80). Note in particular state (b) where the van den Berg domain wall is first destroyed in the applied field, to be replaced by an orthogonal Bryant and Suhl wall (see Fig. 3.25). The process is reversible until the wall touches the sample edge. If the field is reduced after this event (c), a new domain is formed (d) before the wall separates from the edge again to return to the initial state (e). A completely different path is followed if the wall is annihilated in a somewhat higher field (f). Reducing the field from this almost saturated state, a buckling state as in Fig. 5.68 is formed (Fig. 5.80g), followed by irregular intermediate states (h) and often a distinctly different remanent state (i). The blocked buckling state may either decay by the generation of Bloch lines as in extended films or by the tearing off of wall triplets from the sample edge. Sometimes wall triplets are torn off but are not immediately annihilated. They can then be observed in the middle of the element as in (h) in a clearly metastable configuration.

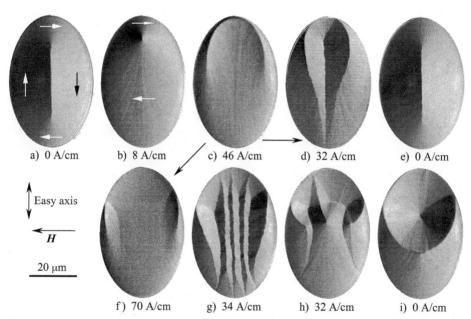

Fig. 5.80 Irreversible hard axis magnetization processes in elliptical thick film elements (Permalloy of 240 nm thickness with longitudinal easy axis). Two series of an increasing and decreasing field are shown, starting from the demagnetized state (a). Although the process in the *top row* is irreversible, the final state (e) is the same as the initial state (a) *if* the wall at (c) did not punch through. In the other case (*bottom row*) the domain pattern passes through a concertina or buckling state (g) as in Fig. 5.68 and ends at a metastable demagnetized state (i), which can only be destroyed in a field of opposite polarity

Such processes play an important role in the hysteresis of thin film elements. The details depend on the nature and strength of the wall-edge interaction.

Under certain circumstances even a concentric, metastable domain state can be obtained after applying hard axis fields to the circular samples of Fig. 5.77 [597] as shown in Fig. 5.81. It should be mentioned that this process, which was quite reproducible in the course of the original investigation [597], could not be reproduced years later on the same samples. A possible explanation might be that strong anisotropic edge stresses that were inferred in [597] had partially relaxed in the course of time, thus modifying the details of the wall-edge interaction.

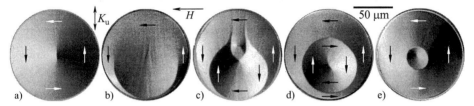

Fig. 5.81 The formation of a concentric *wall* pattern starting at the elementary concentric domain pattern (a) and applying a hard axis field. The strong adherence of the wall to the edge (c) favours a closing of the formed wall [597]

In stronger fields, after saturation along the hard axis, again a system of interacting Néel walls, the buckling state or concertina pattern as in Fig. 5.68 is observed (Fig. 5.80 g). Interactions between the nucleated domain walls and the sample edge lead to marked hysteresis phenomena as demonstrated in Fig. 5.82. A characteristic feature of these complex processes is their irreproducibility, meaning that in spite of an identical magnetic field history the final state is unpredictable.

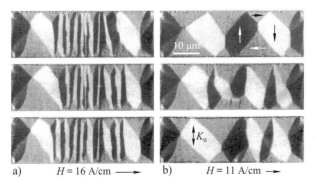

Fig. 5.82 The high-moment states generated coming from saturation along the hard axis (a) decay irreproducibly into largely demagnetized states (b) [999]. (Together with *K. Reber*, Erlangen. Sample: nanocrystalline Fe-Permalloy multilayer of 300 nm thickness, courtesy *W. Bartsch*, Siemens Erlangen)

Similar observations on a pointed thick film element with a transverse easy axis are shown in Fig. 5.83. Again the domain patterns before (a) and after (b) the principal jump connected with the breakdown of the buckling state are shown, this time adding the resulting zero-field states (c). As in the previous figure, complexity of a process leads to irreproducibility, with no possibility to trace its origin in detail.

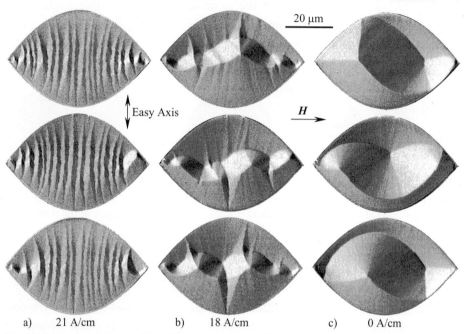

Fig. 5.83 The breakdown of buckling states in a pointed thick film Permalloy element of 240 nm thickness for which the easy axis is oriented perpendicular to the element axis. The particle is first saturated along the hard axis, and the field is then slowly reduced to zero. Three pattern sequences, showing the state before (a) and after (b) the main jump, as well as the remanent state (c) indicate the variability of the process for identical field history

The processes in very thin film elements differ markedly from those in thicker films. The formation of Bloch lines as in Fig. 5.62 is difficult in these films and for this reason the domain walls of the buckling state cannot be converted into cross-tie walls. Often they remain stable even when most of the sample is already remagnetized, so that 360° walls are left over as shown in Fig. 5.84. In other observations (not shown) the buckling state breaks down before the formation of 360° walls, leading immediately to saturation along the external field.

In this sample the 360° walls are not very stable. Even if they are formed as in Fig. 5.84c, they 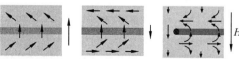 can be annihilated in slightly larger fields by the formation and migration of

Bloch lines. In other, particularly in still thinner films, Bloch lines are more difficult to form and to move, and 360° walls turn out to be remarkably stable.

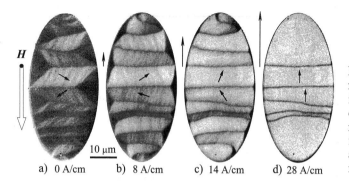

H

10 μm

a) 0 A/cm   b) 8 A/cm   c) 14 A/cm   d) 28 A/cm

Fig. 5.84 The hard axis remagnetization process in an elliptical Permalloy element of 20 nm thickness. Here 360° walls are formed even without any apparent defects or pinning points

(D) Edge Curling Walls. We met complex micromagnetic edge structures, for example in Fig. 5.80, in the form of domain walls attached to an element edge. A simpler, but very interesting configuration is found if a large field is applied perpendicular to a straight film edge.

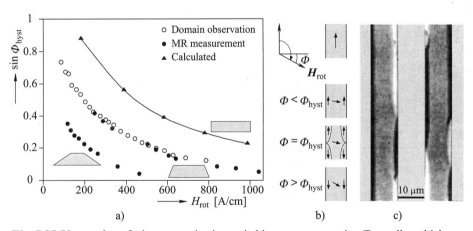

a)   b)   c)

Fig. 5.85 Observation of edge magnetization switching on narrow strips (Permalloy, thickness 35 nm) in a rotating field. The switching field angle [defined in (b)] was measured for different values of the applied field. The data marked by *open circles* were evaluated experimentally from the analysis of observed domain patterns. The switching in high fields cannot be observed directly because the curling zone becomes too narrow. The sign of the curling excursion was therefore inferred from observations in reduced fields after each attempt. The data marked by *black circles* were derived from magnetoresistance measurements on the same samples. The curve marked by triangles was computed by two-dimensional numerical micromagnetics for elements with ideally rectangular cross-sections. (c) Examples of partially switched edge curling zones. Adapted from [1000]

Except for the case of very large fields, the magnetization will try to stay parallel to the edge even if the interior of the element is already saturated parallel to the field. The local de-magnetizing effect at the edge is much larger than in the middle of an element. This specific edge structure 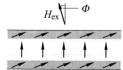 is stable in single films only when a field is present. If the edge is polarized into one direction and if we rotate the field into the opposite direction, an instability will occur at a certain field angle that will depend on the field strength. These effects were studied experimentally and by numerical micro-magnetics in [1000]. Results are shown in Fig. 5.85. The experimentally observed switching angle depends on the edge profile and is smaller than the computed result that applies to an ideal rectangular cross-section. All curves follow the same behaviour, however.

(E) Corner Patterns. A characteristic pattern is often observed in the corners of rectangular elements (Fig. 5.86), which replaces the simple 90° wall expected from the van den Berg construction — the so-called tulip pattern. Figure 5.86a shows a quantitative magneto-optical image of such a pattern. Among the variants in (b) are a simple straight 90° wall, a modified wall coupled to a three-wall edge cluster, a simple wall connected with a complex pattern, as well as various large, small and multiple tulip patterns (bottom row).

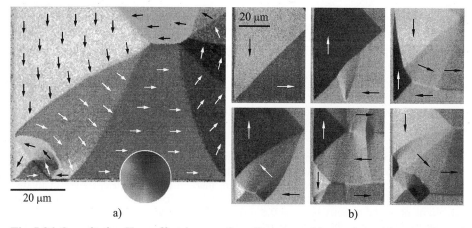

Fig. 5.86 Quantitative Kerr effect image of a tulip pattern (a) together with a number of alternative patterns observed in an equivalent corner (b), ranging from simple and complex 90° wall patterns (*top row*) to complex and multiple tulip patters (*bottom row*). (Together with *J. McCord*, Erlangen; sample: see Fig. 5.77)[1]

---

[1] See Colour Plate

The tulip pattern can be interpreted [591] in terms of a modified van den Berg pattern (see Fig. 3.22b) which is stabilized by Néel wall interaction effects. To see this, follow the magnetization vector along the diagonal across the tulip pattern (dashed line in the sketch), and note the similarity with the buckling state discussed earlier. But patterns of different size along with no tulips at all can be observed after identical demagnetizing treatments (Fig. 5.86 b). This indicates that the tulip pattern is a metastable configuration.

A pinning of the wall edge cluster at imperfections in the element edge appears to be responsible for the stability of the tulip pattern once it has been formed. Although it is often observed experimentally, it has been analysed only qualitatively as far as we are aware.

(F) Particle Interaction Effects. The switching of small particles or elements is influenced by the switching state of near neighbours. Such interactions were studied in detail by transmission electron microscopy on very small, closely packed elements by *Kirk* et al. [1001], as shown in Fig. 5.87. The switching field of the elongate particles depends strongly on the switching state of the neighbours. If one or both neighbours are already switched, a much higher field is needed because the particle is stabilized by the stray field of the neighbours. This has the consequence that second nearest neighbours are preferably switched rather than nearest neighbours, leading to the alternating pattern shown in the picture. Altogether the switching field distribution was found to be strongly widened by the interaction effects.

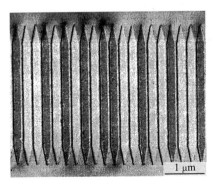

Fig. 5.87 Switching in an interacting array of pointed Permalloy elements of 26 nm thickness. About every second particle has switched in a field along the particle axis, showing lighter contrast. Transmission electron micrograph of the Foucault type. Note the effects of the stray field near the tips which shows up in this technique. (Courtesy *C. Kirk* and *J.N. Chapman*, Glasgow)

Also in larger elements similar phenomena can be observed. In Fig. 5.88 we recorded the equilibrium state in elements of 100 μm length, generated by a decreasing alternating field along the particle axis.  Interestingly, a high remanence state is stabilized for interacting rectangular elements (a) of a certain aspect ratio for which the isolated particle prefers a demagnetized state in equilibrium. This state is favoured by effective flux distribution configurations at the sample ends. Small closure domains are also found in elliptical particles in high-remanence states, but they are much smaller and obviously less effective in reducing the stray field. This may explain why interacting elliptical particles (b) prefer a demagnetized state in equilibrium. The pointed elements (c) were designed with parabolic contours to mimic *ellipsoidal* particles in their cross-sections along the axis. They prefer the saturated state and under the influence of interactions even a regularly alternating arrangement can be observed.

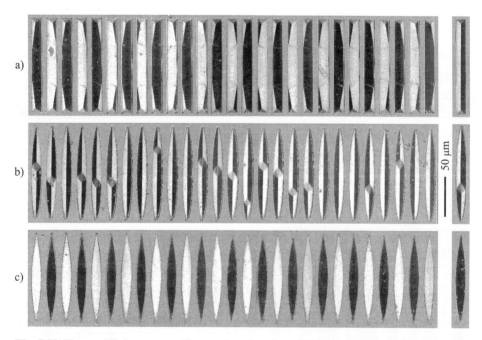

Fig. 5.88 The equilibrium state of interacting elements of rectangular (a), elliptical (b) and pointed (c) Permalloy elements of 240 nm thickness, generated in an alternating field of decreasing amplitude along the particle axis. The equilibrium demagnetized state of the isolated particles is shown on the *right* for comparison

5.5.5 Thick Films with Weak Out-of-Plane Anisotropy

(A) Observation of Dense Stripe Domains. Stripe domains were discovered when searching for the origin of anomalous magnetization curves in certain rather thick Permalloy films [1002]. Instead of the expected low coercivity hysteresis cycle, these samples displayed an awkward behaviour including a "rotatable anisotropy", which means that the permeability in a minor loop depends on the direction of a previous saturation field. High resolution domain observations later revealed the origin of this effect: instead of the expected wide in-plane domains, these samples showed a narrow *stripe* pattern, the period of which is comparable to the film thickness [742, 743, 1003, 1004]. The origin of stripe formation is a weak perpendicular anisotropy usually induced by stress in the evaporated films. According to (3.52c), a uniaxial stress σ_u generates a uniaxial anisotropy $e_{me} = -\frac{3}{2}\lambda_s\sigma_u m_3^2$. The same is true for a planar stress σ_p that leads — after inserting the corresponding stress tensor into (3.52b) — to $e_{me} = \frac{3}{2}\lambda_s\sigma_p m_3^2$. For a negative magnetostriction constant ($\lambda_s < 0$) and tensile stress ($\sigma_p > 0$) this generates a perpendicular easy axis and thus leads, above a critical thickness, to the nucleation of dense stripe domains as discussed in Sect. 3.7.2A.

The best method to observe stripe domains is conventional Fresnel mode Lorentz microscopy. The good spatial resolution and the high sensitivity of this technique make it especially well-suited for the observation of the weak modulation of the stripe domains close to nucleation. Figure 5.89 shows an example of such an observation.

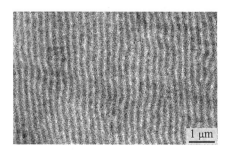

Fig. 5.89 Stripe domains in a Permalloy film of 210 nm thickness made visible by Lorentz microscopy. (Courtesy *I.B. Puchalska*, Meudon)

Magneto-optical observation can be applied if the films are sufficiently thick (remember that at nucleation the stripe width in soft magnetic materials is equal to the film thickness as shown in Fig. 3.109b). Stripes in films thinner than 100 nm that can be easily observed in Lorentz microscopy are beyond the reach of Kerr microscopy. On the other hand, stripe nucleation in thicker films, which occurs at correspondingly larger periods, can be nicely observed magneto-optically. Also the Bitter technique can be used because it is sensitive to small variations in the magnetization.

We show magneto-optical observations of stripe domains. Stripe *switching* which we met in bulk amorphous samples (Fig. 5.49) is an interesting hysteresis phenomenon and is demonstrated for a thick film in Fig. 5.90. While the formation of stripe domains out of the saturated state is a reversible process, the switching of these patterns near zero field is strongly hysteretic.

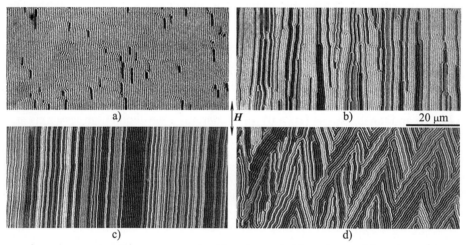

Fig. 5.90 Hysteresis effects connected with stripe switching observed in an amorphous FeSiBCuNb film of 1 μm thickness. (a,b) Progressive switching of stripes approaching the coercive field. (c) Alternating-field demagnetized state. (d) Rotating the stripes in a field perpendicular to the field during demagnetization. (Sample: *G. Henninger*, TU Dresden)

The hallmark of dense stripe domains is the magnetization curve with a steep switching part at small fields, and a linear magnetization rotation part leading to the nucleation field. This curve is shown in Fig. 5.91 for a thick Permalloy film together with observed stripe domain patterns. We see that the magnetization rotation branch is not free of hysteresis, obviously because the stripe period is different, depending on magnetic history. The steep part of the hysteresis is connected with the stripe switching process shown in Fig. 5.90. If the perpendicular anisotropy is much larger, the stripe domains may become invisible in optical microscopy, but the characteristic magnetization curve of stripe domains uniquely indicates their presence.

In Fig. 5.92 the process of stripe nucleation and ripening is shown for a thick film in which near remanence already a three-dimensional modulation of the stripe pattern can be observed, a branching process explained in Fig. 3.142 which one might not expect in a magnetic film. The example establishes a continuous transition between the magnetic domains in thin films and in bulk amorphous material of perpendicular anisotropy.

Usually stripe domains are nucleated along the direction of an in-plane field. In thick films, however, stripes can be forced into the perpendicular orientation by magnetostrictive interactions within the stripe pattern. This effect has been discussed in Fig. 3.123 and can be seen in Fig. 5.48c for the example of a metallic glass sample into which a perpendicular anisotropy has been induced by field annealing.

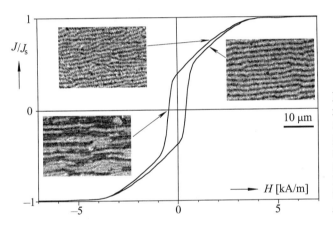

Fig. 5.91 Hysteresis curve measured in a vibrating sample magnetometer, together with some examples of domain patterns observed on the same sample. Permalloy film of 1600 nm thickness. (Together with *K. Reber*, Erlangen)

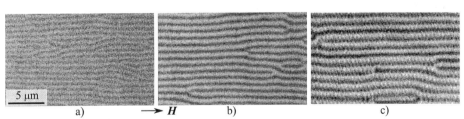

Fig. 5.92 The development from stripe domain nucleation (a) to near remanence band patterns (c). In the latter state a three-dimensional modulation of the bands can be seen, a branching phenomenon as in Sect. 3.7.5G. (Sample: as for Fig. 5.90, 2 μm thickness)

(B) Coexisting Strong and Weak Stripe Domains. A different class of stripe domain patterns is observed in films with *oblique* anisotropy [495, 744–746, 1005]. Uniaxial anisotropies with a tilted axis may develop if a film is evaporated under an oblique angle. The shape and orientation of the crystallites will contribute to the total magnetic anisotropy. Tilted easy axes are also encountered in single-crystal films or platelets depending on crystal orientation — as in the (100)-oriented Ni platelets presented in Sect. 5.5.6B. The anomalous stripe domains found under these circumstances are called *strong* stripe domains because they show a strong contrast in Lorentz microscopy and also in Bitter patterns (see Sect. 3.7.4E).

The distinction between weak and strong stripe domains can be formulated as follows: weak stripe domains develop continuously out of the saturated state along a hard axis with respect to the effective anisotropy. They "remember" their origin by retaining a net magnetization component along the field axis. Strong stripe domains can be best visualized by regular domain models, with the basic domains magnetized along the true easy axis, and closure domains along relative, surface-parallel easy directions [495, 1006]. For perpendicular anisotropy there is no difference between the two: ripening weak stripes continuously converge into a "strong" stripe domain structure. The different situation for tilted anisotropy was demonstrated in Fig. 3.123. In tilted anisotropy samples weak and strong stripe domain configurations form a discontinuous transition, both as a function of thickness and in an applied field for a given thickness. Consequently, the coexistence of both patterns can be observed at certain thicknesses (Fig. 5.93), thus proving their distinct nature.

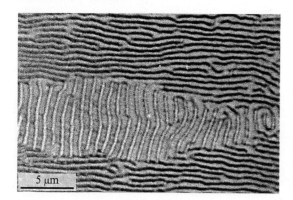

Fig. 5.93 Coexisting weak and strong stripe domains as observed by a dry Bitter pattern technique on an obliquely evaporated $Ni_{80}Co_{20}$ film (0.72 µm thickness). The anisotropy axis is inclined by 27° relative to the sample normal, the relative anisotropy coefficient is $Q = 0.09$ (Courtesy *H. Aitlamine*, Meudon)

A more careful analysis of strong stripe domains has to consider the detailed nature of the walls in the domain structures. As we are dealing with thick films, asymmetric stray-field-free domain wall types are expected in these films. In [1007, 1008] the strong stripe domains were analysed in terms of close-packed arrays of asymmetric Bloch and Néel walls, and good agreement with experimental findings by Lorentz microscopy could be achieved. These concepts were confirmed by rigorous numerical computations [761, 762] which added many details on the complex topology of this domain type.

5.5.6 Cubic Single-Crystal Films and Platelets

Single-crystal films may be prepared by thinning from the bulk, by epitaxial deposition [1009–1011], or by the direct vapour growth of crystal platelets [86, 302, 983, 1012]. The magnetic anisotropy functional of epitaxial films, which are usually under some misfit stress, may be influenced by a resulting magneto-elastic anisotropy; otherwise the magnetic material constants of single-crystal samples should correspond to their bulk counterparts. Beautiful and instructive results were obtained in particular from the magneto-optical analysis of transparent garnet crystals (see for example [210]).

Single-crystal films with one easy direction parallel to the surface do not differ substantially from their polycrystalline, magnetically uniaxial counterparts. Ripple phenomena are almost absent in single-crystal films because of the absence of random crystal anisotropies. In epitaxial films ripple was observed in a very narrow field range around the hard axis [525]. Its origin was attributed to random anisotropies connected with atomic interface structures.

(A) Single-Crystal Films with Two Easy Axes in the Surface. This class, with (100) iron films as the standard example, shows domain patterns that differ markedly from polycrystalline uniaxial thin films, but they also differ in interesting details from their bulk counterparts (Sect. 5.3.1). The most obvious feature is the tendency to use mostly 90° walls instead of the energetically less favourable 180° walls, leading to characteristic diamond patterns. 180° walls are used only to fill gaps in the network of 90° walls.

In Fig. 5.94 observations on epitaxial films are presented for two materials: for pure iron and for Permalloy. Interesting is that in the Permalloy epitaxial film (b) cross-tie walls are observed just as in polycrystalline films, while in the iron film (a) they appear to be absent. Otherwise the internal wall structures of both 90° and 180° walls hardly differ from the corresponding ones of uniaxial films because the anisotropy contribution to the total energy of domain walls in thin films is generally of minor importance as discussed in Sect. 3.6.4.

Figure 5.95 shows domain wall observations for a thickness series of (100)-oriented iron films. In spite of the general preference for 90° walls, 180° walls can be also found, especially around defects or at the sample edges. The 180° walls show the well-known and expected substructures (Bloch lines, cap switches in asymmetric Bloch walls). As in uniaxial, polycrystalline films (Fig. 5.59), the wall width passes a minimum as a function of film thickness, marking the Néel-Bloch transition, which is found in iron at about 50 nm thickness, in contrast to the 90 nm of Permalloy.

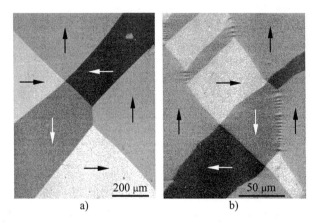

a)   b)

Fig. 5.94 A preference for 90° walls in (100)-oriented epitaxial iron films (thickness 40 nm, courtesy *P. Grünberg*, Jülich) is shown in (a). 180° walls are present in short segments. The same is true for the epitaxial Permalloy film of 50 nm thickness (b), only that the 180° walls display the cross-tie structure. (Sample: courtesy *S.S.P. Parkin*, IBM Almaden)

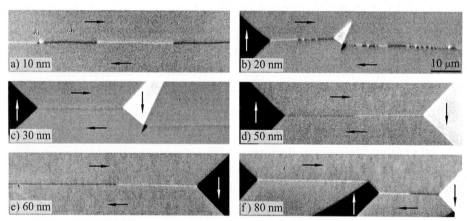

Fig. 5.95 Domain wall observations for iron films of various thicknesses. Subdivided Néel walls with alternating sense of rotation are shown in (a, b). The diffuse appearance of the wall images in (c) and in the left half of (d) indicates submicroscopic cross-tie walls as demonstrated in Fig. 5.96. The asymmetric Bloch walls in (e) and (f) get stronger with increasing film thickness. (Samples: courtesy *P. Grünberg* and *R. Schreiber*, Jülich)

No cross-tie walls show up in Kerr microscopy on single-crystal iron films of any thickness. The reason is simply insufficient resolution of the optical method as was demonstrated by *M. Schneider* et al. [387] by magnetic force microscopy. Iron films of 30 nm thickness do show cross-tie 180° walls

(Fig. 5.96), but their period and width is about 0.1 μm and thus below optical resolution. The cross-tie wall in epitaxial iron films may be considered a benchmark problem for high-resolution domain observation techniques.

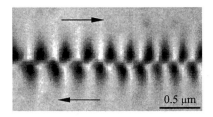

Fig. 5.96 High resolution MFM image of a cross-tie 180° wall in an (100)-oriented iron film of 30 nm thickness. (Courtesy *M. Schneider* [387])

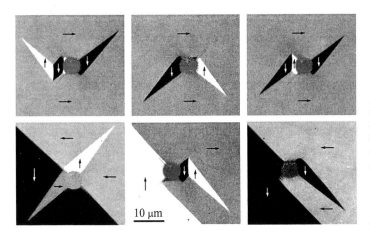

Fig. 5.97 Spike domains formed in the neighbourhood of a non-magnetic inclusion in a single-crystal iron film of 100 nm thickness. (Sample: courtesy *P. Grünberg*)

The effects of holes or non-magnetic inclusions on the domain pattern can be nicely studied in single-crystal films [1013]. Figure 5.97 shows examples of several variations of the famous Néel spikes (Sect. 1.2.3) in iron films, including the interaction of these domains with a 90° wall sweeping over the defect [1014]. One of the earliest visualizations of coercivity mechanisms was the interaction of Néel spike domains with a 180° wall [1015]. This process is demonstrated in Fig. 5.98.

The epitaxial iron films shown in this section were deposited on GaAs single-crystal substrates. A sharp (110) edge of the substrate offers the opportunity to observe the echelon pattern discussed extensively in Sect. 3.7.5C as an example of two-dimensional domain branching. The actually observed patterns are less reproducible and less regular than the theoretically considered patterns (see Fig. 3.133), but all essential features including multiple generation branching can be observed as shown in Fig. 5.99. By using a magneto-optic combination technique (Sect. 2.3.7) with two different objectives, altogether four generations of echelons are shown in this image.

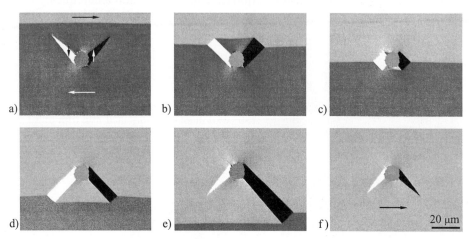

Fig. 5.98 The classical coercivity mechanism of a 180° domain wall interacting with Néel spikes bound to a defect. (Sample as in Fig. 5.97)

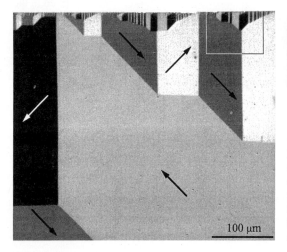

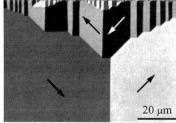

Fig. 5.99 Echelon pattern observed at a [110] boundary of a (001)-oriented iron film of 100 nm thickness. The inset in the main picture (*left*) is shown *above* at higher magnification. (Sample: courtesy *P. Grünberg*, Jülich)

(B) Single-Crystal Films with no Easy Axis in the Surface. Many different types of strong stripe domains can be observed in sufficiently thick single-crystal films if no easy axis lies parallel to the surface [86, 611]. They can again best be understood by domain models that take into account the multiple easy axes. Every conceivable combination of easy directions in the bulk and relative easy directions in the closure domains finds its counterpart in observations, as shown in Fig. 5.100.

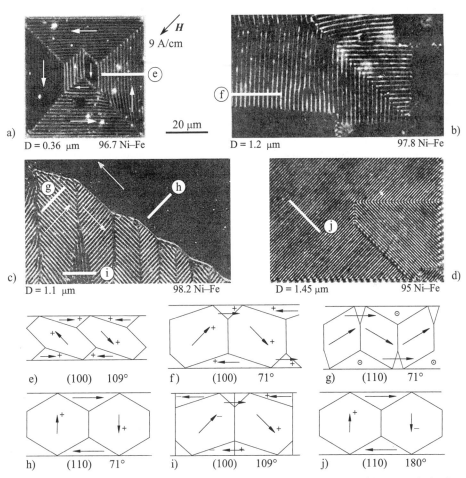

Fig. 5.100 Various stripe-like patterns observed on (100)-oriented Ni-Fe alloy crystal platelets together with plausible domain models. The samples with different Fe content all have a negative anisotropy $K_{c1} < 0$. The cross-section models were constructed assuming that the internal domains are magnetized along the easy $\langle 111 \rangle$ axes, and the closure domains along the relatively easy $\langle 110 \rangle$ axes ($+ -$ signs indicate the third components of the vectors in the cross-sections). The nearly invisible stripes in the *upper part* of (c) [model (h)] can be faintly seen in the original photographs. (Photographs: courtesy *R.W. DeBlois*, General Electric, Schenectady, N.Y. [86])

These observations were done by DeBlois [86] who systematically investigated and classified the different stripe domains for negative-anisotropy materials as a function of platelet thickness and composition. The cross-section domain models were devised in correspondence with DeBlois, but never tested in detail or published. Every cross section is oriented perpendicular to the respective stripe direction as indicated by the white bars in the domain images. The models were inferred from the appearance and the conditions of formation of the different stripe types and from the volume and surface easy directions of the material, relying on the principle of pole avoidance in these low-anisotropy materials as in Sect. 3.7.4D.

The stripe domain patterns in such samples represent a perfect example of *quasi domains* (Sect. 3.4.5). Figure 5.101 demonstrates this aspect by following the movement of quasi-domain walls. Two types of phases can be distinguished in these pictures as shown in (d): primary phases, which differ in their net magnetization, and secondary subphases within the primary quasi-domains that represent the same net magnetization with different internal structures. By changing the field, the primary quasi-domain walls in (c) are displaced relative to (a, b) in a regular and predictable fashion, while the subphase boundaries are not displaced at all or displaced only irregularly.

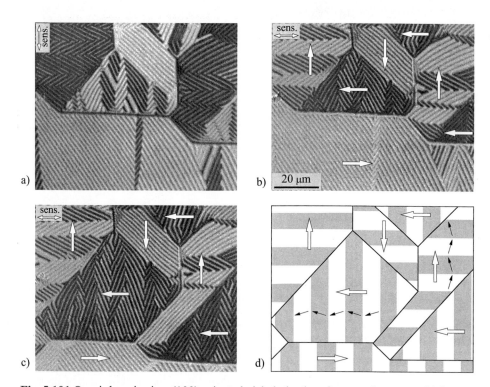

Fig. 5.101 Quasi-domains in a (100)-oriented nickel platelet of some micrometre thickness, shown in (a, b) with two different sensitivity directions. (c) By changing the external field the primary domain walls are displaced. The primary phases of this structure are explained in (d). Platelet grown together with *S. Schinnerling*, IFW Dresden and *W. Habel*, Erlangen

In addition to one-dimensional modulated patterns ("stripe domains"), two-dimensional patterns ("bubble lattices") are also rarely observed. Figure 5.102

shows two examples, one for a similar nickel platelet as in Fig. 5.101, and an example of a garnet film with superimposed cubic and uniaxial anisotropies [1016]. No detailed theory of these micromagnetic configurations has as yet been proposed, but they are clearly distinct from branched domain patterns that can be found for somewhat thicker samples as shown in Fig. 5.102c.

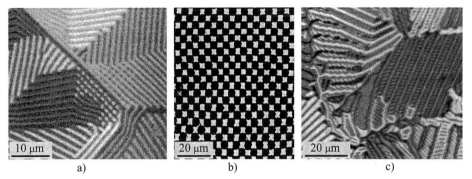

Fig. 5.102 Square domain lattices in (100)-oriented single-crystal films. (a) Nickel platelet in which such patterns are occasionally observed locally. (b) Co-rich garnet film of 5 μm thickness with a strong positive cubic anisotropy and a superimposed perpendicular uniaxial anisotropy. (Courtesy *P. Görnert*, Jena [1016]). (c) Nickel platelet of 13 μm thickness, displaying branched domain patterns as opposed to the two-dimensionally modulated, but unbranched patterns in (a, b)

5.5.7 Double Films

Magnetic double films consist of two ferromagnetic films separated by a non-magnetic layer that modifies or interrupts the exchange interaction. Such sandwich structures must be distinguished from strongly coupled systems, such as ferromagnetic films directly coupled to hard magnetic or antiferromagnetic substrates, and the materials composed of multiple very thin films. For an overview over magnetic phenomena in multilayer systems see [1017].

Three basic types of double films may be distinguished:

- Films with no local coupling between the magnetization directions in the two layers. This condition is met if the non-magnetic layer is free of "pinholes" (i.e. bridges between the magnetic layers) and thicker than 5−10 nm (depending on the nature and perfection of the interlayer).

- Films with a weak ferromagnetic coupling, which favours the parallel orientation in the two films; such a coupling can be due either to quantum-mechanical exchange if the non-magnetic layer is thin enough.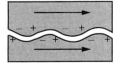

Alternatively, the "orange peel" effect (see sketch) may lead to a ferro-magnetic coupling [1018]. The latter occurs if the interlayer is thin compared to the amplitude of the surface corrugations of the magnetic films [1017].

- Metallic interlayers in the nanometre thickness range may lead to various surprising effects such as an antiferromagnetic coupling between the two magnetic layers [1019], a coupling that oscillates between ferromagnetic and antiferromagnetic depending on the interlayer thickness [1020], and to non-collinear modes of coupling [525].

(A) Decoupled Double Films. In this type of double films the non-magnetic layer is usually more than 10 nm thick, so that direct exchange coupling is reliably suppressed and pinholes are avoided. The properties of such systems are nevertheless surprisingly different from single layers [1021–1024] because most of the stray fields that determine the behaviour of single films can be compensated by corresponding structures in the other film.

Ideally, every magnetic charge in the top layer is matched by an opposite magnetic charge in the bottom layer. Néel walls assume a different, low energy structure without the extended tail of single films. As a consequence, the range of occurrence of Néel walls extends to larger thicknesses for double films and the coercivity is strongly reduced.

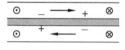

The same principle applies to configurations that would be forbidden in a single film, such as strongly charged walls, or domains meeting an edge head-on. (Partially charged walls are found in *thin* single films; see Sect. 5.5.2B). All these structures are allowed in double films and they have a characteristic width that is determined by anisotropy and stray field energy only. This can best be seen from the explicit calculation of one of these structures.

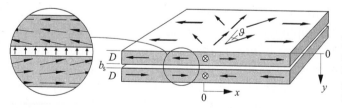

Fig. 5.103 Schematics of a pair of head-on walls in a double film without exchange coupling between the films

We study as an example a pair of head-on walls. Such charged walls would carry a high energy in single films if they were possible at all (see Sect. 3.6.4G). In double films the charges generated in one film can be compensated by opposite charges in the second film. A slight tilting of the magnetization in both layers can displace the charges out of the interior of the two films towards the non-magnetic separation layer, so that a stray field energy is generated only in the gap [1023, 1024].

If the x component of the magnetization depends on x only (Fig. 5.103), the following ansatz fulfils the above-mentioned condition:

$$m_x(x) = \sin \vartheta(x) \ , \quad -\tfrac{\pi}{2} \le \vartheta \le \tfrac{\pi}{2} \ ,$$

$$m_y(x,y) = -y \cos \vartheta(x) \, \mathrm{d}\vartheta/\mathrm{d}x \qquad \text{for } 0 \le y \le D \ ,$$

$$m_y(x,y) = (2D + b_s - y) \cos \vartheta(x) \, \mathrm{d}\vartheta/\mathrm{d}x \qquad \text{for } D + b_s \le y \le 2D + b_s \ ,$$

$$m_z = \sqrt{1 - m_x^2 - m_y^2} \ . \tag{5.2}$$

This magnetization distribution generates a magnetic stray field in the gap that depends primarily on x:

$$\mu_0 H_\mathrm{d}(x)/J_s \cong m_y(x,D) = -D \cos\vartheta(x)\mathrm{d}\vartheta/\mathrm{d}x \ . \tag{5.3}$$

The field is connected with a stray field energy density that may be attributed to the unit volume of the magnetic layers:

$$e_\mathrm{d} = S_\mathrm{d} \cos^2\vartheta \ \vartheta'^2 \ , \quad S_\mathrm{d} = \tfrac{1}{2} K_\mathrm{d} b_s D \ . \tag{5.4}$$

This expression resembles mathematically an exchange stiffness energy, only that the effective stiffness parameter S_d is usually larger than the exchange constant A. Wall structures in double films are therefore rather wide, mostly much wider than the film thickness. We thus might neglect the exchange energy altogether, but some details are better described if we include the exchange stiffness effect at least to lowest order. Keeping the leading ϑ'^2 term and neglecting higher terms containing factors like $y^2 \vartheta'^4$ [1025] leads to:

$$e_x = A(1 + \cos^2\vartheta) \vartheta'^2 \ . \tag{5.5}$$

Finally, we need the anisotropy energy density:

$$e_K = K_\mathrm{u} \cos^2\vartheta \ . \tag{5.6}$$

Because the total energy is a local function, it may be evaluated as in the standard procedure (Sect. 3.6.1D) for planar domain walls. The first integral of the Euler equation corresponding to (3.116) becomes:

$$A(1 + \cos^2\vartheta) \vartheta'^2 + S_\mathrm{d}\cos^2\vartheta \ \vartheta'^2 = K_\mathrm{u} \cos^2\vartheta \ . \tag{5.7}$$

Separating the variables and integrating leads to the solutions shown in Fig. 5.104a. Here the wall profiles are plotted for different values of the relative sizes of A and S_d. For small S_d corresponding to small magnetization or small film thicknesses, we obtain a profile close to the classical Bloch wall profile. For the case of a predominant stray field stiffness constant S_d we get an almost linear wall profile (the profile would be perfectly linear with $\vartheta' = \sqrt{K_u/S_d}$ for negligible exchange stiffness energy).

Also shown in Fig. 5.104b are the profiles of regular Néel walls in double films which differ from head-on walls by the boundary conditions $0 \leq \vartheta \leq \pi$ in (5.2) and by the anisotropy energy expression $e_K = K_u \sin^2 \vartheta$ instead of (5.6). Again the stray-field-dominated profile is much wider than the classical Bloch wall at least in the outer regions of the wall profile. Because the symmetry centre of this wall type is charge free, the maximum slope of the wall profile is independent of stray field effects, scaling as in classical walls with $\sqrt{A/K_u}$. This means that Lilley's wall width (Sect. 3.6.1C), which is based on this slope, is not enhanced, while following any of the integral wall width definitions the 180° walls in double films appear widened. In most cases the characteristic length of domain walls in double films is, however, $\sqrt{S_d/K_u}$ in analogy to $\sqrt{A/K_u}$ for regular Bloch walls.

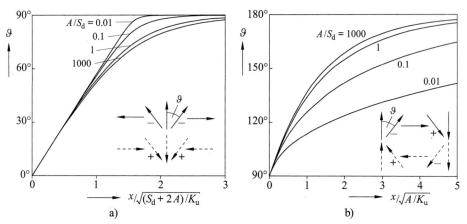

Fig. 5.104 Analytically calculated wall profiles in double layers for different values of the stiffness parameters A and the effective stray field stiffness coefficient S_d for perpendicular orientation [head-on walls; (a)], and parallel, uncharged orientation [regular Néel walls; (b)]. In both cases the charge density is locally compensated by the charges in the second layer (*dashed arrows* in the *inset sketches*)

The advantage of charge compensation between the two layers is so large, however, that, wherever it is possible, it is followed even at the expense of otherwise unnecessary structures such as 0° walls that compensate the charges of 360° walls (see sketch). This compensation is not always possible, however. Figure 5.105 gives an example of an analysis of superimposed Néel walls under the influence of an external field. With the applied field charge compensation becomes impossible, so that long-range stray field interactions lead to the formation of extended tails as in ordinary Néel walls [1026].

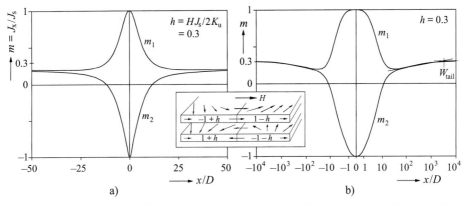

Fig. 5.105 An example in which charge compensation in double layer systems is impossible: Néel wall pairs can be compensated only partially in an applied in-plane field h. There is an unavoidable charge imbalance for $h \neq 0$ (*inset*). In each half wall the net charges $(1-h)$, etc. must be distributed somehow. The imbalance leads to the formation of extended tails that become visible in a quasi-logarithmic plot (b) in which the function $\text{sign}(x) \cdot \log(1 + |x/D|)$ is used as the abscissa. The calculation [1026] was done for films with a weak ferromagnetic coupling (see next section), but the basic features apply also to uncoupled films

A structure for which all these considerations were explicitly confirmed by experimental observations [1027] is that of *edge curling walls* [1024] in patterned double films. An equivalent structure in single films was studied in Sect. 5.5.4D. While these structures in single films were narrow and stable only in a field perpendicular to the edge, they turn out to be wide and stable in the field-free state in double films because the charges of both layers com-

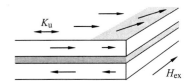

pensate for transverse anisotropy and antiparallel magnetization in the two films. Figure 5.106 shows calculated and experimentally determined profiles in comparison, together with some example pictures of uniform and partially switched edge curling configurations. These configurations and processes are of interest in connection with inductive recording heads.

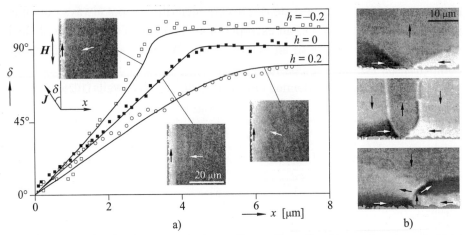

Fig. 5.106 Calculated and measured edge curling wall profiles (a) as a function of a field parallel to the edge, together with observations. No fitting parameters were needed in these calculations [1027]. The pattern in a negative field (*top curve*) is metastable. (b) Increasing the field further the edge curling wall switches in various modes by propagating transitions along the edge. (Sample: 400 nm Permalloy | 20 nm SiO$_2$ | 400 nm Permalloy)

(B) Films with Weak Ferromagnetic Coupling. In uniaxial films with ferromagnetic coupling the *domains* tend to be magnetized in parallel. A remarkable wall structure is observed in such films if they are first magnetized along the hard direction and the field is then reduced towards zero (Fig. 5.107a).

Every wall in the system of interacting walls generated in a hard-axis magnetization loop displays a double contrast [1017, 1028, 1029]. As for the decoupled films treated in (A), also here only Néel walls are to be expected, although the total thickness of the two films in the example (100 nm) would be in the Bloch wall regime for single films already. But in this experiment the "twin walls" (the two Néel walls in the two layers) have the same rotation sense defined by the applied field, so that they cannot ideally compensate each other. Therefore each Néel wall generates a quasi-wall as shown in the calculated profiles of Fig. 5.107e. In the zero-field state (b) the twin wall pattern is metastable with respect to a pair of Néel walls of opposite rotation sense (f). Consequently the transition to the lower-energy "superimposed wall" state has already occurred at some places. It is completed in (c), giving way to twin walls of opposite polarity in (d). Double films thus display an analogous domain wall hysteresis cycle as single-layer films of comparable total thickness (Fig. 5.68), although the micromagnetic details are very different.

Another phenomenon in double films with a ferromagnetic coupling is shown in Fig. 5.108. If the ferromagnetic layers are rather thin (20−30 nm),

360° walls may be generated if a 180° wall sweeps over a strong defect that is able to trap a Bloch line in a similar way as in very thin single layers (Fig. 5.65). In sandwich systems the 360° walls assume an equilibrium zigzag shape, which can be derived theoretically from the orientation dependent specific energy of this wall type, strongly influenced by the degree of charge compensation possible at different orientations [1030, 1031].

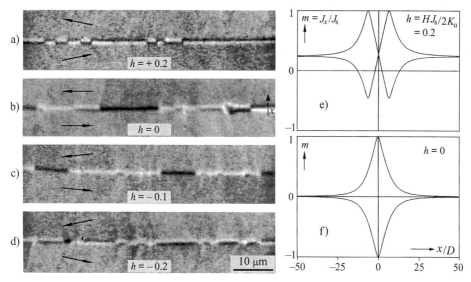

Fig. 5.107 (a) Twin walls in a (50 nm Permalloy | 3 nm carbon | 50 nm Permalloy) sandwich observed after saturation along the hard direction. (b, c) If the field is further reduced, the twin walls are converted into pairs of compensating Néel walls, which is the more stable configuration in zero field. (d) In an opposite field the twin walls reappear. The calculations for the two wall types (e, f) are for the geometrical parameters of the experiment, and for $Q = 0.0005$ and a coupling constant [see (5.8)] of $\kappa_{cp} = 1.8$

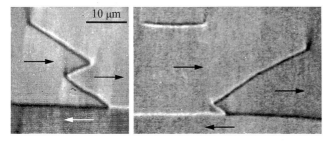

Fig. 5.108 Pinned 360° walls observed in a (20 nm Permalloy | 3 nm carbon | 20 nm Permalloy) sandwich displaying the preferred zigzag shape (a). Free pieces of 360° walls tend to be oriented along the easy axis (b) [1031]

Double layers with perpendicularly oriented easy directions show a remarkable behaviour when they are ferromagnetically coupled [1017]. Their domains resemble those of a single-crystalline cubic material. Figure 5.109

shows a domain pattern on a circular disk of two weakly coupled uniaxial films which displays this feature in a striking way.

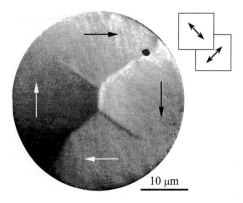

Fig. 5.109 Coupled films with orthogonal easy axes, acting like cubic films with a four-fold axis. The sample consists of two poly-crystalline 30 nm Permalloy films with uni-axial anisotropies, separated by a 5 nm chromium layer that causes a weak "ferromagnetic" orange peel coupling. The contrast is due only to the top layer and reflects, as can be confirmed by a detailed evaluation [1032], the deviation angle calculated in (5.9). (Together with *M. Rührig*, Erlangen)

10 µm

The following calculation demonstrates how an apparently cubic behaviour can be the result of two coupled uniaxial anisotropies. The two uniaxial layers shall be equivalent except that their easy axes are crossed. They shall furthermore be coupled by a Heisenberg-like, "bilinear" interaction [see (3.8)] $e_{\text{coupl}} = C_{\text{bl}}(1-\boldsymbol{m}_1 \cdot \boldsymbol{m}_2)$. If the magnetization vectors in the two layers are in-plane and uniform, they may be characterized by two angles φ and ϑ. Then the total reduced energy can be written as:

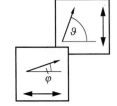

$$e_{\text{tot}} = E_{\text{tot}}/DK_{\text{u}} = \sin^2\varphi + \cos^2\vartheta + \kappa_{\text{cp}}[1 - \cos(\varphi - \vartheta)] , \qquad (5.8)$$

where D is the thickness of the ferromagnetic layers, K_{u} is the uniaxial anisotropy constant and $\kappa_{\text{cp}} = C_{\text{bl}}/DK_{\text{u}}$ is the reduced coupling coefficient that describes a ferromagnetic interaction if $\kappa_{\text{cp}} > 0$. Minimizing the energy with respect to φ and ϑ we obtain as the stable solution for $\kappa_{\text{cp}} > 0$:

$$\vartheta_0 = \tfrac{\pi}{2} - \varphi_0 , \quad \tan(2\varphi_0) = \kappa_{\text{cp}} , \quad e_0 = 1 + \kappa_{\text{cp}} - \sqrt{1 + \kappa_{\text{cp}}^2} . \qquad (5.9)$$

This state carries a net magnetization along the diagonal between the easy directions of the two layers. It is equivalent to three more states for which all angles are rotated by 90°, 180° and 270°. Its energy e_0 is lower than that of the unstable state $\varphi = \vartheta = 0$, the energy of which is 1 in the same units. The system mimics a cubic material with two in-plane orthogonal easy axes. The effective in-plane anisotropy is proportional to $1-e_0$ and decreases with in-creasing coupling strength κ_{cp}. For very strong coupling the two uniaxial anisotropies cancel and the sandwich becomes isotropic.

(C) Antiferromagnetic and Biquadratic Coupling. Antiferromagnetic coupling between two ferromagnetic films was discovered by spectroscopical methods [1019]. It occurs, for example, when an intermediate chromium layer of 1 nm thickness is deposited between two iron layers. Figure 5.110 shows domain phenomena characteristic for such antiferromagnetically coupled films. The observed domains have generally an irregular character because the magnetic moments in the two layers cancel at every point. Without the ordering action of the flux continuity principle domain structures can be quite arbitrary and strongly dependent on magnetic field history. No equilibrium domain patterns can be adjusted in such films.

The local antiferromagnetic exchange coupling has to be distinguished from the case of decoupled films in which an overall antiparallel magnetization arrangement may be preferred for magnetostatic reasons. See the configuration leading to the edge curling wall in Fig. 5.106 as an example.

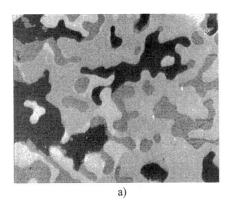 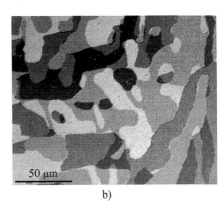

a)   b)

Fig. 5.110 Domains in antiferromagnetically coupled epitaxial iron-chromium-iron sandwich films. (a) Thin films (5 nm Fe) display irregular patch domains. (b) Thick film sandwiches (30 nm Fe) display also straight (horizontal) wall segments within the generally still irregular pattern. (Samples: courtesy *P.Grünberg*, Jülich)

The remarkable phenomenon of an *oscillating* exchange interaction [1020] can best be observed on sandwiches with a wedge-shaped chromium interlayer [1033]. Depending on the chromium thickness an alternating sequence of ferromagnetic and antiferromagnetic coupling is observed, with domain patterns switching between those of Fig. 5.94 and Fig. 5.110 as shown in Fig. 5.111. In between, characteristic transition zones are observed in which the coupling interaction prefers a 90° relative orientation of the two ferromagnetic layers [525]. A phenomenological description derives the canted coupling from a second-order coupling effect, which — instead of the bilinear expression

$m_1 \cdot m_2$ between the magnetization vectors in the two layers [used in (5.8)] — postulates a biquadratic term of the form $(m_1 \cdot m_2)^2$ [525]. Details of the domain patterns and their analysis in these canted coupling transition zones are shown in Fig. 5.112.

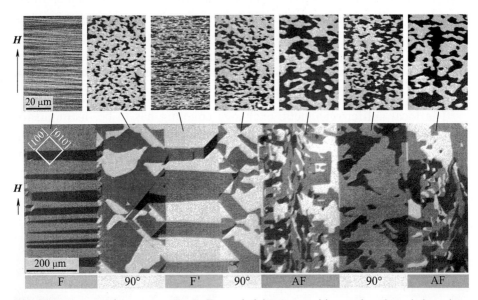

Fig. 5.111 An overview over an Fe-Cr-Fe sandwich system with a wedge-shaped chromium interlayer. The iron layers are 30 nm thick, the chromium interlayer thickness increases from 0 to 0.7 nm from left to right. *F* marks ferromagnetic coupling zones, *AF* antiferromagnetic coupling, *90°* biquadratically coupled zones, *F'* a modified ferromagnetic coupling zone that tends to a 90° arrangement in an applied field along the hard axis. The high-resolution pictures on *top* show the hard axis nucleation patterns typical for the various coupling types: ripple for ferromagnetic coupling, patches for biquadratic and antiferromagnetic coupling, and some transition pattern for the F' zone. See [525] for further details about the interpretation of the coupling zones. Note, however, that this sample shows marked short-period coupling oscillations that were not visible in the less perfect samples studied in [525]. (Sample: courtesy *P. Grünberg*, Jülich)

Figure 5.111 displays another characteristic feature of these exchange-coupled epitaxial sandwich systems: reducing a field from saturation precisely along the hard axis [110], a ripple pattern is formed in ferromagnetically coupled sandwiches

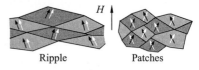

as known from ordinary polycrystalline films. This is not observed in antiferromagnetically and biquadratically coupled films, however. Instead, an irregular patch domain pattern develops in a narrow angular range around the hard axis. This finding [525] can be explained qualitatively by a cancellation of transverse magnetization components as indicated in the sketch. Random step anisotropies at the interface to the single-crystalline substrate are a possible origin of the ripple or patch domain irregularities.

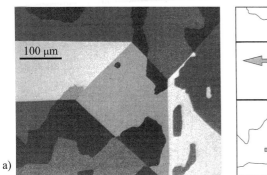
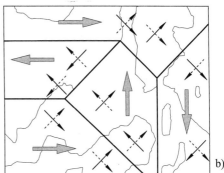

Fig. 5.112 Domains (a) in a double-layer system [Fe (30 nm) | Cr (1.6 nm) | Fe (30 nm)] with a coupling preferring a right angle between the magnetization in the two layers. Together with the cubic anisotropy of the epitaxial iron layers eight different domain types can be constructed (b), which can all be differentiated magneto-optically if the top layer is transparent [*grey arrows*: net magnetization, *black arrows*: magnetization in the top layer, *dashed arrows*: magnetization in the bottom layer]

Direct domain observation is the most reliable method of identifying bi-quadratic coupling phenomena [1034, 1035] although, once established, their strength can be measured by many methods [1036–1040]. An overview over the observed domain and wall patterns in single- and polycrystalline coupled films and their magneto-optical characterization can be found in [155].

The physical origin of the biquadratic coupling effect has been attributed to fluctuations in the coupling, connected with the atomic roughness of the chromium-iron interface [1041]. The notion of a strongly oscillating coupling is supported by the observation of short-range oscillations in the coupling nature on an iron-chromium wedge sandwich deposited on an atomically flat whisker surface [1033]. Other, similar mechanisms may also contribute to a non-collinear coupling effect as reviewed in [527].

Often the coupling effect is weaker than effective coercivities so that a globally measured coupling effect actually represents a mixture of different local configurations. *Daykin* et al. [1042] analysed polycrystalline Co/Cr/Co sandwiches with their newly developed differential phase contrast method in a conventional electron microscope (Sect. 2.4.3B). The resulting image (Fig. 5.113) displays a mixed state of differently aligned regions for a nominally antiferromagnetically coupled sandwich.

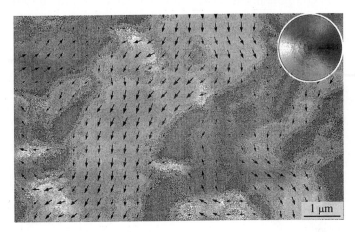

Fig. 5.113 Quantitative transmission electron micrograph of an antiferromagnetically coupled Co (10 nm) | Cr (1 nm)|Co (10 nm) sandwich under the influence of an external field, displaying regions of parallel (coloured, long arrows) and antiparallel (grey, no arrows) alignment. (Courtesy *J.P. Jakubovics*, Oxford) [2]

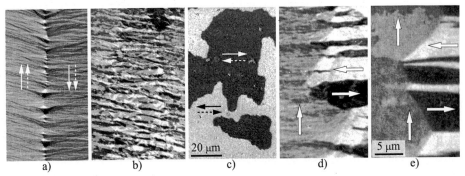

a)   b)   c)   d)   e)

Fig. 5.114 Domains observed on a polycrystalline FeCrFe sandwich with a wedge-shaped chromium interlayer, observed in a hard axis field. After a.c. idealization ripple and patch domains are found in the ferromagnetically (a) and antiferromagnetically coupled parts (c). In between (b) a different pattern is observed, which can be identified by modifying it in an alternating field (d) and comparing it with corresponding single-crystal observations (e) as predominantly 90° coupled. (Together with *M. Rührig*, Erlangen; sample: see [525])

Indications of the occurrence of 90° coupling phenomena in polycrystalline, weakly coupled uniaxial iron sandwiches were presented in [525] and are shown in Fig. 5.114. A biquadratic coupling term can lead to a 90° arrangement only if the uniaxial anisotropy of the polycrystalline films is weak. This condition is fulfilled for iron-chromium-iron sandwich films that were deposited onto glass substrates. Alloy films with their typically stronger induced anisotropy usually suppress 90°-coupling phenomena. A further difficulty is connected with the higher wall motion coercivity of thin polycrystalline films. The coupling nature can only be inferred by comparison after carefully idealizing

[2] See Colour Plate

the domain state in decreasing alternating fields. The nucleation pattern (see Fig. 5.111) and the state in moderate hard-axis fields offer the clearest clues.

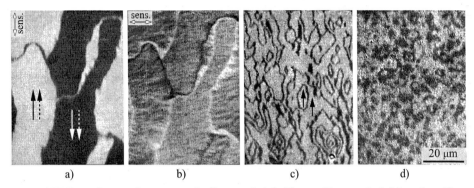

a)   b)   c)   d)

Fig. 5.115 Domains on ferromagnetically coupled (a, b), weakly coupled (c) and antiferromagnetically coupled (d) polycrystalline NiCo (3 nm) / Ru (2—4 nm) / NiCo (3 nm) sandwiches [155]. The 360° walls visible in two different aspects in (a, b) are not an equilibrium feature, but depend on field history. Note the ripple in the domains visible in (b). Also the complex patterns that can be observed in uncoupled (c) and in antiferromagnetically coupled films (d) are non-equilibrium structures. (Samples: courtesy *S.S.P. Parkin*, IBM)

In most cases only ferromagnetic and antiferromagnetic coupling can be distinguished in the domain patterns of polycrystalline sandwiches because of the predominance of uniaxial anisotropy [159, 1043—1045]. Examples are shown in Fig. 5.115 for sputtered sandwiches as they are studied in connection with giant magnetoresistance applications.

5.6 Films with Strong Perpendicular Anisotropy

The films of this section differ from those of Sect 5.5.5 by a predominant perpendicular anisotropy. The material parameter $Q = K_u / K_d$ has to be larger than 1, where K_u is the uniaxial anisotropy constant with the easy axis perpendicular to the surface and K_d is the stray field energy constant $K_d = J_s^2 / 2\mu_0$. Under these conditions a perpendicular magnetization is preferred both for the single-domain state and for the multidomain state. Only in strong in-plane fields or inside domain walls can the horizontal magnetization direction be occupied. In-plane fields lead to dense stripe domains (Sect. 3.7.2), just as in low-anisotropy perpendicular films. But perpendicular or *bias* fields (i.e. fields perpendicular to the film plane, parallel to the easy axis) generate specific patterns, which were discussed theoretically in Sect. 3.7.3.

5.6.1 Extended Domain Patterns

(A) Low-Coercivity Perpendicular Films. Figure 5.116 shows characteristic examples of periodic or quasi-periodic domain patterns observed in low-coercivity perpendicular films: the *maze* pattern (a) obtained after saturation in a perpendicular field, a pattern of parallel *band* domains (b) formed after saturation and subsequent idealization in a tilted field, a *bubble lattice* (c) generated after saturation in a field slightly offset from a hard direction, and a *mixed* pattern (d) obtained after passing the critical point (see Sect. 3.4.4) along the hard direction. The manifold of metastable patterns constitutes a favourite subject of topological [1046−1052] and dynamical [1053−1056] studies.

Remarkably, all patterns in Fig. 5.116 are observed under identical conditions (at zero field). These states (and many more) are (meta-)stable against small perturbations. The occurrence and stability of metastable states is a characteristic feature of the domain structures in perpendicular films. The question of which pattern has the lowest energy under which conditions can best be answered theoretically (Sect. 3.7.3). As the differences in energy between the various patterns are tiny, an experimental decision is almost impossible. Calculations of the *stability limit* of a given metastable pattern, on the other hand, can be checked experimentally.

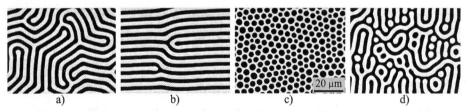

Fig. 5.116 Domain patterns at zero field in a bubble garnet film with a growth-induced perpendicular anisotropy. The patterns were generated after saturation at an angle of 90° (a), 20° (b), 1° (c) and 0° (d) relative to the sample plane

The complex process leading to the mixed bubble pattern in Fig. 5.116d is shown in more detail in Fig. 5.117. It is characterized by a cascade of discontinuous coalescence processes. Starting from a more or less disordered nucleation pattern (a) the final state (f) displays with equal probability black and white band segments and bubbles. The detailed arrangement of the intermediate and the final states depends on additional, secondary anisotropy contributions [609], which are superimposed onto the fundamental perpendicular anisotropy of these films. An orthorhombic anisotropy favouring one of the in-plane hard axis over the other is responsible for the anisotropic domain formation in the bottom row of Fig. 5.117. These additional anisotropies also determine the exact conditions under which the continuous stripe nucleation process (a) is observed. Without an additional perpendicular field this happens only for certain azimuthal angles of the in-plane field. In the absence of an orthorhombic anisotropy the cubic anisotropy of the garnet structure determines six azimuthal directions for the usual (111) sample orientation for which the mixed bubble state is observed. In between these directions pure bubble states of the one or the other polarity as in Fig. 5.116c are generated. Adding a weak perpendicular field can adjust the second-order nucleation condition also for these azimuthal orientations. The perpendicular field necessary for this adjustment is a sensitive tool for the measurement of spurious anisotropies [609].

The nucleation patterns of 5.117a are quite irregular and are obviously influenced by tiny irregularities in the generally highly perfect garnet samples because they form basically in the same configuration after repeating the experiment. Probably only such nucleation studies can reveal these subtle irregularities. The mentioned reproducibility applies to the nucleation pattern only; not to the subsequently developing coarser configurations.

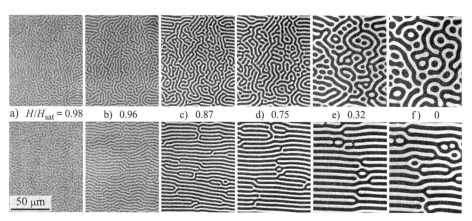

a) $H/H_{sat} = 0.98$ b) 0.96   c) 0.87   d) 0.75   e) 0.32   f) 0

50 µm

Fig. 5.117 From (irregular) stripe domain nucleation (a) to a mixed bubble state formed in a decreasing in-plane field. This garnet film (courtesy *W. Tolksdorf*) shows different secondary anisotropy contributions over the sample, which lead to different domain patterns in the intermediate and final states (*top* and *bottom rows*)

The beautiful patterns developing out of bubble lattices in a perpendicular field parallel to the bubble magnetization have always attracted the attention of investigators [1057, 883]. In such fields the bubbles expand, transforming the former matrix into a network of narrow domains in which each cell represents a former bubble domain. A magnetization cycle starting from a remanent bubble lattice state and ending at a band domain state as shown in Fig. 5.118 demonstrates *configurational hysteresis* phenomena that appear to be independent of crystal lattice defects or imperfections.

The process starts with the collapse of smaller cells and the growth of larger cells. Peculiar is the relative stability of small fivefold cells, so-called bubble traps [755], which persist up to a critical field [slightly above that in Fig. 5.118e]. After their disappearance the network rapidly expands if the field is further enhanced. By reducing it (g–h), the remaining black domains fold to end up in the demagnetized maze pattern (h). Note that the bubbles in the fivefold cells are magnetized parallel to the external field, different from conventional, antiparallel bubbles.

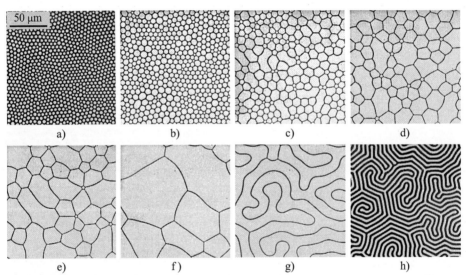

Fig. 5.118 Modification of a bubble lattice in perpendicular fields. The initial "microcrystalline" bubble lattice (a) expands and "melts" under the action of a perpendicular field by the familiar process of the growing of large cells and the collapse of small cells. Remarkably, a certain class of small cells, the five-sided cells, remain metastable up to quite high fields (c–e). After their collapse the network expands rapidly (f). Reducing the field again (g–h) does not lead to a renucleation of bubbles but to a folding of the remaining cell walls into a maze pattern at zero field (h), which appears again "crystallized", but in a completely different configuration compared to (a). (Sample: BiCuGa-modified yttrium iron garnet film of 5 μm thickness; courtesy *W. Tolksdorf*)

Another example of configurational hysteresis is shown in Fig. 5.119. This hysteresis loop begins and ends with maze patterns (a, e) which are not equivalent, however. In (a) the black domains show many branching points, while the white domains are unbranched. Applying a perpendicular field disfavouring the black domains removes the black branching points (c). By reducing the field, the existing domains fold into a maze pattern again, in which now the *white* phase displays branching points (e). The creation and annihilation of branching points is a discontinuous, hysteretic process, which occurs again without any apparent influence of crystal lattice defects.

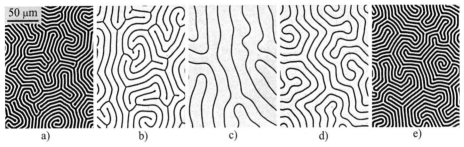

Fig. 5.119 Configurational hysteresis within the maze pattern of Fig. 5.118h. The bifurcations of the black domain phase in (a) are annihilated in increasing fields (b−c). By decreasing the field again, new branching points are nucleated this time in the white domain phase (e)

We explained in the theoretical part (Sect. 3.7.3C) that a transformation between a bubble lattice and a band domain pattern, as well as between band patterns of different periodicity is usually impossible because there are thresholds between these configurations. Such transformations become possible, however, if the respective lattices are not perfect but contain domain dislocations as in Fig. 5.120a. These defects in the magnetic pattern can "climb", thus inserting or removing lattice plains and adjusting the period towards the equilibrium value. Bubbles are a natural source for such dislocations. If a bubble strips out within a band domain array, two dislocations are formed at the ends of the strip segment that replaces the bubble.

Figure 5.120 displays domain pattern conversion processes in a film, in which a movement of dislocations is smoothly possible, although it needs time (in other samples these dislocations tend to be pinned, leading to less regular patterns). If a field almost parallel to the film plane is applied, leading to the nucleation of a bubble lattice, then the pattern, which results after the field is turned off, depends strongly on the speed with which the field is changed. Obviously some thermal activation processes are connected with the process of dislocation movement, the nature of which is unknown. Beautiful dynamic experiments can be performed with such samples, displaying parallels and differences between magnetic and atomic lattices.

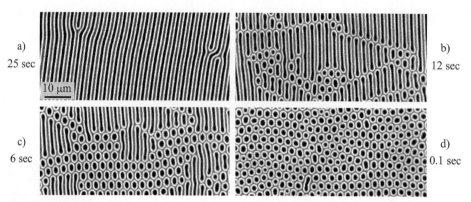

Fig. 5.120 The transition from an initial bubble lattice at high in-plane fields (not shown) to the demagnetized state depends on the field speed in this sample. (a) Slow field change, allowing 25 seconds from the saturating field to zero, generates a near-equilibrium band domain pattern along the field axis with few dislocations. (d) A rapid field change from saturation to zero leads to a metastable bubble lattice with few lattice defects. (b, c) Intermediate speed leaves many dislocations and stacking faults (b), as well as residual bubble lattice islands (c) within the band domain pattern

The particularly defect-free crystal used in Fig. 5.120 also displays a very regular nucleation mode of the dense stripe pattern (Fig. 5.121a) — in contrast to the irregular patterns in Fig. 5.117a. From this second-order phase transition the band domain pattern coarsens by a steady stream of dislocations. Some stages in this process are shown in Fig. 5.121b–d.

Even twin boundaries and small angle boundaries can be generated in the band domain pattern by applying fields of different directions as shown in Fig. 5.122.

To summarize, domains in garnet films offer the opportunity for a transparent study of magnetization processes. Even complex configurational hysteresis phenomena can be analysed in all detail and compared to simulation concepts.

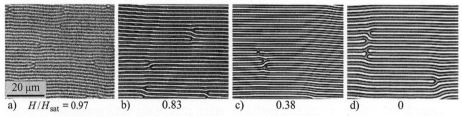

Fig. 5.121 If the in-plane field on the sample of Fig. 5.120 is adjusted to lead to dense stripe domain nucleation (a), the transition to a coarse pattern (d) on reducing the field is found to consist of continuously streaming dislocation climbing processes

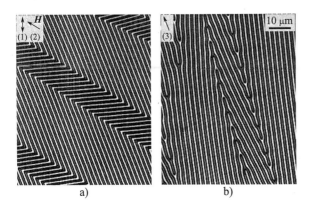

Fig. 5.122 Applying a field at a large angle to the axis of a band pattern leads to the formation of twin boundaries between quasi-domains of a different band axis (a). By reducing the field angle again also small angle boundaries can be formed (b), which consist of a row of dislocations

(B) Perpendicular and Magneto-Optical Recording Media. The regular appearance of the patterns shown in (A) is typical for garnet films and similar materials that are characterized by a high domain wall mobility. Such materials can be used in bubble memories or magneto-optical display or control devices that need this property.

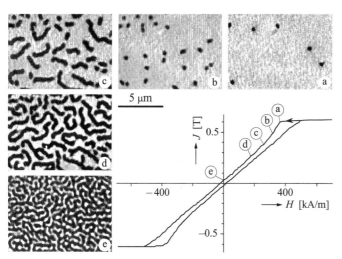

Fig. 5.123 Bubble and band domain growth in a perpendicular recording medium: Co (20 at% Cr) of 622 nm thickness. Starting at saturation in a perpendicular field, oppositely magnetized cylindrical domains are nucleated and then grow into band domains. (Together with *F. Schmidt*, Erlangen)

Domain patterns reminiscent of maze domains are found also on perpendicular recording media as shown in Fig. 5.123. For these applications, materials are used which correspond to bubble films except that a large coercivity, caused by structural features such as the boundaries between the columnar grains,

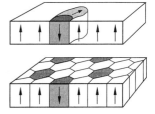

impedes domain wall motion so that information storage becomes possible.

Nevertheless, quasi-periodic domain patterns can be observed in materials optimized for this purpose [162, 98]. Reducing the field from saturation, bubble nucleation and often bubble stripout are observed. We infer a predominantly continuous character because exchange-decoupled columns would tend to be magnetized in a pepper-and-salt-like fashion as sketched [162].

Magneto-optical storage media also have a strong perpendicular anisotropy and rely strongly on domain wall coercivity effects. Usually, these films are thinner than the characteristic length l_c of bubble theory (see Sect. 3.7.3), and for this reason regular periodic domains are not expected. Then only huge domains with irregular, fractal boundaries are observed as demonstrated in Fig. 5.124a. If the films are thicker than the characteristic length, domains developing in a hysteresis loop assume a different character reflecting the equilibrium domain period of maze domains in bubble films (Fig. 5.124b).

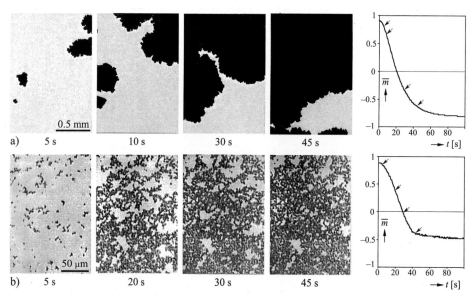

Fig. 5.124 Domains in magneto-optical storage layers developing in the course of time in a perpendicular field close to coercivity. (a) $Tb_{21}Fe_{67}Co_{12}$ of 117 nm thickness. (b) $Tb_{27}Fe_{56}Co_{17}$ of 224 nm thickness. The average magnetization $\overline{m}$ plotted on the *right* is derived from the normalized Kerr signal. (Together with *S. Winkler*, Erlangen)

Characteristic of both types of domain patterns is their global non-equilibrium character. The front of the nucleated reverse domains can be observed to creep slowly into the still saturated area, indicating a thermally activated process. This behaviour is typical for all variants of perpendicular magnetic thin film media.

It must be mentioned that the difference in the domain character demonstrated in Fig. 5.124a,b can also occur in ultrathin films for which magnetostatic effects as expressed by the characteristic length l_c can be excluded as shown by simulation in [1058]. The width of the thermal activation spectrum for domain wall propagation was made responsible for these variations.

5.6.2 Localized Domains (Bubbles)

Applying a sufficiently strong perpendicular field to a film as studied in Sect. 5.6.1A eliminates the extended domain pattern present in the demagnetized state. Before reaching saturation cylindrical domains or "bubbles" are often left over [554], behaving as stable, independent and easily displaceable objects [510]. Bubbles can be individually nucleated, manipulated and annihilated. Figure 5.125 demonstrates the reaction of widely separated bubbles to different applied fields. In uniform perpendicular or "bias" fields they increase or decrease in diameter until they either collapse or "strip out" into band domains again. In strong horizontal fields bubbles are distorted elliptically because of the orientation dependence of domain wall energy.

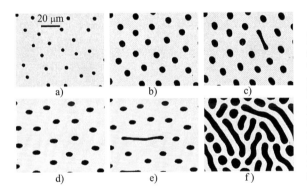

Fig. 5.125 Bubbles in different applied static fields. (a) On the verge of collapse in a strong perpendicular field. (b) Imminent bubble strip-out in a reduced perpendicular field. (c) Bubble strip-out. (d) Strong elliptical distortion in an in-plane field. (e) Strip-out in a higher in-plane field. (f) Remanent state after switching off the field in (c)

In a perpendicular *gradient* field bubbles move laterally into the direction of smaller perpendicular field. The gradients are generated by a current pattern, or by the interaction of a rotating external field with localized soft magnetic thin film elements [510, 751]. Such schemes can be used for various memory and display applications (see Sect. 6.6).

External fields can do more to bubbles than modifying their size and shape. Additional degrees of freedom are hidden in the internal structure of the bubble domain wall. These walls are

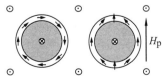

essentially slightly modified Bloch walls. Their detailed structure was discussed in Sect. 3.6.4H. The lowest energy state of a bubble in the absence of an in-plane field is the state with a single rotation sense, i.e. without any Bloch lines (see Sect. 3.6.5A). In a horizontal field a state with two Bloch lines as sketched becomes energetically favoured. A transition between both states is topologically forbidden, however. It needs the generation and motion of a micromagnetic singularity to achieve this transition (see Fig. 3.94). So both states are metastable and can also be used as memory units (see Sect. 6.6.1C).

The different wall states have almost no influence on the static bubble shape. Unless an excessive number of Bloch lines is accumulated in a bubble wall, their static and quasi-static behaviour is only slightly affected. Excessive means that the density of Bloch lines of the same sign exceeds their equilibrium distance $2\pi\sqrt{A/K_d}$ mentioned in Sect. 3.6.5A. Beyond this threshold bubbles get heavily deformed and their collapse field is enhanced. Because they survive in fields in which regular bubbles collapse, they are called "hard bubbles".

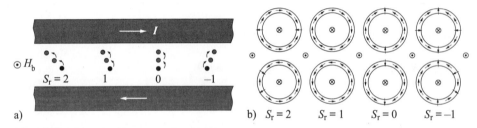

Fig. 5.126 The phenomenon of bubble deflection in a current-induced gradient field (a). The deflection angle depends on the rotation number in the bubble walls. For each of the four cases two examples are shown in (b)

The dynamics of bubble domains represents a highly specialized subject that was covered extensively in the literature [49, 512, 709–712, 751]. For example, bubbles with only a few Bloch lines can be distinguished dynamically. If high speed bubble displacement is tried, bubbles finally do not move to the point of smallest potential energy ("down the field gradient"), but gyromagnetic phenomena become important. Depending on the number and sign of Bloch lines in the domain wall the bubble gets more or less deflected away from the gradient direction as sketched in Fig. 5.126. Hard bubbles with their large rotation number move virtually perpendicular to the gradient. The primary parameter is the *revolution number* S_r of the magnetization in the wall, a topological characteristic that counts the number and sign of rotations of the magnetization vector on a path once around the perimeter of the bubble.

A quantitative description of bubble deflection can be obtained by using *Thiele*'s concept of dynamic reaction forces (3.162).

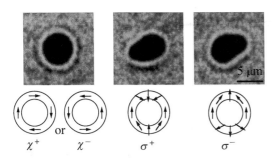

Fig. 5.127 The dynamic expansion of $S_r = 1$ bubbles in a garnet film, observed by stroboscopical Faraday microscopy. The different resulting shapes indicate differences in the wall structure, in particular of the presence or absence of vertical Bloch lines (classified as $\sigma^\pm$- and $\chi^\pm$-states). (Courtesy *L. Zimmermann* and *J. Miltat*, Orsay)

Even further details about bubble wall states can be identified by carefully designed dynamic experiments. Ordinary bubbles with rotation number one can occur in two different rotation senses or "chiralities". In addition, unwinding vertical Bloch line pairs (see Sect. 3.6.5A) can be added without changing the revolution number. Figure 5.127 shows an experiment in which the dynamic reaction in bias field pulses in the presence of additional in-plane fields is stroboscopically detected to distinguish between some of these states. For all details the reader is referred to the literature [49, 1059−1061].

When a bubble collapses, no traces are left according to all experimental evidence. This is somehow surprising for bubbles with a revolution number different from zero, as there is no possible continuous transition between such a bubble state and the uniformly magnetized state. One might think that an exchange-stabilized vortex is left, with a diameter of the order of the exchange length, but a detailed calculation demonstrates that no stable vortex with a finite radius can exist in a ferromagnet. The energy of a cylindrical vortex solution decreases according to micromagnetic continuum theory monotonously to zero radius. Probably singular points are nucleated when the vortex gets very thin, thus opening a path to complete annihilation. Stable vortex solutions were predicted to exist and analysed numerically in special materials without inversion symmetry [546, 1062] but no direct observations of these structures have been reported.

5.6.3 Domain Wall Investigations in Perpendicular Films

In spite of the important role of the domain wall substructures in perpendicular films their direct observation proved to be difficult. The first direct observation of Bloch lines was achieved in the transmission electron microscope by *Grundy* et al. [1063]. This was possible because the polar magnetization in the domains does not interact with the electron beam at perpendicular incidence so that only the walls and their substructure become visible. TEM observations are limited to thin films without their usual substrate, however, and the application of magnetic fields is possible [1064] but not easy in the electron microscope.

An observation with optical means would be preferable because all other domain features in garnet films are conveniently observed with the Faraday effect and magnetic fields of any strength, direction and frequency can be applied. A fundamental difficulty in optical observation is that in perpendicular films the domains generate strong (polar) magneto-optical effects, whereas the interior of a domain wall is magnetized parallel to the surface, generating only weak signals that cannot be separated from domain contrast with conventional means. For a long time only dynamical methods were available for identifying the presence of Bloch lines in optical microscopy [1065−1067].

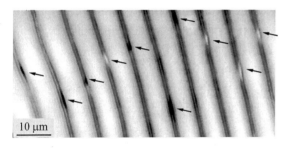

Fig. 5.128 The direct observation of Bloch lines in bubble garnet films of 5 μm thickness by the asymmetric dark field technique. A few of the visible Bloch lines are marked by arrows. (Courtesy *A. Thiaville*, Orsay [1068])

A direct observation of single Bloch lines was finally achieved by *Thiaville* [197, 696, 1069] using a dark field diffraction technique (Fig. 5.128). First a scanning magneto-optical microscope [197] was used for this purpose. Soon after it was demonstrated that equivalent images can be obtained with a conventional microscope [198] equipped for dark field illumination. The precise contrast mechanism was identified later [696]. It turned out that a secondary deformation of the wall connected with the system of magnetic charges accompanying a Bloch line leads to diffraction effects for proper illumination. Many details about the dynamics and transformations of Bloch lines were identified with this technique [1070, 1071] although the method is limited in its applicability, as it needs a certain film thickness of several microns for the mentioned wall deformations to develop. Also the resolution of the method is not satisfactory as it images the wall deformations in the neighbourhood of Bloch lines rather than the Bloch lines themselves. A convenient and high-resolution observation of Bloch lines in perpendicular anisotropy films is still not available.

5.7 Particles, Needles and Wires

In extended thin films no equilibrium single-domain state is expected independent of the material and geometrical parameters. This was discussed in the last section in particular for perpendicular-anisotropy films. The situation changes for samples that are restricted in more than one dimension. Here we deal with thin wires, needles and particles. Such samples are of great practical importance as carriers of information in magnetic recording technology. The details of the transition between multidomain and single-domain behaviour are not well-known because of the small size of the critical diameter that is usually below the resolution limit for optical microscopy. An increased understanding of their behaviour has profited from modern high resolution domain observation techniques and from micromagnetic simulation methods that are particularly suitable for the investigation of small particles as discussed in Sect. 3.3.2.

5.7.1 Observations on High-Anisotropy Uniaxial Particles

Figure 5.129 shows particularly beautiful domain observations on tiny particles of barium ferrite. The patterns are observed in the thermally demagnetized state by a dry colloid technique using a scanning electron microscope [1072, 97]. Such observations may support theoretical studies of the equilibrium domain pattern of small particles, but they cannot contribute to the question of the stability limit of the saturated state, i.e. to the question of coercivity.

Some permanent magnets actually consist of an aggregate of isolated magnetic particles with no exchange interaction between them. Domain observation on such samples can offer valuable insight. Figure 5.130 shows observations for the thermally demagnetized state on the basal plane of an oriented polycrystalline $Nd_2Fe_{14}B$ material. Note first that within the grains the domains in this demagnetized state assume the same branched equilibrium patterns known from single-crystal experiments (Fig. 5.5), indicating a very small value of domain wall pinning forces in this material.

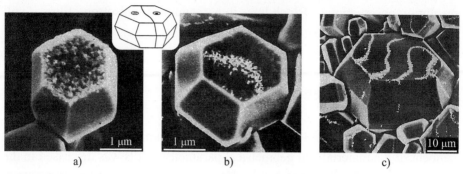

a)   b)   c)

Fig. 5.129 Domains on barium ferrite particles near the critical thickness of single-domain behaviour, made visible by scanning electron microscopy of dried Bitter patterns. In the thermally demagnetized state the particle in (a) shows a single-domain behaviour, while the larger particles in (b, c) are split up into domains. (Courtesy *K. Goto*, Sendai)

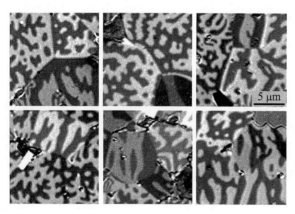

Fig. 5.130 Domains on the basal plane of a $Nd_2Fe_{14}B$ permanent magnet that consists of uncoupled grains. Grain boundaries usually display an antiparallel alignment of the magnetization on either side (*top row*). Exceptions to this rule (*bottom row*) indicate stray field effects to be expected for obliquely oriented grain boundaries. (Sample: courtesy *B. Grieb,* MS Schramberg)

In connection with the small particle aspect we focus on the grain boundaries. If exchange coupling over the grain boundaries would be present, then domains on both sides of these boundaries should always be magnetized parallel. Without exchange coupling at least those grain boundaries that are oriented parallel to the easy axis (perpendicular to the observation plane) should be magnetized antiparallel in the two grains because of the demagnetizing effect. For grain boundaries that are tilted with respect to the easy axis of the material, a parallel alignment would be favoured for magnetostatic reasons.

Observing an antiparallel domain alignment is a clear proof of the absence of full exchange coupling over the grain boundaries because exchange coupling would enforce parallel alignment. Such an antiparallel alignment is indeed observed in several examples (Fig. 5.130, top row), while at other places (bottom row) the domains are aligned in parallel, which is expected, as elaborated, for tilted grain boundaries. The conclusion that in this material the grains are not exchange-coupled is no surprise from the theoretical point of view that could hardly explain a high coercivity in such coarse-grained material otherwise. But a direct confirmation would be difficult with other methods.

5.7.2 Very Small Particles

Particles in the neighbourhood of the single-domain limit (see Fig. 3.32) cannot be analysed experimentally in detail. If a particle is of single-domain character, then this property can be seen by magnetic force microscopy. A somewhat larger particle that is demagnetized remains largely inaccessible to experimental techniques, especially if due to low anisotropy poles are largely avoided. On the other hand, this size regime is the ideal working field for numerical solutions of the micromagnetic equations. With adequate care reliable calculations can be offered for such particles (Fig. 3.21). Some more results are shown in Fig. 5.131. It would certainly be very difficult to verify the detailed three-dimensional configuration of such solutions experimentally.

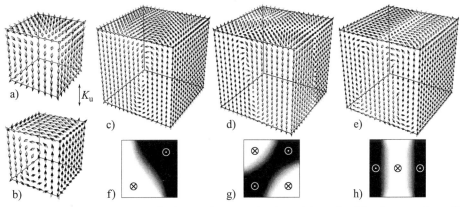

Fig. 5.131 Calculated vector fields for demagnetized particles of cubic shape and small uniaxial anisotropy. While outward views of the patterns (a–e) display complex, three-dimensional configurations, cross-sections showing m_z in the centre planes (f–h) reveal simple internal domain structures: a two-domain state with a twisted wall (c, f), and two different three-domain states (d, g), (e, h). (Courtesy *W. Rave*, IFW Dresden; see [598])

Very small thin film elements can be investigated both with numerical micromagnetics, and with high-resolution experimental techniques. We show examples for both approaches, which display a convincing correspondence between them. Figure 5.132 shows micromagnetically calculated stable and metastable configurations of small thin film elements of different shape. The results are displayed simulating ultrahigh resolution Kerr effect or polarized

electron (SEMPA) images so that the equivalence between the results in submicron particles and those of larger particles (Sect. 5.5.4) becomes apparent. At
least down to the 0.1 μm size range domain patterns in low anisotropy thin
film elements remain largely similar, depending little on size. With numerical
calculations of this type the "single-domain limit" can be calculated [598], the
size for which the uniformly magnetized state becomes energetically favoured.

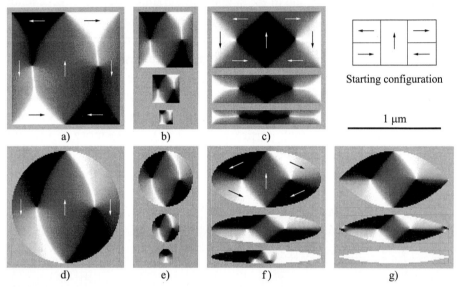

Fig. 5.132 Calculated demagnetized states for thin film elements of different shape, performed
for the material parameters of Permalloy of 20 nm thickness. The starting configuration
was chosen in order to reach a three-domain state in every case, which succeeded except
for the smallest elements in (e–g) (Courtesy *Y. Zheng* and *J.-G. Zhu*, CMU Pittsburgh)

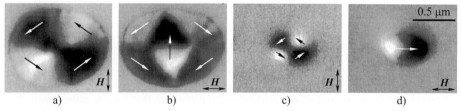

Fig. 5.133 Magnetic force microscopy images of small Co elements of 20 nm thickness.
The larger element shows either a concentric state (a) or a three-domain state (b). The
concentric state can also be observed in the smaller element after applying a field along the
shorter axis; otherwise a single-domain state is observed, which can be recognized by its
black and white charge contrast. (Courtesy *A. Fernandez*, Lawrence Livermore Labs.)

The experimental investigation of elliptical cobalt elements with magnetic
force microscopy [1073] shows results that depend on magnetic history

(Fig. 5.133). After applying a field along the long axis a single-domain state is found in the submicron elements, a state that is stable in larger elements only for more elongate shapes. Closed-flux or domain states are formed after applying a field along the short axis of the elliptical elements. Experimentally, it is difficult to decide which of the states represents the true energy minimum.

The magnetic microstructure of very small particles is of particular interest in the field of geophysics as comprehensively reviewed in [1074]. The basic quantity of interest in this field is the *remanent* magnetization, which is used as an indicator for the alignment of rocks and sediments in the geological past. Of particular interest is the "stable" part of this remanence, the part that is left over after the application of moderate alternating fields (because this stable part is supposed to reflect the originally acquired magnetization of the particle, without "soft" remanence contributions that were acquired perhaps later in geological history). The analysis of observed remanences indicates that single-domain particles which naturally have a high remanence are not sufficiently abundant to explain the observations. Multidomain states such as those shown in Fig. 5.131 have a very low remanence, however, which again appears to be insufficient to explain observations as well as model experiments. This discrepancy led geophysicists to the concept of "pseudo single-domain" (PSD) behaviour [1075]. Many explanations for this phenomenon were offered, but altogether its nature remained elusive [1074, Chap. 12.6].

A possibility that has obviously been overlooked is the influence of particle shape. The low remanence of multidomain states as in Fig. 5.131 is connected with the symmetric shape of the assumed models. Already simple domain theory calculations such as those used in Sect. 3.3.2 for high-anisotropy particles show that *asymmetrical* particles almost always possess a strong, thermodynamically stable remanence as demonstrated in Fig. 5.134. Here a trapezoidal particle is consid- ered for which two mechanisms contribute to an equilibrium $E_{\mathrm{d}}^{\mathrm{min}}$: $m = 0.11$ non-zero remanence: (i) The stray field energy alone favours an asymmetrical position of the domain wall as sketched. Note that sometimes even symmetrical particles show a remanence in thermodynamic equilibrium. An example is the three-domain particle in Fig. 3.19. In asymmetrical particles this is also true for two-domain states. (ii) In asymmetrical particles also the total domain *wall* energy is smaller if the wall position is shifted towards the shorter side, thus enhancing the stray-field-induced remanence for small particles. Altogether, an approximate $1/L$ dependence of the equilibrium remanence seems to result, which is consistent with experiments on artificial model particles

[1076]. It can be expected that a generalization of this argument and proper statistical averaging can account for many of the hitherto unexplained geophysical observations [1077].

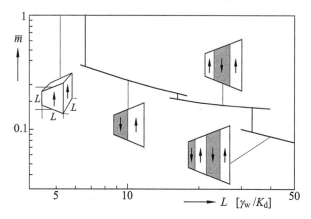

Fig. 5.134 The calculated equilibrium remanence for asymmetrical prismatic high-anisotropy particles as a function of particle size. For a particle of trapezoidal shape the equilibrium positions of one, two and three planar walls were calculated within the framework of domain theory. Transitions are indicated where the energies of the different domain states are equal. (Together with *K. Fabian*, Bremen)

5.7.3 Whiskers

Certainly the most attractive ferromagnetic specimens are magnetic whiskers, tiny single-crystal needles with often perfect surfaces. Whiskers are usually grown by chemical reaction from the vapour phase. The process parameters, such as the temperature and the gas flow, determine the type, size and perfection of the products. The size of whiskers ranges from micrometers to millimetres. Fundamental insight into the laws of magnetic microstructure was first gained from domain observation on whiskers [611, 641, 1078–1082]. Two advantages contributed to this fact: (i) as whiskers can be perfectly oriented and defect-free, the domains can be very simple, and (ii) domain observation is possible from all sides, so that a reliable analysis even of more complex, three-dimensional patterns becomes feasible. A good overview of whisker domain observations can be found in [44]. Figure 5.135 shows some low- and high-resolution Kerr effect examples from thin iron whiskers.

DeBlois [611] succeeded in studying the properties of very thin whiskers that behave like single-domain particles or wires with the techniques of domain observation. Such whiskers are very flexible and they can be bent by the action of a magnetic field. Sudden changes in the curvature indicate the switching of the thin samples. With ingenious experiments he could avoid the influence of the sample ends and thus measure the intrinsic switching field as a function of the whisker thickness.

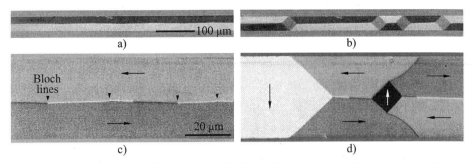

Fig. 5.135 Domain observations on iron whiskers. *Top row*: examples for low-resolution overview images. *Bottom row*: high resolution observations with domain and domain wall contrast. Note the more-or-less periodic occurrence of Bloch lines in (c), and the curved V-lines in (d) (Samples grown and screened by *S. Schinnerling*, IFW Dresden)

The perfect shape of whisker single-crystals also favours the study of field-dependent processes and in particular of the decisive processes at the sample ends. Figure 5.136 resumes the subject of Fig. 5.41; a perfectly simple domain structure (a) in zero field is superimposed by a complex system of supplementary domains employed to distribute the magnetic flux as required by the demagnetizing energy.

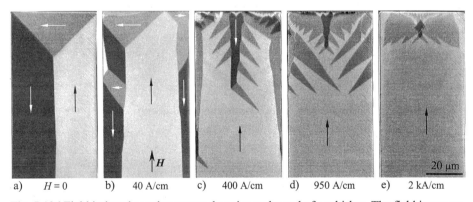

Fig. 5.136 Field induced supplementary domains at the end of a whisker. The field increases from (a) to (e)

5.7.4 Magnetic Wires

The switching behaviour of magnetic wires played a role in the genesis of the domain idea (Sect. 1.2.2). Today they attract renewed interest as sensor devices and pulse generators [1083, 1084]. A rectangular magnetization loop in a field along the wire axis is necessary for all these applications. This requires uniaxial anisotropy that may be generated by an external or an internal stress and the right sign of the magnetostriction constant. Often only the core of the wire shows the switching behaviour; the outer parts produce the necessary stress for the core.

a)

b)

c)

d)

Fig. 5.137 An amorphous wire (Ø 0.127 mm), rapidly quenched into water, shows these domains in the field-free state (a) and in increasing external fields up to about 10 kA/m along the axis (b–d). In much smaller fields the core of the wire "switches", but no trace of this process can be found on the outside. (Sample: courtesy *F.B. Humphrey*, Boston)

In these samples the outer domains do not reflect the switching process [1085]. The longitudinal domains taking part in the switching process can only be made visible by polishing away part of the wire [1086]. The shell is either magnetized perpendicular to the surface as in Fig. 5.137., or circumferentially (see sketch [1087]). In the latter case an attractive "bamboo" pattern [1085, 1088] is sometimes observed as shown in Fig. 5.138. In such samples not only the switching of the core but also the circumferential permeability of the shell can be exploited (see Sect. 6.2.3B).

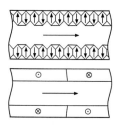

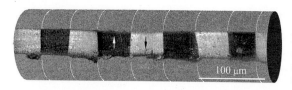

Fig. 5.138 Bamboo domains observed on the central part of an as-quenched amorphous CoSiB wire. (Sample: courtesy *J. Yamasaki*)

5.8 How Many Different Domain Patterns?

Can an overview as the one presented in this chapter be complete? The infinite number of "cases" was emphasized at the beginning (Sect. 5.1.4). But even within well-defined standard cases such as that of bulk uniaxial crystals (Sect. 5.2) a surprising variety of domain patterns can be observed. An increasing degree of branching with increasing plate thickness as in Fig. 5.5 need not be considered as really new, however. Once the principle of branching is understood, new observations for thicker high-anisotropy materials offer no surprises. Unexpected features are observed when the axial symmetry of the pattern is broken by applied in-plane fields (Figs. 5.13 and 5.14) or for tilted surfaces (Fig. 5.8). Also the patterns for low anisotropy uniaxial materials ($Q < 1$) differ in characteristic details (Figs. 2.39 and 5.4). When uniaxial anisotropy is replaced by strongly orthorhombic anisotropy, branched patterns of markedly different character are formed (Fig. 5.9). A combination of these special conditions will most probably yield further unconventional structures.

An even larger variety of domain patterns is observed in bulk cubic crystals (Sect. 5.3). The branching phenomena in cubic crystals are complicated by a transition from two-dimensional to three-dimensional structures as predicted in Sects. 3.7.5D, E and as observed in Figs. 5.22–24. Furthermore, the interplay between different length scales — that of the domain wall width with that of the onset of magnetostrictively dominated phenomena (see Fig. 3.15) — plays an important role. We argued qualitatively, for example, that the complex stress pattern II (Figs. 5.26, 27) is stabilized by its magnetostrictively favourable intermediate domains. As another example, the angular orientation of the cord-like pattern of Fig. 5.23 is known to result from an equilibrium between magnetostrictive energies and orientation-dependent specific wall energy. The interactions between different scales, easy axes, applied fields, surface orientations and external stresses leads to an enormous variety of domain patterns, only a small fraction of which is as yet understood (see Fig. 5.29). Most examples shown in this book in this category are based on the standard sample thickness of electrical steel material around 0.2 mm. The few pictures presented from thicker samples (Fig. 3.147 and 5.24a) suggest that novel domain structures will show up in a systematic search in thick and well-defined samples, waiting for their documentation and analysis.

In electrical steel, external stresses introduce additional, usually uniaxial anisotropies that remain weak compared to the predominant cubic anisotropy so that the easy axes are modified in their energy levels but not in their directions. If the superimposed uniaxial anisotropy becomes comparable in strength with the basic cubic anisotropy, completely new types of magnetic materials may be generated. The unconventional structure of Fig. 5.102b is just one example of this kind.

The many surprises documented in the investigation of amorphous and nanocrystalline materials (Sect. 5.4) were mostly connected with the distributions of internal stresses. As the strength, distribution and orientation of these stresses can be highly variable, new and unexpected domain patterns may show up in every new investigated sample. Some examples connected with partially annealed and crystallized structures were shown in Sect. 5.4.2B.

Domains in thin magnetic films (Sect. 5.5) have been explored for a long time. Still surprising is the large number of metastable solutions in thin film elements of various shapes (Sect. 5.5.4). This feature is also found in the numerical analysis of small particles as shown for a few examples in Fig. 5.131. Neither numerical simulation nor experimental observation has as yet been able to explore the full set of metastable micromagnetic configurations for a given element. Even more complex is the situation for coupled films and multilayers, as demonstrated for a sample with a continuously varying interlayer in Fig. 5.111.

We focused in this chapter on the equilibrium patterns and the magnetization processes in undisturbed, ideal samples, hoping that domain patterns in disturbed, high-coercivity samples can be somehow related to the ideal patterns. The few examples for non-ideal samples (such as Figs. 5.123, 124 for recording media, Figs. 5.113–115 for polycrystalline multilayers) are related to applications to be discussed in the next chapter.

Nevertheless, it is hoped that the presented gallery of domain patterns and magnetization processes in ideal samples can help in the understanding of patterns and processes in disturbed samples or in samples of different symmetry or shape. Every domain pattern that is correctly analysed and understood in one case can serve as a guide to patterns in related cases.

6. The Relevance of Domains

This chapter reviews engineering applications of magnetic materials, and the relevance of magnetic domains in these fields. It connects a discussion of applications with the discussion of magnetic microstructures elaborated in earlier chapters of this book.

6.1 Overview

The role of magnetic microstructures varies strongly between different applications of magnetic materials. In some fields, such as in the cores of electrical machinery (Sect. 6.2.1), domains and domain walls are essential. Electrical machines simply would not work without the easily displaceable domain walls providing the necessary permeability. The same is true for most inductive devices at medium and high frequencies (Sect. 6.2.2–6.2.4). But irregularities in the magnetic microstructure are also the origin of losses and noise (electrical and acoustical) in these devices.

In some applications of magnetic materials domains play no role at all, such as microwave components, nucleation-type permanent magnets (Sect. 6.3.1), and particulate recording media (Sect. 6.4.1A). Other devices would ideally work without any non-uniform magnetic microstructures, but domains are the origin of irregular behaviour if they cannot be suppressed. In these cases, mostly in the field of small sensor and memory elements (Sect. 6.5), domain studies are necessary to understand the conditions of their occurrence and their control. Finally, there are applications in which domain propagation is or was directly put to technical use. These are discussed in Sect. 6.6. Reviews of the application of magnetic materials in general with an emphasis on microscopical mechanisms can be found in [502, 505, 506, 1089–1094].

6.2 Bulk Soft-Magnetic Materials

Inductive devices — from power machinery to radio frequency applications — rely on the high permeability of soft magnetic cores. This extrinsic material property reaches values up to one million times the vacuum value. Domain wall displacement is the primary origin of this extraordinary quality of magnetic materials. In principle, magnetization rotation could also lead to high permeability if anisotropies are uniform and small. Rotational permeability is given by the quantity μ^* discussed in Sect. 3.2.5F. Values of the order 10^2 to 10^4 can be reached in special high-permeability materials. But it is very difficult to prepare a material so that domain wall displacements are suppressed to a degree that magnetization rotation becomes predominant at low frequencies.

In only a few devices, such as thin-film recording heads and special sensors, has a suppression of wall displacement processes in a high-permeability state been achieved by a proper design (see Sect. 6.5). In these devices domain wall processes are of interest as parasitical effects disturbing regular function.

It is interesting to analyse the conditions under which a purely rotational permeability is possible in a multiaxial ferromagnet. Take the case of a positive-anisotropy cubic material like iron. In a soft magnetic material the demagnetized state must consist of a flux-closed configuration that necessarily contains both 180° and 90° walls (the saturated state would not display a high permeability).

A necessary condition for avoiding wall motion would then be that all 180° walls are oriented perpendicular to the working direction, the direction of the applied field.

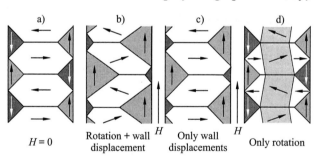

This is not sufficient, however. The unavoidable 90° walls can contribute to the net flux in a variable degree as indicated in the sketch (b,c). But if 90° wall motion is somehow suppressed, an ideal, linear and reversible soft magnetic material based on rotations only would be realized (d). The pinning of the 90° walls could be realized by some induced anisotropy effect as discussed in Sect. 3.6.7. Alternatively, 90° wall motion can be suppressed at very high frequencies, as first observed and analysed by *Mallary* and *Smith* [1095] in thin-film inductive head elements.

Domain wall displacement processes are generally less favourable at high frequencies than magnetization rotation because of the extra ("anomalous") eddy current losses connected with wall motion. The basic mechanism of anomalous eddy current losses was discussed in Sect. 3.6.9 for cores consisting if thin, insulated metal sheets, which is the most common core design. With increasing frequency and for a given sheet thickness, domain wall displacements are more and more suppressed on account of these eddy currents, so that, finally, rotational processes would in fact take over. At such frequencies, however, even classical eddy current losses (3.183) usually become too large, so that a thinner gauge material is preferred. For the thinner material and the same frequency domain wall displacements are less suppressed and may have to be taken into account again.

Grain boundaries in soft magnetic materials must always be fully ferromagnetic, with a complete exchange coupling over the grain interfaces. This feature is not developed consciously, it is the standard property of metallic alloys. The interruption of exchange interactions, as needed in permanent magnets and some recording media, is more difficult to achieve.

In summary, most technical applications of inductive devices are based on the high permeabilities available from domain wall displacement processes. The irregularities connected with domain wall pinning and domain annihilation and creation processes lead to losses in electrical machines, and to noise in communication and sensor devices. If the character of the magnetic microstructure can be improved in the sense of a more regular domain structure, a better performance of inductive devices can be achieved.

6.2.1 Electrical Steels

In the core of transformers, generators, motors and other inductive machinery various iron-based metallic alloys of cubic crystal symmetry are used as reviewed in [1096]. The most important material is iron with up to 3.2 wt% of silicon. Electrically insulated sheets of typically 0.3 mm thickness, but ranging from 0.1 to 0.5 mm depending on application, are either stacked or wound in the cores of these devices. Transformers can be set up so that the magnetic flux alternates along only one, preferred axis of the material. Anisotropic, highly textured material offers the best performance in this case. In rotating machinery the flux necessarily changes its direction during operation, and anisotropic material is of little use. Mostly isotropic, polycrystalline material is therefore used for this purpose. Special textures favouring arbitrary magnetization parallel to the sheet plane have also been developed, their magnetic microstructure is particularly interesting.

The most important quality number for many larger applications of electrical steels is the *loss* per cycle, usually given in watts per kilogram for a given frequency and a given induction amplitude. An overview of this aspect and the available material options can be found in [1097]. *Permeability* in these applications is a secondary quantity. Often a material with smaller losses is preferred even if the permeability is somewhat smaller. The reason is that magnetic losses are the principal contribution to total losses in idling machines — and most transformers are effectively idling about half of the time. In contrast, permeability determines the current needed in the idling state to reach the desired induction level. This current is always much smaller than the current drawn at full load. As the windings must be designed for the full load condition, the losses connected with the no-load magnetizing currents are negligible for reasonable permeabilities.

Core losses belong to the *extrinsic* material properties. They depend strongly on the crystal and magnetic microstructure as will be explained in more detail. A very important intrinsic property is the saturation induction which determines the maximum induction level that can be used in a device. The maximum useful induction level is also limited by anisotropy, another intrinsic property. A favourable texture can increase the magnetizability if it is limited by anisotropy. If the size or weight of a machine is critical, materials with a high magnetizability may be preferred even at the expense of higher losses. This is true, for example, for small motors for which idling losses are negligible compared to other losses. For such applications pure iron ($J_s = 2.17$ T) may be preferred to the standard silicon iron ($J_s = 2.05$T), and in extreme cases even the expensive CoFe ($J_s = 2.4$ T) may be competitive, in particular because its anisotropy is smaller than that of iron.

(A) General Relations for Textured Material. Grain oriented or textured alloys play an important role in the field of electrical steels [1098]. Permeability is used as an efficient means of characterizing such materials. This possibility is based on phase theory, and in particular on the range of field-free magnetization processes ("mode I") which can be reached by domain wall displacement processes only (Sect. 3.4.2C). If the anisotropy field is much larger than the coercivity as is typical for iron-rich alloys, the average magnetization in a field parallel to the sheet plane, which just surpasses wall pinning (coercivity) effects, will correspond essentially to the boundary of the mentioned mode I magnetization processes — the knee in the magnetization curve, or the "polyhedron" as we called it in the context of phase theory. In electrical steels usually a field of 8 A/cm $\cong$ 10 Oe is used to characterize the magnetizability of a material. The *permeability* μ_8 is defined as the total relative permeability at this point. Often simply the reduced average magnetization m, or $J_s m$ at the prescribed field $H = 8$ A/cm are called "permeability".

Let the working and field direction be given by the unit vector a, and let the cubic axes for a given crystal be given by the unit vector tripod (e_1, e_2, e_3). Then, as mentioned in Sect. 3.4.2D, the following formula defines the polyhedron of positive-anisotropy (iron-like) cubic materials:

$$m = \frac{1}{|a \cdot e_1| + |a \cdot e_2| + |a \cdot e_3|} . \tag{6.1}$$

This formula describes the maximum mean magnetization along the working direction under the condition that no net transverse magnetization components are present. Any component of transverse flux is assumed to be absent in this approach because it would lead to demagnetizing effects. The need to compensate magnetization components perpendicular to the sheet plane is obvious from the discussion in Sects. 3.7.1 where the complementary domains were introduced and justified explicitly for this reason. But in-plane net transverse components are also generally unfavourable in a polycrystalline material because of the neighbouring grains. Each grain would prefer a specific in-plane magnetization direction, but the interaction between the grains tends to suppress the individual excursions and favours a common mean flux direction. To first order, therefore, the state characterized by (6.1) will be preferred in all grains. Local fluctuations in the grain arrangement may favour local transverse magnetization excursions. But let us ignore this possibility for the moment. We may then examine the consequences of the phase-theoretical result (6.1). In this state the phase *volumes* of the participating phases are given by:

$$v_i = \frac{|a \cdot e_i|}{|a \cdot e_1| + |a \cdot e_2| + |a \cdot e_3|} , \tag{6.2}$$

where i applies to those three easy directions for which the scalar product with the working direction a is positive. Note that "opposite" magnetization phases as they may occur in actual supplementary domain pattern — such as the surface lancets in the lancet comb pattern (Fig. 3.105) — are ignored. Consider now the case that the easy direction e_1 differs only slightly from the working direction a. Then the cosine $a \cdot e_1$ deviates somewhat from unity. The net magnetization m will deviate markedly from unity anyway because the expressions $|a \cdot e_2|$ and $|a \cdot e_3|$ in the denominator increase with the sine of the misorientation angle. The low-field magnetizability will therefore decrease *linearly* with the misorientation angle.

The predictions from (6.1) agree generally very well with texture measurements as first demonstrated in [1099, 1100] and as reviewed in [726]. This good correlation is used in material development as a convenient texture indicator. The permeability μ_8 characterizes in a single number the usefulness of a texture for a specific magnetic application. A particularly careful analysis was performed in [1101]. The result of the measured permeability as a function of the mean texture deviation angle for various polycrystalline samples is shown in the sketch. This is compared with a predicted curve (6.1) as evaluated with measured average grain orientation distribution functions. The general

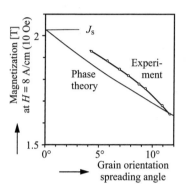

agreement between the prediction of phase theory and measurements is excellent. Deviations show up at very small and probably at very large texture deviation angles, which can both be understood at least qualitatively in terms of observed and expected domains. At large misorientation angles the complex, branched patterns probably can not be constructed with the three most favourable easy directions alone, which are the basis of the prediction (6.1) of phase theory. Also magnetostrictive interactions between the grains as they are discussed in more detail in (C) may be responsible for the reduced permeability at larger texture deviation angles. The discrepancy at small misorientations must be caused by a suppression of supplementary domains. Phase theory loses its validity when the energy of domain walls cannot be neglected any more. Such a case was analysed in Fig. 3.106, where a critical misorientation was calculated, below which the formation of supplementary domains became unfavourable because of the necessary wall energy. It appears plausible that the effect of the domain wall energy causes not only a suppression of supplementary domains below one degree of misorientation, but also leads to a general enhancement of the permeability (resulting from a balance between the stray field energy connected with an imperfect suppression of transverse flux and the energy density of the supplementary domains).

The remaining stray field energy at the critical misorientation angle, when supplementary domains are suppressed, favours a small basic domain spacing. This leads to reduced dynamic losses. For still smaller misorientation angles, wider domains are formed again, with adverse effects on losses. The critical misorientation angle therefore represents a loss optimum for a given material and thickness as will be discussed in more detail in (B).

Magnetic losses, permeabilities, as well as other quantities such as magnetostriction (see Fig. 5.37), are thus intimately connected with magnetic microstructure. This connection becomes particularly apparent in grain oriented transformer material.

(B) Grain Oriented Silicon Iron Sheets. The first transformers were built with ordinary, carbon-containing steel sheets. The disaccommodation connected with carbon diffusion (see sect. 3.6.7B) made the devices useless after a few months, they had to be dismantled and re-annealed. This problem was solved after 1900 with the introduction of carbon-free silicon steel. Losses were halved at the same time by the much increased resistivity of silicon iron compared to pure iron, which reduced eddy current losses of all kinds.

The next big step was the introduction of grain oriented material based on the inventions of *N.P. Goss* [1102] (reviewed in [1098]). Losses were again halved by this technological step. Since then the material has been steadily improved. It continues to be the standard transformer material, exclusively used in all but the smallest electrical transformers. These dramatic and economically important improvements were achieved always in full view of the magnetic microstructure [726, 732, 1103−1105], differing in this aspect from so many other magnetic materials where improvements were developed by trial and error methods, only to be understood microstructurally afterwards.

Grain oriented transformer material consists of large grains of Fe 3 wt% Si oriented within a few degrees deviation from the [001] easy direction along the preferred axis of the material. The surfaces of all grains are (110) surfaces within the same accuracy (forming altogether a [001](110) texture). The preferred axis agrees with the rolling direction during the manufacturing process of the sheet material. The sheets are insulated by thin ceramic layers to avoid eddy currents between the sheets.

The losses in electrical steels can be separated by frequency-dependent measurements as demonstrated in Fig. 3.103. There are two main contributions: (i) Hysteresis losses that are connected with the pinning of domain walls on defects and surface imperfections. Supplementary domains (Sects. 3.7.1 and 5.3.4) contribute decisively to hysteresis losses because they can be attached to the basic walls, impeding their motion (see Fig. 3.104c), and because they must be annihilated and regenerated every time a basic wall sweeps through. (ii) Equally important in transformer material are anomalous eddy current losses. They are connected with domain-wall related eddy current effects (see Sect. 3.6.8) and become important if the basic domain width is larger than the sheet thickness. In general, there is a further important loss mechanism, that of domain rearrangement in the course of magnetization processes. This loss mechanism is largely absent in transformer steel with its well-ordered domain structure. Domain rearrangement losses appear if part of a transformer yoke is not used along the favoured axis as sometimes done in small transformers. Then the complex processes demonstrated in Figs. 5.22 and 5.23 generate strong losses unless the flux amplitude is kept small.

The following procedures led to the present excellent level of loss control.

(i) Better process technologies reduced the mean misorientation and thus hysteresis losses. It would be possible to suppress supplementary domains completely if a misorientation of better than one degree could be achieved (see Fig. 3.106). Such a quasi-single-crystalline material would tend to develop very wide basic domains with correspondingly large anomalous eddy current losses and would therefore be less than optimal, as confirmed experimentally, for example, in [1105].

(ii) For a given, moderate misorientation, supplementary domains can be suppressed by mechanical stresses as discussed in connection with Fig. 5.37. The best practical method of applying a stress consists in using a stress-effective insulation coating. The planar stress σ_p exerted by the coating is for this texture equivalent to a uniaxial stress of $\sigma_u = \frac{1}{2}\sigma_p$ (see Sect. 5.5.5A and [1106] for a more general analysis) and will thus suppress supplementary domains. When the supplementary domains are suppressed, the basic domain spacing is decreased because otherwise the overall stray field energy would rise. This is demonstrated in Fig. 6.1. Here the a.c. demagnetized state of the same material is shown once without and once with stress coating. Without a stress coating, well-oriented grains tend to develop very wide domains as shown in (a). The coating suppresses supplementary domains to a large degree, and, in addition, improves flux continuity along the grains so that even ideally oriented grains assume a small domain width (b) if they are coupled to less well oriented grains.

Interestingly, strong coating stresses turn out to be beneficial only for modern, well-oriented materials as shown for example by detailed measurements in [1107] and reviewed and documented in [1103, 1108]. In materials with stronger misorientation supplementary domains can adjust the flux inside the sheet plane according to the conditions in the individual neighbour grains by deviating from the state defined by (6.2). Putting it in another way: When the average working direction a in (6.1) and (6.2) is adjusted according the needs of each grain, a high permeability can be achieved. Suppressing supplementary domains by stresses in materials with a stronger scatter in grain orientations tends to exclude certain grains and grain clusters from the magnetization process, thus leading to excess losses. The mechanism was proposed as an explanation for the above-mentioned correlation and supported by domain observations in [1109].

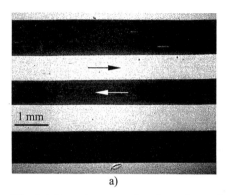
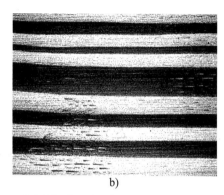

a)   b)

Fig. 6.1 Domains on grain oriented transformer steel. Wide domains in an ideally oriented grain are shown in (a). A stress coating, largely suppressing the supplementary domains, leads to narrow domains even in adjacent ideally oriented grains (b). Obtained by electron backscattering contrast (Sect. 2.5.3) (Courtesy *Y. Matsuo & S. Arai*, Nippon Steel Corp.)

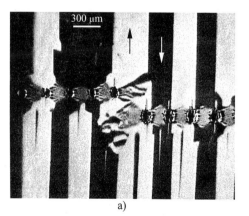
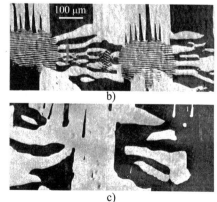

a)   c)

Fig. 6.2 Basic domains interacting with the stress concentrations introduced by laser scribing. (a) Stress pattern observed through the insulation coating with SEM type II contrast (Courtesy *T. Nozawa*, Nippon Steel Corp.). (b, c) High-resolution Kerr images observed after removing the coating, (b) on the front side, where the laser treatment was applied, (c) rear side

(iii) The basic domain width can be artificially reduced by scratching (better "pressing") [332] or by laser scribing [1110]. The stress introduced locally in this way interrupts the basic domains, acting somehow like an artificial grain boundary. The effect of scratching on the domains is shown in Fig. 2.34, the same for a laser-scribed material in Fig. 6.2. This stress-based domain refinement can be applied only to stacked and not to wound transformers because the stress would disappear in the annealing treatment necessary with wound cores. Special "heat-proof" domain control treatments have been developed for the latter case, either based on grooves a few μm deep [1111], or on narrow microcrystalline zones interrupting the flux [1112]. The domain mechanisms connected with these measures are reviewed in [1113]. With artificial domain refinement even ideally oriented material can be favourably used, in contrast to the conclusion in (i).

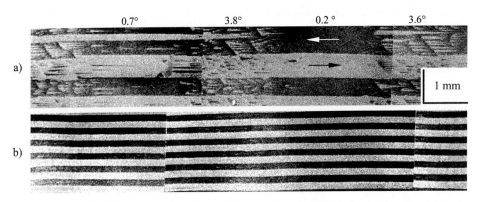

Fig. 6.3 Domains (a) on an oriented transformer sheet in which the grain orientation is periodically modulated as indicated on top of (a) by small-angle boundaries. Applying a tensile stress of 1 kp/mm^2 (b) generates a favourable fine domain pattern. (Courtesy *T. Nozawa*, Nippon Steel Corp.)

The stress pattern introduced by standard laser treatment must cause some hysteresis losses and must contribute to magnetostrictive noise. Conceptually ideal would be the introduction of regularly spaced small-angle boundaries which avoid long-range stresses but generate a precisely adjusted periodic misorientation instead. Very interesting experiments in this direction have been performed. *Sokolov* et al. [1114, 1115] plastically deformed the material, producing with toothed rolls a sinusoidally warped sheet before the secondary recrystallization step. Even if a crystal is ideally oriented at one point in its centre, it is automatically misoriented at its ends because of the bent

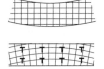

sheet surface. After a flattening heat treatment the crystal orientation relative to the surface is retained, but a stress-free flat shape is reached by the incorporation of small-angle boundaries. *Nozawa* et al. [1116, 1117] carefully evaluated this procedure and combined it with another similar method which generates quasi-single-crystal material by directed recrystallization in a temperature gradient [1118]. Very low losses and regular domain structures can be obtained in this way as demonstrated in Fig. 6.3. But the procedures are difficult to incorporate into commercial production and have so far not produced losses lower than can be achieved by optimized laser domain control methods.

The basic domain width can also be reduced by demagnetization at high frequencies [184, 1119]. To use this effect in practice appears to be difficult.

(iv) Pinning sites are important as a source of hysteresis *and* eddy current losses. Pinning centres can be impurity precipitations in the volume, surface irregularities connected with the formation of the insulation layer, and small-angle boundaries within large grains. Domain wall pinning by the defects leads to irregular domain wall motion. A given flux change must be carried by the unpinned walls alone if others are immobilized by pinning. If all available walls would contribute equally to the flux change, the anomalous eddy current losses assume a minimum value, otherwise they are enhanced. Dynamic domain observations confirmed a much smoother and more uniform wall motion after the surface irregularities were removed by chemical polishing [732]. The effect of this procedure on losses is demonstrated in Fig. 3.103.

(v) The final element in loss reduction techniques is reducing the sheet gauge. By improvements in manufacturing procedures a reduction from 0.3 mm to 0.23 mm has already been achieved without compromising the quality of the texture. The advantages of a reduced gauge are two-fold: Eddy current effects are strongly reduced with sheet thickness, provided that anomalous eddy current losses are not allowed to rise, which can be achieved by artificial domain control as discussed in (iii). The second effect is that supplementary domains need a larger misorientation angle for thinner sheets as shown in Fig. 3.106.

Compressive stresses are detrimental with respect to losses and noise because they induce or at least enhance transverse domains. The same is true for bending because in the bent state about one half of the sheet volume must be under compressive stress. Strong bending causes very complex domain patterns (see Figs. 5.29 and 5.36) and sharply increasing losses. Very slight bending with a radius of curvature beyond 10 m is permitted [333] because the effective tensile stress from the insulation coating (ii) helps in avoiding compressive stresses, which otherwise generate transverse domains and give rise to steeply rising losses. In the art of building good transformers, bending stresses must be carefully avoided, or, as in the case of wound core transformers, they must be removed afterwards by annealing (which amounts to introducing stress-compensating dislocations).

Suppressing supplementary domains also suppresses transformer noise due to magnetostriction. The connection between transverse domains and magneto-striction was demonstrated in Fig. 5.37. Uniaxial tensile stresses and equiv-alently planar stresses suppress the transverse domains and thus remove the magnetostrictive elongation in regular transformer operation. The complex of noise, losses and domains is reviewed in detail in [915].

If all mentioned measures are combined, a loss at 1.7 T and 50 Hz of only 0.35 W/kg seems to be within reach, only about one third of actual values [1104]. All this applies to the regular silicon iron composition that can be produced by cold rolling and annealing processes. Within this process only marginally higher silicon concentrations or other equivalent additions are pos-sible. High-silicon alloys based on completely different, more expensive man-ufacturing procedures are discussed separately in (E).

(C) Non-Oriented Steels. Much less is known about the magnetization pro-cesses of non-oriented silicon iron material commonly used in rotating electrical machines. As discussed in Sect. 5.3.5, the domain patterns of such material are of a heavily branched character for most grains, and only the fact that the branched surface structures in cubic material are confined to a limited near-surface zone as discussed in Sect. 3.7.5C makes a high permeability conceivable, which may rely on the underlying basic domains. According to the arguments of Sects. 3.7.5C, D the depth of the surface pattern will reach as far as about one half of the basic domain width, perhaps 50 μm, depending on many circumstances. Certainly the permanent rearrangement of the branched closure pattern causes a kind of friction for the displacement of the basic domain walls and leads to unavoidable hysteresis losses in non-oriented materials.

The question remains, why acceptable values of permeability and losses are possible at all. Can the basic domains carry the flux all through the sheet consisting of arbitrarily oriented grains? Phase theory predicts in every grain a low-field magnetizability as given by (6.1) from which only the volume of the surface zones filled with branched closure structures has to be subtracted. A possible low-energy scheme compatible with these boundary conditions consists of *folded bands* of basic domains that can be considered as basic "quasi-domains" (Sect. 3.4.5). These bands would in fact be able to carry the flux along the working direction. They ideally consist of the three most favourable easy directions for a given net magnetization invoked in phase theory.

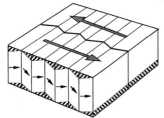 Folded bands with the opposite net magnetization are expected to be separated from the first bands by 180° walls as indicated in the simplified sketch. Certainly, there are many additional degrees of freedom in non-oriented material that can lead to a modification of the simple picture. Some segments of 180° quasi-domain-walls can be replaced by 90° walls as sketched.

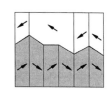 Also the net magnetization direction in the quasi-domains need not follow the global magnetization direction, but can follow local preferences of crystal and stress anisotropy in the individual grain and its environment. Finally, magnetostrictive interactions between the components of the quasi-domains are certainly not negligible in silicon iron material, and may lead to a preference for two-component over three-component quasi-domains (which would entail a deviating net magnetization direction).

If the picture of quasi-domains is basically valid, however, the magnetization mechanism in non-oriented material would again consist primarily of 180° domain wall displacements. An indirect hint in this direction can be found in studies of losses as a function of grain size. While hysteresis losses are found to decrease with increasing grain size because complex grain boundary structures (some examples of which are shown in Fig. 6.4) occupy relatively less volume, the dynamic losses increase with grain size, leading to an optimum grain size [1120] even for non-oriented material with its extremely fine surface domain patterns. Obviously the basic quasi-domains assume larger extensions in larger grains, leading to a rise in anomalous eddy current losses.

Experimentally, the basic domains in these materials can hardly be visualized. One possibility would be X-ray topography (Sect. 2.7.1); but apart from being quite demanding, this technique would not see the decisive 180° walls. In Kerr microscopy the places of 180° wall activity can be somehow "seen" in dynamic observation, and Fig. 6.5 is an attempt to document this activity. In alternating fields of very small amplitude, traces of apparent quasi-domain walls show up as irregular fine lines (b, c). With an increased field, the activity in the coupled surface domain pattern conceals the principal domain wall motion. The observation seems to agree with the proposed concept of basic quasi-domains as the main carriers of the overall alternating flux.

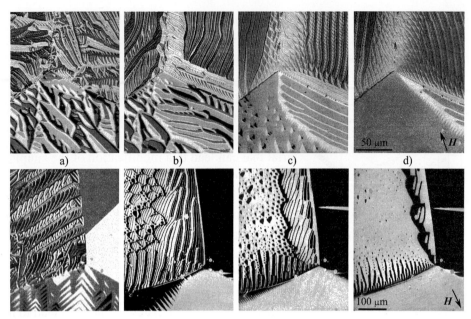

Fig. 6.4 Some observations of domains at grain boundaries in non-oriented silicon iron material in external fields increasing from left to right, leading to an average magnetization of (0, 0.17, 0.26. 0.34) J_s in the top row and similar values in the bottom row. The grain boundaries can be best seen in the highest field (d) and are highlighted there. Note the field-dependent range of grain-boundary-related domain patterns, which extend about 50 μm in (d) and 100 μm in (c)

An interesting feature of non-oriented material is its low remanence that is usually not larger than 50% of saturation. Averaging over independent grains would lead to a remanence beyond 80% for positive cubic anisotropy material, and similar large values would result from phase theory even if branched closure zones are taken into account. The strongly reduced value of the remanence must be due to magnetostrictive interactions between the grains. This conclusion follows already from the empirical fact that high-remanence ("square loop") materials — as they were needed, for example, in ferrite core memories and as they are still applied in non-linear magnetic devices — are obtained exactly when the magnetostriction constant for the easy directions of a material vanishes [1121]. If magnetostriction along the easy axes is not zero, as for most electrical steels, elastic interactions between the grains lead to additional conditions for the allowed phase volumes in each grain. Although a systematic theory of these phenomena appears not to be available, qualitative arguments point in the right direction.

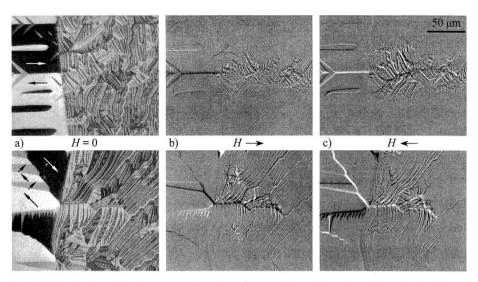

Fig. 6.5 Basic domains in a non-oriented, polycrystalline silicon iron material, indicated with digital difference image techniques. Left: zero field domain patterns near grain boundaries between well-oriented grains and near-(111) grains. For the other pictures these patterns were stored as reference images. With a weak field applied in either direction, the basic domain walls appear as narrow black and white lines. These follow the straight domain boundaries in the well-oriented grains, but can also be seen as broken lines in the misoriented grains. In addition, some reaction of the surface pattern can be seen in the neighbourhood of the traces of the basic domains

Consider a simple two-dimensional example of two elastically and magnetostatically coupled grains in [100] and [110] orientation as sketched. We look for the maximum mean magnetization at zero internal

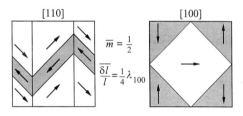

field which is compatible with three postulates: (i) identical average longitudinal magnetizations in both grains, (ii) identical average magnetostrictive strains in both grains, (iii) vanishing average transverse magnetization. This can only be achieved if antiparallel domains are inserted in the [110]-oriented grain to adjust the magnetization, while transverse domains are introduced in the [100]-oriented grain to adjust the strain. In the chosen example the resulting average magnetization of the coupled grains is in fact $\bar{J} = 0.5\,J_{\rm s}$, although the magnetizations of the isolated grains could be $\bar{J} = 0.707\,J_{\rm s}$ and $J = J_{\rm s}$, respectively. In reality, conditions are certainly more complex because strictly identical magnetization and magnetostriction will not be necessary in all grains, allowing perhaps a somewhat larger zero-field magnetization. Indications in this direction were found in domain studies in [1122]. On the other hand, the volume of branched closure domains, which does not contribute to the net magnetization, has to be subtracted, and grains that differ even stronger in their orientation than in the example will have some influence. Perhaps some kind of self-consistent mean-field theory of this phenomenon can be formulated.

Altogether, the presented considerations may explain why acceptable loss values are reached at all. But even in the best non-oriented materials the losses are much larger than in grain oriented material. If for an induction amplitude of 1.5 T at 50 Hz the best available textured materials have losses below 0.6 W/kg, the same value for non-oriented material lies even in the best qualities above 2 W/kg. The reasons are clearly the complex domain re-orientation processes that we met in a more transparent context in Sect. 5.3.2B on magnetizing Goss material along the transverse direction. Similarly dramatic rearrangement processes can be seen in Fig. 6.4 and are shown again in an overview series in Fig. 6.6, selected more for aesthetic value than for the clarity of the displayed processes.

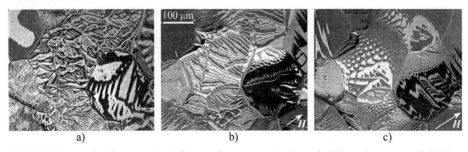

Fig. 6.6 Magnetization process observed on a non-oriented silicon iron material. The volume average magnetization is $(0, 0.39, 0.46) \, J_s$, respectively

The comparison of ordinary losses, measured in a field alternating along a single, preferred axis is not sufficient for a comparison of grain oriented and non-oriented materials. Non-oriented material is used in devices like motors and generators in which the magnetic flux direction also rotates or oscillates in direction. Rotational hysteresis, the energy loss in a constant rotating field, is here an important additional criterion for core materials in all electrical machines except transformers. This loss mechanism increases with increasing induction like ordinary hysteresis loss, but decreases again on approaching saturation. It is mainly caused by domain rearrangement mechanisms that are absent in saturation [1123]. Grain oriented material, while widely superior in ordinary losses when excited along its preferred axis, falls short in the criterion of rotational losses against high-quality non-oriented material [1124]. Figure 6.7 gives an impression of the complex domain reorientation processes responsible for rotational hysteresis losses. Another aspect of rotational magnetization is demonstrated in Fig. 6.8: the magnetization pattern for a fixed field, rotating and stopped at a particular angle, depends on the previous field rotation sense.

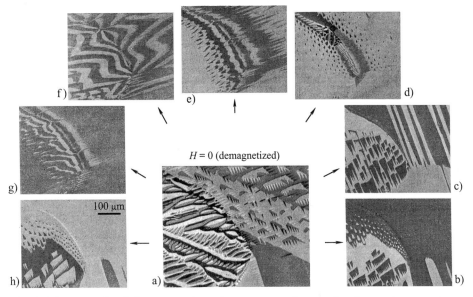

Fig. 6.7 An example of domain reorientation processes that can be observed on rotating a constant field (which leads to about two thirds of saturation) in the sheet plane of a non-oriented material. Such processes are responsible for *rotational hysteresis*

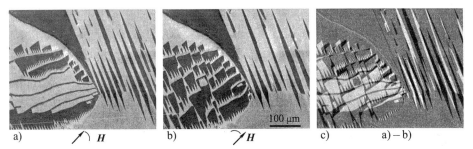

Fig. 6.8 Rotational irreversibility of domain states on a non-oriented material. The same environment as in Fig. 6.7 is shown for a given field orientation after counter-clockwise (a) and clockwise field rotation (b). The difference image (c) indicates the degree of irreversibility

Good quality in non-oriented material is reached by the following measures: (i) In addition to the standard 3% Si also 1% Al is included in the alloy for low-loss qualities. This addition reduces anisotropy and increases the resistivity, leading to an overall improvement in spite of the decreased saturation magnetization. (ii) Avoiding impurities, in particular oxide precipitations, reduces domain wall pinning. On the other hand, precipitations can inhibit grain growth during annealing, making it difficult to adjust the optimum grain size of about 150 µm. (iii) Properly chosen annealing conditions prevent the preferred formation of unfavourable (111)-oriented grains.

In most rotating machines the advantage of grain oriented transformer material cannot be exploited because the flux varies in direction. However, the stator core of large alternators, which is primarily magnetized along the circumference, can be built alternatively with grain oriented and with non-oriented material. The authors of [1124] conclude from model investigations that the use of standard oriented material, and the use of non-oriented material optimized in the indicated sense, lead to practically identical results if all contributions including rotational and higher harmonic loss contributions are included in the analysis. Further improvements in the quality and in the economics of non-oriented material therefore appear to be an important goal.

(D) Cube-Textured Material. The search for a cube-textured material as the "isotropic" analogue to Goss-textured transformer steel has never been abandoned. Early attempts to achieve a (100) [001] texture analogous to the Goss texture [1125, 1126] exploited surface-induced recrystallization. Variations in the rolling and annealing procedures [1127] lead to a fully isotropic (100) [0hk] texture in which the (100) surface is oriented randomly in the sheet surface. Another, almost isotropic texture, consists in a mixture of two orientations which are rotated by ±22.5° as sketched.

Unfortunately, surface-induced crystallization allowed originally only the production of very thin sheets with less than 0.1 mm thickness, and it was found that losses in such a material were even larger than along the preferred direction of comparable Goss-textured material. The dynamic losses in particular were very large. The reason probably lies in domain rearrangement processes which play a more important role in cube-textured than in Goss-textured material with its reduced degrees of freedom. The domain rearrangement processes not only enhance hysteresis losses due to large Barkhausen jumps, they also lead to excessive anomalous eddy current losses because the domain walls have to move multiple paths. Two examples of such field-induced rearrangement processes near grain boundaries are shown in Fig. 6.9 (see [1122] for another study of grain boundary effects). Nevertheless, the losses in cube-textured material are generally lower than those in the transverse direction of Goss material. In applications such as small stacked transformers the material offers advantages in losses at least at high induction levels [1128].

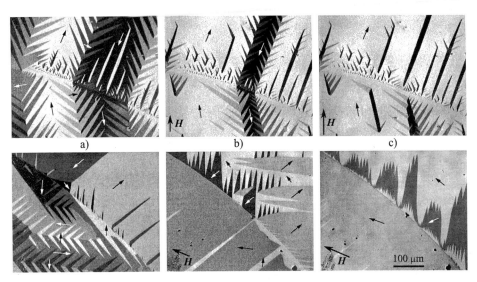

Fig. 6.9 Two examples of domain processes near grain boundaries in a cube-textured material, displaying simple fir tree and spike domains in zero field (a). In applied fields (b, c) some domains expand and shrink in a regular fashion, but others disappear or are newly created in higher fields. (Sample: *S. Arai*, Nippon Steel Corp.)

(E) High-Silicon Alloys. Silicon iron with enhanced silicon levels would offer a number of advantages: At about 6.5 wt% Si the magnetostriction passes through zero. For the same composition the resistivity reaches values almost twice as large as those of standard 3% alloys, and the crystal anisotropy is about halved. In addition, a marked induced anisotropy can be generated by field annealing in these alloys, thus allowing the magnetization curve to be tailored according to the requirements of the device, while this effect is negligible in conventional transformer steel. The only negative feature is a reduced saturation magnetization of less than 1.8 T in high-silicon alloys. The decisive problem that was already mentioned in the introduction is manufacturing this brittle material in the necessary thin sheet shape. Several methods have been demonstrated. Most advanced appears to be a procedure that adds the necessary silicon by a vapour phase reaction [1129, 1130]. In this method conventional grain oriented 3% SiFe is exposed at 1200°C to $SiCl_4$, leading to a deposition of Fe_3Si at the surface which is then allowed to diffuse into the bulk of the sheet. Superior magnetic parameters particularly at enhanced frequencies in the kHz range were demonstrated.

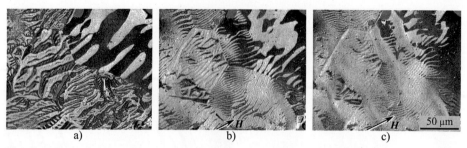

Fig. 6.10 Domains in an increasing field in a non-oriented high-silicon sheet (20 μm thickness), prepared by the method of boron extraction [1131] (Sample: courtesy *S. Roth*, IFW Dresden)

Other possibilities are based on rapid quenching techniques, either directly of the final composition [1132], or supported by a chemical reaction that removes some components of the quenched alloy. The latter possibility was demonstrated in [1131] where the boron was extracted afterwards from standard Si-B-based metallic glasses by annealing in dry hydrogen. Domains on a material prepared with this method are shown in Fig. 6.10. The magnetization typically displays a smoother flow from grain to grain, differing clearly from the appearance of domains in a regular non-oriented material of comparable thickness. This behaviour could be connected either with the absence of stress anisotropy in the zero-magnetostriction alloy, or with the presence of an induced anisotropy in the high-silicon material.

6.2.2 High-Permeability Alloys

Polycrystalline silicon iron material cannot reach a permeability in excess of a few 1000 because of the crystal anisotropy that is only marginally reduced compared to pure iron. Much smaller anisotropies can be reached in NiFe alloys near the 80:20 composition. But also other face-centred cubic alloys down to 30% nickel have interesting properties as reviewed in detail, for example, in [1096, 1133]. Due to the high price of nickel compared to iron the nickel-rich alloys find their market primarily at high frequencies and in applications such as sensors and magnetic shields in which the high permeability is indispensable. Magnetic sensors are reviewed in [1134–1136].

The 80:20 alloys are generically called *Permalloy* materials after an early trademark. If their composition and heat treatment is chosen so that anisotropy and saturation magnetostriction vanish simultaneously, initial permeabilities of more than 100 000 can be obtained. But even in nominally anisotropy- and magnetostriction-free material regular domains with clearly delineated domain walls are observed as demonstrated in Fig. 5.44. The unusual feature of high-permeability alloys is not the absence of domains, but the absence of complex branched surface domains.

Usually, high-permeability materials are prepared with a coarse-grained microstructure, leading to domains *inside* the grains, which are, however, strongly correlated over the grain boundaries as shown in Fig. 5.45a. The alternate magnetic microstructure of fine-grained Permalloy material is shown in Fig. 5.45b. It resembles the domain structure of fine-grained thin films in which the domains extend over the grains, more than conventional soft magnetic materials with large grains. The limit of an ultrafine, nanocrystalline micro-structure for which the effects of crystal anisotropy are smoothed out by exchange stiffness interactions is treated in Sect. 6.2.3B.

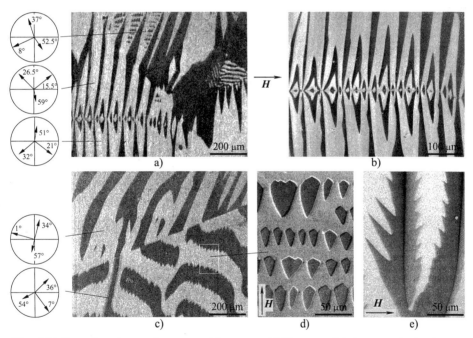

Fig. 6.11 Domains observed on a weakly textured NiFe material (~55% Ni) displaying complex grain-boundary-related patterns (a) in an applied field, which are shown enlarged in (b). In another example of a grain boundary (c) supplementary domains appear in an applied field in the more strongly misoriented grain as shown enlarged in (d, e)

Materials with intermediate Ni contents have a larger saturation magnet-ization and are therefore interesting for certain intermediate frequency power applications. They possess a finite positive crystal anisotropy like iron. At a certain composition near 45% Ni a zero of the magnetostriction constant λ_{100} along the easy axis is found. But more often materials with about 55% Ni are used which have a higher saturation magnetization. A high permeability can be reached in high-saturation NiFe materials if the metallurgical process is optimized to avoid pinning centres for the domain walls. Most observed domain patterns are rather simple as demonstrated in Fig. 6.11. Supplementary domains sometimes appear at grain boundaries (b) and in an applied field (d, e). The material is very stress-sensitive so that stress anisotropy may completely conceal the underlying crystal anisotropy as in Fig. 5.42.

A large variety of NiFe materials with nickel contents between 50 and 65% can be prepared and is offered on the market for various purposes. Among them is a highly perfect cube-textured material in which only very few domain walls separating extremely wide domains are observed [1137]. Excessive anomalous eddy current losses at high frequencies are avoided by a gross deformation of the domain walls towards a surface-parallel configuration under the joint action of intrinsic damping and eddy current effects (see Sect. 3.6.8), as can be inferred from modelling the measured response curves [1138, 1139]. Other varieties of NiFe materials display tailored magnetization loops by exploiting magnetic field annealing or other special textures, or a combination of both [1096, 1133, 1140]. Figure 6.12 shows domains on a core optimized by transverse field annealing for a linear response with low remanence and coercivity. The induced anisotropy is in this material apparently stronger than crystal anisotropy.

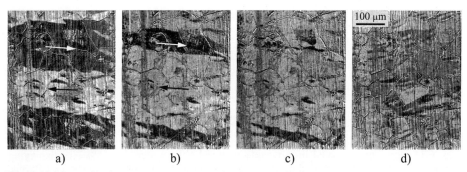

a)   b)   c)   d)

Fig. 6.12 Magnetization process on a wound NiFe core (Ø 40 mm, sheet thickness 0.1 mm) annealed in a transverse field in order to produce a low-remanence, linear hysteresis. (a) Demagnetized state, consisting mainly of extended transverse domains magnetized along the induced anisotropy axis. The grain boundaries are also weakly visible. Magnetizing along the ribbon axis (b–d; average magnetization 0.23, 0.26, 0.91, respectively) causes mainly rotations, accompanied by wall displacements due to slight axis misalignments. Near saturation (d) crystal anisotropy apparently induces grain-orientation-related secondary domains. (Sample: ~$Ni_{55}Fe_{45}$ "Permax F"; courtesy *Ch. Polak*, VAC, Hanau)

6.2.3 Amorphous and Nanocrystalline Alloys

Amorphous magnetic materials ("metallic glasses") are alloys with typically 80% iron, cobalt or nickel, and 20% "metalloids", mostly silicon and boron [1141−1143]. They are quenched from the melt, and if quenched with sufficient rapidity, they are stable in the glass state up to about 400°C, where they crystallize. These substances can be used as magnetic materials both in the amorphous state and in a partially crystallized, "nanocrystalline" state. They compete in applications with high-permeability NiFe alloys and potentially also with iron-based materials for power applications.

(A) **Metallic Glasses for Inductive Devices**. The magnetic quality in the amorphous state is unfavourably influenced by frozen-in stresses stemming from the quenching process. These stresses can be largely, but not completely, relaxed by annealing below the crystallization temperature. While most metallic glasses are ductile in the as-quenched state, they usually become brittle after the stress annealing treatment, which has therefore to be performed after assembly of the device. Metallic glasses are directly produced in a rather thin gauge between 15 and 50 μm (up to 100 μm for selected alloys [1144]). Their resistivity is large, about three times larger than in competing crystalline magnetic alloys. Both properties are favourable in high-frequency applications (into the 100 kHz range). By definition, metallic glasses are free of crystal anisotropy. They are magnetostrictive, however. Highest permeability materials can be prepared by choosing compositions with near-zero magnetostriction constants, which are available in cobalt-rich amorphous alloys.

At power frequencies losses in regular iron-rich metallic glasses are smaller than those of silicon iron material even without a field annealing treatment. When energy prices are high, transformers tend to be used at low induction levels, for example in the range of 1.3 T instead of the standard 1.7 T possible with grain oriented silicon steel [1145]. At this induction level metallic glasses with the typical composition $Fe_{78}Si_{12}B_{10}$ display markedly lower core losses than crystalline material [1146]. This led to the development of wound core distribution transformers based on amorphous ribbons, but not yet to a long-term breakthrough in this application. Several reasons contributed to this state of affairs: (i) Declining energy prices reduced the pressure to reduce power losses. (ii) Metallic glasses suffer from cumbersome side effects such as a low packing factor leading to a larger overall size, extra losses due to incomplete stress relaxation in wound cores, and unavoidable magnetostrictive noise in iron-rich materials. If the small packing factor is enhanced by pressing, interlamellar short-cuts lead to extra eddy current losses [1147]. (iii) The lower losses in metallic glass transformer cores stimulated developments towards loss reduction in conventional silicon iron material as reported in Sect. 6.2.1B. These improvements could only be paralleled in metallic glasses by completely uneconomic measures such as polishing away most of the material [1144]. In spite of the starting advantages of a very thin gauge and a high resistivity, metallic glasses suffer from high anomalous eddy current losses that are difficult to reduce by domain control measures, and they also display strong hysteresis losses due to wall pinning mainly at surface imperfections.

On the other hand, cobalt-rich amorphous materials of the typical composition $Co_{70}Fe_5Si_{10}B_{15}$ can be prepared with zero magnetostriction constant. They compete directly with high-permeability crystalline nickel iron alloys in all applications of inductive devices at elevated frequencies, reaching initial permeabilities in excess of 10^5. An important advantage is their insensitivity to elastic deformations. Metallic glasses also have an excellent yield strength and can be used as flexible magnetic shields [1148], an application which can be realized with no other magnetic material. The same property makes metallic glasses outside the zero-magnetostriction composition excellent materials for all kinds of magneto-mechanical transducers which are reviewed in [1149].

The magnetic domains of magnetostriction-free metallic glasses are characterized by very wide domains even in the as-quenched state. Figure 6.13 shows such a state and demonstrates its insensitivity to mechanical deformation, as compared to a conventional magnetostrictive material.

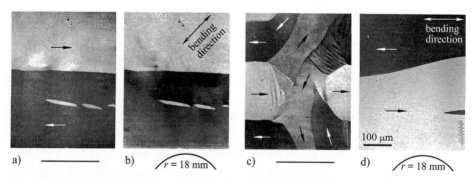

Fig. 6.13 The insensitivity of a magnetostriction-free metallic glass (a, b) with respect to bending the sample, as compared to a magnetostrictive material (c, d). (Samples: Vitrovac 6030 and as-quenched $Fe_{39.5}Co_{39.5}Si_6B_{15}$; courtesy G. *Herzer*, VAC Hanau)

The permeability and in particular the dynamic properties for high frequencies are improved by stress relief annealing and by the introduction of an induced anisotropy that can be produced by annealing in a magnetic field or under uniaxial mechanical load. Wide, irregular magnetic domains (Sect. 5.4.1) can thus be converted into regular, narrow domains with favourable dynamical properties [337, 1150, 1151], as demonstrated in Sect. 5.4.2A. Transverse annealing ideally leads to purely rotational, reversible magnetization as in Fig. 5.51-a–c. Deviations from the ideal behaviour can be studied by domain observations as demonstrated in Fig. 5.51d–f and for the sample edge of a magnetostriction-free material in Fig. 6.14. The maximum rotational permeability obtainable in this way was explored in [928].

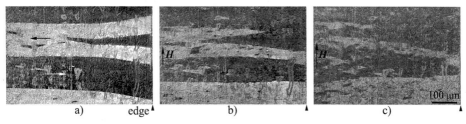

Fig. 6.14 Domains on a transversely annealed wound core of magnetostriction-free amorphous material (dimension: width 6 mm, outer diameter 10 mm). The dull side of the ribbon, which was in contact with the wheel during the quenching process and which displays air pockets and other irregularities, is on the outside of this ring core. Domain observation is thus fairly difficult (without polishing). A zone near the edge of the core shows deviations from the regular transverse domains (a), which lead to unexpected domain rearrangement processes as a field is applied (b, c) in addition to the regular magnetization rotation. The average magnetizations in the core are $(0, 0.75, 0.98)J_s$, respectively. (Sample: courtesy *J. Petzold*, Vacuumschmelze, Hanau)

(B) Sensor Applications of Amorphous Materials. Metallic glasses are used in many forms in sensor devices some of which were already mentioned in (A). In this section we will focus on small-scale field sensors. In addition to the thin sheets discussed in (A) amorphous wires (see Sect. 5.7.4) play an important role in this application field. They are produced with a diameter of about 120 µm by quenching the liquid alloy directly into rotating water. Review [1083] also contains the three principal modes in which such wires (and to some extent also ribbons and films) are used in field sensor applications:

(i) The wire core often switches in a single giant Barkhausen jump. The use of switching wires is reviewed in [1152]. Amorphous wires tend to replace the earlier, less sensitive Wiegand wires (*J.R. Wiegand* 1981) in frequency sensors, revolution counters etc. Magnetization discontinuities can also be observed in specially prepared amorphous ribbons and can be employed in article identification tags [1153–1155].

(ii) The second application field uses magnetostrictive ribbon material and the coupling of a magnetic field with elastic deformations for field sensing. The most successful principle excites a resonant elastic vibration by an alternating field and senses the characteristic ringing after the exciting pulse has been shut off [1156]. Because this mechanical vibration, which is again coupled to a magnetic oscillation, is excited only under resonance conditions, the effects can again be used in article identification.

(iii) The third application principle, which has attracted considerable interest recently, relies on the reversible high permeability of particular metallic glass states. If an alternating current flows through a wire displaying the bamboo domain pattern (Fig. 5.138), the circular magnetic field causes the domain walls to be displaced reversibly. This leads to an inductive reaction enhancing strongly the impedance of the wire. Suppressing the toroidal domain pattern by applying a longitudinal field suppresses the magnetic effect and thus the inductance. The *magneto-impedance* effect turns out to be much more sensitive than competing magnetoresistive effects [1084, 1157–1161]. The working frequencies range from 1 kHz to 1 MHz. The upper limit is given by the suppression of the essential domain wall motions at higher frequencies.

Most domain and wall structures discussed in Sects. 5.4.2A,B, 5.4.3 and 5.7.4 pertained to field sensor application. The variability of sensor materials and application modes asks for dedicated investigations of each single case, making it difficult to develop a general picture.

(C) Nanocrystalline Soft Magnetic Materials. Additional degrees of freedom became available with the development of *nanocrystalline* soft magnetic materials [1162, 1163]. Their production starts with a rapidly quenched amorphous material with a typical composition of $Fe_{73.5}Si_{13.5}B_9Cu_1Nb_3$. The introduction of copper produces many crystallization nuclei, the addition of niobium inhibits grain growth. If annealed at the right temperature between 500 and 600°C the final material consists of crystalline grains of about 10 nm in diameter embedded in an amorphous magnetic matrix. Good soft magnetic properties are closely connected with the strong magnetic coupling between the randomly oriented, sufficiently small crystal grains [1164].

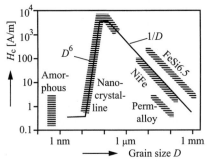

The coercivity shows a characteristic dependence on grain size as sketched, which is supported by theoretical reasoning as well as by experiments [1164]. Very soft magnetic properties are encountered (i) for very large grains, the range of classical high-permeability materials (Sect. 6.2.2), and (ii) for very small grains where the exchange interaction averages over the anisotropic properties of the individual grains. In between, a coercivity maximum is encountered that may be employed in hard magnetic materials (Sect. 6.3.3). The height and the peak position of the $H_c(D)$ curve depend on the intrinsic anisotropy of the crystal grains and on their coupling strength.

In nanocrystalline soft magnetic materials the coupling is mediated by an amorphous ferromagnetic matrix phase. Because the magnetostriction constants of crystallites and matrix phase can be adjusted separately, good soft magnetic properties of the compound material can be obtained easier than for homogeneous metallic glasses. In particular, it is possible to prepare cobalt-free zero-magnetostriction alloys, which is impossible with standard amorphous alloys.

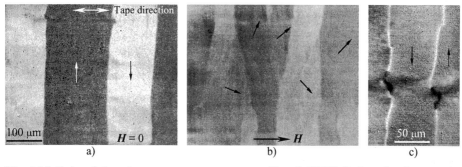

Fig. 6.15 Ordered domain structure on a nanocrystalline FeSiBNbCu-based soft magnetic ring core annealed perpendicular to the tape direction (a). The magnetization process is dominated by magnetization rotation (b) although domain splitting and wall displacement phenomena cannot be neglected. High-resolution Kerr pictures (c) show wide and irregular domain walls, interacting with pinning sites that are probably correlated with air pockets on the rear side formed during quenching. (Sample: courtesy *G. Herzer*, VAC Hanau)

The domain structure of nanocrystalline soft magnetic materials cannot be distinguished at optical resolution from those of metallic glasses as demonstrated in Fig. 5.58a. If the magnetic coupling between the crystal grains is too weak, coercivity rises and this becomes immediately visible in the character of the domain pattern (Fig. 5.58b; [953]). The same connection was demonstrated in [1165, 1166] for another class of nanocrystalline soft magnetic materials with a higher saturation magnetization, zirconium- and niobium-based alloys with the typical composition $Fe_{84}Nb_{3.5}Zr_{3.5}B_8Cu_1$ [1167]. These alloys suffer from the handicap of a strong reactivity of the element zirconium, which requires expensive protective gas processing.

Annealing a nanocrystalline material in an applied field generates an induced anisotropy probably caused by pair ordering in the crystalline phase. A domain state in a transversely annealed nanocrystalline wound core is shown in Fig. 6.15. Similar states can be generated by stress-annealing [954] as in the analogous case of Fig. 5.48e.

Nanocrystalline alloys are primarily used in inductive applications at higher frequencies where they tend to displace both amorphous and Permalloy materials. Even for power applications nanocrystalline materials represent an option for the case of rising energy prices [1167].

6.2.4 Spinel Ferrites

Soft magnetic oxide materials [731, 1168, 1169] are prepared by sintering. Depending on the frequency range they are prepared for, the grain size is chosen smaller or larger. High-permeability manganese-zinc ferrites optimized for a good permeability at moderate frequencies necessarily contain a certain amount of Fe^{2+} ions that share the octahedral lattice sites with regular Fe^{3+} ions. This leads to a conductivity due to electron hopping between both ionic states. The conductivity in the grain boundaries can be reduced by Ca^{2+} and Si^{4+} additions. Titanium doping permits a fine tuning of the anisotropy and, in addition, binds the electrons that otherwise contribute to conductivity [1170]. Suppressing the conductivity reduces not only eddy currents but also after effect and disaccommodation. In the important market of enhanced frequency power supplies, which work in the range of 1–100 kHz, losses and saturation magnetization are the most important quality criteria. For very high frequencies other compositions, such as nickel ferrites, with reduced permeability but enhanced resistivity are used, while the saturation magnetization becomes less significant.

Not much is known about the magnetic microstructure inside polycrysts [507]. But a detailed analysis of the frequency dependence of permeability and losses reveals without doubt that domain wall processes are responsible for the measured behaviour. Initial permeabilities in excess of 10 000 cannot be explained by rotational mechanisms. In addition, ferrites optimized for high permeability are sintered to grain sizes beyond 20 μm, far beyond the single-domain size. Pores left over inside the grains are detrimental to high permeability because they impede domain wall motion, while intergrain pores have little influence [1171] except near saturation. Domain wall motion processes are important in the whole application range of spinel ferrites up to GHz frequencies. Only beyond this point, in the microwave range [1172] where magnetic garnets are preferred to spinel ferrites, the magnetic material is used in saturation and domains and domain wall processes become unimportant.

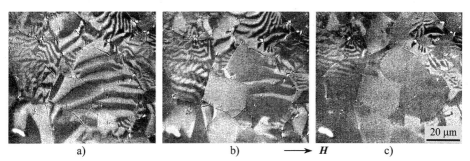

Fig. 6.16 Magnetization process observed on the surface of a piece of hot-pressed manganese-zinc ferrite. (a) Demagnetized state. On applying a field (b, c; average magnetization 0.94, 0.98, respectively) strong reorientation processes are observed. (Sample: courtesy *G. Cuntze*, Exabyte Magnetics, Nürnberg)

Figure 6.16 shows a magnetization process on a sample of polycrystalline hot-pressed manganese-zinc ferrite. This high-permeability material can be used in small components such as video heads. The absence of pores in hot-pressed material facilitates domain observation, although a deeper understanding is hard to derive from surface observations alone.

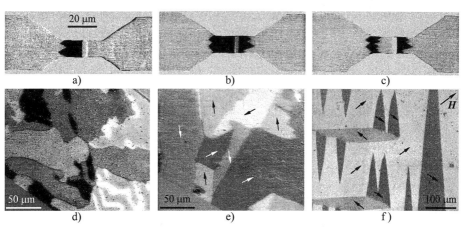

Fig. 6.17 Domains on a single-crystal manganese-zinc ferrite recording head [1173]. The mechanically treated flying surface shows only non-equilibrium domains that depend on field history (a–c). The soft magnetic behaviour becomes visible on the more or less undisturbed surfaces of other parts of the head structure (d, e) and on a perfect, stress-free single crystal [(f); sample courtesy *B. Argyle*, IBM Yorktown Heights]

Ferrites are also used in single-crystal form in recording head applications. Optimizing the crystal orientation is then a possibility to maximize the signal and to minimize the noise of this sensor device. Domain studies have been used to support such efforts ([1174, 1173]; Fig. 6.17).

6.3 Permanent Magnets

Permanent magnets and their coercivity mechanisms are a rich subject quite separate from soft magnetic materials. They are reviewed, for example, in [1175–1179] and in various contributions in [1091] and [1092], in particular in [1180]. Coercivity mechanisms are discussed in detail in [1181, 1182]. The magnetic microstructure of permanent magnets is mostly trivial in the magnetized state: the elementary carriers of the magnetic moment, the more or less independent grains or particles in these materials, are saturated in one or the other easy direction. Interesting is the breakdown of the permanent magnet property, the switching or coercivity mechanism.

Two main classes of permanent magnet materials must be distinguished: Small-particle magnets and anisotropy-based coarse-grained magnets. Small-particle magnets can display simple, uniform switching processes or inhomogeneous switching and pinning of domain walls, which is often hard to decide. In large-grained, sintered magnets two types can be readily separated: nucleation-type magnets and pinning-type magnets. Nucleation-type magnets display no relevant magnetic microstructure in the magnetized state. Only during switching inhomogeneous magnetization states play a role. Pinning-type magnets, on the other hand, rely on the interaction between domain walls and precipitations, which can be studied, for example, by Lorentz microscopy.

The difference between the two classes of large-grained magnets can best be seen in the virgin magnetization curves as sketched [1183]. Starting from the thermally demagnetized state, every grain in a nucleation-type material contains many domain walls that can be displaced easily. This leads to a large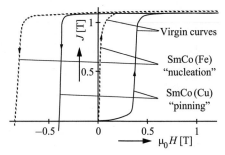
initial permeability. The permanent magnet properties appear only when the initial domain walls are driven out in a large field. The material can be

remagnetized after this step only if new domain walls are nucleated. In contrast, a similar alloy with copper-containing precipitations effectively pins the domain walls, leading to a low initial permeability.

6.3.1 Nucleation-Type Sintered Magnets

The common hexaferrite magnets [731, 1184], as high-quality rare-earth magnets of the $SmCo_5$ [1185, 1186] and the $Nd_2Fe_{14}B$ types [1187, 1188], consist of rather large, highly anisotropic grains in a polycrystalline compound that are prepared in such a way that the switching of one grain has little influence on its neighbours. The coercivity is thus primarily given by some mean switching thresholds of the individual grains. Calculating such switching thresholds needs the tools of (numerical) micromagnetics as discussed in Sect. 3.5.4. Both the initial and the final states of the switching process are essentially saturated in high-coercivity materials, but the switching process itself traverses an inhomogeneous, domain-like state in general. This is clear from the commonly used dimensions of the crystal grains in nucleation-type magnets. While the single-domain size lies generally in the 100 nm range, optimized magnets contain grains in the 10 μm range. Magnetic microstructure is important in this material class in determining the switching process and thus the quality of a magnet. It must be admitted that numerical techniques are not yet able to realistically simulate these switching processes. The particles are too large, they interact mainly magnetostatically with their neighbours in a complex manner, and their detailed shape and surface structure are usually unknown. Numerical simulation can explore possible, not actual mechanisms.

Sometimes the final state after switching is not completely saturated as can be inferred by experiments in an opposite field [1189]. Grains that have just switched, switch back much easier than grains that were fully saturated. Some residual domains must be left over after the switching process, which can be eliminated only in much higher fields than needed for switching. Although the existence of such residual domains is obvious, their location or shape is inaccessible to observation. Their properties can only be inferred from experiments which demonstrate that the coercivity H_c depends on the "saturating" field H_{max} as sketched [1190]. Measurements show that the necessary field for the elimination of the last residual domains may well exceed the saturation field

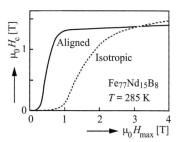

J_s/μ_0 along the easy axis, in particular for isotropic samples or for misoriented grains [1191].

Residual domains are a natural consequence of particle shape and localized stray field concentration as they become evident in thin-film element experiments like in Fig. 5.85. They are not necessarily connected with surface pinning of domain walls. The field necessary for the elimination of residual domains is also unrelated to the nucleation field of new domains. The role of residual domains becomes clearer by analysing a very simple but analytically tractable example which focuses on the intersection between four grains with tilted easy axes (Fig. 6.18, inset). In this situation the vertical grain boundaries carry strong magnetic charges if all grains are saturated along their respective easy axes. A residual domain can remove the charges in the region of strongest stray fields. It turns out to be stable up to high fields, as can be proved by a simple domain theory analysis as in Sect. 3.3.2. The total energy in this approach consists of the demagnetizing energy that is calculated with the tools of Sect. 3.2.5C, the domain wall energy, and the external field energy. A grain boundary coupling is assumed to be absent. Assuming strong anisotropy, we ignore magnetization rotation processes and focus on wall displacement. The result is shown for a specific example in Fig. 6.18 as a function of the size of the residual domain. A stable residual domain is found for most fields, its stability limit increases with the grain orientation angle η_G and reaches at $\eta_G = 45°$ a maximum collapse field for the chosen geometry of $1.364 J_s/\mu_0$. Even larger values can be expected for residual domains that are small along all three dimensions, not only along two dimensions as in the example.

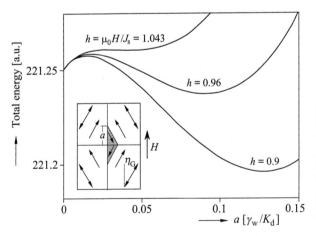

Fig. **6.18** The energy of a residual domain in a polycrystalline, high-anisotropy aggregate as a function of domain size for the grain orientation angle $\eta_G = 30°$ and for different magnetizing fields. $K_d = J_s^2/2\mu_0$, γ_w = specific wall energy

Interestingly, a nucleation field cannot be calculated in this approach. The state without residual domains ($a = 0$) is (meta-)stable in arbitrary opposite fields within rigid domain theory. The calculation of a nucleation field needs a micromagnetic approach [627] and the additional consideration of thermal activation [1192]. Anyway, a number of conclusions can be derived from the simple analysis: (i) The residual domain annihilation process is different from the nucleation process. (ii) Domain wall energy plays a decisive role in the annihilation process. To compare the annihilation field only with demagnetization fields that are proportional to J_s [1190] is unjustified. (iii) The annihilation field can exceed the value J_s/μ_0. This value was formerly considered a boundary for possible annihilation fields in the absence of wall pinning, leading to the wrong conclusion that pinning phenomena necessarily play a role even in samples with high initial permeability [1190, 1193, 1194].

Once a grain is demagnetized in a nucleation-type magnet, the walls can move easily and assume equilibrium positions. This and other states are shown in Fig. 6.19. Further examples for domain patterns in such materials have been shown in Sect. 5.2.1. They are not connected to the permanent magnet function, but may help in offering information about the material parameters, in particular about the domain wall energy as elaborated in Sect. 4.6.3. The uncoupled character of grain boundaries in a nucleation-type permanent magnet material was analysed in detail in Fig. 5.130.

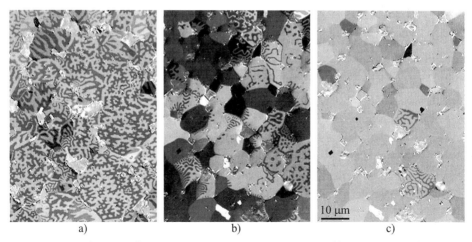

a)   b)   c)

Fig. 6.19 Domains on a sintered NdFeB magnet. (a) The thermally demagnetized state on the basal plane, with the axis of the oriented magnet perpendicular to the viewing surface. (b) The demagnetized state obtained by starting at a fully magnetized state and applying a field slightly larger than coercivity. (c) Remanent state after applying a saturation pulse of 4 T (Sample: corrosion resistant material [1195] courtesy *B. Grieb*, MS Schramberg; sample manipulation: *D. Eckert* and *D. Hinz*, IFW Dresden; see also [1196])

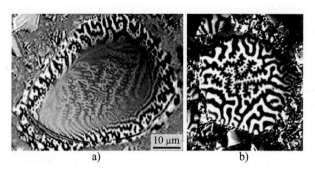

Fig. 6.20 Domains on the basal plane of Sm_2Fe_{17}, monitoring the progress of nitrogenation. The domains in the cores of the grains still display planar anisotropy in (a), and a reduced uniaxial anisotropy in (b) (Courtesy *P.A.P. Wendhausen* and *K.-H. Müller*, IFW Dresden; see [1197])

Some compounds such as Sm_2Fe_{17} are not suited as permanent magnets because their anisotropy constant K_u is negative, leading to a planar anisotropy (see Fig. 5.10). This can be changed by introducing interstitial nitrogen or carbon. The resulting materials combine a high saturation magnetization with large uniaxial anisotropy and are thus attractive potential materials [1198, 1199]. Unfortunately, the interstitials must in most cases be added at the end of the manufacturing process by diffusion at moderate temperatures, because in particular the nitrides are unstable at regular preparation temperatures of the alloys. Domain investigations can serve in monitoring the progress of the diffusion process [1200, 1197, 1201], as shown for two examples in Fig. 6.20.

6.3.2 Pinning-Type Magnets

If the domain walls are narrow due to high intrinsic crystalline anisotropy, the walls may be effectively pinned by a finely dispersed non-magnetic phase. The alloy $Sm_2(CoFe)_{17}$ containing copper-rich Co_5Sm precipitations is an example for a high-quality permanent magnet based on this principle [1202−1204]. Because of the copper contents, the precipitation phase has a reduced magnetization and exchange stiffness constant, so that the domain walls prefer to travel along the platelet-shaped precipitations. No equilibrium domain pattern is expected in pinning-type magnets, the coercivity is determined by the interaction between domain walls and microstructure. The elimination of residual domains plays no role because these domains are pinned like all others, and the initial susceptibility is low. The elementary process of wall–defect interaction is demonstrated in Fig. 6.21 by TEM observations. Very high coercivity can be reached with the addition of zirconium. A detailed analysis of the coercivity mechanism in these advanced alloys can be found in [1205].

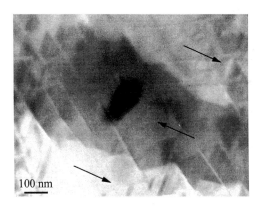

Fig. 6.21 Electron-microscopical observation (using the Foucault technique) of domains and pinning structures in precipitation hardened SmCoFe material. The domain walls get caught by the copper-containing main precipitation phase (visible as tilted thin lines) and assume a zigzag appearance due to this effect. A second set of zirconium-rich precipitations oriented at right angles to the easy axis, only weakly visible in this picture, also helps in the pinning process (Courtesy *J. Fidler*, Vienna)

100 nm

6.3.3 Small-Particle and Nanocrystalline Magnets

(A) General Considerations. Permanent magnets can be prepared also from *single-domain particles*. If the particles are uncoupled or only weakly coupled, the material behaves to first order like an aggregate of independent particles. In principle, the switching field can be calculated micromagnetically again as in the case of nucleation-type materials *if* the particle shape and material parameters are known. The difference is that domain patterns inside single-domain particles are impossible or at least unstable, in contrast to the large grains of nucleation-type magnets. The principal problem in small-particle magnets is their orientation. While magnets consisting of larger particles can be readily oriented in a field before compaction, this becomes increasingly difficult with decreasing particle size. Non-oriented permanent magnets cannot achieve large values of the energy product and can be applied in low-cost bonded materials only.

If the individual particles consist of a low anisotropy material, they must have an elongated shape to achieve high switching fields. We have seen in Sect. 3.3.2C that the single-domain size of elongated particles is much larger than that of isometric particles. For such large particles with their comparatively large volume the danger of thermally activated spontaneous switching (super-paramagnetism) is absent. Elongated single-domain particles near the size of the single-domain limit are therefore thermally stable even if they are completely isolated. If some exchange coupling is present, thermal stability is further enhanced, but it may be more difficult to achieve sufficient coercivity.

High-anisotropy single-domain particles may be isometric, but in this case a certain degree of exchange coupling is preferable to avoid thermal demagnetization. Isolated particles must be larger than a critical size to avoid superparamagnetic effects. This critical size depends on the anisotropy coefficient. Experimental data on this fundamental relation were compiled and measured, for example, by *Luborsky* [1206] and are reproduced in the sketch.

Numerical simulations as discussed in (D) indicate that both coercivity and remanence may be enhanced by the right degree of exchange coupling. In the absence of exchange coupling it becomes difficult to reach acceptable remanence values unless the particles can be oriented. Magnetic forces are not very efficient in small-particle magnet production.

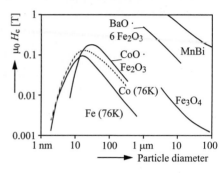

Some other structural mechanism must be used to achieve some orientation as discussed in (C).

Magnetostatic *interaction domains* are expected in all small-particle magnets in the demagnetized state due to the predominance of magnetostatic interactions between the particles, as discussed in Sect. 5.2.3. Examples for interaction domains were shown in Figs. 5.15, 16. In (D) the role of exchange interactions between the particles which are important in some types of small-particle magnets is discussed in more detail.

Small-particle magnet materials are preferred in the important field of resin-bonded magnets which can be easily shaped to any desired form [1207–1209]. The particles that are used for bonded magnets have a diameter of some 100 μm. For small-particle magnet materials these macro-particles consist of many small single-domain grains which are well-protected inside the macro-particles and thus not subject to corrosion, in contrast to single-domain particles obtained by milling coarse-grained sintered magnets.

(B) Classical Small-Particle Magnets. The first nanostructured permanent magnet class and at the same time the first modern magnets were the *Alnico* alloys. They were invented by *T. Mishima* in 1931 and are reviewed in [1210, 1211]. These materials consist of fine filaments of a high-saturation FeCo alloy, embedded into a non-magnetic NiAl matrix. The magnetic needles (or rather very extended threads as demonstrated in [1212]) are formed by spinodal decomposition, a spontaneous solid-state reaction that can be influenced by a magnetic field if the spinodal temperature lies below the Curie point of the magnetic phase. Cobalt-containing alloys fulfil this condition, so that oriented Alnico qualities can be prepared. An expensive variety of this process relies on directional solidification to generate a grain oriented material.

Alnico alloys are still the most temperature-stable magnets and the only magnets that can be used even in the red hot state (up to about 550°C). Their general usage is decreasing, however. The magnetization mechanism is certainly basically a single-domain process, but to what extent this process is modified by interactions between the needles and by inhomogeneous switching is undecided [1181]. Magnetostatic interaction domains in Alnico were shown in Fig. 5.15d. A variant of this magnet group are CrCoFe magnets [1213]. They are based on the same mechanism and reach roughly the same energy product as Alnico. In contrast to Alnico they can be plastically deformed and shaped in an intermediate state before the final heat treatment in which they become magnetically hard and mechanically brittle like most permanent magnets.

Another kind of magnet based on the small-particle principle is the so-called elongated single-domain (ESD) magnet [1206, 1214, 1215], a synthetic magnet that consists of elongated high-saturation FeCo particles embedded in an easily deformable lead matrix. As in the case of Alnico magnets, the particles carry only negligible crystal anisotropy. They have to rely on an elongated, thread- or needle-like shape. These elongated particles can be oriented in a magnetic field within the liquid lead matrix before solidification. At their time ESD magnets were fully competitive and preferred particularly in the field of small precision magnets. This class of magnets fascinated researchers because it appeared to represent an ideal of artificial materials tailored to maximum quality as guided by theory. They became obsolete because of the availability of resin-bonded rare-earth magnets, and also because of their less than ideal and dirty manufacturing method.

(C) High-Anisotropy Nanocrystalline Magnets. A number of techniques can be used to prepare small-particle magnets out of high-anisotropy material precursors such as $Nd_2Fe_{14}B$: rapid quenching and subsequent crystallization [1216], mechanical alloying [1217−1220], and a certain hydrogen-vacuum solid-state reaction that is known as the "hydrogenation disproportionation desorption recombination (HDDR)" process [1221, 1222]. In this method the starting, coarse-grained material is first chemically decomposed by annealing in hydrogen, followed by a regeneration of the hard magnetic phase in vacuum in nanocrystalline form.

All these methods generate a fine powder which then has to be compacted into solid magnets. Three basic methods are available for this step as sketched: resin bonding, hot pressing, and "die-upsetting", an anisotropic hot squeezing process which can generate oriented magnets [1223, 1224]. In variants of the HDDR process an anisotropic powder can be pro-

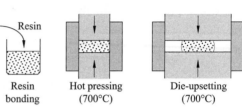

Resin bonding | Hot pressing (700°C) | Die-upsetting (700°C)

duced which can be oriented in a magnetic field [1225−1227]. This unexpected result is not fully understood. Somehow, if certain additives such as cobalt are present, the regenerated nanocrystals remember the orientation of the mother grain. Domain investigations of oriented HDDR magnets [1228, 1229] helped in understanding the microscopic state of this material. The process is a promising candidate for resin-bonded magnets of high-energy product, provided that the production can be controlled in a reproducible way.

(D) Exchange-Enhanced Nanocrystalline Magnets. Also non-oriented nano-crystalline materials remain interesting in light of the discovery that a certain, relatively weak exchange interaction between very small particles can lead to an enhanced remanence without a significant loss in coercivity [1230−1232]. For very small grains the exchange coupling averages over the anisotropy, leading to a soft magnetic behaviour as in Sect. 6.2.3C. Larger grains become partially decoupled, leading to a maximum in the coercivity, while the coupling enhances the remanence above the average $J_r = 0.5 J_s$ of independently oriented uniaxial grains. This mechanism is particularly interesting because it makes grain orientation that is difficult to achieve for very small particles superfluous. A grain size in the nanometre range is indispensable for this method and can be achieved with the methods mentioned in (C). TEM domain and microstructural investigations [902] revealed submicron domains which decreased in scale with decreasing grain size.

Numerical simulations of a simplified model configuration demonstrate the validity of this concept as indicated in the sketch [1233]. In the same grain size range remanence increases monotonously with decreasing grain size in these calculations, in accordance with earlier three-dimensional simulations [1234].

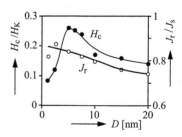

Strong exchange coupling and the material parameters of $Nd_2Fe_{14}B$ were assumed. For the case of reduced coupling the maximum will be shifted to larger grain sizes.

(E) Two-Phase Nanocrystalline Magnets. *Coehoorn* et al. [1235] discovered a further possibility of achieving high remanence in non-oriented small-particle materials. It consists in adding a high-saturation soft magnetic phase, which is strongly exchange coupled to the basic hard magnetic phase if the extension of the soft phase is sufficiently small (< 30 nm). The soft phase formed in the experiments of [1235] was Fe_3B, a compound with a planar anisotropy [1236]. The hard phase is ordinary high-anisotropy $Nd_2Fe_{14}B$, the overall composition can be written as $Nd_4Fe_{77}B_{14}$. Other attempts used pure α-iron as the soft magnetic phase [1237].

The coercivity in the two-phase magnets is dominated by the hard phase as proven in [1238]: it vanishes exactly at the Curie point of $Nd_2Fe_{14}B$, not at that of Fe_3B which is larger. The high remanence in these materials is primarily a consequence of the cubic or planar anisotropy components, which naturally have a high remanence due to their multiple easy axes. Remanence is not enhanced here by exchange effects as in (D), as rightly pointed out and supported by experiments in [1239]. The coupling to the hard magnetic phase is needed to stabilize this remanent state and thus to produce a hard magnetic material.

By carefully controlling the grain size using suitable additives, a material with a competitive quality could be prepared [1240]. It would have the specific advantage of being easy to magnetize, which is important for small magnets of complex shape as they are used in small motors. A problem with this idea appears to be the difficulty to reach reproducibly a sufficient coercivity, while the predicted high remanence values can be readily achieved.

The theoretical claims that this kind of two-phase magnets will be the material of the future with a potential of reaching an energy product of 1 MJ/m^3 [1241, 1242] appear problematic. *Kneller* and *Hawig* [1243] present only qualitative arguments and estimates in which the Zeeman energy term is not explicitly incorporated. More rigorous treatments [1244, 1245] focus on the nucleation field of the soft magnetic phase, which is irrelevant for the hard magnetic behaviour. In experimental verifications of enhanced remanence in hard-soft two-phase materials regularly a so-called *exchange spring* behaviour is observed [1243, 1246−1249]. It means that

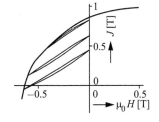

on approaching coercivity the soft phase is reversibly rotated, leading to pronounced reversible recoil loops as sketched [1247]. This characteristic behaviour contradicts the concept of nucleation and switching starting from the fully saturated state. Full-scale numerical simulations [1234] reproduce the measured behaviour, and can help in finding the right dimensions of the soft and hard phases.

6.4 Recording Media

Magnetic recording, or "magnetomotoric data storage", represents by far the most rapidly growing application of magnetic materials. A magnetic pattern is written in narrow tracks into a semi-hard, mostly unstructured magnetic layer, the *recording medium*, when it moves past a writing head. The information is recovered by a reading head. The reading process leaves the written information unaffected. Writing and reading heads can be the same device that is switched between both functions (see Sect. 6.5.1). Recording media are reviewed in [1250–1252].

The classical case of *longitudinal* recording means that the magnetization in the written information lies parallel to the layer surface and parallel to the track direction. In *magneto-optical* recording the magnetic polarization is oriented perpendicular to the layer surface. The same is true for *perpendicular* recording, a technique that is not yet established, but which may ultimately offer the highest recording densities.

The limit for the storage density in magnetic recording will be reached when the independently acting memory elements or particles get magnetically reoriented by thermal activation, meaning that the material becomes superparamagnetic. Thin films allow narrower transition widths and a higher recording density. But very thin films and perfect magnetic isolation of the particles may be disadvantageous for thermal reasons. In several techniques considerable exchange coupling between the particles or grains is allowed and present in optimized media. These media behave to first order like continuous thin films as discussed in Sects. 5.5–6. The traditional predominance of particulate media led to a widespread tendency of using the language of magnetic particles also in the discussion of continuous films, treating regular domains, for example, as "clusters of particles". When this is inadequate, it may lead further developments to wrong directions.

The written pattern carrying the information in a recording medium has nothing to do with equilibrium domain patterns that are the primary subject of this book. However, the recording precision, i.e. the noise connected with the recording pattern, is decisively influenced by micromagnetic conditions. Some media show an increasing noise with increasing recording density. Exploring the origins of noise is an important task of micromagnetic analysis [1253].

6.4.1 Longitudinal Recording

In this most widespread mode of magnetic recording the information elements are oriented in a way that would be avoided in a regular magnetic material. That it can be stable in a recording medium as a metastable state must be a consequence of coercivity. The following elementary consideration applies to both continuous and particulate longitudinal media. The demagnetizing field of a single information element as sketched is:

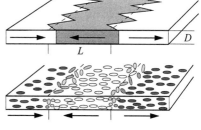

$$H_{d} = 2 D J_{s} / [\mu_{0}(L + D)] \tag{6.3}$$

(see the Néel wall treated in Sect 3.6.4A; the factor 2 stems from the double charge compared to a Néel wall). The information element can only be stable if the coercivity H_{c} is larger than the demagnetizing field H_{d}. For a given medium thickness and a given coercivity, the smallest information element length L and thus the maximum recording density is therefore determined by a self-demagnetizing effect. To increase the recording density either coercivity has to be increased, or a thinner medium has to be chosen. A more detailed analysis of these relations including discussions of the transition profiles can be found in textbooks like [1254, 1255].

(A) Particulate Media. Traditional longitudinal recording media [1250] consist of small, mostly oxidic magnetic particles, surrounded and separated by a plastic binder. The particle size must lie in the single-domain range, with a switching field between some ten to some hundred kA/m. If the basic material has a cubic crystal symmetry, the particles must be elongated to reach the necessary coercivity. Particles consisting of strongly uniaxial material such as barium ferrite may have any shape. To first order, each particle in a particulate medium switches independently, usually in an inhomogeneous fashion, which can be studied with the tools of numerical micromagnetism (see Sect. 3.5) if the shape and material properties of the particle are known. As this is usually not true, model calculations can be used to get an idea about possible switching mechanisms and fields of such particles.

A regular, discrete arrangement of identical isolated particles is claimed to represent the ultimate magnetic recording medium [996, 1256, 1257]. Such particle arrays or *patterned media* can be prepared by optical interference lithography or — as advocated by *Chou* — by mechanically imprinting a pattern on a plastic mould. Highly dense and stable column arrays can be prepared by electroplating into the lithographically defined pattern. Their single-domain switching behaviour can be confirmed with MFM techniques [1073, 1258, 1259]. A practical way of writing and reading on such media is still to be devised [1260]. The principal advantage of patterned media, which use one particle per information bit, is seen in thermal stability. If many, irregularly arranged particles contribute to one bit, the individual particles must be very small, raising the danger of thermal demagnetization if the particles are not coupled by exchange interactions. The presence of weak exchange interactions between the grains in thin-film media may also offer a solution to this problem [1261]. It may avoid the many complications connected with the patterned media concept [1260]. In any case, periodic arrangements of nanometre-size magnetic particles represent a fascinating subject of research [1262–1264].

Ordinary particulate media, as they still dominate in most varieties of tape recording, differ from the ideal of identical, evenly distributed particles. They display a distribution of switching fields which depend on size, shape and interactions of the particles. In recording a narrow switching field distribution is advantageous because it avoids a partial erasure of the written information when the writing head moves on. The detailed analysis of static hysteresis curves is a preferred method for the analysis of these phenomena as reviewed in [1265]. Several characteristic numbers are derived from the various branches of the hysteresis, which characterize the switching and recording behaviour.

Most important are two secondary "remanence" curves, which are defined in the sketch: the saturation remanence curve $J_{sr}(H)$, and the "virgin" or "isothermal" remanence curve $J_{vr}(H)$. The derivative $dJ_{vr}(H)/dH$ is the switching field distribution *if* all particles are isolated, uniaxial and single-domain particles, while the orientation and anisotropy distributions of

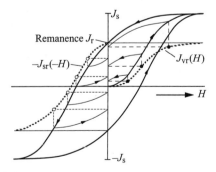

the particles may be arbitrary. Under the same assumption the two curves are not independent, but connected by the Wohlfarth relation [1266]:

$$J_{sr}(H)/J_r = 1 - 2J_{vr}(H)/J_r , \qquad (6.4)$$

which implies that the switching field distribution may be as well derived from the saturation remanence curve $J_{sr}(H)$. Deviations from the Wohlfarth relation, which were first studied by *Henkel* [1267], are attributed to interaction effects between the particles [1268]. But violations of the Wohlfarth relation are expected for multiaxial particles even without interactions.

The same operations on the hysteresis curve can of course be applied to all magnetic materials. This does not mean that they have the same meaning. The success in analysing particulate media in terms of macroscopic hysteresis curves led to a widespread extension of this analysis to other cases, which is not justified. A conventional, exchange-coupled ferromagnet is fundamentally different from an assembly of interacting particles. To describe these deviations by a modified switching field distribution or interpreting continuous media in the language of independent Stoner-Wohlfarth particles is not helpful.

(B) Thin-Film Media. In hard disk storage devices the traditional pigment-type media have been replaced by deposited metallic films. Conventional Permalloy films, as studied for example in Sect. 5.5, cannot be used for this purpose. Thin-film recording media must have a coercivity in the same range as particulate media, which means that they must be somehow heterogeneous microscopically. Nevertheless, films deposited without special precautions are usually of a continuous character, meaning that the exchange coupling between the magnetic grains dominates over demagnetizing effects.

The principal advantage of thin-film media is their smaller demagnetizing field (6.3), which allows a higher recording density. The dense structure with a packing factor of magnetic material close to 1 is also an advantage. The head-on transitions between bits in longitudinal recording are not without problems in longitudinal media, however. The *charged* 180° wall in thin films was discussed in Fig. 5.61 and Fig. 3.83. To distribute the strong magnetic charge of the straight wall, it assumes an equilibrium *zigzag* shape as indicated in the sketch at the beginning of Sect. 6.4.1. The high coercivity of a recording film makes it possible to reduce the zigzag period and to compress the zigzag patterns to a narrow transition zone. But the zigzag transitions tend to be rather irregular in high-coercivity films as was confirmed in Lorentz microscopy images [1269], leading to excessive transition noise [1270–1276].

Noise measurements as a function of the density of flux changes (see sketch [1271]) teach us several interesting lessons: (i) The strongly increasing noise with increasing recording density appears to be a common feature of longitudinal thin-film media. (ii) Particulate media, such as conventional γ-Fe$_2$O$_3$ pigments, do not show this effect. (iii) Perpendicular media like CoCr do not show the enhanced noise at high frequencies either. Conventionally, this finding was interpreted as indicating that vertical CoCr media are of a particulate character. This conclusion is unfounded, however. Regular CoCr perpendicular recording films have a continuous character as first inferred by *Wielinga* et al. [1277] and as definitely demonstrated in [162]

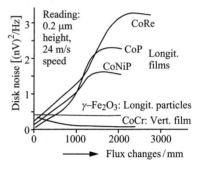

(see also Fig. 5.123). The explanation of this seeming contradiction becomes clear in view of the properties of ordinary, continuous thin films. Zigzag walls are a specific feature of continuous low-anisotropy films with *in-plane* magnetization. No such feature and no accompanying noise is expected in perpendicular films, whether they are particulate or continuous.

Aoi et al. [1278] established an interesting, simpler method to measure the tendency of a medium towards excess noise. Instead of measuring the noise on written tracks, they recorded the noise in partially "d.c. erased" states. After first saturating the medium in one direction, they effectively applied a reverse field by "erasing" with a d.c. writing current of increasing amplitude. For thin-film media, a noise maximum was found when the erasing field was in the neighbourhood of the coercive field. The height of the maximum correlated closely with the high-frequency noise measured on written tracks. The phenomenon is obviously related to the formation of ripple and buckling which also appears in reverse fields of the order of the coercive field (see Sect. 5.5.2–4). Ripple is, like zigzag walls, a characteristic feature of ordinary thin films, and it is sensitive to the irregularities and fluctuations in local anisotropies. The correlation with measured recording noise therefore appears plausible.

Reaching a good signal-to-noise ratio at high recording densities in longitudinal thin-film media required the development of special microstructures. The most transparent success was achieved by depositing the films on a rough chromium underlayer [1279]. In this way the exchange interaction between the grains was geometrically interrupted, leading to a predominantly *particulate* film with a low recording noise at high spatial frequencies. There are various other options to reach the same goal [1252, 1280, 1281]. These alternatives avoid some problems connected with a rough underlayer. They are mostly based on a chemical segregation of chromium in cobalt-based alloys that — at high concentrations and at high deposition temperatures — supplies a non-magnetic grain boundary. Even if the exact mechanism is not clearly established in all cases, there is no doubt that in *longitudinal* recording effectively *particulate* media are the best choice for high density recording. The bit transitions observed in such modified media [1282, 1283] have an irregular appearance, as also confirmed in numerical simulations ([1284]; Fig. 6.22). To still call them zigzag transitions is unjustified. Nevertheless, the irregular bit transition zones observed in particulate media, in which vortex-like configurations can often be distinguished, ultimately limit the recording density. They may be attributed to dipolar forces — as in magnetostatic interaction domains (Sect. 5.2.3) — and to clustering effects [1285].

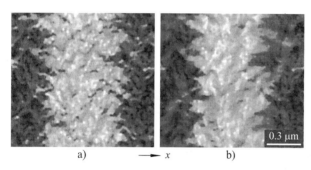

a) $\longrightarrow$ x b)

Fig. 6.22 Simulated bit transitions in longitudinal thin-film media. Without (a) and with strong (b) exchange coupling between grains, the latter displaying distorted zigzags. The grey scale indicates m_x. The original data were smoothed with a Gaussian filter of 1/4 grain size radius. Courtesy *J. Miles*, Manchester [1285]

Interesting is the question whether a *complete* exchange decoupling is necessary and desirable, or if a certain residual coupling enhances the thermal stability of the recording pattern without increasing the noise. Numerical simulations [1284, 1286] will probably be able to answer this question if a reliable inclusion of thermal effects can be achieved. The results of [1284] indicate that zero coupling is better with respect to static transition noise than some specified finite coupling, but insufficient discretization and the lack of more detailed calculations forbids sweeping conclusions such as that *every* exchange coupling is disadvantageous in longitudinal recording. Nevertheless, such statements were widely accepted as guidelines for material development.

(C) Obliquely Evaporated Metal Films. Most magnetic *tapes* are still covered with traditional pigment-type storage layers. The main reason are certainly low cost and the technological reason that the plastic binder of particulate media is compatible with the plastic carrier tape. In high-density video and audio recording another, thinner type of storage layer has been introduced, however, known by the name "metal evaporated (ME) tape" [1287]. It consists of CoNi alloys evaporated in the presence of a considerable amount of oxygen under an oblique angle. This technology is reviewed in [1288–1290]. The layer is rather thick for a thin film (0.1 to 0.3 μm, but much thinner than conventional pigment layers, which is a practical advantage). The process conditions lead to a marked columnar structure, which led analysts to treat these media as basically particulate, with exchange-isolated columns as the elementary magnetic units and an easy axis along the column axis [1291].

Experimental evidence indicates, however, that samples prepared in this way are often characterized by a considerable exchange interaction between the crystallites. Thick films with an oblique anisotropy have *strong stripe domains* as their equilibrium structure (see Sects. 3.7.4E and 5.5.5). Exactly these domains were observed in demagnetized ME samples [349, 1292–1294]. Figure 6.23 shows examples of stripe domain observations obtained with three different techniques.

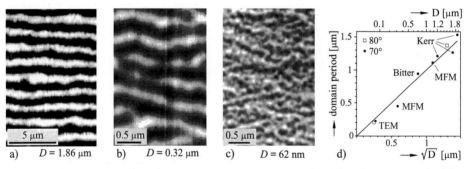

Fig. 6.23 Stripe domains observed in the demagnetized state in obliquely evaporated NiCo films of different thickness. The domains in the thick film of (a) can be seen with the Kerr technique. The thinner film (b) needs magnetic force microscopy. Even in very thin, high coercivity films periodic stripe domains can be made visible in a high voltage electron microscope (c) (in order to emphasize the stripe pattern this image was smoothed with a Gaussian filter of 0.06 μm radius). The classical $\sqrt{D}$ behaviour can be observed throughout the whole thickness range (d). (Courtesy *L. Abelmann* (a, d), Enschede [1295], *W. Rave*, Orsay (b), *I.B. Puchalska* and *J. Jakubovics*, Oxford (c))

Strong stripe domains — a micromagnetic structure which cannot be formed in a particulate medium — carry no or only small net magnetic moments on a scale larger than the stripe period. A numerical simulation of this material, in which a weakening of the exchange interaction between the columns to only 5% of its bulk value was assumed, still resulted in faint, irregular stripe domains with the expected period corresponding to the film thickness [552].

The value of 5% coupling was chosen just as an example to test the algorithm. The actual coupling strength could well be stronger.

Written tracks in this material can be conveniently observed with high-resolution magnetic force microscopy [1290, 1296] as shown in Fig. 6.24. Observations with this technique demonstrate an interesting property: the direction-sensitive performance [1297, 1298] of these media. When writing along the "good" direction, the head field at the trailing edge is better aligned along the tilted columns in the tape, thus leading to sharper and better defined transitions, compared to transitions written along the bad direction. The difference can be clearly seen in the MFM images.

The argument against the use of continuous materials in longitudinal recording, the formation of charged zigzag walls, becomes invalid in this kind of medium. There is no micromagnetic driving force for the formation of zigzag walls within a strong stripe domain pattern. Recording in such films may consist in a modulation of the neutral stripe pattern. In media with strong oxygen contents ["CoNi(O)"], which are often preferred because of their enhanced coercivity, the exchange coupling appears to be suppressed, leading to a predominantly particulate character. According to the arguments presented here this variation of the material would offer no structure-related advantages with respect to transition noise.

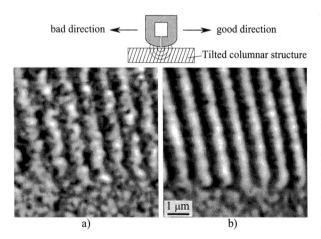

Fig. 6.24 Written tracks on obliquely evaporated NiCo tape observed with magnetic force microscopy, displaying a marked dependence of the result on the writing direction with or against the tilted texture (Courtesy *S. Porthun* and *J.C. Lodder*, Enschede [1290])

6.4.2 Perpendicular Recording

In this recording technique which was pioneered by *Iwasaki* [1299, 1300] the information elements are packed side by side, so that oppositely magnetized elements stabilize each other by the dipolar interaction — in contrast to longitudinal recording for which the dipolar interaction limits the possible information packing density. The same information density can be realized in a perpendicular medium with a much larger film thickness than in longitudinal media of equal coercivity, which is an advantage with respect to wear when the quest for higher recording densities continues. Problems connected with this technique are discussed in [1301].

Perpendicular recording requires specially adapted read and write heads. Not only the radically different "single pole head" advocated by Iwasaki is conceivable, however. An interesting design of an inductive read/write head, which looks essentially like a conventional thin-film head (see Sect. 6.5.1), is shown in the sketch [1302]. In writing, the top yoke is saturated in the thin part and becomes ineffective so that the head operates 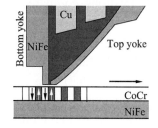 like a single pole head. In reading the "switching head" acts like a sensitive, conventional ring head. Flux closure is achieved by a soft magnetic underlayer under the perpendicular medium, which reduces the effective air gap. Alternatively, magnetoresistive read heads can be used also in perpendicular recording.

Cobalt with some 20% chromium, added to reduce the saturation magnetization and to increase coercivity, has been studied mostly as a perpendicular recording medium. The material is deposited by sputtering, and by proper sputtering conditions a pronounced crystal texture can be achieved, which provides the necessary perpendicular anisotropy. The reduced anisotropy parameter Q is usually smaller than unity, however, so that no extended domain pattern with perpendicular magnetization directions is stable. Although some chromium segregates in the deposition process providing the necessary coercivity, optimized layers appear to be characterized by sufficient exchange coupling to lead to an essentially continuous character of the medium. Band and maze domains are expected and observed under these conditions (see Fig. 5.123). Figure 6.25 shows maze domains in a CoCr layer observed with a high-resolution colloid method [98, 1303, 1304].

In continuous CoCr perpendicular media information is stored as a modulation of the domain pattern in a similar sense as discussed for obliquely evaporated films in Sect. 6.4.1C. Information written in very narrow and dense tracks is displayed by magnetic force microscopy in Fig. 6.26. The written information appears to be embedded in and partially modulated by the natural domain pattern in the environment.

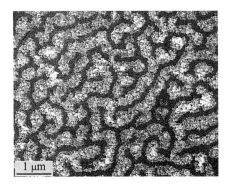

Fig. 6.25 Domains on a perpendicular recording CoCr medium observed with a high-resolution Bitter technique (Courtesy *J. Šimšová*, Prague [1305]),

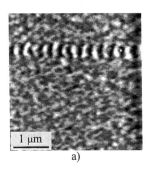

a)

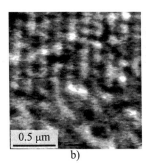

b)

Fig. 6.26 Written tracks and domains on a CoCr perpendicular medium, imaged by magnetic force microscopy (Courtesy *S. Porthun*, Enschede)

Modifying the process conditions leads to a more particulate character of CoCr layers [1306], although domains and domain walls can be observed in a considerable range of coercivities [162, 1307]. For low coercivities regular band domains are easily formed, while high coercivity films display bubble domains only. At a higher chromium content and higher deposition temperatures a complex pattern of chromium segregation can be observed [1308, 1309] that destroys the continuous character of regular CoCr films. No domains are observed in such samples. Another candidate as a perpendicular recording medium are modified barium ferrite particles. This material forms hexagonal platelets with a perpendicular easy axis (see Fig. 5.129). They are easily oriented parallel to the medium surface — a problem consists in their tendency to form agglomerates, which can, however, be solved by applying proper surfactants, as described in [1310]. Which material variant will ultimately offer the best performance is still undecided. A comprehensive critical analysis and review of the perpendicular recording process can be found in [1311]. The authors offer arguments that exchange-coupled columnar grains lead to sharper transitions in perpendicular recording than for the particulate variety. High-density perpendicular media were explored in [1312]. In this study the perpendicular medium was demonstrated to produce lower noise levels than comparable longitudinal media below 0.1 μm bit length. The soft magnetic underlayer can be the source of unwanted disturbances connected with planar domains and domain walls in this film. Probably for this reason only a single perpendicular film was considered in this study.

6.4.3 Magneto-Optical Recording

In this mass storage technique, magnetic heads for reading and writing are replaced by optical setups based on solid-state lasers. Writing relies on locally heating the storage layer by a sharply focused laser beam. In the heated zone the storage layer is then magnetized by an extended magnetic field. Reading uses the magneto-optical (MO) Kerr effect at a weaker laser power. The technology is based on that of the compact disk in basic optics, drive and control, only that MO disks can be arbitrarily rewritten. Reviews of magneto-optical recording can be found in [1313–1317].

(A) Materials and Physics. The storage layer in MO recording consists of a rather thin film with a strong perpendicular anisotropy and a low saturation magnetization at room temperature. Sputtered amorphous TbGdFeCo alloys fulfil these requirements and are used most often. Their ferrimagnetic spin structure leads to a characteristic temperature dependence of the saturation magnetization as sketched [1313]. Near a compensation point T_{comp}, at which the magnetization of the transition metal (Fe, Co) subsystem J_{TM} just balances the magnetization of the rare-earth (Tb, Gd) subsystem J_{RE}, the coercivity rises steeply so that every magnetization pattern is stable even in moderate fields up to several thousand A/cm. If the

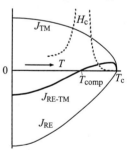

film is heated towards the Curie point (which can be adjusted by the Co content to values of, say, 300°C) the net saturation magnetization rises, while the coercivity gets small, so that a small magnetic field suffices to magnetize the heated spot in the desired direction. Alternatively the spot is heated beyond the Curie point and an even smaller field is then sufficient to write the magnetization during the cooling process.

Domain experiments in high perpendicular fields (Fig. 5.124) leave no doubt about the continuous nature of magneto-optical storage films. The observed domains at room temperature are not equilibrium domain patterns because the film thickness is below the characteristic length $l_c = \gamma_w / 2K_d$ of perpendicular films (γ_w = specific wall energy, $K_d = J_s^2 / 2\mu_0$; see Sect. 3.7.3A). But a pattern written at high temperatures remains frozen in at low temperatures and is separated from its environment by a conventional domain wall. Strong pinning hinders the walls to move at ambient temperatures, so that the information remains reliably stored.

In magneto-optical recording a wall is set into motion when the effective driving field acting on a wall overcomes the wall motion coercivity. The effective driving field consists of the external field, the demagnetizing field and the wall-energy-related effective field. If the applied and the demagnetizing field are known for a circular domain (see Sect. 3.7.3D), then the observation of the onset of wall motion can serve as a method of measuring the specific wall energy: by averaging the onset fields for positive and negative applied fields the coercive effects can be annihilated, leaving as the only unknown the wall energy [1318]. The method works reliably even at enhanced temperatures as long as the domain shape stays circular.

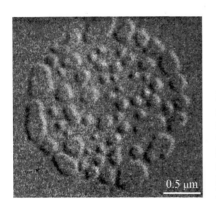

Fig. 6.27 Written spot in a TbFe magneto-optical recording medium, made visible by Fresnel mode transmission electron microscopy. The mark was written in an insufficient magnetic field of only 80 A/cm, resulting in an assembly of tiny subdomains instead of one compact domain needed in recording applications. (Courtesy *C. J. Lin*, IBM Research, Almaden, CA [1319])

Near-equilibrium domains can be generated in magneto-optical storage layers at elevated temperature where coercivity levels are reduced. At the same time the characteristic length may become smaller than the film thickness due to an increased saturation magnetization. The domains observed under these circumstances are useful in characterizing the material and its writing behaviour, as demonstrated in Fig. 5.124. Spontaneous domains may also appear in recording marks written with insufficient power as demonstrated, for example, by TEM observations (Fig. 6.27). The same phenomena can be analysed with magnetic force microscopy as shown for an alternative medium, a Co/Pt multilayer, in [1320].

(B) Direct Overwrite. In the standard MO writing scheme only domains that agree with the constant writing field can be written. Writing new information into a disk therefore requires erasing the old information prior to writing the new one and magnetizing the track to be written first uniformly opposite to the writing field. This operation needs extra time, and to avoid this delay is the purpose of direct overwrite procedures [1321]. Numerous schemes have been devised to achieve a direct overwrite capability in which both polarities can be written in the same step. Conceptually the simplest approach relies in modulating the writing field. This becomes possible when the field is generated by a perpendicular writing head which glides over the MO disk from the "rear" side. Although the writing spot is in this scheme determined by the laser beam and flying heights in the micrometer range are sufficient for this purpose, the addition of a second component on the other side of the disk means an unwanted complication.

Another approach relies on a domain wall induced collapse of previously written domains at high temperatures [1322, 1323]. It has the advantage of using standard materials and heads, but reliable and noise-free operation appears to be difficult to achieve.

The third class of direct overwrite schemes uses multiple, exchange-coupled media with different coercivities and Curie points. The basic design of this approach is indicated in the sketch [1314]. The memory layer has a low Curie point, but a large coercivity at room temperature, while the reference layer has a large Curie point, but a moderate coercivity at room temperature. Writing in opposite directions is achieved by modulating the laser power. At moderate power only the memory layer is heated above the Curie point. After cooling it assumes the magnetization direction from the reference

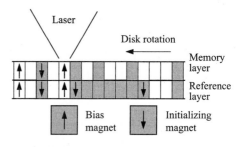

layer which has been magnetized "downward" by an initializing magnet which is mounted ahead of the read/write head. At larger laser power both layers are heated above the Curie point so that on cooling the polarity of the bias field is written. Important in this mechanism is the microstructure and the behaviour of the interface wall separating oppositely magnetized bits in memory and reference layers which have been studied in various papers, including also multiple-layer systems which offer enhanced reliability [1324].

(C) Magnetically Induced Super-Resolution. The storage density of magneto-optic recording is limited by optical resolution. If the numerical aperture of the optical head can be increased or the laser wavelength can be reduced, enhanced recording densities become possible. Another fascinating possibility that relies directly on magnetic microstructure is called "magnetically induced super-resolution" [1325–1327]. In this technique only a small, thermally defined part of the film is activated in writing and in read-out. Since the thermal focus can be smaller than the optical resolution, optical processes that are modulated by thermal conditions can surpass the optical diffraction limit.

To explain the principle of magnetically induced super-resolution we select just one of many schemes [1328] that is based on earlier developments cited there. It consists of two magnetic layers, which are separated by a nonmagnetic spacer layer as sketched. The recording DyFeCo layer is nearly nonmagnetic at room temperature

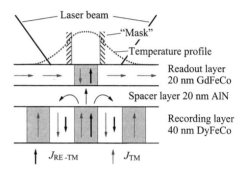

$(T_{comp} = 25°C, T_c = 275°C)$. When heated by the reading laser beam, its net magnetization increases in the hot zone so that it influences the read-out layer by its stray field. The GdFeCo read-out layer $(T_{comp} = 280°C, T_c = 300°C)$ displays only a weak perpendicular anisotropy so that it is magnetized parallel to the film plane at room temperature. Upon heating its net magnetization decreases by the compensation effect, and the anisotropy is then sufficient to turn the magnetization perpendicular to the film plane. Under these conditions the magnetization of the storage layer is locally copied into the read-out layer where it leads to a Faraday effect signal. (In the specific design of [1328] the thicknesses of read-out and spacer layer are optimized with respect to an enhancement of the read-out signal by interference; see Sect. 2.3.4.) Effectively, the temperature profile generates a "mask" inside the laser focus, so that the signal depends only on the heated zone inside the mask. In the example the demonstrated mask diameter is 0.3 μm, much smaller than the laser wavelength. Other schemes manage to generate separate masks in the storage and in the read-out layer, thus achieving still better resolution (for an overview see [1329]). Remarkable is the possibility to expand the domain in the read-out layer by an auxiliary alternating field [1330, 1331] to enhance the signal, which otherwise gets weaker with decreasing bit size. The potential of different variants of magnetically induced super-resolution is explored in [1332].

Magneto-optic recording has intrinsic advantages over conventional recording: it is reliable because the recording head stays well away from the storage layer; information can be erased without any trace of old information; and the media can be conveniently removed and exchanged as in a CD player. For a long time also its storage density exceeded that of conventional recording, with corresponding media price advantages. It appears that by the development of super-resolution, the availability of blue laser diodes and advanced optical systems [1333] the race for the highest storage density can continue.

6.5 Thin-Film Devices

Soft magnetic thin-film elements are used in many applications. Three main effects are used in such devices: (i) The switching or the reversible magnetization of an element leads to an inductive signal in an integrated pick-up loop. (ii) The electrical resistivity of the element itself is measured as it changes with its magnetic state. (iii) The stray field of the element changes on switching and is detected by a different component. We will focus the discussion in this section on computer-related applications. Other applications of magnetic thin films in general-purpose sensors, in thin-film inductors, or in microwave circuits often take advantage of developments in computer technology particularly if miniaturization is required. The general behaviour of thin-film elements was discussed in Sect. 5.5.4. Some of the examples presented there, such as Figs. 5.76 and 5.85 are directly related to sensor applications of thin films.

6.5.1 Inductive Thin-Film Heads in Magnetic Recording

Inductive thin-film heads are used for writing and reading on rigid disk drives. They are fabricated by depositing a sequence of structured magnetic and conductive layers on a rigid substrate. The development and fabrication of this economical technology, which largely replaced traditional ferrite heads (see Fig. 6.17 ; [1334]), are reviewed by *Chiu* et al. [1335].

Thin-film heads consist of a pair of thin-film Permalloy yokes that surround one or more layers of a flat spiral copper coil (see sketch (a) which is based on elements from [1335]). The gap at the tip of the yokes moves with a flying height of less than 0.1 μm over the disk for writing and reading. The yokes are prepared with an easy axis perpendicular to the working axis of the head.

Anisotropy is induced during fabrication by applying a field. Stress anisotropy, connected with a partial relaxation of planar thin-film stresses at the yoke edges, is also important.

A favourable domain pattern in the demagnetized state in the yokes consists of a Landau-Lifshitz-like configuration with closure domains at the lateral edges (as in sketch (a) and in Fig. 6.28a). This configuration can be magnetized largely by rotation, leading to a linear and low-noise reading behaviour.

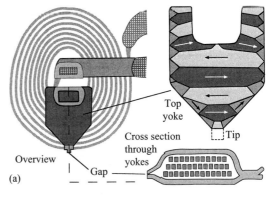

Reading noise was found to be strongly correlated with the domain configuration in the tip, in particular with the volume fraction of longitudinal domains, as indicated in sketch (b) which is based on the material in [1335].

Another domain-related problem occurs when the head has just been used for writing. In writing, the previous pattern is destroyed as the head is saturated at least in some parts. The newly generated pattern formed after the writing process may be metastable. Transitions to a more stable state may occur discontinuously, stimulated thermally or by reading processes, leading

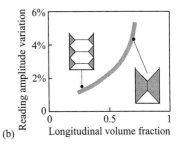

to unwanted noise, known as "popcorn" noise, in the reading signal in some time interval after the write process [1336, 1337]. Figure 6.28 displays such irreversible processes in the upper yoke of a thin-film head using Kerr microscopy (see also Fig. 2.20b). Attempts to control this effect have been reported, for example in [1338], where notches were introduced at the yoke edges of a perpendicular recording head.

Because of the topography of the head yoke, it is difficult to fabricate a head without magnetostriction. In addition, magnetostriction can be used to increase the transverse anisotropy at the pole tip. Temperature transients in connection with the writing process lead to additional inhomogeneous stresses [1339] and therefore necessarily to varying effective anisotropies, which may stimulate domain rearrangement processes. Particularly important in connection with such processes are the neighbourhood of the rear magnetic closure zone and of the pole tip where the yokes are in close contact and therefore most susceptible to stress effects [1340, 1341]. If the domain pattern in these parts of the yokes is unfavourable, the stress transients may lead to domain rearrangement processes and to strong popcorn noise.

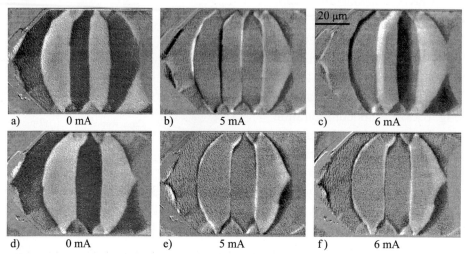

Fig. 6.28 Domains observed on the yoke of a thin-film head. After magnetizing the head with a pulse of 50 mA the metastable configuration (a) is formed. This state is now stored as a reference image. Stimulating it with an alternating field of up to 5 mA (b) generates only small reactions, while with 6 mA alternating field (c) a strong domain rearrangement has been induced. Switching off the a.c. field the state (d) is obtained which is now stored as a new reference image. This state is more stable and not prone to domain rearrangements by small alternating fields (e, f). (Sample: courtesy *H. Grimm*, IBM Mainz)

Another source of noise are domain wall conversion processes [990, 1342]. The films in recording heads are usually several microns thick; they therefore carry asymmetric Bloch and Néel walls (see Figs. 5.68, 70) as shown in Fig. 6.29. Even if the triggering field does not affect the domain pattern as a whole, it may suffice to move, generate and annihilate the transition lines between different wall types and wall orientations. Usually these processes occur discontinuously so that they contribute to noise.

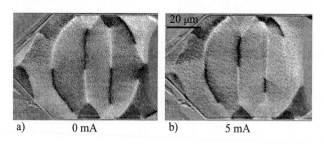

Fig. 6.29 The domain wall substructures in the 180° asymmetric Bloch walls in a recording head. After a small field is applied using the integrated coil, many of the Bloch lines have been displaced

The relatively thick yoke material is usually deposited electrochemically. By varying the deposition parameters the composition and thus the magnetostriction constant of the NiFe alloy can be fine-tuned according to the local distribution of stresses [1341, 1343]. The experimental study of the complex domain structures in inductive heads requires access to both yokes. *Kobayashi* [327] used the depth-dependent backscattering electron contrast method (see also Sect. 2.5.3 and [331]) to explore their domains separately and identified irregularities in the interaction of the two yokes.

At very high operation speeds (beyond some 10^7 Hz) eddy current effects may demand laminated yokes. If the individual layers of these multilayer structures are not only galvanically decoupled, but also magnetically independent, the domain pattern in these multilayers is strongly modified and even simplified compared to single-layer yokes. Note, for example, the absence of closure domains in Fig. 5.106 where they are replaced by edge curling walls. These will certainly be present in multiple films, contributing to noise. They could be avoided by using "closed-edge" laminations [1344] which are difficult to fabricate but would be ideal micromagnetically.

Inductive heads are limited with respect to the maximum recording density [1335] because of two mechanisms which adversely influence the signal/noise ratio on miniaturization: (i) If the recorded bit gets shorter and narrower, the induction signal gets weaker unless this is made up by the number of windings. More windings, however, means a higher resistance and increased thermal noise. (ii) With a reduced track width it becomes more and more difficult to achieve the required transverse domain pattern in the pole tip. Increasing the transverse anisotropy may be difficult and is not an ideal solution because it decreases the permeability which is needed in the reading process.

Because of the low price of inductive heads there is an incentive to stretch the limits of their applicability. This probably requires some fundamental design change. One interesting approach is the fully integrated planar head [1345–1347] which is fabricated with the methods of silicon technology without the additional mechanical polishing step needed for the definition of the pole tip of conventional thin-film heads. The higher symmetry of the planar design leads to a more favourable coupling between coil and yokes. Another approach initiated by *Mallary* [1348, 1349] improves this coupling by feeding the yokes twice or several times around the coil. This multiplies the signal for a given coil and thus enhances the high-density potential of inductive heads, although problems with domain behaviour may have to be overcome as indicated in [1335].

The standard answer to the limitations of the inductive head is the use of magnetoresistive effects in reading, while the writing process still relies on the standard thin-film head design, into which the magnetoresistive sensor is integrated. This approach is discussed in the next section.

6.5.2 Magnetoresistive Sensors

(A) Overview and General Purpose Sensors. General purpose magnetore-sistive (MR) sensors are finding increasing applications in many fields [1350–1352]. They can be fabricated in an integrated fashion, thus offering the potential of a compact and economical design. Usually a Wheatstone bridge is formed by thin-film elements of various shapes. In all magnetoresistive sensor applications uniformly magnetized sensor elements are desired. Domain for-mation determines the stability limit of an element and contributes to noise. The detailed study of magnetic microstructures in sensor elements is therefore essential. The domain studies presented in Fig. 5.76 apply directly to the question of sensitivity and stability of such sensor elements. Another example is shown in Fig. 6.30. Meander-shaped magnetoresistive films are used as components in magnetoresistive sensor bridges of various designs [1350]. Switching processes as a function of field strength can be observed which define the limits of usefulness of these sensors.

Two magnetoresistive effects are used in sensors; they may be active simultaneously in a device. One is the "anisotropic" MR effect, a relatively weak dependence of the resistivity on the angle between current density j and magnetization m. In favourable materials such as nickel iron cobalt films this classical effect amounts to up to 4%. Employed in the compensating bridge mentioned, this change is sufficient to built very compact field sensors that may exceed Hall sensors in sensitivity. The anisotropic MR effect is quadratic in the magnetization components, following a $(j \cdot m)^2$ law. This quadratic dependence on the magnetization direction is mathematically equivalent to the expression for the magnetostrictive elongation (3.51) in an isotropic material.

A second effect was discovered in the course of investigations of exchange coupling between magnetic multilayers [1353, 1354]. If the layers are separated by non-magnetic metallic interlayers that are thinner than the spin diffusion length, and if the adjacent ferromagnetic layers are magnetized in different directions, an additional electron scattering effect is observed. Electrons coming from one layer carry the spin information of the magnetization direction m_1 of their source. If they enter the neighbouring layer with a different magnetization direction m_2, they are strongly scattered, leading to a resistance contribution proportional to $m_1 \cdot m_2$. If a field that aligns the layers magnetically is applied, the extra resistance effect disappears. This heterogeneous or "giant" magnetoresistive effect ("GMR") may be much larger than the "homogeneous", anisotropic effect mentioned above. Values of up to 100% change in resistivity have been reported, and even in the small fields available in most sensor applications enhanced effects have been observed. Example designs of sensors based on this effect are described in [1355–1357].

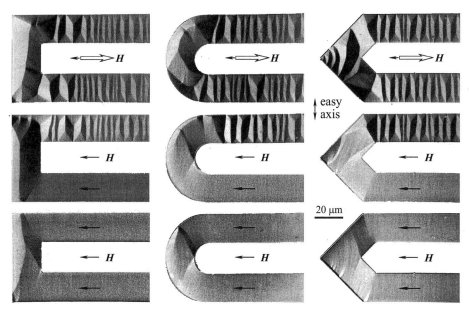

Fig. 6.30 The switching process in the legs of experimental thin-film meander structures, depending on the shape of the U-turns. In these experimental patterns the easy axis is oriented perpendicular to the leg axes. Applying a field along the leg axis, hard axis magnetization processes are observed, as discussed in Sect. 5.5.2C. The buckling state (*top row*) collapses first in one leg (*middle row*) and then in the other (*bottom row*). The residual patterns in the U-turn areas are usually masked from the sensor operation. (Sample: Permalloy single layer films of 40 nm thickness; courtesy *M. Freitag*, Bosch)

(B) Conventional Magnetoresistive Reading Heads. The application of the anisotropic MR effects in reading heads [1334, 1358–1361] offers the advantage that the reading signal is independent of the moving speed of an information carrier. This advantage is important for multi-track tape reading heads. For rigid disk recording a properly designed MR reading head offers advantages in signal level for high-density recording.

Because the $(\boldsymbol{j} \cdot \boldsymbol{m})^2$ dependence of the resistivity of the anisotropic MR effect leads to a quadratic dependence of the signal on the angle between current and magnetization, this angle must be biased to 45° to reach a linear characteristic. Many schemes have been proposed to realize this biasing which are reviewed in [1334]. A favourable configuration for a high-resolution disk head is indicated in the cross-section sketch. It is a *shielded* head, biased by a *soft adjacent layer* (see [1361]). The working principle of these two ideas is quite interesting. The two shields could be the yokes of a thin-film head. They isolate — as indicated by the name — the MR element from outside signals and noise. More important is another function. Because the MR strip is positioned very close to one of the shields, flux that is entering at the reading surface into the MR strip is closed over the shield, so that the

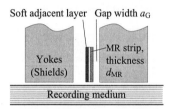

magnetization rotation in the strip is not limited by the demagnetizing field of the isolated strip. It extends, as a more detailed analysis shows, to the decay length $l_{\mathrm{d}} = \sqrt{\frac{1}{2}\mu_{\mathrm{r}}a_{\mathrm{G}}d_{\mathrm{MR}}}$ [1361, 1362], where μ_{r} is the rotational permeability of strip and shield, a_{G} is the gap width, and d_{MR} is the thickness of the MR strip. For $a_{\mathrm{G}} = 100$ nm, $d_{\mathrm{MR}} = 20$ nm and $\mu_{\mathrm{r}} = 3000$ we obtain $l_{\mathrm{d}} = 1.7$ μm for the necessary and useful width of the MR strip.

A further, decisive property of the shielding concept is the high resolution of this design. While an isolated strip reacts to recorded information in a distance which scales with the width of the MR strip, the shielded arrangement is sensitive only to fields entering the MR strip through the gap between the shields.

The soft adjacent layer reacts to the measuring current through the MR strip. If properly adjusted, it is driven almost into saturation. The magnetic charge concentrations at its top and bottom edges generate a field which is strongest at the edges of the MR strip. This inhomogeneous field is favourable to the induction of a uniform biasing in the MR element because the demagnetizing effects of the strip alone are strongest at the edges (see Fig. 5.85).

A necessary requirement for a properly functioning MR head is the absence of domains in the reading area of the MR strip. The possible domains in small rectangular elements were studied by *Decker* and *Tsang* [985, 986]; in this book examples can be found in Figs. 5.78 and 5.82. To avoid these domains, a further biasing is needed, an effective field along the strip length, parallel to the current direction. This can be achieved, for example, by exchange-coupled hard magnetic or antiferromagnetic films outside the sensitive area. A strong effective longitudinal bias field reduces the sensitivity of the sensor, however. An element which is free of domains in the central part without strong longitudinal bias fields is therefore advantageous (see Fig. 5.76 for the analogous arguments in conventional sensor elements).

The MR sensor can be integrated either directly into the gap of the writing head, or, in a *dual head*, adjacent to it, using one of the yokes of the write head as a shield. Because the shields are an integrative part of the sensor magnetic circuit, domain processes in the yokes can enter the reading signal. Longitudinal domains in the pole tip, for example, would offer no permeability and would be ineffective as a shield. Proper domain control in the yokes is therefore not unimportant in integrated heads. The write head component can be simpler than in the case of inductive write/read heads, containing, for example, only one layer of windings because in conventional heads the multiple winding layers are needed to reach an acceptable read signal level. Another source of noise can be domain instabilities in the soft adjacent layer as explored by numerical simulation in [1363].

(C) Giant Magnetoresistance Reading Heads. Reading heads may also exploit the giant magnetoresistive effect mentioned above, offering enhanced sensitivity. Systems exploiting these effects consist of rather thin films. Thin films are in principle desirable because in a narrower gap they can be sensitive to narrower transitions, thus supporting a higher longitudinal recording density. Arguments in favour of GMR read heads are collected in [1364, 1365]. The difficulties in connection with defects and inhomogeneities in such systems were indicated in Figs. 5.114–115.

There are several ways to induce the necessary inhomogeneous magnetization between the components of the multilayer system. One possibility is the use of antiferromagnetic or biquadratic exchange effects (see Sect. 5.5.7C and a recent review [1366]). The drawback of this scheme is that rather large fields are usually needed to overcome exchange coupling. Another way consists in using weakly coupled films, one of which is pinned by a hard magnetic or antiferromagnetic layer, so 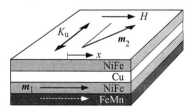 that only the free layer can be rotated in a magnetic field. If the easy axes of the two uniaxial layers are oriented perpendicular to each other as shown in the sketch [1367] and if the two layers are uncoupled, then we have simply $m \sim h$ for the magnetization component in the top layer parallel to the field

($m = J_x/J_s$, $h = H_{ex}J_s/2K_u$). This means the giant magnetoresistive effect, which we have seen in (A) to be proportional to $m_1 \cdot m_2$, is also linear in H up to the saturation field. This ideal characteristic, the conceptual simplicity of this scheme and the potentially large sensitivity have stimulated intense development work. The role of magnetic microstructure in these devices was studied, for example, in [1368] by numerical simulation. In [1369] a less than ideal performance of a model configuration was traced back mainly to a residual orange peel ferromagnetic coupling effect (see p. 487) between the two films, which might be compensated by a weak antiferromagnetic exchange coupling contribution. The various problems with this technique can obviously be overcome [1370], so that it actually reached the market in hard disk drives. Altogether the technique seems to lead the way to the next generations in higher-density recording heads.

6.5.3 Thin-Film Memories

Thin-film elements offer themselves for memory applications because they exist in at least two metastable states like other small particles [1371]. Usually two oppositely near-saturated states are employed between which the element can be switched by pulse fields generated by current loops. The selection of a particular element is performed by a matrix scheme. This kind of random-access, non-volatile memories can serve as the fast central memory of a computer. They were first developed in the sixties to succeed the classical ferrite core memories, sharing with them the inductive read-out method. Eventually, they had to give way to semiconductor memories [1372], although standard, "dynamic" semiconductor memories are volatile and have to be periodically refreshed. Recently, magnetic random-access memories were revived in connection with new magnetoresistive reading methods. Not primarily because they are non-volatile, but because they appear to behave favourably in down-scaling, they are challenging semiconductor memories once more.

(A) Classical Random-Access Thin-Film Memories. In contrast to ferrite core memories in which every core element had to be individually fabricated, tested and assembled, "flat" thin-film memories were fabricated from the beginning in an integrated fashion. They consisted in their simplest form of planar Permalloy elements with a uniaxial in-plane anisotropy. Random access in writing was made possible by the combination of two current systems. The *word line* generates a field along the hard axis preparing the elements along this line to be switched in a reduced field. Switching is then performed by the *bit line*, which generates a field along the easy axis. Only the element that feels the fields from both currents simultaneously is switched. Other elements stay unaffected.

In reading an element is selected in the same way. Voltage pulses are induced in an extra sense loop if the element was magnetized opposite to the bit line field. The reading process, which relies on the full switching of the selected element, is destructive. After reading, the information has to be rewritten which needs extra

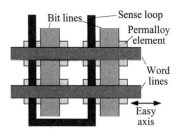

time. Flat thin-film memories would allow non-destructive read-out, but signal levels are too low for reliable operation. Ideally, the element should switch in a fast, coherent fashion in this scheme. In most cases, however, some much slower incoherent switching process is observed (see Fig. 5.64). Fast switching needs strong field pulses and an optimum balance between both field components as analysed, for example, in detail in [1373].

The same principle can be applied to films deposited around thin wires [1374]. In this case rather thick films (~1 μm) can be used, which need not be structured along the wire. The thick films lead to sufficiently large inductive signals, a decisive advantage of the plated wire memory concept, even offering non-

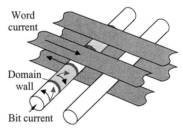

destructive read-out. The written information generates no demagnetizing field in this geometry, a feature that is favourable for stable memory function.

The plated wire memory reached the market for some time [1375, 1372]. The fact that it is not fabricated in a fully integrated way was even seen as an advantage at the time. The wires were plated continuously and immediately tested magnetically. Good segments were cut off and introduced into the memory package. In the case of a malfunction they could be replaced. In comparison with core memory manufacturing this was an advanced procedure which offered a favourable path to good yield.

The problem with all kinds of inductively read random-access thin-film memories was their lack of potential for miniaturization. They could not compete in the race for steadily decreasing scales and cost dominated by semiconductor technology. The reason is twofold: On the one hand, the inductive read-out signals become too small on miniaturization. The other problem was connected with interactions between adjacent memory cells. Hard axis word field pulses that are needed in the reading process lead to unavoidable domain wall conversions. If the bit field intended for one bit extends to the neighbouring bit along the word line, domain walls in these locations may begin to creep (see Sect. 5.5.3A; [1376]), eventually destroying the information. The wall creeping phenomenon sets a limit to the bit density in thin-film memories as long as domain walls are present in these devices.

Thin-film random-access memories attracted renewed interest based on a different read-out principle. If the information is read magnetoresistively, much smaller elements can be employed (see below).

(B) Memories Based on Anisotropic Magnetoresistance Read-Out. The inductive read-out method common to all classical thin-film memory concepts constituted the most important obstacle to their development towards higher density because of their "negative" scaling behaviour: the inductive signal decreases with the volume of the magnetic element. In contrast, the in-plane resistance and also the magnetoresistance of a small element is independent of its size for constant thickness, and the perpendicular resistance (perpendicular to the film plane) increases even with miniaturization. Magnetoresistive read-out methods for magnetic thin-film memories opened important new possibilities.

First concepts of random-access memories with magnetoresistive read-out still used the anisotropic magnetoresistive (AMR) effect (see Sect. 6.5.2A). Here the magnetization has to rotate by 90° relative to the sensing current in order to achieve the maximum magnetoresistive effect. Conventional thin-film elements as analysed in Sect. 5.5.4 do not support stable orthogonal states. Single-domain particles are mostly uniaxial and could store information only in their antiparallel ground states — which do not differ in anisotropic magnetoresistance. Single-crystal cubic films (Sect. 5.5.6) would support orthogonal states, but they are difficult to prepare.

The most interesting representative of this class is the *Y-domain* memory [1377, 1378]. It relies on multi-domain elements of a particular shape. The elements support two types of domain states with about equal energy. One state is partially open and carries a net magnetic moment, the other one corresponds essentially to the stray-field-free ground state of van den Berg (Sect. 3.3.3B). By applying pulse fields H_p of different polarity the element can be switched between the two states as sketched.

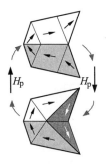

The two states differ in their magnetoresistance for the chosen current path $I(r)$, and after optimizing the geometric parameters the basic memory functions of reading and writing were demonstrated [1378]. The concept has two drawbacks, however: (i) Scaling it down meets limitations when the single-domain size is approached. (ii) Domain observation [1378, 1379] revealed what had to be expected: that the domain states in thin-film elements 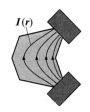 are not unique but highly variable. For example, the "tulip" corner pattern (Fig. 5.86) (connected with domain wall pinning at the element edges) can also be found in the domain images of Y-shaped elements, necessarily disturbing the magnetoresistive read-out signals. Probably for these reasons the Y-domain memory concept was not further pursued.

Other possibilities based on the anisotropic magnetoresistance effect were proposed by *Pohm* et al. [1380–1382]. In these articles double layer elements are used in various schemes. With double layers very small elements can be realized as shall be demonstrated by discussing one interesting variant [1382]. In this scheme the information is stored in edge curling walls. To distinguish the two states an additional field along the element length is applied during reading. As demonstrated in Fig. 5.106, this field will expand the "parallel" edge curling walls of logical "1" and compress the antiparallel edge curling walls of logical "0", thus causing a difference in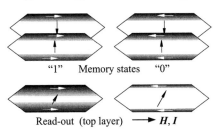

resistance. Although this concept does not allow element dimensions in the single-domain range, very small dimensions (down to 0.25 μm element width) are possible according to micromagnetic simulations.

In any case, the anisotropic magnetoresistance effect is weak (up to 3–4% in FeNiCo films), and only a fraction of the maximum effect can be exploited in the proposed elements. This asks for complex differential or repeated read-out schemes which may be slow. The newly discovered GMR effects behave in principle more favourably in both points: the effects are stronger, and it is possible to exploit the full amplitude of the effect as will be demonstrated. But GMR technologies are difficult to master, while no particular problems are met in the preparation of AMR memory elements. Nevertheless, interest in AMR memory concepts has declined with the advent of GMR effects. Both approaches are critically examined in [1383].

(C) Giant Magnetoresistance Read-Out Memories. In contrast to the AMR effect, the inhomogeneous or "giant" magnetoresistance (see Sect. 6.5.2A) is sensitive to magnetization rotations by 180°. If one element with fixed magnetization is combined with another free element, then the free element has only to be switched by 180° to obtain the maximum magnetoresistive signal. Although the preparation of GMR multilayer systems is much more difficult than that of ordinary thin films, the prospect of high-density magnetic random-access memories (MRAMs) has stimulated worldwide activities. Domains are not required in these devices, the elements work best in the single-domain state, i.e. for small elements. Favourable shapes for a high stability of high-remanence states in larger elements were explored in Fig. 5.76.

The switching of the free elements must be performed by a coincident current scheme as in the classical thin-film memory explained in (A), relying on orthogonal field pulses. If the thin-film elements will finally reach the size range of single-domain particles, they can be described by Stoner-Wohlfarth theory (Sect. 3.5.2A). It demonstrates that a transverse field of 0.5 H_K reduces the switching field to 23% of the value without a transverse field (Fig. 3.45, 46). This is a reliable basis for a coincidence matrix scheme, even if the switching process should not occur in a completely uniform fashion. If at least in the remanent states domain walls are absent, wall creeping effects which tend to destroy stored information in large elements as discussed in (A) have not to be feared in high-density MRAMs. An advantage of magnetoresistive reading schemes is that edge zones of an element which might contain residual domains can be masked from the reading area (see sketch). The potential high switching speed of thin-film elements has been explored for classical thin-film memories [1373].

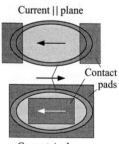

Current || plane

Contact pads

Current ⊥ plane

In the conventional arrangement to measure the magnetoresistance of a thin multilayer element the current flows in the film plane ("CIP"). In principal more favourable would be a scheme in which the current flows perpendicular to the plane ("CPP") through the multilayer stack. The CPP solution needs less space and yields higher values of the relative magnetoresistance. The problem is the low regular resistance in the CPP layout. For conventional GMR sensors with metallic interlayers the CPP method can only be used with specially prepared micro-columns which are hardly conceivable in a memory environment. Two effects have been discovered which combine a large magnetoresistance with an acceptable resistance: In one variant of the giant magnetoresistance effect the electrons travel from one ferromagnet to the other by tunnelling through a thin insulator layer (which is mostly in situ oxidized Al_2O_3) [1384–1386]. The other approach is based on semiconductor interlayers in which hot polarized electrons are transported between the magnetic films [1387, 1388]. However, the applicability of the latter scheme to highly integrated devices has not yet been demonstrated.

Components of GMR or "spin valve" memories in the conventional CIP geometry were reviewed in [1389–1391]. Several authors [1392, 1393] studied the micromagnetics of the needed small elements. First experiments and considerations into the direction of a high-density *tunnelling* memory are reported in [1394–1396]. All GMR devices need very thin magnetic layers. In such films the danger of immobile 360° walls has to be taken into account (see Sect. 5.5.2D). Figure 6.31 shows the top layer of an experimental tunnelling multilayer element with such walls.

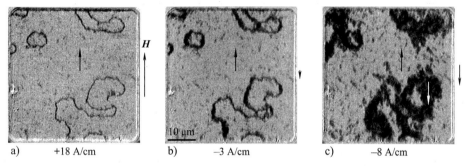

a) +18 A/cm   b) −3 A/cm   c) −8 A/cm

Fig. 6.31 Domains in the top iron film in an experimental tunnelling multilayer arrangement, consisting of the following components: Si | Fe 6 | Cu 30 | Co 1.5 | Cu 1 | Co 1 | Al_2O_3 | Fe 2 | Cu 2 | Cr 3 (thicknesses in nm). The Co/Cu/Co sandwich acts as an "artificial antiferromagnet" which pins the magnetization of the 1 nm Co layer along the *y*-direction. After almost saturating along the same direction (a) 360° walls may be left over in the "free" iron layer. They act in an opposite field (b, c) as nuclei in the remagnetization process. (Sample: courtesy *J. Wecker*, Siemens Erlangen)

6.6 Domain Propagation Devices

The direct and controlled use of domain propagation in memory or other devices always offered a fascinating prospect. Manipulating only magnetic domains without mechanical transport held out the prospect of highly reliable solutions of technical problems that appeared intractable otherwise. Most of the proposed and developed schemes are of historical interest only because other technologies offered more economical solutions after all. It is remarkable that controlled domain displacement does at present not play a role in industrial applications. Practical devices rely mostly on uncontrolled domain processes such as in soft magnetic materials, or on static magnetic structures as in permanent magnets and magnetic recording. The history of domain propagation devices is fascinating anyhow and worth a discussion. Perhaps improved materials and technologies will lead to a revival of these techniques in the future.

6.6.1 Bubble Memories

The basic mechanisms of the formation and stability of cylindrical domains ("bubbles") in soft magnetic perpendicular films were introduced in Sect. 3.7.3 and these domains were demonstrated in Sect. 5.6.2. Bubble domains are generated, shifted, duplicated and read out in a bubble memory that can serve as a fast, fully electronic, non-volatile substitute to magnetic recording devices [510, 751, 1397–1399]. The information carrier in a bubble memory is a low-coercivity, perpendicular film. While many alternatives such as amorphous metallic films [1400] were studied, epitaxial garnet films were the material of choice during most of the period of developing, building and using bubble memories. The smallest bubble diameter possible with this material is about 0.3 μm. Memories of up to 16 Mbit/chip were developed, an integration level that was for many years reached by silicon technology only one or two generations later. Bubble technology acted for a considerable time as a pioneer in the development of lithographic techniques. The latest developments are reviewed in [1401, 1402].

Bubbles are stable in a certain bias field interval. This field is generated by a permanent magnet system and is thus independent of memory activities. As the bubbles are supposed to represent independent information carriers, their interactions should be weak. This means that bubbles should not be used in a densely packed fashion as in a bubble lattice. As a rule of thumb, bubbles in a bubble memory should be separated by about four times their diameter to avoid interference. If bubbles do come closer to each other, their interactions can be exploited in logic functions. While certain functions of this kind were used in memory gates (B), the idea of a fully integrated bubble memory and logic chip [1403 (p. 59–76)] never reached the market. Semiconductor logic is faster and cheaper, and for logic functions the non-volatility of magnetic information storage is irrelevant.

To achieve reliable operation the perpendicular storage film must be covered by a thin in-plane-anisotropy layer, usually generated by ion implantation [1404]. This layer can suppress hard bubbles (which contain many Bloch lines; see Sect. 3.6.5A) because it favours states like that in Fig. 3.94c in the rotating field.

(A) Bubble Transport. Bubbles can be transported in the memory with the help of local magnetic field gradients. Three basic mechanisms of generating these periodic field gradients are known:

(i) Bubbles can be transported by a periodic array of small Permalloy film elements that are deposited in a certain distance on top of the bubble medium. The elements are activated by a rotating in-plane field. The first elements were symmetric and allowed in principle bubble motion in both directions.

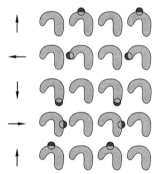

The sketch shows the most advanced asymmetric pattern [1405] that permits, for a given lithographic accuracy, the smallest information cell size. The rotating field (typically of 100 kHz frequency) generates a periodic pattern of magnetic poles, which translates the bubbles by one cell unit for every revolution. The Permalloy elements are used in this technique simply as reversibly magnetizable units. Although they contain domains, they behave largely like elements with an infinite permeability as explained in Sect. 3.3.3B. The domains that are constantly active in the Permalloy elements have never manifested themselves as having an effect on bubble devices. The understanding of soft magnetic thin-film elements was stimulated by the development of bubble technology.

(ii) Bubble transport can be achieved by a pattern of *ion implantation* which modifies the anisotropy of the implanted zone from uniaxial to planar anisotropy. The bubbles are transported in a rotating field along the edge of the implantation zone [1404], guided by charged walls which are formed within the planar-anisotropy implanted drive layer [54] (see sketch). In a rotating field the magnetic

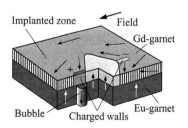

charges in the drive layer change their position, carrying the bubbles along as sketched below. In contrast to Permalloy devices the domain patterns in the control layer play a decisive role in ion-implanted devices.

A Bitter pattern image of charged walls, one of which binds a bubble, is shown in Fig. 2.1a (the charged walls are the short segments close to the unimplanted disk, not the extended "propeller" domains which have no function). The detailed mechanism of charged wall formation and their properties is quite interesting. It turned out that magneto-elastic mechanisms play a decisive role, connected with the stress introduced by implantation and its partial relaxation at the implantation edge [1406–1408]. The advantage of the ion-implantation scheme, which is reviewed in [1409, 1410], is that still smaller cell sizes for a given lithographic

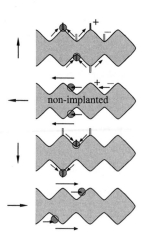

accuracy are possible compared to Permalloy devices. At the point when the development of bubble memories was discontinued, a hybrid ion-implanted-Permalloy memory, in which conventional Permalloy structures were used in the input and output parts of the memory, was planned to enter the market in the next generation [1411]. Such a hybrid device was obviously easier to realize than a conceivable all ion-implanted design [1412]. Although the fully ion-implanted design has the fascinating property that all memory components can be fabricated with a single lithography mask, a successful design was obviously easier to reach with the hybrid approach.

(iii) Bubbles can be transported by the effects of currents flowing in certain overlay patterns. Various possibilities are reviewed in detail in [510 (p. 160–166), 751 (p. 151–161)]. Particularly interesting is the concept of current sheets in which holes of a certain shape can have the same effect as Permalloy elements in a field. Stacking such current sheets relaxes the demands on lithography and adds flexibility in design [1413]. Current access can be faster than rotating field concepts but need a lot of energy. In the course of the evolution of the bubble memory current control was used mostly for special functions as in the transfer gates, not for basic bubble transport.

Every bubble transport system works reliably within a certain range of bias fields and rotating field amplitude. A large *propagation margin* area is favourable for a reliable function of a bubble memory. The propagation margin is always smaller than the stability range of the isolated bubble domains that is shown in Fig. 3.115. In the standard design bubble propagation needs a rotating magnetic field but no currents. When the rotating field is turned off, the bubbles remain stored indefinitely because the necessary bias field is supplied by a permanent magnet.

(B) Memory Functions. The standard design of bubble memories consists mainly of endless *minor loops* in which the bubbles rotate until they are needed. The minor loops are needed only for storage, reading and writing occur in the major loop. A complete bubble chip must contain several special functions in addition to the propagation structures: (i) A bubble generator consisting, for example, of a hair-pin current loop. (ii) Gates transferring or replicating bubbles between minor and major loops. The gates consist of complex overlay patterns interacting with control currents. (iii) A read-out station in which the bubbles are first expanded into band domains and then read magnetoresistively. (iv) Sometimes a special bubble annihilator is added that destroys a bubble when activated by a current pulse. Often bubbles that are not needed any more (for example after reading) are simply guided out of the active area of the memory.

A simple example for the organization of a chip in which two kinds of gates are used is shown in the sketch [1414]. The needed electrical connections are collected on the top of the layout. The transfer gates can move a series of generated bubbles into the minor loops, and they also can extract one bubble from each minor loop in parallel, which are then lead to the detector for read-out. If the replicator is activated, this bubble train can be fed back to the minor loops.

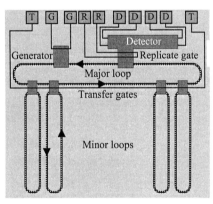

Many variants of this scheme are conceivable and necessary for larger chips.

A function of the bubble memory, which is of general interest, is the bubble detector. Almost from the beginning a magnetoresistive read-out scheme was used in this technology [1415]. Because the anisotropic magnetoresistance is rather weak, the bubbles were expanded into band domains with the help of the "chevron" propagation pattern as shown in the sketch [1416] (the densely packed chevrons effectively reduce the bias field, thus leading to bubble stripout — see Fig. 5.125). In

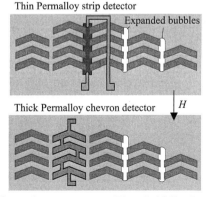

practice much higher expansion factors were used. This is possible because a bubble mass memory needs only one read-out station, which can be accordingly elaborate. Of the two variants shown in the sketch the top one is optimized for sensitivity, while the other one is easy to fabricate, as the same type of Permalloy overlay film is used both for transport and detection.

It is important that the operation margins of all functions have a large overlap with the propagation margin of the memory. Since only few of these functional units are needed on a bubble chip, their design can be permitted to differ strongly from the propagation pattern if a better overlap in the margins can be achieved in this way. It is also very important that the temperature transient occurring when the rotating field is turned on and off is tolerated by all functions reliably. Very detailed design consideration are necessary for the special functions of the bubble memory. We refer to the extensive textbook literature cited at the beginning of this section.

(C) Bloch Line Memories. A new memory system with potentially a much higher density was proposed and explored by *Konishi* [1417, 1418]. It is not based on the presence or absence of bubbles, but on the presence and absence of *Bloch lines* in band or bubble domain walls. As the static properties of bubbles or band domains are hardly influenced by single Bloch lines, the basic domains can be stored in a densely packed pattern. This means that very high storage densities should be achievable with the Bloch line memory concept, but the development has been extremely difficult. One reason is certainly that in contrast to bubbles, Bloch lines can hardly be observed experimentally (see Fig. 5.128 for the only direct, but unfortunately limited method). While the behaviour of bubbles could always be monitored directly, thus guiding the development of complex structures, the same possibility is lacking in the development of the Bloch line memory. Static and dynamic numerical simulations therefore always played a prominent role. References [1066, 1067, 1419–1421] give access to the extensive work in this field.

Band domains are used to carry the Bloch lines. Both are held in position by superimposed longitudinal and transverse structures (trenches, Permalloy strips, ion implanted tracks, etc.). The Bloch lines are written, for example, by an in-plane field acting on an implanted layer as explained in Fig. 3.94. Current pulses lead to dynamical reactions in wall segments and to a lateral displacement of the Bloch lines.

At special stations bubble domains are chopped off from the bands to be read in a conventional bubble detector as described in (B). The chopping process — explained in the sketch below — is the fundamental operation of the Bloch line memory. A current pulse can chop off a bubble from a stripe domain *if* the two domain walls pressed together by the pulse have an unwinding orientation. Otherwise no bubble is generated. The chopping process is therefore sensitive to the wall rotation sense, and thus controlled by the presence or absence of Bloch lines.

Record values of 1 Gbit per chip of non-volatile memory appeared to be in sight with this technology [1422]. The developed prototypes must certainly be counted as remarkable tech-

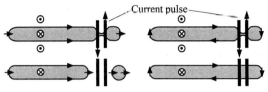

Unwinding walls: Chopping   Winding walls: No chopping

nological achievements, but still no available product is in sight. The basic problem could be the reliability of singular point generation and annihilation. Bloch line pairs may be converted from winding to unwinding by the thermal, dynamic or defect-induced generation of singular points so that they can annihilate, thus destroying the information [1423]. An excess concentration of Bloch lines is found in hard bubbles. The observation that they can decay spontaneously at elevated temperatures [1424, 1425] has not found a consistent explanation. Perhaps a high anisotropy is needed for stable Bloch lines as might be derived from the findings in [1426].

6.6.2 Domain Shift Register Devices

Domains can be propagated also in planar thin films, best along channels defined by a high-coercivity environment [1427, 1428]. As in the bubble memory, domain transport can be done in a rotating field (field access; [1429]), mixed schemes [1430] or by pure current access [1431]. The sketch explains the latter concept, which uses two sets of conductor loops and a two-phase propagation scheme [1431]. The difference to bubble domains is that the domain tips in the propagation channels move in a creeping fashion, stopping at the current position when

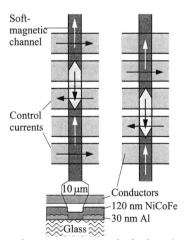

the field is switched off. In contrast, bubbles always assume their local equilibrium shape and position. Reading is done by expanding the domains before reading them, either inductively [1431] or magnetoresistively [1429].

The advantage of this kind of non-volatile mass memory was the use of cheap, conventional thin-film materials, while bubble memory needed expensive epitaxial single-crystal films as the basic material. This applies also to another example of a shift register device which used Bloch line pairs in a thin film as the information unit. Because Bloch lines in thin films are connected with the characteristic cross-tie structure, this scheme is known as the cross tie memory [975]. Domain shift register devices have not been successful in the long run, primarily because of their low speed and their insufficient potential for miniaturization in the planar geometry. Characteristically, the latest offspring of the cross tie memory idea was a magnetoresistively read random-access device [1432] belonging to the category discussed in Sect. 6.5.3A.

6.6.3 Magneto-Optical Display and Sensor Devices

Optical display and switching devices that can be operated electronically at high speed and with high efficiency are still in demand. Many magnetic devices based on the same garnet layers as used in bubble memories were developed and demonstrated their usefulness. A display that is directly based on bubble technology was developed by *Lacklison* [1433]. But in spite of the larger complexity a direct-access display proved to be advantageous because of its better speed [1434, 1435]. A nanosecond light modulator based on a bubble lattice was proposed in [1436]. Even read (and write) heads for recording applications based on bubble films have been proposed as reported in [1403 (p. 123 –125)]. A comprehensive survey of magneto-optical devices can be found in [1437].

The problem with all these approaches is the low efficiency of magneto-optical films in the optical regime. Even with the best magneto-optical materials most of the light in the visible range is still absorbed instead of transmitted. As magneto-optical materials continue to be explored in connection with magneto-optical recording, a breakthrough in efficiency could change the prospect of this device class decisively.

Also magneto-optic indicator and sensor devices belong to this kind of applications. Indicator films were reviewed in Sect. 2.3.12B. Sensor applications for example in current measurements on high-voltage power lines usually do not employ domain effects in their operation. But a special domain-based high-speed sensor has been proposed, based on the magneto-optical detection of domain displacement [1438]. Optical insulators and modulators based on domain-wall-related phase-shift effects are considered in [1439].

6.7 Domains and Hysteresis

Summing up, is there a way to simulate the effects of magnetic domains in the design of devices based on ferromagnetic materials? A realistic simulation of single-domain or near-single-domain particles as they are used in magnetic recording (Sect. 6.4.1), in certain permanent magnets (Sect. 6.3.3), and in new concepts of magnetic memories (Sect. 6.5.3C) is certainly within reach. Unsolved problems under active research exist in the treatment of particle interactions and in the analysis of thermal effects in small particle systems.

All concepts of domain propagation devices (Sect. 6.6) were extensively guided by domain analysis and micromagnetic simulation. Unfortunately, application interest in these concepts has waned and even the magnificent achievements reached in their development could be forgotten when they may be needed again.

Much more complex is the situation in extended soft magnetic materials. Larger thin film elements as they are used in thin-film heads and sensors (Sects. 6.5.1 and 6.5.2) may generate excessive noise because of complex domain patterns and processes. Domain observation helps in controlling such effects by proper adjustment of sample shape and effective anisotropy. Among the applications of bulk soft magnetic materials (Sect. 6.2) grain-oriented transformer steel (Sect. 6.2.1A) stands out. The losses of this material are steadily reduced by studying and controlling the domain structure, which is conveniently observable because of the near-single-crystal nature of this material. The domain patterns and performance of metallic glasses and nanocrystalline materials (Sect. 6.2.3) are also reasonably well understood and under control, suffering from limited possibilities to achieve a well-defined uniform anisotropy.

For all other bulk soft magnetic materials — the non-oriented polycrystalline SiFe, NiFe and ferrite materials treated in Sects. 6.2.1C, 6.2.2 and 6.2.4 — similar possibilities are not available. Observing the domains at the surface of such materials offers only a glimpse at the magnetization processes in the volume. General reasoning as elaborated in Sect. 3.5 and global measurements as discussed in Sect. 2.8 may help in the analysis.

Engineering practice uses, however, different techniques when designing devices based on these complex materials: The measured hysteresis is simulated with the help of mathematical tools, which contain adjustable parameters and functions. These functions are calibrated or "identified" by well-defined magnetic measurements, and within limits the algorithms can then be applied to arbitrary conditions, i.e. to arbitrary sequences of size and direction of the applied field.

These mathematical descriptions of hysteresis phenomena can be grouped in three classes:

(i) The first approach starts from a "particulate" viewpoint, treating every magnetic body as an aggregate of interacting particles or "hysterons". If the distribution of switching and interaction fields of these elementary units is known, the total hysteresis can be calculated. This idea was introduced by *P. Weiss* and *de Freudenreich* [1440], was later elaborated by *Preisach* [1441] and recently reviewed in the textbook by *Mayergoyz* [1442].

(ii) Alternatively, continuous differential equations are used for the description of a general ferromagnet, into which certain measured functions or parameters are inserted. The textbook by *Jiles* [1443], one of the protagonists in this field, gives access to the extensive literature. A treatment of soft ferrites along these lines can be found in [1444].

(iii) The approach of *Bertotti* and his group, which is reviewed in [1445], focuses on dynamic interactions of Barkhausen jumps to describe the frequency dependence of losses in conducting materials, emphasizing stochastic arguments in the analysis of magnetization processes.

All theses approaches use certain arguments and concepts from domain theory. They either postulate identifiable particles or units, which switch under the common action of external, interaction and eddy current fields, or they focus on individual domain walls jumping over a series of obstacles. To what extent the mathematical theories can be reconciled with the complex reality of soft magnetic materials is an open question. Experimental observations like those of Fig. 5.82 suggest that after every major Barkhausen event a completely new domain arrangement is formed, with new instability points that have no relation with the "hysteresis units" present before the rearrangement.

One of the tasks of domain analysis in the future will be to improve the links between microscopic understanding and practically useful global descriptions. Perhaps a new interpretation of the descriptive models is needed, which would then allow the observer to connect domain features and behaviour with favourable global properties.

Colour Plates

1. Quantitative Kerr Microscopy (see page 39)

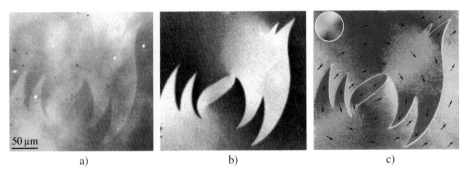

a)   b)   c)

Fig. 2.17 The same domain pattern on an amorphous iron-nickel alloy photographed before (a), after a digital contrast enhancement (b), and after a quantitative evaluation (see Sect. 2.3.8) obtained by combining two digital images (c). In the last case a colour code was used to indicate the magnetization direction at every point in addition to the arrows

2. Quantitative Electron Polarization Microscopy (see page 75)

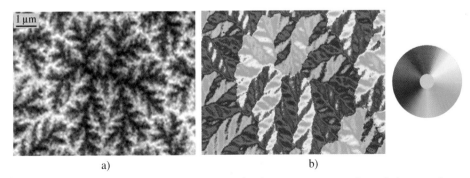

a)   b)

Fig. 2.39 Polar (a) and in-plane (b) magnetization components of a cobalt crystal cut parallel to the basal plane, made visibly with the SEMPA technique. A colour code is used in the representation of the in-plane components (Courtesy *J. Unguris*, NIST)

3. Corner Patterns in Thin Film Elements (see page 474)

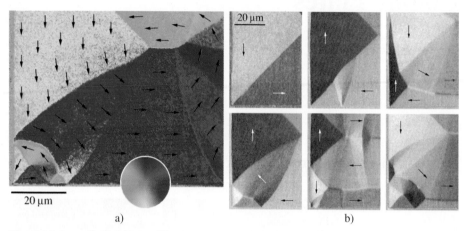

a)   b)

Fig. 5.86 Quantitative Kerr effect image of a tulip pattern (a) together with a number of alternative patterns observed in an equivalent corner (b), ranging from a simple 90°-wall to complex multiple corners. (Together with *J. McCord*, Erlangen)

4. A Weakly Coupled Sandwich System (see page 498)

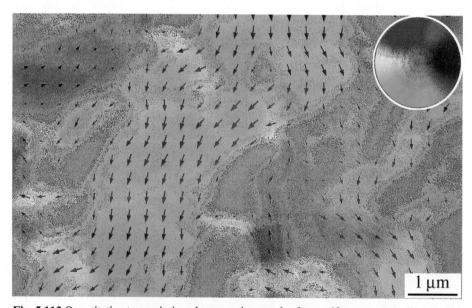

Fig. 5.113 Quantitative transmission electron micrograph of an antiferromagnetically coupled Co (10 nm) | Cr (1 nm) | Co (10 nm) sandwich under the influence of an external field, displaying regions of parallel (coloured) and antiparallel (grey) alignment. (Courtesy *J.P. Jakubovics*, Oxford)

References

1. Introduction

1   L. Pierce Williams: Ampère's electrodynamic molecular model. Contemp. Phys. **4**, 113–123 (1962)
2   P. Langevin: Magnétisme et théorie des électrons. Ann. Chim. Phys. (8) **5**, 70–127 (1905)
    (Magnetism and the theory of electrons)
3   P. Weiss: L'hypothèse du champ moléculaire et la propriété ferromagnétique. J. de Phys. Rad. **6**,
    661–690 (1907) *(The hypothesis of the molecular field and the property of ferromagnetism)*
4   J.D. van der Waals: *Over de continuïteit van den gas- en vloeistoftoestand.*
    Thesis, University of Leiden (1873) *(On the continuity of gas and fluid states)*
5   W. Heisenberg: Zur Theorie des Ferromagnetismus. Z. Phys. **49**, 619–636 (1928)
    (On the theory of ferromagnetism)
6   P. Weiss, G. Foex: *Le Magnétisme* (Armand Colin, Paris, 1926) *(Magnetism)*
7   E.C. Stoner: *Magnetism and Matter* (Methuen, London, 1934)
8   H. Barkhausen: Zwei mit Hilfe der neuen Verstärker entdeckte Erscheinungen.
    Phys. Z. **20**, 401–403 (1919) *(Two phenomena, discovered with the help of the new amplifiers)*
9   M.R. Forrer: Sur les grands phénomènes de discontinuité dans l'aimantation de nickel.
    J. de Phys. **7**, 109 (1926) *(On the great magnetization jump phenomena of nickel)*
10 F. Preisach: Untersuchungen über den Barkhauseneffekt. Ann. Physik **3**, 737–799 (1929)
    (Investigations on the Barkhausen effect)
11 K.J. Sixtus, L. Tonks: Propagation of large Barkhausen discontinuities.
    Phys. Rev. **37**, 930–958 (1931)
12 F. Bloch: Zur Theorie des Austauschproblems und der Remanenzerscheinung der Ferromagnetika.
    Z. Phys. **74**, 295–335 (1932) *(On the theory of the exchange problem and the remanence phenomenon
    of ferromagnets)*
13 N.S. Akulov: Zur Theorie der Magnetisierungskurve von Einkristallen.
    Z. Phys. **67**, 794–807 (1931) *(On the theory of the magnetization curve of single crystals)*
14 R. Becker: Zur Theorie der Magnetisierungskurve. Z. Phys. **62**, 253–269 (1930)
    (On the theory of the magnetization curve)
15 K. Honda, S. Kaya: On the magnetisation of single crystals of iron.
    Sci. Rep. Tohoku Imp. Univ. **15**, 721–754 (1926)
16 R. Becker, W. Döring: *Ferromagnetismus* (Springer, Berlin, 1939) *(Ferromagnetism)*
17 J. Frenkel, J. Dorfman: Spontaneous and induced magnetization in ferromagnetic bodies.
    Nature **126**, 274–275 (1930)
18 W. Heisenberg: Zur Theorie der Magnetostriktion und der Magnetisierungskurve.
    Z. Phys. **69**, 287–297 (1931) *(On the theory of magnetostriction and the magnetization curve)*
19 L. v. Hámos, P.A. Thiessen: Über die Sichtbarmachung von Bezirken verschiedenen ferromagnetischen
    Zustands fester Körper. Z. Phys. **71**, 442–444 (1931) *(Imaging domains of different ferromagnetic
    state in solid materials)*
20 F. Bitter: On inhomogeneities in the magnetization of ferromagnetic materials.
    Phys. Rev. **38**, 1903–1905 (1931)
21 F. Bitter: Experiments on the nature of ferromagnetism. Phys. Rev. **41**, 507–515 (1932)
22 L.D. Landau, E. Lifshitz: On the theory of the dispersion of magnetic permeability in ferromagnetic
    bodies. Phys. Z. Sowjetunion **8**, 153–169 (1935)
23 F. Zwicky: Permanent electric and magnetic moments of crystals. Phys. Rev. **38**, 1722–1781 (1932)
24 E. Lifshitz: On the magnetic structure of iron. J. Phys. USSR **8**, 337–346 (1944)
25 W.C. Elmore: The magnetic structure of cobalt. Phys. Rev. **53**, 757–764 (1938)

26   L. Néel: Les lois de l'aimantation et de la subdivision en domaines élémentaires d'un monocristal de
      fer (II). J. de Phys. (8) **5**, 265–276 (1944) *(The laws of the magnetization and the subdivision into
      elementary domains for an iron single crystal (II))*
27   L. Néel: Effet des cavités et des inclusions sur le champ coercitif.
      Cahiers de Phys. **25**, 21–44 (1944) *(The effect of cavities and inclusions on the coercive field)*
28   H.J. Williams: Direction of domain magnetization in powder patterns.
      Phys. Rev. **71**, 646–647 (1947)
29   C. Kittel: Theory of the structure of ferromagnetic domains in films and small particles.
      Phys. Rev. **70**, 965–971 (1946)
30   C. Kittel: Physical theory of ferromagnetic domains. Rev. Mod. Phys. **21**, 541–583 (1949)
31   W.C. Elmore: Ferromagnetic colloid for studying magnetic structures.
      Phys. Rev. **54**, 309–310 (1938)
32   H.J. Williams, R.M. Bozorth, W. Shockley: Magnetic domain patterns on single crystals of silicon
      iron. Phys. Rev. **75**, 155–178 (1949)
33   W.F. Brown, Jr.: Theory of the approach to magnetic saturation. Phys. Rev. **58**, 736–743 (1940)
34   W.F. Brown, Jr.: *Micromagnetics* (Wiley, New York, 1963)
      [reprinted: R.E. Krieger, Huntingdon N. Y., 1978]
35   W.F. Brown, Jr.: Micromagnetics, domains, and resonance.
      J. Appl. Phys. (Suppl.) **30**, 62S–69S (1959)
36   W.F. Brown, Jr.: Domains, micromagnetics, and beyond, reminiscences and assessments.
      J. Appl. Phys. **49**, 1937–1942 (1978)
37   W. Döring: Mikromagnetismus, in: *Handbuch der Physik*, Vol. 18/2, ed. by S. Flügge
      (Springer, Berlin, Heidelberg, New York, 1966) p. 341–437 *(Micromagnetics)*
38   W. Andrä: Distribution of magnetization in multilayer films. IEEE Trans. Magn. **2**, 560–562 (1966)
39   A. de Simone: Magnetization and magnetostriction curves from micromagnetics.
      J. Appl. Phys. **76**, 7018–7020 (1994)
40   E.D. Dahlberg, J.-G. Zhu: Micromagnetic microscopy and modeling.
      Physics Today **48**.4, 34–40 (1995)
41   A. Mitra, Z.J. Chen, F. Laabs, D.C. Jiles: Micromagnetic Barkhausen emissions in
      2.25 wt% Cr-1wt% Mo steel subjected to creep. Phil. Mag. A **75**, 847–859 (1997)

2. Domain Observation Techniques

Reviews

42   C. Kittel, J.K. Galt: Ferromagnetic domain theory. Solid State Phys. **3**, 437–565 (1956)
43   J.F. Dillon, Jr.: Domains and Domain Walls, in: *Magnetism*, Vol. III,
      ed. by G.T. Rado, H. Suhl (Academic Press, New York, 1963) p. 429–453
44   D.J. Craik, R.S. Tebble: *Ferromagnetism and Ferromagnetic Domains*,
      (North Holland, Amsterdam, 1965)
45   R. Carey, E.D. Isaac: *Magnetic Domains and Techniques for their Observation*,
      (Academic Press, New York, 1966)
46   D.J. Craik: Domain theory and observation. J. Appl. Phys. **38**, 931–938 (1967)
47   W. Andrä: Magnetische Bereiche. Brit. J. Appl. Phys. (J. Phys. D) **1**, 1–16 (1968)
      (Magnetic domains)
48   D.J. Craik: The observation of magnetic domains, in: *Methods of Experimental Physics*, Vol. 11,
      ed. by R.V. Coleman (Academic Press, New York, 1974) p. 675–743
49   A.P. Malozemoff, J.C. Slonczewski: *Magnetic Domain Walls in Bubble Materials*,
      (Academic Press, New York, 1979)
50   R.P. Ferrier: Imaging methods for the study of micromagnetic structure, in: *Noise in Digital Magnetic
      Recording*, ed. by T.C. Arnoldussen, L.L. Nunnelley (World Scientific, Singapore, 1992) p. 141–179
51   R.J. Celotta, J. Unguris, M.H. Kelley, D.T. Pierce: Techniques to measure magnetic domain structures,
      in: *Methods in Materials Research: A Current Protocols Publication* (Wiley, New York, 1999)

Bitter Patterns

52   R. Wolfe, J.C. North: Planar domains in ion-implanted magnetic bubble garnets revealed by Ferrofluid.
      Appl. Phys. Lett. **25**, 122–125 (1974)
53   G.C. Rauch, R.F. Krause, C.P. Izzo, K. Foster, W.O. Bartlett: Enhanced domain imaging
      techniques. J. Appl. Phys. **55**, 2145–2147 (1984)
54   Y.S. Lin, D.B. Dove, S. Schwarzl, C.-C. Shir: Charged wall behavior in 1-μm-bubble implanted struc-
      tures. IEEE Trans. Magn. **14**, 494–499 (1978)

55   W.C. Elmore: Interpretation of ferromagnetic colloid patterns on ferromagnetic crystal surfaces. Phys. Rev **58**, 640–642 (1940)

56   C. Kittel: Theory of the formation of powder patterns on ferromagnetic crystals. Phys. Rev. **76**, 1827 (1949)

57   K. Beckstette, H.H. Mende: Modellrechnungen und Experimente zur Kolloidansammlung bei der Bittertechnik. J. Magn. Magn. Mat. **4**, 326–336 (1977) *(Model calculations and experiments on the colloid accumulation in the Bitter technique)*

58   J.H.P. Watson: Magnetic filtration. J. Appl. Phys. **44**, 4209–4213 (1973)

59   M. Takayasu, R. Gerber, F.J. Friedlaender: The collection of strongly magnetic particles in HGMS. J. Magn. Magn. Mat. **40**, 204–214 (1983)

60   K. Fabian, A. Kirchner, W. Williams, F. Heider, T. Leibl, A. Hubert: Three-dimensional micromagnetic calculations for magnetite using FFT. Geophys. J. Intern. **124**, 89–104 (1996)

61   W.H. Bergmann: Über die Bildung von Bitterstreifen. Z. Angew. Phys. **8**, 559–561 (1956) *(On the formation of Bitter patterns)*

62   J.R. Garrood: Methods of improving the sensitivity of the Bitter technique. Proc. Phys. Soc. **79**, 1252–1262 (1962)

63   C.F. Hayes: Observation of association in a ferromagnetic colloid. J. Colloid Interface Sci. **52**, 239–243 (1975)

64   E.A. Peterson, D.A. Krueger: Reversible, field induced agglomeration in magnetic colloids. J. Colloid Interface Sci. **62**, 24–34 (1977)

65   W.-H. Liao, D.A. Krueger: Theory of large agglomerates in magnetic colloids. J. Colloid Interface Sci. **70**, 564–576 (1979)

66   J.-C. Bacri, D. Salin: Instability of ferrofluid magnetic drops under magnetic field. J. de Phys. Lett. **43**, L649–654 (1982)

67   J.-C. Bacri, D. Salin: Dynamics of the shape transition of a magnetic ferrofluid drop. J. de Phys. Lett. **44**, L415–420 (1983)

68   A.F. Pshenichnikov, I.Y. Shurubor: Stratification of magnetic fluids: Conditions of formation of drop aggregates and their magnetic properties. Bull. Acad. Sci. USSR, Phys. Ser. **51**.6, 40 (1987) *[Izv. Akad. Nauk Fiz. 51 (1987) 1081–1087]*

69   A.O. Tsebers: Thermodynamic stability of magnetofluids. Magneto-Hydrodynamics **18**, 137–142 (1982) *[Magn. Gidrodin. 18.2 (1982) 42–48]*

70   K. Sano, M. Doi: Theory of agglomeration of ferromagnetic particles in magnetic fluids. J. Phys. Soc. Japan **52**, 2810–2815 (1983)

71   A. Cebers: Phase separation of magnetic colloids and concentration domain patterns. J. Magn. Magn. Mat. **85**, 20–26 (1990)

72   G.A. Jones, D.G. Bedfield: Field induced agglomeration in thin films of magnetic fluids. J. Magn. Magn. Mat. **85**, 37–39 (1990)

73   M.I. Shliomis, A.F. Pshenichnikov, K.I. Morozov, I.Y. Shurubor: Magnetic properties of ferrocolloids. J. Magn. Magn. Mat. **85**, 40 (1990)

74   A.F. Pshenichnikov: Equilibrium magnetization of concentrated ferrocolloids. J. Magn. Magn. Mat. **145**, 319–326 (1995)

75   A.O. Ivanov: Phase separation in bidisperse ferrocolloids. J. Magn. Magn. Mat. **154**, 66–70 (1996)

76   H. Bogardus, D.A. Krueger, D. Thompson: Dynamic magnetization in ferrofluids. J. Appl. Phys. **49**, 3422–3429 (1978)

77   H. Pfützner: A new colloid technique enabling domain observations of SiFe sheets with coating at zero field. IEEE Trans. Magn. **17**, 1245–1247 (1981)

78   L.W. MacKeehan: Optical and magnetic properties of magnetic suspensions. Phys. Rev. **57**, 1177–1178 (1940)

79   A. Martinet: Biréfringence et dichroïsme linéaire des ferrofluides sous champ magnétique. Rheol. Acta **13**, 260–264 (1974) *(Birefringence and linear dichroism of ferrofluids in magnetic fields)*

80   G.A. Jones, I.B. Puchalska: The birefringent effects of magnetic colloid applied to the study of magnetic domain structures. Phys. Status Solidi A **51**, 549–558 (1979)

81   G.A. Jones, I.B. Puchalska: Interference colours of colloid patterns associated with magnetic structures. Phil. Mag. B **40**, 89–96 (1979)

82   G.A. Jones, E.T.M. Lacey, I.B. Puchalska: Bitter patterns in polarized light: A probe for microfields. J. Appl. Phys. **53**, 7870–7872 (1982)

83   U. Hartmann, H.H. Mende: The stray-field-induced birefringence of ferrofluids applied to the study of magnetic domains. J. Magn. Magn. Mat. **41**, 244–246 (1984)

84   J. Kranz, A. Hubert, R. Müller: Bitter-Streifen und Bloch-Wände. Z. Phys. **180**, 80–90 (1964) *(Bitter patterns and Bloch walls)*

85   L.F. Bates, P.F. Davis: 'Lozenge' and 'tadpole' domain structures on silicon-iron crystals. Proc. Phys. Soc. B **69**, 1109–1111 (1956)

86 R.W. DeBlois: Ferromagnetic domains in thin single-crystal nickel platelets.
    J. Appl. Phys. **36**, 1647–1658 (1965)

87 I. Khaiyer, T.H. O'Dell: Domain wall observation of Permalloy overlay bars by interference contrast
    technique. AIP Conf. Proc. **29**, 37–38 (1976)

88 G. Nomarski, A.R. Weil: Application a la métallographie des méthodes interferentielles à deux
    ondes polarisées. Rev. Metall. **L11**, 121–134 (1955) *(Application in metallography of interference
    methods based on two polarized waves)*

89 U. Hartmann, H.H. Mende: Observation of Bloch wall fine structures on iron whiskers by a high-
    resolution interference contrast technique. J. Phys. D (Appl. Phys.) **18**, 2285–2291 (1985)

90 B. Wysłocki: Über die Beobachtungsmöglichkeit der magnetischen Pulverfiguren im beliebigen
    Abstand von der Kristalloberfläche. Ann. Physik **13**, 109–114 (1964) *(On the possibilities of powder
    pattern observations at arbitrary distance from the crystal surface)*

91 B. Wysłocki, W. Ziętek: Selective powder-pattern observations of complex ferromagnetic domain
    structures. Acta Phys. Polon. **29**, 223–240 (1966)

92 D.J. Craik: A study of Bitter figures using the electron microscope.
    Proc. Phys. Soc. B **69**, 647–650 (1956)

93 W. Schwartze: Electronenmikroskopische Untersuchung magnetischer Pulvermuster.
    Ann. Physik **19**, 322–328 (1957) *(Electron microscopical investigation of magnetic powder patterns)*

94 D.J. Craik, P.M. Griffiths: New techniques for the study of Bitter figures.
    Brit. J. Appl. Phys. **9**, 279–282 (1958)

95 I.B. Puchalska, H. Jouve, R.H. Wade: Magnetic bubbles, walls, and fine structure in some
    ion-implanted garnets. J. Appl. Phys. **48**, 2069–2076 (1977)

96 K. Goto, T. Sakurai: A colloid-SEM method for the study of fine magnetic domain structures.
    Appl. Phys. Lett. **30**, 355–356 (1977)

97 O. Kitakami, K. Goto, T. Sakurai: A study of the magnetic domains of isolated fine particles of
    Ba ferrite. Jpn. J. Appl. Phys. **27**, 2274–2277 (1988)

98 J. Šimšová, R. Gemperle, J.C. Lodder: The use of colloid-SEM method for domain observation in
    CoCr films. J. Magn. Magn. Mat. **95**, 85–94 (1991)

99 P. Rice, J. Moreland: A new look at the Bitter method of magnetic imaging.
    Rev. Sci. Instrum. **62**, 844–845 (1991)

100 W. Andrä, C. Greiner, W. Schwab: Beobachtung der magnetischen Bereichsstruktur bei Wechsel-
    feldmagnetisierung zur Prüfung der Theorie der Ummagnetisierungsverluste.
    Monatsber. Deut. Akad. Wiss. Berlin **2**, 539–543 (1960) *(Observation of magnetic domain structures
    in alternating fields, examining the theory of core losses)*

101 A. Hubert: Zur Theorie der zweiphasigen Domänenstrukturen in Supraleitern und Ferromagneten.
    Phys. Status Solidi **24**, 669–682 (1967) *(On the theory of two-phase domain structures in supercon-
    ductors and ferromagnets)*

102 W. Andrä: Magnetische Pulvermuster bei höheren Temperaturen. Ann. Physik **3**, 334–339 (1959)
    (Magnetic powder patterns at elevated temperatures)

103 R.R. Birss, P.M. Wallis: Magnetic domains in gadolinium. Phys. Lett. **4**, 313 (1963)

104 L.F. Bates, S. Spivey: Bitter figure observations on gadolinium.
    Brit. J. Appl. Phys. **15**, 705–707 (1964)

105 R. Gemperle, P. Novotný, A. Menorsky: Bitter figure observations on U_3As_4 at low temperatures.
    Phys. Status Solidi A **52**, 587–596 (1979)

106 W. Andrä, E. Schwabe: Eine einfache Methode, magnetische Elementarbezirke mit trockenem Pulver
    sichtbar zu machen. Ann. Physik **17**, 55–56 (1955) *(A simple method to make visible elementary
    magnetic domains)*

107 R.I. Hutchinson, P.A. Lavin, J.R. Moon: A new technique for the study of ferromagnetic domain
    boundaries. J. Sci. Instrum. **42**, 885–886 (1965)

108 U. Essmann, H. Träuble: Comments on a new technique for the study of ferromagnetic domain
    boundaries. J. Sci. Instrum. **43**, 344 (1966)

109 K. Piotrowski, A. Szewczyk, R. Szymczak: A new method for the study of magnetic domains at low
    temperatures. J. Magn. Magn. Mat. **31**, 979–980 (1983)

110 A. Szewczyk, K. Piotrowski, R. Szymczak: A new method of domain structure investigation at
    temperatures below 35 K. J. Phys. D: Appl. Phys. **16**, 687–696 (1983)

111 O. Kitakami, T. Sakurai, Y. Miyashita, Y. Takeno, et al.: Fine metallic particles for magnetic domain
    observations. Jpn. J. Appl. Phys. Pt. 1 **35**, 1724–1728 (1996)

112 T. Sakurai, Y. Shimada: Application of the gas evaporation method to observation of magnetic domains.
    Jpn. J. Appl. Phys., Pt. 1 **31**, 1905–1908 (1992)

113 J. Cernák, P. Macko: The time dependence of particle aggregation in magnetic fluid layers.
    J. Magn. Magn. Mat. **123**, 107–116 (1993)

114 K. Mohri, S.-I. Takeuchi, T. Fujimoto: Domain and grain observation using a colloid technique for
    grain-oriented Si-Fe with coatings. IEEE Trans. Magn. **15**, 1346–1349 (1979)

115 H. Pfützner, C. Bengtsson, A. Leeb: Domain investigations on coated unpolished SiFe sheets. IEEE Trans. Magn. **21**, 2620–2625 (1985)

116 H. Pfützner: Nondestructive rapid investigation of domains and grain boundaries of grain oriented silicon steel. ISIJ International (Iron and Steel Institute of Japan) **29**, 828–835 (1989)

Magneto-Optical Methods

117 H.J. Williams, F.G. Foster, E.A. Wood: Observation of magnetic domains by the Kerr effect. Phys. Rev. **82**, 119–120 (1951)

118 C.A. Fowler, Jr., E.M. Fryer: Magnetic domains on silicon iron by the longitudinal Kerr effect. Phys. Rev. **86**, 426 (1952)

119 J. Kerr: On rotation of the plane of polarization by reflection from the pole of a magnet. Phil. Mag. (5.) **3**, 321–343 (1877)

120 C.A. Fowler, Jr., E.M. Fryer: Magnetic domains in thin films by the Faraday effect. Phys. Rev. **104**, 552–553 (1956)

121 J.F. Dillon, Jr.: Observation of domains in the ferrimagnetic garnets by transmitted light. J. Appl. Phys. **29**, 1286–1291 (1958)

122 R.C. Sherwood, J.P. Remeika, H.J. Williams: Domain behavior in some transparent magnetic oxides. J. Appl. Phys. **30**, 217–225 (1959)

123 M. Faraday: On the magnetization of light and the illumination of magnetic lines of force. Phil. Trans. Royal Soc. (London) **136**, 1–20 (1846)

124 A. Hubert: Beobachtung und Berechnung von magnetischen Bereichsstrukturen auf Siliziumeisen. Z. Angew. Phys. **18**, 474–479 (1965) *(Observation and computation of magnetic domain structures on silicon iron)*

125 E.R. Moog, S.D. Bader: SMOKE signals from ferromagnetic monolayers: p(1×1) Fe/Au(100). Superlattices and Microstructures **1**, 543–552 (1981)

126 S.D. Bader: SMOKE. J. Magn. Magn. Mat. **100**, 440–454 (1991)

127 W. Voigt: Doppelbrechung von im Magnetfeld befindlichem Natriumdampf in der Richtung normal zu den Kraftlinien. Nachr. Kgl. Ges. Wiss. Göttingen, Math.-Phys. Kl. **4**, 355–359 (1898) *(Birefringence of sodium vapour in a magnetic field along a direction perpendicular to the lines of force)*

128 A. Cotton, H. Mouton: Sur les propriétés magnéto-optiques des colloides et des liqueurs hétérogénes. Ann. Chim. Phys. (8.) **11**, 145–203 (1907) *(On magneto-optic properties of colloids and inhomogeneous liquids)*

129 W. Voigt: Magnetooptik, in: *Handbuch der Elektrizität und des Magnetismus*, Vol. IV, ed. by L. Graetz (A. Barth, Leipzig, 1920) p. 667–710 *(Magnetooptics)*

130 M.J. Freiser: A survey of magnetooptic effects. IEEE Trans. Magn. **4**, 152–161 (1968)

131 J.F. Dillon, Jr.: Magneto-optical properties of magnetic crystals, in: *Magnetic Properties of Materials*, ed. by J. Smit (McGraw-Hill, New York, 1971) p. 108–148

132 W. Wettling: Magneto-optics of ferrites. J. Magn. Magn. Mat. **3**, 147–160 (1976)

133 R. Atkinson, P. Lissberger: Sign conventions in magneto-optical calculations and measurements. Appl. Optics **31**, 6076–6081 (1992)

134 G.S. Krinchik, G.M. Nurmukhamedov: Magnetization of a ferromagnetic metal by the magnetic field of light waves. Sov. Phys. JETP **20**, 520–521 (1965) *[Zh. Eksp. Teor. Fiz. 48 (1965) 34–39]*

135 U. Buchenau: Über eine Messung der gyromagnetischen Konstante an Siliziumeisen. Z. Angew. Phys. **21**, 51–57 (1969) *(On a measurement of the gyromagnetic constant on silicon iron)*

136 T. Miyahara, M. Takahashi: The dependence of the longitudinal Kerr magneto-optic effect on saturation magnetization in Ni-Fe films. Jpn. J. Appl. Phys. **15**, 291–298 (1976)

137 J.F. Dillon, Jr.: Magneto-optical properties of magnetic garnets, in: *Physics of Magnetic Garnets*, ed. by A. Paoletti (North Holland, Amsterdam, 1978) p. 379–416

138 B.E. Argyle, E. Terrenzio: Magneto-optic observation of Bloch lines. J. Appl. Phys. **55**, 2569–2571 (1984)

139 R. Schäfer, A. Hubert: A new magnetooptic effect related to non-uniform magnetization on the surface of a ferromagnet. Phys. Status Solidi A **118**, 271–288 (1990)

140 G.S. Krinchik, V.A. Artemev: Magnetooptical properties of Ni, Co, and Fe in the ultraviolet, visible, and infrared parts of the spectrum. Sov. Phys. JETP **26**, 1080–1085 (1968) *[Zh. Eksp. Teor. Fiz. 53 (1990) 1901–1912]*

141 D.B. Dove: Photography of magnetic domains using the transverse Kerr effect. J. Appl. Phys. **34**, 2067–2070 (1963)

142 W. Rave, R. Schäfer, A. Hubert: Quantitative observation of magnetic domains with the magneto-optical Kerr effect. J. Magn. Magn. Mat. **65**, 7–14 (1987)

143 J. Kranz, A. Hubert: Die Möglichkeiten der Kerr-Technik zur Beobachtung magnetischer Bereiche. Z. Angew. Phys. **15**, 220–232 (1963) *(The potential of the Kerr technique for the observation of magnetic domains)*

144 M. Mansuripur: Figure of merit for magneto-optical media based on the dielectric tensor.
     Appl. Phys. Lett. **49**, 19–21 (1986)
145 W.A. Challener: Figures of merit for magneto-optic materials.
     J. Phys. Chem. Solids **56**, 1499–1507 (1995)
146 G. Traeger, L. Wenzel, A. Hubert: Computer experiments on the information depth and the figure of
     merit in magnetooptics. Phys. Status Solidi A **131**, 201–227 (1992)
147 J. Kranz, W. Drechsel: Über die Beobachtung von Weißschen Bereichen in polykristallinem Material
     durch die vergrößerte magnetooptischen Kerrdrehung. Z. Phys. **150**, 632–639 (1958) *(On the observation
     of Weiss domains in polycrystalline material by the enhanced magneto-optical Kerr rotation)*
148 P.H. Lissberger: Kerr magneto-optic effect in nickel-iron films. I. Experimental.
     J. Opt. Soc. Am. **51**, 948–956 (1961)
149 P.H. Lissberger: Kerr magneto-optic effect in nickel-iron films. II. Theoretical.
     J. Opt. Soc. Am. **51**, 957–966 (1961)
150 D.O. Smith: Longitudinal Kerr effect using a very thin Fe film.
     J. Appl. Phys. **36**, 1120–1121 (1965)
151 A. Hubert, G. Traeger: Magnetooptical sensitivity functions of thin film systems.
     J. Magn. Magn. Mat. **124**, 185–202 (1993)
152 L. Wenzel, V. Kamberský, A. Hubert: A systematic first-order theory of magnetooptic diffraction in
     magnetic multilayers. Phys. Status Solidi A **151**, 449–466 (1995)
153 D.E. Fowler: MOKE. Magneto-optic Kerr rotation, in: *Encyclopedia of Materials Characterization*,
     ed. by C.R. Brundle, *et al.* (Butterworth-Heinemann, Boston, MA, 1992) Section 12.3
154 M. Rührig: *Mikromagnetische Untersuchungen an gekoppelten weichmagnetischen Schichten.*
     Thesis, U. Erlangen-Nürnberg (1993) *(Micromagnetic investigations on coupled soft magnetic films)*
155 R. Schäfer: Magneto-optical domain studies in coupled magnetic multilayers.
     J. Magn. Magn. Mat. **148**, 226–231 (1995)
156 G. Pénissard, P. Meyer, J. Ferré, D. Renard: Magneto-optic depth sensitivity to local magnetization
     in a simple ultrathin film structure. J. Magn. Magn. Mat. **146**, 55–65 (1995)
157 J. Ferré, P. Meyer, M. Nyvlt, S. Visnovsky, D. Renard: Magnetooptic depth sensitivity in a simple
     ultrathin film structure. J. Magn. Magn. Mat. **165**, 92–95 (1997)
158 E. Feldtkeller, K.U. Stein: Verbesserte Kerr-Technik zur Beobachtung magnetischer Domänen.
     Z. Angew. Phys. **23**, 100–102 (1967) *(Improved Kerr technique for magnetic domain observation)*
159 J. McCord, H. Brendel, A. Hubert, S.S.P. Parkin: Hysteresis and domains in magnetic multilayers.
     J. Magn. Magn. Mat. **148**, 244–246 (1995)
160 A. Green, M. Prutton: Magneto-optic detection of ferromagnetic domains using vertical illumination.
     J. Sci. Instrum. **39**, 244–245 (1962)
161 A. Hubert: Der Einfluß der Magnetostriktion auf die magnetische Bereichsstruktur einachsiger
     Kristalle, insbesondere des Kobalts. Phys. Status Solidi **22**, 709–727 (1967) *(The influence of magneto-
     striction on the magnetic domain structure of uniaxial crystal, in particular of cobalt)*
162 F. Schmidt, A. Hubert: Domain observations on CoCr layers with a digitally enhanced Kerr microscope.
     J. Magn. Magn. Mat. **61**, 307–320 (1986)
163 H. Boersch, M. Lambeck: Mikroskopische Beobachtung gerader und gekrümmter Magnetisierungs-
     strukturen mit dem Faraday-Effekt. Z. Phys. **159**, 248–252 (1960) *(Microscopical imaging of straight
     and curved magnetization structures by the Faraday effect)*
164 M.H. Kryder, A. Deutsch: A high speed magneto-optic camera system.
     SPIE High speed optical techniques **94**, 49–57 (1976)
165 B.E. Argyle: A magneto-optic microscope system for magnetic domain studies.
     Symposium on Magnetic Materials, Processes and Devices (1990).
     The Electrochemical Society, Electrodeposition Division Vol. **90–8**, p. 85–110
166 N. Bardou, B. Bartenlien, C. Chappert, R. Mégy, *et al.*: Magnetization reversal in patterned Co(0001)
     ultrathin films with perpendicular magnetic anisotropy.
     J. Appl. Phys. **79**, 5848–5850 (1996)
167 D. Treves: Limitations of the magneto-optic Kerr technique in the study of microscopic magnetic
     domain structures. J. Appl. Phys **32**, 358–364 (1961)
168 C.A. Fowler, Jr., E.M. Fryer: Reduction of photographic noise. J. Opt. Soc. Am. **44**, 256 (1954)
169 A. Honda, K. Shirae: Domain pattern measurements using CCD.
     IEEE Trans. Magn. **17**, 3096–3098 (1981)
170 K. Shirae, K. Sugiyama: A CCD image sensor and a microcomputer make magnetic domain observation
     clear and convenient. J. Appl. Phys. **53**, 8380–8382 (1982)
171 F. Schmidt, W. Rave, A. Hubert: Enhancement of magneto-optical domain observation by digital image
     processing. IEEE Trans. Magn. **21**, 1596–1598 (1985)
172 A.B. Smith, W. Goller: New domain configuration in thin-film recording heads.
     IEEE Trans. Magn. **26**, 1331–1333 (1990)

173 J. McCord, A. Hubert: Normalized differential Kerr microscopy - An advanced method for magnetic imaging. Phys. Status Solidi A **171**, 555–562 (1999)

174 E. Kubajewska, A. Maziewski, A. Stankiewicz: Digital image processing for investigation of domain structure in garnet films. Thin Solid Films **175**, 299–303 (1989)

175 P.L. Trouilloud, B. Petek, B.E. Argyle: Methods for wide-field Kerr imaging of small magnetic devices. IEEE Trans. Magn. **30**, 4494–4496 (1994)

176 W. Rave, A. Hubert: Refinement of the quantitative magnetooptic domain observation technique. IEEE Trans. Magn. **26**, 2813–2815 (1990)

177 W. Rave, P. Reichel, H. Brendel, M. Leicht, J. McCord, A. Hubert: Progress in quantitative magnetic domain observation. IEEE Trans. Magn. **29**, 2551–2553 (1993)

178 J. McCord, A. Hubert, G. Schröpfer, U. Loreit: Domain observation on magnetoresistive sensor elements. IEEE Trans. Magn. **32**, 4806–4808 (1996)

179 S. Defoug, R. Kaczmarek, W. Rave: Measurements of local magnetization by Kerr effect on Si-Fe nonoriented sheets. J. Appl. Phys. **79**, 6036–6038 (1996)

180 A. Hubert, R. Schäfer, W. Rave: The analysis of magnetic microstructure. Proc. 5th Symp. Magn. Magn. Mat. (Taipei, 1989). (World Scientific, Singapore 1990) p. 25–42

181 J.P. Jakubovics: Interaction of Bloch-wall pairs in thin ferromagnetic films. J. Appl. Phys. **69**, 4029–4039 (1991)

182 B.E. Argyle, B. Petek, D.A. Herman, Jr.: Optical imaging of magnetic domains in motion. J. Appl. Phys. **61**, 4303–4306 (1987)

183 K. Závěta, Z. Kalva, R. Schäfer: Permeability and domain structure of a nanocrystalline alloy. J. Magn. Magn. Mat. **148**, 390–396 (1995)

184 G.L. Houze, Jr.: Domain-wall motion in grain-oriented silicon steel in cyclic magnetic fields. J. Appl. Phys. **38**, 1089–1096 (1967)

185 W. Drechsel: Eine Anordnung zur photographischen Registrierung von Weißschen Bereichen bei Belichtungszeiten der Größenordnung 1 msec. Z. Phys. **164**, 324–329 (1961) *(A setup for the photographic registration of Weiss domains with exposure times of the order of 1 millisecond)*

186 R.L. Conger, G.H. Moore: Direct observation of high-speed magnetization reversal in films. J. Appl. Phys. **34**, 1213–1214 (1963)

187 B. Passon: Über die Beobachtung ferromagnetischer Bereiche bei Magnetisierung in Wechselfeldern bis zu 20 kHz. Z. Angew. Phys. **25**, 56–61 (1968) *(On the observation of ferromagnetic domains during magnetization in alternating fields up to 20 kHz)*

188 A.P. Malozemoff: Nanosecond camera for garnet bubble domain dynamics. IBM Techn. Discl. Bull. **15**, 2756–2757 (1973)

189 L. Gál, G.J. Zimmer, F.B. Humphrey: Transient magnetic bubble domain configurations during radial wall motion. Phys. Status Solidi A **30**, 561–569 (1975)

190 G.P. Vella-Coleiro: Overshoot in the translational motion of magnetic bubble domains. J. Appl. Phys. **47**, 3287–3290 (1976)

191 M.H. Kryder, P.V. Koeppe, F.H. Liu: Kerr effect imaging of dynamic processes. IEEE Trans. Magn. **26**, 2995–3000 (1990)

192 F.H. Liu, M.D. Schultz, M.H. Kryder: High frequency dynamic imaging of domains in thin film heads. IEEE Trans. Magn. **26**, 1340–1342 (1990)

193 B. Petek, P.L. Trouilloud, B.E. Argyle: Time-resolved domain dynamics in thin-film heads. IEEE Trans. Magn. **26**, 1328–1330 (1990)

194 M.V. Chetkin, S.N. Gadetsky, A.P. Kuzmenko, A.I. Akhutkina: Investigation of supersonic dynamics of domain walls in orthoferrites. Sov. Phys. JETP **59**, 825–830 (1984) *[Zh. Eksp. Teor. Fiz. 86 (1990) 1411–1418]*

195 C.D. Wright, W. Clegg, A. Boudjemline, N.A.E. Heyes: Scanning laser microscopy of magneto-optic storage media. Jpn. J. Appl. Phys. Pt.1 **33**, 2058–2065 (1994)

196 P. Kasiraj, R.M. Shelby, J.S. Best, D.E. Horne: Magnetic domain imaging with a scanning Kerr effect microscope. IEEE Trans. Magn. **22**, 837–839 (1986)

197 A. Thiaville, L. Arnaud, F. Boileau, G. Sauron, J. Miltat: First direct optical evidence of lines in bubble garnets. IEEE Trans. Magn. **24**, 1722–1724 (1988)

198 J. Theile, J. Engemann: Direct optical observation of Bloch lines and their motion in uniaxial garnet films using a polarizing light microscope. Appl. Phys. Lett. **53**, 713–715 (1988)

199 J. Theile, J. Engemann: Determination of twist and charge of Bloch lines by direct optical observation. IEEE Trans. Magn. **24**, 1781–1783 (1988)

200 J. Miltat, A. Thiaville, P.L. Trouilloud: Néel lines structures and energies in uniaxial ferromagnets with quality factor $Q > 1$. J. Magn. Magn. Mat. **82**, 297–308 (1989)

201 S. Egelkamp, L. Reimer: Imaging of magnetic domains by Kerr effect using a scanning optical microscope. Meas. Sci. Technol. **1**, 79–83 (1990)

202 N.A. Heyes, C.D. Wright, W. Clegg: Observations of magneto-optic phase contrast using a scanning laser microscope. J. Appl. Phys. **69**, 5322–5324 (1991)

203 C.D. Wright, N.A.E. Heyes, W. Clegg, E.W. Hill: Magneto-optic scanning laser microscopy. Microscopy and Analysis 34 (March), 23–25 (1995)

204 T.J. Silva, S. Schultz: Non-reciprocal differential detection method for scanning Kerr-effect microscopy. J. Appl. Phys. **81**, 5015–5017 (1997)

205 G.L. Ping, C.W. See, M.G. Somekh, M.B. Suddendorf, J.H. Vincent, P.K. Footner: A fast-scanning optical microscope for imaging magnetic domain-structures. Scanning **18**, 8–12 (1996)

206 G.S. Krinchik, O.M. Benidze: Magneto-optic investigation of magnetic structures under micron resolution conditions. Sov. Phys. JETP **40**, 1081–1087 (1974) *[Zh. Eksp. Teor. Fiz. 67 (1974) 2180–2194]*

207 V.E. Zubov, G.S. Krinchik, A.D. Kudakov: Structure of domain walls in the surface layer of iron single crystals. Sov. Phys. JETP **67**, 2527–2531 (1988) *[Zh. Eksp. Teor. Fiz. 94 (1988) 243–250]*

208 V. Hoffmann, R. Schäfer, E. Appel, A. Hubert, H. Soffel: First domain observations with the magneto-optical Kerr effect on Ti-ferrites in rocks and their synthetic equivalents. J. Magn. Magn. Mat. **71**, 90–94 (1987)

209 J. Basterfield: Domain structure and the influence of growth defects in single crystals of yttrium iron garnet. J. Appl. Phys. **39**, 5521–5526 (1968)

210 V.K. Vlasko-Vlasov, L.M. Dedukh, V.I. Nikitenko: Domain structure of yttrium iron garnet single crystals. Sov. Phys. JETP **44**, 1208–1214 (1976) *[Zh. Eksp. Teor. Fiz. 71 (1976) 2291–2304]*

211 R. Szymczak: Observation of internal domain structure of barium ferrite in infrared. Acta Phys. Polon. A **43**, 571–578 (1973)

212 R. Gemperle, I. Tomáš: Microstructure of thick 180° domain walls. J. Magn. Magn. Mat. **73**, 339–334 (1988)

213 P.R. Alers: Structure of the intermediate state in superconducting lead. Phys. Rev. **105**, 104–108 (1957)

214 W. DeSorbo, W.A. Healy: The intermediate state of some superconductors. Cryogenics **4**, 257–326 (1964)

215 H. Kirchner: High-resolution magneto-optical observation of magnetic structures in superconductors. Phys. Lett. A **26**, 651–652 (1968)

216 H. Kirchner: Ein hochauflösendes magnetooptisches Verfahren zur Untersuchung der Kinematik magnetischer Strukturen in Supraleitern. Phys. Status Solidi A **4**, 531–553 (1971) *(A high resolution magneto-optical method for the investigation of the kinematics of magnetic structures in superconductors)*

217 G. Dietrich, A. Hubert, F. Schmidt: Abbildung und Messung der Streufelder über Magnetköpfen mit Hilfe von Granatschichten, in: *Berichte der Arbeitsgemeinschaft Magnetismus*, Vol. 1 (Verlag Stahleisen, Düsseldorf, 1983) p. 176–180 *(Imaging and measurement of stray fields over magnetic heads using garnet films)*

218 L.A. Dorosinskii, M.V. Indenbom, V.I. Nikitenko, Y.A. Ossipyan, A.A. Polyanskii, V.K. Vlasko-Vlasov: Study of high-T_c superconductors by means of magnetic bubble films, in: *Progress in High Temperature Physics*, Vol. 25 (World Scientific, Singapore, 1990) p. 166–170

219 R. Fichtner, A. Hubert, H. Grimm: The detection of magnetic inclusions in the surface of a non-magnetic ceramic using the critical state of a garnet film. 13. Intern. Conf. Magnetic Films and Surfaces, Extended Abstracts (Glasgow, 1991) p. 423–424

220 L.A. Dorosinskii, M.V. Indenbom, V.I. Nikitenko, Y.A. Ossipyan, A.A. Polyanskii, V.K. Vlasko-Vlasov: Studies of HTSC crystal magnetization features using indicator magnetooptic films with in-plane anisotropy. Physica C **203**, 149–156 (1992)

221 M.R. Koblischka, R.J. Wijngaarden: Magneto-optical investigations of superconductors. Supercond. Sci. Techn. **8**, 199–213 (1995)

222 W. Andrä, K.-H. Geier, R. Hergt, J. Taubert: Magnetooptik für die Materialcharakterisierung. Materialprüfung **36**, 294–297 (1994) *(Magnetooptics for material characterization)*

223 T. Hirano, T. Namikawa, Y. Yamazaki: Bi-YIG magneto-optical coated films for visual applications. IEEE Trans. Magn. **31**, 3280–3282 (1995)

224 R. Schäfer, M. Rührig, A. Hubert: Exploration of a new magnetization-gradient-related magnetooptical effect. IEEE Trans. Magn. **26**, 1355–1357 (1990)

225 A. Thiaville, A. Hubert, R. Schäfer: An isotropic description of the new-found gradient-related magnetooptical effect. J. Appl. Phys. **69**, 4551–4555 (1991)

226 R. Schäfer, M. Rührig, A. Hubert: Magnetooptical domain wall observation at perpendicular incidence. Phys. Status Solidi A **145**, 167–176 (1994)

227 V. Kamberský: The Schäfer-Hubert magnetooptical effect and classical gyrotropy in light-wave equations. J. Magn. Magn. Mat. **104**, 311–312 (1991)

228 L. Wenzel, A. Hubert: Simulating magnetooptic imaging with the tools of Fourier optics. IEEE Trans Magn **32**, 4084–4086 (1996)

229 R. Schäfer: Magnetooptical microscopy for the analysis of magnetic microstructures. 4th Symposium on Magnetic Materials, Processes, and Devices (Chicago, 1995). The Electrochemical Society Vol. **95**–18, p. 300–318

230 R.-P. Pan, H.D. Wei, Y.R. Shen: Optical second-harmonic generation from magnetized surfaces. Phys. Rev. B **39**, 1229–1234 (1989)

231 H.A. Wierenga, M.W.J. Prins, D.L. Abraham, T. Rasing: Magnetization induced optical second harmonic generation: A probe for interface magnetism. Phys. Rev. B **50**, 1282–1285 (1994)

232 W. Hübner: Magneto-optics goes nonlinear. Phys. World **8**.10, 21–22 (1995)

233 T. Rasing: Nonlinear magneto-optics. J. Magn. Magn. Mat. **175**, 35–50 (1997)

234 B. Koopmans, A.M. Janner, H.A. Wierenga, T. Rasing, G.A. Sawatzky, F. van der Woude: Separation of interface and bulk contributions in second-harmonic generation from magnetic and non-magnetic multilayers. Appl. Phys. A **60**, 103–111 (1995)

235 I.L. Lyubchanskii, A.V. Petukhov, T. Rasing: Domain and domain wall contributions to optical second harmonic generation in thin magnetic films. J. Appl. Phys. **81**, 5668–5670 (1997)

236 J. Reif, J.C. Zink, C.-M. Schneider, J. Kirschner: Effects of surface magnetism on optical second harmonic generation. Phys. Rev. Lett. **67**, 2878–2881 (1991)

237 T. Rasing, M. Groot Koerkamp, B. Koopmans, H. van den Berg: Giant nonlinear magneto-optical Kerr effects from Fe interfaces. J. Appl. Phys. **79**, 6181–6185 (1996)

238 V.V. Pavlov, R.V. Pisarev, A. Kirilyuk, T. Rasing: Observation of a transversal nonlinear magneto-optical effect in thin magnetic garnet films. Phys. Rev. Lett. **78**, 2004–2007 (1997)

239 V. Kirilyuk, A. Kirilyuk, T. Rasing: A combined nonlinear and linear magneto-optical microscopy. Appl. Phys. Lett. **70**, 2306–2308 (1997)

Transmission Electron Microscopy

240 M.E. Hale, H.W. Fuller, H. Rubinstein: Magnetic domain observation by electron microscopy. J. Appl. Phys. **30**, 789–790 (1959)

241 H. Boersch, H. Raith: Elektronenmikroskopische Abbildung Weißscher Bezirke in dünnen ferromagnetischen Schichten. Naturwissenschaften **46**, 574 (1959) *(Electron microscopical imaging of Weiss domains in thin ferromagnetic films)*

242 H. Boersch, H. Raith, D. Wohlleben: Elektronenmikroskopische Untersuchung Weißscher Bezirke in dünnen Eisenschichten. Z. Phys. **159**, 388–396 (1960) *(Electron microscopical investigation of Weiss domains in thin iron films)*

243 H.W. Fuller, M.E. Hale: Determination of magnetization distribution in thin films using electron microscopy. J. Appl. Phys. **31**, 238–248 (1960)

244 E. Fuchs: Die Ummagnetisierung dünner Nickeleisenschichten in der schweren Richtung. Z. Angew. Phys. **13**, 157–164 (1961) *(The magnetization of thin nickel iron films along a hard axis)*

245 I.B. Puchalska, R.J. Spain: Observation des mécanismes de l'hystérésis magnétique dans les couches minces. C. R. Acad. Sci. Paris **254**, 72–74 (1962) *(Observation of the magnetic hysteresis mechanisms in thin films)*

246 M.S. Cohen: Magnetic measurements with Lorentz microscopy. IEEE Trans. Magn. **1**, 156–167 (1965)

247 P.J. Grundy, R.S. Tebble: Lorentz electron microscopy. Adv. Phys. **17**, 153–243 (1968)

248 R.H. Wade: Transmission electron microscope observations of ferromagnetic domain structures. J. de Phys. (Coll.) **29**, C2–95–109 (1968)

249 V.I. Petrov, G.V. Spivak, O.P. Pavlyuchenko: Electron microscopy of the magnetic structure of thin films. Sov. Phys. Uspekhi **15**, 66/88–94 (1972) *[Usp. Fiz. Nauk 106 (1972) 229–278]*

250 L. Reimer: *Transmission Electron Microscopy. Physics of Image Formation and Microanalysis*, in: Springer Series in Optical Sciences, Vol. 36 (Springer, Berlin, Heidelberg, New York, 1993)

251 J.N. Chapman: The investigation of magnetic domain structures in thin foils by electron microscopy. J. Phys. D: Appl. Phys. **17**, 623–647 (1984)

252 J.P. Jakubovics: Lorentz microscopy, in: *Handbook of Microscopy*, Vol. 1, ed. by S. Amelinckx, et al. (VCH, Weinheim, New York, 1997) p. 505–514

253 M. Mankos, J.M. Cowley, M.R. Scheinfein: Quantitative micromagnetics at high spatial resolution using far-out-of-focus STEM electron holography. Phys. Status Solidi A **154**, 469–504 (1996)

254 Y. Takahashi, Y. Yajima: Magnetization contrast enhancement of recorded magnetic storage media in scanning Lorentz electron microscopy. Jpn. J. Appl. Phys., Pt. 1 **32**, 3308–3311 (1993)

255 D. Wohlleben: Diffraction effects in Lorentz microscopy. J. Appl. Phys. **38**, 3341–3352 (1967)

256 M.S. Cohen: Wave-optical aspects in Lorentz microscopy. J. Appl. Phys. **38**, 4966–4976 (1967)

257 M. Mansuripur: Computation of electron diffraction patterns in Lorentz electron microscopy of thin magnetic films. J. Appl. Phys. **69**, 2455–2464 (1991)

258 D.C. Hothersall: The investigation of domain walls in thin sections of iron by the electron interference method. Phil. Mag. **20**, 89–112 (1969)

259 D.C. Hothersall: Electronic images of two-dimensional domain walls. Phys. Status Solidi B **51**, 529–536 (1972)

260   J.N. Chapman: The application of iterative techniques to the investigation of strong phase objects in the electron microscope. Phil. Mag. **13**, 85–101 (1965)

261   S. Tsukahara, H. Kawakatsu: Asymmetric 180° domain walls in single crystal iron films. J. Phys. Soc. Japan **32**, 1493–1499 (1972)

262   H.M. Thieringer, M. Wilkens: Abbildung ferromagnetischer Bereiche in einkristallinen dünnen Schichten durch Elektronen-Beugungskontrast. Phys. Status Solidi **7**, K5–8 (1964) *(Imaging ferromagnetic domains in single crystal thin films due to electron diffraction contrast)*

263   J.P. Jakubovics: The effect of magnetic domain structure on Bragg reflection in transmission electron microscopy. Phil. Mag. **10**, 277–290 (1965)

264   J.P. Jakubovics: Application of the dynamical theory of electron diffraction to ferromagnetic crystals. Phil. Mag. **13**, 85–101 (1965)

265   S.J. Hefferman, J.N. Chapman, S. McVitie: In-situ magnetising experiments on small regular particles fabricated by electron beam lithography. J. Magn. Magn. Mat. **83**, 223–224 (1990)

266   G.R. Morrison, J.N. Chapman: STEM imaging with a quadrant detector. Electron Microscopy & Analysis (1982). Inst. of Physics Conf. Ser. Vol. **61**, p. 329–332

267   J.N. Chapman, G.R. Morrison: Quantitative determination of magnetization distributions in domains and domain walls by scanning transmission electron microscopy. J. Magn. Magn. Mat. **35**, 254–260 (1983)

268   N.H. Dekkers, H. de Lang: Differential phase contrast in a STEM. Optik **41**, 452–456 (1974)

269   J.N. Chapman, P.E. Batson, E.M. Waddell, R.P. Ferrier: The direct determination of magnetic domain wall profiles by differential phase contrast electron microscopy. Ultramicroscopy **3**, 203–214 (1978)

270   E.M. Waddell, J.N. Chapman: Linear imaging of strong phase objects using asymmetrical detectors in STEM. Optik **54**, 83–96 (1979)

271   G.R. Morrison, H. Gong, J.N. Chapman, V. Hrnciar: The measurement of narrow domain-wall widths in SmCo₅ using differential phase contrast electron microscopy. J. Appl. Phys. **64**, 1338–1342 (1988)

272   J.N. Chapman, I.R. McFadyen, S. McVitie: Modified differential phase contrast Lorentz microscopy for improved imaging of magnetic structures. IEEE Trans. Magn. **26**, 1506–1511 (1990)

273   J. Zweck, J.N. Chapman, S. McVitie, H. Hoffmann: Reconstruction of induction distributions in thin films from DPC images. J. Magn. Magn. Mat. **104**, 315–316 (1992)

274   I.R. McFadyen, J.N. Chapman: Electron microscopy of magnetic materials. EMSA Bull. **22.2**, 64–76 (1992)

275   J.N. Chapman, R. Ploessl, D.M. Donnet: Differential phase contrast microscopy of magnetic materials. Ultramicroscopy **47**, 331–338 (1992)

276   Y. Takahashi, Y. Yajima: Nonmagnetic contrast in scanning Lorentz electron microscopy of polycrystalline magnetic films. J. Appl. Phys. **76**, 7677–7681 (1994)

277   I.A. Beardsley: Reconstruction of the magnetization in a thin film by combination of Lorentz microscopy and external field measurements. IEEE Trans. Magn. **25**, 671–677 (1989)

278   A.C. Daykin, A.K. Petford-Long: Quantitative mapping of the magnetic induction distribution using Foucault images formed in a transmission electron microscope. Ultramicroscopy **58**, 365–380 (1995)

279   A. Daykin, J.D. Kim, J.P. Jakubovics: A study of cross-tie domain walls in cobalt using small-aperture Foucault imaging. J. Magn. Magn. Mat. **153**, 293–301 (1996)

280   A.C. Daykin, J.P. Jakubovics: Magnetization imaging at high spatial resolution using transmission electron microscopy. J. Appl. Phys. **80**, 3408–3411 (1996)

281   A. Tonomura, T. Matsuda, J. Endo, T. Arii, K. Mihama: Direct observation of fine structure of magnetic domain walls by electron holography. Phys. Rev. Lett. **44**, 1430–1433 (1980)

282   A. Tonomura: Observation of magnetic domain structure in thin ferromagnetic films by electron holography. J. Magn. Magn. Mat. **31**, 963–969 (1983)

283   A. Tonomura: *Electron Holography*, in: Springer Series in Optical Sciences, Vol. 70 (Springer, Berlin, Heidelberg, New York, 1993)

284   H. Lichte: Electron holography methods, in: *Handbook of Microscopy*, Vol. 1, ed. by S. Amelinckx, *et al.* (VCH, Weinheim, New York, 1997) p. 515–536

285   W.J. de Ruijter, J.K. Weiss: Detection limits in quantitative off-axis holography. Ultramicroscopy **50**, 269–283 (1993)

286   T. Hirayama, J. Chen, T. Tanji, A. Tonomura: Dynamic observation of magnetic domains by on-line real-time electron holography. Ultramicroscopy **54**, 9–14 (1994)

287   J. Chen, T. Hirayama, G. Lai, T. Tanji, K. Ishizuka, A. Tonomura: Real-time electron holography using a liquid-crystal panel, in: *Electron Holography*, ed. by A. Tonomura, *et al.* (Elsevier, Amsterdam, 1995) p. 81–102

288   G. Matteucci, G.F. Missiroli, J.W. Chen, G. Pozzi: Mapping of microelectric and magnetic fields with double-exposure electron holography. Appl. Phys. Lett. **52**, 176–178 (1988)

289   M. Mankos, M.R. Scheinfein, J.M. Cowley: Absolute magnetometry at nanometer transverse spatial resolution: Holography of thin cobalt films in a scanning transmission electron microscope. J. Appl. Phys. **75**, 7418–7424 (1994)

290 J.M. Cowley, M. Mankos, M.R. Scheinfein: Greatly defocused, point-projection, off-axis electron holography. Ultramicroscopy 63, 133–147 (1996)

291 M.R. McCartney, P. Kruit, A.H. Buist, M.R. Scheinfein: Differential phase contrast in TEM. Ultramicroscopy 65, 179–186 (1996)

292 J.N. Chapman, A.B. Johnston, L.J. Heyderman, S. McVitie, W.A.P. Nicholson, B. Bormans: Coherent magnetic imaging by TEM. IEEE Trans. Magn. 30, 4479–4484 (1994)

293 A.B. Johnston, J.N. Chapman: The development of coherent Foucault imaging to investigate magnetic microstructure. J. Microsc. Oxford 179, 119–128 (1995)

294 A.B. Johnston, J.N. Chapman, B. Khamsehpour, C.D.W. Wilkinson: In situ studies of the properties of micrometre-sized magnetic elements by coherent Foucault imaging. J. Phys. D: Appl. Phys. 29, 1419–1427 (1996)

295 G. Lai, T. Hirayama, A. Fukuhara, K. Ishizuka, T. Tanji, A. Tonomura: Three-dimensional reconstruction of magnetic vector fields using electron-holographic interferometry. J. Appl. Phys. 75, 4593–4598 (1994)

296 D.G. Streblechenko, M.R. Scheinfein, M. Mankos, K. Babcock: Quantitative magnetometry using electron holography: field profiles near magnetic force microscope tips. IEEE Trans. Magn. 32, 4124–4129 (1996)

297 J.N. Chapman, R.P. Ferrier, L.J. Heyderman, S. McVitie, W.A.P. Nicholson, B. Bormans: Micromagnetics, microstructure and microscopy. EMAG 93 (Liverpool, 1993). Inst. of Phys. Conf. Ser. Vol. 138, p. 1–8

298 T. Suzuki, M. Wilkens: Lorentz-electron-microscopy of ferromagnetic specimens at high voltages. Phys. Status Solidi A 3, 43–52 (1970)

299 R.A. Taylor, J.P. Jakubovics: Observations of magnetic domain structures in thick foils of Ni-Fe-Co-Ti alloys. J. Magn. Magn. Mat. 31, 1001–1004 (1983)

300 J. Dooley, M. De Graef: Energy filtered Lorentz microscopy. Ultramicroscopy 67, 113–131 (1997)

301 R.A. Taylor: Top entry magnetizing stage for the AEI EM7 electron microscope, in: Electron Microscopy 1980, Vol. 4, ed. by P. Bredero, J. van Landuyt (7th European Congress on Electr. Microscopy Foundation, Leiden, 1980) p. 38–41

302 C.G. Harrison, K.D. Leaver: The analysis of two-dimensional domain wall structures by Lorentz microscopy. Phys. Status Solidi A 15, 415–429 (1973)

303 O. Bostanjoglo, T. Rosin: Stroboscopic Lorentz TEM at 100 kV up to 100 MHz. Electron Microscopy 1, 88–89 (1980)

304 I.R. McFadyen: Implementation of differential phase contrast Lorentz microscopy on a conventional transmission electron microscope. J. Appl. Phys. 64, 6011–6013 (1988)

Electron Reflection and Scattering Methods

305 G.V. Spivak, I.N. Prilezhaeva, V.K. Azovtsev: Magnitnii kontrast v electronnom zerkale i nablyudeniye domenov ferromagnetika. Doklady Acad. Nauk SSSR, Fiz. 105, 965–967 (1955) (Magnetic contrast in the electron mirror and the observation of ferromagnetic domains)

306 L. Mayer: Electron mirror microscopy of magnetic domains. J. Appl. Phys. 28, 975–983 (1957)

307 D.J. Fathers, J.P. Jakubovics: Methods of observing magnetic domains by scanning electron microscopes. Physica B 86, 1343–1344 (1977)

308 G.A. Jones: Magnetic contrast in the scanning electron microscope: An appraisal of techniques and their applications. J. Magn. Magn. Mat. 8, 263–285 (1978)

309 D.E. Newbury, D.C. Joy, P. Echlin, C.E. Fiori, J.I. Goldstein: Magnetic contrast in the SEM, in: Advanced Scanning Electron Microscopy and X-Ray Microanalysis (Plenum, New York, London, 1986) p. 147–179

310 K. Tsuno: Magnetic domain observation by means of Lorentz electron microscopy with scanning technique. Rev. Solid State Science 2, 623–656 (1988)

311 J.R. Banbury, W.C. Nixon: The direct observation of domain structure and magnetic fields in the scanning electron microscope. J. Sci. Instrum. 44, 889–892 (1967)

312 D.C. Joy, J.P. Jakubovics: Scanning electron microscope study of the magnetic domain structure of cobalt single crystals. Brit. J. Appl. Phys. (J. Phys. D) 2, 1367–1672 (1969)

313 G.A. Wardly: Magnetic contrast in the scanning electron microscope. J. Appl. Phys. 42, 376–386 (1971)

314 O.C. Wells: Fundamental theorem for type-1 magnetic contrast in the scanning electron microscope SEM. J. Microsc. 131, RP 5–6 (1983)

315 O.C. Wells: Some theoretical aspects of type-1 magnetic contrast in the scanning electron microscope. J. Microsc. 139, 187–196 (1985)

316 J.B. Elsbrock, W. Schroeder, E. Kubalek: Evaluation of 3–dimensional micromagnetic stray fields by means of electron-beam tomography. IEEE Trans. Magn. 21, 1593–1595 (1985)

317 R.P. Ferrier, Y. Liu, J.L. Martin, T.C. Arnoldussen: Electron beam tomography of magnetic recording head fields. J. Magn. Magn. Mat. 149, 387–397 (1995)

318   J. Yin, S. Nomizu, J.-I. Matusda: Reconstruction of three-dimensional magnetic stray fields for magnetic heads using reflection electron beam tomography. J. Phys. D: Appl. Phys. **30**, 1094–1102 (1997)

319   M. Mundschau, J. Romanowicz, J.Y. Wang, D.L. Sun, H.C. Chen: Imaging of ferromagnetic domains using photoelectrons: photoelectron emission microscopy of neodymium-iron-boron ($Nd_2Fe_{14}B$). J. Vac. Sci. Technol. B **14**, 3126–3130 (1996)

320   J. Philibert, R. Tixier: Effets de contraste cristallin en microscopie électronique à balayage. Micron **1**, 174–186 (1969) *(Crystal-related contrast effects in a scanning electron microscope)*

321   D.J. Fathers, J.P. Jakubovics, D.C. Joy: Magnetic domain contrast from cubic materials in the scanning electron microscope. Phil. Mag. **27**, 765–768 (1973)

322   D.J. Fathers, J.P. Jakubovics, D.C. Joy, D.E. Newbury, H. Yakowitz: A new method of observing magnetic domains by scanning electron microscopy. I Theory of image contrast. Phys. Status Solidi A **20**, 535–544 (1973)

323   D.J. Fathers, J.P. Jakubovics, D.C. Joy, D.E. Newbury, H. Yakowitz: A new method of observing magnetic domains by scanning electron microscopy II Experimental confirmation. Phys. Status Solidi A **22**, 609–619 (1974)

324   T. Yamamoto, H. Nishizawa, K. Tsuno: Magnetic domain contrast in backscattered electron images obtained with the scanning electron microscope. Phil. Mag. **34**, 311–325 (1976)

325   O.C. Wells, R.J. Savoy: Magnetic domains in thin-film recording heads as observed in the SEM by a lock-in technique. IEEE Trans. Magn. **17**, 1253–1261 (1981)

326   R.P. Ferrier, S. McVitie, W.A.P. Nicholson: Magnetisation distribution in thin film recording heads by type II contrast. IEEE Trans. Magn. **26**, 1337–1339 (1990)

327   K. Kobayashi: Observation of magnetic domains in thin-film heads by electron microscopy. IEEE Transl. J. Magn. Japan **8**, 595–601 (1993)

328   R.P. Ferrier, S. McVitie: A new method for the observation of type II magnetic contrast. XIIth International Congress for Electron Microscopy (1990). San Francisco Press, p. 764–765

329   K. Tsuno, T. Yamamoto: Observed depths of magnetic domains in high voltage scanning electron microscopy. Phys. Status Solidi A **35**, 437–449 (1976)

330   B. Fukuda, T. Irie, H. Shimanaka, T. Yamamoto: Observation through surface coatings of domain structure in 3% SiFe sheet by a high voltage scanning electron microscope. IEEE Trans. Magn. **13**, 1499–1504 (1977)

331   R.P. Ferrier, S. McVitie: The depth sensitivity of type II magnetic contrast. 49th Annual Meeting Electron Microscopy Society of America (1991). San Francisco Press, p. 766–767

332   T. Nozawa, T. Yamamoto, Y. Matsuo, Y. Ohya: Effects of scratching on losses in 3 percent SiFe single crystals with orientation near (110)[001]. IEEE Trans. Magn. **15**, 972–981 (1979)

333   W. Jillek, A. Hubert: The influence of mechanical stresses on losses and domains of oriented transformer steel. J. Magn. Magn. Mat. **19**, 365–368 (1980)

334   T. Yamamoto, K. Tsuno: Unusual magnetic contrast of domain images obtained in the reflective mode of scanning electron microscopy. Phil. Mag. **34**, 479–484 (1976)

335   D.C. Joy, H.J. Leamy, S.D. Ferris, H. Yakowitz, D.E. Newbury: Domain wall image contrasts in the SEM. Appl. Phys. Lett. **28**, 466–467 (1976)

336   D.J. Fathers, J.P. Jakubovics: Magnetic domain wall contrast in the scanning electron microscope. Phys. Status Solidi A **36**, K13–16 (1976)

337   J.D. Livingston, W.G. Morris: Magnetic domains in amorphous metal ribbons. J. Appl. Phys. **57**, 3555–3559 (1985)

338   O.C. Wells, R.J. Savoy: Type-2 magnetic contrast with normal electron incidence in the scanning electron microscope SEM, in: *Microbeam Analysis*, ed. by D.A. Newbury (San Francisco Press, San Francisco, CA, 1979) p. 17–21

339   L. Pogány, Z. Vértessy, S. Sándor, G. Konczos: Magnetic domain contrast type II detection by pn-junction: a highly effective new method. Proc. XIth Int. Cong. Electron Microscopy (Kyoto, 1986) p. 1737–1738

340   L. Pogany, K. Ramstöck, A. Hubert: Quantitative magnetic contrast – Part I: Experiment. Scanning **14**, 263–268 (1992)

341   G. Chrobok, M. Hofmann: Electron spin polarization of secondary electrons ejected from magnetized europium oxide. Phys. Lett. **57A**, 257–258 (1976)

342   J. Unguris, D.T. Pierce, A. Galejs, R.J. Celotta: Spin and energy analyzed secondary electron emission from a ferromagnet. Phys. Rev. Lett. **49**, 72–76 (1982)

343   H. Hopster, R. Raue, E. Kisker, G. Güntherodt, M. Campagna: Evidence for spin-dependent electron-hole-pair excitations in spin-polarized secondary-electron emission from Ni (110). Phys. Rev. Lett. **50**, 70–73 (1983)

344   D.T. Pierce: Experimental studies of surface magnetism with polarized electrons. Surface Sci. **189**, 710–723 (1987)

345   K. Koike, K. Hayakawa: Observation of magnetic domains with spin-polarized secondary electrons. Appl. Phys. Lett. **45**, 585–586 (1984)

346 K. Koike, K. Hayakawa: Scanning electron microscope observation of magnetic domains using spin-polarized secondary electrons. Jpn. J. Appl. Phys. **23**, L187–188 (1984)

347 K. Koike, H. Matsuyama, K. Hayakawa: Spin-polarized scanning electron microscopy for micromagnetic structure observation. Scanning Microscopy Suppl. **1**, 241–253 (1987)

348 J. Unguris, G. Hembree, R.J. Celotta, D.T. Pierce: Investigations of magnetic microstructure using scanning electron microscopy with spin polarization analysis. J. Magn. Magn. Mat. **54**, 1629–1630 (1986)

349 G.G. Hembree, J. Unguris, R.J. Celotta, D.T. Pierce: Scanning electron microscopy with polarization analysis: High resolution images of magnetic microstructure. Scanning Microscopy Suppl. **1**, 229–240 (1987)

350 M.R. Scheinfein, J. Unguris, M.H. Kelley, D.T. Pierce, R.J. Celotta: Scanning electronic microscopy with polarization analysis (SEMPA). Rev. Sci. Instrum. **61**, 2501–2526 (1990)

351 H.P. Oepen, J. Kirschner: Imaging of magnetic microstructures at surfaces – The scanning electron-microscope with spin polarization analysis. Scanning Microscopy (Chicago) **5**, 1–16 (1991)

352 M.R. Scheinfein, J. Unguris, R.J. Celotta, D.T. Pierce: The influence of the surface on magnetic domain wall microstructure. Phys. Rev. Lett. **65**, 668–670 (1989)

353 T. Kohashi, H. Matsuyama, K. Koike: A spin rotator for detecting all three magnetization vector components by spin-polarized scanning electron microscopy. Rev. Sci. Instrum. **66**, 5537–5543 (1995)

354 M. Haag, R. Allenspach: A novel approach to domain in natural Fe/Ti oxides by spin-polarized scanning electron microscopy. Geophys. Res. Lett. **20**, 1943–1946 (1993)

355 T. VanZandt, R. Browning, M. Landolt: Iron overlayer polarization enhancement technique for spin-polarized electron microscopy. J. Appl. Phys. **69**, 1564–1568 (1991)

356 E. Bauer, W. Telieps: Emission and low energy reflection electron microcospy, in: *Surface and interface characterization by electron optical methods*, ed. by A. Howie, U. Valdré (Plenum, New York, London, 1988) p. 195–233

357 E. Bauer: Low energy electron microscopy. Rep. Progr. Phys. **57**, 895 (1994)

358 E. Bauer: The resolution of the low energy electron reflection microscope. Ultramicroscopy **17**, 51–56 (1985)

359 H. Poppa, E. Bauer, H. Pinkvos: SPLEEM of magnetic surfaces and layered structures. MRS Bull. **20**.10, 38–40 (1995)

360 E. Bauer, T. Duden, H. Pinkvos, H. Poppa, K. Wurm: LEEM studies of the microstructure and magnetic domain structure of ultrathin films. J. Magn. Magn. Mat. **156**, 1–6 (1996)

361 E. Bauer: Spin-polarized low-energy electron microscopy, in: *Handbook of Microscopy*, Vol. 2, ed. by S. Amelinckx, *et al.* (VCH, Weinheim, New York, 1997) p. 751–759

362 T. Duden, E. Bauer: Magnetization wrinkle in thin ferromagnetic films. Phys. Rev. Lett. **77**, 2308–2311 (1996)

363 D.C. Joy, E.M. Schulson, J.P. Jakubovics, C.G. van Essen: Electron channelling pattern from ferromagnetic crystals in the scanning electron microscope. Phil. Mag. **20**, 843–847 (1969)

364 A. Gervais, J. Philibert, A. Rivière, R. Tixier: Contraste de domaines magnétiques dans le fer-silicium observés en microscopie à balayage. Revue Phys. Appl. **9**, 433–441 (1974) *(Contrast of magnetic domains in silicon iron observed in scanning microscopy)*

365 L.J. Balk, D.G. Davies, N. Kultscher: Investigations of Si-Fe transformer sheets by scanning electron acoustic microscopy SEAM. IEEE Trans. Magn. **20**, 1466–1468 (1984)

366 E. Kay, H.C. Siegmann: High resolution dynamic magnetic domain readout. IBM Techn. Discl. Bull. **27**, 317–320 (1984)

367 G.V. Spivak, T.N. Dombrovskaia, N.N. Sedov: The observation of ferromagnetic domain structure by means of photoelectrons. Sov. Phys. Doklady **2**, 120–123 (1957) *[Dokl. Akad. Nauk USSR 113 (1957) 78–81]*

Microscanning Techniques

368 G. Binnig, H. Rohrer: Scanning tunneling microscopy. Surface Science **152**, 17–26 (1985)

369 U. Hartmann: High-resolution magnetic imaging based on scanning probe techniques. J. Magn. Magn. Mat. **157**, 545–549 (1996)

370 G. Binnig, C. Quate, C. Gerber: Atomic force microscope. Phys. Rev. Lett. **56**, 930–933 (1986)

371 Y. Martin, H.K. Wickramasinghe: Magnetic imaging by "force microscopy" with 1000 Å resolution. Appl. Phys. Lett. **50**, 1455–1457 (1987)

372 J.J. Sáenz, N. Garcia, P. Grütter, E. Meyer, *et al.*: Observation of magnetic forces by the atomic force microscope. J. Appl. Phys. **62**, 4293–4295 (1987)

373 P. Grütter, E. Meyer, H. Heinzelmann, L. Rosenthaler, H.-R. Hidber, H.-J. Güntherodt: Application of AFM to magnetic materials. J. Vac. Sci. Technol. A **6**, 279–282 (1988)

374 Y. Martin, D. Rugar, H.K. Wickramasinghe: High resolution magnetic imaging of domains in TbFe by force microscopy. Appl. Phys. Lett. **52**, 244–246 (1988)

375 P. Grütter, T. Jung, H. Heinzelmann, A. Wadas, E. Meyer, H.-R. Hidber, H.-J. Güntherodt:
    10-nm resolution by magnetic force microscopy on FeNdB. J. Appl. Phys. **67**, 1437–1441 (1990)
376 U. Hartmann, T. Göddenhenrich, C. Heiden: Magnetic force microscopy: Current status and future
    trends. J. Magn. Magn. Mat. **101**, 263–270 (1991)
377 P. Grütter, H.J. Mamin, D. Rugar: Magnetic Force Microscopy (MFM),
    in: *Scanning Tunneling Microscopy*, Vol. II, ed. by H.-J. Güntherodt, R. Wiesendanger
    (Springer, Berlin, Heidelberg, New York, 1992) p. 151–207
378 P. Grütter: An introduction to magnetic force microscopy. MSA Bull. **24**, 416–425 (1994)
379 A. Wadas: Magnetic force microscopy, in: *Handbook of Microscopy*, Vol. 2,
    ed. by S. Amelinckx, *et al.* (VCH, Weinheim, New York, 1997) p. 845–853
380 C. Schönenberger, S.F. Alvarado: Understanding magnetic force microscopy.
    Z. Phys. B **80**, 373–380 (1990)
381 C. Schönenberger, S.F. Alvarado, S.E. Lambert, I.L. Sanders: Separation of magnetic and topographic
    effects in force microscopy. J. Appl. Phys. **67**, 7278–7280 (1990)
382 O. Wolter, T. Bayer, J. Greschner: Micromachined silicon sensors for scanning force microscopy.
    KJ. Vac. Sci. Technol. **9**, 1353–1367 (1991)
383 P. Grütter, D. Rugar, H.J. Mamin, G. Castillo, *et al.*: Magnetic force microscopy with batch-fabricated
    force sensors. J. Appl. Phys. **69**, 5883–5885 (1991)
384 P.B. Fischer, M.S. Wei, S.Y. Chou: Ultrahigh resolution magnetic force microscope tip fabricated
    using electron beam lithography. J. Vac. Sci. Technol. B **11**, 2570–2573 (1993)
385 M. Rührig, S. Porthun, J.C. Lodder: Magnetic force microscopy using electron-beam fabricated tips.
    Rev. Sci. Instrum. **65**, 3224–3228 (1994)
386 M. Rührig, S. Porthun, J.C. Lodder, S. McVitie, L.J. Heyderman, A.B. Johnston, J.N. Chapman:
    Electron beam fabrication and characterization of high-resolution magnetic force microscopy tips.
    J. Appl. Phys. **79**, 2913–2919 (1996)
387 M. Schneider, S. Müller-Pfeiffer, W. Zinn: Magnetic force microscopy of domain wall fine structures
    in iron films. J. Appl. Phys. **79**, 8578–8583 (1996)
388 M.S. Valera, A.N. Farley: A high performance magnetic force microscope.
    Meas. Sci. Technol. **7**, 30–35 (1996)
389 T.G. Pokhil, R.B. Proksch: A combined magneto-optic magnetic force microscope study of Co/Pd
    multilayer films. J. Appl. Phys. **81**, 3846–3848 (1997)
390 W. Rave, E. Zueco, R. Schäfer, A. Hubert: Observations on high-anisotropy single crystals using a
    combined Kerr/magnetic force microscope. J. Magn. Magn. Mat. **177**, 1474–1475 (1998)
391 C.D. Wright, E.W. Hill: Reciprocity in magnetic force microscopy.
    Appl. Phys. Lett. **67**, 433–435 (1995)
392 A. Hubert, W. Rave, S.L. Tomlinson: Imaging magnetic charges with magnetic force microscopy.
    Phys. Status Solidi B **204**, 817–828 (1997)
393 W. Rave, L. Belliard, M. Labrune, A. Thiaville, J. Miltat: Magnetic force microscopy analysis of
    soft thin film elements. IEEE Trans. Magn. **30**, 4473–4478 (1994)
394 A. Thiaville, L. Belliard, J. Miltat: Micromagnetism and the interpretation of magnetic force microscopy
    images. Scanning Microsc., accepted, Proc. Conf. "Scanning Microscopy International" Chicago 1997
395 S.L. Tomlinson, A.N. Farley, S.R. Hoon, M.S. Valera: Interactions between soft magnetic samples
    and MFM tips. J. Magn. Magn. Mat. **157**, 557–558 (1996)
396 S.L. Tomlinson, E.W. Hill: Modelling the perturbative effect of MFM tips on soft magnetic thin
    films. J. Magn. Magn. Mat. **161**, 385–396 (1996)
397 J.-G. Zhu, Y. Zheng, X. Lin: Micromagnetics of small size patterned exchange biased Permalloy
    film elements. J. Appl. Phys. **81**, 4336–4341 (1997)
398 L. Folks, R. Street, R.C. Woodward, K. Babcock: Magnetic force microscopy images of high-coercivity
    permanent magnets. J. Magn. Magn. Mat. **159**, 109–118 (1996)
399 S. Foss, R. Proksch, E.D. Dahlberg, B. Moskowitz, B. Walsh: Localized micromagnetic perturbation
    of domain walls in magnetite using a magnetic force microscope.
    Appl. Phys. Lett. **69**, 3426–3428 (1996)
400 S. Foss, E.D. Dahlberg, R. Proksch, B.M. Moskowitz: Measurement of the effects of the localized
    field of a magnetic force microscope tip on a 180° domain wall. J. Appl. Phys. **81**, 5032–5034 (1997)
401 L. Belliard, A. Thiaville, S. Lemerle, A. Lagrange, J. Ferré, J. Miltat: Investigation of the domain
    contrast in magnetic force microscopy. J. Appl. Phys. **81**, 3849–3851 (1997)
402 H.J. Mamin, D. Rugar, J.E. Stern, R.E. Fontana, P. Kasiraj: Magnetic force microscopy of thin
    Permalloy films. Appl. Phys. Lett. **55**, 318–320 (1989)
403 J.R. Barnes, S.J. O'Shea, M.E. Welland: Magnetic force microscopy study of local pinning effects.
    J. Appl. Phys. **76**, 418–423 (1994)
404 M.A. Al-Khafaji, W.M. Rainforth, M.R.J. Gibbs, J.E.L. Bishop, H.A. Davies: The effect of tip type
    and scan height on magnetic domain images obtained by MFM.
    IEEE Trans. Magn. **32**, 4138–4140 (1996)

405 T. Ohkubo, J. Kishigami, K. Yanagisawa, R. Kaneko: Submicron magnetizing and its detection based on the point magnetic recording concept. IEEE Trans. Magn. **27**, 5286–5288 (1991)

406 G.A. Gibson, S. Schultz: Magnetic force microscope study of the micromagnetics of submicrometer magnetic particles. J. Appl. Phys. **73**, 4516–4521 (1993)

407 J.R. Barnes, S.J. O'Shea, M.E. Welland, J.-Y. Kim, J.E. Evetts, R.E. Somekh: Magnetic force microscopy of Co-Pd multilayers with perpendicular anisotropy. J. Appl. Phys. **76**, 2974–2980 (1994)

408 T.G. Pokhil: Domain wall displacements in amorphous films and multilayers studied with a magnetic force microscope. J. Appl. Phys. **81**, 5035–5037 (1997)

409 P.F. Hopkins, J. Moreland, S.S. Malhotra, S.H. Liou: Superparamagnetic magnetic force microscopy tips. J. Appl. Phys. **79**, 6448–6450 (1996)

410 K.L. Babcock, L. Folks, R. Street, R.C. Woodward, D.L. Bradbury: Evolution of magnetic microstructure in high-coercivity permanent magnets imaged with magnetic force microscopy. J. Appl. Phys. **81**, 4438–4440 (1997)

411 R. Proksch, G.D. Skidmore, E.D. Dahlberg, S. Foss, *et al.*: Quantitative magnetic field measurements with the magnetic force microscope. Appl. Phys. Lett. **69**, 2599–2601 (1996)

412 D.W. Pohl, W. Denk, M. Lanz: Optical stethoscopy: Image recording with resolution λ/20. Appl. Phys. Lett. **44**, 651–653 (1984)

413 E. Betzig, J.K. Trautman: Near field optics: Microscopy, spectroscopy, and surface modification beyond the diffraction limit. Science **257**, 189–195 (1992)

414 D. Courjon, M. Spajer: Near field optical microscopy, in: *Handbook of Microscopy*, Vol. 1, ed. by S. Amelinckx, *et al.* (VCH, Weinheim, New York, 1997) p. 83–96

415 E. Betzig, J.K. Trautman, J.S. Weiner, T.D. Harris, R. Wolfe: Polarization contrast in near-field scanning optical microscopy. Appl. Optics **31**, 4563–4568 (1992–1993)

416 C. Durkan, I.V. Shvets, J.C. Lodder: Observation of magnetic domains using a reflection-mode scanning near-field optical microscope. Appl. Phys. Lett. **70**, 1323–1325 (1997)

417 G. Eggers, A. Rosenberger, N. Held, A. Münnemann, G. Güntherodt, P. Fumagalli: Scanning near-field magneto-optic microscopy using illuminated fiber tips. Ultramicroscopy **71**, 249–256 (1998)

418 V.I. Safarov, V.A. Kosobukin, C. Hermann, G. Lampel, J. Peretti: Near-field magneto-optical microscopy. Microsc. Microanal. Microstruct. **5**, 381–388 (1994)

419 M.W.J. Prins, R.H.M. Groeneveld, D.L. Abraham, R. Schad, H. van Kempen, H.W. van Kesteren: Scanning tunneling microscope for magneto-optical imaging. J. Vac. Sci. Technol. B **14**, 1206–1209 (1996)

420 T.J. Silva, S. Schultz: A scanning near-field optical microscope for the imaging of magnetic domains in reflection. Rev. Sci. Instrum. **67**, 715–725 (1996)

421 M. Johnson, J. Clarke: Spin-polarized scanning tunneling microscope: Concept, design, and preliminary results from a prototype operated in air. J. Appl. Phys. **67**, 6141–6152 (1990)

422 R. Wiesendanger, H.J. Güntherodt, G. Güntherodt, R.J. Gambino, R. Ruf: Scanning tunneling microscopy with spin-polarized electrons. Z. Phys. B **80**, 5–6 (1990)

423 R. Wiesendanger, I.V. Shvets, D. Bürgler, G. Tarrach, H.-J. Güntherodt, J.M.D. Coey: Magnetic imaging at the atomic level. Z. Phys. B **86**, 1–2 (1992)

424 I.V. Shvets, R. Wiesendanger, D. Bürgler, G. Tarrach, H.-J. Güntherodt, J.M.D. Coey: Progress towards spin-polarized scanning tunneling microscopy. J. Appl. Phys. **71**, 5489–5499 (1992)

425 R. Jansen, M.C.M.M. van der Wielen, M.W.J. Prins, D.L. Abraham: Progress toward spin sensitive scanning tunneling microscopy using optical orientation in GaAs. J. Vac. Sci. Technol. **12**, 2133–2135 (1994)

426 K. Mukasa, K. Sueoka, H. Hasegawa, Y. Tazuke, K. Hayakawa: Spin-polarized STM and its family. Mater. Sci. Techn. B **31**, 69–76 (1995)

427 R. Allenspach, A. Bischof: Spin-polarized secondary electrons from a scanning tunneling microscope in field emission mode. Appl. Phys. Lett. **54**, 587–589 (1989)

428 B. Kostyshyn, J.E. Brophy, I. Oi, D.D. Roshon, Jr.: External fields from domain walls of cobalt films. J. Appl. Phys. **31**, 772–775 (1960)

429 H. Koehler, B. Kostyshyn, T.C. Ku: A note on Hall probe resolution. IBM Journal **Oct.**, 326–327 (1961)

430 J. Kaczér: A new method for investigating the domain structure of ferromagnetics. Czech. J. Phys. **5**, 239–244 (1955)

431 J. Kaczér, R. Gemperle: Vibrating Permalloy probe for mapping magnetic fields. Czech. J. Phys. **6**, 173–184 (1956)

432 W. Hagedorn, H.H. Mende: A method for inductive measurement of magnetic flux density with high geometrical resolution. J. Phys. E: Sci. Instrum. **9**, 44–46 (1976)

433 H. Pfützner, G. Schwarz, J. Fidler: Computer-controlled domain detector. Jpn. J. Appl. Phys. **22**, 361–364 (1983)

434 A. Oral, S.J. Bending, M. Henini: Scanning Hall probe microscopy of superconductors and magnetic materials. J. Vac. Sci. Technol. B **14**, 1202–1205 (1996)

435 A. Thiaville, L. Belliard, D. Majer, E. Zeldov, J. Miltat: Measurement of the stray field emanating from magnetic force microscope tips by Hall effect microsensors. J. Appl. Phys. **82**, 3182–3191 (1997)

436 R. O'Barr, M. Lederman, S. Schultz: A scanning microscope using a magnetoresistive head as the sensing element. J. Appl. Phys. **79**, 6067–6069 (1996)

437 S.Y. Yamamoto, S. Schultz, Y. Zhang, H.N. Bertram: Scanning magnetoresistance microscopy (SMRM) as a diagnostic for high density recording. IEEE Trans. Magn. **33**, 891–896 (1997)

438 S.Y. Yamamoto, S. Schultz: Scanning magnetoresistance microscopy (SMRM): Imaging with a MR head. J. Appl. Phys. **81**, 4696–4698 (1997)

X-Ray, Neutron and Other Methods

439 A.R. Lang: The projection topograph, a new method in X-ray diffraction topography. Acta Crystallogr. **12**, 249–250 (1959)

440 M. Polcarová, A.R. Lang: X-ray topographic studies of magnetic domain configurations and movements. Appl. Phys. Lett. **1**, 13–15 (1962)

441 M. Polcarová: Applications of X-ray diffraction topography to the study of magnetic domains. IEEE Trans. Magn. **5**, 536–544 (1969)

442 M. Kuriyama, G.M. McManus: X-ray interference fringes and domain arrangements in Fe+3wt.% Si single crystals. Phys. Status Solidi **25**, 667–677 (1968)

443 M. Polcarová, A.R. Lang: On the fine structure of X-ray topography images of 90° ferromagnetic domain walls in Fe-Si. Phys. Status Solidi A **4**, 491–499 (1971)

444 J.E.A. Miltat: Fir-tree patterns. Elastic distortions and application to X-ray topography. Phil. Mag. **33**, 225–254 (1976)

445 M. Polcarová, A.R. Lang: Observation par la topographie aux rayons X des domaines ferromagnétiques dans Fe-3%Si. Bull. Soc. Fr. Minéral. Cristallogr. **91**, 645–652 (1968) *(Observation of ferromagnetic domains in Fe-3%Si by X-ray topography)*

446 M. Schlenker, P. Brissonneau, J.P. Perrier: Sur l'origine du contraste des images de parois de domaines ferromagnétiques par topographie aux rayons X dans le fer silicium. Bull. Soc. Fr. Minéral. Cristallogr. **91**, 653–665 (1968) *(On the origin of X-ray topographical image contrast of ferromagnetic domain walls in silicon iron)*

447 J.E.A. Miltat, M. Kléman: Magnetostrictive displacements around a domain wall junction: Elastic field calculation and application to X-ray topography. Phil. Mag. **28**, 1015–1033 (1973)

448 M. Polcarová, J. Kaczér: X-ray diffraction contrast on ferromagnetic domain walls in Fe-Si single crystals. Phys. Status Solidi **21**, 635–642 (1967)

449 F. Kroupa, I. Vagera: Surface magnetostrictive deformation at 180° Bloch wall. Czech. J. Phys. B **19**, 1204 (1969)

450 H.H. Mende, H. Galinski: Röntgentopographische Untersuchungen von 180°-Blochwänden in Eisenwhiskern. Appl. Phys. **5**, 211–215 (1974) *(X-ray topographic investigations of 180° Bloch walls in iron whiskers)*

451 J. Miltat, M. Kléman: Interaction of moving {110} 90°-walls in Fe-Si single crystals with lattice imperfections. J. Appl. Phys. **50**, 7695–7697 (1979)

452 J. Miltat: Internal strains of magnetostrictive origin: Their nature in the static case and behaviour in the dynamic regime. IEEE Trans. Magn. **17**, 3090–3095 (1981)

453 M. Schlenker, J. Linares-Galvez, J. Baruchel: A spin-related contrast effect. Visibility of 180° ferromagnetic domain walls in unpolarized neutron diffraction topography. Phil. Mag. B **37**, 1–11 (1978)

454 M. Schlenker, J. Baruchel: Neutron techniques for the observation of ferro- and antiferromagnetic domains. J. Appl. Phys. **49**, 1996–2001 (1978)

455 J. Baruchel, M. Schlenker, K. Kurosawa, S. Saito: Antiferromagnetic S-domains in NiO. Phil. Mag. B **43**, 853–868 (1981)

456 G. Schütz, W. Wagner, W. Wilhelm, P. Kienle, R. Zeller, R. Frahm, G. Materlik: Absorption of circularly polarized X rays in iron. Phys. Rev. Lett. **58**, 737–740 (1987)

457 C.T. Chen, F. Sette, Y. Ma, S. Modesti: Soft-X-ray magnetic circular dichroism at the $L_{2,3}$ edges of nickel. Phys. Rev. B **42**, 7262–7265 (1990)

458 C.M. Schneider: Perspectives in element-specific magnetic domain imaging. J. Magn. Magn. Mat. **156**, 94–98 (1996)

459 C.M. Schneider: Soft X-ray photoemission electron microscopy as an element-specific probe of magnetic microstructures. J. Magn. Magn. Mat. **175**, 160–176 (1997)

460 Y. Wu, S.S.P. Parkin, J. Stöhr, M.G. Samant, *et al.*: Direct observation of oscillatory interlayer exchange coupling in sputtered wedges using circularly polarized x rays. Appl. Phys. Lett. **63**, 263–265 (1993)

461 L. Baumgarten, C.-M. Schneider, H. Petersen, F. Schäfers, J. Kirschner: Magnetic X-ray dichroism in core-level photoemission from ferromagnets. Phys. Rev. Lett. **65**, 492–495 (1990)

462 B.P. Tonner: Spin-sensitive magnetic microscopy with circularly polarized X-rays. J. de Phys. IV C9 **4**, 407–414 (1994)

463 B.P. Tonner, D. Dunham: Sub-micron spatial resolution of a micro-XAFS electrostatic microscope with bending magnet radiation: performance assessments and prospects for aberration correction. Nucl. Instrum. Meth. Phys. Res. A **347**, 436–440 (1994)

464 B.P. Tonner, D. Dunham, T. Troubay, J. Kikuma, J. Denlinger: X-ray photoemission electron microscopy: Magnetic circular dichroism imaging and other contrast mechanisms. J. Electr. Spectr. Rel. Phen. **78**, 13–18 (1996)

465 W. Swiech, G.H. Fecher, C. Ziethen, O. Schmidt, et al.: Recent progress in photoemission microscopy with emphasis on chemical and magnetic sensitivity. J. Electr. Spectr. Rel. Phen. **84**, 171–188 (1997)

466 C.M. Schneider: Element specific imaging of magnetic domains in multicomponent thin film systems. J. Magn. Magn. Mat. **162**, 7–20 (1996)

467 T. Kachel, W. Gudat, K. Holldack: Element specific magnetic domain imaging from an antiferromagnetic overlayer system. Appl. Phys. Lett. **64**, 655–657 (1994)

468 F.U. Hillebrecht, T. Kinoshita, D. Spanke, J. Dresselhaus, C. Roth, H.B. Rose, E. Kisker: New magnetic linear dichroism in total photoelectron yield for magnetic domain imaging. Phys. Rev. Lett. **75**, 2224–2227 (1995)

469 Y. Kagoshima, T. Miyahara, M. Ando, J. Wang, S. Aoki: Magnetic domain-specific microspectroscopy with a scanning X-ray microscope using circularly polarized undulator radiation. J. Appl. Phys. **80**, 3124–3126 (1996)

470 P. Fischer, G. Schütz, G. Schmahl, P. Guttmann, D. Raasch: Imaging of magnetic domains with the X-ray microscope at BESSY using X-ray magnetic circular dichroism. Z. Phys. B **101**, 313–316 (1996)

471 P. Fischer, T. Eimüller, G. Schütz, P. Guttmann, G. Schmahl, K. Pruegl, G. Bayreuther: Imaging of magnetic domains by transmission X-ray microscopy. J. Phys. D: Appl. Phys. **31**, 649–655 (1998)

472 S. Stähler, G. Schütz, P. Fischer, M. Knülle, et al.: Distribution of magnetic moments in Co/Pt and Co/Pt/Ir/Pt multilayers detected by magnetic X-ray absorption. J. Magn. Magn. Mat. **121**, 234–237 (1993)

473 K. Futschik, H. Pfützner, A. Doblander, T. Dobeneck, N. Petersen, H. Vali: Why not use magnetotactic bacteria for domain analysis? Physica Scripta **40**, 518–521 (1989)

474 R.P. Blakemore, R.B. Frankel: Magnetic navigation in bacteria. Scient. American **245**.6, 42–49 (1981)

475 B.M. Moskowitz, R.B. Frankel, R.J. Flander, R.P. Blakemore, B.B. Schwartz: Magnetic properties of magnetotactic bacteria. J. Magn. Magn. Mat. **73**, 273–288 (1988)

476 G. Harasko, H. Pfützner, K. Futschik: Domain analysis by means of magnetotactic bacteria. IEEE Trans. Magn. **31**, 938–949 (1995)

477 G. Harasko, H. Pfützner, K. Futschik: On the effectiveness of magnetotactic bacteria for visualizations of magnetic domains. J. Magn. Magn. Mat. **133**, 409–412 (1994)

478 H.-E. Bühler, W. Pepperhoff, W. Schwenk: Zur Frage der Anätzbarkeit magnetischer Bereichsstrukturen auf Kobalt und Kobalt-Eisen-Legierungen. Z. Metallk. **57**, 201–205 (1966) *(On the question of an etchability of magnetic domains on cobalt and cobalt-iron alloys)*

479 D.J. Evans, H.J. Garret: Observation of magnetic domains in NdCo$_5$ using an electro-etching technique. IEEE Trans. Magn. **9**, 197–201 (1973)

480 D.Y. Parpia, B.K. Tanner, D.G. Lord: Direct optical observation of ferromagnetic domains. Nature **303**, 684 (1983)

481 D.G. Lord, V. Elliott, A.E. Clark, H.T. Savage, J.P. Teter, O.D. McMasters: Optical observations of closure domains in Terfenol-D single crystals. IEEE Trans. Magn. **24**, 1716–1718 (1988)

482 A.P. Holden, D.G. Lord, P.J. Grundy: Surface deformations and domains in Terfenol-D by scanning probe microscopy. J. Appl. Phys. **79**, 6070–6072 (1996)

483 F.H. Liu, H.-C. Tong, L. Milosvlasky: Domain structure at the cross sections of thin film inductive recording heads. J. Appl. Phys. **79**, 5895–5897 (1996)

484 A.R. Lang: The early days of high-resolution X-ray topography. J. Phys. D **26**, A1–8 (1993)

485 Y. Chikaura, B.K. Tanner: Evidence of interactions between domain walls and a dislocation bundle in synchrotron X-radiation topographs of iron whisker crystals. Jpn. J. Appl. Phys. **18**, 1389–1390 (1979)

486 J. Miltat: Significance of X-ray imaging techniques in the study of ferro- or ferrimagnets, in: *The Application of Synchrotron Radiation to Problems in Materials Science*, ed. by D.K. Bowen (Daresbury Laboratory, Warrington, 1983) p. 56–65

487 J. Sandonis, J. Baruchel, B.K. Tanner, G. Fillion, V.V. Kvardakov, K.M. Podurets: Coupling between antiferro and ferromagnetic domains in hematite. J. Magn. Magn. Mat. **104**, 350–352 (1992)

488 M. Hochhold, H. Leeb, G. Badurek: Tensorial neutron tomography: A first approach. J. Magn. Magn. Mat. **157**, 575–576 (1995)

489 M. Hochhold, H. Leeb, G. Badurek: Neutron spin tomophraphy: A tool to visualize magnetic domains in bulk materials. J. Phys. Soc. Jpn. **65**, Suppl. A, 292–295 (1996)

490 S. Libovický: Spatial replica of ferromagnetic domains in iron-silicon alloys.
Phys. Status Solidi A **12**, 539–547 (1972)

491 M.T. Rekveldt: Study of ferromagnetic bulk domains by neutron depolarization in three dimensions.
Z. Phys. **259**, 391–410 (1973)

492 O. Schärpf, H. Strothmann: Neutron techniques for magnetic domain and domain wall investigations.
Phys. Scripta **T24**, 58–70 (1988)

493 V.M. Pusenkov, N.K. Pleshanov, V.G. Syromyatnikov, V.A. Ulyanov, A.F. Schebetov: Study of
domain structure of thin magnetic films by polarised neutron reflectometry.
J. Magn. Magn. Mat. **175**, 237–248 (1997)

494 D. Küppers, J. Kranz, A. Hubert: Coercivity and domain structure of silicon-iron crystals.
J. Appl. Phys. **39**, 608–609 (1968)

495 K. Hara: Anomalous magnetic anisotropy of thin films evaporated at oblique incidence.
J. Sci. Hiroshima Univ. Ser. A-II **34**, 139–163 (1970)

496 U. Gonser, H. Fischer: Resonance γ-ray polarimetry, in: *Mössbauer Spectroscopy*, Vol. II,
ed. by U. Gonser (Springer, Berlin, Heidelberg, New York, 1981) p. 98–137

497 T. Zemcik, L. Kraus, K. Závěta: Mössbauer spectroscopy in creep annealed soft amorphous alloys.
Hyperfine Interactions **51**, 1051–1060 (1989)

498 H.D. Pfannes, H. Fischer: The texture problem in Mössbauer spectroscopy.
Appl. Phys. **13**, 317–325 (1977)

499 J.M. Greneche, F. Varret: On the texture problem in Mössbauer spectroscopy.
J. Phys. C: Solid State Phys. **15**, 5333–5344 (1982)

3. Domain Theory

Reviews and Textbooks

500 L.D. Landau, E.M. Lifshitz: *Course of Theoretical Physics*, Vol. IX.2 Theory of the Condensed
State (Pergamon, Oxford, 1980)

501 W.F. Brown, Jr.: *Magnetostatic Principles in Ferromagnetism* (North Holland, Amsterdam, 1962)

502 S. Chikazumi: *Physics of Ferromagnetism* (Clarendon Press, Oxford, 1997)

503 H. Kronmüller: Magnetisierungskurve der Ferromagnetika I, in: *Moderne Probleme der Metallphysik*,
Vol. 2, ed. by A. Seeger (Springer, Berlin, Heidelberg, New York, 1966) p. 24–156
(The magnetization curve of ferromagnets I)

504 H. Träuble: Magnetisierungskurve der Ferromagnetika II, in: *Moderne Probleme der Metallphysik*,
Vol. 2, ed. by A. Seeger (Springer, Berlin, Heidelberg, New York, 1966) p. 157–475
(Magnetization curve of ferromagnets II)

505 R.S. Tebble, D.J. Craik: *Magnetic Materials* (Wiley, London, 1969)

506 D.J. Craik: *Structure and Properties of Magnetic Materials* (Pion, London, 1971)

507 M. Rosenberg, C. Tănăsoiu: Magnetic Domains, in: *Magnetic Oxides*, Vol. 2,
ed. by D.J. Craik (Wiley, London, 1972) p. 483–573

508 I.A. Privorotskii: Thermodynamic Theory of Domain Structures.
Repts. Progr. Phys. **35**, 115–155 (1972)

509 A. Hubert: *Theorie der Domänenwände in Geordneten Medien* (Springer, Berlin, Heidelberg, New
York, 1974) *(Theory of Domain Walls in Ordered Media)*

510 A.H. Bobeck, E. Della Torre: *Magnetic Bubbles* (North Holland, Amsterdam, 1975)

511 I.A. Privorotskii: *Thermodynamic Theory of Domain Structures* (Wiley, New York, 1976)

512 T.H. O'Dell: *Ferromagnetodynamics. The Dynamics of Magnetic Bubbles, Domains and Domain Walls*
(Wiley, New York, 1981)

513 J. Miltat: Domains and domain walls in soft magnetic materials, mostly, in: *Applied Magnetism*,
ed. by R. Gerber, *et al.* (Kluwer, Dordrecht, 1994) p. 221–308

514 A. Aharoni: *Introduction to the Theory of Ferromagnetism* (Clarendon Press, Oxford, 1996)

Energetics of a Ferromagnet

515 P. Asselin, A.A. Thiele: On the field Lagrangians of micromagnetics.
IEEE Trans. Magn. **22**, 1876–1880 (1986)

516 A. Viallix, F. Boileau, R. Klein, J.J. Niez, P. Baras: A new method for finite element calculation of
micromagnetic problems. IEEE Trans. Magn. **24**, 2371–2373 (1988)

517 D.R. Fredkin, T.R. Koehler: Ab initio micromagnetic calculations for particles.
J. Appl. Phys. **67**, 5544–5548 (1990)

518 T. Schrefl, J. Fidler, H. Kronmüller: Nucleation fields of hard magnetic particles in 2D micromagnetic
and 3D micromagnetic calculations. J. Magn. Magn. Mat. **138**, 15–30 (1994)

519 T.R. Koehler, D.R. Fredkin: Finite element methods for micromagnetics.
IEEE Trans. Magn. **28**, 1239–1244 (1992)

520 T. Schrefl, H. Roitner, J. Fidler: Dynamic micromagnetics of nanocomposite NdFeB magnets.
J. Appl. Phys. **81**, 5567–5569 (1997)

521 A.S. Arrott, B. Heinrich, D.S. Bloomberg: Micromagnetics of magnetization processes in toroidal geometries. IEEE Trans. Magn. **10**, 950–953 (1979)

522 A. Hubert, W. Rave: Arrott's ideal soft magnetic cylinder, revisited.
J. Magn. Magn. Mat. **184**, 67–70 (1998)

523 K.M. Tako, T. Schrefl, M.A. Wongsam, R.W. Chantrell: Finite element micromagnetic simulations with adaptive mesh refinement. J. Appl. Phys. **81**, 4082–4084 (1997)

524 P. Grünberg, R. Schreiber, Y. Pang, M.B. Brodsky, H. Sowers: Layered magnetic structures: Evidence for antiferromagnetic coupling of Fe-layers across Cr interlayers.
Phys. Rev. Lett. **57**, 2442–2445 (1986)

525 M. Rührig, R. Schäfer, A. Hubert, R. Mosler, J.A. Wolf, S. Demokritov, P. Grünberg: Domain observations on Fe-Cr-Fe layered structures. Evidence for a biquadratic coupling effect.
Phys. Status Solidi A **125**, 635–656 (1991)

526 P. Bruno: Theory of interlayer magnetic coupling. Phys. Rev. B **52**, 411–439 (1995)

527 J. Slonczewski: Overview of interlayer exchange theory. J. Magn. Magn. Mat. **150**, 13–24 (1995)

528 G. Aubert, Y. Ayant, E. Belorizky, R. Casalengo: Various methods for analyzing data on anisotropic scalar properties in cubic symmetry. Application to magnetic anisotropy of nickel.
Phys. Rev. B **14**, 5314–5326 (1976)

529 R. Gersdorf: Experimental evidence for the X_2 hole pocket in the Fermi surface of Ni from magnetic crystalline anisotropy. Phys. Rev. Lett. **40**, 344–346 (1978)

530 W.P. Mason: Derivation of magnetostriction and anisotropic energies for hexagonal, tetragonal, and orthorhombic crystals. Phys. Rev. **96**, 302–310 (1954)

531 J.C. Slonczewski: Magnetic Annealing, in: *Magnetism*, Vol. I, ed. by G.T. Rado, H. Suhl (Academic Press, New York, 1963) p. 205–242

532 W.J. Carr, Jr.: Secondary effects in ferromagnetism, in: *Handbuch der Physik*, Vol. 18/2, ed. by S. Flügge (Springer, Berlin, Heidelberg, New York, 1966) p. 274–340

533 M.I. Darby, E.D. Isaac: Magnetocrystalline anisotropy of ferro- and ferrimagnetics.
IEEE Trans. Magn. **10**, 259–301 (1974)

534 L. Néel: L'anisotropie superficielle des substances ferromagnétiques.
C. R. Acad. Sci. Paris **237**, 1468–1470 (1953) *(The surface anisotropy of ferromagnetic substances)*

535 U. Gradmann: Magnetic surface anisotropies. J. Magn. Magn. Mat. **54**, 733–736 (1986)

536 H.J. Elmers, T. Furubayashi, M. Albrecht, U. Gradmann: Analysis of magnetic anisotropies in ultrathin films by magnetometry in situ in UHV. J. Appl. Phys. **70**, 5764–5768 (1991)

537 H. Fritzsche, J. Kohlhepp, U. Gradmann: Second- and fourth-order magnetic surface anisotropy of Co(0001)-based interfaces. J. Magn. Magn. Mat. **148**, 154–155 (1995)

538 G.T. Rado: Magnetic surface anisotropy. J. Magn. Magn. Mat. **104**, 1679–1683 (1992)

539 C. Chappert, P. Bruno: Magnetic anisotropy in metallic ultrathin films and related experiments on cobalt films. J. Appl. Phys. **64**, 5736–5741 (1988)

540 F.J.A. den Broeder, W. Hoving, P.H.J. Bloemen: Magnetic anisotropy of multilayers.
J. Magn. Magn. Mat. **93**, 562–570 (1991)

541 W.H. Meiklejohn: Exchange anisotropy – A review. J. Appl. Phys. (Suppl.) **33**, 1328–1335 (1962)

542 I.S. Jacobs, C.P. Bean: Fine Particles, Thin Films and Exchange Anisotropy, in: *Magnetism*, Vol. III, ed. by G.T. Rado, H. Suhl (Academic Press, New York, 1963) p. 271–350

543 A.P. Malozemoff: Mechanisms of exchange anisotropy. J. Appl. Phys. **63**, 3874–3879 (1988)

544 M.J. Carey, A.E. Berkowitz: CoO-NiO superlattices: Interlayer interactions and exchange anisotropy with $Ni_{81}Fe_{19}$. J. Appl. Phys. **73**, 6892–6897 (1993)

545 T.J. Moran, I.K. Schuller: Effects of coupling field strength on exchange anisotropy at Permalloy/CoO interfaces. J. Appl. Phys. **79**, 5109–5111 (1996)

546 A.N. Bogdanov, A. Hubert: Thermodynamical stable magnetic vortex states in magnetic crystals.
J. Magn. Magn. Mat. **138**, 255–269 (1994)

547 P. Rhodes, G. Rowlands: Demagnetizing energies of uniformly magnetized rectangular blocks.
Proc. Leeds Phil. Liter. Soc. **6**, 191–210 (1954)

548 W.H. Press, P.P. Flannery, S.A. Teukolsky, W.T. Vetterly: *Numerical Recipes. The Art of Scientific Computing (Fortran Version)* (Cambridge University Press, Cambridge, 1989)

549 S.W. Yuan, H.N. Bertram: Fast adaptive algorithms for micromagnetics.
IEEE Trans. Magn. **28**, 2031–2036 (1992)

550 D.V. Berkov, K. Ramstöck, A. Hubert: Solving micromagnetic problems - Towards an optimal numerical method. Phys. Status Solidi A **137**, 207–225 (1993)

551   N. Hayashi, K. Saito, Y. Nakatani: Calculation of demagnetizing field distribution based on Fast
      Fourier Transform of convolution. Jpn. J. Appl. Phys. Pt. 1 **35**, 6065–6073 (1996)
552   M. Jones, J.J. Miles: An accurate and efficient 3–D micromagnetic simulation of metal evaporated
      tape. J. Magn. Magn. Mat. **171**, 190–208 (1997)
553   N. Bär, A. Hubert, W. Jillek: Quantitative Untersuchung der Supplement-Domänenstruktur von
      kornorientiertem Elektroblech. J. Magn. Magn. Mat. **6**, 242–248 (1977) *(Quantitative investigation of
      supplementary domain structures on grain oriented electrical steel)*
554   C. Kooy, U. Enz: Experimental and theoretical study of the domain configuration in thin layers of
      $BaFe_{12}O_{19}$. Philips Res. Repts. **15**, 7–29 (1960)
555   M. Fox, R.S. Tebble: The demagnetizing energy and domain structure of a uniaxial single crystal.
      Proc. Phys. Soc. **72**, 765–769 (1958) *[ibd. 73 (1959) 325 correction]*
556   Z. Málek, V. Kamberský: On the theory of the domain structure of thin films of magnetically uniaxial
      materials. Czech. J. Phys. **8**, 416–421 (1958)
557   H.J.G. Draaisma, W.M.J. de Jonge: Magnetization curves of Pd/Co multilayers with perpendicular
      anisotropy. J. Appl. Phys. **62**, 3318–3322 (1987)
558   J. Kaczér, R. Gemperle, M. Zelený, J. Pačes, P. Šuda, Z. Frait, M. Ondris: On domain structure and
      magnetization processes. J. Phys. Soc. Japan **17 B-I**, 530–534 (1962)
559   G. Herzer: Effect of external stresses on the saturation magnetostriction of amorphous Co-based alloys.
      Soft Magnetic Materials 7 (Blackpool, 1985) p. 355–357
560   R.C. O'Handley, S.-W. Sun: Strained layers and magnetoelastic coupling.
      J. Magn. Magn. Mat. **104**, 1717–1720 (1992)
561   R. Koch, M. Weber, K. Thürmer, K.H. Rieder: Magnetoelastic coupling of Fe at high stress investigated
      by means of epitaxial Fe(001) films. J. Magn. Magn. Mat. **159**, L11–16 (1996)
562   W.F. Brown, Jr.: *Magnetoelastic Interactions* (Springer, Berlin, Heidelberg, New York, 1966)
563   M. Kléman, M. Schlenker: The use of dislocation theory in magnetoelasticity.
      J. Appl. Phys. **43**, 3184–3190 (1972)
564   E. Kröner: *Kontinuumstheorie der Versetzungen und Eigenspannungen* (Springer, Berlin, Heidelberg,
      New York, 1958) *(Continuum theory of dislocations and internal stresses)*
565   Z. Jirák, M. Zelený: Influence of magnetostriction on the domain structure of cobalt.
      Czech. J. Phys. B **19**, 44–47 (1969)
566   M. Kléman: *Points, Lines and Walls* (Wiley, Chichester, New York, 1983)
567   M. Kléman: Dislocations, Disclinations and Magnetism, in: *Dislocations in Solids*, Vol. 5,
      ed. by F.R.N. Nabarro (North Holland, Amsterdam, 1980) p. 349–402
568   M. Kléman: Internal stresses due to magnetic wall junctions in a perfect ferromagnet.
      J. Appl. Phys. **45**, 1377–1381 (1974)
569   J.N. Pryor, J.J. Kramer: Stress fields and strain energies associated with closure domains.
      AIP Conf. Proc. **29**, 570–571 (1975)
570   A.K. Head: Edge dislocations in inhomogeneous media. Proc. Phys. Soc. **66**, 793–801 (1953)
571   N.A. Pertsev, G. Arlt: Internal stresses and elastic energy in ferroelectric and ferroelastic ceramics:
      Calculations by the dislocation method. Ferroelectrics **123**, 27–44 (1991)
572   N.A. Pertsev, G. Arlt: Theory of the banded domain structure in coarse-grained ferroelectric ceramics.
      Ferroelectrics **132**, 27–40 (1992)
573   N.A. Pertsev, A.G. Zembilgotov: Energetics and geometry of 90° domain structures in epitaxial ferro-
      electric and ferroelastic films. J. Appl. Phys. **78**, 6170–6180 (1995)
574   K. Fabian, F. Heider: How to include magnetostriction in micromagnetic models of titanomagnetite
      grains. Geophys. Res. Lett. **23**, 2839–2842 (1996)
575   T.L. Gilbert: A Lagrangian formulation of the gyromagnetic equation of the magnetization field.
      Phys. Rev. **100**, 1243 (1955)
576   V.G. Baryakhtar, B.A. Ivanov, A.L. Sukstanskii, E.Y. Melikhov: Soliton relaxation in magnets.
      Phys. Rev. B **56**, 619–635 (1997)
577   F. Hoffmann: Dynamic pinning induced by nickel layers on Permalloy films.
      Phys. Status Solidi **41**, 807–813 (1970)
578   M. Labrune, J. Miltat: Wall structures in ferro/antiferromagnetic exchange-coupled bilayers:
      A numerical micromagnetic approach. J. Magn. Magn. Mat. **151**, 231–245 (1995)
579   A. Hubert: Stray-field-free magnetization configurations. Phys. Status Solidi **32**, 519–534 (1969)

The Origin of Domains

580   W. Williams, D.J. Dunlop: Three-dimensional micromagnetic modelling of ferromagnetic domain
      structure. Nature **337**, 634–637 (1989)
581   M.E. Schabes, H.N. Bertram: Magnetization processes in ferromagnetic cubes.
      J. Appl. Phys **64**, 1347–1357 (1988)

582   M.E. Schabes: Micromagnetic theory of non-uniform magnetization processes in magnetic recording particles. J. Magn. Magn. Mat. **95**, 249–288 (1991)

583   N.A. Usov, S.E. Peschany: Modeling of equilibrium magnetization structures in fine ferromagnetic particles with uniaxial anisotropy. J. Magn. Magn. Mat. **110**, L1–6 (1992)

584   A. Aharoni, J.P. Jakubovics: Cylindrical domains in small ferromagnetic spheres with cubic anisotropy. IEEE Trans. Magn. **24**, 1892–1894 (1988)

585   A. Aharoni, J.P. Jakubovics: Cylindrical magnetic domains in small ferromagnetic spheres with uniaxial anisotropy. Phil. Mag. B **53**, 133–145 (1986)

586   H.A.M. van den Berg: Self-consistent domain theory in soft-ferromagnetic media II. Basic domain structures in thin film objects. J. Appl. Phys. **60**, 1104–1113 (1986)

587   H.A.M. van den Berg, A.H.J. van den Brandt: Self-consistent domain theory in soft-ferromagnetic media III. Composite domain structures in thin-film objects. J. Appl. Phys. **62**, 1952–1259 (1987)

588   H.A.M. van den Berg: Order in the domain structure in soft-magnetic thin-film elements: A review. IBM J. Res. Dev. **33**, 540–582 (1989)

589   H.A.M. van den Berg, D.K. Vatvani: Wall clusters in thin soft ferromagnetic configurations. J. Appl. Phys. **52**, 6830–6839 (1981)

590   P. Bryant, H. Suhl: Thin-film magnetic patterns in an external field. Appl. Phys. Lett. **54**, 2224–2226 (1989)

591   A. Hubert, M. Rührig: Micromagnetic analysis of thin-film elements. J. Appl. Phys. **69**, 6072–6077 (1991)

592   A.S. Arrott, B. Heinrich, A. Aharoni: Point singularities and magnetization reversal in ideally soft ferromagnetic cylinders. IEEE Trans. Magn. **15**, 1228–1235 (1979)

593   A. Hubert: The role of magnetization "swirls" in soft magnetic materials. J. de Phys. (Coll.) **49**, C8–1859–1864 (1989)

594   R. Vlaming, H.A.M. van den Berg: A theory of the three-dimensional solenoidal magnetization configurations in ferro and ferrimagnetic materials. J. Appl. Phys. **63**, 4330–4332 (1988)

595   P. Bryant, H. Suhl: Micromagnetics below saturation. J. Appl. Phys. **66**, 4329 (1989)

596   K.D. Leaver: The synthesis of three-dimensional stray-field-free magnetization distributions. Phys. Status Solidi **27**, 153–163 (1975)

597   M. Rührig, W. Bartsch, M. Vieth, A. Hubert: Elementary magnetization processes in a low-anisotropy circular thin film disk. IEEE Trans. Magn. **26**, 2807–2809 (1990)

598   W. Rave, K. Fabian, A. Hubert: Magnetic states of small cubic particles with uniaxial anisotropy. J. Magn. Magn. Mat. **190**, 332–348 (1998)

Phase Theory

599   R.D. James, D. Kinderlehrer: Theory of magnetostriction with applications to $Tb_xD_{1-x}Fe_2$. Phil. Mag. B **68**, 237–274 (1993)

600   L. Néel: Les lois de l'aimantation et de la subdivision en domaines élémentaires d'un monocristal de fer (I). J. de Phys. Rad. **5**, 241–251 (1944) *(The laws of the magnetization and the subdivision into elementary domains for an iron single crystal (I))*

601   H. Lawton, K.H. Stewart: Magnetization curves for ferromagnetic single crystals. Proc. Roy. Soc. A **193**, 72–88 (1948)

602   R. Pauthenet, G. Rimet: Sur la variation de l'aimantation d'un monocristal de 6 $Fe_2O_3 \cdot PbO$ en fonction de champ. C. R. Acad. Sci. Paris **249**, 656–658 (1959) *(On the field-dependent variation of the magnetization of a magnetoplumbite single crystal)*

603   Y. Barnier, R. Pauthenet, G. Rimet: Sur la variation de l'aimantation d'un monocristal de cobalt en fonction de champ. C. R. Acad. Sci. Paris **252**, 3024–3026 (1961) *(On the field-dependent variation of the magnetization of a cobalt single crystal)*

604   R.R. Birss, B.C. Hegarty, P.M. Wallis: The magnetization process in nickel single crystals II. Brit. J. Appl. Phys. **18**, 459–471 (1967)

605   R.R. Birss, D.J. Martin: The magnetization process in hexagonal ferromagnetic and ferrimagnetic single crystals. J. Phys. C: Sol. State **8**, 189–210 (1975)

606   V.G. Baryakhtar, A.N. Bogdanov, D.A. Yablonskii: The physics of magnetic domains. Sov. Phys. Uspekhi **31**, 810–835 (1988) *[Usp. Fiz. Nauk 156 (1988) 47–92]*

607   A. de Simone: Energy minimizers for large ferromagnetic bodies. Arch. Rat. Mech. Anal. **125**, 99–143 (1993)

608   L.D. Landau, E.M. Lifshitz: *Course of Theoretical Physics*, Vol. V Statistical Physics (Pergamon, Oxford, 1980)

609   A. Hubert, A.P. Malozemoff, J. DeLuca: Effects of cubic, tilted uniaxial, and orthorhombic anisotropies on homogeneous nucleation in a garnet bubble film. J. Appl. Phys. **45**, 3562–3571 (1974)

610   P.W. Shumate, Jr.: Extension of the analysis for an optical magnetometer to include cubic anisotropy in detail. J. Appl. Phys. **44**, 3323–3331 (1973)

611   R.W. DeBlois: Ferromagnetic and structural properties of nearly perfect thin nickel platelets.
      J. Vacuum Sci. Techn. **3**, 146–155 (1965)

Small Particle Switching

612   R. Street, J.C. Wooley: A study of magnetic viscosity. Proc. Phys. Soc. A **62**, 562–572 (1949)
613   L. Néel: Théorie du traînage magnétique de diffusion. J. de Phys. Rad. **13**, 249–263 (1952)
      (Theory of the magnetic after effect by diffusion)
614   C.P. Bean, J.D. Livingston: Superparamagnetism. J. Appl. Phys. (Suppl.) **30**, 120*S*–129*S* (1958)
615   A. Lyberatos, D.V. Berkov, R.W. Chantrell: A method for the numerical simulation of the thermal
      magnetization fluctuations in micromagnetics. J. Phys. Condensed Matter **5**, 8911–8920 (1993)
616   P.J. Thompson, R. Street: Viscosity, reptation and tilting effects in permanent magnets.
  ·   J. Phys. D: Appl. Phys. **30**, 1273–1284 (1997)
617   E.C. Stoner, E.P. Wohlfarth: A mechanism of magnetic hysteresis in heterogeneous alloys.
      Phil. Trans. Roy. Soc. A **240**, 599–644 (1948)
618   W.F. Brown, Jr.: Criterion for uniform micromagnetization. Phys. Rev. **105**, 1479–1482 (1957)
619   E.H. Frei, S. Shtrikman, D. Treves: Critical size and nucleation of ideal ferromagnetic particles.
      Phys. Rev. **106**, 446–455 (1957)
620   J.C. Slonczewski: Theory of magnetic hysteresis in films and its application to computers.
      Research Report RM 003.111.224 (IBM Corp., 1956)
621   A. Aharoni: Magnetization curling. Phys. Status Solidi **16**, 1–42 (1966)
622   A. Holz: Formation of reversed domains in plate-shaped ferrite particles.
      J. Appl. Phys. **41**, 1095–1096 (1970)
623   A.S. Arrott, B. Heinrich, T.L. Templeton, Λ. Aharoni: Micromagnetics of curling configurations in
      magnetically soft cylinders. J. Appl. Phys. **50**, 2387–2389 (1979)
624   H.F. Schmidts, H. Kronmüller: Size dependence of the nucleation field of rectangular ferromagnetic
      parallelepipeds. J. Magn. Magn. Mat. **94**, 220–234 (1991)
625   Y. Uesaka, Y. Nakatani, N. Hayashi: Micromagnetic calculation of applied field effect on switching
      mechanism of a hexagonal platelet particle. Jpn. J. Appl. Phys. Pt. 1 **30**, 2489–2502 (1991)
626   N.A. Usov, Y.B. Grebenshikov, S.E. Peschany: Criterion for stability of a nonuniform micromagnetic
      state. Z. Phys. B **87**, 183–189 (1992)
627   W. Rave, K. Ramstöck, A. Hubert: Corners and nucleation in micromagnetics.
      J. Magn. Magn. Mat. **183**, 328–332 (1998)
628   A. Hubert, W. Rave, K. Ramstöck: Systematic analysis of micromagnetic switching processes.
      J. Magn. Magn. Mat. (in preparation) (1998)
629   K. Ramstöck, W. Hartung, A. Hubert: The phase diagram of domain walls in narrow magnetic strips.
      Phys. Status Solidi A **155**, 505–518 (1996)
630   J. Ehlert, F.K. Hübner, W. Sperber: Micromagnetics of cylindrical particles.
      Phys. Status Solidi A **106**, 239–248 (1988)

Domain Walls

631   E. Dzyaloshinskii: Theory of helicoidal structures in antiferromagnets.
      Sov. Phys. JETP **20**, 665–671 (1965) *[Zh. Eksp. Teor. Fiz. 47 (1990) 992–1002]*
632   Y.A. Izyumov: Modulated, or long-periodic, magnetic structures of crystals.
      Soviet Phys. Uspekhi **27**, 845–867 (1983) *[Usp. Fiz. Nauk 144 (1984) 439–474]*
633   B.A. Lilley: Energies and widths of domain boundaries in ferromagnetics.
      Phil. Mag. (7) **41**, 792–813 (1950)
634   J.P. Jakubovics: Comments on the definition of ferromagnetic domain wall width.
      Phil. Mag. B **38**, 401–406 (1978)
635   W. Döring: Über die Trägheit der Wände zwischen Weißschen Bezirken.
      Z. Naturforschung **3a**, 373–379 (1948) *(On the inertia of walls between Weiss domains)*
636   J. Kaczér: Bloch walls with div $I \neq 0$. J. de Phys. Rad. **20**, 120–123 (1959)
637   A.I. Mitsek, P.F. Gaidanskii, V.N. Pushkar: Domain structure of uniaxial antiferromagnets.
      The problem of nucleation. Phys. Status Solidi **38**, 69–79 (1970)
638   R. Gemperle, M. Zelený: Néel walls in massive uniaxial ferromagnets in an external field.
      Phys. Status Solidi **6**, 839–852 (1964)
639   J.E.L. Bishop: Ruckling: A novel low-loss domain wall motion for (110)[001] SiFe.
      IEEE Trans. Magn. **14**, 248–255 (1976)
640   S. Chikazumi, K. Suzuki: On the maze domain of silicon iron crystal I.
      J. Phys. Soc. Japan **10**, 523–534 (1955)
641   R.W. DeBlois, C.D. Graham, Jr.: Domain observations on iron whiskers.
      J. Appl. Phys. **29**, 931–939 (1958)

642 L. Néel: Énergie des parois de Bloch dans les couches minces.
C. R. Acad. Sci. Paris **241**, 533–536 (1955) *(Bloch wall energy in thin films)*

643 E.E. Huber, Jr., D.O. Smith, J.B. Goodenough: Domain-wall structure in Permalloy films.
J. Appl. Phys. **29**, 294–295 (1958)

644 E. Feldtkeller: Bloch lines in nickel-iron films, in: *Electric and Magnetic Properties of Thin Metallic Layers* (Koninklijke Vlaamse Acad. v. Wetenschapen, Letteren en Schone Kunsten van België, Brussel, 1961) p. 98–110

645 A. Aharoni: Measure of self-consistency in 180° domain wall models.
J. Appl. Phys. **39**, 861–862 (1968)

646 A. Aharoni, J.P. Jakubovics: Self-consistency of magnetic domain wall calculation.
Appl. Phys. Lett. **59**, 369–371 (1991)

647 W.F. Brown, Jr., A.E. LaBonte: Structure and energy of one-dimensional domain walls in ferromagnetic thin films. J. Appl. Phys. **36**, 1380–1386 (1965)

648 R. Collette: Shape and energy of Néel walls in very thin ferromagnetic films.
J. Appl. Phys. **35**, 3294–3301 (1964)

649 R. Kirchner, W. Döring: Structure and energy of a Néel wall. J. Appl. Phys. **39**, 855–856 (1968)

650 A. Holz, A. Hubert: Wandstrukturen in dünnen magnetischen Schichten.
Z. Angew. Phys. **26**, 145–152 (1969) *(Wall structures in thin magnetic films)*

651 A. Hubert: Symmetric Néel walls in thin magnetic films.
Computer Phys. Comm. **1**, 343–348 (1970)

652 H.-D. Dietze, H. Thomas: Bloch- und Néel-Wände in dünnen ferromagnetischen Schichten.
Z. Phys. **163**, 523–534 (1961) *(Bloch and Néel walls in thin ferromagnetic films)*

653 E. Feldtkeller, H. Thomas: Struktur und Energie von Blochlinien in dünnen ferromagnetischen Schichten. Phys. Kondens. Materie **4**, 8–14 (1965) *(Structure and energy of Bloch lines in thin ferromagnetic films)*

654 H. Riedel, A. Seeger: Micromagnetic treatment of Néel walls.
Phys. Status Solidi B **46**, 377–384 (1971)

655 A. Hubert: Interaction of domain walls in thin magnetic films.
Czech. J. Phys. B **21**, 532–536 (1971)

656 Y. Nakatani, Y. Uesaka, N. Hayashi: Direct solution of the Landau-Lifshitz-Gilbert equation for micromagnetics. Jpn. J. Appl. Phys. **28**, 2485–2507 (1989)

657 A.E. LaBonte: Two-dimensional Bloch-type domain walls in ferromagnetic films.
J. Appl. Phys. **40**, 2450–2458 (1969)

658 A. Hubert: Stray-field-free and related domain wall configurations in thin magnetic films II.
Phys. Status Solidi **38**, 699–713 (1970)

659 A.E. LaBonte: *Theory of Bloch-type domain walls in ferromagnetic thin films.*
PhD Thesis, University of Minnesota (1966)

660 A. Aharoni: Two-dimensional model for a domain wall. J. Appl. Phys. **38**, 3196–3199 (1967)

661 A. Hubert: Blochwände in dicken magnetischen Schichten. Z. Angew. Phys. **32**, 58–63 (1971) *(Bloch walls in thick magnetic films)*

662 A. Hubert: Domain wall calculations for thin single crystals. J. de Phys. **32 C1**, 404–405 (1971)

663 S. Höcker, A. Hubert: Theorie der Wandbeweglichkeit in idealen dünnen ferromagnetischen Schichten. Int. J. Magn. **3**, 139–143 (1972) *(Theory of wall mobility in ideal thin ferromagnetic films)*

664 A. Hubert: Domain wall structures in thin magnetic films.
IEEE Trans. Magn. **11**, 1285–1290 (1975)

665 A. Aharoni: Asymmetry in domain walls. Phys. Status Solidi A **18**, 661–667 (1973)

666 J.P. Jakubovics: Analytic representation of Bloch walls in thin ferromagnetic films.
Phil. Mag. **30**, 983–993 (1974)

667 A. Aharoni: A new approach to the structure of domain walls in magnetic films.
IEEE Trans. Magn. **10**, 939–942 (1974)

668 A. Aharoni: Two-dimensional domain walls in ferromagnetic films I-III.
J. Appl. Phys. **46**, 908–916, 1783–1786 (1975)

669 A. Aharoni: Two-dimensional domain walls in ferromagnetic films IV.
J. Appl. Phys. **47**, 3329–3335 (1976)

670 J.P. Jakubovics: Application of the analytical representation of Bloch walls in thin ferromagnetic films to calculations of wall structure with increasing anisotropy. Phil. Mag. **37**, 761–771 (1978)

671 V.S. Semenov: Changes in the structure of Bloch domain walls with increase in depth of film and anisotropy constant. Phys. Met. Metall. **64**.5, 1–8 (1987)
[Fiz. Metal. Metalloved. 64.5 (1987) 837–843]

672 V.S. Semenov: Two-dimensional Bloch and Néel walls in magnetic films. I. 180° domain walls.
Phys. Met. Metall. **71**.2, 61–68 (1991) *[Fiz. Metal. Metalloved. 71.2 (1991) 64–71]*

673 J. Miltat, M. Labrune: An adaptive mesh numerical algorithm for the solution of 2D Néel type walls.
IEEE Trans. Magn. **30**, 4350–4352 (1994)

674 A. Aharoni, J.P. Jakubovics: Structure and energy of 90° domain walls in thin ferromagnetic films. IEEE Trans. Magn. **26**, 2810–2812 (1990)

675 S. Huo, J.E.L. Bishop, J.W. Tucker: Micromagnetic simulation of 90° domain walls in thin iron films. J. Appl. Phys. **81**, 5239–5241 (1997)

676 L. Zepper, A. Hubert: Lorentz-Mikroskopie von Bloch- und Néelwänden in Ni-Fe-Kristallen. J. Magn. Magn. Mat. **2**, 18–24 (1976) *(Lorentz microscopy of Bloch and Néel walls in NiFe crystals)*

677 S.W. Yuan, H.N. Bertram: Domain wall structures and dynamics in thin films. IEEE Trans. Magn. **27**, 5511–5513 (1991)

678 W. Rave, A. Hubert: Micromagnetic calculation of the thickness dependence of surface and interior widths of asymmetrical Bloch walls. J. Magn. Magn. Mat. **184**, 179–183 (1998)

679 F.B. Humphrey, M. Redjdal: Domain wall structure in bulk magnetic material. J. Magn. Magn. Mat. **133**, 11–15 (1994)

680 M.R. Scheinfein, J. Unguris, J.L. Blue, K.J. Coakley, D.T. Pierce, R.J. Celotta, P.J. Ryan: Micromagnetics of domain walls at surfaces. Phys. Rev. B **43**, 3395–3422 (1991)

681 L.A. Finzi, J.A. Hartmann: Wall coupling in Permalloy film pairs with large separation. IEEE Trans. Magn. **4**, 662–668 (1968)

682 A. Hubert: Charged walls in thin magnetic films. IEEE Trans. Magn. **15**, 1251–1260 (1979)

683 A. Hubert: Charged magnetic domain walls under the influence of external fields. IEEE Trans. Magn. **17**, 3440–3443 (1981)

684 M. Labrune, S. Hamzaoui, C. Battarel, I.B. Puchalska, A. Hubert: New type of magnetic domains - the small lozenge type configuration in amorphous thin films. J. Magn. Magn. Mat. **44**, 195–206 (1984)

685 J.C. Slonczewski: Theory of domain wall motion in magnetic films and platelets. J. Appl. Phys. **44**, 1759–1770 (1973)

686 E. Schlömann: Domain walls in bubble films III. J. Appl. Phys. **45**, 369–373 (1974)

687 A. Hubert: Statics and dynamics of domain walls in bubble materials. J. Appl. Phys. **46**, 2276–2287 (1975)

688 H.J. Williams, M. Goertz: Domain structure of Perminvar having a rectangular hysteresis loop. J. Appl. Phys. **23**, 316–323 (1952)

689 J. Kranz, U. Buchenau: Über die Dicke von 180°-Blochwänden auf der Oberfläche von kompaktem Siliziumeisen. IEEE Trans. Magn. **2**, 297–301 (1966) *(On the thickness of 180° Bloch walls on the surface of bulk silicon iron)*

690 L. Schön, U. Buchenau: Observation of Néel lines in silicon iron. Int. J. Magn. **3**, 145–150 (1972)

691 S. Shtrikman, D. Treves: Internal structure of Bloch walls. J. Appl. Phys. (Suppl.) **31**, 147S–148S (1960)

692 E. Feldtkeller: Mikromagnetisch stetige und unstetige Magnetisierungsverteilungen. Z. Angew. Phys. **19**, 530–536 (1965) *(Micromagnetically continuous and discontinuous magnetization distributions)*

693 A. Hubert: Interactions between Bloch lines. AIP Conf. Proc. **18**, 178–182 (1974)

694 Y. Nakatani, N. Hayashi: Computer simulation of two-dimensional vertical Bloch lines by direct integration of Gilbert equation. IEEE Trans. Magn. **23**, 2179–2181 (1987)

695 P. Trouilloud, J. Miltat: Néel lines in ferrimagnetic garnet epilayers with orthorhombic anisotropy and canted magnetization. J. Magn. Magn. Mat. **66**, 194–212 (1987)

696 A. Thiaville, J. Ben Youssef, Y. Nakatani, J. Miltat: On the influence of wall microdeformations on Bloch line visibility in bubble garnets. J. Appl. Phys. **69**, 6090–6095 (1991)

697 K. Ramstöck, A. Hubert, D. Berkov: Techniques for the computation of embedded micromagnetic structures. IEEE Trans. Magn. **32**, 4228–4230 (1996)

698 U. Hartmann, H.H. Mende: Observation of subdivided 180° Bloch wall configurations on iron whiskers. J. Appl. Phys. **59**, 4123–4128 (1986)

699 R. Schäfer, W.K. Ho, J. Yamasaki, A. Hubert, F.B. Humphrey: Anisotropy pinning of domain walls in a soft amorphous magnetic material. IEEE Trans. Magn. **27**, 3678–3689 (1991)

700 H. Bäurich: Berechnung der Energie, Magnetisierungsverteilung und Ausdehnung einer Kreuzbloch-linie. Phys. Status Solidi **16**, K39–43 (1966) *(Calculation of the energy, the magnetization distribution, and the width of a cross Bloch line) [see also Phys. Status Solidi 23 (1967) K137–138]*

701 A. Aharoni: Applications of micromagnetics. CRC Crit. Rev. Sol. State Sci. **2**, 121–180 (1971)

702 W. Döring: Point singularities in micromagnetism. J. Appl. Phys. **39**, 1006–1007 (1968)

703 J. Reinhardt: Gittertheoretische Behandlung von mikromagnetischen Singularitäten. Int. J. Magn. **5**, 263–268 (1973) *(Lattice-theoretical treatment of micromagnetic singularities)*

704 J.C. Slonczewski: Properties of Bloch points in bubble domains. AIP Conf. Proc. **24**, 613–618 (1975)

705 A. Aharoni: Exchange energy near singular points or lines. J. Appl. Phys. **51**, 3330–3332 (1980)

706   M. Toulouse, M. Kléman: Principles of a classification of defects in ordered media.
      J. de Phys. Lett. **37**, L149–151 (1976)
707   A. Hubert: Mikromagnetisch singuläre Punkte in Bubbles. J. Magn. Magn. Mat. **2**, 25–31 (1976)
      (Micromagnetically singular points in bubbles)
708   F.H. de Leeuw, R. van den Doel, U. Enz: Dynamic properties of magnetic domain walls and magnetic
      bubbles. Rep. Prog. Phys. **43**, 689–783 (1980)
709   A.A. Thiele: Steady-state motion of magnetic domains. Phys. Rev. Lett. **30**, 230–233 (1973)
710   A.A. Thiele: Applications of the gyrocoupling vector and dissipation dyadic in the dynamics of mag-
      netic domains. J. Appl. Phys. **45**, 377–393 (1974)
711   J.C. Slonczewski: Theory of Bloch-line and Bloch-wall motion.
      J. Appl. Phys. **45**, 2705–2715 (1974)
712   J.C. Slonczewski: Dynamics of magnetic domain walls. Int. J. Magn. **2**, 85–97 (1972)
713   L.R. Walker: unpublished, reported in J.F. Dillon Jr., Domains and Domain Walls, in: *Magnetism*,
      Vol. III, ed. by G.T. Rado, H. Suhl (Academic Press, New York, 1963) p. 450–454
714   H. Suhl, X.Y. Zhang: Chaotic motions of domain walls in soft magnetic materials.
      J. Appl. Phys. **61**, 4216–4218 (1987)
715   N.L. Schryer, L.R. Walker: The motion of 180° domain walls in uniform dc magnetic fields.
      J. Appl. Phys. **45**, 5406–5421 (1974)
716   V.G. Baryakhtar, B.A. Ivanov, M.V. Chetkin: Dynamics of domain walls in weak ferromagnets.
      Sov. Phys. Usp. **28**, 563–588 (1985) *[Usp. Fiz. Nauk 146 (1985) 417–458]*
717   N. Papanicolaou: Dynamics of domain walls in weak ferromagnets.
      Phys. Rev. B **55**, 12290–12308 (1997)
718   G.W. Rathenau: Time effects in magnetism, in: *Magnetic Properties of Metals and Alloys*,
      ed. by R Bozorth (Am. Soc. Metals, Cleveland, 1959) p. 168–199
719   J.F. Janak: Diffusion-damped domain-wall motion. J. Appl. Phys. **34**, 3356–3362 (1963)
720   H. Kronmüller: *Nachwirkung in Ferromagnetika* (Springer, Berlin, Heidelberg, New York, 1968)
      (After effect in ferromagnets)
721   G.W. Rathenau, G. de Vries: Diffusion, in: *Magnetism and Metallurgy*,
      ed. by A.E. Berkowitz, E. Kneller (Academic Press, New York, 1969) p. 749–814
722   A.F. Khapikov: Domain wall and Bloch line dynamics in a magnet with a magnetic aftereffect.
      Phys. Solid State **36**, 1126–1131 (1994) *[Fiz. Tverd. Tela 36 (1994) 2062–2073]*
723   G.M. Sandler, H.N. Bertram: Micromagnetic simulations with eddy currents of rise time in thin film
      write heads. J. Appl. Phys. **81**, 4513–4515 (1997)
724   H.J. Williams, W. Shockley, C. Kittel: Studies of the propagation velocity of a ferromagnetic domain
      boundary. Phys. Rev. **80**, 1090–1094 (1950)
725   R.H. Pry, C.P. Bean: Calculation of the energy loss in magnetic sheet materials using a domain model.
      J. Appl. Phys. **29**, 532–533 (1958)
726   J.W. Shilling, G.L. Houze, Jr.: Magnetic properties and domain structure in grain-oriented 3% Si-Fe.
      IEEE Trans. Magn. **10**, 195–223 (1974)
727   J.E.L. Bishop: Modelling domain wall motion in soft magnetic alloys.
      J. Magn. Magn. Mat. **41**, 261–271 (1984)
728   J.E.L. Bishop: Eddy current dominated magnetization processes in grain oriented silicon iron.
      IEEE Trans. Magn. **20**, 1527–1532 (1984)
729   J.E.L. Bishop, M.J. Threapleton: An analysis of domain wall ruckling initiated at the points of
      maximum shear-stress on a braket wall in (110)[001] SiFe. J. Magn. Magn. Mat. **40**, 293–302 (1984)
730   J.E.L. Bishop: Steady-state eddy-current dominated magnetic domain wall motion with severe bowing
      and necking. J. Magn. Magn. Mat. **12**, 102–107 (1979)
731   J. Smit, H.P.L. Wijn: *Ferrites* (N.V. Philips, Eindhoven, 1959)
732   T. Nozawa, M. Mizogami, H. Mogi, Y. Matsuo: Domain structures and magnetic properties of
      advanced grain-oriented silicon steel. J. Magn. Magn. Mat. **133**, 115–122 (1994)

Theoretical Analysis of Characteristic Domain Features

733   Y.S. Shur, V.R. Abels: Study of magnetization processes in crystals of silicon-iron.
      Phys. Met. Metall. **6**.3, 167–173 (1958) *[Fiz. Metal. Metalloved. 6 (1958) 556–563]*
734   A. Hubert, W. Heinicke, J. Kranz: Magnetische Oberflächenstrukturen auf Gossblech.
      Z. Angew. Phys. **19**, 521–529 (1965) *(Magnetic surface structures of Goss sheets)*
735   Y.S. Shur, Y.N. Dragoshanskii: The shape of the closure domains inside silicon iron crystals.
      Phys. Met. Metall. **22**.5, 57–63 (1966) *[Fiz. Metal. Metalloved. 22.5 (1966) 702–710]*
736   M. Birsan, J.A. Szpunar, T.W. Krause, D.L. Atherton: Magnetic Barkhausen noise study of domain
      wall dynamics in grain oriented 3% Si-Fe. IEEE Trans. Magn. **32**, 527–534 (1996)
737   M. Imamura, T. Sasaki, A. Saito: Magnetization process and magnetostriction of a four percent
      Si-Fe single crystal close to (110)[001]. IEEE Trans. Magn. **17**, 2479–2485 (1981)

738   S. Arai, A. Hubert: The profiles of lancet-shaped surface domains in iron.
       Phys. Status Solidi A **147**, 563–568 (1995)
739   P.F. Davis: A theory of the shape of spike-like magnetic domains.
       Brit. J. Appl. Phys. (J. Phys. D) **2**, 515–521 (1969)
740   M.W. Muller: Distribution of the magnetization in a ferromagnet.
       Phys. Rev. **122**, 1485–1489 (1961)
741   W.F. Brown, Jr.: Rigorous calculation of the nucleation field in a ferromagnetic film or plate.
       Phys. Rev. **124**, 1348–1353 (1961)
742   R.J. Spain: Dense-banded domain structure in "rotatable anisotropy" Permalloy films.
       Appl. Phys. Lett. **3**, 208–209 (1963)
743   N. Saito, H. Fujiwara, Y. Sugita: A new type of magnetic domain in thin Ni-Fe films.
       J. Phys. Soc. Japan **19**, 421–422 (1964)
744   I.B. Puchalska, R.P. Ferrier: High-voltage electron microscope observation of stripe domains in
       Permalloy films evaporated at oblique incidence. Thin Solid Films **1**, 437–445 (1967/68)
745   R.P. Ferrier, I.B. Puchalska: 360° walls and strong stripe domains in Permalloy films.
       Phys. Status Solidi **28**, 335–347 (1968)
746   E. Tatsumoto, K. Hara, T. Hashimoto: A new type of stripe domains.
       Jpn. J. Appl. Phys. **7**, 176 (1968)
747   R. Allenspach, M. Stampanoni, A. Bischof: Magnetic domains in thin epitaxial Co/Au(111) films.
       Phys. Rev. Lett. **65**, 3344–3347 (1990)
748   L. Hua, J.E.L. Bishop, J.W. Tucker: Simulation of transverse and longitudinal magnetic ripple structures
       induced by surface anisotropy. J. Magn. Magn. Mat. **163**, 285–291 (1996)
749   R.W. Patterson, M.W. Muller: Magnetoelastic effects in micromagnetics.
       Int. J. Magn. **3**, 293–303 (1972)
750   A.N. Bogdanov, D.A. Yablonskii: Theory of the domain structure in ferrimagnets.
       Sov. Phys. Sol. State **22**, 399–403 (1980) *[Fiz. Tverd. Tela 22 (1980) 680–687]*
751   A.H. Eschenfelder: *Magnetic Bubble Technology* (Springer, Berlin, Heidelberg, New York, 1981)
752   J.A. Cape, G.W. Lehman: Magnetic domain structures in thin uniaxial plates with perpendicular
       easy axis. J. Appl. Phys. **42**, 5732–5756 (1971)
753   W.F. Druyvesteyn, J.W.F. Dorleijn: Calculations on some periodic magnetic domain structures;
       consequences for bubble devices. Philips Res. Repts. **26**, 11–28 (1971)
754   P.P. Ewald: Die Berechnung optischer und elektrostatischer Gitterpotentiale. Ann. Physik (4) **64**,
       253–287 (1921) *(The calculation of optical and electrical lattice potentials)*
755   K.L. Babcock, R.M. Westervelt: Elements of cellular domain patterns in magnetic garnet films.
       Phys. Rev. A **40**, 2022–2037 (1989)
756   K.L. Babcock, R. Seshadri, R.M. Westervelt: Coarsening of cellular domain patterns in magnetic garnet
       films. Phys. Rev. A **41**, 1952–1962 (1990)
757   A.A. Thiele: Theory of the static stability of cylindrical domains in uniaxial platelets.
       J. Appl. Phys. **41**, 1139–1145 (1970)
758   W.J. DeBonte: Properties of thick-walled cylindrical magnetic domains in uniaxial platelets.
       J. Appl. Phys. **44**, 1793–1797 (1973)
759   I.A. Privorotskii: Contribution to the theory of domain structures.
       Sov. Phys. JETP **29**, 1145–1152 (1969) *[Zh. Eksp. Teor. Fiz. 56 (1969) 2129–2142]*
760   R. Szymczak: Teoria struktury domenowej jednoosiowych ferromagnetyków. Archiwum Elektro-
       techniki **15**, 477–497 (1966) *(Theory of the domain structure of uniaxial ferromagnetic substances)*
761   M. Labrune, J. Miltat: Micromagnetics of strong stripe domains in NiCo thin films.
       IEEE Trans. Magn. **26**, 1521–1523 (1990)
762   M. Labrune, J. Miltat: Strong stripes as a paradigm of quasi-topological hysteresis.
       J. Appl. Phys. **75**, 2156–2168 (1994)
763   A. Hubert: The calculation of periodic domains in uniaxial layers by a 'micromagnetic' domain model.
       IEEE Trans. Magn. **21**, 1604–1606 (1985)
764   L.D. Landau: On the theory of the intermediate state of superconductors.
       J. Phys. USSR **7**, 99–107 (1943)
765   D.H. Martin: Surface structures and ferromagnetic domain sizes. Proc. Phys. Soc. B **70**, 77–84 (1957)
766   J. Kaczér: On the domain structure of uniaxial ferromagnets.
       Sov. Phys. JETP **19**, 1204–1208 (1964) *[Zh. Eksp. Teor. Fiz. 46 (1964) 1787–1792]*
767   R. Szymczak: A modification of the Kittel open structure. J. Appl. Phys. **39**, 875–876 (1968)
768   R. Bodenberger, A. Hubert: Zur Bestimmung der Blochwandenergie von einachsigen Ferromagneten.
       Phys. Status Solidi A **44**, K7–11 (1977) *(Determining the Bloch wall energy of uniaxial ferromagnets)*
769   R. Carey: Measurement of optimum domain widths in silicon-iron.
       Proc. Phys. Soc. **76**, 567–569 (1960)

770 K. Koike, H. Matsuyama, W.J. Tseng, J.C.M. Li: Fine magnetic domain structure of stressed amorphous metal. Appl. Phys. Lett. **62**, 2581–2583 (1993)

771 L. Hua, J.E.L. Bishop, J.W. Tucker: Simulation of magnetization ripples on Permalloy caused by surface anisotropy. J. Magn. Magn. Mat. **144**, 655–656 (1995)

772 S. Kaya: Pulverfiguren des magnetisierten Eisenkristalls I. Z. Phys. **89**, 796–805 (1934) *(Powder patterns of the magnetized iron crystal I)*

773 S. Kaya: Pulverfiguren des magnetisierten Eisenkristalls II. Z. Phys. **90**, 551–558 (1934)

774 L.F. Bates, F.E. Neale: A quantitative examination of recent ideas on domain structure. Physica **15**, 220–224 (1949)

775 L.F. Bates, F.E. Neale: A quantitative study of the domain structure of single crystals of silicon-iron by the powder pattern technique. Proc. Phys. Soc. A **63**, 374–388 (1950)

776 L.F. Bates, C.D. Mee: The domain structure of a silicon-iron crystal. Proc. Phys. Soc. A **65**, 129–140 (1952)

777 L.F. Bates, G.W. Wilson: A study of Bitter figures on the (110) plane of a single crystal of nickel. Proc. Phys. Soc. A **66**, 819–822 (1953)

778 E.W. Lee: The influence of domain structure on the magnetization curves of single crystals. Proc. Phys. Soc. A **66**, 623–630 (1953)

779 C. Schwink, H. Spreen: Untersuchungen der Bereichsstruktur von Nickeleinkristallen. I. Beobachtungen an Kristallen mit [110] und [100] als Achse. Phys. Status Solidi **10**, 57–74 (1965) *(Investigation of the Domain structure investigations of nickel single crystals. I. Observations on crystals with [110] and [100] axes)*

780 C. Schwink, O. Grüter: ... II. Messungen und Beobachtungen an prismatischen Kristallen mit [110] als Achse. Phys. Status Solidi **19**, 217–229 (1967) *(... II. Measurements and observations on prismatic crystals with [110] axis)*

781 H. Spreen: ... III. Die Abschlußstruktur in zylindrischen Kristallen mit [110] als Achse. Phys. Status Solidi **24**, 413–429 (1967) *(... III. The closure structure in cylindrical crystals with [110] axis)*

782 H. Spreen: ... IV. Die Grundstruktur in zylindrischen Kristallen mit [100] als Achse. Phys. Status Solidi **24**, 431–441 (1967) *(... IV. The basic structure in cylindrical crystals with [110] axis)*

783 D. Krause, H. Frey: Untersuchungen zur Temperaturabhängigkeit der Bereichsstrukturen stabförmiger Nickel-Einkristalle. Z. Phys. **224**, 257–278 (1969) *(Investigations on the temperature dependence of the domain structure of bar-shaped nickel single crystals)*

784 G. Dedié, J. Niemeyer, C. Schwink: Ferromagnetic domain structure of the [110] SiFe crystals. Phys. Status Solidi B **43**, 163–173 (1971)

785 V. Ungemach, C. Schwink: Magnetisierungskurve und Bereichsstruktur zweizähliger CoPd-Einkristalle. Physica **80B**, 381–388 (1975) *(Magnetization curve and domain structure of two-fold CoPd single crystals)*

786 V. Ungemach, C. Schwink: Bereichsgröße von SiFe-Néel-Kristallen in Abhängigkeit von Magnetfeldbehandlung und Temperatur. J. Magn. Magn. Mat. **2**, 167–173 (1976) *(Domain size of SiFe Néel crystals as a function of magnetic field treatment and temperature)*

787 V. Ungemach: Dependence of the domain structure of FeSi single crystals on the frequency of demagnetizing fields. J. Magn. Magn. Mat. **26**, 252–254 (1982)

4. Material Parameters

788 H. Zijlstra: Measurement of Magnetic Quantities, in: *Experimental Methods in Magnetism*, Vol. 2 (North Holland, Amsterdam, 1967)

789 M. Kalvius, R.S. Tebble (Ed.): *Experimental Magnetism*, Vol. I (Wiley, Chichester New York, 1979)

790 P. Weiss: Les propriétés magnétiques de la pyrrhotine. J. de Phys. Rad. **4**, 469–508 (1905) *(The magnetic properties of pyrrhotite)*

791 A.A. Aldenkamp, C.P. Marks, H. Zijlstra: Frictionless recording torque magnetometer. Rev. Sci. Instrum. **31**, 544–546 (1960)

792 F.B. Humphrey, A.R. Johnston: Sensitive automatic torque balance for thin magnetic films. Rev. Sci. Instrum. **34**, 348–358 (1963)

793 G. Maxim: A sensitive torque magnetometer for the measurement of small magnetic anisotropies. J. Sci. Instrum. (J. Phys. E) **2**, 319–320 (1969)

794 Y. Otani, H. Miyajima, S. Chikazumi: High field torque magnetometer for superconducting magnets. Jpn. J. Appl. Phys. **26**, 623–626 (1987)

795 G. Aubert: Torque measurements of the anisotropy of energy and magnetization of nickel. J. Appl. Phys. **39**, 504–510 (1968)

796 L.M. Pecora: The reconstruction of the anisotropy free energy function from magnetic torque data. J. Magn. Magn. Mat. **82**, 57–62 (1989)

797   J.S. Kouvel, C.D. Graham, Jr.: On the determination of magnetocrystalline anisotropy constants from torque measurements. J. Appl. Phys. **28**, 340–343 (1957)

798   G.W. Rathenau, J.L. Snoek: Magnetic anisotropy phenomena in cold rolled nickel-iron. Physica **8**, 555–575 (1941)

799   J.H.E. Griffiths, J.R. MacDonald: An oscillation type magnetometer. J. Sci. Instrum. **28**, 56–58 (1951)

800   U. Gradmann: Struktur und Ferromagnetismus sehr dünner, epitaktischer Ni-Flächenschichten. Ann. Physik **17**, 91–106 (1966) *(Structure and ferromagnetism of very thin epitaxial planar Ni films)*

801   R. Bergholz, U. Gradmann: Structure and magnetism of oligatomic Ni(111)-films on Re(0001). J. Magn. Magn. Mat. **45**, 389–398 (1984)

802   U. Enz, H. Zijlstra: Bestimmung magnetischer Grössen durch Verlagerungsmessungen. Philips Techn. Rundsch. **25**, 112–119 (1963/64) *Philips Techn. Rev. 25 (1963/64) 207*

803   H. Zijlstra: A vibrating reed magnetometer for microscopic particles. Rev. Sci. Instrum. **41**, 1241–1243 (1970)

804   W. Roos, K.A. Hempel, C. Voigt, H. Dederichs, R. Schippan: High sensitivity vibrating reed magnetometer. Rev. Sci. Instrum. **51**, 612–613 (1980)

805   T. Frey, W. Jantz, R. Stibal: Compensating vibrating reed magnetometer. J. Appl. Phys. **64**, 6002–6007 (1988)

806   S. Foner: Versatile and sensitive vibrating-sample magnetometer. Rev. Sci. Instrum. **30**, 548–557 (1959)

807   S. Foner: Further improvements in vibrating sample magnetometer sensitivity. Rev. Sci. Instrum. **46**, 1425–1426 (1975)

808   S. Foner: Review of magnetometry. IEEE Trans. Magn. **17**, 3358–3363 (1981)

809   R.M. Josephs, D.S. Crompton, C.S. Krafft: Vector vibrating sample magnetometry - A technique for three dimensional magnetic anisotropy measurements. Intermag Conf. Digests (Tokyo, 1987) p. CF-05

810   L. Jahn, R. Scholl, D. Eckert: Vibrating sample vector magnetometer coils. J. Magn. Magn. Mat. **101**, 389–391 (1991)

811   J.S. Philo, W.M. Fairbank: High-sensitivity magnetic susceptometer employing superconducting technology. Rev. Sci. Instrum. **48**, 1529–1536 (1977)

812   L. Callegaro, C. Fiorini, G. Triggiani, E. Puppin: Kerr hysteresis loop tracer with alternate driving magnetic field up to 10 kHz. Rev. Sci. Instr. **68**, 2735–2740 (1997)

813   J.C. Suits: Magneto-optical rotation and ellipticity measurements with a spinning analyzer. Rev. Sci. Instrum. **42**, 19–22 (1971)

814   K. Sato: Measurement of magneto-optical Kerr effect using piezo-birefringent modulator. Jpn. J. Appl. Phys. **20**, 2403–2409 (1981)

815   M. Gomi, M. Abe: A new high-sensitivity magneto-optic readout technique and its application to Co-Cr film. IEEE Trans. Magn. **18**, 1238–1240 (1982)

816   R.P. Cowburn, A. Ercole, S.J. Gray, J.A.C. Bland: A new technique for measuring magnetic anisotropies in thin and ultrathin films by magneto-optics. J. Appl. Phys. **81**, 6879–6883 (1997)

817   R.M. Osgood, III, B.M. Clemens, R.L. White: Asymmetric magneto-optic response in anisotropic thin films. Phys. Rev. B **55**, 8990–8996 (1997)

818   K. Postava, H. Jaffres, A. Schuhl, F. Nguyen Van Dau, M. Goiran, A.R. Fert: Linear and quadratic magneto-optical measurements of the spin reorientation in epitaxial Fe films on MgO. J. Magn. Magn. Mat. **172**, 199–208 (1997)

819   S. Shtrikman, D. Treves: On the remanence of ferromagnetic powders. J. Appl. Phys. (Suppl.) **31**, 58*S*–66*S* (1960)

820   K.-H. Müller, D. Eckert, P.A.P. Wendhausen, A. Handstein, S. Wirth, M. Wolf: Description of texture for permanent magnets. IEEE Trans. Magn. **30**, 586–588 (1994)

821   L. Néel: La loi d'approche en a:*H* et une nouvelle théorie de la dureté magnétique. J. de Phys. Rad. **9**, 184–192 (1948) *(The law of the approach to saturation with a:H and a new theory of magnetic hardness)*

822   N.S. Akulov: Über den Verlauf der Magnetisierungskurve in starken Feldern. Z. Phys. **69**, 822–831 (1931) *(On the behaviour of the magnetization curve at high fields)*

823   L. Néel: Relation entre la constante d'anisotropie et la loi d'approche à la saturation des ferromagnétiques. J. de Phys. Rad. **9**, 193–199 (1948) *(A relation between the anisotropy constant and the law of the approach to saturation of ferromagnets)*

824   H.-R. Hilzinger: Stress induced magnetic anisotropy in a non-magnetostrictive amorphous alloy. Proc. 4th Int. Conf. Rapidly Quenched Metals (Sendai, 1981) p. 791–794

825   G. Asti, S. Rinaldi: Singular points in the magnetization curve of a polycrystalline ferromagnet. J. Appl. Phys. **45**, 3600–3610 (1974)

826   R. Grössinger: Pulsed fields: generation, magnetometry and application. J. Phys. D: Appl. Phys. **15**, 1545–1608 (1982)

827   J. Smit, H.G. Beljers: Ferromagnetic resonance absorption in BaFe$_{12}$O$_{19}$, a highly anisotropic crystal.
      Philips Res. Repts **10**, 113–130 (1955)
828   P.E. Seiden: Magnetic Resonance, in: *Magnetism and Metallurgy*, Vol. I,
      ed. by A.E. Berkowitz, E. Kneller (Academic Press, New York, 1969) p. 93–119
829   R.F. Soohoo: *Microwave Magnetics* (Harper & Row, New York, 1985)
830   R. Zuberek, H. Szymczak, R. Krishnan, K.B. Youn, C. Sella: Magnetostriction constant of multilayer
      Ni-C films. IEEE Trans. Magn. **23**, 3699–3700 (1987)
831   C. Kittel: Excitation of spin waves in a ferromagnet by a uniform rf field.
      Phys. Rev. **110**, 1295–1297 (1958)
832   M.H. Seavey, Jr., P.E. Tannenwald: Direct observation of spin-wave resonance.
      J. Appl. Phys. (Suppl.) **30**, 227S–228S (1959)
833   T.G. Phillips, H.M. Rosenberg: Spin waves in ferromagnets. Repts. Prog. Phys. **29**, 285–332 (1966)
834   P.E. Wigen, C.F. Kooi, M.R. Shanabarger, T.D. Rossing: Dynamic pinning in thin-film spin-wave
      resonance. Phys. Rev. Lett. **9**, 206–20 (1962)
835   C.E. Patton: Magnetic excitations in solids. Phys. Reports **103**, 251–315 (1984)
836   J.G. Booth, G. Srinivasan, C.E. Patton, C.M. Srivastava: Spin wave stiffness parameters in lithium-zinc
      ferrites. Solid State Comm. **64**, 287–289 (1987)
837   W.D. Wilber, P. Kabos, C.E. Patton: Brillouin light scattering determination of the spin-wave stiffness
      parameter in lithium-zinc ferrite. IEEE Trans. Magn. **19**, 1862–1864 (1983)
838   J.R. Sandercock, W. Wettling: Light scattering from thermal acoustic magnons in yttrium iron garnet.
      Solid State Comm. **13**, 1729–1732 (1973)
839   R.W. Damon, J.R. Eshbach: Magnetostatic modes of a ferromagnet slab.
      J. Phys. Chem. Solids **19**, 308–320 (1961)
840   S. Demokritov, E. Tsymbal, P. Grünberg, W. Zinn, I.K. Schuller: Light scattering from spin waves
      in thin films and layered systems. J. Phys. Cond. Mat. **6**, 7145–7188 (1994)
841   K. Narita, J. Yamasaki, H. Fukunaga: Measurement of saturation magnetostriction of a thin amorphous
      ribbon by means of small-angle magnetization rotation. IEEE Trans. Magn. **16**, 435–439 (1980)
842   J.M. Barandarián, A. Hernando, V. Madurga, O.V. Nielsen, M. Vázquez, M. Vázquez-López: Tem-
      perature, stress, and structural-relaxation dependence of the magnetostriction in (Co$_{0.94}$Fe$_{0.06}$)$_{75}$Si$_{15}$B$_{10}$
      glasses. Phys. Rev. B **35**, 5066–5071 (1987)
843   O. Song, C.A. Ballentine, R.C. O'Handley: Giant surface magnetostriction in polycrystalline Ni and
      NiFe films. Appl. Phys. Lett. **64**, 2593–2595 (1994)
844   K.V. Vladimirskij: On the magnetostriction of polycrystals.
      C. R. Doklady Acad. Sci. USSR **16**, 10–13 (1943)
845   H.B. Callen, N. Goldberg: Magnetostriction polycrystalline aggregates.
      J. Appl. Phys. **36**, 976–977 (1965)
846   R. Gersdorf: Uniform and non-uniform form effect in magnetostriction.
      Physica **26**, 553–574 (1960)
847   E. de Lacheisserie, J.L. Dormann: An accurate method of the magnetostriction coefficients. Application
      to YIG. Phys. Status Solidi **35**, 925–931 (1969)
848   E. Klokholm: The measurement of magnetostriction in ferromagnetic thin films.
      IEEE Trans. Magn. **12**, 819–821 (1976)
849   A.C. Tam, H. Schroeder: Precise measurements of a magnetostriction coefficient of a thin
      soft-magnetic film deposited on a substrate. J. Appl. Phys. **64**, 5422–5424 (1988)
850   E. du Trémolet de Lacheisserie, J.C. Peuzin: Magnetostriction and internal stresses in thin films: The
      cantilever method revisited. J. Magn. Magn. Mat. **136**, 189 (1994)
851   M. Weber, R. Koch, K.H. Rieder: UHV cantilever beam technique for quantitative measurements of
      magnetization, magnetostriction, and intrinsic stress of ultrathin magnetic films.
      Phys. Rev. Lett. **73**, 1166–1169 (1994)
852   R. Watts, M.R.J. Gibbs, W.J. Karl, H. Szymczak: Finite-element modelling of magnetostrictive
      bending of a coated cantilever. Appl. Phys. Lett. **70**, 2607–2609 (1997)
853   P.M. Marcus: Magnetostrictive bending of a cantilevered film-substrate system.
      J. Magn. Mag. Mat. **168**, 18–24 (1997)
854   M. Ramesh, S. Jo, R.O. Campbell, M.H. Kryder: Submicron bubble garnets with nearly isotropic
      magnetostriction. J. Appl. Phys. **66**, 2508–2510 (1989)
855   R.M. Josephs: Characterization of the magnetic behavior of bubble domains.
      AIP Conf. Proc. **10**, 286–295 (1973)
856   T.G.W. Blake, C.-C. Shir, Y. Tu, E. Della Torre: Effects of finite anisotropy parameter Q in the
      determination of magnetic bubble material parameters. IEEE Trans. Magn. **18**, 985–988 (1982)
857   D. Płusa, R. Pfranger, B. Wysłocki: Dependence of domain width on crystal thickness in YCo$_5$ single
      crystals. Phys. Status Solidi A **92**, 533–538 (1985)
858   P. Salzmann, A. Hubert: Local measurement of magnetic anisotropy in metallic glasses.
      J. Magn. Magn. Mat. **24**, 168–174 (1981)

859 P. Salzmann, W. Grimm, A. Hubert: Anisotropies and domain structures in metallic glasses.
     J. Magn. Magn. Mat. **31**, 1599–1600 (1983)

860 P. Löffler, R. Wengerter, A. Hubert: Experimental and theoretical analysis of "stress pattern" magnetic
     domains in metallic glasses. J. Magn. Magn. Mat. **41**, 175–178 (1984)

861 I.E. Dikshtein, F.V. Lisovskii, E.G. Mansvetova, V.V. Tarasenko: Determination of the anisotropy
     constants of ferrite garnet epitaxial films with different crystallographic orientations by the method of
     phase transitions. Sov. Microelectronics **13**, 176–185 (1984) *[Mikroelektronika 13.4 (1984) 337–347]*

862 M.H. Yang, M.W. Muller: Evidence for non-cubic magnetostriction in epitaxial bubble garnets.
     J. Magn. Magn. Mat. **1**, 251–266 (1976)

863 J. Bindels, J. Bijvoet, G.W. Rathenau: Stabilization of definable domain structures by interstitial atoms.
     Physica B **26**, 163–174 (1960)

864 L. Néel: Nouvelle méthode de mesure de l'énergie des parois de Bloch.
     C. R. Acad. Sci. **254**, 2891–2896 (1962) *(A new method of measuring the Bloch wall energy)*

865 R. Aléonard, P. Brissonneau, L. Néel: New method to measure directly the 180° Bloch wall energy.
     J. Appl. Phys. **34**, 1321–1322 (1963)

866 C. Beatrice, F. Vinai, G. Garra, P. Mazetti: Bloch wall dynamic instability and wall multiplication in
     amorphous ribbons of Metglas 2605SC. J. de Phys. (Coll.) **49**, C8–1323–1324 (1989)

867 J.M. Daughton, G.E. Keefe, K.Y. Ahn, C.-C. Cho: Domain wall energy measurements using narrow
     Permalloy strips. IBM J. Res. Develop. **11**, 555–557 (1967)

868 G.T. Rado, V.J. Folen: Determination of molecular field coefficients in ferrimagnets.
     J. Appl. Phys. **31**, 62–68 (1960)

869 G.F. Dionne: Molecular field coefficients of substituted yttrium iron garnets.
     J. Appl. Phys. **41**, 4874–4881 (1970)

870 G.F. Dionne: Molecular field coefficients of substituted yttrium iron garnets.
     Technical Report 480 (Lincoln Lab., M.I.T., Lexington, Mass., 1970)

871 G.F. Dionne: Molecular field and exchange constants of Gd^{3+} substituted ferromagnetic garnets.
     J. Appl. Phys. **42**, 2142–2143 (1971)

872 G.F. Dionne: Molecular field coefficients of substituted yttrium iron garnets.
     Technical Report 502 (Lincoln Lab., M.I.T., Lexington, Mass., 1974)

873 A. Grill, F. Haberey: Effect of diamagnetic substitutions in $BaFe_{12}O_{19}$ on the magnetic properties.
     Appl. Phys. **3**, 131–134 (1974)

874 T. Holstein, H. Primakoff: Field dependence of the intrinsic domain magnetization of a ferromagnet.
     Phys. Rev. **58**, 1098–1113 (1940)

875 C. Zener: Classical theory of the temperature dependence of magnetic anisotropy energy.
     Phys. Rev. **96**, 1335–1337 (1954)

876 C. Kittel, J.H. van Vleck: Theory of the temperature dependence of the magnetoelastic constants of
     cubic crystals. Phys. Rev. **118**, 1231–1232 (1960)

877 E.R. Callen, H.B. Callen: Static magnetoelastic coupling in cubic crystals.
     Phys. Rev. **129**, 578–593 (1963)

878 A. Hubert, W. Unger, J. Kranz: Messung der Magnetostriktionskonstanten des Kobalts als Funktion der
     Temperatur. Z. Phys. **224**, 148–155 (1969) *(Measurement of the magnetostriction constants of cobalt
     as a function of temperature)*

879 E.P. Wohlfarth: Iron, cobalt and nickel, in: *Ferromagnetic Materials*, Vol. 1–70,
     ed. by E.P. Wohlfarth (North Holland, Amsterdam, 1980) p. 34–43

880 P. Hansen: Magnetic anisotropy and magnetostriction in garnets, in: *Physics of Magnetic Garnets*,
     ed. by A. Paoletti (North Holland, Amsterdam, 1978) p. 56–133 *(Proc. Int. School of Physics "Enrico
     Fermi", Course LXX, L. Como 1977)*

881 H.G. Rassmann, U. Hofmann: Zusammensetzung, Ordnungszustand und Eigenschaften höchstper-
     meabler Nickel-Eisen-Basislegierungen, in: *Magnetismus* (Verl. Grundstoffind., Leipzig, 1967)
     p. 176–198 *(Composition, state of order, and properties of highest permeability nickel-iron-based
     alloys)*

5. Domain Observation and Analysis

Bulk Uniaxial Materials

882 J. Kaczér, R. Gemperle: The thickness dependence of the domain structure of magnetoplumbite.
     Czech. J. Phys. B **10**, 505–510 (1960)

883 J. Kaczér, R. Gemperle: Honeycomb domain structure. Czech. J. Phys. B **11**, 510–522 (1961)

884 B. Wysłocki: Untersuchung der magnetischen Struktur an großen Kobaltkristallen. Phys. Status
     Solidi **3**, 1333–1339 (1963) *(Investigation of the magnetic structure of large cobalt crystals)*

885 B. Wysłocki: Über Remanenzstrukturen, Ummagnetisierungsprozesse und Magnetisierungskurven in grossen Kobalteinkristallen. I. Remanenzstrukturen. Acta Phys. Polon. **27**, 783–797 (1965) *(On remanent structures, magnetization processes and hysteresis loops in large cobalt crystals. I. Remanent structures)*

886 B. Wysłocki: ... III. Magnetisierungskurven. Acta Phys. Polon. **27**, 969–987 (1965) *(... III. Hysteresis loops)*

887 M. Rosenberg, C. Tănăsoiu, V. Florescu: Domain structure of hexagonal ferrimagnetic oxides with high anisotropy field. J. Appl. Phys. **37**, 3826–3834 (1966)

888 B. Wysłocki: Influence of crystal thickness on remanent domain structures in cobalt. I. Acta Phys. Polon. **34**, 327–346 (1968)

889 J. Kaczér: Ferromagnetic domains in uniaxial materials. IEEE Trans. Magn. **6**, 442–445 (1970)

890 J. Unguris, M.R. Scheinfein, R.J. Celotta, D.T. Pierce: The magnetic microstructure of the (0001) surface of hcp cobalt. Appl. Phys. Lett. **55**, 2553–2555 (1989)

891 B.M. Moskowitz, S.L. Halgedahl, C.A. Lawson: Magnetic domains on unpolished and polished surfaces of titanium-rich titanomagnetite. J. Geophys. Res. **93**, 3372–3386 (1988)

892 M. Rosenberg, C. Tănăsoiu, C. Rusu: The domain structure of $Zn_{2-x}Co_xZ$ single crystals. Phys. Status Solidi **6**, 639–650 (1964)

893 A. Hubert: Magnetic domains of cobalt single crystals at elevated temperatures. J. Appl. Phys. **39**, 444–446 (1968)

894 Y.G. Pastushenkov, A. Forkl, H. Kronmüller: Temperature dependence of the domain structure in $Fe_{14}Nd_2B$ single crystals during the spin-reorientation transition. J. Magn. Magn. Mat. **174**, 278–288 (1997)

895 J. Kaczér, R. Gemperle: The rotation of Bloch walls. Czech. J. Phys. B **11**, 157–170 (1961)

896 D.J. Breed, A.B. Voermans: Strip domains in garnet films with large magnetostriction constants under the influence of in-plane fields. IEEE Trans. Magn. **16**, 1041–1043 (1980)

897 E.A. Nesbitt, H.J. Williams: Mechanism of magnetization in Alnico V. Phys. Rev. **80**, 112–113 (1950)

898 W. Andrä: Untersuchung des Ummagnetisierungsvorgangs bei Alnico hoher Koerzitivkraft mit Hilfe der Pulvermustertechnik. Ann. d. Physik **19**, 10–18 (1956) *(Investigation of the remagnetization process of high-coercivity Alnico using the powder pattern technique)*

899 D.J. Craik, E.D. Isaac: Magnetic interaction domains. Proc. Phys. Soc. A **76**, 160–162 (1960)

900 D.J. Craik, R. Lane: Magnetization reversal mechanisms in assemblies of elongated single-domain particles. Brit. J. Appl. Phys. **18**, 1269–1274 (1967)

901 L. Folks, R. Street, R.C. Woodward: Domain structures of die-upset melt-spun NdFeB. Appl. Phys. Lett. **65**, 910–912 (1994)

902 J.N. Chapman, S. Young, H.A. Davies, P. Zhang, A. Manaf, R.A. Buckley: A TEM investigation of the magnetic domain structure in nanocrystalline NdFeB Samples. 13th Int. Workshop on Rare Earth Magnets and their Applications (Birmingham, 1994) p. 95–101

903 W. Rave, D. Eckert, R. Schäfer, B. Gebel, K.-H. Müller: Interaction domains in isotropic, fine-grained $Sm_2Fe_{17}N_3$ permanent magnets. IEEE Trans. Magn. **32**, 4362–4364 (1996)

Bulk Cubic Crystals

904 A. Hubert: Zur Analyse der magnetischen Bereichsstrukturen des Eisens. Thesis, University of Munich (1965) *(On the analysis of magnetic domain structures of iron)*

905 L.V. Kirenskii, M.K. Savchenko, I.F. Degtyarev: On the magnetization process in ferromagnets. Sov. Phys. JETP **37**, 437–441 (1960) *[Zh. Eksp. Teor. Fiz. 37 (1959) 616–619]*

906 P. Brissonneau, M. Schlenker: Domaines de Weiss « en chevrons » sur un monocristal de fer-silicium. C. R. Acad. Sci. Paris **259**, 2089–2092 (1964) *(Chevron type Weiss domains on a silicon iron single crystal)*

907 L.J. Dijkstra, U.M. Martius: Domain pattern in silicon-iron under stress. Rev. Mod. Phys. **25**, 146–150 (1953)

908 Y.S. Shur, V.A. Zaykova: Effect of elastic stresses on the magnetic structure of silicon-iron crystals. Phys. Met. Metall. **6**, 158–166 (1958) *[Fiz. Metal. Metalloved. 6 (1958) 545–555]*

909 W.D. Corner, J.J. Mason: The effect of stress on the domain structure of Goss textured silicon-iron. Brit. J. Appl. Phys. **15**, 709–718 (1964)

910 L.F. Bates, A. Hart: A comparison of the powder patterns on a sample of grain-oriented silicon-iron with those obtained on a single crystal. Proc. Phys. Soc. A **66**, 813–818 (1953)

911 W.S. Paxton, T.G. Nilan: Domain configurations and crystallographic orientation in grain-oriented silicon steel. J. Appl. Phys. **26**, 994–1000 (1955)

912 J.J. Gniewek: Effects of tensile stress on the domain structure in grain-oriented 3.25% silicon steel. J. Appl. Phys. **34**, 3618–3622 (1963)

913   V.A. Zaykova, Y.S. Shur: The shape of silicon iron crystal magnetostriction curves as dependent on the nature of the change in the domain structure on magnetization. Phys. Met. Metall. **18**.3, 31–42 (1964) *[Fiz. Metal. Metalloved. 18.3 (1964) 349–358]*

914   P. Allia, A. Ferro-Milone, G. Montalenti, G.P. Soardo, F. Vinai: Theory of negative magnetostriction in grain oriented 3% SiFe for various inductions and applied stresses. IEEE Trans. Magn. **14**, 362–364 (1978)

915   M. Imamura, T. Sasaki: The status of domain theory for an investigation of magnetostriction and magnetization processes in grain-oriented Si-Fe sheets. Phys. Scripta T **24**, 29–35 (1988)

916   A. Hubert: Magnetische Bereichsstrukturen und Magnetisierungsvorgänge. Z. Angew. Phys. **26**, 35–41 (1969) *(Magnetic domain structures and magnetization processes)*

917   M. Labrune, M. Kléman: Investigation of magnetic domain structures in {111} silicon-iron single crystals. J. de Phys. **34**, 79–89 (1973)

918   R. Szymczak, A. Szewczyk, M. Baran, V.V. Tsurkan: Domain structure in $CuCr_2Se_4$ single crystals. J. Magn. Magn. Mat. **83**, 481–482 (1990)

919   M. Müller, T. Lederer, K.H. Fornacon, R. Schäfer: Grain structure, coercivity and high frequency noise in soft magnetic Fe-81Ni-6Mo alloys. J. Mag. Magn. Mat. **177**, 231–232 (1998)

Amorphous and Nanocrystalline Ribbons

920   A. Hubert: Aus der Schmelze abgeschreckte amorphe Ferromagnetika. J. Magn. Magn. Mat. **6**, 38–46 (1977) *(Amorphous ferromagnets quenched from the melt)*

921   H. Kronmüller, M. Fähnle, M. Domann, H. Grimm, R. Grimm, B. Gröger: Magnetic properties of amorphous ferromagnetic alloys. J. Magn. Magn. Mat. **13**, 53–70 (1979)

922   B. Gröger, H. Kronmüller: Investigation of the domain structure, the Rayleigh region and the coercive field of the amorphous ferromagnetic alloy $Fe_{40}Ni_{40}P_{14}B_6$. J. Magn Magn. Mat **9**, 203–207 (1978)

923   G. Schroeder, R. Schäfer, H. Kronmüller: Magneto-optical investigation of the domain structure of amorphous $Fe_{80}B_{20}$ alloys. Phys. Status Solidi A **50**, 475–481 (1978)

924   A. Veider, G. Badurek, R. Grössinger, H. Kronmüller: Optical and neutron domain structure studies of amorphous ribbons. J. Magn. Magn. Mat. **60**, 182–194 (1986)

925   H.J. Leamy, S.D. Ferris, G. Norman, D.C. Joy, R.C. Sherwood, E.M. Gyorgy, H.S. Chen: Ferromagnetic domain structure of metallic glasses. Appl. Phys. Lett. **26**, 259–260 (1975)

926   J.D. Livingston: Stresses and magnetic domains in amorphous metal ribbons. Phys. Status Solidi A **56**, 637 (1979)

927   T. Egami, P.J. Flanders, C.D. Graham, Jr.: Low-field magnetic properties of ferromagnetic amorphous alloys. Appl. Phys. Lett. **26**, 128–130 (1975)

928   J.D. Livingston, W.G. Morris, T. Jagielinski: Effects of anisotropy on domain structures in amorphous ribbons. IEEE Trans. Magn. **19**, 1916–1918 (1983)

929   W. Grimm, B. Metzner, A. Hubert: Optimized domain patterns in metallic glasses prepared by magnetic field heat treatment and/or surface crystallization. J. Magn. Magn. Mat. **41**, 171–174 (1984)

930   R. Schäfer, M. Rührig, A. Hubert: Loss optimization for iron-rich metallic glasses. Phys. Scripta **40**, 552–557 (1989)

931   O.V. Nielsen, H.J.V. Nielsen: Strain- and field-induced magnetic anisotropy in metallic glasses with positive or negative λ_s. Solid State Comm. **35**, 281–284 (1980)

932   O.V. Nielsen, H.J.V. Nielsen: Magnetic anisotropy in $Co_{73}Mo_2Si_{15}B_{10}$ and $Co_{0.89}Fe_{0.11}$ metallic glasses, induced by stress-annealing. J. Magn. Magn. Mat. **22**, 21–24 (1980)

933   K. Závĕta, L. Kraus, K. Jurek, V. Kambersky: Zig-zag domain walls in creep-annealed metallic glass. J. Magn. Magn. Mat. **73**, 334–338 (1988)

934   K. Závĕta, O.V. Nielsen, K. Jurek: A domain study of magnetization processes in a stress-annealed metallic glass ribbon for fluxgate sensors. J. Magn. Magn. Mat. **117**, 61–68 (1992)

935   J. Yamasaki, T. Chuman, M. Yagi, M. Yamaoka: Magnetization process in hard axis of Fe-Co based amorphous ribbons with induced anisotropy. IEEE Trans. Magn. **33**, 3775–3777 (1997)

936   R. Schäfer, N. Mattern, G. Herzer: Stripe domains on amorphous ribbons. IEEE Trans. Magn. **32**, 4809–4811 (1996)

937   R.H. Smith, G.A. Jones, D.G. Lord: Domain structures in rapidly annealed $Fe_{67}Co_{18}B_{14}Si_1$. IEEE Trans. Magn. **24**, 1868–1870 (1988)

938   J. Sláma, V. Nikulin: Magnetic domains in amorphous FeNiB alloy. Phys. Scripta **39**, 548–551 (1989)

939   P.T. Squire, A.P. Thomas, M.R.J. Gibbs, M. Kuźmiński: Domain studies of field-annealed amorphous ribbon. J. Magn. Magn. Mat. **104**, 109–110 (1992)

940   K. Závĕta, M. Kuźmiński, H.K. Lachowicz, P. Duhaj: Domain structures in "non-magnetostrictive" $Co_{67}Fe_4Cr_7Si_8B_{14}$ wide ribbons. J. de Phys. IV **8**. Pr2, 163–166 (1998)

941   G.S. Krinchik, A.N. Verkhozin: Investigation of the magnetic structure of ferromagnetic substances by a magnetooptical apparatus with micron resolution. Sov. Phys. JETP **24**, 890–894 (1967) [*Zh. Eksp. Teor. Fiz. 51 (1966) 1321–1327*]

942   P.J. Ryan, T.B. Mitchell: Wide domain walls and Bloch lines in Permalloy and Co-Fe films using Kerr effect microscopy. J. Appl. Phys. **63**, 3162–3164 (1988)

943   K. Koike, H. Matsuyama, K. Hayakawa, K. Mitsuoka, *et al.*: Observation of Néel structure walls on the surface of 1.4–μm-thick magnetic films using spin-polarized scanning electron microscopy. Appl. Phys. Lett. **49**, 980–981 (1986)

944   H.P. Oepen, J. Kirschner: Imaging of microstructures at surfaces. J. de Phys. (Coll.) **49**, C8–1853–1857 (1988)

945   T. Göddenhenrich, U. Hartmann, C. Heiden: Investigation of Bloch wall fine structures by magnetic force microscopy. J. Microscopy **152**, 527–536 (1988)

946   M. Schneider, S. Müller-Pfeiffer, W. Zinn: Magnetic force microscopy of domain wall fine structures in iron films. J. Appl. Phys. **79**, 8578–8583 (1996)

947   T.G. Pokhil, B.M. Moskowitz: Magnetic force microscope study of domain wall structures in magnetite. J. Appl. Phys. **79**, 6064–6066 (1996)

948   V.E. Zubov, G.S. Krinchik, S.N. Kuzmenko: The effect of a magnetic field on the near-surface substructure of domain walls in single-crystal iron. Sov. Phys. JETP **72**, 307–313 (1991) [*Zh. Eksp. Teor. Fiz. 99 (1991) 551–561*]

949   U. Hartmann, H.H. Mende: Hysteresis of Néel-line motion and effective width of 180° Blochwalls in bulk iron. Phys. Rev. B **33**, 4777–4781 (1986)

950   K. Pátek, I. Tomáš, P. Bohácek: The process of magnetization of a single Bloch wall with a single Bloch line. J. Magn. Magn. Mat. **87**, 11–15 (1990)

951   V.E. Zubov, G.S. Krinchik, S.N. Kuzmenko: Anomalous coercive force of Bloch points in iron single-crystals. JETP Lett. **51**, 477–480 (1990) [*Pisma Zh. Eksp. Teor. Fiz. 51 (1990) 419–422*]

952   G. Herzer: Nanocrystalline soft magnetic materials. J. Magn. Magn. Mat. **157**, 133–136 (1996)

953   R. Schäfer, A. Hubert, G. Herzer: Domain observation on nanocrystalline material. J. Appl. Phys. **69**, 5325–5327 (1991)

954   L. Kraus, K. Závěta, O. Heczko, P. Duhaj, G. Vlasák, J. Schneider: Magnetic anisotropy in as-quenched and stress-annealed amorphous and nanocrystalline $Fe_{73.5}Cu_1Nb_3Si_{13.5}B_9$ alloys. J. Magn. Magn. Mat. **112**, 275–277 (1992)

Magnetic Films with Low Anisotropy

955   S. Middelhoek: Domain walls in thin Ni-Fe films. J. Appl. Phys. **34**, 1054–1059 (1963)

956   M. Prutton: *Thin Magnetic Films* (Butterworth, Washington D.C., 1964)

957   R.F. Soohoo: *Magnetic Thin Films* (Harper & Row, New York, 1985)

958   W. Schüppel, V. Kamberský: Bereichs- und Wandstrukturen. Phys. Status Solidi **2**, 345–384 (1962) *(Domain and wall structures)*

959   E. Feldtkeller: Review on domains in thin magnetic samples. J. de Phys. (Coll.) **32**, C1–452–456 (1971)

960   E. Feldtkeller, W. Liesk: 360°-Wände in magnetischen Schichten. Z. Angew. Phys. **14**, 195–199 (1962) *(360° walls in magnetic films)*

961   D.O. Smith, E.E. Huber, M.S. Cohen, G.P. Weiss: Anisotropy and inversion in Permalloy films. J. Appl. Phys. **31** (Suppl.) 295S–297S (1960)

962   E. Feldtkeller: Blockierte Drehprozesse in dünnen magnetischen Schichten. Elektron. Rechenanl. **3**, 167–175 (1961) *(Blocked rotation processes in thin magnetic films)*

963   M.S. Cohen: Influence of anisotropy dispersion on magnetic properties of NiFe films. J. Appl. Phys. **34**, 1841–1847 (1963)

964   E. Feldtkeller: Ripple hysteresis in thin magnetic films. J. Appl. Phys. **34**, 2646–2652 (1963)

965   S. Methfessel, S. Middelhoek, H. Thomas: Domain walls in thin Ni-Fe films. IBM Journ. **4**, 96–106 (1960)

966   S. Methfessel, S. Middelhoek, H. Thomas: Nucleation processes in thin Permalloy films. J. Appl. Phys. (Suppl.) **32**, 294S–295S (1961)

967   S. Methfessel, S. Middelhoek, H. Thomas: Partial rotation in Permalloy films. J. Appl. Phys. **32**, 1959–1963 (1961)

968   W. Metzdorf: Vorgänge beim quasistatischen Magnetisieren dünner Schichten. Z. Angew. Phys. **14**, 412–424 (1962) *(Quasistatic magnetization processes in thin films)*

969   H.W. Fuller, H. Rubinstein, D.L. Sullivan: Spiral walls in thin magnetic films. J. Appl. Phys. (Suppl.) **32**, 286S–297S (1961)

970   H. Hoffmann: Theory of magnetization ripple. IEEE Trans. Magn. **4**, 32–38 (1968)

971 K.J. Harte: Theory of magnetization ripple in ferromagnetic films.
     J. Appl. Phys. **39**, 1503–1524 (1968)

972 U. Krey: Die mikromagnetische Behandlung lokaler Störungen mit Hilfe der Greenschen Funktion.
     Phys. Kondens. Materie **6**, 218–228 (1967) *(The micromagnetic treatment of local perturbations using Green's function)*

973 W.F. Brown, Jr.: A critical assessment of Hoffmann's linear theory of ripple.
     IEEE Trans. Magn. **6**, 121–129 (1970)

974 II.A.M. van den Berg, F.A.N. van der Voort: Cluster creation and hysteresis in soft-ferromagnetic thin-film objects. IEEE Trans. Magn. **21**, 1936–1938 (1985)

975 L.J. Schwee, H.R. Irons, W.E. Anderson: The crosstie memory.
     IEEE Trans. Magn. **12**, 608–613 (1976)

976 H.J. Williams, R.C. Sherwood: Magnetic domain patterns on thin films.
     J. Appl. Phys. **28**, 548–555 (1957)

977 M.S. Cohen: Spiral and concentric-circle walls observed by Lorentz microscopy.
     J. Appl. Phys. **34**, 1221–1222 (1963)

978 M. Speckmann, H.P. Oepen, H. Ibach: Magnetic domain structures in ultrathin Co/Au(111): On the influence of film morphology. Phys. Rev. Lett. **75**, 2035–2038 (1995)

979 S. Middelhoek: Domain wall creeping in thin Permalloy films. IBM J. Res. Dev. **6**, 140–141 (1962)

980 H.C. Bourne, Jr., T. Kusuda, C.H. Lin: A study of low-frequency creep in Bloch-wall Permalloy films.
     IEEE Trans. Magn. **5**, 247–252 (1969)

981 J.N. Chapman, G.R. Morrison, J.P. Jakubovics, R.A. Taylor: Determination of domain wall structures in thin foils of a soft magnetic alloy. J. Magn. Magn. Mat. **49**, 277–285 (1985)

982 R. Gemperle: The ferromagnetic domain structure of thin single crystal Fe platelets in an external field. Phys. Status Solidi **14**, 121–133 (1966)

983 R.W. DeBlois: Ferromagnetic domains in single-crystal Permalloy platelets.
     J. Appl. Phys. **39**, 442–443 (1968)

984 E. Huijer, J.K. Watson, D.B. Dove: Magnetostatic effects of I-bars: A unifying overview of domain and continuum results. IEEE Trans. Magn. **16**, 120–126 (1980)

985 S.K. Decker, C. Tsang: Magnetoresistive response of small Permalloy features.
     IEEE Trans. Magn. **16**, 643–645 (1980)

986 C. Tsang, S.K. Decker: The origin of Barkhausen noise in small Permalloy magnetoresistive sensors.
     J. Appl. Phys. **52**, 2465–2467 (1981)

987 E. Huijer, J.K. Watson: Hysteretic properties of Permalloy I-bars.
     J. Appl. Phys. **50**, 2149–2151 (1979)

988 H.A.M. van den Berg, D.K. Vatvani: Wall clusters and domain structure conversions.
     IEEE Trans. Magn. **18**, 880–887 (1982)

989 F.A.N. van der Voort, H.A.M. van den Berg: Irreversible processes in soft-ferromagnetic thin films.
     IEEE Trans. Magn. **23**, 250–258 (1987)

990 D.A. Herman, Jr., B.E. Argyle, B. Petek: Bloch lines, cross ties, and taffy in Permalloy.
     J. Appl. Phys. **61**, 4200–4206 (1987)

991 B.W. Corb: Effects of magnetic history on the domain structure of small NiFe shapes.
     J. Appl. Phys. **63**, 2941–2943 (1988)

992 B.W. Corb: Charging on the metastable domain states of small NiFe rectangles.
     IEEE Trans. Magn. **24**, 2386–2388 (1988)

993 S. McVitie, J.N. Chapman: Magnetic structure determination in small regularly shaped particles using transmission electron microscopy. IEEE Trans. Magn. **24**, 1778–1780 (1988)

994 S. McVitie, J.N. Chapman, S.J. Hefferman, W.A.P. Nicholson: Effect of application of fields on the domain structure in small regularly shaped magnetic particles. J. de Phys. **12**, 1817–1818 (1988)

995 M. Rührig, B. Khamsehpour, K.J. Kirk, J.N. Chapman, P. Aitchison, S. McVitie, C.D.W. Wilkinson: The fabrication and magnetic properties of acicular magnetic nano-elements.
     IEEE Trans Magn. **32**, 4452–4457 (1996)

996 S.Y. Chou: Patterned magnetic nanostructures and quantized magnetic disks.
     Proc. IEEE **85**, 652–671 (1997)

997 C. Coquaz, D. Challeton, J.L. Porteseil, Y. Souche: Memory effects and nucleation of domain structures in small Permalloy shapes. J. Magn. Magn. Mat. **124**, 206–212 (1993)

998 R.D. McMichael, M.J. Donahue: Head to head domain wall structures in thin magnetic strips.
     IEEE Trans. Magn. **33**, 4167–4169 (1997)

999 J. McCord, A. Hubert, A. Chizhik: Domains and hysteresis in patterned soft magnetic elements.
     IEEE Trans. Magn. **33**, 3981–3983 (1997)

1000 R. Mattheis, K. Ramstöck, J. McCord: Formation and annihilation of edge walls in thin-film permalloy strips. IEEE Trans. Magn. **33**, 3993–3995 (1997)

1001 K.J. Kirk, J.N. Chapman, C.D.W. Wilkinson: Switching fields and magnetostatic interactions of thin film magnetic nano-elements. Appl. Phys. Lett. **71**, 539–541 (1997)

1002 M.S. Cohen: Anomalous magnetic films. J. Appl. Phys. **33**, 2968–2980 (1962)

1003 R.J. Spain: Sur une solution stable et périodique du problème de la répartition de l'aimantation spontanée dans une couche mince ferromagnétique. C. R. Acad. Sci. Paris **257**, 2427–2430 (1963) *(On a stable and periodic solution of the problem of spontaneous magnetization distribution in a thin ferromagnetic film)*

1004 N. Saito, H. Fujiwara, Y. Sugita: A new type of magnetic domain structure in negative magnetostriction Ni-Fe films. J. Phys. Soc. Japan **19**, 1116–1125 (1964)

1005 I.B. Puchalska: Stripe domain structure in ferromagnetic films on Ni-Fe. Acta Phys. Polon. **36**, 589–614 (1969)

1006 A.P. Malozemoff, W. Fernengel, A. Brunsch: Observation and theory of strong stripe domains in amorphous sputtered FeB films. J. Magn. Magn. Mat. **12**, 201–214 (1979)

1007 J.N. Chapman, R.P. Ferrier, N. Toms: Strong stripe domains I. Phil. Mag. **28**, 561–579 (1973)

1008 J.N. Chapman, R.P. Ferrier: Strong stripe domains II. Phil. Mag. **28**, 581–595 (1973)

1009 D. Unangst: Über die magnetische Bezirksstruktur dünner Eisen-"Einkristall"-Schichten. Ann. Phys. **7**, 280–301 (1961) *(On the magnetic domain structure of thin iron "single crystal" films)*

1010 H. Sato, R.S. Toth, R.W. Astrue: Checkerboard domain patterns on epitaxially grown single-crystal thin films of iron, nickel, and cobalt. J. Appl. Phys. **34**, 1062–1064 (1963)

1011 H. Sato, R.W. Astrue, S.S. Shinozaki: Lorentz microscopy of magnetic domains of single-crystal films. J. Appl. Phys. **35**, 822–823 (1964)

1012 C.G. Harrison, K.D. Leaver: Micromagnetic structures in single crystal specimens of intermediate thickness studied by Lorentz microscopy. J. Microscopy **97**, 139–145 (1973)

1013 A.F. Smith: Domain wall interactions with non-magnetic inclusions observed by Lorentz microscopy. J. Phys. D: Appl. Phys. **3**, 1044–1048 (1970)

1014 L.F. Bates, D.H. Martin: Ferromagnetic domain nucleation in silicon iron. Proc. Phys. Soc. B **69**, 145–152 (1956)

1015 H.J. Williams, W. Shockley: A simple domain structure in an iron crystal showing a direct correlation with the magnetization. Phys. Rev. **75**, 178–183 (1949)

1016 P. Görnert, Z. Šimša, J. Šimšova, I. Tomáš, R. Hergt, J. Kub: Growth and properties of $Y_{3-u}Pb_u Fe_{5-x-y-z}Co_x Ti_y Ga_z O_{12}$ LPE films. Phys. Status Stolidi A **53**, 297–301 (1986)

1017 A. Yelon: Interactions in multilayer magnetic films, in: *Physics of Thin Films*, Vol. 6, ed. by G. Hass, R.E. Tun (Academic Press, New York, London, 1971) p. 205–300

1018 L. Néel: Sur une nouveau mode de couplage entre les aimantations de deux couches minces ferromagnétiques. Compt. Rend. Acad. Sci. (Paris) **255**, 1676–1681 (1962) *(On a new coupling mode between the magnetization of two ferromagnetic thin films)*

1019 P. Grünberg, R. Schreiber, Y. Pang, U. Walz, M.B. Brodsky, H. Sowers: Layered magnetic structures: Evidence for antiferromagnetic coupling of Fe-layers across Cr interlayers. J. Appl. Phys. **61**, 3750–3752 (1987)

1020 S.S.P. Parkin, N. More, K.P. Roche: Oscillations in exchange coupling and magnetoresistance in metallic superlattice structures: Co/Ru, Co/Cr and Fe/Cr. Phys. Rev. Lett. **64**, 2304–2307 (1990)

1021 S. Middelhoek: Domain-wall structures in magnetic double films. J. Appl. Phys. **37**, 1276–1282 (1966)

1022 J.C. Slonczewski: Structure of domain walls in multiple films. J. Appl. Phys. **37**, 1268–1269 (1966)

1023 J.C. Slonczewski: Theory of domain-wall structure in multiple magnetic films. IBM J. Res. Develop. **10**, 377–387 (1966)

1024 J.C. Slonczewski, B. Petek, B.E. Argyle: Micromagnetics of laminated Permalloy films. IEEE Trans. Magn. **24**, 2045–2054 (1988)

1025 M. Rührig, W. Rave, A. Hubert: Investigation of micromagnetic edge structures of double-layer Permalloy films. J. Magn. Magn. Mat. **84**, 102–108 (1990)

1026 I. Tomáš, H. Niedoba, M. Rührig, G. Wittmann, *et al.*: Wall transitions and coupling of magnetization in NiFe/C/NiFe double films. Phys. Status Solidi A **128**, 203–217 (1991)

1027 M. Rührig, W. Rave, A. Hubert: Investigation of micromagnetic edge structures of double-layer Permalloy films. J. Magn. Magn. Mat. **84**, 102–108 (1990)

1028 F. Biragnet, J. Devenyi, G. Clerc, O. Massenet, R. Montmory, A. Yelon: Interactions between domain walls in coupled films. Phys. Status Solidi **16**, 569–576 (1966)

1029 H. Niedoba, A. Hubert, B. Mirecki, I.B. Puchalska: First direct magneto-optical observations of quasi-Néel walls in double Permalloy films. J. Magn. Magn. Mat. **80**, 379–383 (1989)

1030 L.J. Heyderman, H. Niedoba, H.O. Gupta, I.B. Puchalska: 360° and 0° walls in multilayer Permalloy films. J. Magn. Magn. Mat. **96**, 125–136 (1991)

1031 E. Sanneck, M. Rührig, A. Hubert: A theoretical and experimental study of 360°-walls in magnetically coupled bilayers. IEEE Trans. Magn. **29**, 2500–2505 (1993)

1032 H. Niedoba, L.J. Heyderman, A. Hubert: Kerr domain contrast and hysteresis as a tool for determination of the coupling strength of double soft magnetic films. J. Appl. Phys. **73**, 6362–6364 (1993)

1033 J. Unguris, R.J. Celotta, D.T. Pierce: Observation of two different oscillation periods in the exchange coupling of Fe/Cr/Fe (100). Phys. Rev. Lett. **67**, 140–143 (1991)

1034 J. Unguris, R.J. Celotta, D.T. Pierce: Oscillatory magnetic coupling in Fe/Ag/Fe(100) sandwich structures. J. Magn. Magn. Mat. **127**, 205–213 (1993)

1035 J. McCord, A. Hubert, R. Schäfer, A. Fuss, P. Grünberg: Domain analysis in epitaxial iron-aluminium and iron-gold sandwiches with oscillatory exchange. IEEE Trans. Magn. **29**, 2735–2737 (1993)

1036 C.J. Gutierrez, J.J. Krebs, M.E. Filipkowski, G.A. Prinz: Strong temperature dependence of the 90° coupling in Fe/ Al/ Fe(001) magnetic trilayers. J. Magn. Magn. Mat. **116**, L305–310 (1992)

1037 B. Heinrich, Z. Celinski, J.F. Cochran, A.S. Arrott, K. Myrtle, S.T. Purcell: Bilinear and biquadratic exchange coupling in bcc Fe/Cu/Fe trilayers: Ferromagnetic-resonance and surface magneto-optical Kerr-effect studies. Phys. Rev. B **47**, 5077–5089 (1993)

1038 B. Rodmacq, K. Dumesnil, P. Mangin, M. Hennion: Biquadratic magnetic coupling in NiFe/Ag multilayers. Phys. Rev. B **48**, 3556–3559 (1993)

1039 E.E. Fullerton, K.T. Riggs, C.H. Sowers, S.D. Bader, A. Berger: Suppression of biquadratic coupling in Fe/Cr(001) superlattices below the Néel transition of Cr. Phys. Rev. Lett. **75**, 330–333 (1995)

1040 M.E. Filipkowski, J.J. Krebs, G.A. Prinz, C.J. Gutierrez: Giant near-90° coupling in epitaxial CoFe/Mn/CoFe sandwich structures. Phys. Rev. Lett. **75**, 1847–1850 (1995)

1041 J.C. Slonczewski: Fluctuation mechanism for biquadratic exchange coupling in magnetic multilayers. Phys. Rev. Lett. **67**, 3172–3175 (1991)

1042 A.C. Daykin, J.P. Jakubovics, A.K. Petford-Long: A study of interlayer exchange coupling in a Co/Cr/Co trilayer using transmission electron microscopy. J. Appl. Phys. **82**, 2447–2452 (1997)

1043 R. Schäfer, A. Hubert, S.S.P. Parkin: Domain and domain wall observations in sputtered exchange-biased wedges. IEEE Trans. Magn. **29**, 2738–3740 (1993)

1044 R. Mattheis, W. Andrä, Jr., L. Fritzsch, A. Hubert, M. Rührig, F. Thrum: Magnetoresistance and Kerr effect investigations of Co/Cu multilayers with wedge shaped Cu layers. J. Magn. Magn. Mat. **121**, 424–428 (1993)

1045 P.R. Aitchison, J.N. Chapman, D.B. Jardine, J.E. Evetts: Correlation of domain processes and magneto-resistance changes as a function of field and number of bilayers in Co/Cu multilayers. J. Appl. Phys. **81**, 3775–3777 (1997)

Perpendicular-Anisotropy Films

1046 P. Molho, J.L. Porteseil, Y. Souche, J. Gouzerh, J.C.S. Levy: Irreversible evolution in the topology of magnetic domains. J. Appl. Phys. **61**, 4188–4193 (1987)

1047 K.L. Babcock, R.M. Westervelt: Avalanches and self-organization in the dynamics of cellular magnetic domain patterns. Phys. Rev. Lett. **64**, 2168 (1990)

1048 R.M. Westervelt, K.L. Babcock, M. Vu, R. Seshadri: Avalanches and self-organization in the dynamics of cellular magnetic domain patterns (invited). J. Appl. Phys. **69**, 5436–5440 (1991)

1049 M. Seul, R. Wolfe: Evolution of disorder in magnetic stripe domains II. Hairpins and labyrinth patterns versus branches and comb patterns formed by growing minority component. Phys. Rev. A **46**, 7534–7547 (1992)

1050 M. Seul, L.R. Monar, L. O'Gorman: Pattern analysis of magnetic stripe domains - Morphology and topological defects in the disordered state. Phil. Mag. B **66**, 471–506 (1992)

1051 M. Seul, R. Wolfe: Evolution of disorder in magnetic stripe domains I. Transverse instabilities and disclination unbinding in lamellar patterns. Phys. Rev. A **46**, 7519–7533 (1992)

1052 P. Bak, H. Flyvbjerg: Self-organization of cellular magnetic-domain patterns. Phys. Rev. A **45**, 2192–2200 (1992)

1053 H. Dötsch: Dynamics of magnetic domains in microwave fields. J. Magn. Magn. Mat. **4**, 180–185 (1977)

1054 B.E. Argyle, W. Jantz, J.C. Slonczewski: Wall oscillations of domain lattices in underdamped garnet films. J. Appl. Phys. **54**, 3370–3386 (1983)

1055 F.V. Lisovskii, E.G. Mansvetova, E.P. Nikolaeva, A.V. Nikolaev: Dynamic self-organization and symmetry of the magnetic-moment distribution in thin films. JETP **76**, 116–127 (1993) *[Zh. Eksp. Teor. Fiz. 103 (1993) 213–233]*

1056 F.V. Lisovskii, E.G. Mansvetova, C.M. Pak: Scenarios for the ordering and structure of self-organized two-dimensional domain blocks in thin magnetic films. JETP **81**, 567–578 (1995) *[Zh. Eksp. Teor. Fiz. 108 (1995) 1031–1051]*

1057 J. Kaczér, R. Gemperle: Remanent structure of magnetoplumbite. Czech. J. Phys. B **10**, 614 & 624a (1960)

1058 A. Kirilyuk, J.Ferré, V. Grolier, J.P. Jamet, D. Renard: Magnetization reversal in ultrathin ferromagnetic films with perpendicular anisotropy. J. Magn. Magn. Mat. **171**, 45–63 (1997)

1059 B.E. Argyle, J.C. Slonczewski, P. Dekker, S. Maekawa: Gradientless propulsion of magnetic bubble domains. J. Magn. Magn. Mat. **2**, 357–360 (1976)

1060 T. Kusuda, S. Honda, S. Hashimoto: Nondestructive sensing of wall-state transition and a dynamic bubble collapse experiment. IEEE Trans. Magn. **17**, 2415–2422 (1981)

1061 L. Zimmermann, J. Miltat: Instability of bubble radial motion associated with chirality changes. J. Magn. Magn. Mat. **94**, 207–214 (1991)

1062 A.N. Bogdanov, A. Hubert: The properties of isolated magnetic vortices. Phys. Status Solidi B **186**, 527–543 (1994)

1063 P.J. Grundy, D.C. Hothersall, G.A. Jones, B.K. Middleton, R.S. Tebble: The formation and structure of cylindrical magnetic domains in thin cobalt crystals. Phys. Status Solidi A **9**, 79–88 (1972)

1064 T. Suzuki: Direct observation of the movement of vertical Bloch lines in Ho-Co sputter-deposited films. Jpn. J. Appl. Phys. **22**, L493–495 (1983)

1065 T.M. Morris, G.J. Zimmer, F.B. Humphrey: Dynamics of hard walls in bubble garnet stripe domains. J. Appl. Phys. **47**, 721–726 (1976)

1066 T. Suzuki, H. Asada, K. Matsuyama, E. Fujita, et al.: Chip organization of Bloch line memory. IEEE Trans. Magn. **22**, 784–789 (1986)

1067 P. Pougnet, L. Arnaud, M. Poirier, L. Zimmermann, M.H. Vaudaine, F. Boileau: Characteristics and performance of a 1Kbit Bloch line memory prototype. IEEE Trans. Magn. **27**, 5492–5497 (1991)

1068 A. Thiaville, F. Boileau, J. Miltat, L. Arnaud: Direct Bloch line optical observation. J. Appl. Phys. **63**, 3153–3158 (1988)

1069 A. Thiaville, J. Miltat: Néel lines in the Bloch walls of bubble garnets and their dark-field observation. J. Appl. Phys. **68**, 2883–2891 (1990)

1070 A. Thiaville, J. Miltat: Experimenting with Bloch points in bubble garnets. J. Magn. Magn. Mat. **104**, 335–336 (1992)

1071 K. Pátek, A. Thiaville, J. Miltat: Horizontal Bloch lines and anisotropic-dark-field observations. Phys. Rev. B **49**, 6678–6688 (1994)

Particles, Needles and Wires

1072 K. Goto, M. Ito, T. Sakurai: Studies on magnetic domains of small particles of barium ferrite by colloid-SEM method. Jpn. J. Appl. Phys. **19**, 1339–1346 (1980)

1073 A. Fernandez, P.J. Bedrossian, S.L. Baker, S.P. Vernon, D.R. Kania: Magnetic force microscopy of single-domain cobalt dots patterned using interference lithography. IEEE Trans. Magn. **32**, 4472–4474 (1996)

1074 D.J. Dunlop, Ö. Özdemir: *Rock Magnetism. Fundamentals and Frontiers*, (Cambridge University Press, Cambridge, 1997)

1075 F.D. Stacey, S.K. Banerjee: *The Physical Principles of Rock Magnetism*, (Elsevier, Amsterdam, 1974)

1076 D.J. Dunlop: Magnetism in rocks. J. Geophys. Res. B **100**, 2161–2174 (1995)

1077 K. Fabian, A. Hubert: Shape-induced pseudo single domain remanence. Geophys. J. Intern. (in preparation) (1998)

1078 R.V. Coleman, G.G. Scott: Magnetic domain patterns on single-crystal iron whiskers. Phys. Rev. **107**, 1276–1280 (1957)

1079 R.V. Coleman, G.G. Scott: Magnetic domain patterns on iron whiskers. J. Appl. Phys. **29**, 526–527 (1958)

1080 J. Kaczér, R. Gemperle, Z. Hauptman: Domain structure of cobalt whiskers. Czech. J. Phys. **9**, 606–612 (1959)

1081 C.A. Fowler, Jr., E.M. Fryer, D. Treves: Observation of domains in iron whiskers under high fields. J. Appl. Phys. **31**, 2267–2272 (1960)

1082 C.A. Fowler, Jr., E.M. Fryer: Domain structure in iron whiskers as observed by the Kerr method. J. Appl. Phys. (Suppl.) **32**, 296S–297S (1961)

1083 P.T. Squire, D. Atkinson, M.R.J. Gibbs, S. Atalay: Amorphous wires and their application. J. Magn. Magn. Mat. **132**, 10–21 (1994)

1084 M. Vázquez, A. Hernando: A soft magnetic wire for sensor applications. J. Phys. D **29**, 939–949 (1996)

1085 J. Yamasaki, F.B. Humphrey, K. Mohri, H. Kawamura, H. Takamure, R. Mälmhäll: Large Barkhausen discontinuities in Co-based amorphous wires with negative magnetostriction. J. Appl. Phys. **63**, 3949–3951 (1988)

1086 M. Soeda, M. Takajo, J. Yamasaki, F.B. Humphrey: Large Barkhausen discontinuities of die-drawn Fe-Si-B amorphous wire. IEEE Trans. Magn. **31**, 3877–3879 (1995)

1087 D. Atkinson, P.T. Squire, M.R.J. Gibbs, S.N. Hogsdon: Implications of magnetic and magnetoelastic measurements for the domain structure of FeSiB amorphous wires. J. Phys. D **27**, 1354–1362 (1994)

1088 K. Mohri, F.B. Humphrey, K. Kawashima, K. Kimura, M. Mizutani: Large Barkhausen and Matteucci effects in FeCoSiB, FeCrSiB, and FeNiSiB amorphous wires. IEEE Trans. Magn. **26**, 1789–1791 (1990)

6. The Relevance of Domains

1089 R.M. Bozorth: *Ferromagnetism* (Van Nostrand, Princeton, 1951)
1090 B.D. Cullity: *Introduction to Magnetic Materials* (Addison-Wesley, Reading MA, 1972)
1091 E.P. Wohlfarth, K.H.J. Buschow (Ed.): *Ferromagnetic Materials*, Vol. 1–11
        (North Holland, Amsterdam, 1980f)
1092 J. Evetts (Ed.): *Concise Encyclopedia of Magnetic & Superconducting Materials*,
        in: Advances in Materials Science and Engineering (Pergamon, Oxford, New York, 1992)
1093 J.P. Jakubovics: *Magnetism and Magnetic Materials* (The Intitute of Materials, London, 1994)
1094 R. Gerber, C.D. Wright, G. Asti (Ed.): *Applied Magnetism* (Kluwer, Dordrecht, 1994)

Bulk Soft Magnetic Materials

1095 M. Mallary, A.B. Smith: Conduction of flux at high frequencies by a charge-free magnetization
        distribution. IEEE Trans. Magn. **24**, 2374–2376 (1988)
1096 G.Y. Chin, J.H. Wernick: Soft magnetic metallic materials, in: *Ferromagnetic Materials*, Vol. 2,
        ed. by E.P. Wohlfarth (North Holland, Amsterdam, 1980) p. 55–188
1097 A.J. Moses: Development of alternative magnetic core materials and incentives for their use.
        J. Magn. Magn. Mat. **112**, 150–155 (1992)
1098 C.D. Graham, Jr.: Textured Magnetic Materials, in: *Magnetism and Metallurgy*, Vol. 2,
        ed. by A.E. Berkowitz, E. Kneller (Academic Press, New York, 1969) p. 723–748
1099 K. Foster, J.J. Kramer: Effect of directional orientation on the magnetic properties of cube-oriented
        magnetic sheets. J. Appl. Phys. (Suppl.) **31**, 233*S*–234*S* (1960)
1100 M.F. Littmann: Structures and magnetic properties of grain-oriented 3.2% silicon-iron.
        J. Appl. Phys. **38**, 1104–1108 (1967)
1101 W.M. Swift, W.R. Reynolds, J.W. Shilling: Relationship between statistical distribution of grain
        orientations and B_{10} in polycrystalline (110)[001] 3% Si-Fe sheet.
        AIP Conf. Proc. **10**, 976–980 (1973)
1102 N.P. Goss: New development in electrical strip steels characterized by fine grain structure approaching
        the properties of a single crystal. Trans. Amer. Soc. Metals **23**, 511–544 (1935)
1103 T. Yamamoto, T. Nozawa: Recent development of high-permeability grain-oriented silicon steels,
        in: *Recent Magnetics for Electronics*, ed. by Y. Sakurai (OHM, Tokyo, 1983) p. 227–243
1104 Y. Ushigami, H. Masui, Y. Okazaki, Y. Sugi, N. Takahashi: Development of low-loss grain-oriented
        silicon steel. J. Mater. Eng. Perform. **5**, 310–315 (1996)
1105 T. Nozawa, M. Mizogami, H. Mogi, Y. Matsuo: Magnetic properties and dynamic domain behavior
        in grain-oriented 3% Si-Fe. IEEE Trans. Magn. **32**, 572–589 (1996)
1106 W. Grimm, W. Jillek, A. Hubert: Messung und Auswirkungen der von den Isolierschichten auf
        orientiertes Transformatorenblech übertragenen Spannungen. J. Magn. Magn. Mat. **9**, 225–228 (1978)
        (Measurement and effect of stresses generated by insulation layers on oriented transformer steel)
1107 T. Yamamoto, T. Nozawa: Effects of tensile stress on total loss of single crystals of 3 % silicon-iron.
        J. Appl. Phys. **41**, 2981–2984 (1970)
1108 J.W. Shilling, W.G. Morris, M.L. Osborn, P. Rao: Orientation dependence of domain wall spacing
        and losses in 3-percent Si-Fe single crystals. IEEE Trans. Magn. **14**, 104–111 (1978)
1109 A. Hubert: Some comments on the problem of the "ideal" oriented transformer steel.
        Soft Magnetic Materials 3 (Bratislava, 1977) p. 291–295
1110 T. Iuchi, S. Yamaguchi, T. Ichiyama, N. Nakamura, T. Ishimoto, K. Kuroki: Laser processing for
        reducing core loss of grain oriented silicon steel. J. Appl. Phys. **53**, 2410–2412 (1982)
1111 M. Yabumoto, H. Kobayashi, T. Nozawa, K. Hirose, N. Takahashi: Heatproof domain refining method
        using chemically etched pits on the surface of grain-oriented 3% Si-Fe.
        IEEE Trans. Magn. **23**, 3062–3064 (1987)
1112 H. Kobayashi, K. Kuroki, E. Sasaki, M. Iwasaki, N. Takahashi: Heatproof domain refining method
        using combination of local strain and heat treatment for grain oriented 3% Si-Fe.
        Physica Scripta **T24**, 36–41 (1988)
1113 T. Nozawa, Y. Matsuo, H. Kobayashi, K. Iwayama, N. Takahashi: Magnetic properties and domain
        structures in domain refined grain-oriented silicon steel. J. Appl. Phys. **63**, 2966–2970 (1988)
1114 B.K. Sokolov, V.V. Gubernatorov, V.A. Zaykova, Y.N. Dragoshansky: Influence of substructure
        redistribution on electromagnetic losses of transformer steel. Phys. Met. Metall. **44**.3, 53–58 (1977)
        [Fiz. Metal. Metalloved. 44 (1977) 517–522]
1115 V.A. Zaykova, N.K. Yesina, Y.N. Dragoshanskiy, V.F. Tiunov, B.K. Sokolov, V.V. Gubernatorov:
        Orientation and structure dependencies of the electromagnetic losses of locally deformed single crystals
        of Fe-3% Si. Phys. Met. Metall. **48**.3, 57–65 (1979) *[Fiz. Metal. Metalloved. 48 (1979) 520–529]*

1116 T. Yamamoto, T. Nozawa, T. Nakayama, Y. Matsuo: The influence of grain-orientation on 180° domain wall spacing in (110) [001] grain-oriented 3% Si-Fe with high permeability. J. Magn. Magn. Mat. **31**, 993–996 (1983)

1117 T. Nozawa, M. Yabumoto, Y. Matsuo: Studies of domain refining of grain-oriented silicon steel. Soft Magnetic Materials 7 (Blackpool, 1985). Wolfson Centre of Magnetics Techn., Cardiff. p. 131–136

1118 T. Nozawa, T. Nakayama, Y. Ushigami, T. Yamamoto: Production of single-crystal 3% silicon-iron sheet with orientation near (110)[001]. J. Magn. Magn. Mat. **58**, 67–77 (1986)

1119 T.R. Haller, J.J. Kramer: Observation of dynamic domain size variation in a silicon-iron alloy. J. Appl. Phys. **41**, 1034–1035 (1970)

1120 K. Honma, T. Nozawa, H. Kobayashi, Y. Shimoyama, I. Tachino, K. Miyoshi: Development of non-oriented and grain-oriented silicon steel. IEEE Trans. Magn. **21**, 1903–1908 (1985)

1121 H.P.J. Wijn, E.W. Gorter, C.J. Esveldt, P. Geldermans: Bedingungen für eine rechteckige Hystereseschleife bei Ferriten. Philips Techn. Rundschau **16**, 124–134 (1954) *(Conditions for a rectangular hysteresis loop in ferrites)*

1122 J.W. Shilling: Domain structure during magnetization of grain-oriented 3% Si-Fe as a function of applied tensile stress. J. Appl. Phys. **42**, 1787–1789 (1971)

1123 J. Zbroszczyk, J. Drabecki, B. Wysłocki: Angular-distribution of rotational hysteresis losses in Fe-3.25%Si single crystals with orientations (001) and (011). IEEE Trans. Magn. **17**, 1275–1282 (1981)

1124 K. Matsumura, B. Fukuda: Recent development of non-oriented electrical steel sheets. IEEE Trans. Magn. **20**, 1533–1538 (1984)

1125 F. Assmus, R. Boll, D. Ganz, F. Pfeifer: Über Silizium-Eisen mit Würfeltextur. I. Magnetische Eigenschaften. Z. Metallk. **48**, 341–343 (1957) *(On cube-textured silicon iron. I. Magnetic properties)*

1126 D. Ganz: Über die Richtungsabhängigkeit magnetischer Eigenschaften von 3%-Silizium-Eisen-Blechen mit Goss- und Würfeltextur. Z. Angew. Phys. **14**, 313–322 (1962) *(On the directional dependence of magnetic properties of 3% silicon iron sheets with Goss and cube texture)*

1127 K. Detert: Untersuchungen über eine neue Art der Sekundärrekristallisation in Fe-3%Si-Legierungen. Acta Metall. **7**, 589–598 (1959) *(Investigations on a new kind of secondary recrystallization in Fe-3%Si alloys)*

1128 S. Arai, M. Mizokami, M. Yabumoto: The magnetic properties of doubly oriented 3% Si-Fe and its application. J. Appl. Phys. **67**, 5577–5579 (1990)

1129 S.L. Ames, G.L. Houze, Jr., W.R. Bitler: Magnetic properties of textured silicon-iron alloys with silicon contents in excess of 3.25 %. J. Appl. Phys. **38**, 1577–1578 (1969)

1130 T. Yamaji, M. Abe, Y. Takada, K. Okada, T. Hiratani: Magnetic properties and workability of 6.5% silicon steel sheet manufactured in continuous CVD siliconizing line. J. Magn. Magn. Mat. **133**, 187–189 (1994)

1131 S. Roth: Effect of annealing in hydrogen on composition and properties of rapidly quenched Fe-Co-B and Fe-B-Si alloy ribbons. Mat. Sci. Eng. A **226**, 111–114 (1997)

1132 N. Tsuya, K.I. Arai, K. Ohmori, H. Shimanaka, T. Kan: Ribbon-form silicon-iron alloy containing around 6% silicon. IEEE Trans. Magn. **16**, 728–733 (1980)

1133 F. Pfeifer, C. Radeloff: Soft magnetic Ni-Fe and Co-Fe alloys — Some physical and metallurgical aspects. J. Magn. Magn. Mat. **19**, 190–207 (1980)

1134 R. Boll, K. Overshott (Ed.): *Magnetic Sensors,* in: Sensors. A Comprehensive Survey, ed. by W. Göpel, *et al.*, Vol. 5 (VCH, Weinheim, 1989)

1135 J.E. Lenz: A review of magnetic sensors. Proc. IEEE **78**, 973–989 (1990)

1136 K. Mohri: Magnetic materials for new sensors — Sensor magnetics, in: Magnetic Materials: Microstructure and Properties (Anaheim, 1991). MRS Symp. Proc. Vol. **232**, p. 331–342

1137 T. Bonnema, F.J. Friedlaender, L.F. Silva: Magnetic domain observations in 50-percent Ni-Fe tapes. IEEE Trans. Magn. **4**, 431–434 (1968)

1138 D.S. Rodbell, C.P. Bean: Influence of pulsed magnetic fields on the reversal of magnetization in square-loop metallic tapes. J. Appl. Phys. **26**, 1318–1323 (1955)

1139 F.J. Friedlaender: Flux reversal in magnetic amplifier cores. AIEE Trans. I (Comm. & Electr.) **75**, 268–278 (1956)

1140 R. Boll: Soft magnetic metals and alloys. Mat. Sci. Tech. **3B**, 401–450 (1993)

1141 F.E. Luborsky: Amorphous ferromagnets, in: *Ferromagnetic Materials*, Vol. 1, ed. by E.P. Wohlfarth (North Holland, Amsterdam, 1980) p. 451–529

1142 R. Hasegawa: Amorphous magnetic materials – A history. J. Magn. Magn. Mat. **100**, 1–12 (1991)

1143 F.E. Luborsky, M.R.J. Gibbs: Metallic Glasses, in: *Concise Encyclopedia of Magnetic and Superconducting Materials*, ed. by J. Evetts (Pergamon, Oxford, 1992) p. 314–319

1144 T. Sato: Further improvement of core loss in amorphous alloys. J. Mater. Eng. Perform. **2**, 235–240 (1993)

1145 K. Foster, M.F. Littmann: Factors affecting core losses in oriented electrical steels at moderate inductions. J. Appl. Phys. **57**, 4203–4208 (1985)

1146 T. Sato: Core materials for power devices, in: *Recent Magnetics for Electronics*, ed. by Y. Sajurai (OHM, Tokyo, 1980) p. 151–168

1147 Y. Okazaki: Loss deterioration in amorphous cores for distribution transformers. J. Magn. Magn. Mat. **160**, 217–222 (1996)

1148 L.I. Mendelsohn, E.A. Nesbitt, G.R. Bretts: Glassy metal fabric: A unique magnetic shield. IEEE Trans. Magn. **12**, 924–926 (1976)

1149 G. Hinz, H. Voigt: Magnetoresistive sensors, in: *Magnetic Sensors*, Vol. 5, ed. by R. Boll, K. Overshott (VCH, Weinheim, 1989) p. 97–152

1150 H. Fujimori, H. Yoshimoto, H. Morita: Anomalous eddy current loss in amorphous magnetic thin sheet and its improvement. IEEE Trans. Magn. **16**, 1227–1229 (1980)

1151 H.J. de Wit, M. Brouha: Domain patterns and high-frequency magnetic properties of amorphous metal ribbons. J. Appl. Phys. **57**, 3560–3562 (1985)

1152 G. Rauscher, C. Radeloff: Wiegand and pulse wire sensors, in: *Magnetic Sensors*, Vol. 5, ed. by R. Boll, K. Overshott (VCH, Weinheim, 1989) p. 315–339

1153 K.-H. Shin, C.D. Graham, Jr., P.Y. Zhou: Asymmetric hysteresis loops in Co-based ferromagnetic alloys. IEEE Trans. Magn. **28**, 2772–2774 (1992)

1154 C.D. Graham: Magnetic materials in the fight against crime. Magnews (Spring 1993) 8–10

1155 C.K. Kim, R.C. O'Handley: Development of a pinned wall sensor using cobalt-rich, near-zero magnetostrictive amorphous alloys. Metall. Mater. Trans. A **28**, 423–434 (1997)

1156 C.K. Kim, R.C. O'Handley: Development of a magnetoelastic resonant sensor using iron-rich, non-zero magnetostrictive amorphous alloys. Metall. Mat. Trans. A **27**, 3203–3213 (1996)

1157 K. Mohri, T. Kohzawa, K. Kawashima, H. Yoshida, L.V. Panina: Magneto-inductive effect (MI effect) in amorphous wires. IEEE Trans. Magn. **28**, 3150–3152 (1992)

1158 K. Mohri, K. Kawashima, T. Kohzawa, H. Yoshida: Magneto-inductive element. IEEE Trans. Magn. **29**, 1245–1248 (1993)

1159 L.V. Panina, K. Mohri, K. Bushida, M. Noda: Giant magneto-impedance and magneto-inductive effects in amorphous alloys. J. Appl. Phys. **76**, 6198–6203 (1994)

1160 R.S. Beach, A.E. Berkowitz: Sensitive field- and frequency-dependent impedance spectra of amorphous FeCoSiB wire and ribbon. J. Appl. Phys. **76**, 6209–6213 (1994)

1161 K.V. Rao, F.B. Humphrey, J.L. Costa-Krämer: Very large magneto-impedance in amorphous soft magnetic wires. J. Appl. Phys. **76**, 6204–6208 (1994)

1162 Y. Yoshizawa, Y. Oguma, K. Yamauchi: New Fe-based soft magnetic alloys composed of ultrafine grain structure. J. Appl. Phys. **64**, 6044–6046 (1988)

1163 G. Herzer: Nanocrystalline soft magnetic alloys, in: *Handbook of Magnetic Materials*, Vol. 10, ed. by K.H.J. Buschow (Elsevier, Amsterdam, 1997) p. 415–462

1164 G. Herzer: Grain size dependence of coercivity and permeability in nanocrystalline ferromagnets. IEEE Trans. Magn. **26**, 1397–1402 (1990)

1165 A. Makino, T. Hatanai, Y. Yamamoto, N. Hasegawa, A. Inoue, T. Masumoto: Magnetic domain structure correlated with the microstructure of nanocrystalline Fe-M-B (M=Zr,Nb) alloys. Sci. Rep. Res. Inst. Tohoku Univ. A **42**, 107–113 (1996)

1166 M. Müller, H. Grahl, N. Mattern, U. Kühn: Crystallization behaviour, structure and magnetic properties of nanocrystalline FeZrNbBCu alloys. Mater. Sci. Eng. A **226**, 565–568 (1997)

1167 A. Makino, T. Hatanai, A. Inoue, T. Masumoto: Nanocrystalline soft magnetic Fe-M-B (M = Zr, Hf, Nb) alloys and their applications. Mater. Sci. Eng. A **226**, 594–602 (1997)

1168 P.I. Slick: Ferrites for non-microwave applications, in: *Ferromagnetic Materials*, Vol. 2, ed. by E.P. Wohlfarth (North Holland, Amsterdam, 1980) p. 189–241

1169 R.C. Sundahl, Jr., Y.S. Kim: Ferrites, soft, in: *Concise Encyclopedia of Magnetic and Superconducting Materials*, ed. by J. Evetts (Pergamon, Oxford, 1992) p. 135–141

1170 T.G.W. Stijntjes, J. Klerk, A. Broese van Groenau: Permeabilities and conductivity of Ti-substituted MnZn ferrites. Philips Res. Repts. **25**, 95–107 (1970)

1171 E. Röss: Magnetic properties and microstructure of high-permeability Mn-Zn ferrites. Int. Conf. Ferrites (Japan, 1970) p. 203–209

1172 J. Nicholas: Microwave ferrites, in: *Ferromagnetic Materials*, Vol. 2, ed. by E.P. Wohlfarth (North Holland, Amsterdam, 1980) p. 243–296

1173 R. Schäfer, B.E. Argyle, P.L. Trouilloud: Domain studies in single-crystal ferrite MIG heads with image-enhanced, wide-field Kerr microscopy. IEEE Trans. Magn. **28**, 2644–2646 (1992)

1174 M. Kaneko, K. Aso: Magneto-optical observation of magnetic domains in rubbing surface of a ferrite head. IEEE Transl. J. Magn. Japan **5**, 225–231 (1990)

Permanent Magnets

1175 E.P. Wohlfarth: Permanent magnetic materials, in: *Magnetism*, Vol. III, ed. by G.T. Rado, H. Suhl (Academic Press, New York, 1963) p. 351–393

1176 K.J. Strnat: Modern permanent magnets for applications in electro-technology.
      Proc. IEEE **78**, 923–946 (1990)

1177 G. Asti, M. Solzi: Permanent magnets, in: *Applied Magnetism*, ed. by R. Gerber, *et al.*
      (Kluwer, Dordrecht, 1994) p. 309–375

1178 H.R. Kirchmayr: Permanent magnets and hard magnetic materials.
      J. Phys. D: Appl. Phys. **29**, 2763–2778 (1996)

1179 J.M.D. Coey: Permanent magnetism. Solid State Comm. **102**, 101–105 (1997)

1180 K.H.J. Buschow: Magnetism and processing of permanent magnet materials, in: *Handbook of Magnetic Materials*, Vol. 10, ed. by K.H.J. Buschow (Elsevier, Amsterdam, 1997) p. 463–593

1181 J.D. Livingston: A review of coercivity mechanisms. J. Appl. Phys. **52**, 2544–2548 (1981)

1182 J.D. Livingston: Microstructure and properties of rare earth magnets. Report 86CRD159
      (General Electric Corp. Res. Dev., 1986)

1183 A. Menth, H. Nagel, R.S. Perkins: New high-performance permanent magnets based on rare earth-transition metal compounds, in: *Annual Review of Materials Science*, Vol. 8,
      ed. by R.A. Huggins, *et al.* (Annual Review Inc., Palo Alto, CA, 1978) p. 21–47

1184 H. Stäblein: Hard ferrites and plastoferrites, in: *Ferromagnetic Materials*, Vol. 3,
      ed. by E.P. Wohlfarth (North Holland, Amsterdam, 1982) p. 441–602

1185 G. Hoffer, K. Strnat: Magnetocrystalline anisotropy of YCo_5 and Y_2Co_{17}.
      IEEE Trans. Magn. **2**, 487–489 (1966)

1186 K.J. Strnat, R.M.W. Strnat: Rare earth-cobalt permanent magnets.
      J. Magn. Magn. Mat. **100**, 38–56 (1991)

1187 M. Sagawa, S. Fujimura, N. Togawa, H. Yamamoto, Y. Matsuura: New material for permanent magnets on a base of Nd and Fe. J. Appl. Phys. **55**, 2083–2087 (1984)

1188 J.F. Herbst, J.J. Croat: Neodymium-iron-boron permanent magnets.
      J. Magn. Magn. Mat. **100**, 57–78 (1991)

1189 J.J. Becker: Observation of magnetization reversal in cobalt-rare-earth particles.
      IEEE Trans. Magn. **5**, 211–214 (1969)

1190 K.-D. Durst, H. Kronmüller, H. Schneider: Magnetic hardening mechanisms in Fe-Nd-B type permanent magnets. 5th Int. Symp. on Magnetic Anisotropy and Coercivity in Rare Earth-Transition Metal Alloys (Bad Soden, 1987). DPG Vol. **II**, p. 209–225

1191 E. Adler, P. Hamann: A contribution to the understanding of coercivity and its temperature dependence in sintered $SmCo_5$ and $Nd_2Fe_{14}B$ magnets. 4th Int. Symp. Magn. Anisotr. Coercivity Rare Earth-Trans. Metal Alloys (Dayton (Ohio), 1985). Univ. of Dayton p. 747–760

1192 D. Givord, P. Tenaud, T. Viadieu: Coercivity mechanisms in ferrite and rare earth transition metal sintered magnets (SmCo5, Nd-Fe-B). IEEE Trans. Magn. **24**, 1921–1923 (1988)

1193 S. Chikazumi: Mechanism of high coercivity in rare-earth permanent magnets.
      J. Magn. Magn. Mat. **54**, 1551–1555 (1986)

1194 G.C. Hadjipanayis, A. Kim: Domain wall pinning versus nucleation of reversed domains in R-Fe-B magnets. J. Appl. Phys. **63**, 3310–3315 (1988)

1195 B. Grieb: New corrosion resistant materials based on neodym-iron-boron.
      IEEE Trans. Magn. **33**, 3904–3906 (1997)

1196 D. Eckert, K.H. Müller, P. Nothnagel, J. Schneider, R. Szymczak: Magnetization curves of thermal ac-field and dc-field demagnetized NdFeB magnets. J. Magn. Magn. Mat. **83**, 197–198 (1990)

1197 J. Zawadzki, P.A.P. Wendhausen, B. Gebel, A. Handstein, D. Eckert, K.-H. Müller: Kerr microscopy observation of carbon diffusion profiles in $Sm_2Fe_{17}C_x$. J. Appl. Phys. **76**, 6717–6719 (1994)

1198 J.M.D. Coey, H. Sun: Improved magnetic properties by treatment of iron-based rare earth intermetallic compounds in ammonia. J. Magn. Magn. Mat. **87**, L251–254 (1990)

1199 H. Sun, Y. Otani, J.M.D. Coey: Gas-phase carbonation in R_2Fe_{17}.
      J. Magn. Magn. Mat. **104**, 1439–1440 (1991)

1200 T. Mukai, T. Fujimoto: Kerr microscopy observation of nitrogenated Sm_2Fe_{17} intermetallic compounds.
      J. Magn. Magn. Mat. **103**, 165–173 (1992)

1201 J. Hu, B. Hofmann, T. Dragon, R. Reisser, *et al.*: Carbonation process and domain structure in $Sm_2Fe_{17}C_x$ compounds prepared by gas-solid interaction. Phys. Status Solidi A **148**, 275–282 (1995)

1202 J.D. Livingston, D.L. Martin: Microstructure of aged (Co, Cu, Fe)$_7$Sm magnets.
      J. Appl. Phys. **48**, 1350–1354 (1977)

1203 K.-D. Durst, H. Kronmüller, W. Ervens: Investigation of the magnetic properties and demagnetization processes of an extremely high coercive $Sm(Co,Fe,Cu,Zr)_{7.6}$ permanent magnet. II The coercivity mechanism. Phys. Status Solidi A **108**, 705–719 (1988)

1204 B.Y. Wong, M. Willard, D.E. Laughlin: Domain wall pinning sites in $Sm(CoFeCuZr)_x$ magnets.
      J. Magn. Mag. Mat. **169**, 178–192 (1997)

1205 M. Katter: Coercivity calculation of $Sm_2(Co,Fe,Cu,Zr)_{17}$ magnets. J. Appl. Phys. **83**, 6721–6723 (1998)

1206 F.E. Luborsky: Development of elongated particle magnets.
      J. Appl. Phys. (Suppl.) **32**, 171S–183S (1961)

1207   J.J. Croat: Current status and future outlook for bonded neodymium permanent magnets.
       J. Appl. Phys. **81**, 4804–4809 (1997)

1208   J.M.D. Coey, K. O'Donnell: New bonded magnet materials. J. Appl. Phys. **81**, 4810–4815 (1997)

1209   J. Ormerod, S. Constatinides: Bonded permanet magnets: Current status and future opportunities.
       J. Appl. Phys. **81**, 4816–4820 (1997)

1210   K.J. de Vos: Alnico Permanent Magnet Alloys, in: *Magnetism and Metallurgy*, Vol. 1,
       ed. by A.E. Berkowitz, E. Kneller (Academic Press, New York, 1969) p. 473–512

1211   R.A. McCurrie: The structure and properties of Alnico permanent magnet alloys, in: *Ferromagnetic
       Materials*, Vol. 3, ed. by E.P. Wohlfarth (North Holland, Amsterdam, 1982) p. 107–188

1212   H. Stäblein: Dauermagnetwerkstoffe auf Alnico-Basis. Techn. Mitteilungen Krupp **29**, 101–110 (1971)
       (Alnico-based permanent magnet materials)

1213   H. Kaneko, M. Homma, K. Nakamura: New ductile permanent magnet of Fe-Cr-Co system.
       AIP Conf. Proc. **5**, 1088–1092 (1971)

1214   L.I. Mendelsohn, F.E. Luborsky, T.O.Paine: Permanent-magnet properties of elongated single-domain
       iron particles. J. Appl. Phys. **26**, 1274–1280 (1955)

1215   R.B. Falk: A current review of Lodex permanent magnet technology.
       J. Appl. Phys. **37**, 1108–1112 (1966)

1216   J.J. Croat: Magnetic hardening of Pr-Fe and Nd-Fe alloys by melt-spinning.
       J. Appl. Phys. **53**, 3161–3169 (1982)

1217   L. Schultz, J. Wecker, E. Hellstern: Formation and properties of NdFeB prepared by mechanical
       alloying and solid-state reaction. J. Appl. Phys. **61**, 3583–3585 (1987)

1218   L. Schultz, K. Schnitzke, J. Wecker: Mechanically alloyed isotropic and anisotropic Nd-Fe-B magnetic
       material. J. Appl. Phys. **64**, 5302–5304 (1988)

1219   L. Schultz, K. Schnitzke, J. Wecker, M. Katter, C. Kuhrt: Permanent magnets by mechanical alloying.
       J. Appl. Phys. **70**, 6339–6344 (1991)

1220   P.G. McCormick, J. Ding, E.H. Feutrill, R. Street: Mechanically alloyed hard magnetic materials.
       J. Magn. Magn. Mat. **157**, 7–10 (1996)

1221   T. Takeshita, R. Nakayama: Magnetic properties and microstructure of the NdFeB magnet powder
       produced by hydrogen treatment. 10th Int. Workshop on Rare Earth Magnets and their Applications
       (Kyoto, 1989) Vol. **1**, p. 551–557

1222   O. Gutfleisch, I.R. Harris: Fundamental and practical aspects of hydrogenation, disproportionation,
       desorption and recombination process. J. Phys. D: Appl. Phys. **29**, 2255–2265 (1996)

1223   R.W. Lee, E.G. Brewer, N.A. Schaffel: Processing of neodymium-iron-boron melt-spun ribbons to
       fully dense magnets. IEEE Trans. Magn. **21**, 1958–1963 (1985)

1224   Y. Nozawa, K. Iwasaki, S. Tanigawa, M. Tokunaga, H. Harada: Nd-Fe-B die-upset and anisotropic
       bonded magnets. J. Appl. Phys. **64**, 5285–5289 (1988)

1225   T. Takeshita, R. Nakayama: Magnetic properties and microstructures of the Nd-Fe-B magnet powders
       produced by the hydrogen treatment-(III). 11th Int. Workshop on Rare Earth Magnets and their
       Applications (Pittsburgh, 1990). Carnegie Mellon Univ. Vol. **1**, p. 49–71

1226   R. Nakayama, T. Takeshita: Nd-Fe-B anisotropic magnet powders by the HDDR process.
       J. Alloys Comp. **193**, 259–261 (1993)

1227   C. Short, P. Guegan, O. Gutfleisch, O.M. Ragg, I.R. Harris: HDDR processes in $Nd_{16}Fe_{76-x}Zr_xB_8$ alloys
       and the production of anisotropic magnets. IEEE Trans. Magn. **32**, 4368–4370 (1996)

1228   M. Uehara, T. Tomida, H. Tomizawa, S. Hirosawa, Y. Maehara: Magnetic domain structure of aniso-
       tropic $Nd_2Fe_{14}B$-based magnets produced via the hydrogenation, decomposition, desorption and recom-
       bination (HDDR) process. J. Magn. Magn. Mat. **159**.3, L304–308 (1996)

1229   P. Thompson, O. Gutfleisch, J.N. Chapman, I.R. Harris: Domain studies in thin sections of HDDR pro-
       cessed Nd-Fe-B-type magnets by TEM. J. Magn. Magn. Mat. **177**, 978–979 (1998)

1230   R.W. McCallum, A.M. Kadin, G.B. Clemente, J.E. Keem: High performance isotropic permanent
       magnet based on Nd-Fe-B. J. Appl. Phys. **61**, 3577–3579 (1987)

1231   G.B. Clemente, J.E. Keem, J.P. Bradley: The microstructural and compositional influence upon
       HIREM behavior of $Nd_2Fe_{14}B$. J. Appl. Phys. **64**, 5299–5301 (1988)

1232   H. Kronmüller: Micromagnetism of hard magnetic nanocrystalline materials.
       Nanostruct. Mater. **6**, 157–168 (1995)

1233   W. Rave, K. Ramstöck: Micromagnetic calculation of the grain size dependence of remanence and
       coercivity in nanocrystalline permanent magnets. J. Magn. Magn. Mat. **171**, 69–82 (1997)

1234   R. Fischer, T. Schrefl, H. Kronmüller, J. Fidler: Grain-size dependence of remanence and coercive
       field of isotropic nanocrystalline composite permanent magnets.
       J. Magn. Magn. Mat. **153**, 35–49 (1996)

1235   R. Coehoorn, D.B. de Mooij, J.P.W.B. Duchateau, K.H.J. Buschow: Novel permanent magnetic
       materials made by rapid quenching. J. de Phys. (Colloque) **49**, C8-669–670 (1989)

1236 W. Coene, F. Hakkens, R. Coehoorn, D.B. de Mooij, C. de Waard, J. Fidler, R. Grössinger: Magne-tocrystalline anisotropy of Fe_3B, Fe_2B and $Fe_{1.4}Co_{0.6}B$ as studied by Lorentz electron microscopy, singular point detection and magnetization measurements. J. Magn. Magn. Mat. **96**, 189–196 (1991)

1237 J. Wecker, K. Schnitzke, H. Cerva, W. Grogger: Nanostructured Nd-Fe-B magnets with enhanced remanence. Appl. Phys. Lett. **67**, 563–565 (1995)

1238 D. Eckert, K.-H. Müller, A. Handstein, J. Schneider, R. Grössinger, R. Krewenka: Temperature dependence of the coercive force in $Nd_4Fe_{77}B_{19}$. IEEE Trans. Magn. **26**, 1834–1836 (1990)

1239 K.-H. Müller, D. Eckert, A. Handstein, M. Wolf, S. Wirth, L.M. Martinez: Viscosity and magnetization processes in annealed melt-spun $Nd_4Fe_{77}B_{19}$. 8th Int. Symp. on Magnetic Anisotropy and Coercivity in RE-TM Alloys (Birmingham, 1994), p. 179–187

1240 S. Hirosawa, H. Kanekiyo, M. Uehara: High-coercivity iron-rich rare-earth permanent magnet material based on (Fe, Co)$_3$B-Nd-M (M = Al, Si, Cu, Ga, Ag, Au). J. Appl. Phys. **73**, 6488–6490 (1993)

1241 R. Skomski, J.M.D. Coey: Nucleation field and energy product of aligned two-phase magnets – Progress towards the '1 MJ/m^3' magnet. IEEE Trans. Magn. **29**, 2860–2862 (1993)

1242 R. Skomski: Aligned two-phase magnets: Permanent magnetism of the future? J. Appl. Phys. **76**, 7059–7064 (1994)

1243 E.F. Kneller, R. Hawig: The exchange spring magnet: A new material principle for permanent magnets. IEEE Trans. Magn. **27**, 3588–3600 (1991)

1244 R. Skomski, J.M.D. Coey: Giant energy product in nanostructured two-phase magnets. Phys. Rev. B **48**, 15812–15816 (1993)

1245 T. Leineweber, H. Kronmüller: Micromagnetic examination of exchange coupled ferromagnetic nano-layers. J. Magn. Magn. Mat. **176**, 145–154 (1997)

1246 J. Schneider, D. Eckert, K.-H. Müller, A. Handstein, H. Mühlbach, H. Sassik, H.R. Kirchmayr: Mag-netization processes in $Nd_4Fe_{77}B_{19}$ permanent magnetic materials. Materials Lett. **9**, 201–203 (1990)

1247 S. Hirosawa, A. Kanekiyo: Exchange-coupled permanent magnets based on α-Fe/$Nd_2Fe_{14}B$ nano-crystalline composite. 13th Int. Workshop on Rare Earth Magnets and their Applications (Birmingham, 1994) p. 87–94

1248 E.H. Feutrill, P.G. McCormick, R. Street: Magnetization behavior in exchange-coupled $Sm_2Fe_{14}Ga_3C_2$/α-Fe. J. Phys. D: Appl. Phys. **29**, 2320–2326 (1996)

1249 I. Panagiotopoulos, L. Withanawasam, G.C. Hadjipanayis: Exchange spring behavior in nanocomposite hard magnetic materials. J. Magn. Magn. Mat. **152**, 353–358 (1996)

Recording Media

1250 E. Köster, T.C. Arnoldussen: Recording media, in: *Magnetic Recording*, Vol. I, ed. by C.D. Mee, E.D. Daniel (McGraw-Hill, New York, 1987) p. 98–243

1251 G. Bate: Magnetic recording materials since 1975. J. Magn. Magn. Mat. **100**, 413–424 (1991)

1252 K.E. Johnson: Fabrication of low noise thin-film media, in: *Noise in Digital Magnetic Recording*, ed. by T.C. Arnoldussen, L.L. Nunnelley (World Scientific, Singapore, 1992) p. 7–63

1253 T.C. Arnoldussen, L.L. Nunnelley (Ed.): *Noise in Digital Magnetic Recording* (World Scientific, Singapore, 1992)

1254 C.D. Mee, E.D. Daniel (Ed.): *Magnetic Recording*, Vol. I (McGraw-Hill, New York, 1987)

1255 H.N. Bertram: *Theory of Magnetic Recording* (Cambridge University Press, Cambridge, 1994)

1256 S.Y. Chou, M.S. Wei, P.R. Krauss, P.B. Fischer: Single-domain magnetic pillar array of 35 nm diameter and 65 Gbits/in^2 density for ultrahigh density quantum magnetic storage. J. Appl. Phys. **76**, 6673–6675 (1994)

1257 R.M.H. New, R.F.W. Pease, R.L. White: Lithographically patterned single-domain cobalt islands for high-density magnetic recording. J. Magn. Magn. Mat. **155**, 140–145 (1996)

1258 M. Löhndorf, A. Wadas, G. Lütjering, D. Weiss, R. Wiesendanger: Micromagnetic properties and magnetization switching of single domain Co dots studied by magnetic force microscopy. Z. Phys. B **101**, 1–2 (1996) *[Erratum: Z. Phys. B 102 (1997) 289]*

1259 S.Y. Chou, P.R. Krauss, L. Kong: Nanolithographically defined magnetic structures and quantum magnetic disk. J. Appl. Phys. **79**, 6101–6106 (1996)

1260 R.L. White, R.M.H. New, R.F.W. Pease: Patterned media: A viable route to 50 Gbit/in^2 and up for magnetic recording? IEEE Trans. Magn. **33**, 990–995 (1997)

1261 P.-L. Lu, S.H. Charap: Thermal instability at 10 Gbit/in^2 magnetic recording. IEEE Trans. Magn. **30**, 4230–4232 (1994)

1262 J.F. Smyth, S. Schultz, D.R. Fredkin, D.P. Kern, *et al.*: Hysteresis in lithographic arrays of Permalloy particles: Experiment and theory. J. Appl. Phys. **69**, 5262–5266 (1991)

1263 R.M.H. New, R.F.W. Pease, R.L. White: Submicron patterning of thin cobalt films for magnetic storage. J. Vac. Sci. Technol. B **12**, 3196–3201 (1994)

1264 M.A.M. Haast, J.R. Schuurhuis, L. Abelmann, J.C. Lodder, T.J. Popma: Reversal mechanism of submicron patterned CoNi/Pt multilayers. IEEE Trans. Magn. **34**, 1006–1008 (1998)

1265 R.W. Chantrell, K. O'Grady: Magnetic characterization of recording media.
      J. Phys. D **25**, 1–23 (1992)

1266 E.P. Wohlfarth: Relations between different modes of acquisition of the remanent magnetization of
      ferromagnetic particles. J. Appl. Phys. **29**, 595–596 (1958)

1267 O. Henkel: Remanenzverhalten und Wechselwirkungen in hartmagnetischen Teilchenkollektiven.
      Phys. Status Solidi **7**, 919–929 (1964) *(Remanence behaviour and interactions of hard magnetic
      particle assemblies)*

1268 E.P. Wohlfarth: A review of the problem of fine-particle interactions with special reference to magnetic
      recording. J. Appl. Phys. **35**, 783–790 (1964)

1269 H.C. Tong, R. Ferrier, P. Chang, J. Tzeng, K.L. Parker: The micromagnetics of thin film disk recording
      tracks. IEEE Trans. Magn. **20**, 1831–1832 (1984)

1270 R.A. Baugh, E.S. Murdock, B.R. Natarajan: Measurement of noise in magnetic media.
      IEEE Trans. Magn. **19**, 1722–1724 (1983)

1271 N.R. Belk, P.K. George, G.S. Mowry: Noise in high performance thin-film longitudinal recording
      media. IEEE Trans. Magn. **21**, 1350–1355 (1985)

1272 N.R. Belk, P.K. George, G.S. Mowry: Measurement of the intrinsic signal-to-noise ratio for high-
      performance rigid recording media. J. Appl. Phys. **59**, 557–563 (1986)

1273 T.C. Arnoldussen, H.C. Tong: Zigzag transition profiles, noise, and correlation statistics in highly
      oriented longitudinal film media. IEEE Trans. Magn. **22**, 889–891 (1986)

1274 Y.-S. Tang, L. Osse: Zig-zag domains and metal film disk noise.
      IEEE Trans. Magn. **23**, 2371–2373 (1987)

1275 K. Tang, M.R. Visokay, C.A. Ross, R. Ranjan, T. Yamashita, R. Sinclair: Lorentz transmission electron
      microscopy study of micromagnetic structures in real computer hard disks.
      IEEE Trans. Magn. **32**, 4130–4132 (1996)

1276 J.C. Lodder: Magnetic recording hard disk thin film media, in: *Handbook of Magnetic Materials*,
      Vol. 11.2, ed. by K.H.J. Buschow (Elsevier, Amsterdam, 1998) p. 291–405

1277 T. Wielinga, J.C. Lodder, J. Worst: Characteristics of rf-sputtered CoCr films.
      IEEE Trans. Magn. **18**, 1107–1109 (1982)

1278 H. Aoi, M. Saitoh, N. Nishiyama, R. Tsuchiya, T. Tamura: Noise characteristics in longitudinal thin-
      film media. IEEE Trans. Magn. **22**, 895–897 (1986)

1279 T. Yogi, G.L. Gorman, C. Hwang, M.A. Kakalec, S.E. Lambert: Dependence of magnetics, micro-
      structures and recording properties on underlayer thickness in CoNiCr/Cr media.
      IEEE Trans. Magn. **24**, 2727–2729 (1988)

1280 B.R. Natarajan, E.S. Murdock: Magnetic and recording properties of sputtered Co-P/Cr thin film
      media. IEEE Trans. Magn. **24**, 2724–2726 (1988)

1281 M.F. Doerner, T. Yogi, D.S. Parker, S. Lambeth, B. Hermsmeier, O.C. Allegranza, T. Nguyen:
      Composition effects in high density CoPtCr media. IEEE Trans. Magn. **29**, 3667–3669 (1993)

1282 P.S. Alexopoulos, R.H. Geiss: Micromagnetic and structural studies of sputtered thin-film recording
      media. IEEE Trans. Magn. **22**, 566–569 (1986)

1283 P.S. Alexopoulos, I.R. McFadyen, I.A. Beardsley, T.A. Nguyen, R.H. Geiss: Micromagnetics of longi-
      tudinal recording media, in: *Science and Technology of Nanostructured Magnetic Materials*,
      ed. by G.C. Hadjipanayis, G.A. Prinz (Plenum Press, New York, 1991) p. 239–247

1284 J.-G. Zhu: Micromagnetic modeling of thin film recording media, in: *Noise in Digital Magnetic Recor-
      ding*, ed. by T.C. Arnoldussen, L.L. Nunnelley (World Scientific, Singapore, 1992) p. 181–232

1285 J.J. Miles, M. Wdowin, J. Oakley, B.K. Middleton: The effect of cluster size on thin film media
      noise. IEEE Trans. Magn. **31**, 1013–1024 (1995)

1286 H.N. Bertram, J.-G. Zhu: Fundamental magnetization processes in thin-film recording media.
      Sol. State Phys.: Adv. Res. Appl. Suppl. **46**, 271–371 (1992)

1287 K. Shinohara, H. Yoshida, M. Odagiri, A. Tomaga: Columnar structure and some properties of metal-
      evaporated tape. IEEE Trans. Magn. **20**, 824–826 (1984)

1288 H.J. Richter: An approach to recording on tilted media. IEEE Trans. Magn. **29**, 2258–2265 (1993)

1289 H.J. Richter: An analysis of magnetization processes in metal evaporated tape.
      IEEE Trans. Magn. **29**, 21–33 (1993)

1290 S.B. Luitjens, S.E. Stupp, J.C. Lodder: Metal evaporated tape — State of the art and prospects.
      J. Magn. Magn. Mat. **155**, 261–265 (1996)

1291 B.K. Middleton, J.J. Miles, S.R. Cumpson: Models of metal evaporated tape.
      J. Magn. Magn. Mat. **155**, 266–272 (1996)

1292 I.B. Puchalska, A. Hubert, S. Winkler, B. Mirecki: Strong stripe domains in 80Co20Ni obliquely
      deposited films. IEEE Trans. Magn. **24**, 1787–1789 (1988)

1293 H. Aitlamine, L. Abelmann, I.B. Puchalska: Induced anisotropies in NiCo obliquely deposited films
      and their effect on magnetic domains. J. Appl. Phys. **71**, 353–361 (1992)

1294 T. Kohashi, H. Matsuyama, K. Koike, T. Takayama: Observation of domains in obliquely evaporated
      Co-CoO films by spin-polarized scanning electron microscopy. J. Appl. Phys. **81**, 7915–7921 (1997)

1295 L. Abelmann: Oblique evaporation of Co80Ni20 films for magnetic recording.
      PhD Thesis, University of Twente, The Netherlands (1994)
1296 S. Porthun, L. Abelmann, C. Lodder: Magnetic force microscopy of thin film media for high density
      magnetic recording. J. Magn. Magn. Mat. **182**, 238–273 (1998)
1297 N. Nouchi, H. Yoshida, K. Shinohara, A. Tomago: Analysis due to vector magnetic field for
      recording characteristic of metal evaporated tape. IEEE Trans. Magn. **22**, 385–387 (1986)
1298 G. Krijnen, S.B. Luitjens, R.W. de Bie, J.C. Lodder: Correlation between anisotropy direction and
      pulse shape for metal evaporated tape. IEEE Trans. Magn. **24**, 1817–1819 (1988)
1299 S.-I. Iwasaki, K. Takemura: An analysis for the circular mode of magnetization in short wavelength
      recording. IEEE Trans. Magn. **11**, 1173–1175 (1975)
1300 S.-I. Iwasaki: Perpendicular magnetic recording – Evolution and future.
      IEEE Trans. Magn. **20**, 657–662 (1984)
1301 W. Cain, A. Payne, M. Baldwinson, R. Hempstead: Challenges in the practical implementation of
      perpendicular magnetic recording. IEEE Trans. Magn. **32**, 97–102 (1996)
1302 H. Schewe, D. Stephani: Thin-film inductive heads for perpendicular recording.
      IEEE Trans. Magn. **26**, 2966–2971 (1990)
1303 J. Šimšová, R. Gemperle, J.C. Lodder, J. Kaczér, L. Murtinová, S. Saic, I. Tomáš: Domain period
      determination in CoCr films. Thin Solid Films **188**, 43–56 (1990)
1304 J. Šimšová, R. Gemperle, J. Kaczér, J.C. Lodder: Stripe and bubble domains in CoCr films.
      IEEE Trans. Magn. **26**, 30–32 (1990)
1305 J. Šimšová, V. Kamberský, R. Gemperle, J.C. Lodder, W.J.M.A. Geerts, B. Otten, P. ten Berge:
      Domain structure of Co-Cr films on minor loops. J. Magn. Magn. Mat. **101**, 196–198 (1991)
1306 H. Hoffmann: Thin-film media (CoCr films with perpendicular anisotropy).
      IEEE Trans. Magn. **22**, 472–477 (1986)
1307 J.C. Lodder, D. Wind, G.E. v. Dorssen, T.J.A. Popma, A. Hubert: Domains and magnetic reversals
      in CoCr. IEEE Trans. Magn. **23**, 214–216 (1987)
1308 Y. Maeda, M. Takahashi: Compositional microstructures within Co-Cr film grains.
      J. Magn. Soc. Japan **13**, 673–678 (1989)
1309 Y. Maeda, M. Takahashi: Selective chemical etching of latent compositional microstructures in
      sputtered Co-Cr films. Jpn. J. Appl. Phys. **29**, 1705–1710 (1990)
1310 D.E. Speliotis, J.P. Judge, W. Lynch, J. Burbage, R. Keirsted: Magnetic and recording characterization
      of Ba-ferrite particulate rigid disk media. IEEE Trans. Magn. **29**, 3625–3627 (1993)
1311 W. Andrä, H. Danan, R. Mattheis: Theoretical aspects of perpendicular magnetic recording media.
      Phys. Status Solidi A **125**, 9–55 (1991)
1312 Y. Hirayama: Development of high resolution and low noise single-layered perpendicular recording
      media for high density recording. IEEE Trans. Magn. **33**, 996–1001 (1997)
1313 M.H. Kryder: Magneto-optical storage materials. Ann. Rev. Mater. Sci. **23**, 411–436 (1993)
1314 M.H. Kryder: Magnetic information storage, in: *Applied Magnetism*, ed. by R. Gerber, *et al.*
      (Kluwer, Dordrecht, 1994) p. 39–112
1315 R. Carey, D.M. Newman, B.W.J. Thomas: Magneto-optic recording.
      J. Phys. D **28**, 2207–2227 (1995)
1316 M. Mansuripur: *The Physical Principles of Magneto-Optical Recording*
      (Cambridge University Press, Cambridge, 1995)
1317 T. Suzuki: Magneto-optic recording materials. MRS Bull. **21**.9, 42–47 (1996)
1318 D. Raasch, J. Reck, C. Mathieu, B. Hillebrands: Exchange stiffness constant and wall energy density
      of amorphous GdTb–FeCo films. J. Appl. Phys. **76**, 1145–1149 (1994)
1319 J.C. Suits, R.H. Geiss, C.J. Lin, D. Rugar, A.E. Bell: Observation of laser-written magnetic domains
      in amorphous TbFe films by Lorentz microscopy. J. Appl. Phys. **61**, 3509–3513 (1987)
1320 H.W. van Kesteren, A.J. den Boef, W.B. Zeper, J.H.M. Spruit, B.A.J. Jacobs, P.F. Carcia: Scanning
      magnetic force microscopy on Co/Pt magneto-optical disks. J. Appl. Phys. **70**, 2413–2422 (1991)
1321 K. Tsutsumi, T. Fukami: Direct overwrite in magneto-optic recording.
      J. Magn. Magn. Mat. **118**, 231–247 (1993)
1322 H.-P.D. Shieh, M.H. Kryder: Magneto-optic recording materials with direct overwrite capability.
      Appl. Phys. Lett. **49**, 473–474 (1986)
1323 H.-P.D. Shieh, M.H. Kryder: Operating margins for magneto-optic recording materials with direct
      overwrite capability. IEEE Trans. Magn. **23**, 171–173 (1987)
1324 D. Mergel: Magnetic interface walls under applied magnetic fields.
      J. Appl. Phys. **70**, 6433–6435 (1991)
1325 M. Kaneko, K. Aratani, M. Ohta: Multilayered magneto-optical disk for magnetically induced super-
      resolution. Jpn. J. Appl. Phys. **31**, 568–575 (1992)
1326 T.D. Milster, C.H. Curtis: Analysis of superresolution in magneto-optic data storage devices.
      Appl. Optics **31**, 6272–6279 (1992)

1327 Y. Murakami, A. Takahashi, S. Terashima: Magnetic super-resolution.
      IEEE Trans. Magn. **31**, 3215–3220 (1995)

1328 J. Hirokane, A. Takahashi: Magnetically induced superresolution using interferential in-plane mag-
      netization readout layer. Jpn. J. Appl. Phys. Pt. 1 **35**, 5701–5704 (1996)

1329 K. Takahashi, K. Katayama: The influence of the rear mask on copying recorded marks for magnetically
      induced superresolution disks. Jpn. J. Appl. Phys. Pt. 1 **36**, 6329–6338 (1997)

1330 H. Awano, S. Ohnuki, H. Shirai, N. Ohta, A. Yamaguchi, S. Sumi, K. Torazawa: Magnetic domain
      expansion readout for amplification of an ultra high density magneto-optical recording signal.
      Appl. Phys. Lett. **69**.27, 4257–4259 (1996)

1331 X. Ying, K. Shimazaki, H. Awano, M. Yoshihiro, H. Watanabe, N. Ohta, K.V. Rao: Magnetic expansion
      readout of 0.2 µm packed domains on a magneto-optical disk with an in-plane magnetized readout
      layer. Appl. Phys. Lett. **72**, 614–616 (1998)

1332 M. Birukawa, N. Miyatake, T. Suzuki: MSR high density recording.
      IEEE Trans. Magn. **34**, 438–443 (1998)

1333 I. Ichimura, S. Hayashi, G.S. Kino: High-density optical recording using a solid immersion lens.
      Appl. Opt. **36**, 4339–4348 (1997)

Thin Film Devices

1334 F. Jeffers: High-density magnetic recording heads. Proc. IEEE **74**, 1540–1586 (1986)

1335 A. Chiu, I. Croll, D.E. Heim, R.E. Jones, Jr, *et al.*: Thin-film inductive heads.
      IBM J. Res. Dev. **40**, 283–300 (1996)

1336 F.H. Liu, M.H. Kryder: Dynamic domain instability and popcorn noise in thin film heads.
      J. Appl. Phys. **75**, 6391–6393 (1994)

1337 F.H. Liu, M.H. Kryder: Inductance fluctuation, domain instability and popcorn noise in thin film
      heads. IEEE Trans. Magn. **30**, 3885–3887 (1994)

1338 H. Muraoka, Y. Nakamura: Artificial domain control of a single-pole head and its read/write perfor-
      mance in perpendicular magnetic recording. J. Magn. Magn. Mat. **134**, 268–274 (1994)

1339 K.B. Klaassen, J.C.L. van Peppen: Delayed relaxation in thin-film heads.
      IEEE Trans. Magn. **25**, 3212–3214 (1989)

1340 P.L. Trouilloud, B.E. Argyle, B. Petek, D.A. Herman, Jr.: Domain conversion under high frequency
      excitation in inductive thin film heads. IEEE Trans. Magn. **25**, 3461–3463 (1989)

1341 P.V. Koeppe, M.E. Re, M.H. Kryder: Thin film head domain structures versus Permalloy composition:
      Strain determination and frequency response. IEEE Trans. Magn. **28**, 71–75 (1992)

1342 B.E. Argyle, B. Petek, M.E. Re, F. Suits, D.A. Herman , Jr.: Bloch line influence on wall motion
      response in thin film heads. J. Appl. Phys. **64**, 6595–6597 (1988)

1343 O. Shinoura, T. Koyanagi: Magnetic thin film head with controlled domain structure by electroplating
      technology. Electrochim. Acta **42**, 3361–3366 (1997)

1344 J.C. Slonczewski: Micromagnetics of closed–edge laminations.
      IEEE Trans. Magn. **26**, 1322–1327 (1990)

1345 P. Deroux-Dauphin, J.P. Lazzari: A new thin film head generation IC head.
      IEEE Trans. Magn. **25**, 3190–3193 (1989)

1346 J.-P. Lazzari: Planar silicon heads / conventional thin-film heads recording behavior comparisons.
      IEEE Trans. Magn. **32**, 80–83 (1996)

1347 W. Cain, A. Payne, G. Qiu, D. Latev, *et al.*: Achieving 1 Gbit/in^2 with inductive recording heads.
      IEEE Trans. Magn. **32**, 3551–3553 (1996)

1348 M.L. Mallary, S. Ramaswamy: A new thin film head which doubles the flux through the coil.
      IEEE Trans. Magn. **29**, 3832–3836 (1993)

1349 M.L. Mallary, L. Dipalma, K. Gyasi, A.L. Sidman, A. Wu: Advanced multi-via heads.
      IEEE Trans. Magn. **30**, 287–290 (1994)

1350 U. Dibbern: Magnetoresistive sensors, in: *Magnetic Sensors*, Vol. 5, ed. by R. Boll, K. Overshott
      (VCH, Weinheim, 1989) p. 341–380

1351 F. Rottmann, F. Dettmann: New magnetoresistive sensors: Engineering and applications.
      Sensors & Actuators **25**, 763–766 (1991)

1352 D.J. Mapps: Magnetoresistive sensors. Sensors & Actuators A **59**, 9–19 (1997)

1353 M.N. Baibich, J.M. Broto, A. Fert, F. Nguyen Van Dau, *et al.*: Giant magnetoresistance of
      (001)Fe/(001)Cr magnetic superlattices. Phys. Rev. Lett. **61**, 2472–2475 (1988)

1354 G. Binasch, P. Grünberg, F. Saurenbach, W. Zinn: Enhanced magnetoresistance in layered magnetic
      structures with antiferromagnetic interlayer exchange. Phys. Rev. B **39**, 4828–4830 (1989)

1355 J.K. Spong, V.S. Speriosu, R.E. Fontana, M.M. Dovek, T.L. Hylton: Giant magnetoresistive spin
      valve bridge sensor. IEEE Trans. Magn. **32**, 366–371 (1996)

1356 H.A.M. van den Berg, W. Clemens, G. Gieres, G. Rupp, W. Schelter, M. Vieth: GMR sensor scheme
      with artificial antiferromagnetic subsystem. IEEE Trans. Magn. **32**, 4624–4626 (1996)

1357 H. Yamane, J. Mita, M. Kobayashi: Sensitive giant magnetoresistive sensor using ac bias magnetic field. Jpn. J. Appl. Phys. Pt. 2 **36**, L1591–L1593 (1997)

1358 R.P. Hunt: A magnetoresistive readout transducer. IEEE Trans. Magn. **7**, 150–154 (1971)

1359 D.A. Thompson, L.T. Romankiw, A.F. Mayadas: Thin film magnetoresistors in memory, storage, and related applications. IEEE Trans. Magn. **11**, 1039–1050 (1975)
D.A. Thompson: Magnetoresistive transducers in high-density magnetic recording. AIP Conf. Proc. **24**, 528–533 (1975)

1360 C.D. Mee, E.D. Daniel: Recording heads, in: *Magnetic Recording*, Vol. I, ed. by C.D. Mee, E.D. Daniel (McGraw-Hill, New York, 1987) p. 244–336

1361 F.B. Shelledy, J.L. Nix: Magnetoresistive heads for magnetic tape and disk recording. IEEE Trans. Magn. **28**, 2283–2288 (1992)

1362 R.I. Potter: Digital magnetic recording theory. IEEE Trans. Magn. **10**, 502–508 (1974)

1363 J.-G. Zhu, D.J. O'Connor: Stability of soft-adjacent-layer magnetoresistive heads with patterned exchange longitudinal bias. J. Appl. Phys. **81**, 4890–4892 (1997)

1364 J.A. Brug, T.C. Anthony, J.H. Nickel: Magnetic recording head materials. MRS Bull. **21**.9, 23–27 (1996)

1365 J.A. Brug, L. Tran, M. Bhattacharyya, J.H. Nickel, T.C. Anthony, A. Jander: Impact of new magneto-resistive materials on magnetic recording heads. J. Appl. Phys. **79**, 4491–4495 (1996)

1366 S. Demokritov: Biquadratic interlayer coupling in layered magnetic systems. J.Phys. D: Appl. Phys. **31**, 925–941 (1998)

1367 W. Folkerts, J.C.S. Kools, T.G.S.M. Rijks, R. Coehoorn, *et al.*: Application of giant magnetoresistive elements in thin film tape heads. IEEE Trans. Magn. **30**, 3813–3815 (1994)

1368 T.R. Koehler, M.L. Williams: Micromagnetic simulation of 10 GB/in^2 spin valve heads. IEEE Trans. Magn. **32**.5, 3446–3448 (1996)

1369 J. McCord, A. Hubert, J.C.S. Kools, J.J.M. Ruigrok: Domain observations on NiFeCo/Cu/NiFeCo-sandwiches for giant magnetoresistive sensors. IEEE Trans. Magn. **33**, 3984–3986 (1997)

1370 B.A. Gurney, V.S. Speriosu, D.R. Wilhoit, H. Lefakis, R.E. Fontana, Jr., D.E. Heim, M. Dovek: Can spin valves be reliably deposited for magnetic recording applications? J. Appl. Phys. **81**, 3998–4003 (1997)

1371 M.S. Blois, Jr.: Preparation of thin magnetic films and their properties. J. Appl. Phys. **26**, 975–980 (1955)

1372 S. Middelhoek, P.K. George, P. Dekker: *Physics of Computer Memory Devices* (Academic Press, London, New York, 1976)

1373 K.-U. Stein: Kohärente und inkohärente Drehung bei der Impulsummagnetisierung dünner Nickel-eisenschichten. Z. Angew. Phys. **20**, 36–46 (1965) *(Coherent and incoherent rotation in pulse magnetization reversal of thin nickel iron films)*

1374 T.R. Long: Electrodeposited memory elements for a nondestructive memory. J. Appl. Phys. (Suppl.) **31**, 123S–124S (1960)

1375 J.S. Mathias, G.A. Fedde: Plated-wire technology: A critical review. IEEE Trans. Magn. **5**, 728–751 (1969)

1376 S. Middelhoek, D. Wild: Review of wall creeping in thin magnetic films. IBM J. Res. Dev. **11**, 93–105 (1967)

1377 D.S. Lo, G.J. Cosimini, L.G. Zierhut, R.H. Dean, M.C. Paul: A Y-domain magnetic thin film memory element. IEEE Trans. Magn. **21**, 1776–1778 (1985)

1378 G.J. Cosimini, D.S. Lo, L.G. Zierhut, M.C. Paul, R.H. Dean, K.J. Matysik: Improved Y-domain magnetic film memory elements. IEEE Trans. Magn. **24**, 2060–2067 (1988)

1379 G.P. Cameron, W.J. Eberle: Characterization of chevron-shaped Permalloy magnetic memory elements by four magnetic imaging techniques. Scanning **13**, 419–428 (1991)

1380 A.V. Pohm, C.S. Comstock: 0.75, 1.25 and 2.0 μm wide M-R transducers. J. Magn. Magn. Mat. **54**, 1667–1669 (1986)

1381 A.V. Pohm, J.M. Daughton, C.S. Comstock, H.Y. Yoo, J. Hur: Threshold properties of 1, 2 and 4 μm multilayer M-R memory cells. IEEE Trans. Magn. **23**, 2575–2577 (1987)

1382 A.V. Pohm, J.M. Daughton, K.E. Spears: A high output mode for submicron m-r memory cells. IEEE Trans. Magn. **28**, 2356–2358 (1992)

1383 J.M. Daughton: Magnetoresistive memory technology. Thin Solid Films **216**, 162–168 (1992)

1384 M. Julliere: Tunneling between ferromagnetic films. Phys. Lett. **54A**, 225–226 (1975)

1385 T. Miyazaki, N. Tezuka: Giant magnetic tunneling effect in Fe/Al$_2$O$_3$/Fe junction. J. Magn. Magn. Mat. **139**, L231–L234 (1995)

1386 J.S. Moodera, L.R. Kinder, T.M. Wong, R. Meservey: Large magnetoresistance at room temperature in ferromagnetic thin film tunnel junctions. Phys. Rev. Lett. **74**, 3273–3276 (1995)

1387 D.J. Monsma, J.C. Lodder, T.J.A. Popma, B. Dieny: Perpendicular hot electron spin-valve effect in a magnetic field sensor: The spin valve transistor. Phys. Rev. Lett. **74**, 5260–5263 (1995)

1388   D.J. Monsma, R. Vlutters, T. Shimatsu, E.G. Keim, R.H. Mollema, J.C. Lodder: Development of the spin-valve transistor. IEEE Trans. Magn. **33**, 3495–3499 (1997)

1389   J.L. Brown, A.V. Pohm: 1-MB memory chip using giant magnetoresistive memory cells. IEEE Trans Comp., Pack. & Man. Techn. **17**, 373–378 (1994)

1390   D.D. Tang, P.K. Wang, V.S. Speriosu, S. Le, K.K. Kung: Spin valve RAM cell. IEEE Trans. Magn. **31**, 3206–3208 (1995)

1391   K. Nordquist, S. Pendharkar, M. Durlam, D. Resnick, *et al.*: Process development of sub-0.5 μm nonvolatile magnetoresistive random access memory arrays. J. Vac. Sci. Techn. B **15**, 2274–2278 (1997)

1392   Y. Zheng, J.-G. Zhu: Micromagnetics of spin valve memory cell. IEEE Trans. Magn. **32**, 4237–4239 (1996)

1393   J.O. Oti, S.E. Russek: Micromagnetic simulations of magnetoresistive behavior of sub-micrometer spin-valve MRAM devices. IEEE Trans. Magn. **33**, 3298–3300 (1997)

1394   W.J. Gallagher, S.S.P. Parkin, Y. Lu, X.P. Bian, *et al.*: Microstructured magnetic tunnel junctions. J. Appl. Phys. **81**, 3741–3746 (1997)

1395   J.M. Daughton: Magnetic tunneling applied to memory. J. Appl. Phys. **81**, 3758–3763 (1997)

1396   Z.G. Wang, D. Mapps, L.N. He, W. Clegg, D.T. Wilton, P. Robinson, Y. Nakamura: Feasibility of ultra-dense spin-tunneling random-access memory. IEEE Trans. Magn. **33**, 4498–4512 (1997)

Domain Propagation Devices

1397   H. Jouve (Ed.): *Magnetic Bubbles* (Academic Press, New York, 1986)

1398   A.H. Bobeck, H.E.D. Scovil: Magnetic bubbles. Scient. American **224**.6, 78–90 (1971)

1399   T.H. O'Dell: Magnetic-bubble domain devices. Rep. Progr. Phys. **49**, 589–620 (1986)

1400   P. Chaudhari, J.J. Cuomo, R.J. Gambino: Amorphous metallic films for bubble domain applications. IBM J. Res. Develop. **17**, 66–68 (1973)

1401   Y. Sugita: Magnetic bubble memories. Solid state file utilizing micromagnetic domains, in: *Physics and Engineering Applications of Magnetism*, Vol. 92, ed. by N. Miura, Y. Ishikawa (Springer, Berlin, Heidelberg, New York, 1991) p. 231–259

1402   R. Suzuki: Recent development in magnetic-bubble memory. Proc. IEEE **74**, 1582–1590 (1986)

1403   H. Chang (Ed.): *Magnetic Bubble Technology: Integrated Circuit Magnetics for Digital Storage and Processing* (IEEE Press, New York, 1975)

1404   R. Wolfe, J.C. North, W.A. Johnson, R.R. Spiwak, L.J. Varnerin, R.F. Fischer: Ion implanted patterns for magnetic bubble propagation. AIP Conf. Proc. **10**, 339–343 (1973)

1405   S. Orihara, T. Yanase: Field access Permalloy devices, in: *Magnetic Bubbles*, ed. by H. Jouve (Academic Press, New York, 1986) p. 137–213

1406   S.C.M. Backerra, W.H. de Roode, U. Enz: The influence of implantation-induced stress gradients in magnetic bubble layers. Philips J. Res. **36**, 112–121 (1980)

1407   Y. Hidaka, H. Matsutera: Charged wall formation mechanism in ion-implanted contiguous disk bubble devices. J. Appl. Phys. **39**, 116–118 (1981)

1408   A. Hubert: Micromagnetics of ion-implanted garnet layers. IEEE Trans. Magn. **20**, 1816–1821 (1984)

1409   T.J. Nelson, D.J. Muehlner: Circuit design and properties of patterned ion-implanted layers for field access bubble devices, in: *Magnetic Bubbles*, ed. by H. Jouve (Academic Press, New York, 1986) p. 215–290

1410   R. Suzuki: Recent development in magnetic-bubble memory. Proc. IEEE **74**, 1582–1590 (1986)

1411   Y. Sugita, R. Suzuki, T. Ikeda, T. Tacheuki, *et al.*: Ion-implanted and Permalloy hybrid magnetic bubble memory devices. IEEE Trans. Magn. **22**, 239–246 (1986)

1412   Y.S. Lin, G.S. Almasi, G.E. Keefe, E.W. Pugh: Self-aligned contiguous-disk chip using 1 μm bubbles and charged wall functions. IEEE Trans. Magn. **15**, 1642–1647 (1979)

1413   K. Matsuyama, H. Urai, K. Yoshimi: Optimum design consideration for dual conductor bubble devices. IEEE Trans. Magn. **19**, 111–119 (1983)

1414   H. Chang: *Magnetic Bubble Technology* (Marcel Dekker, New York, Basel, 1978)

1415   G.S. Almasi, G.E. Keefe, Y.S. Lin, D.A. Thompson: Magnetoresistive detector for bubble domains. J. Appl. Phys. **42**, 1268–1269 (1971)

1416   A.H. Bobeck, I. Danylchuk, F.C. Rossol, W. Strauss: Evolution of bubble circuits processed by a single mask level. IEEE Trans. Magn. **9**, 474–480 (1973)

1417   S. Konishi: A new ultra-high-density solid state memory: Bloch line memory. IEEE Trans. Magn. **19**, 1838–1840 (1983)

1418   Y. Hidaka, K. Matsuyama, S. Konishi: Experimental confirmation of fundamental functions for a novel Bloch line memory. IEEE Trans. Magn. **19**, 1841–1843 (1983)

1419   F.B. Humphrey, J.C. Wu: Vertical Bloch line memory. IEEE Trans. Magn. **5**, 1762–1766 (1985)

1420 Y. Maruyama, T. Ikeda, R. Suzuki: Primary operation of R/W gate for Bloch line memory devices. IEEE Transl. J. Magn. Japan **4**, 730–740 (1989)

1421 J.C. Wu, R.R. Katti, H.L. Stadler: Major line operation in vertical Bloch line memory. J. Appl. Phys. **69**, 5754–5756 (1991)

1422 S. Konishi, K. Matsuyama, I. Chida, S. Kubota, H. Kawahara, M. Ohbo: Bloch line memory, an approach to gigabit memory. IEEE Trans. Magn. **20**, 1129–1134 (1984)

1423 L. Zimmermann, J. Miltat, P. Pougnet: Stability of information in Bloch line memories. IEEE Trans. Magn. **27**, 5508–5510 (1991)

1424 B.S. Han, R. Dahlbeck, Y. Yuan, J. Engemann: On the mechanism of the critical temperature for the break-down of VBL chains. J. Magn. Magn. Mat. **104**, 305–306 (1992)

1425 B.S. Han: Behavior of vertical Bloch-line chains of hard domains in garnet bubble films. J. Magn. Magn. Mat. **100**, 455–468 (1991)

1426 K. Fujimoto, Y. Maruyama, R. Imura: Dependence of read/write operations on uniaxial anisotropy constant in Bloch line memories. IEEE Trans. Magn. **33**, 4469–4474 (1997)

1427 R.J. Spain: Controlled domain tip propagation. Part I. J. Appl. Phys. **37**, 2572–2583 (1966)

1428 R.J. Spain, H.I. Javits: Controlled domain tip propagation. Part II. J. Appl. Phys. **37**, 2584–2593 (1966)

1429 R. Spain, M. Marino: Magnetic film domain-wall motion devices. IEEE Trans. Magn. **6**, 451–463 (1970)

1430 C.P. Battarel, R. Morille, A. Caplain: Increasing the density of planar magnetic domain memories. IEEE Trans. Magn. **19**, 1509–1513 (1983)

1431 H. Deichelmann: Ferromagnetic domain memories, in: *Digital Memory and Storage*, ed. by W.E. Proebster (Vieweg, Braunschweig, 1978) p. 239–245

1432 R.F. Hollman, M.C. Mayberry: Switching and signal characteristics of a zigzag-shaped crosstie RAM cell. IEEE Trans. Magn. **23**, 245–249 (1987)

1433 D.E. Lacklison, G.B. Scott, A.D. Giles, J.A. Clarke, R.F. Pearson, J.L. Page: The magnetooptic bubble display. IEEE Trans. Magn. **13**, 973–981 (1977)

1434 B. Hill, K.P. Schmidt: Fast switchable magneto-optic memory-display components. Philips J. Res. **33**, 211–225 (1978)

1435 B. Hill: x-y addressing methods for iron-garnet display components. IEEE Trans. Electr. Dev. **27**, 1825–1834 (1980)

1436 E.I. Nikolaev, A.I. Linnik, V.N. Sayapin: Magnetic bubble dynamics in an iron garnet film working as a spatial light modulator. Tech. Phys. **39**, 580–583 (1994) *[Zh. Tekh. Fiz. 64.6 (1994) 113–120]*

1437 J.-P. Castera: Magneto-optical devices, in: *Encyclopedia of Applied Physics*, Vol. 9, ed. by G.L. Trigg (VCH, Weinheim, New York, 1994) p. 229–244

1438 H. Hauser, F. Haberl, J. Hochreiter, M. Gaugitsch: Measurement of small distances between light spots by domain wall displacements. Appl. Phys. Lett. **64**, 2448–2450 (1994)

1439 A.F. Popkov, M. Fehndrich, M. Lohmeyer, H. Dötsch: Nonreciprocal TE-mode phase shift by domain walls in magnetooptic rib waveguides. Appl. Phys. Lett. **72**, 2508–2510 (1998)

Domains and Hysteresis

1440 P. Weiss, J. de Freudenreich: Étude de l'aimantation initiale en fonction de la température. Archives des Sciences Physiques et Naturelles (Genève) **42**, 449–470 (1916) *(Study of the initial magnetization as a function of temperature)*

1441 F. Preisach: Über die magnetische Nachwirkung. Z. Phys. **94**, 277–302 (1935) *(On magnetic aftereffect)*

1442 I.D. Mayergoyz: *Mathematical Models of Hysteresis* (Springer, Berlin, Heidelberg, New York, 1991)

1443 D.C. Jiles: *Introduction to Magnetism and Magnetic Materials* (Chapman & Hall, London, 1995)

1444 M. Esguerra: Computation of minor hysteresis loops from measured major loops. J. Magn. Magn. Mat. **157**, 366–368 (1995)

1445 G. Bertotti: *Hysteresis in Magnetism* (Academic Press, New York, 1998)

Textbooks and Review Articles

Aharoni, A.: *Introduction to the Theory of Ferromagnetism* (Clarendon Press, Oxford, 1996)

Arnoldussen, T.C. and L.L. **Nunnelley** (Ed.): *Noise in Digital Magnetic Recording* (World Scientific, Singapore, 1992)

Bate, G.: Magnetic recording materials since 1975. J. Magn. Magn. Mat. 100, 413–424 (1991)

Becker, R., W. **Döring**: *Ferromagnetismus* (Springer, Berlin, 1939) *(Ferromagnetism)*

Bertram, H.N.: *Theory of Magnetic Recording* (Cambridge University Press, Cambridge, 1994)

Bertotti, G.: *Hysteresis in Magnetism* (Academic Press, New York, 1998)

Bobeck, A.H. and H.E.D. **Scovil**: Magnetic bubbles. Scient. American 224.6, 78–90 (1971)

Bobeck, A.H. and E. **Della Torre**: *Magnetic Bubbles* (North Holland, Amsterdam, 1975)

Boll, R. and K. **Overshott** (Ed.): *Magnetic Sensors,* in: Sensors. A Comprehensive Survey, ed. by W. Göpel, *et al.*, Vol. 5 (VCH, Weinheim, 1989)

Bozorth, R.M.: *Ferromagnetism* (Van Nostrand, Princeton, 1951)

Brown, W.F., Jr.: *Magnetostatic Principles in Ferromagnetism* (North Holland, Amsterdam, 1962)

Brown, W.F., Jr.: *Micromagnetics* (Wiley, New York, 1963) *[reprinted: R.E. Krieger, Huntingdon N. Y., 1978]*

Carey, R. and E.D. **Isaac**: *Magnetic Domains and Techniques for their Observation,* (Academic Press, New York, 1966)

Castera, J.-P.: Magneto-optical devices, in: *Encyclopedia of Applied Physics*, Vol. 9, ed. by G.L. Trigg (VCH, Weinheim, New York, 1994) p. 229–244

Celotta, R.J., J. **Unguris**, M.H. **Kelley** and D.T. **Pierce**: Techniques to measure magnetic domain structures, in: *Methods in Materials Research: A Current Protocols Publication* (Wiley, New York, 1999)

Chang, H. (Ed.): *Magnetic Bubble Technology: Integrated Circuit Magnetics for Digital Storage and Processing* (IEEE Press, New York, 1975)

Chapman, J.N.: The investigation of magnetic domain structures in thin foils by electron microscopy. J. Phys. D: Appl. Phys. 17, 623–647 (1984)

Chikazumi, S.: *Physics of Ferromagnetism* (Clarendon Press, Oxford, 1997)

Craik, D.J. and R.S. **Tebble**: *Ferromagnetism and Ferromagnetic Domains* (North Holland, Amsterdam, 1965)

Craik, D.J.: *Structure and Properties of Magnetic Materials* (Pion, London, 1971)

Craik, D.J.: The observation of magnetic domains, in: *Methods of Experimental Physics*, Vol. 11, ed. by R.V. Coleman (Academic Press, New York, 1974) p. 675–743

Cullity, B.D.: *Introduction to Magnetic Materials* (Addison-Wesley, Reading MA, 1972)

Döring, W.: Mikromagnetismus, in: *Handbuch der Physik*, Vol. 18/2, ed. by S. Flügge (Springer, Berlin, Heidelberg, New York, 1966) p. 341–437 *(Micromagnetics)*

Dunlop, D.J. and Ö. **Özdemir**: *Rock Magnetism. Fundamentals and Frontiers*, (Cambridge University Press, Cambridge, 1997)

Evetts, J. (Ed.): *Concise Encyclopedia of Magnetic & Superconducting Materials,* in: Advances in Materials Science and Engineering (Pergamon, Oxford, New York, 1992)

Gerber, R., C.D. **Wright** and G. **Asti** (Ed.): *Applied Magnetism* (Kluwer, Dordrecht, 1994)

Grütter, P., H.J. **Mamin** and D. **Rugar**: Magnetic force microscopy (MFM), in: *Scanning Tunneling Microscopy*, Vol. II, ed. by H.-J. Güntherodt and R. Wiesendanger (Springer, Berlin, Heidelberg, New York, 1992) p. 151–207

Hubert, A.: *Theorie der Domänenwände in Geordneten Medien* (Springer, Berlin, Heidelberg, New York, 1974) *(Theory of Domain Walls in Ordered Media)*

Jakubovics, J.P.: *Magnetism and Magnetic Materials* (The Intitute of Materials, London, 1994)

Jiles, D.C.: *Introduction to Magnetism and Magnetic Materials* (Chapman & Hall, London, 1995)

Jouve, H. (Ed.): *Magnetic Bubbles* (Academic Press, New York, 1986)

Kalvius, M. and R.S. **Tebble** (Ed.): *Experimental Magnetism*, Vol. I (Wiley, Chichester New York, 1979)

Kittel, C.: Physical theory of ferromagnetic domains. Rev. Mod. Phys. 21, 541–583 (1949)

Kittel, C. and J.K. **Galt**: Ferromagnetic domain theory. Solid State Phys. 3, 437–565 (1956)

Kronmüller, H. and H. **Träuble**: Magnetisierungskurve der Ferromagnetika, in: *Moderne Probleme der Metallphysik*, Vol. 2, ed. by A. Seeger (Springer, Berlin, Heidelberg, New York, 1966) p. 24–475 *(The magnetization curve of ferromagnet)*

Kryder, M.H.: Magneto-optical storage materials. Ann. Rev. Mater. Sci. 23, 411–436 (1993)

Landau, L.D. and E.M. **Lifshitz**: *Course of Theoretical Physics*, Vol. IX.2: Theory of the Condensed State (Pergamon, Oxford, 1980)

Lichte, H.: Electron holography methods, in: *Handbook of Microscopy*, Vol. 1, ed. by S. Amelinckx, *et al.* (VCH, Weinheim, New York, 1997) p. 515–536

Malozemoff, A.P. and J.C. **Slonczewski**: *Magnetic Domain Walls in Bubble Materials*, (Academic Press, New York, 1979)

Mansuripur, M.: *The Physical Principles of Magneto-Optical Recording* (Cambridge University Press, Cambridge, 1995)

Mayergoyz, I.D.: *Mathematical Models of Hysteresis* (Springer, Berlin, Heidelberg, New York, 1991)

Mee, C.D. and E.D. **Daniel** (Ed.): *Magnetic Recording*, Vol. I (McGraw-Hill, New York, 1987)

Middelhoek, S., P.K. **George** and P. **Dekker**: *Physics of Computer Memory Devices* (Academic Press, London, New York, 1976)

Newbury, D.E., D.C. **Joy**, P. **Echlin**, C.E. **Fiori** and J.I. **Goldstein**: Magnetic contrast in the SEM, in: *Advanced Scanning Electron Microscopy and X-Ray Microanalysis* (Plenum, New York, London, 1986) p. 147–179

O'Dell, T.H.: *Ferromagnetodynamics. The Dynamics of Magnetic Bubbles, Domains and Domain Walls* (Wiley, New York, 1981)

O'Dell, T.H.: Magnetic-bubble domain devices. Rep. Progr. Phys. 49, 589–620 (1986)

Privorotskii, I.A.: *Thermodynamic Theory of Domain Structures* (Wiley, New York, 1976)

Rado, G.T. and H. **Suhl** (Ed.): *Magnetism* (Academic Press, New York, 1963)

Reimer, L.: *Transmission Electron Microscopy. Physics of Image Formation and Microanalysis*, in: Springer Series in Optical Sciences, Vol. 36 (Springer, Berlin, Heidelberg, New York, 1993)

Rosenberg, M. and C. **Tǎnǎsoiu**: Magnetic Domains, in: *Magnetic Oxides*, Vol. 2, ed. by D.J. Craik (Wiley, London, 1972) p. 483–573

Smit, J. and H.P.L. **Wijn**: *Ferrites* (N.V. Philips, Eindhoven, 1959)

Soohoo, R.F.: *Microwave Magnetics* (Harper & Row, New York, 1985)

Stacey, F.D. and S.K. **Banerjee**: *The Physical Principles of Rock Magnetism* (Elsevier, Amsterdam, 1974)

Tebble, R.S. and D.J. **Craik**: *Magnetic Materials* (Wiley, London, 1969)

Tonomura, A.: *Electron Holography*, in: Springer Series in Optical Sciences, Vol. 70 (Springer, Berlin, Heidelberg, New York, 1993)

Tsuno, K.: Magnetic domain observation by means of Lorentz electron microscopy with scanning technique. Rev. Solid State Science 2, 623–656 (1988)

Weiss, P. and G. **Foex**: *Le Magnétisme* (Armand Colin, Paris, 1926) *(Magnetism)*

Wohlfarth, E.P. and K.H.J. **Buschow** (Ed.): *(Handbook of) Ferromagnetic Materials*, Vol. 1–11 (North Holland/Elsevier, Amsterdam, 1980–1998 f)

Zijlstra, H.: Measurement of magnetic quantities, in: *Experimental Methods in Magnetism*, Vol. 2 (North Holland, Amsterdam, 1967)

References

Symbols

ω_{opt}	Optimum reduced domain width	337
ω_{res}	Resonance frequency	374
ω_L	Light frequency	50
Ω	Angular defect	145
Ω_w	Wall angle	239

P

φ	Magnetization angle	112
φ^*	Switching magnetization angle	204
φ_0	Equilibrium magnetization angle	207
φ_{ap}	Optical amplitude penetration function	34
φ_e	Electron phase	55
φ_p	Azimuth angle at peak velocity	276
φ_∞	Asymptotic magnetization angle	218
Φ_d	Stray field potential	119
Φ_q	Anisotropy type angle	192
Φ_{tip}	Tip scalar potential	80
φ_K	Kerr rotation	29
ψ	Wall orientation angle	141
ψ_p	Polarizer angle relative to p-axis	28
p	Reduced period	307
$\boldsymbol{p}$	Elastic lattice distortion	134
p_c	Colloid particle concentration	14
$\boldsymbol{p}^{cp}$	Compensating distortion	140
$\boldsymbol{p}_e$	Compatible magnetostrictive distortion	139
p_{kin}	Kinetic potential	272
p_{opt}	Optimum domain period	350
P	Period	133
P_x	Function (abbreviation)	122

Q

q	Numerical coefficient	211
q_c	Particle magnetic moment	14
q_e	Electron charge	53
Q	Reduced anisotropy material constant	120

Q_{cr}	Maximum Q-value for stripe domains	301
Q_V	Magneto-optical parameter	25

R

ρ	Reduced particle radius	211
ρ_c	Reduced critical radius	211
ρ_{ms}	Relative magnetostrictive self energy	233
$\boldsymbol{r}$	Position vector	11
r_a	Elastic anisotropy ratio	379
r_p	Parallel reflectivity (intensity)	33
r_{SN}	Signal to noise ratio	31
R	Radius	174
R_c	Critical radius	211
R_p	Parallel amplitude reflection coefficient	28
R_F	Faraday amplitude	27
R_K	Kerr amplitude	27
R_N	Regularly reflected amplitude	26

S

σ	Conductivity	286
σ^0	Quasi-plastic stress tensor	137
σ^{bal}	Balancing stress tensor	140
$\bar{\sigma}$	Average stress tensor	223
σ^{cp}	Compensating stress tensor	140
σ_{ex}	External stress tensor	148
σ^{form}	Stress tensor of the form effect	380
σ_p	Planar stress	477
σ_s	Magnetic surface charge density	119
σ_{sample}	Sample surface charge	80
σ_u	Uniaxial stress	477
s	Derivative function	249
$\boldsymbol{s}$	Inverse magneto-elastic tensor	142
s_b	Bubble spacing	311
$\boldsymbol{s}_b$	Electron beam direction	55
$\boldsymbol{s}^{ch}$	Charge pattern set	127
s_{mo}	Relative magneto-optical signal	30
S	Shearing function	248
S_{bc}	Function for bubble collapse	314

Acronyms

Author Index

Subject Index

Druck: Strauss Offsetdruck, Mörlenbach
Verarbeitung: Schäffer, Grünstadt